T0205307

Lecture Notes in Computer Science 14075

Founding Editors

Gerhard Goos
Juris Hartmanis

Editorial Board Members

Elisa Bertino, *Purdue University, West Lafayette, IN, USA*
Wen Gao, *Peking University, Beijing, China*
Bernhard Steffen, *TU Dortmund University, Dortmund, Germany*
Moti Yung, *Columbia University, New York, NY, USA*

The series Lecture Notes in Computer Science (LNCS), including its subseries Lecture Notes in Artificial Intelligence (LNAI) and Lecture Notes in Bioinformatics (LNBI), has established itself as a medium for the publication of new developments in computer science and information technology research, teaching, and education.

LNCS enjoys close cooperation with the computer science R & D community, the series counts many renowned academics among its volume editors and paper authors, and collaborates with prestigious societies. Its mission is to serve this international community by providing an invaluable service, mainly focused on the publication of conference and workshop proceedings and postproceedings. LNCS commenced publication in 1973.

Jiří Mikyška · Clélia de Mulatier ·
Maciej Paszynski · Valeria V. Krzhizhanovskaya ·
Jack J. Dongarra · Peter M. A. Sloot
Editors

Computational Science – ICCS 2023

23rd International Conference
Prague, Czech Republic, July 3–5, 2023
Proceedings, Part III

Springer

Editors
Jiří Mikyška
Czech Technical University in Prague
Prague, Czech Republic

Clélia de Mulatier
University of Amsterdam
Amsterdam, The Netherlands

Maciej Paszynski
AGH University of Science and Technology
Krakow, Poland

Valeria V. Krzhizhanovskaya
University of Amsterdam
Amsterdam, The Netherlands

Jack J. Dongarra
University of Tennessee at Knoxville
Knoxville, TN, USA

Peter M. A. Sloot
University of Amsterdam
Amsterdam, The Netherlands

ISSN 0302-9743 ISSN 1611-3349 (electronic)
Lecture Notes in Computer Science
ISBN 978-3-031-36023-7 ISBN 978-3-031-36024-4 (eBook)
https://doi.org/10.1007/978-3-031-36024-4

© The Editor(s) (if applicable) and The Author(s), under exclusive license
to Springer Nature Switzerland AG 2023, corrected publication 2023

This work is subject to copyright. All rights are reserved by the Publisher, whether the whole or part of the material is concerned, specifically the rights of translation, reprinting, reuse of illustrations, recitation, broadcasting, reproduction on microfilms or in any other physical way, and transmission or information storage and retrieval, electronic adaptation, computer software, or by similar or dissimilar methodology now known or hereafter developed.
The use of general descriptive names, registered names, trademarks, service marks, etc. in this publication does not imply, even in the absence of a specific statement, that such names are exempt from the relevant protective laws and regulations and therefore free for general use.
The publisher, the authors, and the editors are safe to assume that the advice and information in this book are believed to be true and accurate at the date of publication. Neither the publisher nor the authors or the editors give a warranty, expressed or implied, with respect to the material contained herein or for any errors or omissions that may have been made. The publisher remains neutral with regard to jurisdictional claims in published maps and institutional affiliations.

This Springer imprint is published by the registered company Springer Nature Switzerland AG
The registered company address is: Gewerbestrasse 11, 6330 Cham, Switzerland

Preface

Welcome to the 23rd annual International Conference on Computational Science (ICCS - https://www.iccs-meeting.org/iccs2023/), held on July 3–5, 2023 at the Czech Technical University in Prague, Czechia.

In keeping with the new normal of our times, ICCS featured both in-person and online sessions. Although the challenges of such a hybrid format are manifold, we have always tried our best to keep the ICCS community as dynamic, creative, and productive as possible. We are proud to present the proceedings you are reading as a result.

ICCS 2023 was jointly organized by the Czech Technical University in Prague, the University of Amsterdam, NTU Singapore, and the University of Tennessee.

Standing on the Vltava River, Prague is central Europe's political, cultural, and economic hub.

The Czech Technical University in Prague (CTU) is one of Europe's largest and oldest technical universities and the highest-rated in the group of Czech technical universities. CTU offers 350 accredited study programs, 100 of which are taught in a foreign language. Close to 19,000 students are studying at CTU in 2022/2023. The Faculty of Nuclear Sciences and Physical Engineering (FNSPE), located along the river bank in Prague's beautiful Old Town (Staré Mesto) and host to ICCS 2023, is the only one in Czechia to offer studies in a broad range of fields related to Nuclear Physics and Engineering. The Faculty operates both fission (VR-1) and fusion (GOLEM Tokamak) reactors and hosts several cutting-edge research projects, collaborating with a number of international research centers (CERN, ITER, BNL-STAR, ELI).

The International Conference on Computational Science is an annual conference that brings together researchers and scientists from mathematics and computer science as basic computing disciplines, as well as researchers from various application areas who are pioneering computational methods in sciences such as physics, chemistry, life sciences, engineering, arts, and humanitarian fields, to discuss problems and solutions in the area, identify new issues, and shape future directions for research.

Since its inception in 2001, ICCS has attracted increasingly higher-quality attendees and papers, and this year is not an exception, with over 300 participants. The proceedings series have become a primary intellectual resource for computational science researchers, defining and advancing the state of the art in this field.

The theme for 2023, "**Computation at the Cutting Edge of Science**", highlights the role of Computational Science in assisting multidisciplinary research. This conference was a unique event focusing on recent developments in scalable scientific algorithms; advanced software tools; computational grids; advanced numerical methods; and novel application areas. These innovative novel models, algorithms, and tools drive new science through efficient application in physical systems, computational and systems biology, environmental systems, finance, and others.

ICCS is well known for its excellent lineup of keynote speakers. The keynotes for 2023 were:

- **Helen Brooks**, United Kingdom Atomic Energy Authority (UKAEA), UK
- **Jack Dongarra**, University of Tennessee, USA
- **Derek Groen**, Brunel University London, UK
- **Anders Dam Jensen**, European High Performance Computing Joint Undertaking (EuroHPC JU), Luxembourg
- **Jakub Šístek**, Institute of Mathematics of the Czech Academy of Sciences & Czech Technical University in Prague, Czechia

This year we had 531 submissions (176 to the main track and 355 to the thematic tracks). In the main track, 54 full papers were accepted (30.7%); in the thematic tracks, 134 full papers (37.7%). A higher acceptance rate in the thematic tracks is explained by the nature of these, where track organizers personally invite many experts in a particular field to participate in their sessions. Each submission received at least 2 single-blind reviews (2.9 reviews per paper on average).

ICCS relies strongly on our thematic track organizers' vital contributions to attract high-quality papers in many subject areas. We would like to thank all committee members from the main and thematic tracks for their contribution to ensuring a high standard for the accepted papers. We would also like to thank *Springer, Elsevier,* and *Intellegibilis* for their support. Finally, we appreciate all the local organizing committee members for their hard work in preparing for this conference.

We are proud to note that ICCS is an A-rank conference in the CORE classification.

We hope you enjoyed the conference, whether virtually or in person.

July 2023

Jiří Mikyška
Clélia de Mulatier
Maciej Paszynski
Valeria V. Krzhizhanovskaya
Jack J. Dongarra
Peter M. A. Sloot

Organization

The Conference Chairs

General Chair

Valeria Krzhizhanovskaya — University of Amsterdam, The Netherlands

Main Track Chair

Clélia de Mulatier — University of Amsterdam, The Netherlands

Thematic Tracks Chair

Maciej Paszynski — AGH University of Science and Technology, Poland

Scientific Chairs

Peter M. A. Sloot — University of Amsterdam, The Netherlands | Complexity Institute NTU, Singapore

Jack Dongarra — University of Tennessee, USA

Local Organizing Committee

LOC Chair

Jiří Mikyška — Czech Technical University in Prague, Czechia

LOC Members

Pavel Eichler — Czech Technical University in Prague, Czechia
Radek Fučík — Czech Technical University in Prague, Czechia
Jakub Klinkovský — Czech Technical University in Prague, Czechia
Tomáš Oberhuber — Czech Technical University in Prague, Czechia
Pavel Strachota — Czech Technical University in Prague, Czechia

Thematic Tracks and Organizers

Advances in High-Performance Computational Earth Sciences: Applications and Frameworks – IHPCES

Takashi Shimokawabe, Kohei Fujita, Dominik Bartuschat

Artificial Intelligence and High-Performance Computing for Advanced Simulations – AIHPC4AS

Maciej Paszynski, Robert Schaefer, Victor Calo, David Pardo, Quanling Deng

Biomedical and Bioinformatics Challenges for Computer Science – BBC

Mario Cannataro, Giuseppe Agapito, Mauro Castelli, Riccardo Dondi, Rodrigo Weber dos Santos, Italo Zoppis

Computational Collective Intelligence – CCI

Marcin Maleszka, Ngoc Thanh Nguyen

Computational Diplomacy and Policy – CoDiP

Michael Lees, Brian Castellani, Bastien Chopard

Computational Health – CompHealth

Sergey Kovalchuk, Georgiy Bobashev, Anastasia Angelopoulou, Jude Hemanth

Computational Modelling of Cellular Mechanics – CMCM

Gabor Zavodszky, Igor Pivkin

Computational Optimization, Modelling, and Simulation – COMS

Xin-She Yang, Slawomir Koziel, Leifur Leifsson

Computational Social Complexity – CSCx

Vítor V. Vasconcelos. Debraj Roy, Elisabeth Krüger, Flávio Pinheiro, Alexander J. Stewart, Victoria Garibay, Andreia Sofia Teixeira, Yan Leng, Gabor Zavodszky

Computer Graphics, Image Processing, and Artificial Intelligence – CGIPAI

Andres Iglesias, Lihua You, Akemi Galvez-Tomida

Machine Learning and Data Assimilation for Dynamical Systems – MLDADS

Rossella Arcucci, Cesar Quilodran-Casas

MeshFree Methods and Radial Basis Functions in Computational Sciences – MESHFREE

Vaclav Skala, Samsul Ariffin Abdul Karim

Multiscale Modelling and Simulation – MMS

Derek Groen, Diana Suleimenova

Network Models and Analysis: From Foundations to Complex Systems – NMA

Marianna Milano, Pietro Cinaglia, Giuseppe Agapito

Quantum Computing – QCW

Katarzyna Rycerz, Marian Bubak

Simulations of Flow and Transport: Modeling, Algorithms, and Computation – SOFTMAC

Shuyu Sun, Jingfa Li, James Liu

Smart Systems: Bringing Together Computer Vision, Sensor Networks and Machine Learning – SmartSys

Pedro Cardoso, Roberto Lam, Jânio Monteiro, João Rodrigues

Solving Problems with Uncertainties – SPU

Vassil Alexandrov, Aneta Karaivanova

Teaching Computational Science – WTCS

Angela Shiflet, Nia Alexandrov

Reviewers

Zeeshan Abbas
Samsul Ariffin Abdul Karim
Tesfamariam Mulugeta Abuhay
Giuseppe Agapito
Elisabete Alberdi
Vassil Alexandrov
Nia Alexandrov
Alexander Alexeev
Nuno Alpalhão
Julen Alvarez-Aramberri
Domingos Alves
Sergey Alyaev
Anastasia Anagnostou
Anastasia Angelopoulou
Fabio Anselmi
Hideo Aochi
Rossella Arcucci
Konstantinos Asteriou
Emanouil Atanassov
Costin Badica
Daniel Balouek-Thomert
Krzysztof Banaś
Dariusz Barbucha
Luca Barillaro
João Barroso
Dominik Bartuschat
Pouria Behnodfaur
Jörn Behrens
Adrian Bekasiewicz
Gebrail Bekdas
Mehmet Belen
Stefano Beretta
Benjamin Berkels
Daniel Berrar
Piotr Biskupski
Georgiy Bobashev
Tomasz Boiński
Alessandra Bonfanti

Carlos Bordons
Bartosz Bosak
Lorella Bottino
Roland Bouffanais
Lars Braubach
Marian Bubak
Jérémy Buisson
Aleksander Byrski
Cristiano Cabrita
Xing Cai
Barbara Calabrese
Nurullah Çalık
Victor Calo
Jesús Cámara
Almudena Campuzano
Cristian Candia
Mario Cannataro
Pedro Cardoso
Eddy Caron
Alberto Carrassi
Alfonso Carriazo
Stefano Casarin
Manuel Castañón-Puga
Brian Castellani
Mauro Castelli
Nicholas Chancellor
Ehtzaz Chaudhry
Théophile Chaumont-Frelet
Thierry Chaussalet
Sibo Cheng
Siew Ann Cheong
Lock-Yue Chew
Su-Fong Chien
Marta Chinnici
Bastien Chopard
Svetlana Chuprina
Ivan Cimrak
Pietro Cinaglia

Noélia Correia
Adriano Cortes
Ana Cortes
Anna Cortes
Enrique Costa-Montenegro
David Coster
Carlos Cotta
Peter Coveney
Daan Crommelin
Attila Csikasz-Nagy
Javier Cuenca
António Cunha
Luigi D'Alfonso
Alberto d'Onofrio
Lisandro Dalcin
Ming Dao
Bhaskar Dasgupta
Clélia de Mulatier
Pasquale Deluca
Yusuf Demiroglu
Quanling Deng
Eric Dignum
Abhijnan Dikshit
Tiziana Di Matteo
Jacek Długopolski
Anh Khoa Doan
Sagar Dolas
Riccardo Dondi
Rafal Drezewski
Hans du Buf
Vitor Duarte
Rob E. Loke
Amir Ebrahimi Fard
Wouter Edeling
Nadaniela Egidi
Kareem Elsafty
Nahid Emad
Christian Engelmann
August Ernstsson
Roberto R. Expósito
Fangxin Fang
Giuseppe Fedele
Antonino Fiannaca
Christos Filelis-Papadopoulos
Piotr Frąckiewicz

Alberto Freitas
Ruy Freitas Reis
Zhuojia Fu
Kohei Fujita
Takeshi Fukaya
Wlodzimierz Funika
Takashi Furumura
Ernst Fusch
Marco Gallieri
Teresa Galvão Dias
Akemi Galvez-Tomida
Luis Garcia-Castillo
Bartłomiej Gardas
Victoria Garibay
Frédéric Gava
Piotr Gawron
Bernhard Geiger
Alex Gerbessiotis
Josephin Giacomini
Konstantinos Giannoutakis
Alfonso Gijón
Nigel Gilbert
Adam Glos
Alexandrino Gonçalves
Jorge González-Domínguez
Yuriy Gorbachev
Pawel Gorecki
Markus Götz
Michael Gowanlock
George Gravvanis
Derek Groen
Lutz Gross
Tobias Guggemos
Serge Guillas
Xiaohu Guo
Manish Gupta
Piotr Gurgul
Zulfiqar Habib
Yue Hao
Habibollah Haron
Mohammad Khatim Hasan
Ali Hashemian
Claire Heaney
Alexander Heinecke
Jude Hemanth

Marcin Hernes
Bogumila Hnatkowska
Maximilian Höb
Rolf Hoffmann
Tzung-Pei Hong
Muhammad Hussain
Dosam Hwang
Mauro Iacono
Andres Iglesias
Mirjana Ivanovic
Alireza Jahani
Peter Janků
Jiri Jaros
Agnieszka Jastrzebska
Piotr Jedrzejowicz
Gordan Jezic
Zhong Jin
Cedric John
David Johnson
Eleda Johnson
Guido Juckeland
Gokberk Kabacaoglu
Piotr Kalita
Aneta Karaivanova
Takahiro Katagiri
Mari Kawakatsu
Christoph Kessler
Faheem Khan
Camilo Khatchikian
Petr Knobloch
Harald Koestler
Ivana Kolingerova
Georgy Kopanitsa
Pavankumar Koratikere
Sotiris Kotsiantis
Sergey Kovalchuk
Slawomir Koziel
Dariusz Król
Elisabeth Krüger
Valeria Krzhizhanovskaya
Sebastian Kuckuk
Eileen Kuehn
Michael Kuhn
Tomasz Kulpa
Julian Martin Kunkel

Krzysztof Kurowski
Marcin Kuta
Roberto Lam
Rubin Landau
Johannes Langguth
Marco Lapegna
Ilaria Lazzaro
Paola Lecca
Michael Lees
Leifur Leifsson
Kenneth Leiter
Yan Leng
Florin Leon
Vasiliy Leonenko
Jean-Hugues Lestang
Xuejin Li
Qian Li
Siyi Li
Jingfa Li
Che Liu
Zhao Liu
James Liu
Marcellino Livia
Marcelo Lobosco
Doina Logafatu
Chu Kiong Loo
Marcin Łoś
Carlos Loucera
Stephane Louise
Frederic Loulergue
Thomas Ludwig
George Lykotrafitis
Lukasz Madej
Luca Magri
Peyman Mahouti
Marcin Maleszka
Alexander Malyshev
Tomas Margalef
Osni Marques
Stefano Marrone
Maria Chiara Martinis
Jaime A. Martins
Paula Martins
Pawel Matuszyk
Valerie Maxville

Pedro Medeiros
Wen Mei
Wagner Meira Jr.
Roderick Melnik
Pedro Mendes Guerreiro
Yan Meng
Isaak Mengesha
Ivan Merelli
Tomasz Michalak
Lyudmila Mihaylova
Marianna Milano
Jaroslaw Miszczak
Dhruv Mittal
Miguel Molina-Solana
Fernando Monteiro
Jânio Monteiro
Andrew Moore
Anabela Moreira Bernardino
Eugénia Moreira Bernardino
Peter Mueller
Khan Muhammad
Daichi Mukunoki
Judit Munoz-Matute
Hiromichi Nagao
Kengo Nakajima
Grzegorz J. Nalepa
I. Michael Navon
Vittorio Nespeca
Philipp Neumann
James Nevin
Ngoc-Thanh Nguyen
Nancy Nichols
Marcin Niemiec
Sinan Melih Nigdeli
Hitoshi Nishizawa
Algirdas Noreika
Manuel Núñez
Joe O'Connor
Frederike Oetker
Lidia Ogiela
Ángel Javier Omella
Kenji Ono
Eneko Osaba
Rongjiang Pan
Nikela Papadopoulou

Marcin Paprzycki
David Pardo
Anna Paszynska
Maciej Paszynski
Łukasz Pawela
Giulia Pederzani
Ebo Peerbooms
Alberto Pérez de Alba Ortíz
Sara Perez-Carabaza
Dana Petcu
Serge Petiton
Beata Petrovski
Toby Phillips
Frank Phillipson
Eugenio Piasini
Juan C. Pichel
Anna Pietrenko-Dabrowska
Gustavo Pilatti
Flávio Pinheiro
Armando Pinho
Catalina Pino Muñoz
Pietro Pinoli
Yuri Pirola
Igor Pivkin
Robert Platt
Dirk Pleiter
Marcin Płodzień
Cristina Portales
Simon Portegies Zwart
Roland Potthast
Małgorzata Przybyła-Kasperek
Ela Pustulka-Hunt
Vladimir Puzyrev
Ubaid Qadri
Rick Quax
Cesar Quilodran-Casas
Issam Rais
Andrianirina Rakotoharisoa
Célia Ramos
Vishwas H. V. S. Rao
Robin Richardson
Heike Riel
Sophie Robert
João Rodrigues
Daniel Rodriguez

Marcin Rogowski
Sergio Rojas
Diego Romano
Albert Romkes
Debraj Roy
Adam Rycerz
Katarzyna Rycerz
Mahdi Saeedipour
Arindam Saha
Ozlem Salehi
Alberto Sanchez
Ayşin Sancı
Gabriele Santin
Vinicius Santos Silva
Allah Bux Sargano
Azali Saudi
Ileana Scarpino
Robert Schaefer
Ulf D. Schiller
Bertil Schmidt
Martin Schreiber
Gabriela Schütz
Jan Šembera
Paulina Sepúlveda-Salas
Ovidiu Serban
Franciszek Seredynski
Marzia Settino
Mostafa Shahriari
Vivek Sheraton
Angela Shiflet
Takashi Shimokawabe
Alexander Shukhman
Marcin Sieniek
Joaquim Silva
Mateusz Sitko
Haozhen Situ
Leszek Siwik
Vaclav Skala
Renata Słota
Oskar Slowik
Grażyna Ślusarczyk
Sucha Smanchat
Alexander Smirnovsky
Maciej Smołka
Thiago Sobral

Isabel Sofia
Piotr Sowiński
Christian Spieker
Michał Staniszewski
Robert Staszewski
Alexander J. Stewart
Magdalena Stobinska
Tomasz Stopa
Achim Streit
Barbara Strug
Dante Suarez
Patricia Suarez
Diana Suleimenova
Shuyu Sun
Martin Swain
Edward Szczerbicki
Tadeusz Szuba
Ryszard Tadeusiewicz
Daisuke Takahashi
Osamu Tatebe
Carlos Tavares Calafate
Andrey Tchernykh
Andreia Sofia Teixeira
Kasim Terzic
Jannis Teunissen
Sue Thorne
Ed Threlfall
Alfredo Tirado-Ramos
Pawel Topa
Paolo Trunfio
Hassan Ugail
Carlos Uriarte
Rosarina Vallelunga
Eirik Valseth
Tom van den Bosch
Ana Varbanescu
Vítor V. Vasconcelos
Alexandra Vatyan
Patrick Vega
Francesc Verdugo
Gytis Vilutis
Jackel Chew Vui Lung
Shuangbu Wang
Jianwu Wang
Peng Wang

Katarzyna Wasielewska
Jarosław Wątróbski
Rodrigo Weber dos Santos
Marie Weiel
Didier Wernli
Lars Wienbrandt
Iza Wierzbowska
Maciej Woźniak
Dunhui Xiao
Huilin Xing
Yani Xue
Abuzer Yakaryilmaz
Alexey Yakovlev
Xin-She Yang
Dongwei Ye
Vehpi Yildirim
Lihua You

Drago Žagar
Sebastian Zając
Constantin-Bala Zamfirescu
Gabor Zavodszky
Justyna Zawalska
Pavel Zemcik
Wenbin Zhang
Yao Zhang
Helen Zhang
Jian-Jun Zhang
Jinghui Zhong
Sotirios Ziavras
Zoltan Zimboras
Italo Zoppis
Chiara Zucco
Pavel Zun
Karol Życzkowski

Contents – Part III

Computational Collective Intelligence

Balancing Agents for Mining Imbalanced Multiclass
Datasets – Performance Evaluation 3
    Joanna Jedrzejowicz and Piotr Jedrzejowicz

Decision Tree-Based Algorithms for Detection of Damage in AIS Data 17
    Marta Szarmach and Ireneusz Czarnowski

The Role of Conformity in Opinion Dynamics Modelling with Multiple
Social Circles 33
    Stanisław Stępień, Jarosław Jankowski, Piotr Bródka,
    and Radosław Michalski

Impact of Input Data Preparation on Multi-criteria Decision Analysis
Results 48
    Jarosław Wątróbski, Aleksandra Bączkiewicz, and Robert Król

Computational Diplomacy and Policy

Automating the Analysis of Institutional Design in International
Agreements 59
    Anna Wróblewska, Bartosz Pieliński, Karolina Seweryn,
    Sylwia Sysko-Romańczuk, Karol Saputa, Aleksandra Wichrowska,
    and Hanna Schreiber

Modeling Mechanisms of School Segregation and Policy Interventions:
A Complexity Perspective 74
    Eric Dignum, Willem Boterman, Andreas Flache, and Mike Lees

An Approach for Probabilistic Modeling and Reasoning of Voting Networks ... 90
    Douglas O. Cardoso, Willian P. C. Lima, Guilherme G. V. L. Silva,
    and Laura S. Assis

An Analysis of Political Parties Cohesion Based on Congressional Speeches ... 105
    Willian P. C. Lima, Lucas C. Marques, Laura S. Assis,
    and Douglas O. Cardoso

Computational Health

Comparative Study of Meta-heuristic Algorithms for Damage Detection
Problem 123
 Kamel Belhadj, Najeh Ben Guedria, Ali Helali, Omar Anis Harzallah,
 Chokri Bouraoui, and Lhassane Idoumghar

Estimation of the Impact of COVID-19 Pandemic Lockdowns on Breast
Cancer Deaths and Costs in Poland Using Markovian Monte Carlo
Simulation 138
 Magdalena Dul, Michal K. Grzeszczyk, Ewelina Nojszewska,
 and Arkadiusz Sitek

Machine Learning for Risk Stratification of Diabetic Foot Ulcers Using
Biomarkers 153
 Kyle Martin, Ashish Upadhyay, Anjana Wijekoon, Nirmalie Wiratunga,
 and Stewart Massie

A Robust Machine Learning Protocol for Prediction of Prostate Cancer
Survival at Multiple Time-Horizons 162
 Wojciech Lesiński and Witold R. Rudnicki

Supervised Machine Learning Techniques Applied to Medical Records
Toward the Diagnosis of Rare Autoimmune Diseases 170
 Pedro Emilio Andrade Martins, Márcio Eloi Colombo Filho,
 Ana Clara de Andrade Mioto, Filipe Andrade Bernardi,
 Vinícius Costa Lima, Têmis Maria Félix, and Domingos Alves

Handwriting Analysis AI-Based System for Assisting People
with Dysgraphia 185
 Richa Gupta, Deepti Mehrotra, Redouane Bouhamoum,
 Maroua Masmoudi, and Hajer Baazaoui

Universal Machine-Learning Processing Pattern for Computing
in the Video-Oculography 200
 Albert Śledzianowski, Jerzy P. Nowacki, Konrad Sitarz,
 and Andrzej W. Przybyszewski

RuMedSpellchecker: Correcting Spelling Errors for Natural Russian
Language in Electronic Health Records Using Machine Learning
Techniques 213
 Dmitrii Pogrebnoi, Anastasia Funkner, and Sergey Kovalchuk

Named Entity Recognition for De-identifying Real-World Health Records
in Spanish 228
Guillermo López-García, Francisco J. Moreno-Barea, Héctor Mesa,
José M. Jerez, Nuria Ribelles, Emilio Alba, and Francisco J. Veredas

Discovering Process Models from Patient Notes 243
Rolf B. Bänziger, Artie Basukoski, and Thierry Chaussalet

Knowledge Hypergraph-Based Multidimensional Analysis for Natural
Language Queries: Application to Medical Data 250
Sana Ben Abdallah Ben Lamine, Marouane Radaoui,
and Hajer Baazaoui Zghal

Coupling Between a Finite Element Model of Coronary Artery Mechanics
and a Microscale Agent-Based Model of Smooth Muscle Cells Through
Trilinear Interpolation 258
Aleksei Fotin and Pavel Zun

Does Complex Mean Accurate: Comparing COVID-19 Propagation
Models with Different Structural Complexity 270
Israel Huaman and Vasiliy Leonenko

Multi-granular Computing Can Predict Prodromal Alzheimer's Disease
Indications in Normal Subjects 278
Andrzej W. Przybyszewski

Accounting for Data Uncertainty in Modeling Acute Respiratory
Infections: Influenza in Saint Petersburg as a Case Study 286
Kseniya Sahatova, Aleksandr Kharlunin, Israel Huaman,
and Vasiliy Leonenko

A Web Portal for Real-Time Data Quality Analysis on the Brazilian
Tuberculosis Research Network: A Case Study 300
Victor Cassão, Filipe Andrade Bernardi, Vinícius Costa Lima,
Giovane Thomazini Soares, Newton Shydeo Brandão Miyoshi,
Ana Clara de Andrade Mioto, Afrânio Kritski, and Domingos Alves

Use of Decentralized-Learning Methods Applied to Healthcare:
A Bibliometric Analysis 313
Carolina Ameijeiras-Rodriguez, Rita Rb-Silva, Jose Miguel Diniz,
Julio Souza, and Alberto Freitas

Computational Modelling of Cellular Mechanics

Simulating Initial Steps of Platelet Aggregate Formation in a Cellular
Blood Flow Environment 323
  Christian J. Spieker, Konstantinos Asteriou, and Gabor Zavodszky

Estimating Parameters of 3D Cell Model Using a Bayesian Recursive
Global Optimizer (BaRGO) 337
  Pietro Miotti, Edoardo Filippi-Mazzola, Ernst C. Wit, and Igor V. Pivkin

A Novel High-Throughput Framework to Quantify Spatio-Temporal
Tumor Clonal Dynamics 345
  Selami Baglamis, Joyaditya Saha, Maartje van der Heijden,
  Daniël M. Miedema, Démi van Gent, Przemek M. Krawczyk,
  Louis Vermeulen, and Vivek M Sheraton

Computational Optimization, Modelling and Simulation

Expedited Metaheuristic-Based Antenna Optimization Using EM Model
Resolution Management 363
  Anna Pietrenko-Dabrowska, Slawomir Koziel, and Leifur Leifsson

Dynamic Core Binding for Load Balancing of Applications Parallelized
with MPI/OpenMP 378
  Masatoshi Kawai, Akihiro Ida, Toshihiro Hanawa, and Kengo Nakajima

Surrogate-Assisted Ship Route Optimisation 395
  Roman Dębski and Rafał Dreżewski

Optimization of Asynchronous Logging Kernels for a GPU Accelerated
CFD Solver 410
  Paul Zehner and Atsushi Hashimoto

Constrained Aerodynamic Shape Optimization Using Neural Networks
and Sequential Sampling 425
  Pavankumar Koratikere, Leifur Leifsson, Slawomir Koziel,
  and Anna Pietrenko-Dabrowska

Optimal Knots Selection in Fitting Degenerate Reduced Data 439
  Ryszard Kozera and Lyle Noakes

A Case Study of the Profit-Maximizing Multi-Vehicle Pickup and Delivery
Selection Problem for the Road Networks with the Integratable Nodes 454
  Aolong Zha, Qiong Chang, Naoto Imura, and Katsuhiro Nishinari

Symbolic-Numeric Computation in Modeling the Dynamics
of the Many-Body System TRAPPIST 469
  Alexander Chichurin, Alexander Prokopenya, Mukhtar Minglibayev,
  and Aiken Kosherbayeva

Transparent Checkpointing for Automatic Differentiation of Program
Loops Through Expression Transformations 483
  Michel Schanen, Sri Hari Krishna Narayanan, Sarah Williamson,
  Valentin Churavy, William S. Moses, and Ludger Paehler

Performance of Selected Nature-Inspired Metaheuristic Algorithms Used
for Extreme Learning Machine 498
  Karol Struniawski, Ryszard Kozera, and Aleksandra Konopka

Simulation–Based Optimisation Model as an Element of a Digital Twin
Concept for Supply Chain Inventory Control 513
  Bożena Mielczarek, Maja Gora, and Anna Dobrowolska

Semi-supervised Learning Approach to Efficient Cut Selection
in the Branch-and-Cut Framework 528
  Jia He Sun and Salimur Choudhury

Efficient Uncertainty Quantification Using Sequential Sampling-Based
Neural Networks 536
  Pavankumar Koratikere, Leifur Leifsson, Slawomir Koziel,
  and Anna Pietrenko-Dabrowska

Hierarchical Learning to Solve PDEs Using Physics-Informed Neural
Networks 548
  Jihun Han and Yoonsang Lee

SPMD-Based Neural Network Simulation with Golang 563
  Daniela Kalwarowskyj and Erich Schikuta

Low-Cost Behavioral Modeling of Antennas by Dimensionality Reduction
and Domain Confinement 571
  Slawomir Koziel, Anna Pietrenko-Dabrowska, and Leifur Leifsson

Real-Time Reconstruction of Complex Flow in Nanoporous Media: Linear
vs Non-linear Decoding 580
  Emmanuel Akeweje, Andrey Olhin, Vsevolod Avilkin,
  Aleksey Vishnyakov, and Maxim Panov

Model of Perspective Distortions for a Vision Measuring System
of Large-Diameter Bent Pipes 595
Krzysztof Borkowski, Dariusz Janecki, and Jarosław Zwierzchowski

Outlier Detection Under False Omission Rate Control 610
Adam Wawrzeńczyk and Jan Mielniczuk

Minimal Path Delay Leading Zero Counters on Xilinx FPGAs 626
Gregory Morse, Tamás Kozsik, and Péter Rakyta

Reduction of the Computational Cost of Tuning Methodology
of a Simulator of a Physical System 641
*Mariano Trigila, Adriana Gaudiani, Alvaro Wong, Dolores Rexachs,
and Emilio Luque*

A Hypergraph Model and Associated Optimization Strategies for Path
Length-Driven Netlist Partitioning 652
Julien Rodriguez, François Galea, François Pellegrini, and Lilia Zaourar

Graph TopoFilter: A Method for Noisy Labels Detection
for Graph-Structured Classes 661
Artur Budniak and Tomasz Kajdanowicz

Dynamic Data Replication for Short Time-to-Completion in a Data Grid 668
Ralf Vamosi and Erich Schikuta

Detection of Anomalous Days in Energy Demand Using Leading Point
Multi-regression Model 676
Krzysztof Karpio and Piotr Łukasiewicz

Correction to: Coupling Between a Finite Element Model of Coronary
Artery Mechanics and a Microscale Agent-Based Model of Smooth
Muscle Cells Through Trilinear Interpolation C1
Aleksei Fotin and Pavel Zun

Author Index 685

Computational Collective Intelligence

Balancing Agents for Mining Imbalanced Multiclass Datasets – Performance Evaluation

Joanna Jedrzejowicz[1] and Piotr Jedrzejowicz[2]([⊠])

[1] Institute of Informatics, Faculty of Mathematics, Physics and Informatics,
University of Gdansk, 80-308 Gdansk, Poland
joanna.jedrzejowicz@ug.edu.pl
[2] Department of Information Systems, Gdynia Maritime University, 81-225 Gdynia,
Poland
p.jedrzejowicz@umg.edu.pl

Abstract. The paper deals with mining imbalanced multiclass datasets. The goal of the paper is to evaluate the performance of several balancing agents implemented by the authors. Agents have been constructed from 5 state-of-the-art classifiers designed originally for mining binary imbalanced datasets. To transform binary classifiers into multiclass ones, we use the one-versus-one (OVO) approach making use of the collective decision taken by the majority voting. The paper describes our approach and provides detailed description of the respective balancing agents. Their performance is evaluated in an extensive computational experiment. The experiment involved multiclass imbalanced datasets from the Keel imbalanced datasets repository. Experiment results allowed to select best performing balancing agents using statistical tools.

Keywords: Imbalanced data · Multiclass datasets · Dataset balancing

1 Introduction

Multiclass imbalanced data mining is a challenging task in the field of machine learning and data mining. It refers to the scenario where a dataset contains multiple classes, but the number of instances in each class is significantly imbalanced. This can occur when one or more classes dominate the dataset, while the instances of other classes are scarce.

Imbalanced datasets can pose problems for machine learning algorithms, as they may not be able to accurately classify the minority classes due to the lack of sufficient training data. Additionally, in multiclass imbalanced data mining the mutual relationships between classes are complex and hence difficult to identify [24]. As a result, traditional machine learning algorithms tend to perform poorly on multiclass imbalanced datasets, leading to poor prediction accuracy and biased models.

To address these issues, various approaches have been proposed in the literature. As has been observed by [16], algorithms for dealing with multiclass

© The Author(s), under exclusive license to Springer Nature Switzerland AG 2023
J. Mikyška et al. (Eds.): ICCS 2023, LNCS 14075, pp. 3–16, 2023.
https://doi.org/10.1007/978-3-031-36024-4_1

problems can be broadly categorized into binarization approaches and ad hoc solutions. According to [7], binarization aims at decomposing the M-class problem into M(M-1)/2 binary subproblems (OVO one-versus-one) or M binary subproblems (OVA one-versus-all). There has been a significant amount of research in the field of multiclass imbalanced data mining over the past few decades. Some of the key techniques that have been proposed in the literature include:

1. Undersampling: This approach involves reducing the number of instances in the majority classes, such that the resulting dataset is more balanced (see for example [2,20]).
2. Oversampling: This approach involves increasing the number of instances in the minority classes, such that the resulting dataset is more balanced (see for example [1,16,18,23]).
3. Cost-sensitive learning: This approach involves modifying the loss function of a machine learning algorithm such that it takes into account the relative importance or cost of misclassifying different classes (see for example [17,27]).
4. Ensemble methods: Methods such as bagging, boosting, and stacking have been shown to be effective in handling imbalanced datasets (see for example [6,9,10,22,26]).
5. Algorithm level methods: Dedicated methods adapted to multiclass imbalance (see for example [5,12,19]).

The main advantage of undersampling is that it is simple and fast to implement, but it can also lead to the loss of important information from the majority classes. The main advantage of oversampling is that it can help improve the performance of machine learning algorithms on minority classes, but it can also lead to overfitting if the synthetic samples are not generated carefully. Another potential drawback, as pointed out in [16], is that classic oversampling algorithms consider only information from the minority class neglecting information from the majority classes. Cost-sensitive learning suffers often from the lack of information on the relative importance of misclassifying different classes. Ensemble methods can combine the predictions of multiple classifiers to improve the overall performance. Their performance relies on diversity between classifiers involved which is not always easy to achieve.

In this paper we evaluate the performance of 5 balancing agents. Balancing numbers of the majority and minority classes examples is one of a key approaches in constructing classifiers able to deal with imbalanced datasets. The idea is to preprocess training dataset to maximize the performance of a classifier used to classify data with unknown class labels. Balancing can be based on oversampling, undersampling or both. Balancing numbers of the majority and minority classes examples is one of a key approaches in constructing classifiers able to deal with imbalanced datasets. The idea is to preprocess training dataset to maximize the performance of a classifier used to classify data with unknown class labels.

All of the discussed agents have roots in binary imbalanced data classification methods and all have been implemented by us in the form of a software agents using the OVO approach to make them suitable for solving multiclass problems. The goal of the paper is to evaluate their performance. The proposed software agents are goal driven, reactive, autonomous, and display collaborative behaviors in the following sense:

- They try to balance majority and minority classes examples to improve the classification performance.
- They adapt to various imbalance ratio in the training datasets.
- They are independent of the classification algorithm used.
- They reach a final decision through comparing classification decisions of base classifiers and selecting the final outcome basing on the majority vote paradigm.

The list of the original algorithms implemented as agents for balancing multiclass imbalanced datasets follows:

- Adaptive synthetic sampling approach for imbalanced learning ADASYN-M [11].
- Local distribution-based adaptive minority oversampling LAMO-M [25].
- Combined synthetic oversampling and undersampling technique CSMOUTE-M [14].
- Feature-weighted oversampling approach FWSMOTE-M [21].
- Dominance-based oversampling approach DOMIN-M [13].

ADASYN has been selected as one of the most often applied oversampling techniques. The remaining algorithms have been selected as they are known to outperform many "classic" balancing techniques. Besides, they are based on relatively newly published concepts, and all use information from not only the minority class but also from the majority classes.

The rest of the paper is constructed as follows: Sect. 2 contains a description of the discussed agents. Section 3 contains agent performance evaluation based on computational experiment results. The final section contains conclusions and ideas for future research.

2   Balancing Agents

Assume that $D \subset X \times Y$ is a multiclass training dataset with samples $(\mathbf{x}, y)$ where $\mathbf{x}$ is an instance (datarow) and y is the class identity label associated with it.

The algorithm applied to learn the best classifier for D uses one-vs-one (OVO) method. For each pair of classes from Y, the dataset D is filtered, resulting in a subset with data from two classes. If it is imbalanced, an agent modifies it performing oversampling and/or undersampling. The modified set is used to generate the best possible classifier for the two classes. Finally, all the generated classifiers are merged to perform majority vote on data from the testing set. The OVO approach makes use of collective decision taken through a voting procedure. Algorithm 1 shows the pseudo-code for the proposed approach.

As mentioned before, the agents used to balance datasets are: ADASYN-M, LAMO-M, FWSMOTE-M, DOMIN-M, CSMOUTE-M. They are briefly described.

Algorithm 1: Schema of the approach

Input: Multiclass dataset $D = Train \cup Test$, threshold α, agent $\mathcal{G}$
Output: values of performance metrics for D

```
/* learning                                                              */
```
1   $geneList \leftarrow \emptyset$;
2   **foreach** *pair of classes* $c_1, c_2 \in Y$ **do**
```
       /* filtering training dataset to classes c₁,c₂                    */
```
3      $T(c_1, c_2) \leftarrow \{(\mathbf{x}, y) \in Train : y = c_1 \vee y = c_2\}$;
4   **if** *imbalance ratio of* $T(c_1, c_2)$ *is above* α **then**
5   use agent $\mathcal{G}$ to transform $T(c_1, c_2)$
6   **end**
7   apply GEP to $T(c_1, c_2)$ to obtain gene g;
8   merge g to $geneList$;
9   **end**
```
   /* testing                                                            */
```
10 **foreach** $(\mathbf{x}, y) \in Test$ **do**
11   **foreach** $g \in geneList$ **do**
12   apply g to $\mathbf{x}$, compare with y and store the result;
13   **end**
14 **end**
15 **return** *performance metrics as defined in 3.2*

The first four agents perform undersampling of the majority subset using Algorithm 2. The idea is to find the centroid of minority data and keep in the majority subset only those closest to the centroid.

Oversampling is specific for each agent type. SMOTE [4], which is an oversampling method, extending minority set via interpolation, adds elements of type $x + \lambda \cdot (z - x)$ for any minority example x, z its K-neighbor and random $\lambda \in (0,1)$(K is a parameter). ADASYN-M agent uses adaptive sampling approach introduced in [11] to extend the minority subset, using the weighted distribution of minority examples by generating with SMOTE new minority instances whose number is proportional to the proportion of K-neighbors which are in the majority subclass.

In case of LAMO-M agent, using the method introduced in [25] (Local distribution-based Adaptive Minority Oversampling), differently than in ADASYN, not all data from the minority subset are used in generation of new synthetic data. Two steps are performed:

- using two parameters k_1, k_2 defining respectively the number of neighbors for minority and majority instances, the distribution of instances is inspected and sampling seeds identified: first instances from majority subset which appear in k_1 neighborhood of minority instances are identified and sampling seeds are those minority instances which are in k_2 neighborhood of any from the first set,
- synthetic minority instances are generated from sampling seeds using interpolation.

Algorithm 2: Undersampling with minority class centroid

Input: data from majority class $majC$, data from minority class $minC$,
parameter s - size of reduced majority class.

Output: reduced majority class $redMaj \subset majC$ of size s.

1 calculate centroid CN of $minC$
2 define distances of CN to majority instances
3 $DIST \leftarrow \{dist(x, CN) : x \in majC\}$
4 sort $DIST$ in ascending order $SORT = \{d_1, \ldots, d_{|majC|}\}$
5 keep in reduced majority class instances whose distances are in the initial s
   segment of $DIST$
6 $redMaj \leftarrow \{x \in majC : dist(x, CN) \le d_s\}$
7 **return** $redMaj$

FWSMOTE-M uses the algorithm introduced in [21] which applies a method based on SMOTE where the importance of attributes is weighted by Fisher score making use of difference of attribute means in each of two classes; the weights are used when calculating distances in interpolation.

In case of DOMIN-M which is based on our method suggested in [13] the relation of domination among instances is introduced and using the genetic algorithm (GA) in subsequent iteration steps the minority subset is oversampled with non-dominated members of GA population. The relation of domination uses two criteria. Assume majority objects $majC$, minority objects $minC$ fixed. The first criterion makes use of an approach suggested in [15] for oversampling strategies based on calculating real-valued potential of each instance. The potential is defined by a radial basis function based on a set of majority objects $majC$, minority objects $minC$ and parameter γ representing the spread of the function. For an instance x the potential is defined as:

$$\phi(x, majC, minC, \gamma) = \sum_{y \in majC} \exp^{-(\frac{dist(x,y)}{\gamma})^2} - \sum_{y \in minC} \exp^{-(\frac{dist(x,y)}{\gamma})^2}$$

For any two instances x, y we write:

$$x \prec_1 y \iff \phi(x, majC, minC, \gamma) < \phi(y, majC, minC, \gamma)$$

The second criterion makes use of an average distance of an instance to 25% of nearest neighbors from the majority instances. For a fixed instance x and fixed majority dataset $majC$, let $\{x_1, \ldots, x_n\}$ stand for the 25% of nearest neighbors from $majC$. Define:

$$distMajority(x, majC) = \sum_{i=1}^{n} dist(x, x_i)/n$$

$$x \prec_2 y \iff distMajority(x, majC) < distMajority(y, majC)$$

Finally, x dominates y iff

$$x \prec y \iff x \prec_1 y \, \& \, x \prec_2 y$$

The genetic algorithm starts with random population of instances and fitness defined as level of domination. After each iteration members with lowest fitness are merged into the oversampled minority set. Details are in [13].

Agent CSMOUTE is balancing using the algorithm introduced in [14]. For oversampling SMOTE is used, and undersampling is performed in the following steps: for randomly selected majority instance x and its random k-neighbor z, both are deleted from majority set and the new interpolated instance $x+\lambda\cdot(z-x)$ is introduced; this procedure is repeated to reach the proper size of the majority subset.

3   Computational Experiment

3.1   Experiment Plan

Experiment involved multiple class imbalanced datasets from the Keel Dataset Repository as shown in Table 1.

Table 1. Datasets used in the reported experiment. source [3]

Dataset	#Attr	#Inst	#Clas	IR	Dataset	#Attr	#Inst	#Clas	IR
Balance	4	625	3	5.88	New Thyroid	5	215	3	4.84
Contraceptive	9	1473	3	1.89	Pageblocks	10	548	5	164
Dermatology	34	336	6	5.55	Penbased	15	1100	10	1.95
Ecoli	7	336	8	71.50	Shuttle	9	2175	5	853
Glass	9	214	6	8.44	Thyroid	21	720	3	36.94
Hayes-Roth	4	132	3	1.70	Wine	13	178	3	1.50
Lymphography	18	148	4	40.50	Yeast	8	1484	10	23.15

Each of the discussed balancing agents has been used in the experiment to produce synthetic minority examples followed by applying the binary GEP classifier under the OVO scheme to obtain the confusion matrix from which values of the performance measures have been calculated using formula from 3.2, that is geometric mean (Gmean), mean of recall values (M.Rec.) shown as (1), accuracy (Acc.), index kappa (Kappa) shown as (2) and area under the roc curve (MAUC) shown as (3). To obtain the average values we used 5-CV scheme repeated 6 times.

Gene Expression Programming (GEP) technique, introduced by [8] combines the idea of genetic algorithms and genetic programming and makes use of a population of genes. Each gene is a linear structure divided into two parts. The first part, the head, contains functions and terminals while the second part, the tail, contains only terminals. For this study terminals are of type $(oper; attr; const)$, where the value of $const$ is in the range of attribute $attr$ and $oper$ is a relational operator from $\{<, \leq, >, \geq, =, \neq\}$. Functions are from the set

$\{AND, OR, NOT, XOR, NOR\}$. For a fixed instance x from the dataset, the value $g(x)$ of a gene g is boolean and thus a gene can be treated as a binary classifier.

In the reported experiment, for all considered balancing agents and datasets, the GEP classifier has been used with the following parameter value settings: population size - 100; number of iterations - 200; probabilities of mutation, RIS transposition, IS transposition, 1-point and 2-point recombination - 0.5, 0.2, 0.2, 0.2, 0.2, respectively. For selection the roulette wheel method has been used. Parameter values for oversampling agent algorithms have been set as in original papers describing implementation for the binary classification task.

3.2   Performance Measures

To define classifier performance measures used in the experiments, assume that the dataset contains data from k classes. The elements of confusion matrix $C = \{c_{ij} : i, j \leq k\}$, where c_{ij} describes the number of instances that were predicted as class i but belonged to class j, allow to define for each class $m \leq k$:

- $TP_m = c_{mm}$ - the number of true positives (examples of class m which were classified correctly),
- $FP_m = \sum_{i=1,i\neq m}^{k} c_{mi}$ - the number of false positives (examples that were wrongly assigned to class m),
- $TN_m = \sum_{i=1,i\neq m}^{k} \sum_{j=1,j\neq m}^{k} c_{ij}$ - the number of true negative predictions regarding class m,
- $FN_m = \sum_{i=1,i\neq m}^{k} c_{im}$ - the number of false negatives for class m.

The Precision and Recall for class m are defined as:

$$Precision_m = \frac{TP_m}{TP_m + FP_m}, \ Recall_m = \frac{TP_m}{TP_m + FN_m}$$

and used for the measure $Gmean$ and average accuracy which is the arithmetic mean of recall values of all the classes $Mean - recall$:

$$Gmean = (\prod_{i=1}^{k} Recall_i)^{\frac{1}{k}}, \ M.Rec = \frac{1}{k} \sum_{i=1}^{k} Recall_i \qquad (1)$$

As noted in [22], for multiclass classification $Kappa$ measure is less sensitive to class distribution than $Accuracy$:

$$Kappa = \frac{n \sum_{i=1}^{k} TP_i - ABC}{n^2 - ABC}, \ Accuracy = \frac{1}{n} \sum_{i=1}^{k} TP_i \qquad (2)$$

where n is the size of the dataset and $ABC = \sum_{i=1}^{k}(TP_i + FP_i)(TP_i + FN_i)$. The average AUC (area under the ROC curve) is defined as:

$$MAUC = \frac{1}{k(k-1)} \sum_{i\neq j}^{k} AUC(i,j) \qquad (3)$$

where $AUC(i,j)$ is the area under the curve for the pair of classes i and j.

3.3   Experiment Results

In Table 2 computational experiment results averaged over 30 runs produced using 5 cross-validation scheme for 6 times, and further on, averaged over 14 considered datasets, are shown. In Figs. 1, 2, 3, 4 and 5 the above results are shown in the form of Box & Whiskers plots.

Table 2. Average values of performance measures obtained in the experiment.

Performance measure	DominM	CsmouteM	FWSmoteM	AdasynM	LamoM
Accuracy	**0.559**	0.362	0.548	0.522	0.355
Kappa Index	**0.393**	0.257	0.345	0.362	0.190
Balanced recall	**0.607**	0.520	0.555	0.588	0.431
MAUC	0.345	0.315	0.329	**0.352**	0.258
Gmean	**0.584**	0.419	0.483	0.548	0.410

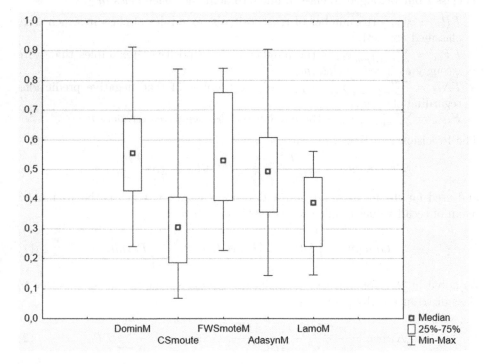

Fig. 1. Box & Whisker plot of average accuracies obtained in the reported experiment.

To evaluate the results shown in Table 2 and Figs. 1, 2, 3, 4 and 5 we have performed the Friedman ANOVA by ranks test for each of the considered performance measures. The null hypothesis for the procedure is that the different agents produced statistically similar results i.e. produced samples drawn from

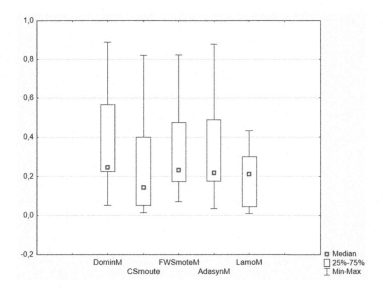

Fig. 2. Box& Whisker plot of average kappa indexes obtained in the reported experiment.

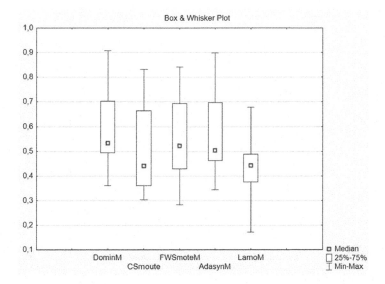

Fig. 3. Box & Whisker plot of average balanced recall values obtained in the reported experiment.

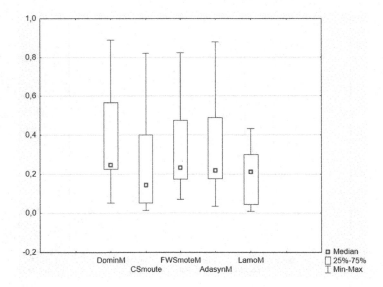

Fig. 4. Box & Whisker plot of average Kappa indexes obtained in the reported experiment.

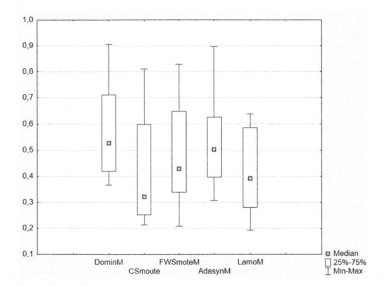

Fig. 5. Box & Whisker plot of average Gmean values obtained in the reported experiment.

the same population, or specifically, populations with identical medians. As it is shown in Table 3 summarizing the above test results, the null hypothesis for results measured using each of the considered performance measure, should be rejected at the significance level of 0.05. The Kendall concordance coefficient calculated for results produced using each of performance measures shows a fair agreement in the rankings of the variables among cases. The above findings tell us that there are statistically significant differences in the performance of the considered agents.

Table 3. Summary of the Friedman ANOVA test results

Measure	Chi Sqr	p-value	Conc. C
Accuracy	32.58700	0.0000	0.58193
Kappa	36.17204	0.0000	0.64593
M. Rec	34.42140	0.0000	0.54324
MAUC	31.63880	0.0000	0.52731
Gmean	28.97491	0.0001	0.51741

Table 4. Wilcoxon matched pair test results

Measure	Compared agents	T	Z	p-value
Accuracy	**DOMIN-M vs. FWSMOTE-M**	**38.00000**	**0.91026**	**0.36269**
Accuracy	DOMIN-M vs. ADASYN-M	5.00000	2.98188	0.00287
Accuracy	DOMIN-M vs. CSMOUTE-M	0.00000	3.29577	0.00098
Accuracy	DOMIN-M vs. LAMO-M	2.00000	3.17021	0.00152
Kappa	DOMIN-M vs. ADASYN-M	5.00000	2.98188	0.00287
Kappa	DOMIN-M vs. FWSMOTE-M	12.00000	2.54245	0.01101
Kappa	DOMIN-M vs. CSMOUTE-M	0.00000	3.29577	0.00098
Kappa	DOMIN-M vs. LAMO-M	0.00000	3.17980	0.00147
M.Rec	DOMIN-M vs. FWSMOTE-M	10.00000	2.83981	0.00451
M.Rec	DOMIN-M vs. ADASYN-M	6.00000	3.06699	0.00216
M.Rec	DOMIN-M vs. CSMOUTE-M	0.00000	3.40777	0.00066
M.Rec	DOMIN-M vs. LAMO-M	0.00000	3.29577	0.00098
MAUC	**DOMIN-M vs. ADASYN-M**	**40.00000**	**1.13592**	**0.25599**
MAUC	DOMIN-M vs. FWSMOTE-M	3.00000	3.10744	0.00189
MAUC	DOMIN-M vs. CSMOUTE-M	5.00000	2.98188	0.00287
MAUC	DOMIN-M vs. LAMO-M	0.00000	3.29577	0.00098
Gmean	**DOMIN-M vs. ADASYN-M**	**18.00000**	**1.92186**	**0.05462**
Gmean	DOMIN-M vs. FWSMOTE-M	8.00000	2.79355	0.00521
Gmean	DOMIN-M vs. CSMOUTE-M	0.00000	3.29577	0.00098
Gmean	DOMIN-M vs. LAMO-M	6.00000	2.91911	0.00351

To gain better knowledge of the performance of the considered balancing agents we have carried out a series of pairwise comparisons using the Wilcoxon matched pairs tests. The null hypothesis in such case states that results produced by two different agents are drawn from samples with the same distribution. Test results are summarized in Table 4.

Data from Table 4 allow drawing the following observations valid at the significance level of 0.05:

- For Accuracy measure, DOMIN-M and FWSMOTE perform statistically equally well.
- For the Accuracy measure, DOMIN-M outperforms statistically ADASYN-M, CSMOUTE-M, and LAMO-M.
- For the Kappa index measure, DOMIN-M outperforms statistically all the remaining agents.
- For the Mean-Recall measure, DOMIN-M outperforms statistically all the remaining agents.
- For the MAUC measure, DOMIN-M and ADASYN-M perform statistically equally well.
- For the MAUC measure, DOMIN-M outperforms statistically FWSMOTE-M, CSMOUTE-M, and LAMO-M.
- For the Gmean measure, DOMIN-M and ADASYN-M perform statistically equally well.
- For the Gmean measure, DOMIN-M outperforms statistically FWSMOTE-M, CSMOUTE-M, and LAMO-M.

4   Conclusions

The paper contributes by proposing a set of balancing agents able to deal with mining multiclass imbalanced datasets. The proposed agents are based on several state-of-the-art binary classifiers and use OVO (One-versus-One) strategy to deal with the multiclass problems. Agents can be used as stand-alone classifiers or serve as components (base classifiers) in ensembles of classifiers. The goal of the paper was to evaluate the performance of the considered agents when mining multiclass imbalanced datasets. The computational experiment has shown that among considered approaches the DOMIN-M assures the best performance no matter which performance measure was used. DOMIN-M agent performance is closely followed by that of the ADASYN-M and FWSMOTE-M agents. The above three agents should be considered as a promising option when looking for a tool for mining multiclass imbalanced datasets.

Future research will concentrate on the experimental study of the proposed agent's performance in different ensemble architectures including boosting, bagging, and stacking. Another promising direction of research would concentrate on undersampling part of balancing strategies. Using more advanced techniques like PCA or metaheuristic based techniques could be advantageous.

References

1. Abdi, L., Hashemi, S.: To combat multi-class imbalanced problems by means of over-sampling and boosting techniques. Soft. Comput. **19**(12), 3369–3385 (2015)
2. Agrawal, A., Viktor, H.L., Paquet, E.: SCUT: multi-class imbalanced data classification using SMOTE and cluster-based undersampling. In: 2015 7th International Joint Conference on Knowledge Discovery, Knowledge Engineering and Knowledge Management (IC3K), vol. 01, pp. 226–234 (2015)
3. Alcalá-Fdez, J., et al.: KEEL: a software tool to assess evolutionary algorithms for data mining problems. Soft Comput. **13**(3), 307–318 (2009)
4. Chawla, N.V., Bowyer, K.W., Hall, L.O., Kegelmeyer, W.P.: SMOTE: synthetic minority over-sampling technique. J. Artif. Intell. Res. **16**, 321–357 (2002)
5. Díaz-Vico, D., Figueiras-Vidal, A.R., Dorronsoro, J.R.: Deep MLPs for imbalanced classification. In: 2018 International Joint Conference on Neural Networks, IJCNN 2018, Rio de Janeiro, Brazil, 8–13 July 2018, pp. 1–7. IEEE (2018)
6. Fernandes, E.R.Q., de Carvalho, A.C.P.L.F., Yao, X.: Ensemble of classifiers based on multiobjective genetic sampling for imbalanced data. IEEE Trans. Knowl. Data Eng. **32**(6), 1104–1115 (2020)
7. Fernández, A., del Jesus, M.J., Herrera, F.: Hierarchical fuzzy rule based classification systems with genetic rule selection for imbalanced data-sets. Int. J. Approx. Reason. **50**(3), 561–577 (2009)
8. Ferreira, C.: Gene expression programming: a new adaptive algorithm for solving problems. Complex Syst. **13**(2) (2001)
9. Haixiang, G., Yijing, L., Yanan, L., Xiao, L., Jinling, L.: BPSO-Adaboost-KNN ensemble learning algorithm for multi-class imbalanced data classification. Eng. Appl. Artif. Intell. **49**, 176–193 (2016)
10. Hastie, T.J., Rosset, S., Zhu, J., Zou, H.: Multi-class adaboost. Stat. Its Interface **2**, 349–360 (2009)
11. He, H., Bai, Y., Garcia, E.A., Li, S.: ADASYN: adaptive synthetic sampling approach for imbalanced learning. In: IEEE International Joint Conference on Neural Networks (IEEE World Congress on Computational Intelligence), IJCNN 2008, pp. 1322–1328 (2008)
12. Hoens, T.R., Qian, Q., Chawla, N.V., Zhou, Z.-H.: Building decision trees for the multi-class imbalance problem. In: Tan, P.-N., Chawla, S., Ho, C.K., Bailey, J. (eds.) PAKDD 2012. LNCS (LNAI), vol. 7301, pp. 122–134. Springer, Heidelberg (2012). https://doi.org/10.1007/978-3-642-30217-6_11
13. Jedrzejowicz, J., Jedrzejowicz, P.: Bicriteria oversampling for imbalanced data classification. In: Knowledge-Based and Intelligent Information & Engineering Systems: Proceedings of the 26th International Conference KES-2022. Procedia Computer Science, vol. 207C, pp. 239–248. Elsevier (2022)
14. Koziarski, M.: CSMOUTE: combined synthetic oversampling and undersampling technique for imbalanced data classification. In: International Joint Conference on Neural Networks, IJCNN 2021, Shenzhen, China, 18–22 July 2021, pp. 1–8. IEEE (2021)
15. Koziarski, M.: Potential anchoring for imbalanced data classification. Pattern Recognit. **120**, 108114 (2021)
16. Koziarski, M., Krawczyk, B., Wozniak, M.: Radial-based oversampling for noisy imbalanced data classification. Neurocomputing **343**, 19–33 (2019)
17. Krawczyk, B.: Cost-sensitive one-vs-one ensemble for multi-class imbalanced data. In: 2016 International Joint Conference on Neural Networks, IJCNN 2016, Vancouver, BC, Canada, 24–29 July 2016, pp. 2447–2452. IEEE (2016)

18. Li, Q., Song, Y., Zhang, J., Sheng, V.S.: Multiclass imbalanced learning with one-versus-one decomposition and spectral clustering. Expert Syst. Appl. **147**, 113152 (2020)
19. Lin, M., Tang, K., Yao, X.: Dynamic sampling approach to training neural networks for multiclass imbalance classification. IEEE Trans. Neural Netw. Learn. Syst. **24**(4), 647–660 (2013)
20. Liu, X.Y., Wu, J., Zhou, Z.H.: Exploratory undersampling for class-imbalance learning. IEEE Trans. Syst. Man Cybern. Part B (Cybern.) **39**(2), 539–550 (2009)
21. Maldonado, S., Vairetti, C., Fernández, A., Herrera, F.: FW-SMOTE: a feature-weighted oversampling approach for imbalanced classification. Pattern Recognit. **124**, 108511 (2022)
22. Rodríguez, J.J., Díez-Pastor, J., Arnaiz-González, Á., Kuncheva, L.I.: Random balance ensembles for multiclass imbalance learning. Knowl. Based Syst. **193**, 105434 (2020)
23. Sáez, J.A., Krawczyk, B., Wozniak, M.: Analyzing the oversampling of different classes and types of examples in multi-class imbalanced datasets. Pattern Recognit. **57**, 164–178 (2016)
24. Wang, S., Yao, X.: Multiclass imbalance problems: analysis and potential solutions. IEEE Trans. Syst. Man Cybern. Part B (Cybern.) **42**(4), 1119–1130 (2012)
25. Wang, X., Xu, J., Zeng, T., Jing, L.: Local distribution-based adaptive minority oversampling for imbalanced data classification. Neurocomputing **422**, 200–213 (2021)
26. Yijing, L., Haixiang, G., Xiao, L., Yanan, L., Jinling, L.: Adapted ensemble classification algorithm based on multiple classifier system and feature selection for classifying multi-class imbalanced data. Knowl. Based Syst. **94**, 88–104 (2016)
27. Zhang, Z.L., Luo, X.G., García, S., Herrera, F.: Cost-sensitive back-propagation neural networks with binarization techniques in addressing multi-class problems and non-competent classifiers. Appl. Soft Comput. **56**, 357–367 (2017)

Decision Tree-Based Algorithms for Detection of Damage in AIS Data

Marta Szarmach[1]([✉]) [ID] and Ireneusz Czarnowski[2] [ID]

[1] Department of Marine Telecommunications, Gdynia Maritime University,
ul. Morska 81-87, 81-225 Gdynia, Poland
`m.szarmach@we.umg.edu.pl`
[2] Department of Information Systems, Gdynia Maritime University,
ul. Morska 81-87, 81-225 Gdynia, Poland
`i.czarnowski@umg.edu.pl`

Abstract. Automatic Identification System (AIS) is a system developed for maritime traffic monitoring and control. The system is based on an obligatory automatic exchange of information transmitted by the ships. The Satellite-AIS is a next generation of AIS system based on a satellite component that allows AIS to operate with a greater range. However, due to technical limitations, some AIS data collected by the satellite component are damaged, which means that AIS messages might contain errors or unspecified (missing) values. Thus, the problem of reconstruction of the damaged AIS data needs to be considered for improving performance of the Satellite-AIS in general. The problem is still open from research point of view. The aim of the paper is to compare selected decision tree based algorithms for detecting the damaged AIS messages. A general concept of detection of the damaged AIS data is presented. Then, the assumption and results of the computational experiment are reported, together with final conclusions.

Keywords: Damaged data detection · Multi-label classification · AIS data analysis · Decision Tree-based algorithms

1 Introduction

AIS (Automatic Identification System) has been developed to support the monitoring and control of maritime traffic,as well as collision avoidance. AIS messages can also be used for tracking vessels. An example of a system based on AIS messages is a vessel traffic services (VTS) system. Based on AIS data, VTS is a source of information of position of ships and other traffic data. The information provided by AIS can be classified as static (including vessel's identificafition number, MMSI) and dynamic (including ship's position, speed, course, and so on) [1].

The Satellite-AIS is a next generation of AIS system based on satellite component for extending range of AIS operationality. Due to technical limitations, resulting from lack of synchronization while receiving AIS packets from two or

Supported by Gdynia Maritime University.

© The Author(s), under exclusive license to Springer Nature Switzerland AG 2023
J. Mikyška et al. (Eds.): ICCS 2023, LNCS 14075, pp. 17–32, 2023.
https://doi.org/10.1007/978-3-031-36024-4_2

more so-called terrestrial cells at the same time, a problem of packet collision exists. Packet collision leads to the problem of the satellite component processing the AIS data that can be incomplete (missing) or incorrect, i.e. damaged. Of course, not all data is damaged, but the damage can be eliminated and not forwarded in an easy way.

When the damage of the data is detected in a form of missing data (i.e. a some part of AIS message being lost), the aim can be to predict proper values for missing fields and an elimination of data gaps. When the damage of the data manifests itself in the incorrectness of the data, the AIS data may be corrected. However, at first, this incorrectness must be detected. In literature, the problem of incorrect data detection is often discussed as a anomaly detection problem [2]. In this paper we called it as a data damage detection. One approach to this is to use statistical tools. Alternative approach is to use machine learning methods to the considered detection. Finally, at the last stage, the reconstruction of the damaged data is carried out.

This paper focuses on only one of these stages, i.e. the damage detection stage. A general concept for damaged AIS data detection (based on pre-clustering) is presented. After detecting of which AIS messages are damaged, the aim is to detect which parts of those messages are damaged and require correction. The main aim of the paper is to propose and evaluate the performance of decision tree-based algorithms, implemented to detect possible incorrections in individual parts of the AIS message. The role of these classifiers is to assign proper labels to each part of the AIS message, and thus indicate the damaged ones. Therefore, the problem of damage detections in AIS message fields is considered as a problem of multi-label classification[1].

The discussed approach is based on an ensemble of the decision tree models used for damage detection. Three different decision tree models have been compared, i.e. single Decision Tree, Random Forest and XGBoost [4,5]. The decision tree based algorithms were chosen for evaluation because of their relatively low computational complexity, nevertheless, they are also powerful and popular tool for classification and prediction, being competitive to others [6]. On the other hand, the problem under consideration is related to its physical and engineering solution, where the mentioned attributes matter. The paper also extends research results presented before in [7].

The remainder of this paper is organized as follows. In the next section, there is a literature review regarding the considered topic, then the problem formulation is included. In Sect. 4, a general concept for the proposed approach is presented, with details on the process of anomaly detection and damage detection in the separate parts of the AIS message. Computational experiment results on evaluating the performance of the proposed method, together with their discussion, are included in Sect. 5. Finally, Sect. 6 concludes this paper.

---

[1] The multi-label classification problem and a review of different approaches for solving this kind of machine learning problem have been discussed in [3].

2   Related Works

The topic of anomaly detection in AIS data is not entirely new in scientific literature. In most of works that can be found, existing frameworks consider "anomalies" as parts of ships' trajectories that do not fit to the expected trend of vessels' behaviour within a given area [8,9]. The ships' typical trajectories are often extracted by analysing AIS data recorded during a long observation period: one way is to cluster collected trajectory points with the use of available clustering algorithms, such as DBSCAN [9], the other requires finding a specific trajectory points in a given area (called waypoints), where ships typically turn, speed up, etc [8,10]—trajectories can be then expressed as egdes of graphs whose vertices are waypoints.

Other approaches for detecting anomalous trajectory points utilize neural networks [11,12] or statistical models, like Bayesian model based on Gaussian processes in [13].

However, to the best of our knowledge, there are few frameworks that would reconstruct the AIS messages based on a single ship trajectory (in contrast to those mentioned above, relying on identyfing trajectory trends) and detect damages in AIS messages no matter whether the ship is following the trend or not.

There are also works that focus on the reconstruction of incomplete AIS data. What can be noticed is that deep learning is also widely seen in this task. For example, in paper [14] a classic neural network is used, in [15]—a convolutional U-Net, in [16]—recurrent neural network. Nonetheless, as mentioned before, there is still a need for developing algorithms for AIS data reconstruction that are as accurate as deep learning approaches, but requires less computational power and relatively small amount of time to execute—it is important for the ships not to wait long for the reconstructed data, otherwise they may collide.

3   Problem Formulation

Let T_i be a ship trajectory, described by a set of vectors $T_i^{t_m}$. Each $T_i^{t_m}$ represents ith ship's trajectory point observed in time (observation step) t_m, that can be expressed as [7]:

$$T_i^{t_m} = [x_{i1}^{t_m}, x_{i2}^{t_m}, x_{i3}^{t_m}, \ldots, x_{iN}^{t_m}], \tag{1}$$

where:

- $m = 1, ..., M$ and M denotes the number of received AIS messages during a given observation time,
- N denotes the number of features of AIS message,
- $x_{in}^{t_m}$ represents nth feature of an AIS message from ith ship received in time t_m.

As it has been formulated in [1], typical data (features) of AIS message are as follows: message ID, repeat indicator, ship's ID (MMSI), navigational status, rate of turns, speed over ground, position accuracy, longitude, latitude, course over ground, true heading, time stamp, special manoeuvre indicator. They have either static or dynamic character.

Therefore, a trajectory of the ith ship can be defined as:

$$T_i = \{T_i^{t_1}, T_i^{t_2}, T_i^{t_3}, \ldots, T_i^{t_M}\}. \tag{2}$$

The data received by the AIS satellite may contain errors. After decoding the data, such a state of affairs will mean that the message could be damaged (totally or partially) or incomplete. Thus, the data received from ith ship (i.e. $T_i^{t_m}$ or elements of the vector, $x_{in}^{t_m}$) may be incorrect (or undefined), which can be expressed in the following way:

$$\exists_{t_m} T_i^{t_m}, \text{ that is missing/incorrect} \tag{3}$$

or

$$\exists_{n:n=1\ldots N} \; x_{in}^{t_m} \text{ , that is missing/incorrect.} \tag{4}$$

In general, the authors' interest is to reconstruct each T_i, when the ith ship's trajectory points are partially damaged, i.e. there are incorrect or missing values within the vector $T_i^{t_m}$. Therefore, the reconstruction of missing or incorrect AIS data can be defined as a correction of detected incorrect/missing (damaged) messages, to the point where finally the following condition is fulfilled:

$$\forall T_i(\nexists T_i^{t_m} \vee \nexists x_{in}^{t_m}), \text{ that are not missing/incorrect.} \tag{5}$$

The important step in the reconstruction of missing or incorrect AIS data is a detection of the damaged AIS data. The problem of detection of the damaged AIS data (which is the main aim of the paper) can be considered as a classification problem. In that case, machine learning tools can be helpful in this process.

Considering the above, we can therefore associate each of the ship's trajectory point with a label c, where c is an element of a finite set of decision classes $C = \{0, 1\}$, where its elements are called "false" or "true", respectively. A true value means that the message is damaged, otherwise, it means that the message is proper (correct, no damage). Whereas, $|C|$ is equal to 2, the problem of detection of the damaged data is an example of a binary classification problem.

Thus, the one considered problem is to find a model (classifier) that describes and distinguishes data class for AIS messages and predicts whether ship's trajectory point is damaged or not. From the formal point of view and a considered binary classification problem, classification is the assignment of a class $c \in C$ to each trajectory point $T_i^{t_m}$.

The problem of detection of the damaged AIS data can be also considered with respect to the multi-label classification problem. In such case, each of ship's trajectory point is associated with a vector of labels, where each item in the vector refers to one feature of the ship's trajectory point. When the label is set to true, it means that the corresponding feature is damaged, otherwise, it is considered proper.

Thus, let now $C = [c_1, \ldots, c_N]$ be a vector of labels, where each element $c_{n:n=1,\ldots,N} \in \{0, 1\}$. The elements of C are related with elements of vector $T_i^{t_m}$.

Then, in case of any element x_{in}^{tm} being damaged, a respective c_n is set to be 1, otherwise to 0. It also means that the vector of labels can have one or more than one true values, and in the extreme case all true values, which means that all features (fields) of the AIS message are damanged. Finally, the role of a classification model (classifier) is to assign a vector of labels C to each trajectory point T_i^{tm}.

4 Proposed Approach for Damage Detection

4.1 Anomaly Detection

To successfully reconstruct the damaged fragments of AIS data, it is necessary to first find the AIS messages and their parts that actually require correction—this is exactly the purpose of the anomaly detection stage.

Algorithm 1. Framework of AIS data reconstruction using machine learning

---

Require: D—dataset;
    K (optional, according to the chosen clustering algorithm)—number of clusters (equal to the number of individual vessels appearing in a dataset)
 1: **begin**
 2:   $D_c = D$
 3:   Map messages from D into K clusters
      (1 cluster = messages from 1 ship) using a selected clustering algorithm.
 4:   Let $D_1, \ldots, D_K$ denote the obtained clusters and let $D = D_1 \cup D_2 \cup \ldots \cup D_K$.
 5:   **for all** $i = 1, \ldots, K$ **do**
 6:      Search for anomalies (potentially damaged messages) in D_i.
 7:      Let $\hat{D}_i$ denote cluster with detected anomalies.
 8:      Predict the correct values of damaged fields for $\tilde{D}_i$.
 9:      Update D_c using $\tilde{D}_i$.
10:   **end for**
11:   **return** D_c—corrected dataset.
12: **end**

To accomplish this, during the very first step of the reconstruction, a clustering stage is provided. During the clustering stage, the analysed AIS data is divided into groups, such that (ideally) one group consists of messages from one and only one vessel. DBSCAN algorithm [17] was used in this task. Although AIS messages contain a field with the transmitting ships' identifier (which is called MMSI, Maritime Mobile Service Identity) we decided not to sort the data accoring to the MMSI because, as other fields in AIS messages, this value might contain errors and clustering algorithms should recognize the similarity between messages from the same vessel despite those errors or mark the message as clearly outlying. Only then, when the individual trajectories have been distinquished, they can be further processed to find datapoints that somehow do not fit to the rest (we will call them outliers, which in other words are potentially damaged messages) and reconstructed.

The desired algorithm, as mentioned before, should require low computational complexity and manage to detect damages in AIS messages no matter whether the ship is acting typically or not.

We focused to find anomalies in MMSI, navigational status, speed over ground, longitude, latitude, course over ground and special manouvre indicator fields (unfortunately, the data that we was working on contains mostly default values in rate of turns and true heading fields). Also, we decided not to reconstruct values in fields such as message type, repeat indicator, position accuracy, since they describe the message itself rather than vessel's trajectory.

The whole proposed reconstruction algorithm (including the damage detection stage) is presented as Algorithm 1. In next sections, the two-step process of anomaly detection part (including searching for both standalone and proper, multi-element clusters) is discussed.

4.2   Detection of Damage in Standalone Clusters

Background. To find AIS messages that are potentially outliers, it is advisable to search for messages that after the clustering stage became a part of standalone, 1-element clusters. If a clustering algorithm decided to place those messages that far from any other points, it might suggest that the fields inside such message contain anomalous values, i.e. are damaged.

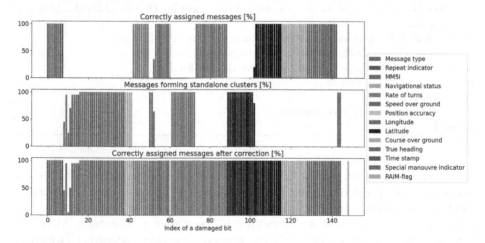

Fig. 1. The motivation behind the analysis of standalone clusters for AIS data damage detection

Indeed, an auxilary test that was conducted confirms this logic. For each bit, we randomly selected 20 messages, artificially corrupted the given bit (swapped its value from 0 to 1 or from 1 to 0) and repeat the clustering stage. It turned out that for some bits, its corrruption makes the message become a 1-element cluster j, as can be seen in Fig. 1—those fields are mainly ships' identifier MMSI, navigational status, special manouvre indicator and most significant bits from speed over ground, longitude and latitude. Therefore, when a standalone, 1-element cluster is found, we can assume that the message in it is an outlier that should be further examined.

However, it is not enough to find a corrupted message $\hat{T}_j^1$—we would also like to know which part (or parts) of it is actually damaged. In order to accomplish this, the origin of such message must be established—in other words, the message must be assigned to a right cluster together with other messages that were transmitted by the same ship. We decided to use a k nearest neighbour classifier [18] for this task. The index of a cluster assigned to a message by the clustering algorithm becomes a classifying label and then $k = 5$ messages closest to the outlying one are found—the cluster that most of messages found this way is assigned to becomes the new group for the outlying message.

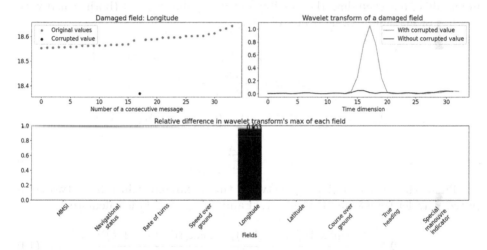

Fig. 2. The intuition behind the usage of waveform transform in AIS data damage detection

After the origin of an outlying message is established, its trajectory can be evaluated and the fields that contain wrong values can be detected. If we take a look at a waveform consisting of consecutive values from a given AIS message field, i.e. how the values from a given field change over time (Fig. 2), we can see that the (exemplary) outlying value changes much from the rest and the whole waveform starts to resemble a specific mathematic function called a wavelet, basically speaking, in the form of a sudden value change. A wavelet $\psi(t)$ is a function that meets the following criterium [19]:

$$\int_{-\infty}^{\infty} \frac{|\hat{\psi}(\omega)|^2}{\omega} \, d\omega < \infty \qquad (6)$$

where $\hat{\psi}(\omega)$ is a Fourier transform of a wavelet $\psi(t)$. Wavelet transform [19] is a transformation that measures how much a given signal $f(t)$ is similar to a wavelet $\psi(t)$. Wavelet transform can be described as [19]:

$$F_\psi(a,b) = \frac{1}{\sqrt{a}} \int_{-\infty}^{\infty} f(t) \cdot \psi \left(\frac{t-b}{a} \right) \, dt \tag{7}$$

where:

- $F_\psi(a,b)$ is a wavelet transform of a signal $f(t)$ computed for wavelet $\psi(t)$,
- t is time,
- a denotes a scale parameter,
- b is a time-shift parameter.

A special algorithm for detecting anomalies in AIS message fields, based on wavelet transform, is proposed. For each analysed field, we compute the wavelet transform (based on a Morlet wavelet) for two waveforms: $\Delta\hat{w}_{in}$ consisting of differences between the consecutive values in a given field (8) and Δw_{in}, very much alike, but excluding the possibly outlying value $\hat{x}_{jn}^{t_m}$ (9) (both scaled with the use of maximum value from $\Delta\hat{w}_{in}$):

$$\Delta\hat{w}_{in} = [x_{in}^{t_2} - x_{in}^{t_1}, \quad x_{in}^{t_3} - x_{in}^{t_2}, \quad \dots \quad x_{in}^{t_M} - x_{in}^{t_{M-1}}]$$

$$\Delta\hat{w}_{in} = \frac{\Delta\hat{w}_{in}}{\max(\Delta\hat{w}_{in})} \tag{8}$$

$$\Delta w_{in} = [x_{in}^{t_2} - x_{in}^{t_1}, \quad \dots \quad x_{in}^{t_M} - x_{in}^{t_{M-1}}]^{x_{in}^{t_m} \neq \hat{x}_{jn}^{t_m}}$$

$$\Delta w_{in} = \frac{\Delta w_{in}}{\max(\Delta\hat{w}_{jn})} \tag{9}$$

Then, the relative difference between the maximum values from two computed wavelet transforms $\hat{W}_\psi^{in}(a=1,b)$ and $W_\psi^{in}(a=1,b)$ is calculated:

$$\Delta\psi_{in} = \frac{|\max(\hat{W}_\psi^{in}(a=1,b)) - \max(W_\psi^{in}(a=1,b))|}{\max(\hat{W}_\psi^{in}(a=1,b))} \tag{10}$$

The higher the difference $\Delta\psi_{in}$, the more impact to the wavelet transform was given by introducing the potentially damaged value $\hat{x}_{jn}^{t_m}$, indicating higher chance for this value to require further correction.

Based on the same logic, the relative difference of standard deviation of values from the two given waveforms $\hat{w}_{in}$ and w_{in} can be calculated:

$$\hat{w}_{in} = [x_{in}^{t_1}, \quad x_{in}^{t_2}, \quad \dots \quad x_{in}^{t_M}]$$

$$w_{in} = [x_{in}^{t_1}, \quad x_{in}^{t_2}, \quad \dots \quad x_{in}^{t_M}]^{x_{in}^{t_m} \neq \hat{x}_{jn}^{t_m}}$$

$$\Delta\sigma_{in} = \frac{|\sigma_{\hat{w}_{in}} - \sigma_{w_{in}}|}{\sigma_{\hat{w}_{in}}} \tag{11}$$

Again, the higher the difference $\Delta\sigma_{in}$, the more change to the distribution of the values from a given field n was introduced by the damaged value $\hat{x}_{jn}^{t_m}$.

Algorithm 2. Detection of damage in AIS data

Require: D_1, D_2, ... D_K — subsets of the AIS dataset (groups obtained during the clustering phase);

N — number of fields in AIS messages.

1: **begin**
2:    **for** $j = 1, 2, ... K$ **do**
3:      **if** $|D_j| == 1$ **then**
4:        Add index of T_j^1 to *idx_list*.
5:        Run k-NN algorithm to find group i ($i \neq j$) — the one that the message T_j^1 should be assigned to.
6:        Add i to *i_list*.
7:        **for** $n = 1, 2, ... N$ **do**
8:          Compute the wavelet transform $\hat{W}_\psi^{in}$ of normalized sequence of differences between the consecutive values of field n in group i.
9:          Compute the wavelet transform W_ψ^{in} of the same sequence, excluding the value from the potentially damaged message T_i^1.
10:        Compute $\Delta\psi_{in}$, the relative difference between the maximum values of both wavelet transforms.
11:        Compute $\Delta\sigma_{in}$, the relative difference between the standard deviation of values of field n from group i and those values excluding the one from the potentially damaged message.
12:        Run classification algorithm to classify vector $[\Delta\psi_{in}, \Delta\sigma_{in}]$.
13:        **if** classification result $==$ 'field n is damaged' **then**
14:          Add n to *n_list*.
15:        **end if**
16:        **end for**
17:      **end if**
18:      **for** $n = 1, 2, ... N$ **do**
19:        **if** $n = 2, 3, 12$ **then**
20:          Run Isolation Forest to detect outlying values from nth field and jth ship.
21:        **else**
22:          **for** $m = 1, 2, ... M$ **do**
23:            Create vectors consisting of field values (longitude, latitude, speed over ground, course over ground, timestamp) from previous, given and next message.
24:            Run classification algorithm to classify those vectors.
25:          **end for**
26:        **end if**
27:        **if** classification/anomaly detection result $==$ 'field n is damaged' **then**
28:          Add index of T_j^{tm} to *idx_list*.
29:          Add j to *i_list*.
30:          Add n to *n_list*.
31:        **end if**
32:      **end for**
33:    **end for**
34:    **return** *idx_list* — list of indices of AIS messages that require correction,
     i_list — list of indices of groups that those messages should be assigned to,
     n_list — list of indices of AIS message fields that require correction.
35: **end**

Multi-label Field Classification. For each element of $T_i^{t_m}$, 2-element vectors $[\Delta\psi_{in}, \Delta\sigma_{in}]$ is computed using the equations presented in the previous section. Based on the values of the 2-element vectors it is possible to predict whether a given field $x_{jn}^{t_m}$ from $T_i^{t_m}$ is damaged or not. The higher the values $[\Delta\psi_{in}, \Delta\sigma_{in}]$, the more likely the field n can be considered anomalous (damaged). The prediction process is here associated with solving multi-label classification problem. For each analysed field, independent classifier is build.

4.3 Detection of Damage Inside Proper Clusters

After examining the AIS messages that formed standalone clusters, also messages from proper, multi-element clusters must be checked—the damage might be so slight that the clustering algorithm manages to assign the corrupted message along with other messages from the same vessel.

For dynamic data: speed over ground, longitude and latitude fields, we created vectors consisting of values from the current, previous and next AIS message from the following fields: longitude, latitude, speed over ground, course over ground, timestamp: $[x_{i7}^{t_m} - x_{i7}^{t_{m-1}}, x_{i7}^{t_{m+1}} - x_{i7}^{t_m}, x_{i8}^{t_m} - x_{i8}^{t_{m-1}}, x_{i8}^{t_{m+1}} - x_{i8}^{t_m}, x_{i5}^{t_{m-1}}, x_{i5}^{t_m}, x_{i5}^{t_{m+1}}, x_{i9}^{t_{m-1}}, x_{i9}^{t_m}, x_{i9}^{t_{m+1}}, t_{m+1} - t_m, t_m - t_{m-1}]$. For course over ground, $\arctan\left(\frac{x_{i8}^{t_{m+1}} - x_{i8}^{t_m}}{x_{i7}^{t_{m+1}} - x_{i8}^{t_m}}\right)$ and $\arctan\left(\frac{x_{i8}^{t_m} - x_{i8}^{t_{m-1}}}{x_{i7}^{t_m} - x_{i8}^{t_{m-1}}}\right)$ was put instead of values related to speed, longitude and latitude differences to help the algorithm deal with the trigonometric functions that field relies on. To some of those fields in a training data, we introduced an artificial damage, those were labelled as 1 ("damage"), to some we did not (labelled 0, "no damage"). Here, we again used the dedicated classifier (such as Decision Tree, Random Forest or XGBoost) to make a prediction of whether a given field is corrupted or not.

For static fields, values from a given nth field and given ith ship (x_{i2}, x_{i3}, x_{i12}) were analysed by Isolation Forest [20] to distinguish anomalies.

The entire proposed damage detection process in presented as Algorithm 2.

5   Computational Experiment

5.1   Overview

The Goal of the Experiment. The computational experiment focuses on validation of the performance of the tree-based algorithms, i.e. Decision Tree, Random Forest and XGBoost—as base classifiers in the ensemble approach, in proposed machine learning based framework for detecting damaged AIS messages and their fields.

Environment. The framework and all algorithms have been implemented in Python programming language using Sci-kit learn [20] and XGBoost [5] libraries in Visual Studio Code.

Quality Metrics. The following quality measures [20] have been used in order to evaluate the performance of the proposed approach for AIS data damage detection: recall (percentage of detected positive instances among all truly positive instances) and precision (percentage of true positive instances among those classified as positive).

Hyperparameters. During some initial tests, we found the optimal hyperparameters for our Random Forest [7] for analysing standalone clusters: *max_depth*, indicating how deep the tree can be, was set to 5, and *n_estimators*, indicating the number of trees in our Forest, was set to 15 (for XGBoost classifier, those parameter are 3 and 15, respectively).

Also the values of hyperparameters for analysing proper clusters have been discovered: for Random Forest *max_depth* = 15 and *n_estimators* = 15; for XGBoost *max_depth* = 7 and *n_estimators* = 15; as a tradeoff between good performance on validation set (in the sense of classification F1 score) and overfitting to the training set. Note: only for the speed over ground field the *max_depth* was set to 12 and 5, respectively.

For Desision Trees, the same *max_depths* were used as for Random Forests (as mentioned earlier).

Data. In this experiment, data from a real, operational AIS was used. Recorded AIS messages of types 1–3 (those that carry the information regarding the vessels' movement) were divided into 3 datasets (Fig. 3):

1. 805 messages from 22 vessels from the area of Gulf of Gdańsk,
2. 19 999 messages from 524 vessels from the area of Gibraltar,
3. 19 999 messages from 387 vessels from the area of Baltic Sea.

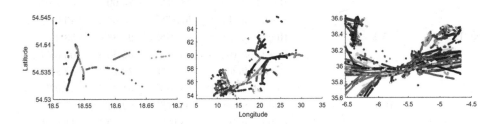

Fig. 3. Visualisation of vessels' trajectiories from each dataset used in this experiment

Each dataset was later divided into training (50%), validation (25%) and test (25%) sets. For each, a matrix X was build, serving as an input to the machine learning algorithms and consisting of the following fields of AIS messages type 1–3: longitude, latitude, MMSI, navigational status, speed over ground, course over ground, true heading (not used in anomaly detection phase), special manouvre indicator. In clustering stage, some features being the identifiers (MMSI, navigational status, special manouvre indicator) were additionally one-hot-encoded and the whole data was standarized.

5.2   Results

Performance of the Proposed Method. In the first part of the experiment, we checked how well our framework of detecting false values in AIS data works (in terms of the quality metrics described earlier).

Table 1. Performance of the proposed method in AIS data damage detection

Algorithm	Damaged messages	Metric	1. dataset	2. dataset	3. dataset
Decision Tree	5%	Recall (mess.)	90.77%	91.00%	93.51%
		Precision (mess.)	05.78%	06.02%	05.50%
		Recall (field)	58.46%	69.38%	60.60%
		Precision (field)	55.73%	49.82%	50.89%
	10%	Recall (mess.)	90.74%	91.20%	92.82%
		Precision (mess.)	11.84%	11.83%	10.87%
		Recall (field)	60.74%	69.24%	61.17%
		Precision (field)	50.89%	47.67%	45.23%
Random Forest	5%	Recall (mess.)	95.38%	91.68%	93.35%
		Precision (mess.)	06.01%	06.20%	05.73%
		Recall (field)	72.31%	73.74%	64.00%
		Precision (field)	58.46%	49.93%	49.00%
	10%	Recall (mess.)	93.70%	92.20%	92.88%
		Precision (mess.)	12.29%	12.06%	11.32%
		Recall (field)	71.85%	74.00%	64.99%
		Precision (field)	51.87%	48.04%	46.07%
XGBoost	5%	Recall (mess.)	92.31%	91.88%	93.43%
		Precision (mess.)	06.56%	06.50%	05.68%
		Recall (field)	71.54%	72.76%	63.80%
		Precision (field)	58.08%	51.71%	49.97%
	10%	Recall (mess.)	91.85%	92.24%	92.72%
		Precision (mess.)	13.67%	12.60%	11.21%
		Recall (field)	71.85%	73.30%	64.77%
		Precision (field)	53.48%	49.92%	46.30%

The performance was examined in two scenarios: when 5% or 10% of messages were artificially corrupted (2 randomly chosen bits of randomly chosen messages of given amount were swapped) for each of the three available datasets. The corrupted bits were chosen among the features of our interest. The test was executed 10 times for each percentage and the mean value of the following metrics were calculated: recall and precision in detecting messages, recall and precision in detecting fields. During the test, Decision Tree, Random Forest and XGBoost were examined. The results are presented in Table 1.

It can be noticed that a single Decision Tree underperforms in most cases, while both enseble methods give better and similar results. The recall (of both message and field detection) looks promising, however, we find precision values slightly unsatisfying. It seems like the proposed method marks more messages as "damaged" that it should (high false positive rate), but when it does, it predicts only a few fields to be corrupted in those messages. In fact, this is not that big issue since it is more important to detect all damaged messages (their appearance in the dataset might eventually result in two ships colliding) than bother with the false positive instances (which would be handled during the last, prediciton stage). What also can be noticed is that the increasing of number of damaged messages has only a little impact on the performance.

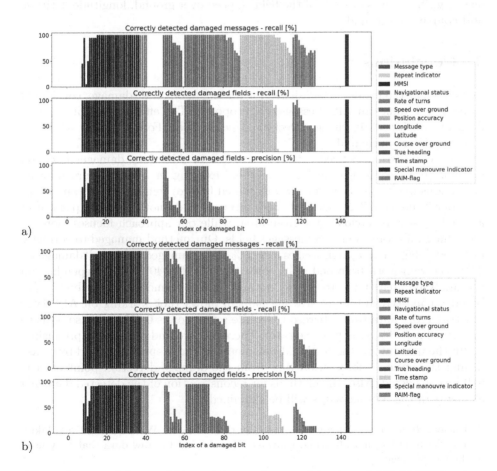

Fig. 4. Impact of damaged bit position on damage detection: a) RF, b) XGBoost

Impact of the Damaged Bit Position. In another part of the experiment, the impact of where exactly the damaged bit is on the performance of the proposed method of AIS data damage detection was examined.

While iterating over each of the AIS position report bit (excluding the bits related to fields that we decided not to investigate), the value of that exact bit was swapped in 20 randomly selected messages (only from the first dataset) and the performance (in the form of recall and precision) of detecting that the selected message and the field corresponding to the damaged bit were indeed corrupted was calculated.

The results are presented on Fig. 4. While using Random Forest, the average recall of detecting corrupted AIS messages was 92.26%, the average recall and precision of detecting fields were 78.98% and 61.28%, respectively. For XGBoost, the same results are 89.87%, 81.42% and 61.47%, respectively. It can be noticed that the precision is slightly lower than the recall (a part that still can be improved). Moreover, the proposed method struggled to detect damage placed in least significant bits of some of the fields (speed over ground, longitude, latitude and course over ground).

6   Conclusions

In this paper the framework designed for finding anomalies (damage, considered as incorrect values in AIS messages) is proposed. The effectiveness of usage of decision tree based algorithms (Decision Tree, Random Forest, XGBoost) in this framework was evaluated.

The comparison between the effectiveness of AIS data damage detection method proposed in this paper and others existing in the literature may be ambiguous due to the fact that (as mentioned before) some existing frameworks consider "anomalies" as parts of trajectories that do not fit to the pattern of a usual vessels' behaviour in a given area, while our approach focuses more on identifing false values in AIS messages fields. Our method managed to correctly mark 90%–95% damaged messages and 60%–74% damaged fields as '"damaged". TREAD framework proposed in [8] correctly identifies 40%–80% (depending on the analysed area) trajectories as "normal", while framework described in [2], based on masked autoregressive normalizing flows (not working on AIS data, but trajectory data from Microsoft GeoLife set) receives 0.055–0.6 false positive rate (for fixed 80% true positive rate). Therefore, we find our results promising.

In the near future, the further development of the proposed method by establishing the optimal observation time from the damage detection point of view will be considered. Also, the methods for reconstruction of the detected damaged elements of the AIS messages will be examined.

Acknowledgements. The authors specially thank Marcin Waraksa and Prof. Jakub Montewka from Gdynia Maritime University for sharing the raw data that they used in their experiment.

References

1. International Telecommunications Union, Recommendation ITU-R M.1371-5. https://www.itu.int/dms_pubrec/itu-r/rec/m/R-REC-M.1371-5-201402-I!!PDF-E.pdf. Accessed 10 Jan 2023
2. Dias, M.L.D., Mattos, C.L.C., da Silva, T.L.C., de Macedo J.A.F., Silva, W.C.P.: Anomaly detection in trajectory data with normalizing flows. In: 2020 International Joint Conference on Neural Networks (IJCNN), pp. 1–8. IEEE, Glasgow (2020)
3. Wang, R., Kwong, S., Wang, X., Yuheng, J.: Active k-label sets ensemble for multi-label classification. Pattern Recogn. **109**, 107583 (2021)
4. Ho, T.K.: The random subspace method for constructing decision forests. IEEE Trans. Pattern Anal. Mach. Intell. **20**(8), 832–844 (1998)
5. Chen, T., Guestrin, C.: XGBoost: a scalable tree boosting system. In: KDD-16 Proceedings, pp. 785–794. ACM, New York (2016)
6. Rossbach, P.: Neural Networks vs. Random Forests – Does it always have to be Deep Learning? https://blog.frankfurt-school.de/de/neural-networks-vs-random-forests-does-it-always-have-to-be-deep-learning/. Accessed 31 Jan 2023
7. Szarmach, M., Czarnowski, I.: Multi-label classification for AIS data anomaly detection using wavelet transform. IEEE Access **10**, 109119–109131 (2022)
8. Pallotta, G., Vespe, M., Bryan, K.: Vessel pattern knowledge discovery from AIS data: a framework for anomaly detection and route prediction. Entropy **15**, 2218–2245 (2013)
9. Lei, B., Mingchao, D.: A distance-based trajectory outlier detection method on maritime traffic data. In: 2018 4th International Conference on Control, Automation and Robotics, pp. 340–343. IEEE, Auckland (2018)
10. Kontopoulos, I., Varlamis, I., Tserpes, K.: Uncovering hidden concepts from AIS data: a network abstraction of maritime traffic for anomaly detection. In: Tserpes, K., Renso, C., Matwin, S. (eds.) MASTER 2019. LNCS (LNAI), vol. 11889, pp. 6–20. Springer, Cham (2020). https://doi.org/10.1007/978-3-030-38081-6_2
11. Singh, S.K., Heymann, F.: Machine learning-assisted anomaly detection in maritime navigation using AIS data. In: 2020 IEEE/ION Position. Location and Navigation Symposium (PLANS), pp. 832–838. IEEE, Portland (2020)
12. Zhongyu, X., Gao, S.: Analysis of vessel anomalous behavior based on bayesian recurrent neural network. In: 2020 IEEE 5th International Conference on Cloud Computing and Big Data Analytics, pp. 393–397. IEEE, Chengdu (2020)
13. Kowalska, K., Peel, L.: Maritime anomaly detection using Gaussian Process active learning. In: 2012 15th International Conference on Information Fusion, pp. 1164–1171. IEEE, Singapore (2012)
14. Zhang, Z., Ni, G., Xu, Y.: Trajectory prediction based on AIS and BP neural network. In: 2020 IEEE 9th Joint International Information Technology and Artificial Intelligence Conference, pp. 601–605. IEEE, Chongqing (2020)
15. Li, S., Liang, M., Wu, X., Liu, Z., Liu, R.W.: AIS-based vessel trajectory reconstruction with U-Net convolutional networks. In: 2020 IEEE 5th International Conference on Cloud Computing and Big Data Analytics, pp. 157–161. IEEE, Chengdu (2020)
16. Jin, J., Zhou, W., Jiang, B.: Maritime target trajectory prediction model based on the RNN network. In: Liang, Q., Wang, W., Mu, J., Liu, X., Na, Z., Chen, B. (eds.) Artificial Intelligence in China. LNEE, vol. 572, pp. 334–342. Springer, Singapore (2020). https://doi.org/10.1007/978-981-15-0187-6_39

17. Ester, M., Kriegel, H.-P., Sander, J., Xu, X.: A density-based algorithm for discovering clusters in large spatial databases with noise. In: KDD-1996 Proceedings, pp. 226–231. AAAI, Portland (1996)
18. Altman, N.S.: An introduction to kernel and nearest-neighbor nonparametric regression. Am. Stat. **46**(3), 175–185 (1992)
19. Debnath, L.: Wavelet transform and their applications. PINSA-A **64**(6), 685–713 (1998)
20. Pedregosa, F., et al.: Scikit-learn: machine learning in python. J. Mach. Learn. Res. **12**, 2825–2830 (2011)

The Role of Conformity in Opinion Dynamics Modelling with Multiple Social Circles

Stanisław Stępień[1] , Jarosław Jankowski[2] , Piotr Bródka[1] ,
and Radosław Michalski[1(✉)]

[1] Department of Artificial Intelligence, Wrocław University of Science and
Technology, 50-370 Wrocław, Poland
`229852@student.pwr.edu.pl`, `{piotr.brodka,radoslaw.michalski}@pwr.edu.pl`
[2] Department of Computer Science and Information Technology, West Pomeranian
University of Technology, 70-310 Szczecin, Poland
`jjankowski@wi.zut.edu.pl`

Abstract. Interaction with others influences our opinions and behaviours. Our activities within various social circles lead to different opinions expressed in various situations, groups, and ways of communication. Earlier studies on agent-based modelling of conformism within networks were based on a single-layer approach. Contrary to that, in this work, we propose a model incorporating conformism in which a person can share different continuous opinions on different layers depending on the social circle. Afterwards, we extend the model with more components that are known to influence opinions, e.g. authority or openness to new views. These two models are then compared to show that only sole conformism leads to opinion convergence.

Keywords: opinion spread · opinion formation · conformism · multilayer networks

1 Introduction

We tend to believe that we have a single opinion on a given topic. Yet, one can say that we also manifest different behaviour and opinions when interacting with others from each social circle we are a part of. This is because our expressed opinions are a function of our honest worldview, the type of group that we are exposing it to, and multiple other factors [21]. This leads to the existence of our multiple Is, where each is slightly different from the other and dependent on local social pressures. This type of behaviour is not considered a symptom of any social disorder but is a natural slight adjustment of our expression to the setting we are currently in. Moreover, this does not apply to all of us, as some non-conformists always present their inherent opinion. Most people, however, typically employ some level of conformism or adaptation of opinions.

© The Author(s), under exclusive license to Springer Nature Switzerland AG 2023
J. Mikyška et al. (Eds.): ICCS 2023, LNCS 14075, pp. 33–47, 2023.
https://doi.org/10.1007/978-3-031-36024-4_3

This work focuses on investigating which effects are observable when we model opinion dynamics in a setting in which we belong to multiple social circles but also expose conformism and other social phenomena, such as sociability, openness, or obeying authorities. In order to represent different social circles, we reached for the framework of multilayer networks that serve this purpose well. The main objective of this work is to demonstrate the outcomes of a simulation of a continuous opinion spread process in a setting in which we model multiple social circles as separate layers of a multilayer social network and also incorporate the aspect of conformism in such a way that individuals are capable of exposing different opinions at different layers. The agent-based simulations led to a better understanding of how continuous opinions fluctuate and whether this leads to differences in the opinions of individuals related to a variety of contexts they expose them. To make the model even more realistic, we also incorporate to it the influence of authority, openness to different views and sociability.

This work is organised as follows. In the next section, we present the related work, mostly underlining conformism, as it is the base of our model. Next, in Sect. 3, we present the conceptual framework, where we also refer to all other aspects the model incorporates (influence of authority, sociability and openness to new views). Section 4 is devoted to presenting two experimental scenarios: solely with conformism (Sect. 4.1) and with all the model components (Sect. 4.2). In Sect. 5, we draw conclusions and present future work directions.

2   Related Work

Various aspects of human activity take place among others and the need for popularity or acceptance can be one of the social targets. Social scientists explore how it leads to conformity due to the fact that not following accepted norms or standards can lead to penalization of individuals due to their altitudes [1]. From another perspective, conformism can be positioned within social learning strategies, and an approach to copying the majority is based on conformist bias [11]. Conformist transmission takes a vital role in effective social learning strategies [20]. The selection of social learning strategy is based on cognitive abilities, social status, and cultural background. Conformism can slow the diffusion of innovations occurring in the population with low frequency [8]. They can be lost in conditions with conformist bias. In terms of awareness and, for example, epidemics, a high fraction of conformists within the network can lower efforts for epidemics suppression because of the focus on defection and the impact of anti-vaccination altitudes [19]. From the perspective of cascade behaviours, sequences of individual decision and information cascades conformism can be supported by rewarding conformity institutions or at the community level [10].

To better understand the mechanisms of conformism, several attempts were made to model the impact of conformism on social influence and information spreading. For example, an algorithm for analysing and detecting influence and conformism within social networks was proposed in [14]. The authors focused on a context-aware approach, and the method is based on the extraction of topic-based sub-graphs, with each sub-graph assigned to specific topics. Each node can be characterised by different influence and conformism measures for various issues. Positive signs represent support, while harmful distrust and opposition. Influence and conformity indices are computed for each node. The empirical study was performed on Epinions and Twitter datasets. Influence within networks was calculated using conformism based on positive and negative relations. Another work emphasised the lack of conformity factors within influence maximisation algorithms [15]. Typical methods focus on influence parameters but are conformity-unaware. The authors proposed a greedy approach with integrated influence and conformity for effective seed selection. The critical element is the conformity-aware cascade model and its extension to the context-aware model with contextual information used. The method is based on partitioning of the network into sub-networks and computations within components. Influence and conformity indices are computed using the earlier presented method [14].

The need for integrating conformity parameters in spreading models was emphasised in the SIS model applied to cultural trait transmission [24]. The conformist influence was added as a conformity function resulting in an additional weight to transmission rates. Conformity bias was represented by a sigmoidal shape covering the tendency to follow the majority. Conformism was also studied for the SIR model, and a set of infected neighbours is added to overall probability and is integrated with transmission rate [18]. Conformity rate was also integrated within the social opinion formation model for vaccination decision-making with a probability of converting nodes to a social opinion of their neighbours implemented [26].

3   Conceptual Framework

3.1   Multilayer Networks

Conformity as an act of matching or changing one behaviour to adjust it to others' beliefs, opinions, attitudes, etc. [4] is hard to study using a simple one-layer network as they cannot reflect the full complexity of different behaviours and environments we are in. Every day we immerse ourselves in different social circles like family, friends, coworkers, colleagues from the basketball team, online friends and so on. Towards each of those cycles, we might present different 'masks' to be accepted, look better, feel comfortable, etc. For example, we might present our true opinion to our family and friends while we only pretend to accept the opinion of our coworkers because we are afraid that our true opinion won't be accepted in this social circle. On the other hand, people we trust, like family or friends, are more likely to influence us to truly change our beliefs than our colleagues from the basketball team.

All these situations are next to impossible to model using a one-layer network, thus in this study, we have decided to use a multilayer networks [2,3,6] which allows us to model different interactions and behaviour in various social circles. The multilayer network is defined as $M = (N, L, V, E)$ [12], where:

- N is a not empty set of actors $\{n_1, ..., n_n\}$,
- L is a not empty set of layers $\{l_1, ..., l_l\}$,
- V is a not empty set of nodes, $V \subseteq N \times L$,
- E is a set of edges $(v_1, v_2) : v_1, v_2 \in V$, and if $v_1 = (n_1, l_1)$ and $v_2 = (n_2, l_2) \in E$ then $l_1 = l_2$.

The example of a multilayer network is presented in Fig. 1. This network contains: six actors $\{n_1, n_2, n_3, n_4, n_5, n_6\}$, two layers $\{l_1, l_2\}$, ten nodes $\{v_1 = (n_1, l_1)$, $v_2 = (n_2, l_1), v_3 = (n_3, l_1), v_4 = (n_4, l_1), v_5 = (n_5, l_1), v_6 = (n_1, l_2), v_7 = (n_2, l_2)$, $v_8 = (n_3, l_2), v_9 = (n_4, l_2), v_{10} = (n_6, l_2)\}$, and ten edges $\{(v_1, v_2), (v_1, v_5), (v_2, v_5), (v_2, v_3), (v_2, v_4), (v_6, v_9), (v_6, v_{10}), (v_7, v_8), (v_7, v_{10}), (v_8, v_9)\}$. To align with the naming convention used in related works, we often use the term *agent* instead of *actor*. However, both terms' meaning is the same in our paper.

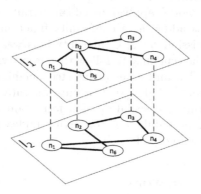

Fig. 1. An example of a multilayer network

3.2 Continuous Opinions

Most opinion dynamics models derived from physics related to spin models treat opinion assigned to nodes in a discrete space with a binary representation [7]. It is adequate when selecting only one of the options, e.g., *pro* or *contra*, *convinced* or *not convinced*. However, in reality, it is also possible to express attitudes toward specific situations to some extent. Growth of confidence can be modelled in a continuous process taking into account gradual changes with early models presented in [5,9]. They assume that opinion is adjusted together with each contact and the ability to hear thoughts from others. In [9] opinion of each agent evolves towards the average opinion of neighbouring agents. Other mechanics are based on random pairwise encounters leading to possible compromise [5]. Continuous models were extended towards bounded confidence with interaction limited to agents with close opinions [17]. External opinions are taken into account only if the difference from the own opinion of the agent is lower than a certain bound of confidence. Other extensions include the ability to model heterogeneous confidence thresholds, disagreement, the wisdom of the crowd, or integrate a hybrid approach based on the degree assigned to discrete opinions [23]. Mechanics of continuous opinions in the proposed model and multilayer setup will be independent for each layer and interactions within it. For each layer, the value of the agent's opinion before and after the interaction with each agent will be stored and will be used for conformity measure computation.

3.3 Authority

Authority factors in how one may consider other individuals' credentials to be greater or lesser than these of ourselves [4]. Such a view of how knowledgeable another person may be as compared to us affects how likely we are to heed this person's influence [13]. To factor this into the model, a single parameter has been introduced, describing how authoritative a given agent is going to be when interacting with his or her neighbourhood. Real-world social influence is often dependent on how the influencing party is perceived by the influenced one. This is often the case when a person is interacting with a specialist or a professional in a given area. In such situations, one is much more prone to subduing or even completely reversing his or her personal view of the given matter if the person or people he or she spoke to are considered as of great authority.

This model's value of the parameter describing authority will be a floating point number ranging from 0.0 to 1.0, where 0.0 will signify a person without any authority whatsoever. This would refer to a real-life situation where if met with such a person, one would offhandedly disregard this person's opinion on the matter as extremely uninformed and gullible. People tend to distrust or sometimes even ridicule information obtained from such individuals; we could revoke, for example, a situation where an inexperienced person (e.g. a child or a person whose opinion we consider to be outdated) tries to argue his or her opinion against a person who is much more acquainted with the matter of argument or holds a position of respect (e.g. a scientist or a courtroom judge). Of course, the degree to which one considers a person to be authoritative varies between individuals, and the one who could be considered to be of great knowledge and authority by Person A could just as well be considered to be of close-to-none authority by Person B. However, since the experiments will not focus on how one's authority changes, the model will not consider this factor, and each person will be given their own constant value of this parameter which will not later change throughout the course of the experiment.

During an interaction between two agents, the algorithm will have to take into account both the authorities of the two individuals in order to decide which one is considered to be more authoritative, as well as the average value of this parameter from one's neighbourhood, as to verify how a given individual compares with the rest of agents in one's neighbourhood [25]. During each interaction, a result of Eq. 1 will be one of the factors deciding how this interaction will affect the agent's opinion.

$$A_{bl} < \bar{A}_{Nl} \tag{1}$$

where A_{bl} denotes the value of agent B's opinion in the layer l of the network and $\bar{A}_{Nl}$ is an average opinion of all of the agent A's neighbours in this layer. This check will determine whether the person who interacts with the given agent may be considered reliable and trustworthy or should rather be dismissed as unqualified depending on how close their opinion is to the agent. This may either make them consolidate more with the status quo of their neighbours or even remove them further away from it.

3.4 Conformity

Depending on one's inclinations toward conforming behaviours influences of other people differ in effectiveness. This behaviour is predicated on multiple factors, such as one's tendency to be subservient, amicable or simply obedient. One may distinguish two main types of influences which cause people to behave in a more conforming way: normative influence for situations when an individual would like to fit in with his or her group and informational influence for situations when one believes the group is better informed than they are [22].

As it is described in Asch's experiment, one is likely to alter its opinion on a given matter if other people around express different views, even if the individual knows this information is not true. To factor these into the model, each agent will be assigned a random float value from 0.0 to 1.0, where 0.0 denotes a person being practically immune to the influence of their environment and not afraid of displaying their own honest opinion regardless of how others view it. This could relate to an example of a very confident person who follows their chosen ideal exclusively because of its merit, not being affected by whether it is frowned upon by others. On the other end of the spectrum, a person with a value of 1.0 will most of the time conceal their true view on the matter and adjust the opinions that they express to others to what appears to be the most appropriate in the given situation. The rest of the spectrum will describe different degrees to which one is willing to alter a genuine opinion to adjust to the group.

3.5 Openness to New Views

Having accounted for one's conformity when interacting with others and their opinions, the next step is to include the influence of one's cognitive bias. Even though there have not been too many models dealing with this problem, some noteworthy works have been created, like the one by Longzhao Liu's team [16], in which agents determine who belongs to their in-group or out-group based on the other neighbour's similarity in opinion and based on that a likeliness of interaction between the two is being determined.

As a product of this bias, most individuals tend to downplay or downright disregard any data or opinions which disagree with their prior held beliefs [27] while acting in a more accepting way towards new ideas which are closer to their personal opinion. This problem is especially interesting in the context of a network of multiple layers in which one agent may already be altering his or her opinions to fit in more with a specific group.

The degree to which one is antagonistic or accepting towards views that stray from his or her own is predicated on a number of sociological and psychological factors, such as the specificity of one's community, gullibility, dogmatism, etc. We have decided to parameterise these factors with the use of a single coefficient which can be more broadly addressed as the *openness* and assign each of the actors in the model with their own attribute, which value will be randomly set to a float from the scope of [0.0, 1.0] where the value of zero would denote a person who is extremely close-minded and displays a very strong cognitive bias which causes her or him to ignore opinions which are even remotely different than their own and a value of one would indicate a person who does not discriminate even these opinions which are on the other end of the spectrum from their own.

With this parameter, it will now be possible to include the influence of peer pressure in the model by comparing the difference between the opinions of two agents to the value of this parameter. Therefore an exemplary interaction between agent A and agent B will call out the formula as shown in Eq. 2.

$$|O_{bl} - O_{al}| < \bar{\theta}_{Nl} \tag{2}$$

where O_{al} and O_{bl} are consecutively agent's A and agent's B opinions in the layer l of the network and $\bar{\theta}_{Nl}$ is the value of the average openness of agent's A neighbours. This equation may be interpreted as agent A assessing how acceptable a new opinion presented to them will be with their peers from the social circle where the interaction took place. The result of this inequality is going to be one of the two factors (next to the comparison of authority) deciding how this interaction is going to affect the agent's opinion.

3.6 Sociability

How often one interacts with others in different spheres of his or her life is a result of a variety of factors such as, e.g. one's character or social situation and background. It is not within the scope of this system to model these complex interactions and instead, we have decided to garner them into a single coefficient describing the likeliness of the given agent to form connections with the others and later interact with them.

Similarly to the *Openness* and *Authority* parameters, *Sociability* will also be assigned to be a floating point number from within the scope of [0.0, 1.0], where 0.0 will signify an individual who is extremely solitary and avoids all the interactions with their neighbours and 1.0 an agent who interacts with almost all of their neighbours in each step of the simulation.

4 Simulations

4.1 Conformism in Multiple Social Circles

Introduction. This experiment is designed to illustrate the operation of the multilayer social network. It employs a relatively simple formula that does not use the other parameters and instead utilises only agents' conformity measures to decide how social influence affects one's opinions.

This experiment consists of one thousand agents whose conformity and opinions have been initialised to be random floating point values from within their discussed scopes: [0.0, 1.0] for Conformity and [−1.0, 1.0] for each of the presented opinions in the layers. In this simulation the only factor deciding how one's opinion is influenced by the interaction with other actors is the agent's conformity parameter as described in the Formula 3 ensures that conformity is a factor deciding to what degree a person will adopt others' opinion when exposed to it:

$$V(O_{al}, O_{bl}) = (1 - C_{al}) \times O_{al} + C_{al} \times O_{bl} \tag{3}$$

where O_{aln} denotes the new value of agent's a opinion at the layer l after the interaction with agent b and similarly, O_{al} denotes value of agent's opinion before this interaction and O_{bl} stands for agent's b opinion in the given layer. C_a denotes the value of agent's a conformity measure.

This formula entails that an agent whose conformity would be of a maximal value (equal to 1) would always fully adopt the opinion of a person he or she last spoke to. Similarly, an agent with the minimal value of this attribute would not alter his or her opinion no matter how many interactions took place.

After each interaction, one's personal opinion is updated as well. The algorithm just assigns it to be an average of all the opinions displayed in the other layers of the simulation.

Overview of the Experiment. This experiment consists of a thousand steps of simulation with a thousand agents. Firstly, a multilayer network has been constructed. This one, similarly to others consists of five layers: 'personal worldview', 'workplace', 'personal life', 'social media 1' and 'social media 2'. Secondly, all of the experiment agents were initialised; this included assigning each individual with randomised values of the parameters, transferring their population into the network in the form of nodes and creating connections between the agents' nodes in each of the dimensions of the network. To obtain a relatively realistic number of particular agent's neighbours for each of the layers, the random parameter responsible for creating a connection between two individuals has been set to 0.05% chance. A set of agent's neighbours, once set at the beginning of the experiment, is not going to be altered throughout its duration. After these initial conditions have been created, redetermined number of simulation steps is going to take place. In each step, for each of the layers, each of the agents may interact with his or her neighbours. Sets of one's neighbours may vary between the consecutive layers. Each of these interactions has a 5% chance of taking place. Therefore in each of the steps, each agent may hold between 0 and $l \times n$ interactions (l being a number of layers and n denoting one's number of neighbours in a given layer of the network).

Results of the Experiment. The experiment illustrates how this version of the model operates and how the agents' opinions are being affected by the simplified

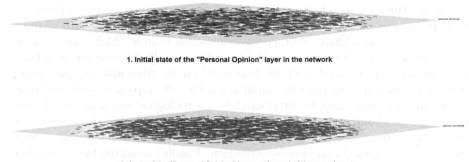

1. Initial state of the "Personal Opinion" layer in the network

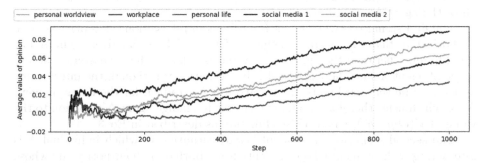

2. State of the "Personal Opinion" layer at the end of the experiment

Fig. 2. Comparison of the state of the 'Personal Opinion' layer before (top of the figure) and after (bottom of the figure) the course of the experiment.

social influence algorithm. As it is shown in Fig. 2, the initial distribution of opinions has been quite random. However, throughout the run of the simulation, agents, as a group, have become more aligned in their personal opinions even in the remaining layers (which are not shown in this figure). This proves the fact that, as expected, agents' opinions seem to coincide more than before the experiment began.

Fig. 3. A graph illustrating changes in average values of opinions for each of the multiplex network layers.

The initial random distribution of the opinions was randomised - causing the average value of opinion (which can be any float value between −1.0 and 1.0) to oscillate close to the value of zero. However, as can be observed in Fig. 3, these values began to change quite drastically in the initial phases of the experiment. Before the 200th step of the simulation, all except one of the social media layers seem to converge at zero.

In the later phases of the experiment, opinions appear to very slowly and steadily be rising, it is easy to predict that this trend will continue until all of the opinion layers would finally obtain one, common value after which point, no more modifications to one's opinion would be possible, as all of his or her neighbours from the given layer would be of the same mind on the given issue and no amount of interactions could further alter opinion's value. Discrepancies between layers, visible in this graph may be attributed to a topological separateness of some agents group, who in random process at the beginning might have been somewhat isolated from the wider population and because of that skew the overall value of the layer's average opinion in either direction. The line denoting the value of the personal opinion remains in the middle of the graph lines, as its value is always assigned as a simple average of the agent's opinions from other layers.

4.2   Conformism and Other Factors

Introduction. This run of the experiment utilises all of the parameters described before. Therefore formula responsible for calculating how one's opinion will be affected (marked as $O_{al_{new}}$) by the interaction between two agents is more complex. When agents A and B interact, depending on how their parameters rank against each other, one of the four results may take place. In the Eq. 4 each of the cases is reliant on the values of A_b(authority of the agent B), $\bar{A}_{Nl}$ (average authority of all of the agent A's neighbours), O_{al} (agent A's opinion in the layer l), O_{bl} (agent B's opinion in the layer l) and $\bar{\theta}_{Nl}$ (average openness of all of the agent A's neighbours).

The first scenario is when agent B with whom interaction takes place is more authoritative than the average member of the neighbourhood of agent A but the difference between the opinions of A and B is larger than the average acceptance of agent A's neighbourhood. In this scenario even though the interlocutor appears to be very credulous, their opinion is too radical for the liking of agent A's environment, therefore model will choose to change agent A's opinion to the value of $V(\bar{O}_{Nl}, O_{bl})$, where V is a function as described in Eq. 3.

The second scenario is a straightforward situation in which individual A is interacting with an agent whose credentials are both of high authority and whose opinions are not too different from the other opinions in the neighbourhood. This will cause the agent to simply adopt this opinion as their own in this layer.

The third case denotes a situation in which another individual is both untrustworthy when it comes to their authority, as well as radically different in opinion from the rest of one's neighbourhood. This denotes a situation in which agent A would find it desirable to stray even further away from agent B's opinion and therefore, with the use of the V function, would move toward the value of either 1.0 or −1.0, depending on which of these will move them further away from agent B's opinion. This kind of interaction may potentially cause some agents to paradoxically stray further away from the average value of their neighbourhood's opinion in what could be called a radicalisation process.

The fourth scenario will describe a situation in which agent B's opinion is not too different from the neighbourhood's average, however, agent B is not too authoritative him or herself in which case agent A will simply consolidate their opinion with the rest of the neighbourhood by using the V function.

$$O_{al_{new}} = \begin{cases} V(\bar{O}_{Nl}, O_{bl}) & \text{,if } A_b > \bar{A}_{Nl} \text{ and } |O_{bl} - O_{al}| > \bar{\theta}_{Nl} \\ O_{bl} & \text{,if } A_b > \bar{A}_{Nl} \text{ and } |O_{bl} - O_{al}| < \bar{\theta}_{Nl} \\ V(O_{al}, 1 \times sgn(O_{al} - O_{bl})) & \text{,if } A_b < \bar{A}_{Nl} \text{ and } |O_{bl} - O_{al}| > \bar{\theta}_{Nl} \\ V(O_{al}, \bar{O}_{Nl}) & \text{,if } A_b < \bar{A}_{Nl} \text{ and } |O_{bl} - O_{al}| < \bar{\theta}_{Nl} \end{cases}$$

$$(4)$$

Overview of the Experiment. The experiment has taken place over $n = 1000$ steps and included a population of 1000 agents. Similarly to the previous one, all of the values have been initialised to random values from within the scope of each consecutive coefficient at the beginning of the simulation.

Results of the Experiment. Throughout the run, most of the more radical opinions held by the agents have been somewhat moderated which is perhaps best visible in the comparison between network 1 - a network at the beginning of the experiment and network 2 - network after **1000** steps of the simulation in Fig. 4. Bright green and red colour marks these opinions which are closer to either **1.0** or **−1.0** and therefore are more radical, whereas darker shades describe opinions which are closer to the value of **0.0**. As one may easily notice, the second image is mostly devoid of these brighter spots. Nevertheless, the distribution of red and green colours shows only subtle changes (mostly in the 'Personal opinion' layer) between these two images. Moreover, in the image on the right side, nodes marked with the magenta colour indicate agents whose opinions have shifted by at least 1.0 downwards throughout the duration of the experiment and with the cyan colour those, which, similarly, have shifted significantly upwards.

As the result of the effect new coefficients bring into the model, as it is visible in Fig. 5.1, the average measure of opinions tended to fluctuate quite intensely when compared with the first experiment in which their values appeared to slowly rise, with minimal oscillation (as illustrated in Fig. 3). There is no trend towards convergence within any of the network layers. Since every interaction between any two neighbours occurs randomly (the likeliness of it taking place is predicated upon one's sociability parameter), more visible fluctuations were due to relatively rare occurrences in which a comparatively remote or unsociable agent or a group of agents happened to interact with the rest of the group and therefore skewing their average opinion. Worth noting is the fact that almost all of such fluctuations came to be as a result of more than one of such unlikely interactions (as may be deduced from the multi-step shape of their slopes). These were however quite rare as compared with the others and none lead to a new, long-standing trend in the shape of the graph.

The stability of opinions in this approach, as compared to a simpler model from Experiment 1, may also be observed in Fig. 5.2. Here, the standard deviation of actors' opinions has reliably remained under the value of **0.5** for the

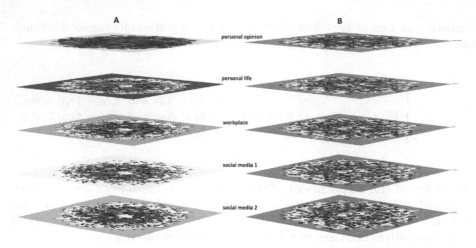

Fig. 4. Graphical representation of the multiplex network at the beginning of the simulation (network A) and after the experiment (network B).

entirety of the experiment. This lack of volatility is also reminiscent of the first experiment's results where the model's values did not take long to converge and settle on a value close to zero to synchronously drift towards one later on. What is different here is how large the value of convergence remains as compared to the previous simulation. This is an effect of how new mechanics take into the account parameters such as authority and openness to new views. These cause the system to form some sub-groups of agents who refuse to have their opinion changed by the current set of their neighbours in any further capacity and unless a new neighbour with perhaps a bigger authority or views a little bit closer to their own is introduced into their neighbourhood.

To elaborate on the idea of altering this balance of opinions, a set of simulations was run which included injections of new agents into the system while the experiment was still in progress. What at first has been a small population of **25%** of the original size, was then injected at step **500** with another **25%** of this number and then, at step **750**, again with the remaining **50%**. This could be interpreted as a real-life occurrence where for example a large new population of individuals is introduced to the local community as a result of, e.g. a merger between two companies which caused people to work with a larger number of co-workers. Here however this injection of new individuals took place in each of the layers simultaneously and, similarly as before, all the links between individuals were predicated on one's sociability. In Fig. 5.3 it is well visible how the average value of opinion has been affected after each of the consecutive injections of new agents into the simulation. They both caused the value of the opinions to become less spread out. Initial averages were oscillating quite violently at first but after the introduction of the remainder of the population, they lost some of that volatility and returned to the state of greater balance, with values more akin to these in Fig. 5.1.

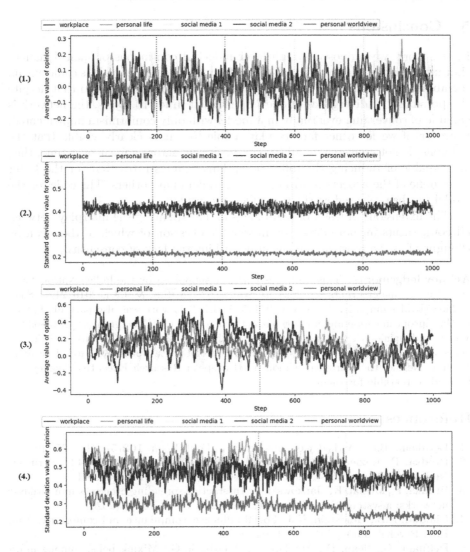

Fig. 5. From the top: **1)** A graph illustrating changes in the average value of opinions in the consecutive layers throughout the run of the second experiment. **2)** Graph showing values of the standard deviation of opinions throughout the second experiment. **3)** Graph of the average value of opinion throughout the run of the simulation with the injections of new agents during the run of the experiment. **4)** Graph of the measure of the standard deviation of agents' opinions during the run of the experiment with the injections of new agents.

Fig. 5.4 illustrates this trend even more clearly, as values of the standard deviation of opinions in each layer gradually converge towards lower value as new agents are being added into the population. With every new set injected into the simulation, the distribution of opinions agglomerates at a lower magnitude.

46   S. Stępień et al.

5 Conclusions

In this work, we proposed a multilayer model of continuous opinion dynamics that incorporates multiple factors known from the literature, such as conformism, sociability or authority.. By conducting the simulations, we showed that despite the possibility of divergence of one's opinions at different layers, the network is capable of converging eventually in a model with only conformism implemented. However, if we let other factors play a role too, it is clearly visible that the opinions do not converge on all layers, but if new actors appear over time, they are capable of reducing the differences. This demonstrates that for the latter case, none of the aspects is strong enough to dominate others. The code for the model has ben released as the Code Ocean capsule[1].

The extensions of this work will focus on investigating the interplay between all components in more detail to answer the question of which of these can be thought of as the most important in disturbing the lack of consensus.

Acknowledgements. This work was partially supported by the Polish National Science Centre, under grants no. 2021/41/B/HS6/02798 (S.S. & J.J.), 2016/21/D/ST6/02408 (R.M.) and 2022/45/B/ST6/04145 (P.B.). This work was also partially funded by the European Union under the Horizon Europe grant OMINO (grant no. 101086321). Views and opinions expressed are however those of the author(s) only and do not necessarily reflect those of the European Union or the European Research Executive Agency. Neither the European Union nor European Research Executive Agency can be held responsible for them.

References

1. Bernheim, B.D.: A theory of conformity. J. Polit. Econ. **102**(5), 841–877 (1994)
2. Bródka, P., Kazienko, P.: Multilayer Social Networks, pp. 1408–1422. Springer, New York (2018). https://doi.org/10.1007/978-1-4614-7163-9_239-1
3. Bródka, P., Musial, K., Jankowski, J.: Interacting spreading processes in multilayer networks: a systematic review. IEEE Access **8**, 10316–10341 (2020)
4. Cialdini, R.B., Goldstein, N.J.: Social influence: compliance and conformity. Ann. Rev. Psychol. **55**(1), 591–621 (2004)
5. Deffuant, G., Neau, D., Amblard, F., Weisbuch, G.: Mixing beliefs among interacting agents. Adv. Complex Syst. **3**(01n04), 87–98 (2000)
6. Dickison, M.E., Magnani, M., Rossi, L.: Multilayer Social Networks. Cambridge University Press, Cambridge (2016)
7. Galam, S.: Minority opinion spreading in random geometry. Eur. Phys. J. B-Condens. Matter Complex Syst. **25**, 403–406 (2002)
8. Grove, M.: Evolving conformity: conditions favouring conformist social learning over random copying. Cogn. Syst. Res. **54**, 232–245 (2019)
9. Hegselmann, R., et al.: Opinion dynamics and bounded confidence models, analysis, and simulation. J. Artif. Societies Social Simul. **5**(3) (2002)
10. Hung, A.A., Plott, C.R.: Information cascades: replication and an extension to majority rule and conformity-rewarding institutions. Am. Econ. Rev. **91**(5), 1508–1520 (2001)

[1] The code of the model: https://doi.org/10.24433/CO.1006113.v1.

11. Kendal, R.L., Boogert, N.J., Rendell, L., Laland, K.N., Webster, M., Jones, P.L.: Social learning strategies: bridge-building between fields. Trends Cogn. Sci. **22**(7), 651–665 (2018)
12. Kivelä, M., Arenas, A., Barthelemy, M., Gleeson, J.P., Moreno, Y., Porter, M.A.: Multilayer networks. J. Complex Netw. **2**(3), 203–271 (2014)
13. Lewis, G.C.: An essay on the influence of authority in matters of opinion. JW Parker (1849)
14. Li, H., Bhowmick, S.S., Sun, A.: Casino: towards conformity-aware social influence analysis in online social networks. In: Proceedings of the 20th ACM International Conference on Information and Knowledge Management, pp. 1007–1012 (2011)
15. Li, H., Bhowmick, S.S., Sun, A., Cui, J.: Conformity-aware influence maximization in online social networks. VLDB J. **24**(1), 117–141 (2015)
16. Liu, L., Wang, X., Chen, X., Tang, S., Zheng, Z.: Modeling confirmation bias and peer pressure in opinion dynamics. Front. Phys. **9**, 649852 (2021)
17. Lorenz, J.: Continuous opinion dynamics under bounded confidence: a survey. Int. J. Mod. Phys. C **18**(12), 1819–1838 (2007)
18. Lu, Y., Geng, Y., Gan, W., Shi, L.: Impacts of conformist on vaccination campaign in complex networks. Physica A: Stat. Mech. Appl. **526**, 121124 (2019)
19. Miyoshi, S., Jusup, M., Holme, P.: Flexible imitation suppresses epidemics through better vaccination. J. Comput. Soc. Sci. **4**(2), 709–720 (2021). https://doi.org/10.1007/s42001-021-00105-z
20. Muthukrishna, M., Morgan, T.J., Henrich, J.: The when and who of social learning and conformist transmission. Evol. Human Behav. **37**(1), 10–20 (2016)
21. Noelle-Neumann, E.: The spiral of silence a theory of public opinion. J. Commun. **24**(2), 43–51 (1974)
22. Asch, S.E.: Studies in the principles of judgements and attitudes: II. determination of judgements by group and by ego standards. J. Social Psychol. **12**(2), 433–465 (1940)
23. Sîrbu, A., Loreto, V., Servedio, V.D.P., Tria, F.: Opinion dynamics: models, extensions and external effects. In: Loreto, V., et al. (eds.) Participatory Sensing, Opinions and Collective Awareness. UCS, pp. 363–401. Springer, Cham (2017). https://doi.org/10.1007/978-3-319-25658-0_17
24. Walters, C.E., Kendal, J.R.: An sis model for cultural trait transmission with conformity bias. Theor. Popul. Biol. **90**, 56–63 (2013)
25. Walton, D.: Appeal to Expert Opinion: Arguments from Authority. Penn State Press, University Park (2010)
26. Xia, S., Liu, J.: A computational approach to characterizing the impact of social influence on individuals' vaccination decision making. PloS One **8**(4), e60373 (2013)
27. Zollo, F., et al.: Debunking in a world of tribes. PLOS ONE **12**(7), 1–27 (2017)

Impact of Input Data Preparation on Multi-criteria Decision Analysis Results

Jarosław Wątróbski[1,2]($\boxtimes$) (iD), Aleksandra Bączkiewicz[1]($\boxtimes$) (iD), and Robert Król[1] (iD)

[1] Institute of Management, University of Szczecin, ul. Cukrowa 8, 71-004 Szczecin, Poland
{jaroslaw.watrobski,aleksandra.baczkiewicz}@usz.edu.pl
[2] National Institute of Telecommunications, ul. Szachowa 1, 04-894 Warsaw, Poland

Abstract. Multi-criteria decision analysis (MCDA) methods support stakeholders in solving decision-making problems in an environment that simultaneously considers multiple criteria whose objectives are often conflicting. These methods allow the application of numerical weights representing the relevance of criteria and, based on the provided decision matrices with performances of alternatives, calculate their scores based on which rankings are created. MCDA methods differ in their algorithms and can calculate the scores of alternatives given constructed reference solutions or focus on finding compromise solutions. An essential initial step in many MCDA methods is the normalization procedure of the input decision matrix, which can be performed using various techniques. The possibility of using different normalization techniques implies getting different results. Also, the imprecision of the data provided by decision-makers can affect the results of MCDA procedures. This paper investigates the effect of normalizations other than the default on the variability of the results of three MCDA methods: Additive Ratio Assessment (ARAS), Combined Compromise Solution (CoCoSo), and Technique for Order of Preference by Similarity to Ideal Solution (TOPSIS). The research demonstrated that the normalization type's impact is noticeable and differs depending on the explored MCDA method. The results of the investigation highlight the importance of benchmarking different methods and techniques in order to select the method that gives solutions most robust to the application of different computing methods supporting MCDA procedures and input data imprecision.

Keywords: Multi-criteria decision analysis · MCDA · Normalization · Input data preparation

1 Introduction

Multi-criteria decision analysis (MCDA) methods support decision-making processes for problems that require simultaneous consideration of multiple conflicting criteria. Currently, many MCDA methods are available that differ in their

© The Author(s), under exclusive license to Springer Nature Switzerland AG 2023
J. Mikyška et al. (Eds.): ICCS 2023, LNCS 14075, pp. 48–55, 2023.
https://doi.org/10.1007/978-3-031-36024-4_4

algorithms, which scale the input data differently and determine the best solution. Differences in MCDA methods obviously produce different results, which can cause confusion among stakeholders [6]. The importance of this problem is evidenced by the fact that published studies of the impact of normalization techniques on the results of MCDA methods can be found in the literature. For example, Jafaryeganeh et al. studied the impact of four normalization methods (linear, vector, minimum-maximum, and logarithmic) for the WSM, TOPSIS, and ELECTRE methods for a case study of ship internal layout design selection [7]. Vafaei et al. studied the effect of Max, Min-Max, Sum, and Vector on the MCDA Simple Additive Weighting results for the supplier selection case study [12,13]. As can be noted, there have been several attempts to formulate methods for evaluating the most appropriate normalization techniques for decision-making problems, which indicate obtaining different MCDA results depending on the chosen normalization techniques [15]. However, they are characterized by a lack of consistency and a robust evaluation framework that takes into account aspects such as the repeatability of the test, the universality of the problem domain, and the impact of the structure of decision problems in the form of different dimensions of the matrix representing the problem.

The initial stage of most MCDA methods is a normalization of a decision matrix containing performance values of considered alternatives regarding criteria assessment [1]. Normalization of the decision matrix plays a significant role in MCDA methods [16]. With this procedure, it is possible to use data provided in different units and for criteria with different objectives without requiring additional action regarding preprocessing data by the decision maker [3]. There are many normalization techniques, and among the most popular used in MCDA procedures are minimum-maximum, sum, maximum, vector, and linear normalizations [4]. The original algorithms of many MCDA methods have normalization methods assigned to them by their authors. For example, vector normalization is recommended for Technique for Order of Preference by Similarity to Ideal Solution (TOPSIS), minimum-maximum normalization is advised for Multi-Attributive Border Approximation area Comparison (MABAC) [8] and Combined Compromise Solution (CoCoSo) [17], linear normalization is one of the stages of Weighted Aggregated Sum Product Assessment (WASPAS) [11] and COmbinative Distance-based ASsessment (CODAS) [14], sum normalization is recommended for Additive Ratio Assessment (ARAS) [5]. Normalization techniques must also be suited for a given decision-making problem. Not every type of normalization can be applied to data containing negative or zero values due to the nature of the mathematical operations required [1]. In such cases, the original normalization technique for a given method must be replaced by another. For instance, minimum-maximum normalization is adequate for data including negative and zero values. However, such a procedure can lead to variability in the results. Variability of values in the decision matrix may occur not only due to different normalization methods but also when decision-makers provide imprecise data, which can also affect the results.

In this paper, the authors present a benchmarking procedure as a numerical experiment that makes it possible to evaluate selected MCDA methods concerning their robustness to changes in values in the decision matrix caused by applying different normalization methods. In addition, the numerical experiment makes it possible to identify the normalization methods that, for a given MCDA method, cause the greatest and least variability in rankings compared to the original normalization. The procedure presented can be useful when it is necessary to choose a different normalization technique than the original one due to the data. Besides, the procedure can facilitate the identification of the MCDA method that is most resistant to data variability in the case of awareness of input data imprecision.

2   Methodology

Three MCDA methods were selected for this research, including TOPSIS [9], ARAS [5], and CoCoSo [17]. All three methods use a different normalization of the decision matrix in the initial step. In evaluating alternatives, the ARAS method determines the utility of each considered option relative to the ideal solution, which effectively supports the prioritization of alternatives. In the original algorithm of this method, the decision matrix is normalized using sum normalization [5]. The TOPSIS method is also based on reference solutions, except that it evaluates alternatives concerning their distance from the ideal and anti-ideal solution using the Euclidean distance metric. In the original version of the TOPSIS algorithm, vector normalization is used to normalize the decision matrix [9]. The CoCoSo method also uses normalization of the decision matrix, but it evaluates alternatives differently from the TOPSIS and ARAS methods. CoCoSo considers a combination of compromise approaches. Its algorithm includes an integrated simple additive weighting and exponentially weighted product model and can provide comprehensive compromise solutions [17].

Since the main focus of this paper is decision matrix normalization techniques, the basics and mathematical formulas of the particular normalization methods investigated in this paper are presented below. The fundamentals and formulas of the MCDA methods investigated are provided in Supplementary material in an open-source repository made available by authors on GitHub [2]. The Supplementary material also explains normalization methods applied in this research, namely minimum-maximum, maximum, sum, linear, and vector, together with mathematical formulas describing them. The research was conducted using a Python 3 script implemented in the Visual Studio Code environment based on the pseudocode provided in Listing 1 using MCDA methods and supporting techniques from the author's pyrepo-mcda Python 3 library [16]. GitHub also provides software in Python 3 implemented for procedures performed in this paper.

Numerical Experiment. Listing 1 demonstrates a research algorithm in the form of pseudocode employed for the numerical experiment performed in this paper. The experiment was conducted for each considered MCDA method:

ARAS, CoCoSo, and TOPSIS in an iterative procedure involving 1000 iterations. The convergence of the rankings obtained using the original normalization
for the given method with the rankings received using the four alternative normalization techniques was examined.

Algorithm 1. Research algorithm for benchmarking normalization methods.

1: *iterations* ← 1000
2: *list_of_matrix_sizes* ← *matrix_sizes*
3: **for** i = 1 to iterations **do**
4:     **for** s in list_of_matrix_sizes **do**
5:         *matrix* ← *generate_random_matrix*(*s, s*)
6:         *types* ← *determined_criteria_types*
7:         *weights* ← *generate_weights*(*matrix*)
8:         *rank_ref* ← *mcda_method*(*matrix, weights, types, default_normalization*)
9:         *normalizations* ← *list_of_normalizations*
10:         *result* ← *empty_list*()
11:         **for** normalization in normalizations **do**
12:             *rank* ← *mcda_method*(*matrix, weights, types, normalization*)
13:             *result.append*(*correlation*(*rank_ref, rank*))
14:         **end for**
15:         *save_result*(*result*)
16:     **end for**
17: **end for**

The convergence of rankings was determined using two rank correlation coefficients: the Weighted Spearman rank correlation coefficient r_w [9] and the Spearman rank correlation coefficient r_s [10]. The procedure was repeated for different
dimensions of decision matrices $\{5 \times 5, 8 \times 8, 11 \times 11, 14 \times 14, 17 \times 17, 20 \times 20\}$ filled
with random values in the range from 1 to 100. The criteria weights were deter-
mined by the CRITIC (Criteria Importance Through Inter-criteria Correlation)
method [11]. Types of all criteria were set as profit.

3   Results

This section presents the results of numerical experiments investigating the effect
of using alternative normalization techniques on the outcomes of three MCDA
methods: ARAS, CoCoSo, and TOPSIS. The correlation results of the compared
rankings are shown in the graphs in Fig. 1 for the ARAS method, Fig. 2 for the
CoCoSo method, and Fig. 3 for the TOPSIS method.

Figure 1 shows the results of comparisons of ARAS rankings obtained using
sum normalization as in its original algorithm with the results obtained with linear, maximum (Max), minimum-maximum (Minmax), and vector normalizations.
It can be observed that for the matrix dimensions examined, the results obtained
using vector normalization converge most closely with sum normalization. In contrast, the lowest values of convergence were obtained for minimum-maximum

normalization. The values of both correlation coefficients are high and, in most cases, range from 0.8 to 1. It implies that the ARAS method shows high resilience to changes in the value of the decision matrix caused, for example, by the normalization technique chosen for data preprocessing. The correlation values are also high when the dimensions of the decision matrices are increased, which means that an increase in the complexity of the decision problem does not reduce the convergence of the obtained results.

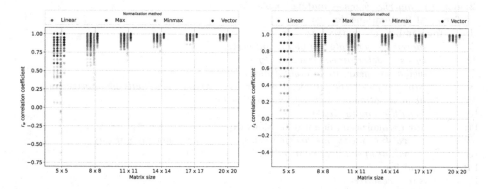

Fig. 1. Comparison of sum normalization results with alternative normalizations for ARAS.

Figure 2 shows the results of comparisons of CoCoSo rankings obtained using minimum-maximum normalization as recommended in the original algorithm, with the results obtained using linear, maximum, sum, and vector normalizations.

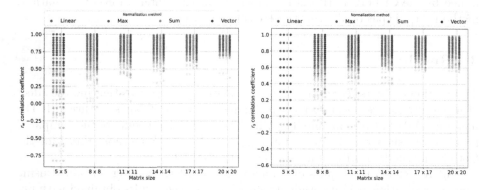

Fig. 2. Comparison of minimum-maximum normalization results with alternative normalizations for CoCoSo.

In this case, the correlation values for the compared rankings are lower than for the experiment conducted for the ARAS method. It indicates that the CoCoSo method is more sensitive to changes in the input data caused by other normalization methods, resulting in noticeable changes in the results. For CoCoSo, the level of divergence of the compared rankings is comparable to the alternative normalizations included in the experiment.

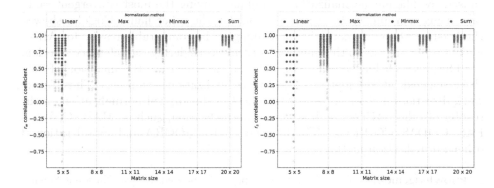

Fig. 3. Comparison of vector normalization results with alternative normalizations for TOPSIS.

Figure 3 displays the results of the experiment conducted with the TOPSIS method. In this case, rankings obtained using the vector normalization suggested in the original TOPSIS algorithm were compared with those obtained using alternative normalizations. The received correlation values for the compared rankings are similar to the experiment for the ARAS method, but the values of the r_w and r_s coefficients are slightly lower. As for the ARAS method, the highest correlation values were registered for rankings generated using sum and vector normalizations. On the other hand, the lowest correlations occurred for rankings obtained using minimum-maximum and vector normalizations.

4   Conclusions

The multitude of MCDA methods and supporting techniques cause decision makers to often wonder which method will be most suitable for solving considered decision problems. These doubts are justified because applying different computing techniques affects the results obtained differently. This influence is evident from the initial stages of MCDA methods, including providing performance data by decision-makers and data normalization. Therefore, an important role in MCDA procedures is played by benchmarking, demonstrating the impact of different methods on the variability of results. The research presented in this paper showed that the use of different normalization techniques causes variability in the MCDA results obtained, which differs depending on the normalization technique and MCDA method. The main implication of the presented research is a universal framework that enables experiments to test the impact of different normalization techniques applied to different MCDA methods and different complexity of data structure using two objective correlation metrics. Such a framework can be applied to analogous studies with different parameters depending on the objectives of the researchers. An experiment considering ARAS, CoCoSo, and TOPSIS showed that the ARAS method was the most resilient to changes caused by different normalization techniques, giving the most convergent rankings using sum and vector normalization. On the other hand, CoCoSo showed the least resistance to normalization change. The comparable results for ARAS and TOPSIS are due to the similar algorithms considering the reference solution.

This research has some limitations, among which are the inclusion of only three selected MCDA methods and five normalization techniques. Thus, future work directions include exploring the effect of the applied normalization on the results of other MCDA methods using normalization, such as MABAC, WAS-PAS, CODAS, COPRAS, MOORA, and MULTIMOORA and consideration of other normalization methods.

Acknowledgements. This research was partially funded by the National Science Centre 2022/45/B/HS4/02960 (J.W., A.B.).

References

1. Bielinskas, V., Burinskienė, M., Podviezko, A.: Choice of abandoned territories conversion scenario according to MCDA methods. J. Civil Eng. Manag. **24**(1), 79–92 (2018). https://doi.org/10.3846/jcem.2018.303
2. Bączkiewicz, A.: Software for the benchmarking influence of different normalization techniques on MCDA results (2023). https://github.com/energyinpython/MCDA-normalizations-benchmark
3. Bouraima, M.B., Qiu, Y., Stević, Ž, Marinković, D., Deveci, M.: Integrated intelligent decision support model for ranking regional transport infrastructure programmes based on performance assessment. Expert Syst. Appl. **222**, 119852 (2023). https://doi.org/10.1016/j.eswa.2023.119852
4. Cinelli, M., Kadziński, M., Gonzalez, M., Słowiński, R.: How to support the application of multiple criteria decision analysis? let us start with a comprehensive taxonomy. Omega **96**, 102261 (2020). https://doi.org/10.1016/j.omega.2020.102261
5. Goswami, S.S., Behera, D.K.: Solving material handling equipment selection problems in an industry with the help of entropy integrated COPRAS and ARAS MCDM techniques. Process Integr. Optim. Sustainabil. **5**(4), 947–973 (2021). https://doi.org/10.1007/s41660-021-00192-5
6. Güneri, B., Deveci, M.: Evaluation of supplier selection in the defense industry using q-rung orthopair fuzzy set based EDAS approach. Expert Syst. Appl. **222**, 119846 (2023). https://doi.org/10.1016/j.eswa.2023.119846
7. Jafaryeganeh, H., Ventura, M., Guedes Soares, C.: Effect of normalization techniques in multi-criteria decision making methods for the design of ship internal layout from a Pareto optimal set. Struct. Multidisc. Optim. **62**(4), 1849–1863 (2020). https://doi.org/10.1007/s00158-020-02581-9

8. Pamučar, D., Ćirović, G.: The selection of transport and handling resources in logistics centers using Multi-Attributive Border Approximation area Comparison (MABAC). Expert Syst. Appl. **42**(6), 3016–3028 (2015). https://doi.org/10.1016/j.eswa.2014.11.057

9. Rana, H., Umer, M., Hassan, U., Asgher, U., Silva-Aravena, F., Ehsan, N.: Application of fuzzy TOPSIS for prioritization of patients on elective surgeries waiting list-a novel multi-criteria decision-making approach. Decis. Mak. Appl. Manag. Eng. **6**(1), 603–630 (2023). https://doi.org/10.31181/dmame060127022023r

10. Sajjad, M., Sałabun, W., Faizi, S., Ismail, M., Wątróbski, J.: Statistical and analytical approach of multi-criteria group decision-making based on the correlation coefficient under intuitionistic 2-tuple fuzzy linguistic environment. Expert Syst. Appl. **193**, 116341 (2022). https://doi.org/10.1016/j.eswa.2021.116341

11. Tuş, A., Aytaç Adalı, E.: The new combination with CRITIC and WASPAS methods for the time and attendance software selection problem. OPSEARCH **56**(2), 528–538 (2019). https://doi.org/10.1007/s12597-019-00371-6

12. Vafaei, N., Ribeiro, R.A., Camarinha-Matos, L.M.: Selection of normalization technique for weighted average multi-criteria decision making. In: Camarinha-Matos, L.M., Adu-Kankam, K.O., Julashokri, M. (eds.) DoCEIS 2018. IAICT, vol. 521, pp. 43–52. Springer, Cham (2018). https://doi.org/10.1007/978-3-319-78574-5_4

13. Vafaei, N., Ribeiro, R.A., Camarinha-Matos, L.M.: Assessing normalization techniques for simple additive weighting method. Procedia Comput. Sci. **199**, 1229–1236 (2022). https://doi.org/10.1016/j.procs.2022.01.156

14. Wątróbski, J., Bączkiewicz, A., Król, R., Sałabun, W.: Green electricity generation assessment using the CODAS-COMET method. Ecol. Indic. **143**, 109391 (2022). https://doi.org/10.1016/j.ecolind.2022.109391

15. Wątróbski, J., Bączkiewicz, A., Sałabun, W.: New multi-criteria method for evaluation of sustainable RES management. Appl. Energy **324**, 119695 (2022). https://doi.org/10.1016/j.apenergy.2022.119695

16. Wątróbski, J., Bączkiewicz, A., Sałabun, W.: pyrepo-mcda - reference objects based MCDA software package. SoftwareX **19**, 101107 (2022). https://doi.org/10.1016/j.softx.2022.101107

17. Yazdani, M., Zarate, P., Kazimieras Zavadskas, E., Turskis, Z.: A combined compromise solution (CoCoSo) method for multi-criteria decision-making problems. Manag. Decis. **57**(9), 2501–2519 (2019). https://doi.org/10.1108/MD-05-2017-0458

Computational Diplomacy and Policy

Automating the Analysis of Institutional Design in International Agreements

Anna Wróblewska[1]([✉]) [iD], Bartosz Pieliński[2]([✉]) [iD], Karolina Seweryn[1,3] [iD],
Sylwia Sysko-Romańczuk[4] [iD], Karol Saputa[1] [iD], Aleksandra Wichrowska[1],
and Hanna Schreiber[2] [iD]

[1] Faculty of Mathematics and Information Science, Warsaw University
of Technology, Koszykowa 75, 00-662 Warsaw, Poland
`anna.wroblewska1@pw.edu.pl`
[2] Faculty of Political Science and International Studies, University of Warsaw,
Krakowskie Przedmieście 26/28, 00-927 Warsaw, Poland
`b.pielinski@uw.edu.pl`
[3] NASK - National Research Institute, Kolska 12, 01-045 Warsaw, Poland
`karolina.seweryn@nask.pl`
[4] Faculty of Management, Warsaw University of Technology, Ludwika Narbutta 85,
02-524 Warsaw, Poland

Abstract. This paper explores the automatic knowledge extraction of formal institutional design - norms, rules, and actors - from international agreements. The focus was to analyze the relationship between the visibility and centrality of actors in the formal institutional design in regulating critical aspects of cultural heritage relations. The developed tool utilizes techniques such as collecting legal documents, annotating them with Institutional Grammar, and using graph analysis to explore the formal institutional design. The system was tested against the 2003 UNESCO Convention for the Safeguarding of the Intangible Cultural Heritage.

Keywords: Natural Language Processing · Information Extraction · Institutional Design · Institutional Grammar · Graphs · UNESCO · Intangible Cultural Heritage

1 Introduction

Solving contemporary interlinked and complex problems requires international cooperation. This cooperation is expressed through international agreements establishing norms, rules, and actors to facilitate collaboration between nations. Though digitization processes have facilitated access to official documents, the sheer volume of international agreements makes it challenging to keep up with the number of changes and understand their implications. The formal institutional design expressed in legal texts is not easily comprehensible. This is where computational diplomacy with NLP models comes into play. They hold

© The Author(s), under exclusive license to Springer Nature Switzerland AG 2023
J. Mikyška et al. (Eds.): ICCS 2023, LNCS 14075, pp. 59–73, 2023.
https://doi.org/10.1007/978-3-031-36024-4_5

the potential to analyze vast amounts of text effectively, but the challenge lies in annotating legal language [11]. Our paper and accompanying prototype tool aims to facilitate the analysis of institutional design in international agreements.

The authors annotated the legal text using Institutional Grammar (IG), a widely used method for extracting information from written documents [7]. IG offers a comprehensive analysis of the rules that regulate a particular policy, revealing the relationships between rules, reducing complexity, and identifying key actors involved in the policy. Despite being relatively new, IG has achieved high standardization [7] through close collaboration between political/international relations and computer scientists. As such, the prototype aligns well with the intersection of these two areas of scientific research.

The growing popularity of IG in public policy studies has led to the development of several tools helping to upscale the usage of IG. There is the IG Parser created by Christopher Frantz[1] , the INA Editor created by Amineh Ghorbani (https://ina-editor.tpm.tudelft.nl/), and attempts have been made to develop software for the automatization of IG annotations [22]. The first tool is handy in manually annotating legal texts, and the second is beneficial in semi-manually transforming IG data into a network representation of rules. The third one shows the potential of IG parsing automatization. However, each tool focuses only on one aspect of the multilevel process of transforming a large corpus of legal text into a graph. The tool presented in this paper is more than a single element – it chains together three separate modules that automate the whole process of working with IG. Although IG allows relatively easy annotation of legal text, the data produced at the end is raw. It provides no information on institutional design expressed in legal documents. This is where graphs come to the rescue.

Graphs are the most popular way of expressing the composition of rules identified by IG. Several published studies have already proved the utility of such an approach [9,15,19]. It makes it possible to look at elements of rules (e.g., addressees of rules or targets of action regulated by rules) as nodes connected by an edge representing the rule. If an individual rule can be modeled as an edge, then the whole regulation could be seen as a graph or a network. This conceptualization allows the introduction of measures used in network analysis to study specific policy design, making the entire field of institutional design more open to quantitative methods and analysis.

We use a graph representation of a formal institutional design to describe the position of actors in a given institutional arrangement. To achieve this aim, we use conventional network metrics associated with centrality. Using the UNESCO Convention for the Safeguarding of the Intangible Cultural Heritage (the 2003 UNESCO Convention) as our use case, we focus on the actors' location in the network. We wish to address a research issue discussed in the International Public Administration (IPA) literature: the role of people working in international organizations versus states and other institutional institutions actors [6]. Researchers are interested in the degree that international civil servants are

---

[1] https://s.ntnu.no/ig-parser-visual.

autonomous in their decisions and to what extent they can shape the agenda of international organizations. Particularly interesting is the position of treaty secretariats – bodies created to manage administrative issues related to a specific treaty [2,10]. We wanted to learn about the Secretariat's position concerning other actors described by the Convention, both in data coming from IG and the Convention's network representation.

Parsing legal texts by way of IG requires our tool to consist of three technical components: (1) a scraper, (2) a tagger, and (3) a graph modeler. Thanks to this, our prototype can scrape many legal documents from the Internet. Then, it can use the IG method, NLP techniques, and IG-labelled legal regulations to generate hypergraph representation and analyze inter-relations between crucial actors in formal institutional design. The aim is to use IG as an intermediate layer that makes it possible to transform a legal text into a graph which is then analyzed to learn about the characteristics of the institutional design expressed in the text. A rule, or to use IG terminology, an institutional statement, is represented here as an edge connecting actors and objects mentioned in the statement. This approach makes it possible to model a legal regulation text as a graph and analyze institutional design through indicators used in network analysis. The approach has already been implemented in research papers [8,15,18,19].

Our purpose while designing this prototype was to build an effective system based on existing technologies. All of the system's sub-modules are derived from well-known techniques (scraping, OCR, text parsing, graph building); what is innovative is how they are adjusted and coupled.[2] To our knowledge, no other system takes input regulations distributed throughout the Internet and produces an output graph representation of institutional design. The biggest challenge in the prototype design was adjusting NLP methods for the specificity of IG. Thus, in the following use case, we concentrate on this challenge. We show how to build a graph of inter-relations between important objects and actors based on information extracted using IG from a single but crucial document.

For this paper, our prototype was tested on the 2003 UNESCO Convention. This document was chosen for three reasons. Firstly, the Convention text is relatively short and well-written, which allows for the first test to be one where the IG tagger works on a document that does not breed unnecessary confusion on the syntactic and semantic levels. Secondly, we have involved an expert on the Convention in our project, allowing us to check our analysis against her expertise.[3] Thirdly, the Convention is considered to be one of the most successful international documents, almost universally ratified (181 States Parties as of April 2023). The document is analyzed to compare the position of institutional actors mentioned in the document.

The following sections present our current achievements leading to our research goal. Section 2 highlights the current achievements in working with the IG method, developed independently, and sketches out our current research ideas. We then describe our prototype and its main modules (Sect. 3). Our use

---

[2] The source code of our analysis: https://github.com/institutional-grammar-pl.
[3] The expert is Hanna Schreiber.

case – modeling the 2003 UNESCO Convention – is presented in Sect. 4. The paper concludes with Sect. 5.

2   Institutional Grammar and Graphs

Our approach oscillates around building the semantic network from texts [4,20]. In our use case, the challenge is to extract meaningful information from the legal text and visualize the extracted inter-related information. The most crucial part of our system is IG, employed as an interlayer between the regulation text and its graph representation.

IG was used as the layer because it was designed as a schema that standardizes and organizes information on statements coming from a legal text. The statements are understood in IG as bits of institutional information. They have two main functions: to set up prominent institutional actors (organizations, collective bodies, organizational roles, etc.) and to describe what actions those actors are expected, allowed, or forbidden to perform (see Table 1). Therefore, IG makes it possible: (1) to identify how many statements are written into a sentence; (2) to categorize those statements; (3) to identify links between them; (4) to identify animated actors and inanimate objects regulated by the statements; and (5) to identify relations between actors and objects defined by the statements.

Table 1. IG components depending on statement type based on [7, pp. 10–11].

Regulative	Description	Constitutive	Description
Attribute (A)	The addressee of the statement	Constituted Entity (E)	The entity being defined
Aim (I)	The action of the addressee regulated by the statement	Constitutive Function (F)	A verb used to define (E)
Deontic (D)	An operator determines the discretion or constraint associated with (I)	Modal (M)	An operator determining the level of necessity and possibility of defining (E)
Object (B)	The action receiver described by (I). There are two possible receivers: Direct or Indirect	Constituting Properties (P)	The entity against which (E) is defined
Activation Condition (AC)	The setting to which the statement applies	Activation Condition (AC)	The setting to which the statement applies
Execution Constraint (EC)	Quality of action described by (I)	Execution Constraint (EC)	Quality of (F)

Figure 1a presents an example of a regulative statement from the 2003 UNESCO Convention (Article 23 par. 1). In this statement, "State Party" is an entity whose actions are regulated; therefore, it is annotated as Attribute.

An action regulated explicitly in the statement is "submit," and this element is identified as Aim. The statement informs us through the Deontic "may" that the action "submit" is allowed to be performed by a State Party. However, the statement also tells through the Execution Constraint what kind of action of submission is allowed – "through an online form." From the Activation Condition "once a year," we also learn when a State Party is entitled to the action "submit." Another piece of information provided by the statement is the kind of target directly affected by the regulated activity – it is a Direct Object "request" that has the property "for financial assistance." The statement also identifies who is obliquely affected by the State Party's action – this is the Indirect Object of "the Committee."

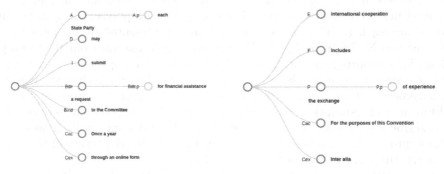

(a) A regulative statement: "Once a year, each State Party may submit a request for financial assistance to the Committee through an online form"

(b) A constitutive statement: "For the purpose of this Convention, international cooperation includes, inter alia, the exchange of experience"

Fig. 1. Examples of parsed statements. The IG Parser generated the tree (https://s.ntnu.no/ig-parser-visual).

Another statement, a part of Article 19 par. 1, from the 2003 UNESCO Convention, is a simple example of a constitutive statement (Fig. 1b). Here, the statement defines one of the essential properties of "international cooperation" that is indicated as a Constituted Entity. The statement informs that international cooperation "includes" (Function) a specific type of exchange (Constituting Property) – the exchange "of experience" (Constituting Entity property). We also learn that this definition of international cooperation applies only in a specific context – inside the institutional setting described by the 2003 UNESCO Convention. The Activation Condition provides information about the context in which the definition applies: "for the purposes of this Convention."

The application of IG makes it possible to extract standardized data on rules from legal texts. However, the IG schema is not associated with any popular way of data analysis, which is challenging to utilize in research. Therefore, researchers have been searching for ways to transform IG data into more popular data formats. The most promising direction is graph theory. Statements can be seen as edges that connect nodes represented by Attributes, Objects, Constituted Entities, and Constituting Properties. In this regard, a set of edges –statements constitutes a graph representing an individual regulation. This approach allows for using popular network analysis indicators to characterize the institutional setting developed by the regulation. A legal act is a list prescribed by the legal relations between different objects and actors. Three approaches have been developed to transform IG data into graph data. The first is represented by T. Heikkila and C. Weible [9]. IG is used here to identify actors in a variety of legal documents. They are linked in a single document and across many regulations. The aim is to identify a network of influential actors in a particular policy and use network analysis to characterize their associations. Heikkila and Weible used their method to study the regulation of shale gas in Colorado and found that the institutional actors governing this industry create a polycentric system. The second approach to IG and network analysis is represented by research by O. Thomas and E. Schlager [18,19]. Here, data transformation from IG to graph allows for comparing the intensity with which different aspects of institutional actors' cooperation are regulated. The authors used this approach to study water management systems in the USA to link IG data with graph representation. It is built around assumptions that network analysis based on IG makes it possible to describe institutional processes written into legal documents in great detail. This approach was also used in a study on the climate adaptation of transport infrastructures in the Port of Rotterdam [15].

In contrast, our approach is more actor-centered than the previous ones (see Sect. 4 for more details). As stated in [5], entity mentions may provide more information than plain language in some texts. We believe that actors are crucial in analyzing the legal acts in our use case. We aim to compare the positions of prominent actors in a network of rules and use indicators developed by network analysis to describe the place of institutional actors in the formal design of the UNESCO Convention. We also confront these metrics with IG data to determine if an actor's position in a rules network correlates with its place in IG statements.

Information structured in a graph, with nodes representing entities and edges representing relationships between entities, can be built manually or using automatic information extraction methods. Statistical models can be used to expand and complete a knowledge graph by inferring missing information [16]. There are different technologies and applications of language understanding, knowledge acquisition, and intelligent services in the context of graph theory. This domain is still in *statu nascendi* and includes the following key research constructs: graph representation and reasoning; information extraction and graph construction; link data, information/knowledge integration and information/knowledge graph storage management; NLP understanding, semantic computing, and information/knowledge graph mining as well as information/knowledge graph application [14]. In the last decade, graphs in various social science and human life fields have become a source of innovative solutions [13,14,24].

Interactions do not usually occur in an institutional vacuum; they are guided and constrained by agreed-on rules. It is also essential to understand the parameters that drive and constrain them. Hence, the design of institutional behaviors can be measured through Networks of Prescribed Interactions (NPIs), capturing patterns of interactions mandated by formal rules [18]. Social network analysis offers considerable potential for understanding relational data, which comprises the contacts, ties, and connections that relate one agent to another and are not reduced to individual agent properties. In network analysis, relations express the linkages between agents. That analysis consists of a body of qualitative measures of network structure. Relational data is central to the principal concerns of social science tradition, emphasizing the investigation of the structure of social action. The structures are built from relations, and the structural concerns can be pursued through the collection and analysis of relational data [12]. Our approach uses social network analysis powered by institutional grammar.

3 Our Prototype

Our prototype's workflow starts with setting up a crawler to collect legal documents from websites of interest. Its functionalities mirror the stages of work with legal acts: (1) retrieving documents from different internet resources and (2) selecting, for further analysis, only documents relevant for a specific IG-related research task (filtered by customizable defined rules or keywords). Then, (3) legal texts are prepared, pre-processed, and parsed with IG. The output – selected IG components – are then (4) refined and transformed into a graph where nodes represent institutional actors and objects, and the statements containing them are expressed as edges (relations). In this process, we also incorporate quality checks and super-annotations with an IG and domain expert in policy design. Finally, we consider institutional design research questions that can be answered using quantitative analysis based on the generated graph. Figure 2 shows this process of annotating legal documents by incorporating our automated tools in the prototype. The first significant challenge regarding building the above system is the semantic understanding of legal policy texts. To automate the process of legal text analysis, we utilize Institutional Grammar as an intermediate layer for text

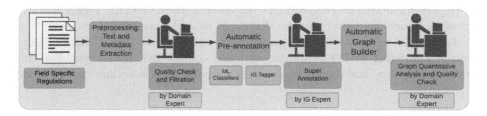

Fig. 2. The analysis process of legal acts with our system support and manual work of IG human expert and refinement of a domain expert.

interpretation and extraction of crucial entities. Thus, in our study, we incorporate IG as a text-semantic bridge. On the one hand, IG is a tool for describing and analyzing relations between institutional actors more systematically and structured than natural language texts. However, there are still open issues we are currently addressing with the manual work of policy design researchers. Firstly, there is still a need to define some IG statements more precisely, e.g., context data like Activation Conditions entities. Some IG entities are not defined precisely and are not structuralized as a graph structure or formal ontology with classes and logical rules between them. Such structurally ordered and precisely defined data (a graph/ an ontology) can further be used, for example, for defining rules of institutions and network analysis and reasoning [17]. We have yet to manually address atomic statement extractions and classification (regulative vs. constitutive) and their transformations into active voice.

Our prototype's second challenge is transforming the IG format with the extracted entities into a graph structure and format, which is a proper data structure for automated analysis. Under the label of "social network analysis," social science has developed significant expertise in expressing information about particular elements of our reality in computer-readable data structures to analyze the interrelation between entities. This task is accomplished using IG as an interlayer between a legal text and mainly graph-based metrics expressing interrelations between entities.

3.1 IG Tagger

Institutional statements have two types – regulative and constitutive – with a few distinct IG tags. In IG, constitutive statements serve defining purposes, while regulative statements describe how actors' behaviors and actions should be regulated. For this task – i.e., the differentiation of statement types – we trained a classifier based on TF-IDF with 70 of the most relevant (1–3)-grams and Random Forest to distinguish between these types (our train set contained 249 statements). The model's F1-score is 92% on the test dataset (84 observations out of 382).[4]

Then in our IG tagger, the first tagging stage, each word in the statement gets an annotation containing a lemma, a part of speech tag, morphological features, and relation to other words (extracted with Stanza package [21]). Due to different IG tags in regulative and constitutive statements, the automatic tagger has two different algorithms based on rules dedicated to each type of statement.

Table 2 shows overall tagger performance tested on the 2003 UNESCO Convention consisting of 142 regulative and 240 constitutive statements. For this analysis, predicted and correct tags were mapped before the evaluation: (A, prop)

---

[4] The model is used in the tagger's repository (https://github.com/institutional-grammar-pl/ig-tagger) and the experiments are included in a separate repository: https://github.com/institutional-grammar-pl/sentence-types.

to (A), (B, prop) to (B) for regulative ones, and (E, prop) to (E), (P, prop) to (P) for constitutive statements (prop means property). Applied measures were determined based on the accuracy of the classification of the individual words in a sentence. The best results are achieved in recognizing Aim, Deontic, Function, and Modal components – F1-score of over 99%. These tags are precisely defined. As we can see in Table 2, the constitutive statements and some tags of regulative statements are still a big challenge that should be defined more precisely and solved with machine learning-based text modeling.

Table 2. Results of IG Tagger on regulative and constitutive statements.

Layer	Component	F1 score	Precision	Recall
Regulative	Attribute	0.51	0.62	0.56
	Object	0.55	0.67	0.61
	Deontic	1.0	0.96	0.98
	Aim	0.99	0.86	0.92
	Context	0.53	0.73	0.62
	Overall	0.71	0.69	0.68
Constitutive	Entity	0.79	0.64	0.71
	Property	0.57	0.78	0.66
	Function	0.82	0.82	0.82
	Modal	1.00	0.83	0.90
	Context	0.09	0.02	0.03
	Overall	0.58	0.57	0.56

3.2 Graph of Entities Extraction

We extracted hypergraph information from our analyzed documents to get insights into a given document. The hypergraph vertices are the essential entities – actors (e.g., institutions) and objects (e.g., legal documents). The statements (in the analyzed documents) represent connections (edges) in this structure.

Let $H = (V, E)$ be a hypergraph, where verticles V represent actors and objects appearing in the document. $E = (e_1, ..., e_k), e_i \subset V \forall_{i \in 1,...,k}$ means hyperedges that are created where objects appear together in one statement from the analyzed legal documents. We chose hypergraphs, not graphs, because relations are formed by more than one vertex.

After constructing the hypergraph, our analysis employed various graph measures, including the hypergraph centrality proposed in [1]. We can also use other graph metrics to help describe information and relationships between entities and define the interrelations and expositions of particular actors.

4   Use Case and Impact

Our use case is the 2003 UNESCO Convention, which establishes UNESCO listing mechanisms for the safeguarding of intangible cultural heritage, including the Representative List of the Intangible Cultural Heritage of Humanity, the List of Intangible Cultural Heritage in Need of Urgent Safeguarding, and the Register of Good Practices in Safeguarding of the Intangible Cultural Heritage [3,23].

As mentioned previously (Sect. 1), we aim to study the institutional design of an international convention by comparing the actors' positions in the network of rules described by the convention. Following other researchers who transformed IG data into network data [9,15,18], we use network metrics to describe the locations of actors in the institutional settings produced by the 2003 UNESCO Convention. We outline the Convention's institutional design using data from its text explored using IG and network analysis.

In addition, we compare metrics from network analysis with metrics from IG data. The comparison is the first step in understanding if an actor's position in a set of statements is a good indicator of their location in the institutional setting expressed as a graph. Looking at it from another angle, we aim to empirically verify to what extent indicators derived from IG analysis are good predictors of indicators related to network analysis. We want to determine whether future studies must process IG-derived data into data suitable for network analysis.

In our study, the position is considered from the perspective of two measures - visibility and centrality - referring to two sets of data: IG data and IG data converted into network data. *Visibility* of an actor is defined as the level of the directness with which institutional statements regulate the actor's actions. In the context of IG, this relates to the institutional statement components in which the actor is mentioned. It can be measured on an ordinal scale built around the actor's place in a statement (see Table 3). An actor mentioned in the most prominent elements of a statement (Attribute or Direct Object) has a higher score than an actor mentioned in concealed parts of statements (for example, properties of Objects).

Table 3. Actors' visibility scale in constitutive (CS) and regulative (RS) texts.

Actor:	Weight
As Attribute in RS or Constituted Property in CS	6
As Direct Object in RS or Constituting Entity in CS	5
As Indirect Object in RS	4
In properties of Attribute or Constituted Property	3
In properties of Direct Object or Constituting Entity	2
In properties of Indirect Object	1

Actors are ranked by their measures of directness. Each actor is assigned a weight depending on the class (see the ranking in Table 3). Thus, we can define a *visibility* measure as:

$$visibility = \sum_{c \in \{1,...,C\}} w_c * \frac{n_c}{N},$$

where N - number of statements, n_c - occurrence in class c, w_c - rank of class c.

Centrality is associated with actors' positions in a network of statements. This analysis assumes an actor can be mentioned in less prominent IG components. However, at the same time, it could have a central role in the formal institutional setting created by a legal document. It is "involved" in many statements but is usually mentioned in subsidiary IG elements. Each statement is seen as an edge, and actors are depicted as nodes independent of their association with IG components. A set of all edges forms a hypergraph mapping the regulation. The measure of centrality in the hypergraph is computed for each actor. Then, the actors are ranked by their measures of centrality. Throughout this article, centrality denotes closeness centrality $C(u)$ defined as

$$C(u) = \frac{n-1}{\sum_{v \neq u, v \in U} d(u, v)},$$

where U is a set of vertices in the graph, $|V| = n$ and $d(u, v)$ denotes the shortest path distance between vertices u and v.

Based on the above operationalizations, the research question that can be formulated for the exemplary case is: What is the relation of visibility versus centrality in the case of the Secretariat?

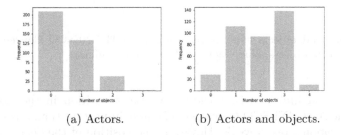

(a) Actors.   (b) Actors and objects.

Fig. 3. Histogram of mentions of actors and objects in statements which refer to the number of connected nodes in the hypergraph (hyperedges).

Figure 3 presents how many actors occur in one statement. The baseline centrality approach analyzes the occurrence of actors together within one sentence. As shown in Fig. 3a, usually, only one actor appears within one atomic sentence. Therefore, the analysis was expanded to consider the presence of actors and objects such as a report, individual or group. Figure 3b illustrates that this approach has much greater potential. Then, finally, Fig. 4 presents our hypergraph extracted from the 2003 UNESCO Convention. Based on our analysis, we can answer the research question stating that the Secretariat belongs to a group of actors with low levels of visibility but relatively high normalized values of their centrality – see Fig. 5. This group comprises non-governmental organizations, the Intergovernmental Committee, Communities, Groups, and Individuals (CGIs), and the Secretariat. However, the Secretariat has the highest level of visibility in the group. It appears in more essential IG elements than the rest of the group.

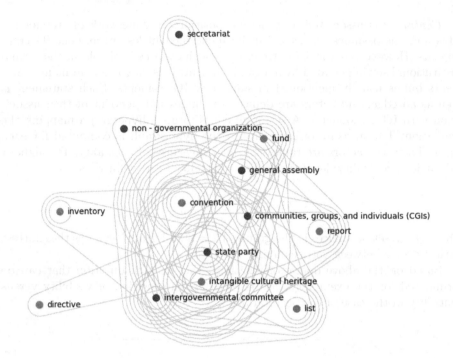

Fig. 4. Actors and objects hypergraph from the 2003 UNESCO Convention. Note: green dots denote objects, blue – actors. (Color figure online)

This observation indicates that the Secretariat's position in the institutional design of the 2003 UNESCO Convention is more important than a brief analysis of the Convention's text reveals. However, the position of the Secretariat is not as undervalued from the perspective of visibility as the position of others in the group. The issue of undervaluation leads us to the second observation – State Party is the only overexposed actor in the 2003 Convention.

State Party is the only actor with a visibility score higher than its centrality score. However, we can only talk about a relative "overexposure" of the State Party – it has, after all, the highest centrality score among all actors. We can tell that states that are parties to the Convention are crucial actors in the formal institutional design of the 2003 UNESCO Convention. This observation probably reflects the fact that international conventions are usually formulated in a manner that, above all, regulates the actions of actors being parties to international conventions – the states themselves.

The third observation from our case study relates to the relationship between pure IG analysis and network analysis based on data provided by IG. The indicators coming from raw IG are poor proxies for network analysis ones. This observation shall encourage building systems that treat IG only as an interlayer for developing data on a formal institutional design from legal texts.

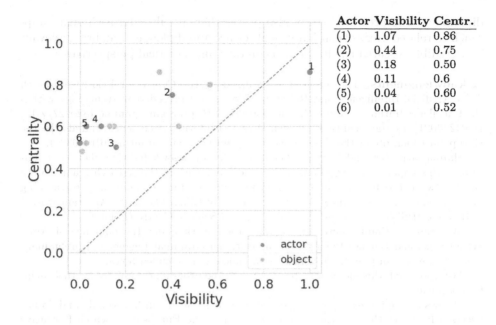

Actor	Visibility	Centr.
(1)	1.07	0.86
(2)	0.44	0.75
(3)	0.18	0.50
(4)	0.11	0.6
(5)	0.04	0.60
(6)	0.01	0.52

Fig. 5. Visibility vs Centrality (closeness centrality), where (1) is State Party, (2) – General Assembly, (3) – Secretariat, (4) – Intergovernmental Committee, (5) – Communities, Groups, and Individuals (CGIs), and (6) is Non-governmental organization. Note: green dots denote objects, blue – actors. (Color figure online)

5   Conclusions

This paper addresses the challenge of easily comprehending international agreements to solve global issues. It presents a prototype tool for efficiently extracting formal institutional design – norms, rules, and actors – from collected legal documents. The tool employs Institutional Grammar as a bridge between legal text and a graph representation of the institutional design, enabling practical NLP analysis of a complex legal text. The tool was tested in the 2003 UNESCO Convention. Its analysis reveals the position of the Convention Secretariat and other actors within the institutional design using the measures of visibility and centrality. The results indicate that the Secretariat holds average importance and is part of a group of relatively hidden actors in the Convention's institutional design.

The developed approach shows an analytical and research potential for optimizing and effectively designing the processes of formalizing any new organizations. The study demonstrates the potential of combining NLP tools, Institutional Grammar, and graph analysis to facilitate a deeper understanding of formal institutional design in international agreements. The resulting graph representation provides a clear, easily-comprehensible representation of the institutional structure and holds promise for advancing the field of computational diplomacy. The presented study also proves that a truly interdisciplinary app-roach to related areas of computational science and diplomacy can bring relevant and credible research results from theoretical and practical perspectives.

Acknowledgments. The research was funded by the Centre for Priority Research Area Artificial Intelligence and Robotics of Warsaw University of Technology within the Excellence Initiative: Research University (IDUB) program (grant no 1820/27/Z01/POB2/2021), by the Faculty of Mathematics and Information Science, and also by the European Union under the Horizon Europe grant OMINO (grant no 101086321).

Hanna Schreiber and Bartosz Pieliński wish to acknowledge that their contribu-tion to this paper was carried out within the framework of the research grant Sonata 15, "Between the heritage of the world and the heritage of humanity: researching international heritage regimes through the prism of Elinor Ostrom's IAD framework," 2019/35/D/HS5/04247 financed by the National Science Centre (Poland).

We want to thank many Students for their work under the guidance of Anna Wróblewska and Bartosz Pieliński on Institutional Grammar taggers and preliminary ideas of the system (e.g. [25]), which we modified and extended further.

We also thank the anonymous reviewers and the program committee for their help-ful comments.

(Views and opinions expressed are however those of the author(s) only and do not necessarily reflect those of the European Union or the European Research Executive Agency. Neither the European Union nor European Research Executive Agency can be held responsible for them.)

References

1. Aksoy, S.G., Joslyn, C., Ortiz Marrero, C., Praggastis, B., Purvine, E.: Hypernet-work science via high-order hypergraph walks. EPJ Data Sci. **9**(1), 1–34 (2020). https://doi.org/10.1140/epjds/s13688-020-00231-0
2. Bauer, M.W., Ege, J.: Bureaucratic autonomy of international organizations' sec-retariats. J. Eur. Public Policy **23**(7), 1019–1037 (2016)
3. Blake, J., Lixinski, L.: The 2003 UNESCO Intangible Heritage Convention: A Commentary. Oxford University Press, Cambridge (2020)
4. Celardo, L., Everett, M.G.: Network text analysis: a two-way classification app-roach. Int. J. Inf. Manag. **51**, 102009 (2020)
5. Dai, H., Tang, S., Wu, F., Zhuang, Y.: Entity mention aware document represen-tation. Inf. Sci. **430**, 216–227 (2017)
6. Eckhard, S., Ege, J.: International bureaucracies and their influence on policy-making: a review of empirical evidence. J. Eur. Public Policy **23**(7), 960–978 (2016)
7. Frantz, C.K., Siddiki, S.: Institutional grammar 2.0: a specification for encoding and analyzing institutional design. Public Adm. **99**(2), 222–247 (2021)
8. Ghorbani, A., Siddiki, S., Mesdaghi, B., Bosch, M., Abebe, Y.: Institutional com-pliance in flood risk management: a network analysis approach (2022)
9. Heikkila, T., Weible, C.M.: A semiautomated approach to analyzing polycentricity. Environ. Policy Govern. **28**(4), 308–318 (2018)
10. Jörgens, H., Kolleck, N., Saerbeck, B.: Exploring the hidden influence of inter-national treaty secretariats: using social network analysis to analyse the Twitter debate on the 'Lima Work Programme on Gender'. J. Eur. Public Policy **23**(7), 979–998 (2016)

11. Libal, T.: A meta-level annotation language for legal texts. In: Dastani, M., Dong, H., van der Torre, L. (eds.) CLAR 2020. LNCS (LNAI), vol. 12061, pp. 131–150. Springer, Cham (2020). https://doi.org/10.1007/978-3-030-44638-3_9
12. Liu, B.: Social Network Analysis, pp. 269–309. Springer, Heidelberg (2011)
13. Lüschow, A.: Application of graph theory in the library domain-building a faceted framework based on a literature review. J. Libr. Inf. Sci **54**(4), 558–577 (2021)
14. Majeed, A., Rauf, I.: Graph theory: a comprehensive survey about graph theory applications in computer science and social networks. Inventions **5**(1), 10 (2020)
15. Mesdaghi, B., Ghorbani, A., de Bruijne, M.: Institutional dependencies in climate adaptation of transport infrastructures: an institutional network analysis approach. Environ. Sci. Policy **127**, 120–136 (2022)
16. Nickel, M., Murphy, K., Tresp, V., Gabrilovich, E.: A review of relational machine learning for knowledge graphs. Proc. IEEE **104**(1), 11–33 (2016)
17. Olivier, T., Scott, T.A., Schlager, E.: Institutional design and complexity: protocol network structure in response to different collective-action dilemmas. In: Fischer, M., Ingold, K. (eds.) Networks in Water Governance. PSWGPP, pp. 267–293. Springer, Cham (2020). https://doi.org/10.1007/978-3-030-46769-2_10
18. Olivier, T.: How do institutions address collective-action problems? bridging and bonding in institutional design. Polit. Res. Q. **72**(1), 162–176 (2019)
19. Olivier, T., Schlager, E.: Rules and the ruled: understanding joint patterns of institutional design and behavior in complex governing arrangements. Policy Stud. J. **50**(2), 340–365 (2022)
20. Paranyushkin, D.: Infranodus: generating insight using text network analysis. In: The World Wide Web Conference, WWW 2019, p. 3584–3589. Association for Computing Machinery, New York (2019)
21. Qi, P., Zhang, Y., Zhang, Y., Bolton, J., Manning, C.D.: Stanza: a python natural language processing toolkit for many human languages. In: ACL: System Demonstrations, pp. 101–108. ACL (2020)
22. Rice, D., Siddiki, S., Frey, S., Kwon, J.H., Sawyer, A.: Machine coding of policy texts with the institutional grammar. Public Adm. **99**, 248–262 (2021)
23. Schreiber, H.: Intangible cultural heritage and soft power - exploring the relationship. Int. J. Intangible Herit. **12**, 44–57 (2017)
24. Spyropoulos, T.S., Andras, C., Dimkou, A.: Application of graph theory to entrepreneurship research: new findings and a new methodology framework. KnE Social Sci. **5**(9), 139–156 (2021)
25. Wichrowska, A.E., Saputa, K.: System of a full-text database of legislative documents and information created during the COVID-19 pandemic with a module for detection of IG objects (2021). https://repo.pw.edu.pl/info/bachelor/WUTea29ae51c3164edcba3d5016e9d38e9e/

Modeling Mechanisms of School Segregation and Policy Interventions: A Complexity Perspective

Eric Dignum[1]([⊠]) [iD], Willem Boterman[2] [iD], Andreas Flache[3] [iD],
and Mike Lees[1] [iD]

[1] Computational Science Lab, University of Amsterdam, Amsterdam, Netherlands
e.p.n.dignum@uva.nl
[2] Urban Geographies, University of Amsterdam, Amsterdams, Netherlands
[3] Department of Sociology, Rijksuniversiteit Groningen, Groningen, Netherlands

Abstract. We revisit literature about school choice and school segregation from the perspective of complexity theory. This paper argues that commonly found features of complex systems are all present in the mechanisms of school segregation. These features emerge from the interdependence between households, their interactions with school attributes and the institutional contexts in which they reside. We propose that a social complexity perspective can add to providing new generative explanations of resilient patterns of school segregation and may help identifying policies towards robust school integration. This requires a combination of theoretically informed computational modeling with empirical data about specific social and institutional contexts. We argue that this combination is missing in currently employed methodologies in the field. Pathways and challenges for developing it are discussed and examples are presented demonstrating how new insights and possible policies countering it can be obtained for the cases of primary school segregation in the city of Amsterdam.

Keywords: complex systems · policy · school segregation · agent-based modeling

1 Introduction

In many educational systems, levels of school segregation are still substantial to high. Meaning that children with different characteristics such as race [43], ethnicity [10], income levels [24], educational attainment [7] and ability [50] cluster together in different schools, which is widely associated with existing inequalities and their reproduction [51]. Even despite the wealth of knowledge about the problem and the many policy interventions that have been proposed to counteract it, segregation continues to plague educational systems globally and hence proves a hard problem to solve [8].

© The Author(s), under exclusive license to Springer Nature Switzerland AG 2023
J. Mikyška et al. (Eds.): ICCS 2023, LNCS 14075, pp. 74–89, 2023.
https://doi.org/10.1007/978-3-031-36024-4_6

Research in the field of (social) complexity has highlighted how interactions between components of a system, can cause an unanticipated and possibly unintended social outcome such as segregation at the macro-level due to choices individuals make [11]. Anticipating such dynamics and developing policies that could prevent or mitigate them is a notoriously hard problem to solve. Moreover, if the focus of research lies primarily on isolated individual choices or on macro-level characteristics, these explanations miss potentially important effects of interactions [34]. Similarly for empirical studies in the field of school segregation, the focus has mainly been on individual-level decision making or macro-level patterns. For example through parent interviews [2], discrete choice analysis [14], changes in levels and trends of segregation [43], or associations of segregation levels with characteristics of neighborhoods or municipalities [33]. Additionally, the factors in school segregation are often treated as individual variables that do not affect each other [42]. Ignoring such interactions can lead to a lack of understanding of the mechanisms underlying school segregation and could result in ineffective policies. This could be an explanation of why school segregation is still a robust and resilient phenomenon in society even with counteracting policies. This is substantiated by recent theoretical work, utilizing complexity-inspired methodologies, that substantial school segregation can even result from relatively tolerant individuals due to their interactions in the system [20,45,48]. However, these stylized models have limited applicability to reality and methodologies inspired by complexity have been, in our view, underused in the field of school segregation.

Therefore, this paper argues that the dynamics of school choice constitute a complex system not only in principle, but also in light of theoretical and empirical research in the field. The dynamics of school choice at the micro-level (e.g., household) on the one hand, and macro-level contextual factors (e.g., school compositions, quality, institutional aspects) on the other, are interdependent and interacting elements of a broader complex system. These interactions result in the emergent patterns of school segregation at the macro-level. This system, we contend, is characterized by the adaptive behavior of individual actors who respond to changing contextual conditions and simultaneously influence these conditions by the choices they make. Therefore, the behavior of the system as a whole (e.g., school segregation) is difficult to infer from the analysis of individual components in isolation [34]. Common methodologies in this field often take a reductionist approach, treating the macro-level as the sum of its isolated individual parts. We elaborate and exemplify how Agent-Based Models (ABMs) can overcome these limitations, allowing for the explicit modeling of interactions between the components conditional on their specific educational context [17,40]. We identify pathways and challenges for how to incorporate ABM in future empirical research and show examples of how it can lead to new insights and possible policies countering primary school segregation in the city of Amsterdam.

2   Features of Complex Systems

Complex systems consist of numerous, typically heterogeneous, elements operating within an environment, whose behavior is defined by a set of local rules. Interactions can occur directly between components or via their environment and are not restricted to be physical, but can also be thought of as exchanging energy or information. The state of an element can be influenced by the current and/or previous states of possibly all other elements in the system. Although there is no universal definition of complexity, there is more consensus on features arising in complex systems, which are described in the next sections.

2.1   Emergence

A complex system shows emergent behavior in the sense that the system exhibits properties and dynamics that are not observed in the individual elements of the system. Typically, the quantity of interest is some system level emergent property, which might be lost by isolating the individual elements. Emergence makes complex systems hard to reduce and study from the isolated components, the interactions of the elements are crucial to the system dynamics [1]. The interactions is what establish the link between the micro- and macro-level and it is the understanding of this relationship that is often of importance. Examples of emergent phenomena are the forming of coalitions in societies [17] and swarming behavior in animals [37].

2.2   Self-organization, Adaptation and Robustness

Where emergence is focused on new, sometimes unexpected, measures and structures that arise from the interaction of individual components, self-organization stresses adaptive and dynamic behavior that leads to more structure or order without the need for external control. The resulting organization is decentralized and is formed by the collection of individual components. These elements are able to adapt their behavior by learning and react to the environment and other elements [16]. This typically creates robust structures that are able to adapt and repair and can make it difficult to control complex systems using traditional global steering mechanisms. Self-organization and emergence can exist in isolation, yet a combination is often present in complex systems [19]. The idea of robustness is the ability of a system to resist perturbations or its ability to function with a variety of different inputs/stimuli. The closely related concept of resilience, a specific form of adaptation, is the ability of the system to continue to function by adapting or changing its behavior in the face of perturbations.

2.3   Feedback, Non-linearity and Tipping Points

The interactions between the elements and the environment and adaptation imply that the components can respond to and thereby change the properties of the system. This generally leads to feedback loops, which can create complex causality mechanisms and nonlinear responses [30]. These feedback loops can have (de)stabilizing effects [36] and can take place at different time scales. Non-linear systems have the property that the change in the output of the system does not scale proportionally with the change of input of the system. This makes it hard to predict the behavior of the system, the same small perturbation that caused small responses may at some point lead to drastic changes at the macro-level. Such non-linearity can be triggered by feedback and other mechanisms [39], which might even lead to tipping points where systems can shift—abruptly—from one state to another, such as virus outbreaks, stock market crashes, and collapsing states [46]. In the case of segregation, the relocation of one individual or a small group can increase the proportion of the majority in the new neighborhood/school. This in turn could lead to individuals of the minority group to leave, which decreases their number even more, resulting in more moves and so on. This can trigger a whole cascade where a small effect can tip a neighborhood/school to become homogeneous [15, 45].

2.4 Path-dependency

Path-dependency explains how decisions at a given point in time are constrained by the decisions made in the past or by events that occurred leading up to that particular moment, although past events may no longer be relevant. This implies that once a decision has been made, by simply making that decision, you make it hard/impossible to change or re-make that decision in the future. A form of path-dependency may appear in self-fulfilling prophecies [13] or in politics, where a certain sequence or timing of political decisions, once introduced, can be almost impossible to reverse.

3 Complexity in the Mechanisms of School Segregation

This section extracts general mechanisms underlying school segregation from the existing literature and connects them to the previously discussed features of complex systems. Although the factors in school segregation are highly context-specific, different educational systems exhibit similarities through which the dynamics of school segregation operate similarly [8]. These factors are discussed in separate sections, however, it should be stressed that they are intrinsically linked and interact with each other. The described dynamics are found in most, if not all, educational systems, but relevant importance of mechanisms can differ substantially depending on the context, which is crucial for eventual understanding.

3.1 Distance

Numerous studies find that, irrespective of household characteristics or educational system, most children attend a school close to their home [8]. Hence, residential segregation (partially) portrays itself in the population of schools. However, socio-economically advantaged subgroups of parents are found to be more willing/able to travel further or even move neighborhoods for more favorable school characteristics [14], increasing school segregation and complicating the dynamics. The latter might be more prominent in educational systems with geographic assignment mechanisms, forcing households to either accept their neighborhood schools or *adapt* and opt-out of the system. This establishes a mutual relationship (*feedback*) between school characteristics and residential segregation.

3.2   School Profile

In various educational systems, schools can differ on their religious foundation, pedagogical principles, curricula, or status. These profiles are important in choice of school, but they might attract only certain groups of parents and thereby increasing segregation between schools [10]. For example, in the Netherlands, highly-educated parents are found to be more attracted to schools with a certain pedagogy than their counterparts [5]. Moreover, existing residential segregation might make schools more inclined to adopt a certain profile [25], connecting the location of a school and the profile, resulting in possible interactions between these factors.

3.3   School Quality

Although an ambiguous concept, quality could refer to a school's added value to academic achievement, but also to its climate, order, and discipline [9]. Nevertheless, parents have a strong preference for proxies of quality, such as the academic performance of schools, and even more so if the family is advantaged (e.g., education, income) [26]. However, academic performance might say more about a school its ability to attract a better performing student population via profile, gatekeeping or its location, rather than measuring its "added value" or actual quality [10]. Perceived high-quality schools could increase house prices in the neighborhood, making it more likely that only high-income households can afford living close [41], affecting residential segregation and school segregation in turn (via distance preferences).

3.4   School Assignment

Many cities employ school allocation mechanisms. If geographically restricted, these could strengthen the link between residential and school segregation [8], but can also lead to specific types of households opting out and attend non-public schools for example. Additionally, advantaged households are suggested to be better aware of registration deadlines, accompanying assignment mechanisms and their consequences [23]. This can lead to disadvantaged parents registering late more often, resulting in less choice or lower rankings. Another form of assignment is early ability tracking, where most European countries, the US and the UK group students together based on ability [50] or use selection criteria. Although implementations differ, [47] find that tracking and selecting increase inequality with respect to socio-economic status. Moreover, grouping children based on ability is itself already segregation, but also strongly overlaps with ethnicity and social class.

3.5   Gatekeeping

Gatekeeping or cream skimming can be summarized as the (un)intentional and (in)formal use of selective criteria by schools. High tuition fees, voucher systems, waiting lists, catchment areas, advising children about different schools or organizing education in such a way that particular children do not feel at home, are all reported gatekeeping practices [28]. Hence, schools can also *adapt* their behavior by reacting to the system, creating additional interactions between households and schools. Many systems offer alternatives to public education. Private schools, exclusively funded by school fees, hybrid schools, where the state also partly funds the privately organized schools and charter/magnet schools also play a substantial role and hence should be mentioned here [44]. However, their effect on school segregation can be attributed to already described factors. Schools in this sector, depending on funding, often have substantially more autonomy. This could lead to highly specialized school profiles,—perceived—high quality (attracting a high-ability population), gatekeeping practices (e.g., high fees) and allocation mechanisms [8].

3.6   Social Network

The decision of which school to attend could also be influenced by interactions through social networks, preschool groups or school visits for example [6]. Social networks can provide (mis)information on school choice and act as a platform for social comparison, where people trust the opinions of high-status individuals in their choice of school [2]. As one tends to be friends with similar people, these networks tend to be socially and geographically structured [35]. Hence, opinions or factual information about schools are not equally accessible to everyone in the network and could enforce certain group-specific preferences [31], leading to *feedback loops, non-linearity* and *robustness*. For example, high-status schools could be more accessible to those with better connections and, in turn, can offer status and connections to those who attend them [27].

3.7   School Composition

Another important determinant, irrespective of context, is that people are more attracted to schools attended by larger proportions of their own group and avoid schools with large shares of other groups [8]. This behavior has obvious effects on school segregation and could induce self-reinforcing processes. Empirically, this is demonstrated in phenomena such as White flight, where households of a particular group affect each other and opt out of schools and/or neighborhoods to avoid undesirable school composition/quality [18]. If schools are already segregated, a preference for attending a school with similar pupils will reinforce future patterns of school segregation. These *feedback mechanisms* make segregation both *robust* and *path-dependent* (i.e., historical school segregation affects future school choices). Furthermore, school closure/founding might lead to *non-linear* effects, where the new choices of the movers induce children at other schools to move. This can result in some systems going from an integrated to a segregated state (*tipping*). Homophily, the tendency to associate with similar others [35] can be an explanation for this mechanism, but the composition of a school could also emerge from the profile it propagates, as particular types of people might be attracted to certain profiles, or due to the projection of existing residential segregation in schools. Another way in which school composition affects school choice is because parents, apart from official, "cold knowledge", also rely on social network-based information "hot knowledge", [2] to assess school quality. It has been demonstrated that (perceptions of) school composition, circulated by parents is used as a proxy for quality of a school [3].

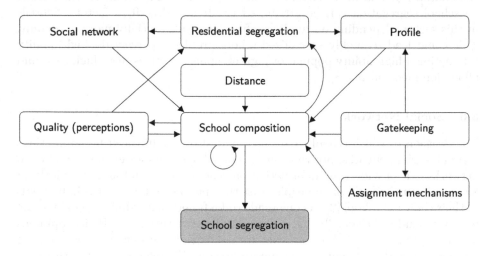

Fig. 1. Mechanisms underlying school segregation and their interactions.

4   Modeling School Segregation Dynamics

We have argued that school choice and school composition form a complex system that is hard to understand, let alone predict or be influenced by policy without a good understanding of the interactions between its various levels and components. From a complexity perspective, a suitable model of school choice dynamics thus needs to contain descriptions of the micro- and macro-level, as well as of the interactions between levels and their components. Interviews, document analysis, and surveys are all methods used to study school choice and are useful for obtaining an overview of the factors decision makers and experts consider important in a concrete case. However, it is often hard to quantify the relative importance of factors and how these vary across different types of people in the system under study. Complementary, using observed choices, discrete choice models are able to quantify the effect that individual, school and institutional characteristics have on individual choice behavior [34]. Nevertheless, common assumptions are that decision makers operate independently of each other, have full knowledge of the system and are rationally maximizing their utility. These methods, separately or combined, provide valuable and necessary information, as one needs to have input for how to model the individual elements in the system. However, a remaining challenge is to model the interactions within and between the micro- and macro-level simultaneously [40].

4.1   Agent-Based Models

Agent-Based Models (ABM) have long been used to model complex systems, but recently also to model the dynamics of school choice. ABMs are algorithmic, aiming to describe dynamics of systems in terms of the processes or algorithms through which each of the individual elements of the systems changes its state (e.g., school choice), responding to perceived inputs from its local environment. At the simplest level, an ABM consists of a system of agents, an environment and the relationships between them [4]. These agents are often autonomous, yet interdependent, and behave according to certain rules, where these rules can vary from simplistic to very sophisticated. Individual level decision rules can be simple heuristics, be specified by discrete choice models, neural networks or evolutionary algorithms as behavioral processes. This allows agents, interactions between them, or with the environment to be modeled explicitly, including assumptions describing their learning and adaptation to changing situational conditions. This approach becomes more important if analyzing average behavior is not enough, because a system is composed of heterogeneous substructures such as in clustered social networks or heterogeneous spatial patterns shaping the spatial distribution and accessibility of schools and the composition of their residential environments. Further, *adaptation* can occur when schools open or close, or households move due to changing demographics or residential patterns [13]. Moreover, *feedback effects* and *adaptation* potentially *emerge* from the rules and *interactions* governing agents' (inter)actions. ABM also allow to simulate multiple generations choosing schools, which could be used in analyzing *robustness, resilience, path-dependency* and *tipping points* with respect to school segregation. Also, agents can all be different to incorporate heterogeneity (i.e., varying household preferences or resources) and the environment allows for the explicit modeling of space, such as cities with infrastructure. These models have been applied in a stylized manner for school choice already, to link theoretical behavior rules to aggregate patterns [20,45,48] and more data-driven models to test system interventions or component behavior [21,38,49]. However, their usage in the field have been limited so far. To emphasize what ABM can add to existing methodologies, two examples are provided on how both stylized and more empirically calibrated ABM can improve our understanding of school segregation.

4.2   An Alternative Explanation for Excess School- Relative to Residential Segregation

In various educational systems, the level of school segregation is consistently higher than that of residential segregation [8,51]. One hypothesis is that parents might want to live in a diverse neighborhood, but when it comes to their children, they are less tolerant with respect to school compositions leading to less diverse schools [18]. However, residential and school segregation are not distinct emergent processes, they are intertwined as school choices are partially dependent on residential segregation (i.e., via distance preferences). Hence, this phenomenon can maybe also emerge, not because people really want it (intolerance), but due to *feedback loops* and *nonlinear* effects of interacting processes.

To study this, [20] create a stylized ABM where households are assumed to belong to one of two groups and choose neighborhoods based on composition and schools based on a trade-off between composition and distance. For composition preferences, households are assumed to have an optimal fraction of their own group ($t_i \in [0, 1]$) in a school/neighborhood, giving a utility of 1 (linearly increasing from 0) and receive only ($M \in [0, 1]$) if it is completely homogeneous (Fig. 2b), as research suggests many parents actually want some diversity in schools [9]. Schools are chosen by weighting ($\alpha \in [0, 1]$) distance and school composition, where 0 means only distance matters and 1 means only composition. Households strive to maximize their residential utility first and after that their school utility (given residential location). Importantly, households have the same composition preferences for neighborhoods as for schools, to test the alternative scenario. For a more detailed description of the model, the reader is referred to [20].

Unexpectedly, the level of residential- and school segregation increase when the optimal fraction increases or the penalty for homogeneity decreases (Fig. 3). However, given the exact same composition preference for neighborhoods as for schools, it is surprising that school segregation is consistently higher than residential segregation. This provides an alternative explanation of why schools are often more segregated than neighborhoods, not because of extra intolerance but due to compounding *feedback* and *nonlinear* effects. First, neighborhoods segregate more than expected based on individual preferences and, given distance preferences, households would prefer to attend the nearest school. However, some schools have a composition that likely deviates from the optimal one, due to residential segregation. Thus, at least one group starts to move out and travel further for a more favorable composition, triggering a whole cascade and increasing the level of school segregation above the level that would be expected without residential patterns.

4.3   An ABM of Amsterdam Primary School Choice

However, this stylized example has limited applicability to reality. In an more empirically calibrated model, [21] approximate residential locations of low- (blue)

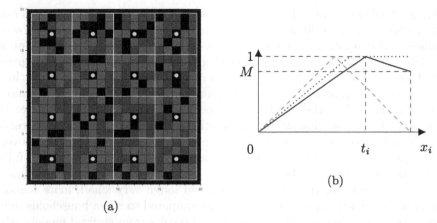

(a)

(b)

Fig. 2. (a) The environment of the ABM of [20], only 16 neighborhoods (white) and schools (yellow) are portrayed for visualization purposes. Households belong to one of two groups, either blue or red. (b) Single-peaked utility function. For the blue line, the agent obtains the maximum utility at t_i and only M if the neighborhood or school is homogeneous with respect to their own group. The green (dashed) and red line (dotted) show alternative values for t_i and M.

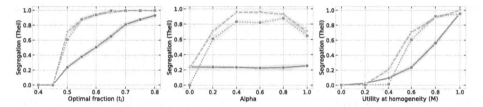

Fig. 3. The level of residential segregation (blue), school segregation with residential patterns (orange, dashed) and without (green, dotted). The plotted level of segregation is the average over 30 model runs and the bands represent the 2.5% and 97.5% percentiles. When the optimal fraction (t_i) is varied, $\alpha = 0.2, M = 0.6$, otherwise $t_i = 0.5$, fixed values of other parameters that are not varied can be found in Table 1 (Color figure online) of [20].

or high-income (red) households (Fig. 4). In Amsterdam, you get priority if you apply at one of you eight closest primary schools, which around 86% of pupils ultimately does [12]. As this percentage is quite high, this experiment models this priority-scheme as a strict geographic assignment mechanism: households are only allowed to consider their $n \in \{1, 2, 4, 8\}$ closest schools. Within this subset, households are assumed to be indifferent from a distance perspective and only composition matters (following the same function as in 2b). While this might be a strong assumption, 50% of both groups still have their eight closest schools within 1.9 km [21]. Hence, within their feasible choice set, households strive to find schools with their "optimal" composition.

Figure 5 indicates that school segregation is shown to be affected by both residential segregation and limiting the amount of schools households are allowed to consider. If everyone is assigned their closest school, residential segregation is fully determining school choices. Interestingly, as soon as households are given more choice (2, 4 or 8 schools) and they require at least a slight majority of their own group ($t \geq 0.5$), school segregation increases drastically. Segregation almost exclusively rises when the number of schools is increased, even for moderate preferences. Hence, in this model, more school choice leads to more segregation, which is also a consistent finding in empirical research [8,51]. Moreover, at some point more choice does not mean anything anymore (schools are fully segregated). Note that in the part of the parameter space where households strive to be a minority ($t < 0.5$, $M < 0.4$), although very small, more schools actually allow for less segregation to emerge compared to when households only consider the closest schools, resonating earlier results from stylized models [45]. This experiment suggests the final segregation that emerges is a delicate balance between preferred and locally available compositions (i.e., within "reasonable" distance). Hence, understanding how households consider distance and composition is vitally important to understand the potential for segregation in the city and also offering the number of priority schools can have significant effects on the resulting segregation levels. If one prefers more own group and residential segregation is strong: more schools might not help to desegregate, but if one prefers a smaller fraction of similar households and residential segregation is strong, then more choice could lead to less segregation. Note that context is still important, in an additional experiment with five income groups instead of two, no group will (realistically) have more than 50% in a school, hence the transitions in similar plots already happen at lower values values of the optimal fraction. This also stresses the importance of finding out what groups households feel they belong to and what they consider to be others. Also, constraining households to their closest school might have the adverse affect of inducing particular households (high-income) to move neighborhoods, increasing residential- and school segregation, which is not modeled here.

5   Challenges

While ABM are theoretically able to model the described underlying mechanisms of school segregation and can simulate (hypothetical) what-if scenarios, they also present several challenges. One major challenge is the difficulty in determining the appropriate level of detail and granularity for agent behavior and interactions. Specifying these rules at a fine-grained level can result in models that are computationally intractable and difficult to interpret, while overly simplified rules may not accurately reflect the complexity of real-world social phenomena [13]. Moreover, obtaining data to calibrate and validate the model can pose issues. Data on the micro-level (household) can be hard to gather or raise privacy and ethical concerns for example. Additionally, for actual decisions and/or policy interventions it is necessary that model assumptions, mechanisms

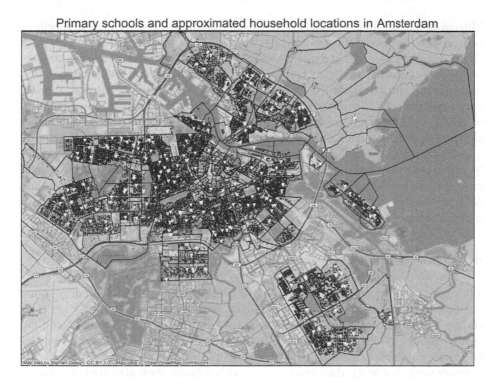

Fig. 4. Q1 households are in blue, Q5 in red, yellow triangles are the schools. (Color figure online)

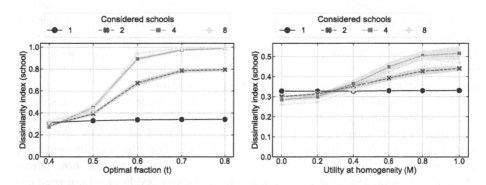

Fig. 5. The level of school segregation when varying the optimal fraction (t) and utility at homogeneity (M). $t = 0.5, M = 0.6$, when they are not varied. Estimates are the means of 10 model runs, error bands are one standard deviation.

and outcomes correspond with empirical observations, allowing validation and possible refutation of hypotheses generated by an ABM [40]. Specifying interactions between agents may require making assumptions about how information is transmitted and interpreted, which can be difficult to validate empirically.

Moreover, traditional estimation/calibration methods often require a specification of the likelihood function, which are almost impossible to write down for ABM. Existing studies have estimated parameters using different methodologies, before incorporating them in their ABM. However, likelihood-free inference provides techniques to calibrate ABM directly, but these can be computationally very expensive [22]. Finally, due to the nonlinear, stochastic, and dynamic nature of ABM, model output may be highly sensitive to initial conditions, and small changes in parameters can produce vastly different outcomes. Hence uncertainty quantification is an important aspect as well. However, these caveats should not be mistaken as argument that ABM is impractical to use in empirical research. It only shows that using ABM fruitfully requires adherence to methodologies and best practices for model analysis, calibration and validation which are increasingly developed in the field of computational social science [32].

6   Conclusion

This paper has argued that school choice exhibits all features commonly found in complex systems and therefore qualifies as one [29]. Households base their choice on information/perception of school properties (e.g., distance, composition, profile, quality). As their actual decision can change these properties it possibly influences choices of current/future generations (e.g., *feedback*, *path dependence*). Not only do they interact with schools, but also with their geographical/institutional environment (e.g., residential patterns, assignment mechanisms) and each other (e.g., school compositions, social network). Schools are players as well, depending on their level of autonomy they can impact affordability, school profiles, and their own admission policies. This indicates *feedback mechanisms* that can possibly lead to *non-linear* effects and *tipping points* that are hard to anticipate from the analysis of isolated individual components. School segregation might even be a sub-optimal outcome, as there is evidence of parents preferring more diverse schools than they actually attend [9].

Hence, to model the mechanisms behind school segregation one needs to model the individual households in the system, but also how they interact, influence macro-level components and feedback into individual choices. We argue that ABM can adhere to this need [40]. Depending on the specific question, these models can take a more stylized approach and allow to test how (theoretical) behavioral rules could be related to particular levels of school segregation or to provide alternative explanations for observed phenomena. The models can also be more data-driven, for example how primary school segregation in Amsterdam can emerge from relatively tolerant households with distance constraints and residential segregation. This can be extended by augmenting them with large-scale empirical data about socio-geographic contexts and properties of individual households and schools. This is important to validate model outcomes about household behavior (i.e., using micro-level data) and for testing whether the assumptions about which mechanisms lead to observed levels of school segregation explain real-life behavior and macro-level patterns [13,32]. However, determining the appropriate level of detail for these models involve a trade-off with computational tractability. Additionally ABM can be very sensitivity to initial conditions where small changes can lead to very different outcomes.

Keeping these challenges in mind, future modeling attempts in the domain of school choice and school segregation, could make more use of a social complexity approach such as ABM [21,49]. We believe that this offers a way forward to provide generative explanations of the mechanisms that lead to resilient patterns of school segregation and a pathway towards understanding policies that can help achieve more robust levels of school integration [13,32,40].

References

1. Anderson, P.W.: More is different. Science **177**(4047), 393–396 (1972)
2. Ball, S.J., Vincent, C.: "i heard it on the grapevine":'hot'knowledge and school choice. British journal of Sociology of Education **19**(3), 377–400 (1998)
3. Billingham, C.M., Hunt, M.O.: School racial composition and parental choice: new evidence on the preferences of white parents in the united states. Sociol. Educ. **89**(2), 99–117 (2016)
4. Bonabeau, E.: Agent-based modeling: methods and techniques for simulating human systems. Proc. Natl. Acad. Sci. **99**(suppl 3), 7280–7287 (2002)
5. Borghans, L., Golsteyn, B.H., Zölitz, U.: Parental preferences for primary school characteristics. The BE J. Econ. Anal. Policy **15**(1), 85–117 (2015)
6. Bosetti, L.: Determinants of school choice: understanding how parents choose elementary schools in alberta. J. Educ. Policy **19**(4), 387–405 (2004)
7. Boterman, W., Musterd, S., Pacchi, C., Ranci, C.: School segregation in contemporary cities: socio-spatial dynamics, institutional context and urban outcomes. Urban Stud. **56**(15), 3055–3073 (2019)
8. Boterman, W., Musterd, S., Pacchi, C., Ranci, C.: School segregation in contemporary cities: socio-spatial dynamics, institutional context and urban outcomes. Urban Stud. **56**(15), 3055–3073 (2019)
9. Boterman, W.R.: Dealing with diversity: middle-class family households and the issue of 'black'and 'white'schools in amsterdam. Urban Stud. **50**(6), 1130–1147 (2013)
10. Boterman, W.R.: The role of geography in school segregation in the free parental choice context of dutch cities. Urban Studies p. 0042098019832201 (2019a)
11. Boudon, R.: The Unintended Consequences of Social Action. Palgrave Macmillan UK, London (1977). https://doi.org/10.1007/978-1-349-04381-1
12. Breed Bestuurlijk Overleg: Placement of children in Amsterdam primary schools 2020-2021. https://bboamsterdam.nl/wat-we-doen/toelatingsbeleid/uitvoering-van-stedelijk-toelatingsbeleid-in-cijfers/ (2021) (Accessed July 2022)
13. Bruch, E., Atwell, J.: Agent-based models in empirical social research. Sociological Methods Res. **44**(2), 186–221 (2015)
14. Candipan, J.: Choosing schools in changing places: Examining school enrollment in gentrifying neighborhoods. Sociology of Education p. 0038040720910128 (2020)
15. Card, D., Mas, A., Rothstein, J.: Tipping and the dynamics of segregation. Q. J. Econ. **123**(1), 177–218 (2008)
16. Cioffi-Revilla, C.: Introduction to computational social science. Springer, London and Heidelberg (2014)

17. Conte, R., et al.: Manifesto of computational social science. The European Physical Journal Special Topics **214**(1), 325–346 (2012)
18. Cordini, M., Parma, A., Ranci, C.: 'white flight'in milan: school segregation as a result of home-to-school mobility. Urban Stud. **56**(15), 3216–3233 (2019)
19. De Wolf, T., Holvoet, T.: Emergence versus self-organisation: different concepts but promising when combined. In: Brueckner, S.A., Di Marzo Serugendo, G., Karageorgos, A., Nagpal, R. (eds.) ESOA 2004. LNCS (LNAI), vol. 3464, pp. 1–15. Springer, Heidelberg (2005). https://doi.org/10.1007/11494676_1
20. Dignum, E., Athieniti, E., Boterman, W., Flache, A., Lees, M.: Mechanisms for increased school segregation relative to residential segregation: a model-based analysis. Comput. Environ. Urban Syst. **93**, 101772 (2022)
21. Dignum, E., Boterman, W., Flache, A., Lees, M.: A data-driven agent-based model of primary school segregation in amsterdam [working paper] (2023)
22. Dyer, J., Cannon, P., Farmer, J.D., Schmon, S.: Black-box bayesian inference for economic agent-based models. arXiv preprint arXiv:2202.00625 (2022)
23. Fong, K., Faude, S.: Timing is everything: late registration and stratified access to school choice. Sociol. Educ. **91**(3), 242–262 (2018)
24. Gutiérrez, G., Jerrim, J., Torres, R.: School segregation across the world: has any progress been made in reducing the separation of the rich from the poor? J. Econ. Inequal. **18**(2), 157–179 (2020)
25. Haber, J.R.: Sorting schools: A computational analysis of charter school identities and stratification (2020)
26. Hastings, J.S., Kane, T.J., Staiger, D.O.: Parental Preferences And School Competition: Evidence From A Public School Choice Program. Tech. rep, National Bureau of Economic Research (2005)
27. Holme, J.J.: Buying homes, buying schools: School choice and the social construction of school quality. Harv. Educ. Rev. **72**(2), 177–206 (2002)
28. Ladd, H.F.: School vouchers: a critical view. J. Econ. Perspect. **16**(4), 3–24 (2002)
29. Ladyman, J., Lambert, J., Wiesner, K.: What is a complex system? Eur. J. Philos. Sci. **3**(1), 33–67 (2013)
30. Ladyman, J., Wiesner, K.: What is a complex system? Yale University Press (2020)
31. Lauen, L.L.: Contextual explanations of school choice. Sociol. Educ. **80**(3), 179–209 (2007)
32. León-Medina, F.J.: Analytical sociology and agent-based modeling: is generative sufficiency sufficient? Sociol Theor. **35**(3), 157–178 (2017)
33. Logan, J.R., Minca, E., Adar, S.: The geography of inequality: why separate means unequal in american public schools. Sociol. Educ. **85**(3), 287–301 (2012)
34. Mäs, M.: Interactions. In: Research Handbook on Analytical Sociology. Edward Elgar Publishing (2021)
35. McPherson, M., Smith-Lovin, L., Cook, J.M.: Birds of a feather: homophily in social networks. Ann. Rev. Sociol. **27**(1), 415–444 (2001)
36. Miller, J.H., Page, S.E.: Complex adaptive systems: An introduction to computational models of social life. Princeton University Press (2009)
37. Mitchell, M.: Complexity: A guided tour. Oxford University Press (2009)
38. Mutgan, S.: Free to Choose?: Studies of Opportunity Constraints and the Dynamics of School Segregation. Ph.D. thesis, Linköping University Electronic Press (2021)
39. Nicolis, G., Nicolis, C.: Foundations of complex systems: emergence, information and prediction. World Scientific (2012)
40. Page, S.E.: What sociologists should know about complexity. Ann. Rev. Sociol. **41**, 21–41 (2015)

41. Pearman, F.A., Swain, W.A.: School choice, gentrification, and the variable significance of racial stratification in urban neighborhoods. Sociol. Educ. **90**(3), 213–235 (2017)
42. Perry, L.B., Rowe, E., Lubienski, C.: School segregation: theoretical insights and future directions (2022)
43. Reardon, S.F., Owens, A.: 60 years after brown: trends and consequences of school segregation. Ann. Rev. Sociol. **40**, 199–218 (2014)
44. Renzulli, L.A.: Organizational environments and the emergence of charter schools in the united states. Sociol. Educ. **78**(1), 1–26 (2005)
45. Sage, L., Flache, A.: Can ethnic tolerance curb self-reinforcing school segregation? a theoretical agent based model. J. Artif. Societies Social Simulation **24**(2), 2 (2021). https://doi.org/10.18564/jasss.4544, http://jasss.soc.surrey.ac.uk/24/2/2.html
46. Scheffer, M.: Critical transitions in nature and society, vol. 16. Princeton University Press (2009)
47. Schmidt, W.H., Burroughs, N.A., Zoido, P., Houang, R.T.: The role of schooling in perpetuating educational inequality: an international perspective. Educ. Res. **44**(7), 371–386 (2015)
48. Stoica, V.I., Flache, A.: From schelling to schools: a comparison of a model of residential segregation with a model of school segregation. J. Artif. Soc. Soc. Simul. **17**(1), 5 (2014)
49. Ukanwa, K., Jones, A.C., Turner, B.L., Jr.: School choice increases racial segregation even when parents do not care about race. Proc. Natl. Acad. Sci. **119**(35), e2117979119 (2022)
50. Van de Werfhorst, H.G.: Early tracking and social inequality in educational attainment: educational reforms in 21 european countries. Am. J. Educ. **126**(1), 65–99 (2019)
51. Wilson, D., Bridge, G.: School choice and the city: Geographies of allocation and segregation. Urban Studies, pp. 1–18 (2019)

An Approach for Probabilistic Modeling and Reasoning of Voting Networks

Douglas O. Cardoso[2]([✉])[iD], Willian P. C. Lima[1][iD], Guilherme G. V. L. Silva[1], and Laura S. Assis[1][iD]

[1] Celso Suckow da Fonseca Federal Center of Technological Education, Rio de Janeiro, RJ, Brazil
{willian.lima,guilherme.lourenco}@aluno.cefet-rj.br,
laura.assis@cefet-rj.br
[2] Smart Cities Research Center, Polytechnic Institute of Tomar, Tomar, Portugal
douglas.cardoso@ipt.pt

Abstract. This work proposes a methodology for a sounder assessment of centrality, some of the most important concepts of network science, in the context of voting networks, which can be established in various situations from politics to online surveys. In this regard, the network nodes can represent members of a parliament, and each edge weight aims to be a probabilistic proxy for the alignment between the edge endpoints. In order to achieve such a goal, different methods to quantify the agreement between peers based on their voting records were carefully considered and compared from a theoretical as well as an experimental point of view. The results confirm the usefulness of the ideas herein presented, and which are flexible enough to be employed in any scenario which can be characterized by the probabilistic agreement of its components.

Keywords: Node Centrality · Weighted Networks · Data Summarization · Social Computing · Voting Networks

1 Introduction

Information Systems can combine tools from various branches of knowledge, such as Data Science, enabling the discovery of interesting insights into complex systems from data respective to these [1,13]. In this regard, a central aspect is the proper use of theoretical models which soundly fit available data leading to meaningful predictions. The trending importance of interpretability in such a context has been deservedly indicated in the recent literature [25]. Likewise, this has also happened in the more specific scenario of techniques related to complex networks: for example, spectral graph theory [34], random graph models [32], and social network analysis [3]. This work aims to contribute to the literature similarly while focusing on probabilistic networks modeling and analysis [16,18, 19,30,31].

The importance of such modeling is evidenced by the wide range of concrete problems which can be approached from this point of view. This could be

© The Author(s), under exclusive license to Springer Nature Switzerland AG 2023
J. Mikyška et al. (Eds.): ICCS 2023, LNCS 14075, pp. 90–104, 2023.
https://doi.org/10.1007/978-3-031-36024-4_7

expected given the ubiquity of uncertainty and chance in all aspects of life. Such diversity also implies some challenges, considering the goal of developing tools whose use is not limited to a single probabilistic setting but ideally are flexible enough to help make better sense of most of these occasions. With that in mind, in this work a bottom-up strategy was carried out, from the reconsideration of basic probabilistic network features up to methods and results which rely on this fresher look.

Notwithstanding the just indicated ideas, since our work is oriented towards mining concrete systems modeled as complex networks, we decided to present the proposed methodology concomitantly with a use case regarding relationships between members of a political congress [14,15,26,27] based on their voting records [8,9,11]. Therefore, in a rough description, the target of this paper is to show how to draw interesting conclusions from such political data, assessing the alignment between peers, which enables a multifaceted centrality analysis built on a solid yet straightforward mathematical foundation. And as a completing aspect, empirical results derived from applying this methodology on real data of the lower house of the Brazilian National Congress are reported and discussed.

The remainder of this paper is organized as follows. Section 2 surveys the theoretical background that substantiates this work. Section 3 addresses the related works and also highlights the contributions of this work in view of the existing literature. The proposed methodology is described in Sect. 4, while Sect. 5 presents its experimental evaluation and the respective discussion of the results. Some concluding remarks are stated in Sect. 6.

2 Fundamental Concepts

Many real-world circumstances can be represented using graph theory, popularly depicted as a diagram composed of a set of points and lines connecting pairs of this set. A simple undirected *graph* $G(V, E)$ consists of a set V of n *nodes*, and a set E of m *edges*. Each edge in a graph is specified by a pair of nodes $\{u, v\} \in E$, with $u \in V$, $v \in V$. A node u is *adjacent* to node v if there is an edge between them. The degree of a node is the number of edges that are incident upon the node. As an extension, in a *weighted graph* each edge $e_i \in E, i = 1, \cdots, m$ has a value $w(e_i) \in \mathbb{R}$ associated to it. This kind of graph can be denoted as $G = (V, E, w)$, where $w : \binom{V}{2} \to \mathbb{R}$.

A graph whose set of nodes equals the union of two disjoint sets (i.e., $V = T \cup U$, and $T \cap U = \varnothing$) such that any two nodes in the same set are not adjacent is called *bipartite graph, two-mode graph*, or *bigraph* [35]. A *projection* of a bipartite graph $G(T \cup U, E)$ is a graph whose nodes are a subset of T or U, and whose edges represent the existence of a common neighbor between its endpoints in G. Edges in a projection of a bipartite graph can be weighted in numerous ways, e.g. considering the absolute or relative number of shared neighbors of their endpoints in the original graph [6], or using a context-specific method [28].

Given a representation of a real-world system as a network (i.e., a concrete analog of a graph, which is an abstract structure), it is common to examine the characteristics and structural properties of the network. Many properties can be associated with graph nodes, such as distance and centrality. A certain awareness of the individual system elements' importance may be obtained by measures of how "central" the corresponding vertex is in the network. The search for "communities" and analogous types of indefinite "clusters" within a system may be addressed as a graph partitioning problem [21]. Some important definitions for this sort of network characterization are presented next.

Measures of centrality are designed to quantify notions of importance and thereby facilitate answering some questions related to the network. A common notion of a *central* node is a node that can easily reach many other nodes in the graph through its edges. There is an immense number of different centrality measures that have been proposed over the years [12]. This subsection discusses two classic measures on which the proposed methodology was developed: degree and strength.

The Degree Centrality [17] of a node is possibly the simplest centrality measure and refers to the number of incident edges upon this node, i.e., its own degree. From this point of view, the higher the degree, the more central the node is. This measure better represents influence from a local perspective on the graph. This is a straightforward yet effectual measure: in various contexts nodes with high degrees are ruled as notably central by other measures [5]. The Degree Centrality of the node u can be established as shown in Eq. (1).

$$C_D(u) = |\{e \in E : u \in e\}| . \tag{1}$$

This measure was originally proposed for unweighted graphs. Afterwards, some generalizations were developed to contemplate valued networks. Barrat et al. [4] extended the notion of degree centrality of a node and called it the node *strength*, which was defined by the sum of the weights of all edges incident in it. Equation (2) presents the calculation of the strength of a given node u.

$$s_u = C_D^w(u) = \sum_{e \in E : u \in e} w(e) . \tag{2}$$

3   Literature Review

As stated in Sect. 1, probabilistic networks have a broad and rich history of applications. One of the oldest elements of such a collection is the work of Frank [16], which regards finding shortest paths in graphs whose edge weights are random variables instead of deterministic values. Sevon et al. [31] considered a similar problem but in graphs whose edge weights regard the probability of interaction between nodes. This setting was also explored by Potamias et al. [30], who proposed an approach to answering k-nearest neighbors queries in graphs as such. More recently, Fushimi et al. [18] used Monte Carlo simulations on the same kind of graph to establish *connected centrality* for suggesting the positioning of evacuation facilities against natural disasters. And regarding social networks, the study of probabilistic ego networks can be mentioned [19].

In the political science context, network analysis has been recognized as an invaluable tool for the development of research at the highest level [33]. Several studies have been conducted based on the relational fundamentals of politics, giving rise to different political networks models and methodologies to infer and validate contextual features [9]. As an example, Lee and Bearman [22] used that platform for arguing that political and ideological isolation and homogenization of U.S. population have never been so strong, according to data obtained through a survey.

Alternatively, studies concerning networks whose nodes are members of a political congress are very popular, even considering those which are not based on voting records: co-sponsorship was already used in this regard [14,15,27], while other dyadic actions as co-authoring and co-attendance were discussed [26]. These networks are commonly studied because it signals peer endorsement very explicitly and publicly. Networks induced by co-sponsorship of bills in legislative bodies have been used to examine, e.g., homophily [7] and characterize ideological evolution over time [20].

Lee et al. [24] presented a study of the community structure in time-dependent legislation co-sponsorship networks in the Peruvian Congress, which is compared with the networks of the US Senate. A multilayer representation of temporal networks is employed, and a multilayer modularity maximization is used to detect communities in these networks. How much the legislators tend to form ideological relationships with members of the opposite party is measured by Andris et al. [2]. The authors quantify the level of cooperation or lack thereof between Democrat and Republican party members in the U.S. House from 1949 to 2012.

A study about the Brazilian Congress network to investigate the relationships between the donations received by congressmen elected and their voting behaviors during next two years is presented in a paper by Bursztyn et al. [9]. Two networks are built and analyzed, the donation network, and the voting network. In both networks, the vertices are the congressmen. The homophily and cohesion of congressmen are investigated. The results indicate that regions exhibit stronger homophily than political parties in the donation network, while this trend is opposite for the voting network.

Using data from Twitter and other sources (e.g., roll-call vote data), Peng et al. [29] examined how and why the members of congress connect and communicate with one another on Twitter and also what effects such connection and communication have on their vote behavior. This study shows a high degree of partisan homogeneity and the homophily effect in social network research. A network with voting data from the Brazilian Deputies Chamber is determined in the work of Brito et al. [8]. A methodology for studying the evolution of political entities over time is presented. Although a multiparty political system characterizes the Brazilian Chamber of Deputies, the results obtained reveal that the expected plurality of ideas did not occur.

The network of relations between parliament members according to their voting behavior regarding political coalitions and government alliances was analyzed by Dal Maso et al. [11]. Existing tools for complex networks were used to exploit such a network and assist in developing new metrics of party polarization, coalition internal cohesion, and government strength. The methodology presented was applied to the Chamber of Deputies of the Italian Parliament. The results made it possible to characterize the heterogeneity of the governing coalition and the specific contributions of the parties to the government's stability over time.

4   Methodology

In this work, the starting point of the proposed methodology is establishing a bipartite network relating members of parliament (MPs) to ballots in which they participated. This network is unweighted, but each edge is marked with the option chosen for the single vote it represents, e.g.: *(i)* yes, *(ii)* no, *(iii)* obstruction, *(iv)* abstention. It is then possible to project an all-MPs network, whose links indicate at least one joint participation in the ballots considered, regardless of the options taken. For the sake of simplicity, the bipartite network is assumed to be connected, implying the same property for its projection.

4.1   Edge Weighting

The edges of this projected network can be weighted for a deeper assessment of the relationships between MPs. An interesting approach in this regard is to model the empirical probability of agreement, i.e., voting likewise, or sharing a common interest in some political topics. With this in mind, let a matrix $V_{m \times n}$ regard the participation of m MPs in n ballots, such that $v_{i,j}$ may represent the option chosen by the i-th MP in the j-th ballot, or it indicates that the MP did not take part in this ballot ($v_{i,j} = None$). Also let $H_i = \{j : v_{i,j} \neq None\}$ be the set of ballots in which the i-th MP indeed participated, and $O_i = \{(j, v_{i,j}) : j \in H_i\}$ be the set of pairs representing options chosen by an MP i in the respective ballots.

Then, the *agreement* between MPs x and y could be assessed according to ratio between of the number of votes they had in common and the number of ballots in which both participated, as shown in Eq. (3), which is hereinafter referred to as *co-voting agreement*. Consider $\alpha > 0$ a parameter for additive smoothing of the Jaccard Index-like computation of $A(x,y)$ [10,23]: by default, it was used in this work $\alpha = 1$.

$$A(x,y) = \frac{|O_x \cap O_y| + \alpha}{|O_x \cup O_y| + 2\alpha} , \qquad (3)$$

A projected network of MPs whose edge weights are defined by this method is referred to as a Direct Agreement Network (DAN). The edge weights in the DAN have a positive interpretation as a proxy to voting alignment. Figure 1 provides an example of a voting network and the DAN which results from its projection as described.

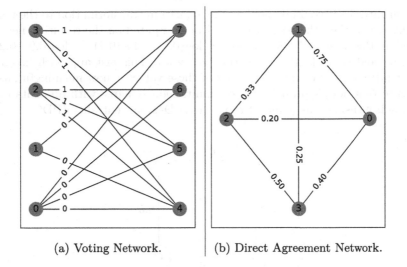

(a) Voting Network. | (b) Direct Agreement Network.

Fig. 1. Hypothetical voting network (left) and its respective DAN (right), considering $\alpha = 1$. Nodes 0 to 3 represent voters, while nodes 4 to 7 represent polls. In (a), each edge implies the participation of the voter in one of its ends on the poll in its other end, by voting according to the edge label (in this example, 0 or 1). In (b), each edge implies that the voters it connects took part of the same poll at least once.

4.2 Probabilistic Centrality

Now consider evaluating node strength in this network: weights summation is inherent to such computation. However, this is questionable since there is no guarantee that the events the edges denote are mutually exclusive: the probabilistic understanding of such sums is unclear, as they could become even greater than 1.

The substitution of the probabilities by their logarithms may allow to counter these problems. This transformation also avoids the semantic loss previously imposed by weights summation. Adding log-probabilities is directly related to multiplying probabilities, which in turn can point to the joint probability of independent events occurrence. Although the independence of agreement events cannot be assured a priori, its assumption can be seen as a modeling simplification with no negative consequences observed so far.

The logarithm transformations can be conveniently employed to compute more insightful centrality indexes. The Strength can be computed for all DAN nodes so that it could be considered reasonable to affirm that the MP/node with the highest strength is very well-connected to its neighbors. However, comparing these values can be tricky: in the toy DAN shown in Fig. 2, nodes A and C have the same strength (equal to 1.1), although the weight distribution over their edges differs.

Alternatively, it could be applied the logarithm transformation to the respective DAN, and then the strengths could be computed on the resulting network. Hence, in this case, cologDAN: $s_A = \log(0.9) + \log(0.2) = \log(0.9 \cdot 0.2) = \log(0.18)$, and $s_C = \log(0.30)$, so that $s_C > s_A$. This last approach has a significant advantage over the first. Each of these values is now meaningful, as the logarithm of a node's probability to simultaneously agree with all of its neighbors. Such measure was named Probabilistic Degree Centrality (PDC).

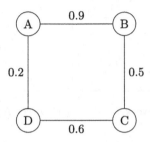

Fig. 2. A toy DAN. The edge weights regard the probability of agreement between the voters represented by the nodes.

4.3 Political Contextualization

The just introduced centrality indexes are flexible enough to be used for any network whose edge weights regard the chance of agreement, or some similar concept, between the nodes they connect. These are far from being limited to voting scenarios. However, in which context this work is focused allows specific interpretations and uses of some of these indexes as well as concepts used for their definition. The insights in this regard are detailed next.

First of all, in a broad sense, it is considered how the idea of node centrality provides a concrete and useful perspective of the probabilistic political networks in question. The most central node of a network is often regarded as the most important of all, for example, the one closest to its peers, the strongest influencer, or the one whose removal would result in the greatest loss of data traffic efficiency. Although it is tempting to consider the most central MP as the greatest leader of the house, there is no support to such hypothesis since this could only be reasonably stated if the agreement between peers was assessed in a directed fashion, which is not the case.

Despite this negative, a sensible characterization of the MPs according to their centralities is still possible. One perspective which enables reaching this goal is that of summarization, where the MP with whom there is a greater chance of an agreement with all others can serve as an archetype of the house members. Thus, it is processing a voting profile that resembles a medoid of collecting all items of this kind. Moreover, the same principles can be employed in the identification of the member of a party that better represents it.

Going a little further in the just indicated direction, if known in advance the orientation of a central MP in a ballot, this information can be used as a predictor of its result. How confident one can be about such prediction can be conveniently evaluated since the outputs of the centrality measures proposed are probabilistically sound. These values can also objectively assess how cohesive the entire group of MPs or even a subset is. Thereby enabling compare different parties in a given period as well as the house in different periods, for example.

5   Experimental Evaluation

This section presents and discusses practical results obtained from the application of the proposed concepts to synthetic and naturally-produced data. This aimed at analyzing the proposed methodology under controlled circumstances, enabling a clearer perception of its features, as well as observing its behavior when subject to the idiosyncrasies which are inherent to real applications.

5.1   Synthetic Data

This first collection of experiments aimed at validating the idea which inspired all subsequent developments which were realized: taking into account the probabilistic structure of a voting network instead of ignoring it enables a more insightful analysis of its properties. In order to confirm this statement it was then hypothesized that the regular assessment of strength (Eq. (2)) in a DAN would lead to a node ranking which could be notably dissimilar to that produced by its logDAN counterpart. This was tested as follows.

First, it was established a simple model to generate random voting networks. Its parameters were the number of polls to consider, the number of parties, the number of members of each party, the probability of attendance of the voters to polls, and the probability of loyalty of the voters to their parties: that is, in every poll, each of the voters can only be absent, or be loyal by voting in its own party, or be independent by voting in any party randomly. This model was used to produce numerous artificial voting networks, each of which was transformed to a DAN in order to compute the regular and probabilistic node centrality rankings, whose similarity was at last evaluated using the Kendall's τ non-parametric correlation coefficient [23].

To put the intended results into perspective, other methods for the projection of bipartite networks were also employed, despite the fact that they rely solely on the topological configuration of its input graph and but ignore attributes of its edges, as votes. Targeting to mimic the consideration of such attributes, instead of applying these methods on the original voting network, whose edges regarded every type of vote, they were applied on alternative versions of these in which only the edges of votes to party #1 were preserved. This way the coincidence between node neighborhoods, a concept which all these methods share, could be better used for assessing voting alignment. The rivals methods considered were:

weighted projection, Jaccard-based overlap projection, and maximum overlap ratio projection [6].

Figure 3 illustrates the results obtained considering 2 parties, each with 100 members, 100 polls, and varied scenarios with respect to the loyalty and attendance probabilities. In each scenario, a total of 100 random voting networks were generated and processed, targeting to ensure the statistical stability of the reported averages. It can be noticed that the proposed methodology (top-left subfigure) reflected more clearly than its counterparts the variation of both model parameters in question, exhibiting a smooth gradient. Moreover, it is possible to reason that as loyalty is diminished, what makes parties more irrelevant as voting becomes purely random, the distinction between regular and probabilistic degree centrality also vanishes.

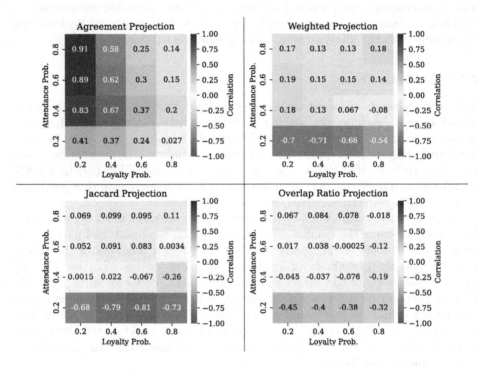

Fig. 3. Average Correlation between regular and probabilistic degree centrality rankings for synthetic data: 2 parties, 100 members per party, 100 polls. Compared to its alternatives, the proposed Agreement Projection better reflects variations in parameters of randomly generated voting networks.

As a final remark regarding artificial data, Fig. 4 presents the results of the proposed methodology in a setting which is similar to the one just described but now with 3 parties. The aforementioned rival methods are unable to handle this case since opting to keep only the edges regarding the votes of a single party means discarding the votes of the other two parties. Our methodology not only can handle this case, as evidenced by the same sensitiveness and smoothness in the results, but any number of different types of votes.

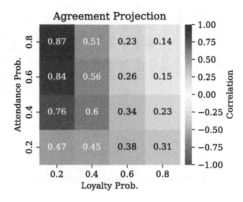

Fig. 4. Average Correlation between regular and probabilistic degree centrality rankings for synthetic data: 3 parties, 100 members per party, 100 polls. Agreement Projection can once again capture data peculiarities, even in this scenario which its counterparts are unable to properly represent by definition.

5.2   Real Data

These tests rely on publicly available records of Brazil's Chamber of Deputies, the lower house of the National Congress of the country: results from open votes that happened during 2021. First, a descriptive analysis of the dataset is provided, displaying some general information for the portrait of the context at hand. Then the network-based modeling and its developments are carefully reported.

As a start, it is presented next a broad description of the dataset focusing on its size. Figure 5 depicts the total number of votes respective to each of the 25 parties featured in the Congress ("S.PART." regards independent, unaffiliated deputies), which were declared in a total of 956 opportunities. Figure 6 also presents the total number of votes but according to its type: there is a total of 5 types, although two of those clearly dominate the distribution.

At last, Fig. 7 displays the overall correlation between regular and probabilistic degree centrality node rankings of DANs produced from voting networks resulting from all votes grouped on a monthly basis. It also displays the correlation considering scenarios limited to some chosen pairs of parties: PSL and PT are the more numerous parties of the Congress, and represent the right-wing and

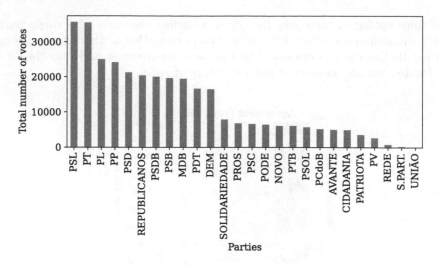

Fig. 5. Number of votes by each party.

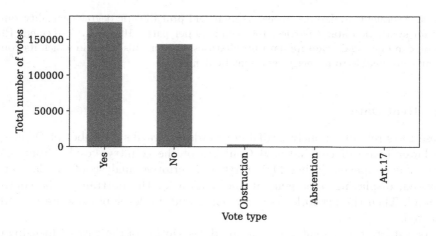

Fig. 6. Number of votes of each type. "Art. 17" is a vote type reserved to the president of the house, who can vote only in exceptional occasions.

left-wing spectrum, respectively; NOVO and PSOL are also from the right-wing and left-wing spectrum respectively but are smaller, more cohesive, and more ideologically-oriented than the first two mentioned.

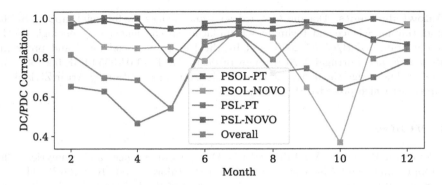

Fig. 7. Correlations between node rankings defined by their respective regular and probabilistic degree centralities, considering members of all parties or selected pairs of them, as well as monthly intervals. The behavior of the entire congress resembles that of the PSL-PT pair, which are the poles to the right and to the left of the political spectrum, respectively. The PSL-NOVO and PSOL-PT pairs are highly correlated, which is coherent with their ideological alignment.

As it can be observed, the results of the PSL-PT pair resemble those of the entire congress ("Overall"), which is consistent with the fact that these parties encompass the greatest number of MPs. Moreover, the correlation of this pair is in general smaller than that of all parties together, evidencing an antagonism in a higher level than that which is employed in general in the house. It is also interesting to confirm the ideological alignment of the PSOL-PT and PSL-NOVO pairs: the high correlation during the entire year indicates that they work with minimal discordance, in a similar fashion to the scenarios of random voting discussed in the previous subsection.

6   Conclusion

The inference and analysis of relationships in a political context are undeniably valuable, not only for politicians themselves but for the population in general, contributing for a better understanding of the social landscape in which we are all inserted. The employment of methods from network science in this regard is not new. However, generic approaches sometimes fail to take into account the specificities of the scenario in which they are used.

We believe the probabilistic centrality indexes proposed in this work directly tackle this issue, providing the means to a deeper and more reliable understanding of voting networks. Moreover, the proposed methodology is flexible enough to be used on other systems with similar basic attributes, such as having the alignment of its components as random events. As a future work, a comparison of parliaments of different countries based on the proposed methodology could lead to interesting results and discussions. Another idea that could be explored is to use such the proposed ideas in the context of recommender systems, considering the agreement between customers given their behavioral records.

Acknowledgements. Douglas O. Cardoso acknowledges the financial support by the Foundation for Science and Technology (Fundação para a Ciência e a Tecnologia, FCT) through grant UIDB/05567/2020, and by the European Social Fund and programs Centro 2020 and Portugal 2020 through project CENTRO-04-3559-FSE-000158. The authors acknowledge the motivational and intellectual inspiration by Artur Ziviani (*in memoriam*) which preceded this work.

References

1. Agarwal, R., Dhar, V.: Editorial-Big Data, Data Science, and Analytics: The Opportunity and Challenge for IS Research. Inform. Syst. Res. **25**(3), 443–448 (Sep 2014). https://doi.org/10.1287/isre.2014.0546, https://pubsonline.informs.org/doi/10.1287/isre.2014.0546

2. Andris, C., Lee, D., Hamilton, M.J., Martino, M., Gunning, C.E., Selden, J.A.: The rise of partisanship and super-cooperators in the u.s. house of representatives. PLoS One **10**(4), 127–144 (2016). https://doi.org/doi.org/10.1371/journal.pone.0123507

3. Atzmueller, M.: Towards socio-technical design of explicative systems: Transparent, interpretable and explainable analytics and its perspectives in social interaction contexts information. In: Nalepa, G.J., Ferrández, J.M., Palma-Méndez, J.T., Julián, V. (eds.) Proceedings of the 3rd Workshop on Affective Computing and Context Awareness in Ambient Intelligence (AfCAI 2019), Universidad Politécnica de Cartagena, Spain, November 11–12, 2019. CEUR Workshop Proceedings, vol. 2609. CEUR-WS.org (2019)

4. Barrat, A., Barthelemy, M., Pastor-Satorras, R., Vespignani, A.: The architecture of complex weighted networks. Proc. Natl. Acad. Sci. **101**(11), 3747–3752 (2004)

5. Borgatti, S.P., Everett, M.G.: A graph-theoretic perspective on centrality. Soc. Netw. **28**(4), 466–484 (2006). https://doi.org/10.1016/j.socnet.2005.11.005

6. Borgatti, S.P., Halgin, D.S.: Analyzing Affiliation Networks. In: The SAGE Handbook of Social Network Analysis, pp. 417–433. SAGE Publications Ltd, 1 Oliver's Yard, 55 City Road, London EC1Y 1SP United Kingdom (2014). https://doi.org/10.4135/9781446294413.n28

7. Bratton, K.A., Rouse, S.M.: Networks in the legislative arena: how group dynamics affect cosponsorship. Legislative Stud. Quart. **36** (2011). https://doi.org/doi-org.ez108.periodicos.capes.gov.br/10.1111/j.1939-9162.2011.00021.x

8. Brito, A.C.M., Silva, F.N., Amancio, D.R.: A complex network approach to political analysis: Application to the brazilian chamber of deputies. PLOS ONE **15**(3), 1–21 (03 2020). https://doi.org/10.1371/journal.pone.0229928

9. Bursztyn, V.S., Nunes, M.G., Figueiredo, D.R.: How brazilian congressmen connect: homophily and cohesion in voting and donation networks. J. Complex Netw. **8**(1) (2020). https://doi.org/10.1093/comnet/cnaa006

10. Cardoso, D.O., França, F.M.G., Gama, J.: WCDS: a two-phase weightless neural system for data stream clustering. N. Gener. Comput. **35**(4), 391–416 (2017). https://doi.org/10.1007/s00354-017-0018-y

11. Dal Maso, C., Pompa, G., Puliga, M., Riotta, G., Chessa, A.: Voting behavior, coalitions and government strength through a complex network analysis. PLOS ONE **9**(12), 1–13 (12 2015). https://doi.org/10.1371/journal.pone.0116046

12. Das, K., Samanta, S., Pal, M.: Study on centrality measures in social networks: a survey. Soc. Netw. Anal. Min. **8**(1), 1–11 (2018). https://doi.org/10.1007/s13278-018-0493-2

13. Dhar, V.: Data science and prediction. Commun. ACM **56**(12), 64–73 (2013). https://doi.org/10.1145/2500499
14. Fowler, J.H.: Connecting the congress: a study of cosponsorship networks. Polit. Anal. **14**(4), 456–487 (2006). https://doi.org/10.1093/pan/mpl002
15. Fowler, J.H.: Legislative cosponsorship networks in the US house and senate. Soc. Netw. **28**(4), 454–465 (2006). https://doi.org/10.1016/j.socnet.2005.11.003
16. Frank, H.: Shortest Paths in Probabilistic Graphs. Operations Research **17**(4), 583–599 (Aug 1969). https://doi.org/10.1287/opre.17.4.583, publisher: INFORMS
17. Freeman, L.C.: Centrality in social networks conceptual clarification. Social Netw. **1**(3), 215–239 (1978)
18. Fushimi, T., Saito, K., Ikeda, T., Kazama, K.: A new group centrality measure for maximizing the connectedness of network under uncertain connectivity. In: Aiello, L.M., Cherifi, C., Cherifi, H., Lambiotte, R., Lió, P., Rocha, L.M. (eds.) Complex Networks and Their Applications VII - Volume 1 Proceedings The 7th International Conference on Complex Networks and Their Applications COMPLEX NETWORKS 2018, Cambridge, UK, December 11–13, 2018. Studies in Computational Intelligence, vol. 812, pp. 3–14. Springer (2018). https://doi.org/10.1007/978-3-030-05411-3_1
19. Kaveh, A., Magnani, M., Rohner, C.: Defining and measuring probabilistic ego networks. Soc. Netw. Anal. Min. **11**(1), 1–12 (2020). https://doi.org/10.1007/s13278-020-00708-w
20. Kirkland, J.H., Gross, J.H.: Measurement and theory in legislative networks: the evolving topology of congressional collaboration. Social Netw. **36** (2014). https://doi.org/dx.doi.org/10.1016/j.socnet.2012.11.001
21. Kolaczyk, E.D.: Statistical Analysis of Network Data: Methods and Models. 1 edn. (2009). https://doi.org/10.1007/978-0-387-88146-1
22. Lee, B., Bearman, P.: Political isolation in america. Network Science, pp. 1–23 (02 2020). https://doi.org/10.1017/nws.2020.9
23. Lee, L.: Measures of distributional similarity. In: Proceedings of the 37th annual meeting of the Association for Computational Linguistics on Computational Linguistics, pp. 25–32. ACL '99, Association for Computational Linguistics, USA (Jun 1999). https://doi.org/10.3115/1034678.1034693
24. Lee, S.H., Magallanes, J.M., Porter, M.A.: Time-dependent community structure in legislation cosponsorship networks in the congress of the republic of peru. J. Complex Netw. **5**(1), 127–144 (2016). https://doi.org/doi-org.ez108.periodicos.capes.gov.br/10.1093/comnet/cnw004
25. Murdoch, W.J., Singh, C., Kumbier, K., Abbasi-Asl, R., Yu, B.: Definitions, methods, and applications in interpretable machine learning. Proc. Natl. Acad. Sci. **116**(44), 22071–22080 (2019). https://doi.org/10.1073/pnas.1900654116
26. Neal, Z.: The backbone of bipartite projections: inferring relationships from co-authorship, co-sponsorship, co-attendance and other co-behaviors. Soc. Netw. **39**, 84–97 (2014). https://doi.org/10.1016/j.socnet.2014.06.001
27. Neal, Z.P.: A sign of the times? weak and strong polarization in the U.S. congress, 1973–2016. Soc. Netw. **60**, 103–112 (2020). https://doi.org/10.1016/j.socnet.2018.07.007
28. Newman, M.E.: Scientific collaboration networks. ii. shortest paths, weighted networks, and centrality. Phys. Rev. **64**(1), 016132 (2001)
29. Peng, T.Q., Liu, M., Wu, Y., Liu, S.: Follower-followee network, communication networks, and vote agreement of the u.s. members of congress. Commun. Res. **43**(7), 996–1024 (2014). https://doi.org/doi-org.ez108.periodicos.capes.gov.br/10.1177/0093650214559601

30. Potamias, M., Bonchi, F., Gionis, A., Kollios, G.: k-nearest neighbors in uncertain graphs. Proc. VLDB Endow. **3**(1), 997–1008 (2010). https://doi.org/10.14778/1920841.1920967
31. Sevon, P., Eronen, L., Hintsanen, P., Kulovesi, K., Toivonen, H.: Link Discovery in Graphs Derived from Biological Databases. In: Data Integration in the Life Sciences, pp. 35–49. Springer, Berlin, Heidelberg (Jul 2006). https://doi.org/10.1007/11799511_5
32. Tsur, O., Lazer, D.: On the interpretability of thresholded social networks. In: Proceedings of the Eleventh International Conference on Web and Social Media, ICWSM 2017, Montréal, Québec, Canada, May 15–18, 2017, pp. 680–683. AAAI Press (2017)
33. Ward, M.D., Stovel, K., Sacks, A.: Network analysis and political science. Annu. Rev. Polit. Sci. **14**(1), 245–264 (2011). https://doi.org/10.1146/annurev.polisci.12.040907.115949
34. Zenil, H., Kiani, N.A., Tegnér, J.: Numerical investigation of graph spectra and information interpretability of eigenvalues. In: Guzman, F.M.O., Rojas, I. (eds.) Bioinformatics and Biomedical Engineering - Third International Conference, IWBBIO 2015, Granada, Spain, April 15–17, 2015. Proceedings, Part II. Lecture Notes in Computer Science, vol. 9044, pp. 395–405. Springer (2015). https://doi.org/10.1007/978-3-319-16480-9_39
35. Zhang, X., Wang, H., Yu, J., Chen, C., Wang, X., Zhang, W.: Polarity-based graph neural network for sign prediction in signed bipartite graphs. World Wide Web **25**(2), 471–487 (Mar 2022). https://doi.org/10.1007/s11280-022-01015-4, https://link.springer.com/10.1007/s11280-022-01015-4

An Analysis of Political Parties Cohesion Based on Congressional Speeches

Willian P. C. Lima[1], Lucas C. Marques[1], Laura S. Assis[1],
and Douglas O. Cardoso[2]($\boxtimes$)

[1] Celso Suckow da Fonseca Federal Center of Technological Education,
Rio de Janeiro, RJ, Brazil
{willian.lima,lucas.custodio}@aluno.cefet-rj.br, laura.assis@cefet-rj.br
[2] Smart Cities Research Center, Polytechnic Institute of Tomar, Tomar, Portugal
douglas.cardoso@ipt.pt

Abstract. Speeching is an intrinsic part of the work of parliamentarians, as they expose facts as well as their points of view and opinions on several subjects. This article details the analysis of relations between members of the lower house of the National Congress of Brazil during the term of office between 2011 and 2015 according to transcriptions of their house speeches. In order to accomplish this goal, Natural Language Processing and Machine Learning were used to assess pairwise relationships between members of the congress which were then observed from the perspective of Complex Networks. Node clustering was used to evaluate multiple speech-based measures of distance between each pair of political peers, as well as the resulting cohesion of their political parties. Experimental results showed that one of the proposed measures, based on aggregating similarities between each pair of speeches, is superior to a previously established alternative of considering concatenations of these elements relative to each individual when targeting to group parliamentarians organically.

Keywords: Politics · Social Networks · Natural Language Processing · Machine Learning

1 Introduction

Recent research works were based on the relational foundations of politics, giving rise to different models and methodologies to explore such a subject [3,6,7,14,32]. In this context, a familiar scenario is that of a National Congress of any country wherein debates happen as an inherent aspect of the nation's legislative process. Our work proposes a methodology for knowledge discovery based on the affinity between members of the congress estimated from the speeches given in those debates. Characterizing interactions between parliamentarians is an interesting activity for democracy, as it may provide evidence of suspicious associations and conflicts of interest. Targeting such challenges, we

© The Author(s), under exclusive license to Springer Nature Switzerland AG 2023
J. Mikyška et al. (Eds.): ICCS 2023, LNCS 14075, pp. 105–119, 2023.
https://doi.org/10.1007/978-3-031-36024-4_8

focused on the lower house of the bicameral National Congress of Brazil, known as the Chamber of Deputies, which is more dynamic and nuanced than those of some other countries [17].

For this task, the concept of complex networks can be used [4], since it is a very common way of representing relationships between individuals [13]. A network can be defined as a graph in which there is a set of nodes representing the objects under study, and a set of edges connecting these nodes. The edges represent an existing relationship between two nodes according to the context in which they are inserted [24]. In this work, a complete graph was considered in which each node represents a deputy, and the weights associated with the edges that connect each pair of nodes represent the similarities between their political positions which were orally stated.

From the creation of the network using deputies and their speeches, it is possible to analyze the cohesion of the parties to which they belong, as members of ideological groups tend to have similar discourses [8,11,15,25]. The fact that each deputy explicitly belongs to a single political party allows a more objective assessment of the relationship between congressional peers. This can be based on verbal statements of their positions and ideas. Therefore, party cohesion can be estimated according to the panorama across the network of deputies, through quantitative aggregation of the conformity of their speeches. This examination is performed considering speeches of deputies belonging to the same party and that of deputies from different parties. This targeted not only to evaluate the association of deputies to parties but also to highlight phenomena as secessions within parties.

Multiple Natural Language Processing (NLP) approaches were considered to quantify the weights of the edges in the deputies network, based on the cosine similarity between the *TF-IDF (term frequency-inverse document frequency)* vectors regarding their speeches. Our main contributions are: *(i)* establishing measures of distance between deputies based on their speeches; *(ii)* use Agglomerative Hierarchical Clustering to estimate the quality of such measures; *(iii)* evaluate party cohesion from a complex networks perspective.

The remainder of this paper is organized as described next. Section 2 examines the theoretical base that underlies this work. Section 3 includes works related to the problem of measuring textual similarity. The proposed methodology is described in Sect. 4. The results obtained through the computational experiments performed and their respective discussion are discussed in Sect. 5. Finally, Sect. 6 presents conclusions and future work.

2   Theoretical Reference

This section presents the necessary information to understand the developed research properly. First, some concepts of Natural Language Processing are introduced, which serve as a basis for the proposed methods. Subsequently, graph and complex network concepts are presented, and a validation method of clustering is reported.

2.1   Natural Language Processing

Natural Language Processing (NLP) is the branch of Artificial Intelligence that allows machines to interpret human language combining computational linguistics with statistical models, allowing computers to process human language. NLP uses various pre-processing and text representation techniques aimed at computation to realize automatic text cognition [9].

To model the data and enable an artificial intelligence algorithm a better understand the text and make better associations, it is necessary to perform a pre-processing that abstract and structure the language, ideally preserving only what is the relevant information. The following pre-processing techniques were used in this work [23]: tokenization, stemming, and stop words removal.

A commonly used method to represent the text is to generate a matrix, in which each row is a vector that refers to one of the documents under analysis and each column is relative to a token. A token is a sequence of contiguous characters that play a certain role in a written language, such as words or even parts of them. Each matrix entry is calculated using a widely used measure called *TF-IDF* [5]. It is a manner of determining the piece of content quality based on an established expectation of what an in-depth piece of content contains. This is a statistical measure that is intended to indicate the importance of a word in a document in relation with a collection of documents or a linguistic corpus. The purpose of using *TF-IDF* is to reduce the impact of tokens that occur in a large majority or in very few documents, being of little use to differentiate them. The formula used to calculate the *TF-IDF* for a term t of a document d in a set of documents is given by Eq. (1), where $\mathrm{tf}(t, d)$ is the frequency of the term t in the document d.

$$\mathrm{TFIDF}(t, d) = \mathrm{tf}(t, d) \cdot \mathrm{idf}(t) \tag{1}$$

The $\mathrm{idf}(t)$ is calculated as shown in Eq. (2), where n is the total number of documents, in the document set e $\mathrm{df}(t)$ is the frequency of documents where t occurs. Document frequency is the number of documents in the dataset that contain the term t.

$$\mathrm{idf}(t) = \log\left(\frac{n}{\mathrm{df}(t)}\right) + 1 \tag{2}$$

For two elements of any vector space, their dot product is proportional to the cosine of the angle between them. Such a value can be interpreted as an indication of similarity between these vectors on a scale that varies, if they are unitary, from -1 (diametrically opposite) to 1 (coincident). Using the matrix generated by *TF-IDF*, we can use cosine similarity to measure the similarity between pairs of documents, a measure that can vary from 0 to 1 given that all entries are non-negative. For the calculation of distances between documents, the similarity complement is used. Since a and b are distinct vectors of the matrix *TF-IDF* the distance between these vectors is given by Eq. (3):

$$d(a,b) = 1 - (\langle a,b \rangle)/(||a|| \cdot ||b||) \tag{3}$$

2.2   Clustering

There are several ways to perform a clustering, and one of them is through hierarchical agglomerative clustering methods [31]: initially each group contains a single element, and iteratively, the two most similar groups are concatenated until, in the end, the number of groups is equal to the number of parties. To define the clustering criterion of the agglomerative cluster, the hyper-parameter linkage is used. The linkage criterion determines which distance to consider between the observation sets. The algorithm will merge the cluster pairs considering the minimization of this criterion. The following types of linkage were considered in this work: average, complete, and single.

When organizing the collection items into clusters, there are two criteria that can be used to evaluate the resulting cluster: *i)* homogeneity (*h*) and *ii)* completeness (*c*). Both depend on the establishment *a priori* of a cluster considered ideal for the data, indicating the respective class for each item. A clusterization has a maximum homogeneity value if all its clusters contain only samples that are originally members of the same class, without merging elements that should be separate. Completeness, on the other hand, is maximum if all samples that are members of a given class are elements of the same cluster, and have not been improperly allocated to different groups.

In some ways, cluster homogeneity and completeness are competing. The trivial clustering, in which each item is in a unitary cluster, has maximum homogeneity but minimum completeness, while this scenario is reversed if all items are placed in a single group. The *V-Measure* [27] measures the success of the criteria of homogeneity and completeness in a combined way, as can be seen in Eq. (4):

$$V_\beta = ((1 + \beta)hc)/(\beta h + c) \tag{4}$$

The *V-Measure* is a measure calculated as the harmonic mean of the values of homogeneity and completeness. These values can be adjusted to assign different weights to the contributions of homogeneity or completeness through the parameter $\beta \geq 0$. Usually, a default value of $\beta = 1$ is used. If $\beta \in [0,1[$, the homogeneity has greater relevance in the weighting. However, if $\beta \in (1, \infty)$ the completeness has greater weight in the calculation. In the case of $\beta = 1$, there is a balance between the two criteria. The closer the result of V-Measure is to 1, the better the performance.

A measure called *silhouette coefficient* can be used to validate the consistency of the clustering process. It is calculated for each item, based on the average distance of it to other items which belong to the same cluster (*a*) and the minimum average distance of it to elements of another cluster (b), as shown in Eq. (5). The closer to 1 is the average silhouette, the better was the clustering process.

$$silhouette = \frac{(b-a)}{max(a,b)} \qquad (5)$$

In complex networks, the clustering coefficient measures the degree to which graph nodes tend to cluster. In its most basic definition, considering a given node u, the clustering coefficient represents the relative frequency of triangles formed by u and its neighbors. There are several manners to perform such an evaluation for weighted graphs [28]. One of them is based on the geometric mean of the edges weights of each triangle, as described in Eq. (6).

$$ca_u = \frac{\sum\limits_{v,z \in N^u} (\hat{w}_{uv}.\hat{w}_{uz}.\hat{w}_{vz})^{1/3}}{deg(u).(deg(u)-1))}, \qquad (6)$$

where $deg(u)$ is the degree of node u. Nodes v and z are neighbors of u, thus N^u is the neighborhood of the node u. $\hat{w}_{ab}$ is the weight of the edge (a,b). Evidence suggests that in most real-world networks, and especially in social networks, nodes tend to form strongly connected groups characterized by a relatively high density of loops. This probability inclines to be greater than the average probability of a loop being randomly established between two nodes [29].

3   Related Works

Historically, political scientists have devoted much attention to the role that political institutions and actors play in a variety of phenomena in this context. Recently, however, there has been a strong trend towards a point of view based on the relational foundations of politics, giving rise to various political networks and methodologies to characterize their structures [20].

Some of the most recent works related to this topic use different information to associate parliamentarians with each other. For example, voting, campaign donations, and participation in events. In [7] complex networks are used to assess the relationship between donations received by parliamentarians elected in 2014 and their voting behavior during 2015 and 2016. The authors examine the homophily and cohesion of parliamentarians in networks created concerning their political parties and constituencies.

Determining the thematic profile of federal deputies, through the processing of texts obtained from their speeches and propositions is held in [16]. This work presents natural language processing techniques used to analyze the speeches of deputies. Such speeches include the removal of words with little semantic meaning (stop words), terms reduction into their morphological roots (stemming), computational representation of texts (bag-of-words), and uses of the Naive Bayes model for the classification of discourses and propositions.

The challenge of relating the deputies, by the similarity of their votes, using the votes in which the parliamentarians participate was addressed in [6].

This approach can be applied in a variety of political scenarios, as most current legislative processes have votes for the approval of the proposals.

A network technique approach to relating different portals of news according to the ideological bias of the published news is presented in [2]. The paper uses hyperlinks to develop an automatic classification method in order to analyze the relationship between the network's structural characteristics and political bias. That is if the ideology of portals is reflected in the properties of the networks that model citations (hyperlinks) among them. In the works [33] and [12] NLP and machine learning techniques were used to determine news bias. They use overtly political texts for a fully automatic assessment of the political tendency of online news sites.

Another major area of research is those focused on social network analysis, as approached by [15], which used tweets to automatically verify the political and socio-economic orientation of Chilean news portals. Sentiment analysis techniques were applied in [11] to assess the relationship between the opinion of Twitter users (text in Portuguese) and the elections final result. The same occurs for other countries and languages [25], considering studies on identification of political positioning and ideology through textual data.

Although a greater abundance of works on NLP are focused on supervised learning, there are also those in the literature on this topic aimed at unsupervised learning. For example, there are works [13,30] on methods of grouping textual objects, where they are automatically organized into similar groups and a comparative study of grouping algorithms applied to the clustering of such objects is carried out. In [18], a modeling of topics of speeches in the European Union parliament is carried out.

4   Methodology

In this section, we describe a baseline approach to the proposed study based on previous works, the proposed method in this work to overcome the presented previous literature, and the measure used to evaluate the considered approaches.

4.1   Preliminary Approach

To define the distance between parliamentarians based on their speeches, the User-as-Document (UaD) [10] model was used. This consists of concatenating all the speeches of a parliamentarian, thus forming a single document that represents him. Then, the similarity between two parliamentarians is calculated using the similarity between their respective documents.

To create these documents, two NLP pre-processing techniques are used: *(i)* removal of stop-words and other tokens considered less relevant, and *(ii)* stemming [22]. The list of stop-words was based on the list available in Portuguese in the package NLTK [21] with the addition of very common words in the political context, such as the sra , v. exa among others. For the execution of stemming, the module of NLTK SnowballStemmer was applied to the corpus. After this step, the documents are represented in a vector way using *TF-IDF* and the similarities between them are calculated by the cosine similarity.

An illustration that exemplifies a network where the nodes represent parliamentarians, the edges are determined by the similarity between them according to their discourses is presented in Fig. 1. The vectors that goes along with each vertex represent the speeches of each parliamentarian. The speeches which are classified by subject/idea through the colors, and the values are the number of speeches given of each type. The calculation of similarities is nothing more than the dot product of these vectors. It is carried out by multiplying the number of speeches of a deputy by that of another deputy. This multiplication occurs only among the number of the same type discourses, that is, situated in the same coloring. The sum of these multiplications results in the weight value of each edge.

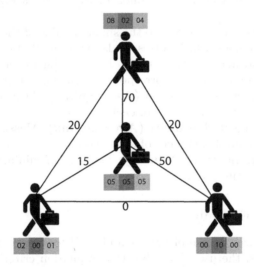

Fig. 1. Illustration of the similarity evaluation between parliamentarians through their speeches.

4.2   Proposed Approach

In the context of parliamentarians' speeches, the use of the UaD approach can cause some difficulties in calculating distances. One reason is the mischaracterization of the number of documents in the corpus, which directly reflects on the *TF-IDF* calculation. Another related problem is the terms crossing from different discourses, which can lead to an exaggerated extrapolation in the parliamentarian characterization.

An alternative approach to assessing the alignment between two parliamentarians is to calculate the distance between each pair of their speeches, and then aggregate that collection of distances into a single value that represents the distance between parliamentarians. In this way, the distribution of tokens in the generated corpus would be more realistic compared to the one corresponding achieved from the UaD approach, employing the NLP techniques used in the preliminary approach. Furthermore, the impact of crossing different terms between discourses can be minimized using different aggregation methods.

It is possible to apply several aggregation methods to summarize in a single measure the distances between the pairs of speeches of any two parliamentarians. In this work, six types of aggregation are proposed. All of them are premised on the formation of a complete bipartite graph determined in such a way that each vertex corresponds to a discourse, and each partition of vertices corresponds to a parliamentarian. Each edge weight represents the distance between the discourses on which it falls. The aggregation types are described next:

1. **Average of distances**: It uses the average of all values distances between pairs of discourses, which corresponds to the edges' average weight of the bipartite graph.
2. **Minimum distances**: It refers to the smallest value of the distance between each pair of discourses, which corresponds to the smallest of the edge weights.
3. **Maximum distances**: It refers to the largest value of the distance between each pair of discourses, which corresponds to the largest of the edge weights.
4. **Average of shortest distances (AverageMin)**: Measure relative to the average of the smallest distances of each speech.
5. **Average of longest distances (AverageMax)**: Measure relative to the average of the greatest distances of each speech.
6. **Minimal spanning tree (MinST)**: The sum of minimum spanning tree edges of the bipartite graph.

4.3   Validation of Results

In order to legitimize the use of graphs and NLP to determine organic groups of parliamentarians, the use of an objective evaluation criterion was idealized. Thus, it would be possible to identify the optimal configuration of the aggregation type and the pre-processing hyper-parameters. This criterion would ideally indicate how well a clustering reflects the prevailing party structure. However, it should be noted that this does not preclude such clustering from including evidence that contrasts with the aforementioned structure, due to the similarity (or dissimilarity) between parliamentarians.

In this work, the clusters of parliamentarians are defined according to their two-by-two distances, and the number of groups is defined by the number of parties among the parliamentarians, resulting in 21 clusters. Given these conditions, an agglomerative hierarchical clustering approach was used, using pre-defined similarities instead of coordinates of points whose distance could be evaluated [19].

For the construction of a cluster applying agglomerative hierarchical clustering, three types of criteria applicable to the algorithm to define the clusters were used in this work: *(i) single linkage, (ii) average linkage* and *(iii) complete linkage.* From the agglomerative hierarchical clustering, it is possible to assess the quality of clustering, by comparing the membership of parliamentarians to clusters with the membership of parliamentarians to parties. To obtain the clustering quality assessment value, the V-Measure was used. Figure 2 presents a flowchart indicating the steps performed to obtain the evaluation measure.

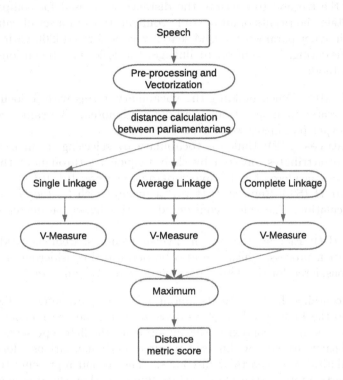

Fig. 2. Flowchart with each step taken to obtain the quality assessment measure of the clusters generated from the applied methodology.

5   Experimental Evaluation

In this section, experiments related to the questions raised in the previous section are detailed. The results of these experiments were obtained using the 8,378 speeches of the 627 active federal deputies during the 54^a House legislature, which spans the years from 2011 to 2015. Each congressman was treated as a party member to which he belonged when he was elected, disregarding possible party changes during the term of office. The parliamentary data used in this work are open data, available in the Chamber of Deputies Open Data Service [1]. All experiments were performed on the Google Colaboratory platform free of charge.

5.1   Hyper-parameter Optimization and Correlation Analysis

Hyper-parameters fine-tuning can help achieving to simultaneously reflect the current party structure as well as evidence that contrasts with that structure. Using the combinations of hyper-parameters for vectorization based on the *TF-IDF* statistic and the agglomerative hierarchical clustering algorithm, different versions of the method to measure the distance were used for comparison. In order to obtain the results of all possible configurations in a set of values specified for each hyper-parameter, a grid search was performed [26], with 120 value settings considering, in addition to linkage, the following dimensions, for the *TF-IDF* method:

- **max_df** (MD): When building the vocabulary, terms with a document frequency greater than the given threshold are ignored. The values considered for this hyper-parameter were: $2^0, 2^{-1}, \cdots, 2^{-5}$;
- **max_features** (MF): Builds a vocabulary by selecting no more than this number of attributes, ordered by their frequencies throughout the corpus. The values considered for this hyper-parameter were: $10^1, 10^2, \cdots, 10^5$;
- **sublinear_tf** (ST): Replaces tf (term frequency) with $1 + log(tf)$ in the *TF-IDF* calculation. The values considered for this hyper-parameter were: *Yes* and *No*;
- **use_idf** (UI): Enables the use of inverse-document-frequency (idf) in the vectorization process, which can also be performed considering only *tf*. The values considered for this hyper-parameter were: *Yes* and *No*.

Table 1 contains the best results among all hyper-parameter configurations, according to the highest values of V-Measure, which add *i)* homogeneity and *ii)* completeness. In the context of this experiment, the first represents the minimization of party diversity within each group of parliamentarians inferred based on the similarities between their discourses. The second represents the preservation of party unity in the obtained grouping, so that the fragmentation of originally related parliamentarians is avoided. Most of the results presented use the distance aggregation approach proposed here (Sect. 4.2): the best configuration used the Average of distances; only one of the top seven is based on the User-as-Document approach.

Looking at the hyper-parameters, approaches, and aggregation methods individually, it is possible to see that the option $max\_df = 2^{-4}$ generally has the best performance among the evaluated values. This can be interpreted as an indication that there are many relatively frequent and non-discriminating terms in the analyzed corpus. The max_features parameter for the method that calculates the distance between each pair of deputies obtains better results with the maximum amount of attributes $\geq 10^4$, while the User-as-Document approach obtained your best result using a maximum of 10^3 attributes. In general, better results were achieved using *True* as an argument for both sublinear_tf and use_idf parameters.

To finalize the comparison of distance measures between parliamentarians, Fig. 3 presents the correlation matrix of the considered measures. This matrix is

Table 1. Best hyperparameter settings found according to the highest values of V-Measure.

Average	V-measure	Linkage	max_df	max_features	sublinear_tf	use_idf
Average	0,2264	average	2^{-4}	10^5	No	Yes
UaD	0,2156	average	2^{-4}	10^3	No	Yes
AverageMax	0,2015	complete	2^{-3}	10^5	Yes	Yes
MinST	0,1950	complete	2^{-4}	10^4	Yes	No
AverageMin	0,1659	average	2^{-4}	10^4	Yes	Yes
Maximum	0,1635	complete	2^0	10^5	Yes	Yes
Minimum	0,1598	complete	2^{-4}	10^2	Yes	No

colored like a heat map for better visualization. Kendall's τ was used as correlation statistic, i.e. a non-parametric correlation measure, that is able to capture even non-linear relationships between random variables, making the analysis more flexible and robust. According to the presented matrix, the segregation of the UaD measure from the others is evident, also it can be observed the Average is positioned as an intermediary between this first and the remaining five measures. Such an organization is consistent with the definition of the assessed measures. However, attention is drawn to the fact that the strongest correlation is between the alternatives AGMin and AverageMax, despite the two having principles that do not appear to be similar.

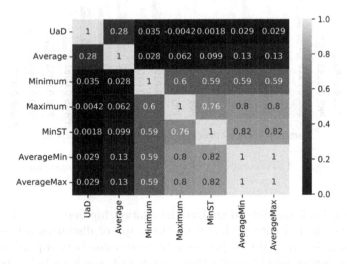

Fig. 3. Correlation matrix between the measures used to estimate the similarity between parliamentarians.

5.2   Party Cohesion Analysis

Figure 4 depicts party cohesion based on the parliamentary distance measure in its already detailed optimal configuration (Avarage). The order of the parties is arranged from left to right from the highest to the lowest clustering coefficient of the subgraph induced by the vertices relative to parliamentarians of each party. On the ordinate axis, the values of the clustering coefficient are shown together with the party acronym. The clustering coefficient evaluates the degree to which the graph vertices tend to colligate, considering the distance between the vertices of the same group. Parties whose parliamentarians have speeches at small distances from each other will have larger clustering coefficients. Smaller clustering coefficients indicate greater distances between speeches by parliamentarians from a given party. The red bars represent the party size in terms of the parliamentarians' number, while the blue bars characterize the entropy present in the distribution of parliamentarians of a party in clusters. An entropy value equal to zero means that all the members of a party were clustered into the same cluster. As the members of a party are dispersed into more groups the entropy increases.

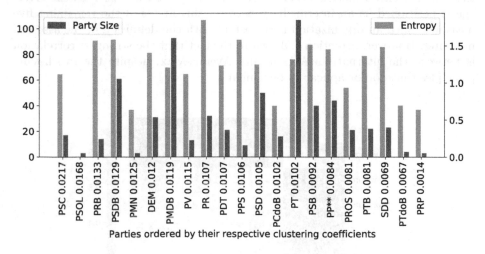

Fig. 4. Assessment of similarity between parliamentarians.

From Fig. 4 it is possible to analyze the relationship between party size, clustering coefficient and entropy based on the graph of distances between parliamentarians. It is expected that parties with greater numbers of parliamentarians have higher entropy and lower clustering coefficient, resulting in less party cohesion, as there is a greater chance of divergences between parliamentarians, given the large number. It is possible to observe that there are cases where this expectation is confirmed, as in the second party with the best clustering coefficient, the PSOL, whose number of parliamentarians is small, and its entropy is all small and clustering coefficient is larger. Other examples of parties that also follow this expectation are the PT, PSB and PP, where the number of parliamentarians is large, and entropy is also large and clustering coefficient is lower. Some parties do not correspond to this behavior, such as the PRP and the PTdoB, which have a very small number of parliamentarians and a relatively high entropy and lower clustering coefficient. These results suggest great distances between the speeches of parliamentarians that belong to the same party.

The PMDB and PSDB, parties with large numbers of parliamentarians, are examples of parties with a relatively high value of clustering coefficient, and with a high value of entropy, indicating small distances between parliamentarians of a single party, and also a dispersion of parliamentarians from one party to other parties. This means that the process of clustering spread these parties' parliamentarians into several other clusters, despite the similarity between the speeches of their parliamentarians on average. This can be observed as evidence of the existence of cohesive and distant wings in these parties.

6   Conclusion

Analyzing the relations between deputies and parties is an interesting activity for democracy, because it provides data about deputies and possible conflicts of interest. In this work, the relationship and party cohesion of parliamentarians of the Deputies Chamber active during the 54^a legislature, which comprises the years between 2011 and 2015, according to their speeches, was analyzed. Complex networks were used to model the relationships between deputies. An unsupervised machine learning approach, named agglomerative hierarchical clustering, was used for the analysis of party cohesion.

The proposed methodology assesses the affinity between any pair of parliamentarians based on the discourses sets of each of them. For this, a complete bipartite graph was considered. In this graph, the vertices represent the parliamentarians' speeches, divided into two partitions according to the parliamentarian to which they correspond. The edges are weighted according to the respective speeches' textual similarity. Affinity can then be calculated by aggregating these weights according to one of the six alternatives presented and considered for this purpose. The presented methodology obtained superior results compared to the reference measure previously established in the literature. In the meantime, the aggregation method that obtained the best results was the one related to the average of the similarities of each pair of speeches, considering any two parliamentarians.

In the future, we intend to use other Natural Language Processing techniques to obtain different measures in addition to *TF-IDF*. In the next study, we want to consider word embedding, since *TF-IDF* is a simpler technique that may not consider the nuances of language as synonyms, for example. We also intend to test different methods to compute the measures of aggregation using algorithms in graphs. Such algorithms can lead to measures that better reflect the distances between parliamentarians based on their speeches.

Acknowledgements. Douglas O. Cardoso acknowledges the financial support by the Foundation for Science and Technology (Fundação para a Ciência e a Tecnologia, FCT) through grant UIDB/05567/2020, and by the European Social Fund and programs Centro 2020 and Portugal 2020 through project CENTRO-04-3559-FSE-000158.

References

1. Dados abertos da câmara dos deputados. https://dadosabertos.camara.leg.br/swagger/api.html. (Accessed 17 Aug 2021)
2. Aires, V., da Silva, A., Nakamura, F., Nakamura, E.: An evaluation of structural characteristics of networks to identify media bias in news portals. In: Proceedings of the Brazilian Symposium on Multimedia and the Web, pp. 225–232 (2020)
3. Aref, S., Neal, Z.P.: Identifying hidden coalitions in the US house of representatives by optimally partitioning signed networks based on generalized balance. Sci. Rep. **11**(1), 19939 (2021)
4. Barabási, A.L., Pósfai, M.: Network science. Cambridge University Press, Cambridge, United Kingdom (2016), oCLC: ocn910772793
5. Beel, J., Gipp, B., Langer, S., Breitinger, C.: Research-paper recommender systems: a literature survey. Int. J. Digit. Libr. **17**(4), 305–338 (2016)
6. Brito, A.C.M., Silva, F.N., Amancio, D.R.: A complex network approach to political analysis: application to the brazilian chamber of deputies. PLoS ONE **15**(3), e0229928 (2020)
7. Bursztyn, V.S., Nunes, M.G., Figueiredo, D.R.: How brazilian congressmen connect: homophily and cohesion in voting and donation networks. J. Complex Netw. **8**(1), cnaa006 (2020)
8. Caetano, J.A., Lima, H.S., Santos, M.F., Marques-Neto, H.T.: Using sentiment analysis to define twitter political users' classes and their homophily during the 2016 american presidential election. J. Internet Serv. Appl. **9**(1), 1–15 (2018)
9. Camacho-Collados, J., Pilehvar, M.T.: On the role of text preprocessing in neural network architectures: An evaluation study on text categorization and sentiment analysis. arXiv preprint arXiv:1707.01780 (2017)
10. Cossu, J.V., Labatut, V., Dugué, N.: A review of features for the discrimination of twitter users: application to the prediction of offline influence. Soc. Netw. Anal. Min. **6**(1), 25 (2016)
11. Cristiani, A., Lieira, D., Camargo, H.: A sentiment analysis of brazilian elections tweets. In: Anais do VIII Symposium on Knowledge Discovery, Mining and Learning, pp. 153–160. SBC (2020)
12. Dallmann, A., Lemmerich, F., Zoller, D., Hotho, A.: Media bias in german online newspapers. In: Proceedings of the 26th ACM Conference on Hypertext & Social Media, pp. 133–137 (2015)
13. Dias, M., Braz, P., Bezerra, E., Goldschmidt, R.: Contextual information based community detection in attributed heterogeneous networks. IEEE Lat. Am. Trans. **17**(02), 236–244 (2019)
14. Doan, T.M., Gulla, J.A.: A survey on political viewpoints identification. Online Social Netw. Media **30**, 100208 (2022)
15. Elejalde, E., Ferres, L., Herder, E.: The nature of real and perceived bias in chilean media. In: Proceedings of the 28th ACM Conference on Hypertext and Social Media, pp. 95–104 (2017)
16. Fernandes, M.S.: Tenho dito: uma aplicação para análise de discursos parlamentares utilizando técnicas de processamento de linguagem natural (2017)

17. Gomes Ferreira, C.H., Murai Ferreira, F., de Souza Matos, B., Marques de Almeida, J.: Modeling Dynamic Ideological Behavior in Political Networks (2019)
18. Greene, D., Cross, J.P.: Exploring the political agenda of the european parliament using a dynamic topic modeling approach. Polit. Anal. **25**(1), 77–94 (2017)
19. Jain, A.K., Murty, M.N., Flynn, P.J.: Data clustering: a review. ACM Comput. Surv. (CSUR) **31**(3), 264–323 (1999)
20. Lee, S.H., Magallanes, J.M., Porter, M.A.: Time-dependent community structure in legislation cosponsorship networks in the congress of the republic of peru. J. Complex Netw. **5**(1), 127–144 (2016)
21. Loper, E., Bird, S.: Nltk: The natural language toolkit. CoRR cs.CL/0205028 (2002)
22. Lovins, J.B.: Development of a stemming algorithm. Mech. Transl. Comput. Linguist. **11**(1–2), 22–31 (1968)
23. Méndez, J.R., Iglesias, E.L., Fdez-Riverola, F., Díaz, F., Corchado, J.M.: Tokenising, stemming and stopword removal on anti-spam filtering domain. In: Marín, R., Onaindía, E., Bugarín, A., Santos, J. (eds.) CAEPIA 2005. LNCS (LNAI), vol. 4177, pp. 449–458. Springer, Heidelberg (2006). https://doi.org/10.1007/11881216_47
24. Metz, J., Calvo, R., Seno, E.R., Romero, R.A.F., Liang, Z., et al.: Redes complexas: conceitos e aplicações. (2007)
25. Pastor-Galindo, J., Zago, M., Nespoli, P., Bernal, S.L., Celdrán, A.H., Pérez, M.G., Ruipérez-Valiente, J.A., Pérez, G.M., Mármol, F.G.: Spotting political social bots in twitter: a use case of the 2019 spanish general election. IEEE Trans. Netw. Serv. Manage. **17**(4), 2156–2170 (2020)
26. Pedregosa, F.: Scikit-learn: Machine learning in Python. J. Mach. Learn. Res. **12**, 2825–2830 (2011)
27. Rosenberg, A., Hirschberg, J.: V-measure: A conditional entropy-based external cluster evaluation measure. In: Proceedings of the 2007 Joint Conference On Empirical Methods In Natural Language Processing And Computational Natural Language Learning (EMNLP-CoNLL), pp. 410–420 (2007)
28. Saramäki, J., Kivelä, M., Onnela, J.P., Kaski, K., Kertesz, J.: Generalizations of the clustering coefficient to weighted complex networks. Phys. Rev. E **75**(2), 027105 (2007)
29. Watts, D.J., Strogatz, S.H.: Collective dynamics of 'small-world'networks. Nature **393**(6684), 440–442 (1998)
30. Wives, L.K.: Um estudo sobre agrupamento de documentos textuais em processamento de informações não estruturadas usando técnicas de" clustering" (1999)
31. Yim, O., Ramdeen, K.T.: Hierarchical cluster analysis: comparison of three linkage measures and application to psychological data. Quant. Methods Psychol. **11**(1), 8–21 (2015)
32. Zhang, X., Wang, H., Yu, J., Chen, C., Wang, X., Zhang, W.: Polarity-based graph neural network for sign prediction in signed bipartite graphs. World Wide Web **25**(2), 471–487 (2022)
33. Zhitomirsky-Geffet, M., David, E., Koppel, M., Uzan, H.: Utilizing overtly political texts for fully automatic evaluation of political leaning of online news websites. Online Information Review (2016)

Computational Health

Comparative Study of Meta-heuristic Algorithms for Damage Detection Problem

Kamel Belhadj[1,2,3]($\boxtimes$), Najeh Ben Guedria[1], Ali Helali[1], Omar Anis Harzallah[3], Chokri Bouraoui[1], and Lhassane Idoumghar[2]

[1] Laboratory of Mechanics of Sousse (LMS), University of Sousse, National Engineering School of Sousse, Sousse, Tunisia
kamel.belhadj@eniso.u-sousse.tn
[2] University of Haute-Alsace, IRIMAS-UHA, 68093 Mulhouse, France
lhassane.idoumghar@uha.fr
[3] University of Haute-Alsace, LPMT-UHA, UR, 4365, F-68093 Mulhouse, France
Omar.harzallah@uha.fr

Abstract. This study presents a comprehensive comparative analysis of several meta-heuristic optimization algorithms for solving the damage detection problem in concrete plate structures. The problem is formulated as a bounded single objective optimization problem. The performance and efficiency of the algorithms are compared under various scenarios using noise-contaminated data. The results show that these meta-heuristics are powerful methods for global optimization and are suitable for solving the damage detection problem. The study compares the performance of these algorithms in: (1) identifying the location and extent of damaged elements, and (2) robustness to noisy data. The proposed meta-heuristic algorithms show promise for solving the damage detection problem. Particularly, the GSK-ALI, MRFO, and Jaya algorithms demonstrate superior performance compared to the other algorithms in identifying damaged elements within concrete plate structures.

Keywords: Structure Health Monitoring (SHM) · Damage detection · Plates · Meta-Heuristic · Flexibility matrix

1 Introduction

The safety of structures can be threatened by damage, which is a significant concern for maintaining their integrity. To address this issue, structural health monitoring (SHM) is commonly employed to gather vast amounts of data using wireless sensors, signal processing technology, and artificial intelligence. However, analysing and assessing the condition of a structure based on this data can be a complex task. Damage detection is crucial in structural condition assessment, as it helps to identify damage, determine its severity, and estimate the remaining life of a structure [1–3]. It is important to note that structural damage may not always be predictable, which is why early detection of damage is crucial for fast and efficient repairs, and to ensure the safety and serviceability

© The Author(s), under exclusive license to Springer Nature Switzerland AG 2023
J. Mikyška et al. (Eds.): ICCS 2023, LNCS 14075, pp. 123–137, 2023.
https://doi.org/10.1007/978-3-031-36024-4_9

of the structure [4]. Non-destructive vibration studies that display structural dynamic characteristics behaviour, such as frequency response functions (FRFs) and modal properties, are typically used to identify structural damage. Since these features are related to physical structural properties, changes in these properties can be used to infer damage, under the assumption that changes in physical structural properties lead to modifications in dynamic structural properties [5]. Damage detection in structures typically relies on analysing the vibrational responses of the structure [6]. However, small differences in these responses can be difficult to detect due to variations in environmental conditions such as temperature, wind, and rain [6]. Many methods have been proposed in the literature to address this issue and improve the practical applicability of damage detection techniques. These methods can be used for both structural health monitoring and early detection of damage [7–12]. An extensive review of methods for the detection of damage is available in [3, 13].

Numerous approaches have been put forward to tackle the problem of identifying damage in structures, employing a wide range of techniques. Several of these methods are associated with particular characteristics of the structure, such as its modal strain energy [14], mode shape derivative [15, 16], natural frequency response [17], wavelet transform [18, 19], and residual force vector [20, 21]. These features are typically employed as indicators of damage in the structure and furnish valuable insights into the site and the magnitude of the damage.

The presented study aims to evaluate the effectiveness of several meta-heuristic methods for solving the damage detection problem in structures. The problem is modelled as a bounded single objective optimization problem, and the performance and the efficiency of the algorithms are compared under various scenarios using noisy data. The meta-heuristic algorithms studied in this research are inspired by the collective intelligence of social animals or insects, such as herds, birds, or fish, and include well-known algorithms such as Particle Swarm Optimization (PSO) [22], Artificial Bee Colony (ABC) [23], Differential Evolution (DE) [24], and Teaching-Learning-Based Optimization (TLBO) [25].

In recent years, there has been a sustained interest in swarm-based methods, and several advanced swarm intelligence methods have been developed. These methods are known for their excellent computing performance and have been applied in a wide range of fields such as Mechanical Engineering [26], Aerospace Engineering [27], Structural Design [28], Automotive Industry [29], Civil Engineering [30], to examine the performance of these algorithms, a numerical simulation of different scenarios with noisy data in a concrete plate structure is performed. The results of this study will be of interest to researchers in the field of structural health monitoring, as they will help identify the most effective and efficient meta-heuristic algorithms for solving the damage detection problem.

The article is organized as follows: In Sect. 2, the problem of damage detection is explained theoretically. The algorithmic concepts used in the study are briefly outlined in Sect. 3. The results are then presented and discussed in Sect. 4. Finally, the findings are presented in Sect. 5, as well as prospective research interests.

2 Structure of Damage Detection Problem

2.1 Damage Detection Modelling

The simplest expression of the damage detection problem is by applying a linear equation of motion representing the undamped free vibration [31].

$$[M][\ddot{x}] + [K][x] = 0 \tag{1}$$

where [x] is the displacement vector, [K] and [M] represent respectively the stiffness and the mass matrix. In this case, the equation of motion can be expressed as follows:

$$x(t) = \phi_i u_i(t),$$
$$u_i(t) = A_i \cos(\omega_i t - \theta_i), \tag{2}$$

where ϕ_i and ω_i represent the *ith* mode shape and modal frequency, respectively, u_i is a displacement time variation given by the harmonic excitation, θ_i stands for the *ith* angle of phase, A_i denotes the *ith* constant associated with the *ith* mode shape. Replacing Eq. (2) into (1), this gives:

$$u_i(t)\left(-\omega_i^2[M]\phi_i + [K]\phi_i\right) = 0 \tag{3}$$

However, the eigenvalue formula used to represent the vibrational mode characteristic of a healthy plate structure is expressed as follow:

$$\left([K] - \omega_i^2[M]\right)\phi_i = 0 \tag{4}$$

In the literature, there are several methods for modeling a damaged plate structure, such as a cracked model [32]. However, these methods can increase the complexity of the simulation and may not be effective for studying the structural performance response in optimization analyses. As a result, a commonly used method for modeling damage to structural elements in optimization problems is to reduce the stiffness of element. The global stiffness matrix of the structure is the sum of the intact and affected stiffness matrices, and can be represented mathematically as follows:

$$K = \sum_{e=1}^{nele} (1 - a_e)K_e \tag{5}$$

where K_e corresponds to the stiffness matrix of the element *eth*, *nele* indicates the number of elements and a_e corresponds to the damage ratio $\in [0, 1]$ representing the degree of damage of the elements, where 0 means a healthy element and 1 meaning that the element is fully damaged.

2.2 The Objective Function Based on Modal Flexibility

According to [33], structural damage leads to a reduction in stiffness and an increase in flexibility of a structure. This means that any changes to the flexibility matrix can be considered as an indication of structural damage and can provide further information about the damage's site and severity. However, previous research as cited in [34] has shown that using the flexibility matrix is more effective for identifying damage compared to other methods. In the current study, the objective function for evaluating damage in concrete structures is based on the difference between the flexibility matrix calculated from a numerical model and the one obtained from a measured model, which is used to compare the two flexibility matrices and assess the damage.

$$f(x) = \frac{\left\| F^{\exp} - F^{ana}(x) \right\|_{Fro}}{\left\| F_j^{\exp} \right\|_{Fro}}, \; x = (x_1, ..., x_n) \in [0, 1]^n \quad (6)$$

F^* is the flexibility matrix expressed as follows:

$$F^* = \sum_{i=1}^{n \, mod} \frac{1}{(\omega_i^*)^2} \phi_i^* (\phi_i^*)^T \quad (7)$$

where ω_i^* and ϕ_i^* represent the *ith* natural frequency and its associated mode shape, the superscripts exp and ana refer to the damaged model and the analytic model, the number of modes is *nmod*, the design vector of variable for the damage extent of n elements is x, and $\|.\|_{Fro}$ present the Frobenius norm of a matrix. However, in this study, we use generated data which were obtained by numerical simulations of damage scenarios on a structural model. To improve the generalizability of our results, we also applied data augmentation techniques, such as adding noise and changing the location of the damage, on the generated data.

3   Instruction to the Optimization Method

Real-world engineering problems often present a wide range of complex optimization challenges. To address these issues, metaheuristic algorithms can be employed as they are user-friendly and do not rely on gradient information. This study delves into various techniques for addressing the problem of damage detection, and comparisons of different algorithms are also presented.

3.1   Differential Evolution DE

The differential evolutionary algorithm (DE) [24] utilizes a combination of individuals from the same population, including the parent, to generate new candidates. The DE algorithm only selects candidates that are superior to their parents. Due to its straightforward design and minimal number of control settings, the DE has been successful in many real-world applications and has been used to find optimal solutions in various complex optimization problems.

3.2 Particle Swarm Optimization PSO

This method is a type of stochastic optimization technique and evolutionary algorithm that was developed by James Kennedy and Russ Eberhart in 1995 [22], to solve difficult computational optimization problems. It has been widely used in various optimization problems and research projects since its introduction. The approach is based on the concept of swarm theory, which is inspired by the behaviour of swarms of birds, fish, etc. The swarm theory is a population-based evolution algorithm where each swarm or particle represents a specific decision. The position of the particle is updated through its velocity vector and aims to reach the optimal solution.

3.3 Teaching Learning Based Optimization TLBO

The TLBO method, introduced by Rao et al. in 2011 [25], is a population-based optimization algorithm that is based on the teaching-learning methodology. It relies on a population of learners, where the optimization procedure is carried out through two distinct phases. The initial phase, also known as the "teaching stage", involves learners acquiring knowledge from a teacher. Subsequently, the "learning phase" takes place, during which learners interact to optimize problem-solving. In this approach, the group of learners represents diverse variable configurations employed to tackle optimization issues. These different configurations are handled as different subjects available for the learners, and their performance is evaluated through a fitness estimation value. The best solution among the entire population is the "teacher" in this algorithm. The TLBO approach uses the concept of a population of learners, where the different configurations of variables are considered as individuals in the population, and the optimization is carried out through the interactions among the learners, using the teaching and learning methodology. The optimization process is divided into two phases, teacher phase and learner phase, where the best solution in the population is considered to be the teacher.

3.4 Artificial Bee Colony ABC

The ABC algorithm was introduced by Karaboga (2005) [23] as a Swarm Intelligent Metaheuristic method to solve optimization problems. The inspiration for this algorithm comes from the neighbourhood honeybee behaviour while searching for food sources in the wild in a colony. The artificial bees present is classified into three basic categories: worker bees, observer bees and scout bees. There are an equal number of worker bees and scout bees as there are food sources, and each individual bee is related and associated to each source of food. The worker bees look up the food sources in memory. Then, they exchange their data with the spectator bees. The spectator bees will be waiting inside the hive and decide which of the food sources to choose. As a result, the most favourable food sources tend to be more likely to be chosen. Scout bees are changed from few worker bees that abandoned food sources and looked for new sources.

3.5 Harmony Search HS

Harmony Search (HS) [35] is a metaheuristic optimization algorithm that was first proposed by Zong Woo Geem in 2001. It is inspired by the process of improvisation in music, where a musician searches for a perfect harmony by adjusting the pitch and playing time of musical notes. In the HS algorithm, a set of potential solutions (referred to as "harmonies") is represented as a set of decision variables. The algorithm uses a set of heuristic rules to generate new harmonies and to update existing ones. The objective function, which represents the desired harmony, is used to evaluate the quality of each harmony. The algorithm continues to iterate and update the harmonies until a satisfactory solution is found or a stopping criterion is met. HS has been shown to be effective in finding global optimal solutions and is relatively easy to implement.

3.6 Sparrow Search Algorithm SSA

The Sparrow Search Algorithm SSA [36] is an optimization algorithm that employs swarm-based techniques, drawing inspiration from the hunting behaviour of sparrows. It models the behaviour of birds in a flock, where specific members, known as "generators," lead the search for food while others, referred to as "followers," trail behind. To find the best solution to a problem, the algorithm mimics this process by utilizing a group of individuals that search for the optimal solution. The SSA algorithm uses a mechanism called "detection and early alert" to identify and avoid suboptimal solutions. This is done by selecting certain individuals from the population to act as "scouts" and explore different parts of the search space. If a scout detects a suboptimal solution, it "flies away" and searches for a new solution.

3.7 Gaining Sharing Knowledge-Based Algorithm GSK-Ali

The GSK algorithm (Gaining Sharing Knowledge-based Algorithm) is an optimization algorithm that is based on the human process of acquiring and sharing knowledge. The algorithm consists of two phases: the junior phase and the senior phase. In the junior phase, initial solutions are generated by using different methods such as randomization or heuristics. Then, in the senior phase, the solutions are transferred and interoperate with other solutions generated by the algorithm. This allows the algorithm to explore different parts of the search space and identify a global optimal solution. Several variations of the GSK algorithm have been developed to adapt it to specific types of problems and to enhance its performance [37, 38].

3.8 Manta Ray Foraging Optimization MRFO

The Manta Ray Foraging Optimization (MRFO) [39] algorithm is a metaheuristic optimization algorithm that is inspired by the foraging behaviour of manta rays. Manta rays are known for their ability to efficiently search for food in their environment, by using a combination of random exploration and directed search. The MRFO algorithm simulates this behaviour by using a population of individuals, called "manta rays", to explore the search space and identify an optimal solution. Each manta ray is characterized by its own search strategy and step-size, which are updated during the optimization process. The algorithm uses a combination of random exploration and directed search, to efficiently explore the search space and identify the global optimal solution.

3.9 Pathfinder Algorithm PFA

The Pathfinder algorithm is a novel metaheuristic optimization algorithm that is inspired by the survival strategies employed by animal groups in nature. It was first proposed by Yapici and Cetinkaya in 2019 [40]. The algorithm is based on the division of group members into two distinct roles: leaders and followers. Leaders are responsible for exploring new territories in the search space and identifying potential solutions. Followers rely on the guidance of the leaders to locate these solutions. As the search progresses, individuals may switch roles based on their relative proficiency in each. This allows for a greater degree of adaptability in the search process, as individuals can switch between exploring and exploiting the search space. Unlike other swarm intelligence algorithms, the PFA algorithm does not impose any specific limitations on the population size or the number of leaders and followers. This allows the algorithm to adapt to different optimization problems and search spaces.

4 Experimental Results

4.1 Test Setup

This section is dedicated to evaluating the efficiency of all the methods for identifying damage through a numerical simulation of a concrete plate. The assessment of the algorithm's ability to accurately detect and quantify damaged elements will be displayed by testing it on finite element models (FEM) of a concrete plate exposed to three different damage cases. The resilience of these methods is also examined by incorporating noisy data. Noise levels of 3% on frequencies and 0.15% on mode shapes are incorporated in the simulation, as outlined in [41].

$$f_i^{noise} = (1 + \eta_f (2.rand[0.1] - 1))f_i \tag{10}$$

$$\varphi_{ij}^{noise} = (1 + \eta_m (2.rand[0.1] - 1))|\varphi_{ij}| \tag{11}$$

It's important to keep in mind that the matrices used to determine the flexibility are resolute through the use of numerical models and experimental data simulations. Only the displacement of the plate degrees of freedom in the transverse direction are considered. Additionally, the process of identifying damage is repeated 20 times per case and the averages are calculated. Table 1 provides the settings parameters for the listed algorithms, corresponding to the optimal parameter values determined in their original research papers. The optimization algorithms are designed to stop under two specific conditions. First, the process stops when the maximum number of iteration (I_{max}) is reached, second, the process can also be stopped if the value of the objective function of the best individual reaches an extremely weak or 0 value (threshold = 10^{-8}). Both above-mentioned conditions will indicate the end of this process. The FEM for the plate and all algorithms were developed using the MATLAB programming software. The simulation and numerical solutions were carried out on a powerful personal computer featuring an Intel(R) Core (TM) i7-8750H CPU @ 2.20 GHz, 12GB of Random-Access Memory (RAM), and running on a 64-bit version of Windows 10.

Table 1. Parameter setting of the algorithms.

Algorithm	Np	Max_{ietr}	Parameters Setting
MDE	30	500	F Cr
TLBO	30	500	Teaching Factors 1 or 2
ABC	30	500	Food number 15 Inertia weight 0.7
JAYA	30	500	r_1, r_2 random number between [0, 1]
GSK-ALI	30	500	$K_r = 0.9$, $K_f = 0.5$ $P = 0.1$, $K = 10$
HS	30	500	Par = 0.4 HMCR = 0.8
MRFO	30	500	α, β random generated and subject to iteration number
PFA	30	500	α, β random number in the range of [1, 2] possibility of switch 0.8

4.2 Numerical Result

An examination of a concrete plate is conducted in this example, which was first used in [42]. The geometric and material properties of the plate outlined in Table 2. The plate is rigidly fixed along its four edges and discretized using quadrilateral finite elements with four nodal points, as depicted in Fig. 1. Three damage cases are evaluated with different site, as outlined in Fig. 1. The specifics of the damaged elements and the corresponding damage ratios are presented in Table 3. The identification of damage is carried out utilizing only the initial three frequencies and their associated mode shapes for all cases.

Table 2. The Concrete plate parameters

Settings / Unit	Value
Length (Lx, Ly)/m	2
Thickness (t)/m	0.15
Young's modulus (E) /GPa	$2e^{10}$
Poisson ratio (v)	0.20
Mass density (ρ)	2400

As part of this study, a Kruskal-Wallis test was performed at the 0.05 level of significance, a non-parametric statistical test, followed by a Mann-Whitney test to compare the results of the algorithms. Table 5 shows the *p-values* obtained from this statistical analysis, and utilized to specify whether there is a significant difference between the algorithms, respectively, with a 95% confidence level.

Table 3. Damaged Elements Cases

Cases	Element No	Damage rate
1	27	0.20
2	43	0.20
3	67	0.20

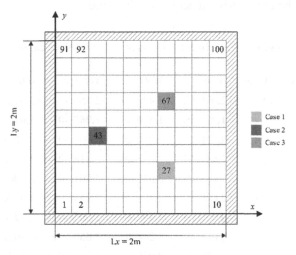

Fig. 1. Discretised a plate with three damage cases.

The statistical result compares the performance of several algorithms, identified by their acronyms (ABC, GSK-ALI, HS, JAYA, MDE, MRFO, PFA, and TLBO). Table 4 provides information on the dimension of the problem (Dim), as well as the best value, worst, average, the standard deviation, while the errors refer to the discrepancies between the average damage ratios obtained by the algorithms and the actual damage ratio. The table is divided into three cases with different site of the damage, each representing the results for a 10 dimension of the problem. It appears that the table is comparing the performance of these algorithms across three different experiments, represented by the three cases, it is also comparing the effectiveness of different algorithms and how they perform in several scenarios of damage in presence of noisy data.

Comparing these algorithms, the performance of the algorithms varies depending on the detection of damage taking into consideration the presence of noise. The best results for each algorithm are not always consistent across the location of the damage. As an illustration, the ABC algorithm records a best result of 2.09E-01 in the initial case, while it is 1.85E-01 in the second set of findings and 1.97E-01 in the third case of damage. Additionally, the majority of algorithms display a low standard deviation, indicating that the outcomes remain consistent during multiple iterations. Furthermore, GSK-ALI and TLBO exhibit superior performance in the first two scenarios with best values of 1.98E-01, 2.01E-01, 2.02E-01, and 2.29E-01 respectively. In the third case, it seems that GSK-ALI is performing better with the best value of 2.00E-01 compared to the actual damage ratio. Overall, it can be seen that ABC and MRFO have similar performance with an average of 2.153e-01 and 2.044e-01 respectively.

The GSK-ALI technique displays a commendable performance, recording a slightly elevated mean score of 1.965e-01 and a standard deviation of 1.044e-02. Conversely, the HS algorithm yields the poorest results, with an average value of 2.708e-01, while the MDE has the highest standard deviation of 3.706e-02. However, some algorithms such as ABC, MDE, and JAYA have a relatively high standard deviation, which indicates that their results may be less consistent across multiple runs.

Table 4. Experimental results of test of the three cases

Algorithms	Dim	Case	Best	Worst	average	Std	Error%
ABC	10	1	2,09E-01	2,40E-01	2.153e-01	3.614e-02	7.65
GSK-ALI			**1,98E-01**	**1,87E-01**	**1.965e-01**	**1.044e-02**	**1.75**
HS			2,32E-01	2,95E-01	2.708e-01	2.729e-02	15
JAYA			2,54E-01	2,40E-01	2.218e-01	1.926e-02	27
MDE			2,43E-01	3,02E-01	2.501e-01	3.706e-02	35.4
MRFO			2,09E-01	2,40E-01	2.044e-01	1.979e-02	2.2
PFA			2,09E-01	2,40E-01	2.095e-01	2.772e-02	4.75
TLBO			2,02E-01	2,24E-01	2.055e-01	2.808e-02	2.75
ABC	10	2	1,85E-01	3,03E-01	2.244e-01	3.724e-02	22
GSK-ALI			**2,01E-01**	1,90E-01	**1.999e-01**	1.625e-02	**0.05**
HS			2,23E-01	2,58E-01	2.418e-01	2.013e-02	20.9
JAYA			2,02E-01	2,23E-01	2.257e-01	**1.249e-02**	28.5
MDE			1,90E-01	2,97E-01	2.271e-01	2.780e-02	35.5
MRFO			2,04E-01	**2,01E-01**	2.044e-01	2.601e-02	2.2
PFA			2,04E-01	1,76E-01	2.014e-01	1.597e-02	0.7
TLBO			2,29E-01	2,21E-01	2.078e-01	1.509e-02	3.9
ABC	10	3	1,97E-01	2,79E-01	2.233e-01	3.102e-02	11.65
GSK-ALI			**2,00E-01**	**2,09E-01**	2.053e-01	**1.221e-02**	2.65
HS			2,54E-01	2,21E-01	**2.039e-01**	2.748e-02	**1.95**
JAYA			2,54E-01	2,21E-01	2.422e-01	1.730e-02	21.1
MDE			2,49E-01	2,93E-01	2.225e-01	3.076e-02	12.5
MRFO			2,56E-01	2,83E-01	2.593e-01	2.072e-02	29.65
PFA			2,48E-01	**2,09E-01**	2.469e-01	2.150e-02	23.45
TLBO			2,48E-01	2,54E-01	2.524e-01	1.907e-02	26.2

Table 5. The p-values obtained by the comparison of the Kruskal-Wallis test followed by Mann-Whitney test

Algorithms	*p-value*	analysis
ABC VS GSK-ALI	2.068e-02	Algorithms differ significantly
ABC VS HS	3.846e-02	Algorithms differ significantly
ABC VS JAYA	1.413e-01	Algorithms differ significantly
ABC VS MDE	1.393e-01	Algorithms differ significantly
ABC VS MRFO	7.958e-01	No significant difference
ABC VS PFA	9.705e-01	No significant difference
ABC VS SSA	9.411e-01	No significant difference
ABC VS TLBO	8.187e-01	No significant difference
GSK-ALI VS HS	1.491e-06	Algorithms differ significantly
GSK-ALI VS JAYA	4.489e-08	Algorithms differ significantly
GSK-ALI VS MDE	4.636e-05	Algorithms differ significantly
GSK-ALI VS MRFO	7.958e-03	Algorithms differ significantly
GSK-ALI VS PFA	7.957e-03	Algorithms differ significantly
GSK-ALI VS SSA	2.052e-03	Algorithms differ significantly
GSK-ALI VS TLBO	4.636e-03	Algorithms differ significantly
HS VS JAYA	2.311e-01	Algorithms differ significantly
HS VS MDE	6.897e-01	No significant difference
HS VS MRFO	5.942e-02	Algorithms differ significantly
HS VS PFA	2.975e-02	Algorithms differ significantly
HS VS SSA	2.150e-02	Algorithms differ significantly
HS VS TLBO	4.513e-02	Algorithms differ significantly
JAYA VS MDE	6.952e-01	No significant difference
JAYA VS MRFO	2.398e-01	Algorithms differ significantly
JAYA VS PFA	8.235e-02	Algorithms differ significantly
JAYA VS SSA	5.942e-02	Algorithms differ significantly
JAYA VS TLBO	1.646e-01	Algorithms differ significantly
MDE VS MRFO	2.226e-01	Algorithms differ significantly
MDE VS PFA	1.023e-01	Algorithms differ significantly
MDE VS SSA	9.049e-02	Algorithms differ significantly
MDE VS TLBO	2.282e-01	Algorithms differ significantly
MRFO VS PFA	7.172e-01	No significant difference

(continued)

Table 5. *(continued)*

Algorithms	*p-value*	analysis
MRFO VS SSA	7.788e-01	No significant difference
MRFO VS TLBO	9.941e-01	No significant difference
PFA VS SSA	9.528e-01	No significant difference
PFA VS TLBO	7.618e-01	No significant difference
SSA VS TLBO	8.130e-01	No significant difference

Based on the comparison criteria of the best values, mean, standard deviation, and the error, it can be concluded that GSK-ALI outperforms its competitors. It is noteworthy, that in real-world applications, where the damage detection problem involves high complexity and dimensionality, there are time constraints. Therefore, GSK-ALI is preferred due to its efficient in detection of the damaged elements in a concrete plate structure.

5 Conclusions

In this study, a comprehensive evaluation of seven meta-heuristic algorithms is presented in order to determine the optimal detection of damage in a concrete plate structure. The algorithms were applied to three different scenarios of damage site, and a thorough comparison of the experimental results was conducted. The results indicate that GSK-ALI, MRFO, and Jaya algorithms exhibit superior performance when compared to the other algorithms. Notably, GSK-ALI demonstrates the best performance in most of cases, as evidenced by the highest best, mean, and standard deviation values and the errors refer to the discrepancies between the average damage ratios obtained by the algorithms and the actual damage ratio. These findings can be utilized by researchers to generalize these selected approaches to other classes of damage detection problems. Future research will focus on the implementation of an improved version of the GSK-ALI algorithm to solve damage detection issues in different types of structures under complex scenario and the investigation of the performance of GSK-ALI for multi-objective functions in the context of optimal sensor placement OSP for damage detection in structure health monitoring.

Acknowledgment. The authors would like to express their appreciation to IRIMAS-UHA Lab for providing the resources and support necessary for this research. We would also like to extend our gratitude to Sir Lhassane Idoumghar for his valuable insights, guidance, and expertise throughout the course of this project.

References

1. Wei Fan et Pizhong Qiao: Vibration-based damage identification methods: a review and comparative study. Struct. Health Monit., **10**(1), 83–111 (2011) https://doi.org/10.1177/147 5921710365419

2. Yan, Y.J., Cheng, L., Wu, Z.Y., Yam, L.H.: Development in vibration-based structural damage detection technique. Mech. Syst. Signal Process. **21**(5), 2198–2211 (2007), https://doi.org/10.1016/j.ymssp.2006.10.002

3. Carden, E.P., Fanning, P.: Vibration based condition monitoring: a review., Struct. Health Monit., **3**(4), 355–377 (2004) https://doi.org/10.1177/1475921704047500

4. Joshuva et, A., Sugumaran, V.: A comparative study of bayes classifiers for blade fault diagnosis in wind turbines through vibration signals, p. 23 (2017)

5. Alkayem, N.F., Cao, M., Zhang, Y., Bayat, M., Su, Z.: Structural damage detection using finite element model updating with evolutionary algorithms: a survey. Neural Comput. Appl. **30**(2), 389–411 (2017). https://doi.org/10.1007/s00521-017-3284-1

6. Farrar, C.R., Doebling, S.W., Nix, D.A.: Vibration–based structural damage identification. Philos. Trans. R. Soc. Lond. Ser. Math. Phys. Eng. Sci. **359**(1778), 131–149 (2001) https://doi.org/10.1098/rsta.2000.0717

7. Balmès, E., Basseville, M., Mevel, L., Nasser, H.: handling the temperature effect in vibration monitoring of civil structures: a combined subspace-based and nuisance rejection approach. IFAC Proc. **39**(13), 611–616 (2006) https://doi.org/10.3182/20060829-4-CN-2909.00101

8. Bernal, D.: Kalman filter damage detection in the presence of changing process and measurement noise. Mech. Syst. Signal Process. **39**(12), 361–371, août 2013, https://doi.org/10.1016/j.ymssp.2013.02.012

9. Döhler, M., Hille, F.: Subspace-Based Damage Detection on Steel Frame Structure Under Changing Excitation. In: Wicks, A. (ed.) Structural Health Monitoring, Volume 5. CPSEMS, pp. 167–174. Springer, Cham (2014). https://doi.org/10.1007/978-3-319-04570-2_19

10. Döhler, M., Mevel, L., Hille, F.: Subspace-based damage detection under changes in the ambient excitation statistics. Mech. Syst. Signal Process. **45**(1), 207–224, mars (2014) https://doi.org/10.1016/j.ymssp.2013.10.023

11. Döhler, M., Mevel, L.: Subspace-based fault detection robust to changes in the noise covariances. Automatica, **49**(9) 2734–2743, sept. (2013) https://doi.org/10.1016/j.automatica.2013.06.019

12. Döhler, M., Hille, F., Mevel, L., Rücker, W.: Structural health monitoring with statistical methods during progressive damage test of S101 Bridge. Eng. Struct. **69**, 183–193, juin (2014) https://doi.org/10.1016/j.engstruct.2014.03.010

13. Doebling, S.W., Farrar, C.R., et Prime, M.B.: A Summary Review of Vibration-Based Damage Identification Methods. Shock Vib. Dig. **30**(2), 91–105, mars (1998) https://doi.org/10.1177/058310249803000201

14. Ručevskis, S., Chate, A.: Identification in a plate-like structure using modal data. Aviation **17**(2), 45–51 (2013) https://doi.org/10.3846/16487788.2013.805863

15. Navabian, N., Bozorgnasab, M., Taghipour, R., Yazdanpanah, O.: Damage identification in plate-like structure using mode shape derivatives. Arch. Appl. Mech. **86**(5), 819–830 (2015). https://doi.org/10.1007/s00419-015-1064-x

16. Moreno-García, P., Dos Santos, J.A., Lopes, H.: A new technique to optimize the use of mode shape derivatives to localize damage in laminated composite plates. Compos. Struct., **108**, 548–554 (2014) https://doi.org/10.1016/j.compstruct.2013.09.050

17. Yang, Z., Chen, X., Yu, J., Liu, R., Liu, Z., He, Z.: A damage identification approach for plate structures based on frequency measurements. Nondestruct. Test. Eval. **28**(4), 321–341 (2013) https://doi.org/10.1080/10589759.2013.801472

18. Katunin, A.: 953. Vibration-based damage identification in composite circular plates using polar discrete wavelet transform. VOLUME, **15**(1) 9

19. Cao, M.S., Xu, H., Bai, R.B., Ostachowicz, W., Radzieński, M., Chen, L.: Damage characterization in plates using singularity of scale mode shapes. Appl. Phys. Lett. **106**(12) 121906, mars (2015) https://doi.org/10.1063/1.4916678

20. Eraky, A., Saad, A., Anwar, A.M., Abdo, A.: Damage detection of plate-like structures based on residual force vector. HBRC J. **12**(3), 255–262, (2016) https://doi.org/10.1016/j.hbrcj. 2015.01.005

21. Yun, G.J., Ogorzalek, K.A., Dyke, S.J., Song, W.: A parameter subset selection method using residual force vector for detecting multiple damage locations. Struct. Control Health Monit. **17**(1), 4867 (2010) https://doi.org/10.1002/stc.284

22. Kennedy, J., Eberhart, R.C.: A discrete binary version of the particle swarm algorithm. In: 1997 IEEE International Conference on Systems, Man, and Cybernetics. Computational Cybernetics and Simulation, Orlando, FL, USA: IEEE, 1997, pp. 4104–4108. https://doi.org/10.1109/ ICSMC.1997.637339

23. Karaboga, D.: An idea based on honey bee swarm for numerical optimization, p. 10

24. Storn, R.: Differential Evolution – A simple and efficient heuristic for global optimization over continuous spaces. Differ. Evol. **11**(4) 19 (1997)

25. Rao, R.V., Savsani, V.J., Vakharia, D.P.: Teaching–learning-based optimization: a novel method for constrained mechanical design optimization problems. Comput.-Aided Des., **43**(3), 303-315, mars (2011) https://doi.org/10.1016/j.cad.2010.12.015

26. Abderazek, H., Ferhat, D., Ivana, A.: Adaptive mixed differential evolution algorithm for bi-objective tooth profile spur gear optimization. Int. J. Adv. Manufact. Technol. **90**(5–8), 2063–2073 (2016). https://doi.org/10.1007/s00170-016-9523-2

27. Acar, E., Haftka, R.T.: Reliability Based Aircraft Structural Design Pays Even with Limited Statistical Data, p. 19

28. Kamjoo, V., Eamon, C.D.: Reliability-based design optimization of a vehicular live load model. Eng. Struct. **168**, 799–808, août (2018) https://doi.org/10.1016/j.engstruct.2018. 05.033

29. Youn, B.D., Choi, K.K., Yang, R.-J., Gu, L.: Reliability-based design optimization for crashworthiness of vehicle side impact. Struct. Multidiscip. Optim. **26**(3–4), 272–283 (2004). https://doi.org/10.1007/s00158-003-0345-0

30. Spence, S.M., Gioffrè, M.: Large scale reliability-based design optimization of wind excited tall buildings. Probabilistic Eng. Mech., **28,** 206–215, avr.(2012) https://doi.org/10.1016/j. probengmech.2011.08.001

31. Marwala, T.: Finite-element-model Updating Using Computional Intelligence Techniques. London: Springer London (2010). https://doi.org/10.1007/978-1-84996-323-7

32. Fang, J., Wu, C., Rabczuk, T., Wu, C., Sun, G., Li, Q.: Phase field fracture in elasto-plastic solids: a length-scale insensitive model for quasi-brittle materials. Comput. Mech. **66**(4), 931–961 (2020). https://doi.org/10.1007/s00466-020-01887-1

33. Yang, Q.W., Liu, J.K.: Damage identification by the eigenparameter decomposition of structural flexibility change. Int. J. Numer. Methods Eng. **78**(4), 444-459 (2009) https://doi.org/ 10.1002/nme.2494

34. Li, J., Li, Z., Zhong, H.,Wu, B.: Structural Damage Detection Using Generalized Flexibility Matrix and Changes in Natural Frequencies. AIAA J. **50**(5), 1072–1078, mai (2012) https:// doi.org/10.2514/1.J051107

35. Kang, J., Zhang, W.: Combination of Fuzzy C-Means and Particle Swarm Optimization for Text Document Clustering. In: Advances in Electrical Engineering and Automation, A. Xie et X. Huang, Éd., in Advances in Intelligent and Soft Computing, vol. 139. Berlin, Heidelberg: Springer Berlin Heidelberg, 2012, p. 247–252. https://doi.org/10.1007/978-3-642-27951-5_37

36. Gharehchopogh, F.S., Namazi, M., Ebrahimi, L., Abdollahzadeh, B.: Advances in Sparrow Search Algorithm: A Comprehensive Survey , Arch. Comput. Methods Eng., août (2022) https://doi.org/10.1007/s11831-022-09804-w

37. Agrawal, P., Ganesh, T., Oliva, D., Mohamed, A.W.: S-shaped and V-shaped gaining-sharing knowledge-based algorithm for feature selection. Appl. Intell., **52**(1), 81–112 (2022) https://doi.org/10.1007/s10489-021-02233-5

38. Agrawal, P., Ganesh, T., Mohamed, A.W.: A novel binary gaining–sharing knowledge-based optimization algorithm for feature selection. Neural Comput. Appl. **33**(11), 5989–6008 (2020). https://doi.org/10.1007/s00521-020-05375-8

39. Zhao, W., Zhang, Z., Wang, L.: Manta ray foraging optimization: An effective bio-inspired optimizer for engineering applications. Eng. Appl. Artif. Intell., **87**(C), (2020) https://doi.org/10.1016/j.engappai.2019.103300

40. Yapici, H., Cetinkaya, N.: A new meta-heuristic optimizer: Pathfinder algorithm . Appl. Soft Comput., **78**, 545–568, mai (2019) https://doi.org/10.1016/j.asoc.2019.03.012

41. Dinh-Cong, D., Vo-Duy, T., Nguyen-Minh, N., Ho-Huu, V., Nguyen-Thoi, T.: A two-stage assessment method using damage locating vector method and differential evolution algorithm for damage identification of cross-ply laminated composite beams. Adv. Struct. Eng., **20**(12), (2017) https://doi.org/10.1177/1369433217695620

42. Dinh-Cong, D., Vo-Duy, T., Ho-Huu, V., Nguyen-Thoi, T.: Damage assessment in plate-like structures using a two-stage method based on modal strain energy change and Jaya algorithm. Inverse Probl. Sci. Eng., **27**(2), 166–189, févr. (2019) https://doi.org/10.1080/17415977.2018.1454445

Estimation of the Impact of COVID-19 Pandemic Lockdowns on Breast Cancer Deaths and Costs in Poland Using Markovian Monte Carlo Simulation

Magdalena Dul[1,2], Michal K. Grzeszczyk[1](✉) (iD), Ewelina Nojszewska[2](iD), and Arkadiusz Sitek[3](iD)

[1] Sano Centre for Computational Medicine, Cracow, Poland
m.grzeszczyk@sanoscience.org
[2] Warsaw School of Economics, Warsaw, Poland
[3] Massachusetts General Hospital, Harvard Medical School, Boston, MA, USA
asitek@mgh.harvard.edu

Abstract. This study examines the effect of COVID-19 pandemic and associated lockdowns on access to crucial diagnostic procedures for breast cancer patients, including screenings and treatments. To quantify the impact of the lockdowns on patient outcomes and cost, the study employs a mathematical model of breast cancer progression. The model includes ten different states that represent various stages of health and disease, along with the four different stages of cancer that can be diagnosed or undiagnosed. The study employs a natural history stochastic model to simulate the progression of breast cancer in patients. The model includes transition probabilities between states, estimated using both literature and empirical data. The study utilized a Markov Chain Monte Carlo simulation to model the natural history of each simulated patient over a seven-year period from 2019 to 2025. The simulation was repeated 100 times to estimate the variance in outcome variables. The study found that the COVID-19 pandemic and associated lockdowns caused a significant increase in breast cancer costs, with an average rise of 172.5 million PLN (95% CI [82.4, 262.6]) and an additional 1005 breast cancer deaths (95% CI [426, 1584]) in Poland during the simulated period. While these results are preliminary, they highlight the potential harmful impact of lockdowns on breast cancer treatment outcomes and costs.

Keywords: Breast Cancer · Costs · Markov Model · Covid Lockdowns

1 Introduction

The COVID-19 pandemic impacted the lives of people around the world. To slow down the spread of the disease, many countries introduced lockdown restrictions

---

M. Dul and M. K. Grzeszczyk —Authors contributed equally.

© The Author(s), under exclusive license to Springer Nature Switzerland AG 2023
J. Mikyška et al. (Eds.): ICCS 2023, LNCS 14075, pp. 138–152, 2023.
https://doi.org/10.1007/978-3-031-36024-4_10

in form of banning gatherings, limiting outdoor trips and canceling public events [23]. While lockdowns positively influenced the pandemic progression (decreased doubling time) [25] or even environment [6], the negative impact on mental health [1], physical fitness [49], dietary habits [8] and other important aspects of our lives are evident. In this work we analyze the effect of pandemic lockdowns on breast cancer care in Poland.

Breast cancer is the most frequent cause of cancer deaths among women [48] and is a high burden to public finance. There is an estimated 2.3 million women diagnosed with breast cancer and 685,000 deaths globally in 2020 [46]. The direct cause of breast cancer is unknown, but there exist a number of risk factors like obesity, late menopause or alcohol use [22]. Since there are few to no symptoms at the early stage of breast cancer, many countries introduced screening programs in the form of free mammography procedures to support the early detection of the disease [47]. COVID lockdowns resulted in restricted access to healthcare [19] which consequently reduced the number of diagnosed and treated breast cancer patients.

In this paper, we present a Markov Model-based approach to the Monte Carlo simulation of breast cancer disease progression. The nodes of the model are different states or cancer stages that the subject can be in at a specific point in time. The probabilities of transitions between states are computed based on the existing literature and empirical experiments. We use this method to conduct 100 repetitions of seven-year-long simulations on 1% of the total women population in Poland. In the simulation, we consider the direct costs (medicines, surgeries, chemotherapy), indirect costs (premature death, absenteeism, disability) of breast cancer and statistics of the number of subjects in all states. We conduct two types of experiments. First, we perform the simulation taking into consideration the impact of COVID lockdowns on the accessibility of public healthcare, screening programs and treatment. Secondly, we conduct the simulation as if there was no pandemic. We extrapolate results to the population of the entire country.

The main contributions of this paper are as follows:

1. We present a Markov Model-based simulation of the progression of breast cancer.
2. We analyze the impact of COVID lockdowns on mortality and healthcare costs using a comparison of simulations conducted on the population of Polish women with and without the simulated effect of pandemic.
3. We provide a publicly available code to simulate the progression of breast cancer: https://github.com/SanoScience/BC-MM.

The rest of the paper is structured as follows. In Sect. 2, we describe the existing methods for the simulation of disease progression and present our approach to breast cancer modeling based on Markov Models in Sect. 3. We show the results of simulations with and without the effects of pandemic and discuss how COVID-19 impacted breast cancer patients and the costs of the disease in Sect. 4 and conclude in Sect. 5.

2 Related Work

In this section, we describe works presented in the literature related to the investigation of the impact of the COVID-19 pandemic on healthcare, modeling the progression of diseases and analysis of disease costs in public finance.

2.1 The Impact of COVID-19 Pandemic on Healthcare

Since the beginning of the pandemic, researchers have been concerned about the possible, negative side effects of lockdowns [41]. Paltrinieri et al. [33] reported that 35.1% lifestyle survey participants encountered worsening of physical activity during lockdowns. Similar concerns were presented by Tsoukos and Bogdanis [49] who described lower body fitness, poorer agility tests results and increased body mass index in adolescent students as the effects of a five-month lockdown. The negative impact of lockdowns does not end on the deterioration of physical fitness. Mental health is one of the factors that suffered the worst during the pandemic. Adams et al. [1] discussed a significant decrease in self-reported mental health in the United States. The self-harm incidents due to stress related to COVID-19 were reported in India [40]. Cases of depression, anxiety and post-traumatic stress disorders were expected to rise in Sub-Saharan Africa [43].

During the pandemic, access to healthcare, especially related to the treatment of other diseases was limited [19]. Many global healthcare resources were reallocated to prevent and treat coronavirus infections [14]. The expected results of the depletion of healthcare resources were the increase of COVID-19 and all-cause mortality [38]. Additionally, more than 28 million surgeries were expected to be canceled in the UK due to lockdowns [15]. In most cases, those were operations for benign diseases, however, the effect cannot be neglected. Jiang et. al [21] described examples of the co-epidemics of COVID-19 and other infectious diseases and potential negative effects on the treatment of non-communicable and chronic diseases.

Concerns regarding the impact of COVID-19 on the treatment of diseases give a justified basis for the analysis the influence of lockdowns on breast cancer prevalence and costs. As reported by Gathani et al. [18], there was a steep decrease in the number of referrals for suspected breast cancer (28% lower) and breast cancer diagnosis (16% lower) in the UK in 2020. Yin et al. [50] describe the decline in the usage number of 3 services (breast imaging, breast surgery and genetics consultation) in the early stages of the pandemic. In [17], the pauses in screening programs that occurred in various countries were described, and disease modeling was mentioned as one of the possible approaches to analyse the repercussions of COVID-19 on breast cancer. In this paper, we analyse the impact of those radical changes in breast cancer diagnosis and treatment on the costs of breast cancer in public finance.

2.2 Modelling Progression of the Disease

There are multiple methods for developing disease models for the purpose of conducting simulations [24]. One of the approaches to stochastic process modeling (like the progression of the chronic disease) and economic impact analysis is the utilization of Markov Modelling [11]. In such a graph model, nodes represent stages of the disease and edges the probabilities of moving from one state to another. For instance, Liu *et al.* [26] presented the Continuous-Time Hidden Markov Model for Alzheimer's disease progression. In [37], a multi-state semi-Markov model was used to investigate the impact of type 2 diabetes on the co-occurrence of cardiovascular diseases.

Markov Model can be successfully utilized to conduct an analysis of breast cancer. Momenzadeh *et al.* [29] used hidden Markov Model to predict the recurrence of breast cancer, while Pobiruchin *et al.* [35] presented a method for Markov Model derivation out of real-world datasets (cancer registry's database). Breast cancer modeling was also used to investigate the decline in screening, delays in diagnosis and delays in treatment during COVID-19 pandemic in the USA [3]. Alagoz *et al.* [3] developed three models representing each issue. The conducted simulation exposed that there is a projected excess of breast cancer deaths by 2030 due to the pandemic. In this paper, we present the results of the experiments conducted with Monte Carlo simulation based on the Markov Model of breast cancer progression in Poland.

2.3   Costs of Breast Cancer Care

The analysis of disease costs for public finance is a difficult task as there are different methods that could be used and various types of costs that have to be taken into consideration [12]. Costs in pharmacoeconomics can be divided into four categories: direct medical costs, direct non-medical costs, indirect costs and intangible costs [34]. Direct medical costs are the easiest to determine. They include the costs of medicines, diagnostic tests, hospital visits etc. Direct non-medical costs are costs mainly related to the treatment of the patient, but not having a medical basis. Another group of costs are indirect costs. They are mainly related to the loss of productivity associated with the patient's illness or death. The intangible costs are the costs associated with pain, suffering, fatigue and anxiety associated with the disease, as well as side effects of treatment such as nausea. They are difficult to estimate and measure, as they are mainly related to the patient's feelings [39]. In this paper, we take into consideration direct and indirect costs only.

Depending on the methodology the calculated costs may highly vary (e.g. $US20,000 to $US100,000 of the per-patient cost) [12]. Different studies analyse different types of costs, making it difficult to compare them. In [7], the mathematical model of continuous tumor growth and screening strategies was applied for the evaluation of screening policies. In [16], cost-effectiveness studies were used to estimate the costs of breast cancer screening per year of life saved to be $13,200-$28,000. The total cost of breast cancer in the USA was estimated to be $3.8 billion. Blumen *et al.* [10] conducted a retrospective study to compare the treatment costs by tumor stage and type of service. The analysis was undertaken on a population selected from the commercial claims database which facilitated the study as costs-related data was directly available.

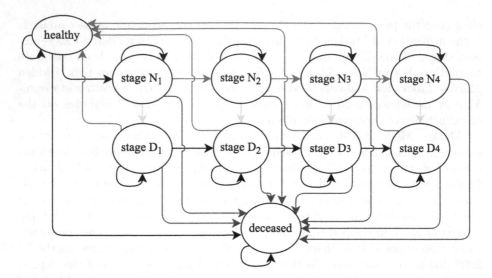

Fig. 1. Breast cancer Markov Model with 10 states. Blue arrows indicate transitions between different stages of breast cancer, yellow ones show the diagnosis of breast cancer, green arrows indicate that the patient was healed and red represent the death event related to breast cancer. (Color figure online)

3   Methodology

In this section, we describe the Markov Model used for the simulation of breast cancer progression. We define the types and values of costs used and provide details on the parameters used in simulations. For more details on the actual algorithmic implementation refer to the source code available at https://github.com/SanoScience/BC-MM.

Breast Cancer Markov Model. We use the Monte Carlo simulation based on the Markov Model to describe the course of breast cancer disease. The time horizon of the analysis is divided into equal time increments, called Markov cycles. During each cycle, the patient can transition from one state to another. Arrows connecting two different states indicate allowed transitions. We applied values of those transitions using clinical information, derived from previous studies or estimated empirically. The arrows from the state to itself indicate that the patient may remain in that state during cycles [44]. Transition probabilities are used to estimate the proportion of patients who will transfer from one state to another. The probability of an event occurring at a constant rate (r) during time (t) can be expressed by the equation:

$$p = 1 - e^{-rt} \qquad (1)$$

In Fig. 1, we present the Markov Model describing the progression of breast cancer. There are ten states in our model: *healthy*, four states describing a nondiagnosed person with breast cancer at four stages [5] of the disease ($stageN_i$, where $i \in \{1, 2, 3, 4\}$), four states for diagnosed stages ($stageD_i$, where $i \in \{1, 2, 3, 4\}$) and *deceased*. *Deceased* is a terminal stage of our model and if a subject reaches this state its simulation is terminated. We follow The American Joint Committee on Cancer which defines four stages of breast cancer [5].

Simulation. We assume that each Markov cycle is equal to one week. Breast cancer is a disease that develops for many years, however, the longer the simulated period is, the less reliable are the results due to assumptions made about the future. Therefore, we set the number of cycles to 364, corresponding to 7 years (assuming that every year has 52 weeks). This period allows us to measure the long-term effects of the COVID-19 pandemic. We set the beginning of the simulation to January 1st 2019 so that the simulation stabilizes (reaches equilibrium) before the 2020 year year with a pandemic. We conduct two types of simulations, one taking into account COVID-19 lockdowns, and one which assumes that there was no effect of COVID-19 on breast cancer treatment and progression. We assume that lockdowns in Poland were lasting from March 2020 till the beginning of March 2021. We repeat 100 times each type and average the collected results.

In the simulation, we take into account malignant breast cancer only (C50 ICD-10 code). Stage 0 of breast cancer (which has a different ICD-10 code) often has a 100% 5-year survival rate. Thus, we omit this stage in the analysis as it should not have a significant impact on costs and survival. We simulate the breast cancer progression in women as this sex accounts for most of the cases of the disease. We conduct computation on 1% of the representative women population in Poland with the age distribution according to Table 1 - after the end of simulations, we multiply results by 100. The minimum age of simulated patients is set to 25 because below this age the occurrence of breast cancer is rare (Table 2). We increase the age of each simulated person every 52 cycles.

To find the number of women diagnosed with breast cancer in 2019 we compute a linear trend line based on the available data. There were 143,911, 151,831, 158,534, 166,031, and 174,005 patients with breast cancer in 2010, 2011, 2012, 2013, and 2014 respectively [31]. According to Agency for Health Technology Assessment and Tariff System in Poland, those numbers rose to 227,784 and 242,838 in 2016 and 2017 [2]. The projected trend line indicated that in 2018 and 2019 there were 247,013 and 263,590 women with this disease in Poland. Taking into consideration the distribution of cancer stages among women with the diagnosed disease in the UK [13] and the distribution of age (Table 1) we derive the number of diagnosed women in Poland by cancer stage and by age (Table 3). In addition, we estimate the number of undiagnosed people. We assume that breast cancer would only be detected using mammography and follow-up diagnostic regimen, and around 71% of patients show up at this procedure [42]. In 2019 the number of mammography tests was 1.041 million with 19620 cases detected (2%). The number of people who fell ill in 2019 was 263,590. This is 2% of the 71% of people who appeared on mammograms. On this basis, the remaining people who did not show up on the mammogram and would have a

Table 1. The age distribution of Polish women above 25 years in 2019 [45].

Age	Number of women	Percentage
25–29	1,233,777	8%
30–34	1,436,161	10%
35–39	1,596,757	11%
40–44	1,502,164	10%
45–49	1,294,636	9%
50–54	1,142,203	8%
55–59	1,234,478	8%
60–64	1,460,924	10%
65–69	1,359,815	9%
70–74	1,013,368	7%
75–79	640,118	4%
80–84	581,529	4%
85+	583,545	4%
Total	15,079,475	100%

Table 2. The distribution of patients diagnosed with breast cancer in 2019 in Poland [36].

Age	0–24	25–29	30–34	35–39	40–44	45–49	50–54	55–59	60–64	65–69	70–74	75–79	80–84	85+
#	8	79	292	697	1,252	1,591	1,832	2,120	2,970	3,377	2,039	1,442	1,107	814
%	0.0	0.4	1.5	3.6	6.4	8.1	9.3	10.8	15.1	17.2	10.4	7.3	5.6	4.1

positive result can be calculated. They are 2% of the remaining 29% that should come for a mammogram. The estimated number of people is 108,427. Using this number and information about the percentage of patients in a specific stage, we calculate the number of undiagnosed patients in stages II, III and IV (Table 3). The same strategy for stage I destabilizes the model. Thus, we set the number of undiagnosed patients in the first stage to the same value as for those diagnosed in the first stage in 2019. We make this assumption due to the fact that people in stage I very often do not have symptoms yet and the model.

State Transition Probabilities. We derive the following state transition probabilities:

1. $P(healthy \rightarrow stageN_1)$ - the probability of developing breast cancer,
2. $P(healthy \rightarrow deceased)$ - the probability of non-cancer related death,
3. $P(stageN_i \rightarrow stageN_{i+1})$ - the probability of cancer stage increase,
4. $P(stageN_i \rightarrow stageD_i)$ - the probability of cancer diagnosis,
5. $P(stageN_i \rightarrow deceased)$ - the probability of breast cancer death,
6. $P(stageD_i \rightarrow stageD_{i+1})$ - the probability of cancer stage increase,
7. $P(stageD_i \rightarrow healthy)$ - the probability of healing,
8. $P(stageD_i \rightarrow deceased)$ - the probability of breast cancer death.

Table 3. The projected distribution of diagnosed and undiagnosed women with breast cancer in 2019 in Poland.

Age \ Stage	Diagnosed				Undiagnosed			
	I	II	III	IV	I	II	III	IV
25–29	461	441	102	57	461	181	42	24
30–34	1,705	1,630	377	211	1,705	670	155	87
35–39	4,069	3,891	900	504	4,069	1,600	370	208
40–44	7,308	6,989	1,617	906	7,308	2,875	665	373
45–49	9,287	8,881	2,055	1,151	9,287	3,653	845	474
50–54	10,694	10,227	2,366	1,326	10,694	4,207	973	545
55–59	12,375	11,834	2,738	1,534	12,375	4,868	1,126	631
60–64	17,337	16,579	3,836	2,150	17,337	6,820	1,578	884
65–69	19,713	18,851	4,361	2,444	19,713	7,754	1,794	1,005
70–74	11,902	11,382	2,633	1,476	11,902	4,682	1,083	607
75–79	8,417	8,050	1,862	1,044	8,417	3,311	766	429
80–84	6,462	6,179	1,430	801	6,462	2,542	588	330
85+	4,752	4,544	1,051	589	4,752	1,869	432	242

To simulate the effects of covid lockdowns we modify three transition probabilities. The probability of cancer diagnosis is decreased because of lockdowns and restricted access to healthcare. The probability of breast cancer-related death is increased due to a lack of proper healthcare assistance, and the probability of healing is decreased due to poorer treatment during the COVID-19 pandemic. Numerically we implement the models as follows.

We assume **probability of developing breast cancer** is only dependent on women's age. We set this probability in the following manner - age (probability; probability in one cycle): 20 (0.1%; 0.0002%), 30 (0.5%; 0.001%), 40 (1.5%; 0.0029%), 50 (2.4%; 0.0046%), 60 (3.5%; 0.0067%), 70 (4.1%; 0.0079%), 80 (3.0%; 0.0058%) [4]. We define the **probability of non-cancer related death** according to the life tables from 2019 [45]. There are multiple resources defining the Progression-Free Survival parameter which is a time during which the patient lives with the disease but their state is not worsening. For example, Haba-Rodriguez *et al.* [20] state that the median of PFS varies between 4 to 18 months. Thus, we empirically set the **probability of cancer stage increase for diagnosed patient** to $p = k(1 - e^{-\lambda t})$ where k is 0.0009, t is the number of cycles and λ is 10, 15, 20 or 25 depending on the stage of the disease. It is difficult and highly uncertain to assess the progression of breast cancer in an undiagnosed patient as no data exist that describe those transitions. Therefore, we define the **probability of cancer stage increase for undiagnosed women** in the same manner as in the case of diagnosed cases and set λ to 20, 25, 30 or 35 depending on the stage which is a reasonable approximation.

We define the **probability of healing** based on the 5-year Disease Free Survival (DFS) parameter which changes depending on the cancer stage [32]. The

5-year DFS in 2019 was 0.987 (stage I), 0.873 (stage II), 0.52 (stage III), and 0.037 (stage IV). We decrease those values during lockdowns by 19% (due to a 19% decrease in hospitalizations [30]) to 0.801 (stage I), 0.708 (stage II), 0.422 (stage III), and 0.03 (stage IV) and set the probability of healing to $p = k(1 - e^{-\lambda t})$ where k is set to $\frac{1}{3}$ and λ is computed based on the 5-year DFS. The **probability of death for diagnosed patient** is computed from the 5-year survival rate which indirectly provides the probability of death within 5 years. Taking into consideration the 5-year survival in stages I, II, III and IV of 0.975, 0.856, 0.44, 0.23 [32], we compute λ parameter of the probability of death in cycle $\leq t$ ($p(x \leq t) = 1 - e^{-\lambda t}$) to be 0.0061, 0.0302, 0.1642 and 0.2939 respectively. For covid simulation according to predictions that the mortality rate might increase by 9.6% [27] the λ parameter is set to 0.0056, 0.0344, 0.1903, 0.3715 for every stage. The 3-month delay in cancer diagnosis may increase the chances of premature death by 12% [9]. We, therefore, find the **probability of death for undiagnosed patient** by increasing the 5-year death probability for diagnosed patients and compute λ for those probabilities equal to 0.0061, 0.0349, 0.1989, 0.3932 depending on the stage.

In 2019, approximately 7% of all women aged 25 and over had a mammogram. The situation changed in 2020 when the number of mammograms decreased by over 305,000, which was a decrease of about 29%. We assume that malignant breast cancer can only be detected by mammography and diagnostic follow-ups. The newly detected cases in 2019 (19,620) accounted for 2% of all mammograms performed. In 2020, the number of detected cases decreased due to the COVID-19 pandemic. The average annual growth rate of new cases of breast cancer is 3%. This means that around 20,209 new cases of cancer should have been diagnosed in 2020. We assume that the percentage of positive mammograms did not change, which will bring the number of detected cases to about 13,873. This is a difference of about 6,000 cases compared to what should have been detected. About 12.5% of mammograms are thought to have a false-negative result and data shows that only 71% of all women show up for the examination. In 2020, the number of these women probably decreased even more (by 29%). Therefore, we define the **probability of diagnosis** in one year as $p = \frac{l_{pos}}{l_{pos} + l_{fneg} + l_{nm}}$ where l_{pos} is the number of positive mammography cases, l_{fneg} is the number of false negative mammography and l_{nm} is the number of women that should have taken part in the mammography and would have had positive results. Thus, this probability for 2019 is 12.4% and 11.6% during lockdowns.

Costs of Breast Cancer in Poland. We collect two types of costs during simulation: direct costs and indirect costs. We divide the latter into indirect costs related to premature death and other indirect costs related to absenteeism, presenteeism, absenteeism of informal caregivers, presenteeism of informal carers, disability etc.

We derive direct per-person, per-stage costs in the following way. We estimate the total direct costs for 2019 based on the estimated number of breast cancer patients and direct costs in 2010–2014 years [31] to be 846,653 thousand PLN.

We follow the distribution of the stage-specific costs in [28]. We compute that direct per-stage yearly costs, based on the number of patients in every stage, in the simulation are: stage I (25% of total costs, 207,560 thousand PLN, 1881 PLN per person), stage II (38%, 325,108 thousand PLN, 3,185 PLN per person), stage III (25%, 215,076 thousand PLN, 5,573 PLN per person), stage IV (12%, 98,909 thousand PLN, 7,869 PLN per person). We add those costs (divided by the number of cycles in one year) for every diagnosed woman in each cycle of the simulation.

We compute indirect death costs by averaging the per-person death costs related to breast cancer patients in 2010–2014 years [31]. The average value of premature death cost is 123,564 PLN. We add this value every time a simulated person enters *deceased* state from one of the breast cancer stages. We estimate other indirect costs in the same way. The average value of indirect per-patient cost in 2010–2014 years is 13,159 PLN. We add this value (divided by the number of cycles in year) for every patient in the $stageD_i$ state in every cycle.

Experimental Setup. We develop the Monte Carlo simulation with Python 3.10 programming language. We conduct all simulations on a 1.80 GHz Intel Core i7 CPU and 16 GB RAM. The execution of all simulations took 3 h to complete.

4   Results and Discussion

Costs of Breast Cancer. In Table 4, we present changes in average direct and indirect costs over the 7-years period. In all cases, the total costs incurred in the absence of lockdowns are smaller than the ones that resulted from the simulation with the COVID-19 pandemic. However, the only statistically significant difference (p-value < 0.001) is in the case of indirect costs related to premature death. This is reasonable because breast cancer is a long-lasting disease and the costs of treatment or patient care did not change drastically due to lockdowns. On the other hand, delayed diagnoses and surgeries resulted in more premature deaths. The impact of the pandemic is also reflected in the total costs of breast cancer (Table 5). the pandemic resulted in an increase in breast cancer costs of 172.5 million PLN (average total costs with covid - average total costs without covid) with 95% confidence interval (CI) of [82.4, 262.6]. The difference between total costs with and without lockdowns is statistically significant (p-value < 0.001). The positive influence of lockdowns on the progression of the pandemic should not be neglected. However, the presented results suggest that lockdowns had a negative impact on overall disease treatment, both socially and economically.

Breast Cancer with and Without COVID-19 Pandemic. Simulations also showed that there was a significant difference in the number of women deaths due to COVID-19. On average, during 7 years, 60,052 women died taking into consideration lockdowns. This number would be smaller by 1005 deaths if the pandemic did not occur (95% CI [426, 1584]). Year-by-year visualization

Table 4. Direct and indirect simulated costs (in million PLN) of breast cancer in Poland with and without COVID-19 pandemic.

year	DIRECT NO COVID	COVID	INDIRECT_DEATH NO COVID	COVID	INDIRECT_OTHER NO COVID	COVID
2019	763,4	764,4	625,79	625,93	3297,14	3302,13
2020	770,7	771,7	827,93	875,106	3317,26	3321,28
2021	775,10	775,9	990,107	996,113	3337,36	3338,34
2022	777,11	777,12	1087,118	1119,120	3352,40	3355,44
2023	777,11	779,13	1186,116	1197,121	3367,43	3372,50
2024	775,12	779,14	1275,120	1270,119	3377,49	3387,51
2025	774,13	777,15	1306,150	1340,30	3389,53	3399,56
TOTAL	5411,55	5422,64	7296,246	7420,267	23437,213	23474,236

Table 5. Total simulated costs (in million PLN) of breast cancer in Poland.

year	TOTAL NO COVID	COVID
2019	4686,77	461,91
2020	4915,91	4967,107
2021	5101,116	5109,118
2022	5216,133	5251,122
2023	5329,122	5348,135
2024	5428,130	5435,133
2025	5469,170	5516,140
TOTAL	36144,316	36317,333

of deaths is presented in Fig. 2. It can be noticed that delayed diagnoses and poorer treatment resulted in an overall increase in the number of deaths. The long-term effects will be visible in years to come. Figure 2 depicts also the average number of cases of diagnosed breast cancer in Poland. There is a sharp decline in the number of diagnoses between covid and no covid simulations in 2020. The delayed diagnoses resulted in an increased probability of complications and death. In the following years, the trend was reversed and more of the delayed cases were diagnosed in covid simulation. However, the inefficient healthcare system is not capable of making up for the lost diagnostic time.

Limitations. Our study is subject to limitations. First, the model and simulations presented are only approximations of the real-world situation, and therefore, the results should be interpreted with caution. Second, the impact of the COVID-19 pandemic on the costs associated with breast cancer is complex and rapidly evolving, and our study provides only a snapshot of the situation at the time of the analysis. Third, in order to build the model, we had to make several assumptions and rely on estimates or information from countries other

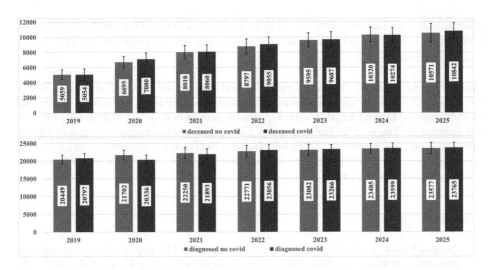

Fig. 2. The average number of breast cancer-related deaths (top) and the average number of breast cancer diagnoses (bottom) with and without lockdowns.

than Poland due to a lack of national data. Access to healthcare and treatment may vary across different countries, and this may have resulted in overestimated or underestimated data in our model. Therefore, our findings should be interpreted in the context of these limitations and further research is needed to validate and expand our results. To account for the uncertainty around the course of the tumor, empirical fitting of transition probabilities was necessary. This is because, upon diagnosis, patients are immediately referred for treatment, leaving no research data on the disease's development. Furthermore, the study assumes that people with cancer did not directly die from coronavirus infection, but those at an advanced stage of the disease may have had a higher risk of succumbing faster after being infected with the pathogen. It is also worth noting that the model does not consider potential changes in healthcare policies or treatment protocols during the pandemic, which could have affected breast cancer care costs and patient outcomes. Despite these limitations, the study provides valuable insights into the potential impact of the pandemic on breast cancer care costs, and its findings could be beneficial to healthcare policymakers, clinicians, and researchers. Nevertheless, more research is necessary to confirm and expand the results presented in this study.

5 Conclusion

In this study, we have used a Monte Carlo simulation approach and a Markov Model to analyze the effects of COVID-19 lockdowns on the costs and mortality of breast cancer in Poland. Our findings indicate a significant negative impact on breast cancer treatment, resulting in increased costs and higher mortality rates. Although these findings are preliminary, they offer important insights for future discussions on strategies that could be employed during future pandemics. As

part of our ongoing research, we plan to conduct a sensitivity analysis of model
parameters and expand our analysis to estimate the impacts of lockdowns on
other diseases.

Acknowledgements. This publication is partly supported by the European Union's
Horizon 2020 research and innovation programme under grant agreement Sano No.
857533 and the International Research Agendas programme of the Foundation for Polish Science, co-financed by the European Union under the European Regional Development Fund. We would like to thank Tadeusz Satława, Katarzyna Tabor, Karolina
Tkaczuk, and Maja Więckiewicz from Sano for help and their initial contribution in
the development of the model.

References

1. Adams-Prassl, A., Boneva, T., Golin, M., Rauh, C.: The impact of the coronavirus
lockdown on mental health: evidence from the united states. Econ. Policy **37**(109),
139–155 (2022)
2. Agencja Oceny Technologii Medycznych i Taryfikacji: Kompleksowa diagnostyka i
leczenie nowotworów piersi (breast unit). https://bipold.aotm.gov.pl/assets/files/
zlecenia_mz/2017/033/RPT/WS.434.3.2017_bcu.pdf (2018) [Accessed 25 Jan
2023]
3. Alagoz, O., et al.: Impact of the covid-19 pandemic on breast cancer mortality in
the us: estimates from collaborative simulation modeling. JNCI: J. National Cancer
Inst. **113**(11), 1484–1494 (2021)
4. American Cancer Society: Breast Cancer Facts & Figures (2019)
5. Amin, M.B., et al.: AJCC cancer staging manual, vol. 1024. Springer (2017)
6. Arora, S., Bhaukhandi, K.D., Mishra, P.K.: Coronavirus lockdown helped the environment to bounce back. Sci. Total Environ. **742**, 140573 (2020)
7. Baker, R.D.: Use of a mathematical model to evaluate breast cancer screening
policy. Health Care Manag. Sci. **1**(2), 103–113 (1998)
8. Bennett, G., Young, E., Butler, I., Coe, S.: The impact of lockdown during the
covid-19 outbreak on dietary habits in various population groups: a scoping review.
Front. Nutr. **8**, 626432 (2021)
9. Bish, A., Ramirez, A., Burgess, C., Hunter, M.: Understanding why women delay
in seeking help for breast cancer symptoms. J. Psychosom. Res. **58**(4), 321–326
(2005)
10. Blumen, H., Fitch, K., Polkus, V.: Comparison of treatment costs for breast cancer,
by tumor stage and type of service. Am. Health Drug Benefits **9**(1), 23 (2016)
11. Briggs, A., Sculpher, M.: An introduction to markov modelling for economic evaluation. Pharmacoeconomics **13**(4), 397–409 (1998)
12. Campbell, J.D., Ramsey, S.D.: The costs of treating breast cancer in the us. Pharmacoeconomics **27**(3), 199–209 (2009)
13. Cancer Research UK: Breast cancer survival statistics. https://www.cancerresear
chuk.org/health-professional/cancer-statistics/statistics-by-cancer-type/breast-
cancer/survival#heading-Zero Accessed 25 Jan 2023
14. Chudasama, Y.V., et al.: Impact of covid-19 on routine care for chronic diseases: a
global survey of views from healthcare professionals. Diabetes Metabolic Syndrome:
Clin. Res. Rev. **14**(5), 965–967 (2020)

15. Collaborative, C.: Elective surgery cancellations due to the COVID-19 pandemic: global predictive modelling to inform surgical recovery plans. British J. Surg. **107**(11), 1440–1449 (06 2020). https://doi.org/10.1002/bjs.11746
16. Elixhauser, A.: Costs of breast cancer and the cost-effectiveness of breast cancer screening. Int. J. Technol. Assess. Health Care **7**(4), 604–615 (1991)
17. Figueroa, J.D., et al.: The impact of the covid-19 pandemic on breast cancer early detection and screening. Prev. Med. **151**, 106585 (2021)
18. Gathani, T., Clayton, G., MacInnes, E., Horgan, K.: The covid-19 pandemic and impact on breast cancer diagnoses: what happened in england in the first half of 2020. Br. J. Cancer **124**(4), 710–712 (2021)
19. Goyal, D.K., et al.: Restricted access to the nhs during the covid-19 pandemic: Is it time to move away from the rationed clinical response? The Lancet Regional Health-Europe 8 (2021)
20. de la Haba-Rodríguez, J., Aranda, et al.: Time-to-progression in breast cancer: A stratification model for clinical trials. Breast **17**(3), 239–244 (2008)
21. Jiang, P., Klemeš, J.J., Fan, Y.V., Fu, X., Bee, Y.M.: More is not enough: a deeper understanding of the covid-19 impacts on healthcare, energy and environment is crucial. Int. J. Environ. Res. Public Health **18**(2), 684 (2021)
22. Key, T.J., Verkasalo, P.K., Banks, E.: Epidemiology of breast cancer. Lancet Oncol. **2**(3), 133–140 (2001)
23. Koh, D.: Covid-19 lockdowns throughout the world. Occup. Med. **70**(5), 322–322 (2020)
24. Kopec, J.A., et al.: Validation of population-based disease simulation models: a review of concepts and methods. BMC Public Health **10**(1), 1–13 (2010)
25. Lau, H., et al.: The positive impact of lockdown in wuhan on containing the covid-19 outbreak in china. J. Travel Med. (2020)
26. Liu, Y.Y., Li, S., Li, F., Song, L., Rehg, J.M.: Efficient learning of continuous-time hidden markov models for disease progression. Adv. Neural Inform. Process. Syst. **28**, 3599–3607 (2015)
27. Maringe, C., et al.: The impact of the covid-19 pandemic on cancer deaths due to delays in diagnosis in england, uk: a national, population-based, modelling study. Lancet Oncol. **21**(8), 1023–1034 (2020)
28. Mittmann, N., et al.: Health system costs for stage-specific breast cancer: a population-based approach. Curr. Oncol. **21**(6), 281–293 (2014)
29. Momenzadeh, M., Sehhati, M., Rabbani, H.: Using hidden markov model to predict recurrence of breast cancer based on sequential patterns in gene expression profiles. J. Biomed. Inform. **111**, 103570 (2020)
30. Narodowy Fundusz Zdrowia: https://statystyki.nfz.gov.pl/Benefits/1a Accessed 15 Aug 2022
31. Nojszewska, E., Bodnar, L., Łyszczarz, B., Sznurkowski, J., Śliwczyński, A.: Ocena strat ekonomicznych i kosztów leczenia nowotworów piersi, szyjki macicy i jajnika w Polsce. Fundacja Instytut Innowacyjna Gospodarka (2016)
32. Nowikiewicz, T., et al.: Overall survival and disease-free survival in breast cancer patients treated at the oncology centre in bydgoszcz-analysis of more than six years of follow-up. Contemporary Oncology/Współczesna Onkologia **19**(4), 284–289 (2015)
33. Paltrinieri, S., et al.: Beyond lockdown: the potential side effects of the sars-cov-2 pandemic on public health. Nutrients **13**(5), 1600 (2021)
34. Petryszyn, P.W., Kempiński, R., Michałowicz, J., Poniewierka, E.: Non-medical costs of colonoscopy. Gastroenterology Review/Przegląd Gastroenterologiczny **9**(5), 270–274 (2014)

35. Pobiruchin, M., Bochum, S., Martens, U.M., Kieser, M., Schramm, W.: A method for using real world data in breast cancer modeling. J. Biomed. Inform. **60**, 385–394 (2016)
36. Polish National Cancer Registry: https://onkologia.org.pl/pl/raporty Accessed 15 Aug 2022
37. Ramezankhani, A., Azizi, F., Hadaegh, F., Momenan, A.A.: Diabetes and number of years of life lost with and without cardiovascular disease: a multi-state homogeneous semi-markov model. Acta Diabetol. **55**(3), 253–262 (2018)
38. Randolph, H.E., Barreiro, L.B.: Herd immunity: understanding covid-19. Immunity **52**(5), 737–741 (2020)
39. Rascati, K.: Essentials of pharmacoeconomics. Lippincott Williams & Wilkins (2013)
40. Sahoo, S., Rani, S., Parveen, S., Singh, A.P., Mehra, A., Chakrabarti, S., Grover, S., Tandup, C.: Self-harm and covid-19 pandemic: An emerging concern-a report of 2 cases from india. Asian J. Psychiatr. **51**, 102104 (2020)
41. Schippers, M.C.: For the greater good? the devastating ripple effects of the covid-19 crisis. Front. Psychol. **11**, 577740 (2020)
42. Screening & Immunisations Team, NHS Digital: Breast Screening Programme, England 2017–18 (2018)
43. Semo, B.w., Frissa, S.M.: The mental health impact of the covid-19 pandemic: implications for sub-saharan africa. Psychol. Res. Behav. Manage. **13**, 713 (2020)
44. Sonnenberg, F.A., Beck, J.R.: Markov models in medical decision making: a practical guide. Med. Decis. Making **13**(4), 322–338 (1993)
45. Statistics Poland: https://bdl.stat.gov.pl/bdl/start Accessed 15 Aug 2022
46. Sung, H. et al.: Global cancer statistics 2020: Globocan estimates of incidence and mortality worldwide for 36 cancers in 185 countries. CA: Cancer J. Clin. **71**(3), 209–249 (2021)
47. Tabar, L., et al.: Mammography service screening and mortality in breast cancer patients: 20-year follow-up before and after introduction of screening. The Lancet **361**(9367), 1405–1410 (2003)
48. Torre, L.A., et al.: Global cancer statistics, 2012. CA: Cancer J. Clin. **65**(2), 87–108 (2015)
49. Tsoukos, A., Bogdanis, G.C.: The effects of a five-month lockdown due to covid-19 on physical fitness parameters in adolescent students: a comparison between cohorts. Int. J. Environ. Res. Public Health **19**(1), 326 (2022)
50. Yin, K., Singh, P., Drohan, B., Hughes, K.S.: Breast imaging, breast surgery, and cancer genetics in the age of covid-19. Cancer **126**(20), 4466–4472 (2020)

Machine Learning for Risk Stratification of Diabetic Foot Ulcers Using Biomarkers

Kyle Martin$^{(\boxtimes)}$, Ashish Upadhyay , Anjana Wijekoon ,
Nirmalie Wiratunga , and Stewart Massie

School of Computing, Robert Gordon University, Aberdeen, Scotland
{k.martin3,a.upadhyay,a.wijekoon1,n.wiratunga,s.massie}@rgu.ac.uk
https://rgu-repository.worktribe.com/tag/1437329/artificial-intelligence
-reasoning-air

Abstract. Development of a Diabetic Foot Ulcer (DFU) causes a sharp decline in a patient's health and quality of life. The process of risk stratification is crucial for informing the care that a patient should receive to help manage their Diabetes before an ulcer can form. In existing practice, risk stratification is a manual process where a clinician allocates a risk category based on biomarker features captured during routine appointments. We present the preliminary outcomes of a feasibility study on machine learning techniques for risk stratification of DFU formation. Our findings highlight the importance of considering patient history, and allow us to identify biomarkers which are important for risk classification.

Keywords: Diabetic Foot Ulceration · Machine Learning · Biomarkers

1 Introduction

Diabetic Foot Ulcers (DFUs) are a severe complication of Diabetes Mellitus. It is estimated that 15% of diabetic patients will develop a DFU during their lives [12]. Development of a DFU can cause a sharp decline in health and quality of life, often leading to further infection, amputation and death [3,10]. Predicting the likelihood of DFU formation is based on risk stratification.

Risk stratification is crucial for informing the level and regularity of care that a patient should receive. Improper treatment of a DFU can exacerbate patient condition and lead to further health complications, impacting quality of life and increasing cost of treatment [3]. Given medical knowledge of biological markers which act as patient features, clinicians leverage their domain expertise to manually allocate a risk category which describes the likelihood of developing a DFU [1]. Biological markers (henceforth 'biomarkers', as per domain terminology) are physiological features captured during routine medical examinations which contribute to a clinician's understanding of patient condition and their capability to effectively stratify future risk. For example, concentration of albumin in the blood is a recognised indicator of Diabetes and its complications [5]. Of specific interest are biomarkers which describe the onset of peripheral neuropathy (i.e. damaged nerve endings in the extremities [2]).

© The Author(s), under exclusive license to Springer Nature Switzerland AG 2023
J. Mikyška et al. (Eds.): ICCS 2023, LNCS 14075, pp. 153–161, 2023.
https://doi.org/10.1007/978-3-031-36024-4_11

Risk stratification is therefore an expertise-driven task with good potential for automation using machine learning. However, few existing works have used patient health records for this purpose. The authors in [11] produced a review of journal papers describing applications of machine learning algorithms for the diagnosis, prevention and care of DFUs. They surveyed 3,769 papers (reduced to 37 after application of inclusion and exclusion criteria) and found the majority of ML algorithms have been applied to thermospectral or colour images of the foot (29 papers). Only a single reviewed paper examined patient health records. In [4] the authors collated a dataset of 301 patient records from a hospital in India and trained a decision tree for the purposes of explaining amputations caused by DFUs. Key differences to our work include: (a) our work is on a much larger dataset - we apply machine learning algorithms to health records for approximately 27,000 individual patients; (b) our work is targeted towards risk stratification of ulcer formation (predictive) whereas [4] aimed to explain amputation decisions due to DFUs (retrospective); and (c) the works are applied in the context of different health systems.

In [8] the authors describe a method of predicting DFU formation (and subsequent amputation) using national registry data for 246,000 patients in Denmark. Their dataset is formed from socio-economic features, with some knowledge of concurrent health conditions. Though their results are not comparable to those we present here, due to differences in features and task, they emphasise the limitations of high-level data. In their results, they stated that knowledge of DFU history is an important indicator of recurrence. Our findings mirror this, hence the use of historical data to augment the biomarker dataset.

In this paper we present three contributions. Firstly, we compare 3 machine learning algorithms for risk stratification using a large dataset of health records extracted from SCI-Diabetes. As part of this experimentation, we provide a novel comparison of recent and historical features to identify their impact on decision-making. Finally, we identify the contributory power of each feature in our dataset using mutual information values. Results indicate that risk labels are highly dependant on features for detection of peripheral neuropathy.

We structure this paper in the following manner. In Sect. 2 we formalise our methodology and evaluation by introducing the dataset and machine learning task. In Sect. 3 we describe the results of our evaluation, while in Sect. 4 we provide some discussion. Finally, in Sect. 5 we present our conclusions.

2 Methods

The Scottish Care Information - Diabetes (SCI-Diabetes) platform is operated by the National Health Service (NHS) Scotland and contains digital health records of patients with Diabetes. The SCI-Diabetes platform has been used since 2014 to allow sharing of records across multi-disciplinary care teams (i.e. endocrinologist, diabeteologist, etc.) to facilitate treatment of Diabetes patients in Scotland. Records within SCI-Diabetes store information captured during routine healthcare appointments, including results of medical tests and procedures. We describe these features as biomarkers. Biomarker values are recorded for a patient from the time they were first diagnosed with Diabetes until their most recent examination, and are used to predict the progression of associated health concerns. This includes our primary interest; risk of DFU formation in either foot.

We have been given access to a subset of electronic health records from the entire scope of SCI-Diabetes[1]. The SCI-Diabetes Biomarker dataset (henceforth simply the Biomarker Dataset) contains data describing biomarker features for 30,941 unique patients. The dataset contains a mix of recent and historical biomarker information, and the length of patient history varies based upon the number of appointments they have attended or tests they have taken. These biomarkers are of two types: numerical, where values are continuous numbers; and categorical, where values are discrete categories. Each patient is also allocated a risk classification, provided by clinicians based upon National Institute for Health and Care Excellence (NICE) guidelines [6].

To our knowledge, patient SCI-Diabetes records have not previously been used to stratify risk of DFU development using machine learning algorithms. However, researchers have applied machine learning to patient records within SCI-Diabetes to predict whether a patient has Type 1 or Type 2 Diabetes [7]. They found that a simple neural network model could outperform clinicians on this task. This suggests that SCI-Diabetes is a promising data source for us to explore risk stratification using biomarker data.

2.1 Recent and Historical Biomarker Datasets

We derive a machine learning task to classify the risk of a patient developing a DFU based on their biomarker data. As a first step, a data quality assessment was performed and anomalies were corrected. Categorical data was standardised across the dataset. Missing numerical values were infrequent, and only occurred in patients with multiple appointments, so were imputed using the mean of that biomarker value for that patient. Finally, missing class labels were removed - of the initial 30,941 patients, 4,621 have no recorded risk status and thus are dropped. The remaining 26,320 patients are divided into three risk classes: 19,419 Low risk, 4,286 Moderate risk and 2,615 High risk patients respectively.

We differentiate between using only recent biomarkers (basing risk prediction only upon a patient's most recent appointment) and using historical biomarkers (incorporating knowledge of a patient's medical history) to create two distinct datasets. This allows us to compare whether additional historical knowledge allows improved risk stratification for DFU formation. To create the Recent Biomarker dataset, we extracted the biomarker values recorded at a patients' most recent appointment. This resulted in a dataset with 26,320 instances (one for each patient), each of which had 37 features. To create the Historical Biomarker dataset, the Recent dataset was augmented with additional features calculated from historical biomarkers. Records of patient appointments were grouped using their CHI number, then aggregated using the following techniques:

---

[1] Access provided by NHS Data Safe Haven Dundee following ethical approval and completion of GDPR training.

- **Numerical:** Calculating the mean of values from first examination to second most recent examination and subtract this from the most recent examination value. Use of an average allows us to capture a baseline for the patient and standardise variable lengths of patient histories. Using a difference measure allows us to explicitly capture feature changes between appointments.
- **Categorical:** Using the second most recent examination value as the new feature.

When historical features were unavailable (i.e. a patient only had a single appointment), we reused recent feature values. The resulting dataset had 26,320 instances and 47 features - the 37 features of the Recent dataset, and 10 additional features created through our aggregation of historical biomarkers (broken down into 4 numerical and 6 categorical features). Not all features were suitable for aggregation, as some were unalterable (such as Diabetes Type).

We highlight that we have removed a feature describing whether a patient has previously developed a foot ulcer. While we acknowledge the literature suggests that DFU history is an important feature for risk prediction [8], we removed this feature from our task because it is linked with the class label. In SCI-Diabetes, a patient is always allocated a High risk label if they have a history of DFU.

2.2   Machine Learning for Risk Stratification

We now have a multi-class classification problem where the goal is to identify whether a patient is at Low, Moderate or High risk of developing a DFU. We apply three different machine learning algorithms for this purpose:

Logistic Regression (LR). estimates the probability of an event or class using a logit function. For multi-class classification, the prediction based on a one-vs-all method (i.e. the probability of belonging to one class vs belonging to any other class, repeated for each class in the dataset).

Multi-Layer Perceptron (MLP). is a neural network formed of an input layer, multiple hidden layers and an output layer. The goal is to model the decision boundary of different classes by learning an easily separable representation of the data. Within each hidden layer, data is transformed by applying a set of weights and biases to the output of the previous layer (i.e. creating a new representation of the data). The final layer converts this representation to a probability distribution across the range of possible classes (i.e. modelling the decision boundary).

Random Forest (RF). classifier is an ensemble of multiple decision tree classifiers. Decision trees learn by inferring simple decision rules to classify data. Random forest learns an uncorrelated set of decision trees where the output of the forest is an improvement over any individual tree it contains.

These algorithms were selected as they compliment the feature composition of the dataset. RF is well suited for categorical data, as inferred rules are inherently categorical and previous work suggests decision trees perform well in this context [7]. However, conversion from numerical to categorical data reduces granularity. Therefore we also evaluated LR and MLP algorithms, which are well suited towards numerical data.

Table 1. Comparison of ML Models for Risk Stratification of DFU.

Model	Recent		Historical	
	Accuracy (%)	F1 Score (%)	Accuracy (%)	F1 Score (%)
Random Forest	82.57	63.00	**82.98**	**64.34**
Multi-Layer Perceptron	79.28	53.39	78.91	54.65
Logistic Regression	81.12	58.68	79.80	48.54

2.3 Understanding Important Biomarkers for Risk Stratification

Finally, it is desirable to understand biomarkers which contribute to allocation of risk classes. Mutual Information (MI) is a method to calculate dependency between two variables X and Y (see Eq. 1).

$$MI(X;Y) = H(X) - H(X|Y) \tag{1}$$

where $H(X)$ is the entropy for X and $H(X|Y)$ is the conditional entropy for X given Y. Greater MI values indicate greater dependency between two variables, whereas low values indicate independence. We calculate MI between each feature and label in both the Recent and Historical Biomarker datasets.

3 Results

Overall, the results (shown in Table 1) are very promising for this challenging problem, with a peak accuracy of 82.98% and Macro F1 Score of 64.14% respectively. We believe the difference indicates some overfitting to the majority class, suggesting the ML algorithms are capable of recognising low risk patients, but struggle to accurately identify Moderate and High risk patients. RF is the best performing ML algorithm, obtaining the highest accuracy and Macro F1 Score on both datasets. The difference in Macro F1 Score highlights that RF is more capable of correctly predicting minority classes, Moderate and High risk.

Next we examine whether risk stratification of DFU is more accurate if we consider patient history. In the Recent Biomarker dataset, we achieve a peak accuracy of 82.57% and Macro F1 Score of 63% using RF. Using the Historical Biomarker dataset demonstrates a slight increase to 82.98% and 64.14% respectively (the best performing set up from our experiments). However, results on MLP and LR are mixed. For example, the MLP algorithm demonstrates a minor reduction in accuracy on the Historical Biomarker dataset compared with the Recent Biomarker dataset (dropping from 79.28% to 78.91%), but an improved Macro F1 Score (increasing from 53.39% to 54.65%). This suggests a drop in performance on the majority class (Low risk), but an increase in performance in one of the minority classes (Moderate or High risk). The LR algorithm demonstrates a noticeable drop in performance when historical features are included.

Finally, as a proxy for feature importance we calculate MI between each biomarker feature the risk label (see Table 2).

Table 2. MI Values for Features in Recent and Historical Biomarker Datasets.

Feature	MI Recent	Historical
Albumin Concentration	0.79	1.04
Angina	0.7	0.63
Body Mass Index	0	0
CVA Haemorrhagic	0.18	0.13
CVA Non-Haemorrhagic	0	0
CVA Summary	0.24	0.79
CVA Unspecified	0.73	1.03
Coronary Artery Bypass Surgery	0.66	0.05
Current Tobacco Nicotine Consumption Status	0	0
Diabetes Mellitius Sub Type	0.9	0.47
Diabetes Mellitius Sub Type 2	0.3	0.31
Diabetes Mellitius Type	0.29	0.54
Diastolic Blood Pressure	0.57	0.69
Estimated Glomerular Filtration Rate	2.68	2.78
HbA1c	0.08	0.44
Hypertension	0.41	0.96
Ischaemic Heart Disease	0.41	0.23
Maculopathy Left	0.16	0.97
Maculopathy Right	0.02	0.6
Maculopathy Summary	0.18	0.59
Monofilament Left Sites	16.97	16.37
Monofilament Right Sites	16.87	17.41
Myocardinal Infraction	0.57	0.92
Peripheral Pulses Left	8.34	8.3
Peripheral Pulses Right	7.95	7.86
Peripheral Vascular Disease	0.92	0.82
Protective Sensation Left	17.01	17.2
Protective Sensation Right	16.68	17.12
Protective Sensation Summary	19.38	18.94
Retinopathy Left	1.04	0.94
Retinopathy Right	1.21	0.66
Retinopathy Summary	0.93	0.65
Systolic Blood Pressure	0.57	0.27
Total Cholesterol	0.81	0.85
Transient Ischemic Attack	0.52	0
Triglyceride Level	0.44	0
Weight	0.08	0
Historical Numerical Features		
Historical Estimated Glomerular Filtration Rate	–	2.39
Historical Albumin Concentration	–	1.53
Historical Systolic Blood Pressure	–	1.07
Historical Diastolic Blood Pressure	–	0.73
Historical Categorical Features		
Historical Protective Sensation Left	–	16.18
Historical Protective Sensation Right	–	16.06
Historical Peripheral Pulses Left	–	7.8
Historical Peripheral Pulses Right	–	7.57
Historical Monofilament Left Sites	–	15.81
Historical Monofilament Right Sites	–	16.11

4   Discussion

Our results suggest that inclusion of historical features make algorithms more capable of recognising Moderate and High risk patients. However, we observed evidence of all algorithms overfitting, which we suspected was due to class imbalance in the dataset. We applied downsampling to reduce the number of samples in the Low risk class to 5,000 (making class sizes comparable across the three classes). In almost all scenarios, downsampling resulted in decreased accuracy and F1 Score. The drop in F1 Score highlights that additional instances within the Low risk class is contributory to ML algorithms' ability to learn this task.

By applying MI, we find the most important features for model decision-making are derived from tests of foot health (such as protective sensation, peripheral pulses, and monofilament tests), many of which are directly related to tests for peripheral neuropathy. However, these features are mostly categorical in nature - for example, monofilament feature values are calculated based on clinical test thresholds. It would be useful to capture more granular detail about these features, for improved risk stratification. Interestingly, there are several features which are not traditionally associated with foot health which also contribute, specifically albumin concentration, retinopathy and estimated glomerular filtration rate. We suspect that unusual recordings of these features indicate further complications of Diabetes, which would be linked with increasing risk of DFU formation. A recent study suggesting links between retinopathy and peripheral neuropathy [9] supports this idea. Biomarkers such as smoking status or BMI (which can lead to complications in other aspects of Diabetes) are not so relevant to DFU formation. Finally, we note large dependency values using the historical knowledge we have captured from relevant categorical features, and low dependency values on historical numerical features. Even several features which previously showed low importance scores when only considered using knowledge from the most recent appointment (such as systolic and diastolic blood pressure) demonstrated a noticeable increase. This supports the outcome of our experiments; that an effective risk stratification system should be based on evolving knowledge of the patient and their history.

There are several limitations to our study. SCI-Diabetes captures data for patients within Scotland, so our results may not generalise to other healthcare systems, though we highlight that our findings overlap with [8] despite this. We have only tried a subset of possible machine learning algorithms, selected as they compliment the feature composition of the dataset. It would be desirable to compare more algorithms on this task. Finally, while we have performed some initial experimentation to address dataset imbalance and subsequent overfitting of trained algorithms, further strategies in the literature could be investigated (both at the training and evaluation stages of model development). Despite these limitations, we believe our results show good potential for risk stratification of DFU formation using ML algorithms.

5   Conclusions

In this paper, we presented a comparative study of machine learning algorithms for risk stratification of diabetic foot ulceration. Our results have indicated the empirical value of examining historical biomarker features of a patient to stratify this risk. Finally, we have highlighted the importance of biomarker indicators of peripheral neuropathy as contributing to risk categorisation of a patient.

In future work, we plan to incorporate historical and recent biomarkers into a single time-series record for a given patient. This should allow more granular prediction of the evolution of DFU formation risk. Another interesting aspect is addressing the dataset imbalance by examining cost by error class. Finally, we plan to improve capability to explain model decision-making by combining feature importance and counterfactual explainer algorithms to identify contributing factors to DFU formation and which biomarkers to target for treatment.

Acknowledgements. This work was funded by SBRI Challenge: Delivering Safer and Better Care Every Time for Patients with Diabetes. The authors would like to thank NHS Data Safe Haven Dundee for providing access to SCI-Diabetes, and Walk With Path Ltd as a partner on this project.

References

1. Dhatariya, K., et al.: Nhs diabetes guideline for the perioperative management of the adult patient with diabetes. Diabet. Med. **29**(4), 420–433 (2012)
2. Dros, J., Wewerinke, A., Bindels, P.J., van Weert, H.C.: Accuracy of monofilament testing to diagnose peripheral neuropathy: a systematic review. Ann. Family Med. **7**(6), 555–558 (2009)
3. Guest, J.F., Fuller, G.W., Vowden, P.: Diabetic foot ulcer management in clinical practice in the uk: costs and outcomes. International Wound Journal **15**(1), 43–52 (2018). https://doi.org/10.1111/iwj.12816, https://onlinelibrary.wiley.com/doi/abs/10.1111/iwj.12816
4. Kasbekar, P.U., Goel, P., Jadhav, S.P.: A decision tree analysis of diabetic foot amputation risk in indian patients. Front. Endocrinol. **8**, 25 (2017)
5. Li, L., et al.: Serum albumin is associated with peripheral nerve function in patients with type 2 diabetes. Endocrine **50**(2), 397–404 (2015)
6. National Institute for Health and Care Excellence: Diabetic foot problems: prevention and management (NICE Guideline NG19) (2015)
7. Sainsbury, C., Muir, R., Osmanska, J., Jones, G.: Machine learning (neural network)-driven algorithmic classification of type 1 or type 2 diabetes at the time of presentation significantly outperforms experienced clinician classification. In: Diabetic Medicine. vol. 35, pp. 168–168. Wiley (2018)
8. Schäfer, Z., Mathisen, A., Svendsen, K., Engberg, S., Rolighed Thomsen, T., Kirketerp-Møller, K.: Towards machine learning based decision support in diabetes care: A risk stratification study on diabetic foot ulcer and amputation. Front. Med. **7**, 957 (2020)
9. Sharma, V.K., Joshi, M.V., Vishnoi, A.A., et al.: Interrelation of retinopathy with peripheral neuropathy in diabetes mellitus. J. Clin. Ophthalmol. Res. **4**(2), 83 (2016)

10. Snyder, R.J., Hanft, J.R.: Diabetic foot ulcers-effects on qol, costs, and mortality and the role of standard wound care and advanced-care therapies. Ostomy Wound Manage. **55**(11), 28–38 (2009)
11. Tulloch, J., Zamani, R., Akrami, M.: Machine learning in the prevention, diagnosis and management of diabetic foot ulcers: a systematic review. IEEE Access **8**, 198977–199000 (2020)
12. Yazdanpanah, L., Nasiri, M., Adarvishi, S.: Literature review on the management of diabetic foot ulcer. World J. Diabetes **6**(1), 37 (2015)

A Robust Machine Learning Protocol for Prediction of Prostate Cancer Survival at Multiple Time-Horizons

Wojciech Lesiński[1]([✉]) [iD] and Witold R. Rudnicki[1,2,3] [iD]

[1] Institute of Informatics, University of Białystok, Białystok, Poland
w.lesinski@uwb.edu.pl
[2] Computational Centre, University of Białystok, Białystok, Poland
[3] Interdisciplinary Centre for Mathematical and Computational Modelling,
University of Warsaw, Warsaw, Poland

Abstract. Prostate cancer is one of the leading causes of cancer death
in men in Western societies. Predicting patients' survival using clinical
descriptors is important for stratification in the risk classes and selecting
appropriate treatment. Current work is devoted to developing a robust
Machine Learning (ML) protocol for predicting the survival of patients
with metastatic castration-resistant prostate cancer. In particular, we
aimed to identify relevant factors for survival at various time horizons. To
this end, we built ML models for eight different predictive horizons, starting at three and up to forty-eight months. The model building involved
the identification of informative variables with the help of the MultiDimensional Feature Selection (MDFS) algorithm, entire modelling procedure was performed in multiple repeats of cross-validation. We evaluated
the application of 5 popular classification algorithms: Random Forest,
XGBoost, logistic regression, k-NN and naive Bayes, for this task. Best
modelling results for all time horizons were obtained with the help of
Random Forest. Good prediction results and stable feature selection were
obtained for six horizons, excluding the shortest and longest ones. The
informative variables differ significantly for different predictive time horizons. Different factors affect survival rates over different periods, however, four clinical variables: ALP, LDH, HB and PSA, were relevant for all
stable predictive horizons. The modelling procedure that involves computationally intensive multiple repeats of cross-validated modelling, allows
for robust prediction of the relevant features and for much-improved estimation of uncertainty of results.

Keywords: prostate cancer · feature selection · machine learning ·
random forest · xgboost · logistic regression · knn

1 Introduction

Prostate cancer is the sixth most common cancer in the world (in the number
of new cases), the third most common cancer in men, and the most common

© The Author(s), under exclusive license to Springer Nature Switzerland AG 2023
J. Mikyška et al. (Eds.): ICCS 2023, LNCS 14075, pp. 162–169, 2023.
https://doi.org/10.1007/978-3-031-36024-4_12

cancer in men in Europe, North America, and some parts of Africa [7]. Prostate cancer is a form of cancer that develops in the prostate gland. As with many tumours, they can be benign or malignant. Prostate cancer cells can spread by breaking away from a prostate tumour. They can travel through blood vessels or lymph nodes to reach other body parts. After spreading, cancer cells may attach to other tissues and grow to form new tumours, causing damage where they land. Metastatic castrate-resistant prostate cancer (mCRPC) is a prostate cancer with metastasis that keeps growing even when the amount of testosterone in the body is reduced to very low levels. Many early-stage prostate cancers need normal testosterone levels to grow, but castrate-resistant prostate cancers do not.

One of the essential goals of research on prostate cancer is the development of predictive models for patients' survival. The classical approach was proposed by Halabi [9,10] and coworkers, who used proportional hazard models. Halabi's model is based on eight clinical variables: lactate dehydrogenase, prostate-specific antigen, alkaline phosphatase, Gleason sum, Eastern Cooperative Oncology Group performance status, haemoglobin, and the presence of visceral disease. Another notable work, [18], used a joint longitudinal survival-cure model based mainly on PSA level. In Mahapatra et al. [12] survival prediction models were built using DNA methylation data.

More recently, the Prostate Cancer DREAM Challenge was organised in 2015 by DREAM community to improve predictive models [1]. The DREAM Challenges [4] is a non-profit, collaborative community effort comprising contributors from across the research spectrum, including researchers from universities, technology companies, not-for-profits, and biotechnology and pharmaceutical companies. Many independent scientists made survival models and predictions within the Prostate Cancer DREAM Challenge. The challenge resulted in developing multiple new algorithms with significantly improved performance compared to a reference Halabi model, see [8].

Both the reference model and best model developed within DREAM Challenge, are variants of the proportional hazard model. This model has a relatively rigid construction - the influence of variables used for modelling is identical for different predictive horizons. The current study explores an alternative approach, where a series of independent models is developed for different predictive horizons. Identification of informative variables is performed independently at each horizon. This approach allows a more fine-grained analysis of the problem.

What is more, the format of the DREAM Challenges has two significant limitations, namely, inefficient use of data and inefficient evaluation of modelling error. The main problem is the strict division into training and testing sets. Solutions were evaluated based only on test sets' results, and such setup necessarily results in random biases arising due to the data split. Moreover, not all available data is used for the development of the model, and estimates of the error bounds of the model are based on the single data split between the training and validation set.

The current study applies a robust modelling protocol for building a series of predictive models with different time horizons. This leads to obtaining comparable predictions compared to aggregate models from the DREAM Challenge but also allows for detailed analysis of the influence of various factors for survival in different time horizons.

2   Materials and Methods

2.1   Data

The publicly released data from the DREAM Challenge was used for the current study. The data set comprises 1600 individual cases collected in three clinical trials: VENICE [17], MAILSAIL [15] and ENTHUSE [6]. Data contains five groups of variables and two clinical indicators of prostate cancer progression corresponding to the patient's medical history, laboratory values, lesion sites, previous treatments, vital signs, and basic demographic information. The final record for the patient consists of 128 descriptive variables and two decision variables: the patient's status (alive or deceased) and the time of last observation. Based on this data nine binary decision variables were created for different predictive horizons: 3 months, 6 months, 1 year, 18 months, 2, 3, and 4 years.

2.2   Modelling

Machine learning methods often produce models biased towards the training set. In particular, the selection of hyper-parameters of the algorithms and the selection of variables that will be used for modelling can introduce strong biases. What is more - a simple selection of the best-performing model also can lead to a bias. Finally, dividing the data set into training and validation sets involves bias by creating two partitions with negative correlations between fluctuations from the actual averages. To minimize the influence of biases and estimate the variance of the model-building process, the entire modelling procedure was performed within multiple repeats of the cross-validation loop.

A single iteration of our approach is based on the following general protocol:

- Split the data into training and validation set;
- Identify informative variables in the training set;
- Select most informative variables;
- Build model on training set;
- Estimate models on validation set,

This protocol was repeated 150 times for each classifier – time horizon pair (30 iterations of 5-fold cross-validation). The series of predictive models were constructed for various horizons of prediction. A binary classification model of survival beyond this horizon was built at each horizon.

The data set was imbalanced, especially in the short time horizons. While some algorithms, e.g. Random Forest, are relatively robust when dealing with imbalanced data, in many cases, imbalanced data may cause many problems for machine learning methods [11]. The main difficulty lies in finding properties that discern the minority class from the much more numerous majority class. The simple downsampling of the majority class can deal with this. It involves randomly removing observations from the majority class to prevent its signal from dominating the learning algorithm. Downsampling was used for prediction in 3 and 6 months time horizon.

Two measures used to evaluate models' quality are suitable for unbalanced data. Matthews Correlation Coefficient (MCC) [13], MCC measures the correlation of distribution of classes in predictions with actual distribution in the sample. The area under the receiver operating curve (AUROC or AUC) is a global measure of performance that can be applied to any classifier that ranks the binary prediction for objects.

Feature Selection. Identification of informative variables was performed with the help of the Multidimensional Feature Selection (MDFS) algorithm [14, 16]. The method is based on information theory and considers synergistic interactions between descriptive variables. It was developed in our laboratory and implemented in the R package *MDFS*. The algorithm returns binary decisions about variables' relevance and ranking based on Information Gain and p-value.

Classification. For modelling, we used five popular classifiers: **Random Forest** algorithm [2], **XGboost** [3] **k-Nearest Neighbours (k-NN), Logistic regression,** and **Naive Bayes.** Random Forest and XGboost are based on decision trees and work well *out of the box* on most data sets [5]. Logistic regression represents generalized linear models, whereas k-NN is a simple and widely known method based on distances between objects. Finally, Naive Bayes is a simple algorithm that may work well for additive problems. All tests were performed in 30 repeats of the 5-folds cross-validation. Both feature selection and ML model building were performed within cross-validation. All ML algorithms were applied to the identical folds in the cross-validation to ensure fair comparisons.

Table 1. Cross-validated quality of predictions for five classification algorithms at 8 predictive horizons.

time horizon	Random Forest		XGboost		logistic regression		k-NN		Naive Bayes	
	MCC	AUC	MCC	AUC	MCC	AUC	MCC	AUC	MCC	AUC
3 months	0.23	0.66	0.26	0.65	0.17	0.62	0.16	0.57	0.23	0.64
6 months	0.34	0.72	0.30	0.70	0.27	0.68	0.27	0.62	0.23	0.66
9 months	0.43	0.78	0.36	0.74	0.36	0.74	0.33	0.65	0.28	0.71
12 months	0.44	0.79	0.39	0.76	0.40	0.76	0.35	0.66	0.30	0.72
18 months	0.44	0.77	0.38	0.74	0.39	0.75	0.34	0.67	0.28	0.71
24 months	0.45	0.80	0.41	0.77	0.38	0.74	0.40	0.68	0.43	0.76
36 months	0.42	0.77	0.37	0.74	0.32	0.71	0.26	0.62	0.37	0.72
48 months	0.21	0.65	0.24	0.66	0.20	0.61	0.18	0.58	0.17	0.66

3   Results and Discussion

Survival predictions were made for eight different time horizons with the help of the five classifiers mentioned before.

As can be expected, the worst-performing models were built for the shortest and longest horizons, most likely due to a small number of cases in the minority class (non-survivors) for the shortest horizon and an overall small number of non-censored cases for the longest horizon, see Table 1. The Random Forest classifier obtained the best results for all predictive horizons. At the shortest horizon, we obtained $AUC = 0.66$ and $MCC = 0.23$. The quality of predictions increases with an increased horizon. The best results were obtained for horizons between 9 and 24 months, with acceptable results for 6 and 36 months. The prediction quality falls at 48 months horizon back to $AUC = 0.65$ and $MCC = 0.21$. AUC curves for all Random Forest models for all examined time horizons are displayed in Fig. 1.

The XGboost produced slightly worse models than Random Forest. The difference is insignificant for a single horizon but significant when all horizons are considered together. The logistic regression generally produced slightly worse models than XGboost. The two simplest methods produced significantly worse models – differences from the best model were significant for almost all predictive horizons.

Evaluation of feature importance was built based on 30 repeats of 5 folds cross-validation procedure. The cumulative ranking of importance for each period was obtained as a count of occurrences of variables in the sets of the top ten most relevant variables in all repeats of the cross-validation procedure. In cases when the feature selector reports fewer than ten relevant variables, the highest-ranked descriptors were included. The feature selection results were volatile in the shortest and the longest time horizons. In particular, none of the

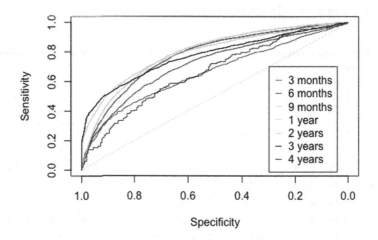

Fig. 1. ROC plots for Random Forest model.

Table 2. Feature selection ranking in given period. Importance for short-, medium-, and long-term prediction was computed as the geometric mean of ranking. If a variable has not been ranked in a given horizon ranking, equal to 20 was used.

Descriptor	Predictive horizons											
	Months						short		medium		long	
	6	9	12	18	24	36	imp.	score	imp.	score	imp.	score
ALP	1	1	1	1	1	3	1	1	1	1	1.7	1
PSA	4	5	5	3	3	8	2.2	2	3.9	3	4.9	5
HB	2	3	3	4	6	5	2.4	3	3.5	4	5.5	6
LDH	3	4	4	2	2	7	3.5	4	2.8	2	3.7	4
ECOG-C	9	9	-	8	10	9	9	6	12.6	10	9.5	8
CCRC	-	8	6	-	7	6	12.6	10	11.0	8	6.5	4
NA	-	7	9	-	-	10	11.8	8	13.4	-	14.1	10
AST	6	2	2	5	8	-	3.5	4	3.2	5	12.6	9
ANALGESICS	-	10	10	-		-	14.1	-	14.1	6	-	-
ALB	5	6	8	-	-	-	5.5	4	12.6	6	-	-
TBILI	8	-	-	-	-	-	12.6	10	-	-	-	-
REGION-C	-	-	-	6	5	1	-	-	11.0	7	2.2	2
MHNEOPLA	7	-	-	-	-	-	11.8	8	-	-	-	-
ALT	10	9	-	-	-	-	9.5	7	-	-	-	-
NUE	-	-	7	7	-	-	-	-	7.0	9	-	-
PLT	-	-	-	9	-	-	-	-	13.4	-	-	-
BMI	-	-	-	10	-	-	-	-	14.1	-	-	-
SMOKE	-	-	-	-	4	2	-	-	-	-	2.8	3
TSTAG-DX	-	-	-	-	9	4	-	-	-	-	6.0	7

variables was present within the top ten variables in all 150 repeats of cross-validation. In comparison, at 12 months horizon, six variables were included in the top ten in all 150 cases. Therefore, the two extreme horizons were removed from further analyses. Other horizons were divided into three groups: short-term (6 and 9 months), medium-term (12 and 18 months) and long-term (24 and 36 months). The results of the feature selection procedure for these horizons are displayed in Table 2

Nineteen variables were included in top ten most relevant variables in six predictive horizons. Only four of them, alkaline phosphatase level (ALP), lactate dehydrogenase level (LDH), prostate-specific antigen (PSA) and haemoglobin level (HB) were important for all predictive horizons. One should note, that these four variables were the most relevant ones for short- and medium-term predictions, and were also quite important for the long-term predictions. In particular ALP was the most relevant variable for all but one predictive horizons. Four other variables, namely ECOG-C, calculated creatinine clearance (CCRC), sodium (NA) and aspartate aminotransferase (AST) were relevant for short-, medium-, and long-term predictions, but with lower ranks. Three variables, ANALGESICS, total bilirubin level (TBILI) and albumin level (ALB) were relevant for the short- and medium-term predictions. Appearance of other sites of neoplasms (MHNEOPLA) and level of alanine transaminase (ALT) are important only for short-term predictions. Region of the world (REGION-C) is relevant for medium- and long-term predictions. Neutrophils (NUE), body mass index (BMI) and platelet count (PLT) appear only in medium prediction. Finally two variables (smoking status (SMOKE) and primary tumor stage score (TSTAG-DX)) are relevant for the medium and long-term predictions only.

What is interesting, the variables corresponding to socioeconomic status and behaviour (REGION-C and SMOKE), while irrelevant for the short- and medium-term predictions become the most important ones for prognosis of the long-term survival, in particular for the 36 months prediction.

These results agree very well with basic medical knowledge. The bad results of medical tests showing the overall health of the patient, such as total bilirubin level or platelet count level are strong indicators of a bad prognosis in the short period but are not very important for long-term prediction. On the other hand presence of other neoplasms may not be an indicator of immediate threat, but are very serious risk factors in the mid- and long-term horizon.

4   Conclusions and Future Works

The approach presented in the current study relies on computationally intensive procedures. The multiple repeats of cross-validation and inclusion of feature selection within cross-validation allow for the removal of biases that are observed for a single division between training and validation set, or even for a single run of cross-validation. What is more, it gives the opportunity to estimate variance that results from both performing feature selection and model building on finite and relatively small samples. For future work, we would like to combine our methods with classical survival models. Finding new datasets also seems like a good idea.

References

1. Abdallah, K., Hugh-Jones, C., Norman, T., Friend, S., Stolovitzky, G.: The prostate cancer dream challenge: a community-wide effort to use open clinical trial data for the quantitative prediction of outcomes in metastatic prostate cancer (2015)
2. Breiman, L.: Random forests. Mach. Learn. **45**, 5–32 (2001)
3. Chen, T., Guestrin, C.: XGBoost: a scalable tree boosting system, pp. 785–794 (2016). https://doi.org/10.1145/2939672.2939785
4. Costello, J., Stolovitzky, G.: Seeking the wisdom of crowds through challenge-based competitions in biomedical research. Clini. Pharmacol. Ther. **93** (2013). https://doi.org/10.1038/clpt.2013.36
5. Fernández-Delgado, M., et al.: Do we need hundreds of classifiers to solve real world classification problems. J. Mach. Learn. Res. **15**(1), 3133–3181 (2014)

6. Fizazi, K., et al.: Phase iii, randomized, placebo-controlled study of docetaxel in combination with Zibotentan in patients with metastatic castration-resistant prostate cancer. J. Clin. Oncol. **31**(14), 1740–1747 (2013)
7. Grönberg, H.: Prostate cancer epidemiology. Lancet **361**(9360), 859–864 (2003)
8. Guinney, J., et al.: Prediction of overall survival for patients with metastatic castration-resistant prostate cancer: development of a prognostic model through a crowdsourced challenge with open clinical trial data. Lancet Oncol. **18**(1), 132–142 (2017). https://doi.org/10.1016/S1470-2045(16)30560-5. https://www.sciencedirect.com/science/article/pii/S1470204516305605
9. Halabi, S., et al.: Prognostic model for predicting survival in men with hormone-refractory metastatic prostate cancer. J. Clin. Oncol. Official J. Am. Soc. Clin. Oncol. **21**, 1232–1237 (2003). https://doi.org/10.1200/JCO.2003.06.100
10. Halabi, S., et al.: Updated prognostic model for predicting overall survival in first-line chemotherapy for patients with metastatic castration-resistant prostate cancer. J. Clin. Oncol. Official J. Am. Soc. Clin. Oncol. **32** (2014). https://doi.org/10.1200/JCO.2013.52.3696
11. He, H., Garcia, E.A.: Learning from imbalanced data. IEEE Trans. Knowl. Data Eng. **21**(9), 1263–1284 (2009)
12. Mahapatra, S., et al.: Global methylation profiling for risk prediction of prostate cancer. Clin. Cancer Res. **18**(10), 2882–2895 (2012). https://doi.org/10.1158/1078-0432.CCR-11-2090
13. Matthews, B.: Comparison of the predicted and observed secondary structure of T4 phage lysozyme. Biochem. Biophys. Acta. **405**(2), 442–451 (1975)
14. Mnich, K., Rudnicki, W.R.: All-relevant feature selection using multidimensional filters with exhaustive search. Inf. Sci. **524**, 277–297 (2020). https://doi.org/10.1016/j.ins.2020.03.024
15. Petrylak, D., et al.: Docetaxel and prednisone with or without lenalidomide in chemotherapy-Naive patients with metastatic castration-resistant prostate cancer (mainsail): a randomised, double-blind, placebo-controlled phase 3 trial. Lancet Oncol. **16**(4), 417–425 (2015)
16. Piliszek, R., et al.: MDFS: multidimensional feature selection in R. R J. (2019). https://doi.org/10.32614/RJ-2019-019
17. Tannock, I.F., et al.: Aflibercept versus placebo in combination with docetaxel and prednisone for treatment of men with metastatic castration-resistant prostate cancer (VENICE): a phase 3, double-blind randomised trial. Lancet Oncol. **14**(8), 760–768 (2013)
18. Yu, M., et al.: Individual prediction in prostate cancer studies using a joint longitudinal survival-cure model. J. Am. Stat. Assoc. **103**(481), 178–187 (2008)

Supervised Machine Learning Techniques Applied to Medical Records Toward the Diagnosis of Rare Autoimmune Diseases

Pedro Emilio Andrade Martins[1]([✉]) [iD], Márcio Eloi Colombo Filho[1] [iD],
Ana Clara de Andrade Mioto[1] [iD], Filipe Andrade Bernardi[1] [iD],
Vinícius Costa Lima[1] [iD], Têmis Maria Félix[2] [iD], and Domingos Alves[1,3] [iD]

[1] Health Intelligence Laboratory, Ribeirão Preto Medical School, Ribeirão Preto, Brazil
pedroemilioam02@usp.br
[2] Medical Genetics Service, Porto Alegre Clinical Hospital, Porto Alegre, Brazil
[3] Department of Social Medicine, Ribeirão Preto Medical School, Ribeirão Preto, Brazil

Abstract. Rare autoimmune diseases provoke immune system malfunctioning, which reacts and damages the body's cells and tissues. They have a low prevalence, classified as complex and multifactorial, with a difficult diagnosis. In this sense, this work aims to support the diagnosis of a rare autoimmune disease using the analysis of medical records from supervised machine learning methods and to identify the models with the best performance considering the characteristics of the available data set. A synthetic database was created with 1000 samples from epidemiological studies identified in the literature, simulating demographic data and symptoms from patient records for the diagnosis of Amyotrophic Lateral Sclerosis (ALS), Multiple Sclerosis (MS), Systemic Lupus Erythematosus (SLE), Crohn's Disease (CD) and Autoimmune Hepatitis (AIH). Data were segmented into training (80%) and test (20%), assigning the diagnosis as the class to be predicted. The models with the highest accuracy were Logistic Regression (84%), Naive-Bayes (82.5%), and Support Vector Machine (79%). Only LR obtained quality values greater than 0.8 in all metrics. SLE had the highest precision and recall for all classifiers, while ALS had the worst results. The LR accuracy model had the best performance with good quality metrics, although it was impossible to predict ALS accurately. Although a real dataset was not used, this work presents a promising approach to support the challenging diagnosis of rare autoimmune diseases.

Keywords: Machine learning · Rare diseases · Medical records

1 Introduction

The concept of autoimmune diseases is derived from the improper functioning of the immune system. The defense cells initiate an erroneous process of attacking the body's substances since it recognizes them as possible antigens. It initiates an inflammatory cascade effect and destruction of the target component. The lymphocyte mistakenly recognizes its molecule as an antigen, continuously stimulating an immune response and directly affecting the functioning of tissues and other body structures [1].

© The Author(s), under exclusive license to Springer Nature Switzerland AG 2023
J. Mikyška et al. (Eds.): ICCS 2023, LNCS 14075, pp. 170–184, 2023.
https://doi.org/10.1007/978-3-031-36024-4_13

Although the pathophysiological procedure of autoimmune diseases and their mechanisms of action is widely recognized, the same causal factor that triggers autoimmune diseases is unknown. However, studies suggest that genetic aspects, hormonal dysregulation, and environmental factors combine in an interdependent way in the manifestation of these diseases, allowing the understanding of these as complex multifactorial [2].

According to the medical literature, it is estimated that about 3% to 5% of the world's population suffers from an autoimmune disease, which is more common in women [3]. Some already documented diseases are Type 1 Diabetes Mellitus, Psoriasis, Multiple Sclerosis, Rheumatoid Arthritis, Systemic Lupus Erythematosus, and others [4].

Diagnosing autoimmune diseases is often complex and difficult to be carried out assertively, given the reaction similarity between the diseases in their initial stages. The distinction between symptoms and laboratory tests is insufficient, requiring a comprehensive investigation of the patient's family health history, medical imaging exams, and the study of previous medical events and symptoms [5].

The medical experience is a predominant factor, directly influencing decision-making in carrying out specific tests and, consequently, in the diagnostic process. In this way, although there are key indicators to guide the possibilities during the investigation, the diagnosis is not fully supported by tests but also by what is most plausible and likely given the combination of the physician's intrinsic knowledge, literature, and information obtained from the patient and the exams.

According to the Brazilian Ministry of Health, rare diseases are categorized as those with a prevalence of up to 65 people per 100.000 inhabitants [6]. Rare autoimmune diseases fit into this scenario according to epidemiological studies in Brazil and other countries [6–10].

Rare autoimmune diseases accentuate the problem regarding its assertive diagnosis due to the similarity between symptoms, the need for different laboratory tests and medical images, and professional experience. Examples such as Amyotrophic Lateral Sclerosis (ALS), Multiple Sclerosis (MS), Systemic Lupus Erythematosus (SLE), Crohn's Disease (CD), and Autoimmune Hepatitis (AIH) accurately represent this complexity, as they denote a costly and time-consuming diagnosis, with a severe prognosis, besides their clinical and epidemiological significance in Brazil.

According to the literature, the most recurrent uses of machine learning and deep learning aimed at autoimmune diseases are for predicting disease progression and diagnosis [11]. Also, the identification of possible biomarkers for the detection and formulation of inhibitory drugs and the prediction of candidate genetic sequences to be interpreted as autoantigens is a growing field [12, 13].

It is reported a wide variation of data types applied to these methods with a main focus on clinical data, especially magnetic resonance images, and genomic data [11]. Few studies have addressed the diagnosis between multiple autoimmune diseases, and these utilized rRNA gene data [14, 15].

No related studies were found regarding the correlation of the main topics as machine learning techniques towards the diagnosis of multiple rare autoimmune diseases and medical records as the main data source. Therefore, this research provides an early comprehension about the subject, with relevance on avoiding the diagnosis delay by possibly identifying features on medical records and guiding diagnosis confirmation tests.

2 Objective

This research aims to study the performance of machine learning models as a decision support tool in the diagnosis of a rare autoimmune disease, based on patient medical records.

3 Methods

3.1 Database Elaboration

The database is an essential component to carry out analyses and inferences toward the objective of this research. A database was developed for the project, simulating data from the patient's medical records due to the lack of relevant public data related to rare autoimmune diseases.

First, it is well established that criteria are needed to build the data set to simulate information about patient records and rare autoimmune diseases more faithfully with reality. The data structure was based on studies and systematic reviews regarding the epidemiology of rare autoimmune disorders [16–24]. Thus, the following parameters were applied:

Epidemiological Studies. For the selection of reviews and epidemiological studies as a reference, the authors put the following reasoning into practice: Limit the search to only five rare autoimmune diseases, namely: ALS, MS, SLE, CD, and AIH, since they have clinical, epidemiological relevance in Brazil; Prioritize epidemiological studies on the Brazilian population; Search for systematic reviews of the global epidemiology.

Sample Set. As these are rare diseases, patient medical records data is highly scarce. Given this, to determine the number of samples (n) to create the database, a systematic review with meta-analysis of Autoimmune Hepatitis was used as a reference [25]. Therefore, the elaborated database has n = 1000.

Variables Selection. From the data that make up the patient's medical records according to the literature, together with those from the epidemiological studies surveyed, the selected attributes were: Patient Identification; Age; Race; Sex; Symptoms, and Diagnosis. The attributes and structure of the dataset are shown in Table 1.

Values Parameters. The following epidemiological data were used to determine how the values within each attribute would be distributed:

Prevalence. Used to confirm the disease as a rare autoimmune disease, specify the frequency of each diagnosis within the database, and verify which rare diseases are more common or less common.

Prevalence by Sex. Used to indicate the frequency of female or male patients by disease (MS, ALS, SLE, CD, AIH).

Table 1. Dataset Attributes.

Attribute name	Meaning	Attribute Type
ID	Patient Identification	Discrete
Sex	Gender	Categorical
Race	Ethnicity	Categorical
Age	Age arranged in time interval	Categorical
Symptom_F	Fatigue Symptom	Boolean
Symptom_M	Stiffness and loss of muscle strength	Boolean
Symptom_W	Loss weight	Boolean
Symptom_GI	Gastrointestinal Symptom	Boolean
Symptom_VD	Visual disorder	Boolean
Symptom_SI	Skin injuries	Boolean
Diagnosis	Rare autoimmune disease diagnosis	Categorical

Prevalence by Race. Used to show the distribution of races by disease.

Graphs of distribution of age groups to the detriment of diagnosis. Used to indicate the percentage of age groups in each disease, applying it to our sample set.

Symptoms. Used to map lists of symptoms for each disease and trace common ones.

Furthermore, for the symptoms' selection, inclusion criteria were established that symptoms must be present in 3 or more diseases, as shown in Table 2: be a frequent symptom; be a generalized symptom; and finally, a patient must present at least two symptoms that match the characteristics of a given disease.

Table 2. Distribution of symptoms.

Symptoms	MS	ALS	SLE	CD	AIH
Symptom_F	Present	Present	Present	Present	Present
Symptom_M	Present	Present	Present	Absent	Absent
Symptom_W	Present	Present	Present	Present	Present
Symptom_GI	Present	Present	Present	Present	Present
Symptom_VD	Present	Present	Present	Present	Absent
Symptom_SI	Absent	Absent	Present	Present	Present

Thus, from these specifications, whose objective is to guarantee the greatest possible basis for the creation of the database, data were created and randomly allocated between the attributes and the class.

Data Processing. With selecting the sample set, it is crucial to process the data to avoid bias and unbalanced values. However, since the authors authored the dataset and it was standardized, there were few steps necessary to process the data.

Among the data pre-processing activities, the "Age" attribute was changed from discrete values to age range intervals to assist in the analysis process as categorical data. The "ID" attribute was removed since it has no significance among the other attributes and may even interfere during the analysis and classification process of the algorithms.

Finally, the dataset was segmented into two groups: the training and test sets. This division was made at random, following the proportion of 80% for the training set and 20% for the test stage of the models. In this scenario, the diagnosis of the autoimmune disease present in the patient's medical record is the "Class" to be predicted by machine learning algorithms.

3.2 Analysis Software

The Weka 3 software, a data mining tool, was chosen to analyze data by applying supervised machine learning algorithms. This software was chosen because it presents classification, regression, clustering, data visualization techniques, and various classification algorithms. Also, Weka 3 has helpful quality metrics for validating the machine learning model, such as the PRC Area, Confusion Matrix, and the F-Score.

3.3 Classification Algorithms

Supervised learning algorithms were used to select classification algorithms, which are already described in the literature for presenting good results in data analysis from patient records and for data related to rare diseases, especially Support Vector Machine and Artificial Neural Networks [26]. KNN, Naive-Bayes, Random Forest and Logistic Regression, has previously been used on MS diagnosis and others neurodegenerative diseases [27, 28].

All algorithms were used with default parameters. No optimal hyperparameters were selected, since this study aims at the initial comparison between well described classifiers.

The algorithms to which the dataset will be applied are shown below.

Support Vector Machine (SVM). Supervised machine learning algorithm applied for classification and regression. It consists of building a set of hyperplanes in an 'n'-dimensional space, which allows a better distinction between classes using the maximum values between each class for its delimitation [29].

KNN. The supervised learning algorithm, known as 'lazy,' is used for dataset classification and regression. A class separation method based on a 'K' parameter selects the closest variables in space by the point-to-point distance (e.g., Euclidean, Hamming,

Manhattan). Given the class of these close variables, the model classifies from the one that appears most in the set within a certain distance [30].

Naive-Bayes (NB). The supervised classifier uses Thomas Bayes' theorems to ensure a simplification approach to the problem, being expressively relevant in real-world scenarios in the health area. This model assumes that all attributes are independent and relevant to the result, generating probabilities and frequency of existing values in comparison with the class to be predicted [31].

Random Forest (RF). Supervised learning algorithm executed for classification and regression problems, being able to use data sets containing continuous and categorical variables. This model uses decision trees in different samples, predicting the classification and regression from most data or its mean [32].

Logistic Regression (LR). The statistical model is used in supervised machine learning, allowing classification. It combines the attributes present in the dataset with mathematical and statistical methods to predict a class. In addition, it allows for analyzing the relationships between the present attributes, such as autocorrelation and multicollinearity, verifying the variables that best explain the expected output [33].

Multilayer Perceptron (MLP). It is an artificial neural network with one or more hidden layers and an indeterminate number of neurons. It consists of non-linear mathematical functions based on the backpropagation technique, which trains neurons by changing synaptic weights at each iteration [34].

3.4   Quality Metrics

First, all models presented the same evaluation metrics as the tables in the results. The most significant measures to qualify the performance of the classifiers were: Precision, Recall, F-score, Matthews Correlation Coefficient (MCC), and PRC Area.

Precision and Recall are fundamental validation statistics, as they are based on the number of true positive (PV) ratings, in other words, how many individuals were correctly categorized according to the class. The F-score is relevant because it synthesizes the Precision and Recall rates from a harmonic mean.

As it deals with diagnosis and differentiation between classes of diseases, the MCC is a highly relevant metric since it only presents a satisfactory result if all categories of the confusion matrix obtain good values, highlighting cases of false negative (FN) and false positive (FP). Finally, the PRC Area allows the graphical analysis of the relationship between precision and sensitivity, being preferable in comparison to the ROC Area because it is an unbalanced data set.

4   Results

4.1   Training Set

First, due to the segmentation of the dataset between the training and test groups, a sample set n = 800 (80%) was used for the training process of the classification algorithms. The distribution between classes in the training set is shown in Table 3. The set was applied to each classifier using cross-validation with K folds = 10, obtaining the results presented in the following sections.

Table 3. Distribution of Classes in Training Set.

Class	Sample	Percentage
MS	117	14.6%
ALS	43	5.4%
SLE	321	40%
CD	206	25.8%
AIH	113	14.2%

4.2   Test Set

The test set was randomly chosen, selecting the first n = 200 (20%) patients. The distribution between classes is shown in Table 4.

Table 4. Distribution of Classes in Test Set.

Class	Sample	Percentage
MS	33	16.5%
ALS	7	3.5%
SLE	79	39.5%
CD	44	22%
AIH	37	18.5%

The test set was applied to each of the previously described models without any parameter changes for validation and verification of the performance with the new data.

4.3   Applied Classifiers in the Test Set

Support Vector Machine (SVM). From inserting new data, the SVM model presented a correct classification of 158 patients (79%) and weighted mean precision equivalent to 0.811. Such values were similar to those obtained during training, 0.783 and 0.790, respectively. Furthermore, the model had an average recall between classes of 0.790 and an F-score of 0.795.

From a class perspective, there was a considerable variation in metrics. In comparison, the SLE PRC Area was 0.970. The others did not have a value greater than 0.650,

Table 5. SVM Detailed Performance by Class.

TP Rate	Precision	Recall	F-Measure	MCC	PRC Area	Class
0.899	1,000	0.899	0.947	0.918	0.970	SLE
0.571	0.667	0.571	0.615	0.604	0.312	ALS
0.773	0.618	0.773	0.687	0.592	0.577	DC
0.606	0.800	0.606	0.690	0.647	0.594	MS
0.784	0.674	0.784	0.725	0.660	0.638	AIH
0.790	0.811	0.790	0.795	0.743	0.737	Weighted Avarage

with a lower value of 0.312 for ALS. As shown in Table 5, other metrics highlight the imbalance between classes.

KNN. The KNN models with K = 3, K = 5, and K = 7 did not show disparity between the metrics, with minimal to no difference when detailed by Class. Therefore, as seen in Table 6, we only focused on the comparison between the overall performance of this models.

The three models demonstrated an accuracy of 73.5% (n = 147), with more excellent recall for KNN-7 and KNN-5 (0.735) compared to KNN-3 (0.730). The KNN-7 provided slightly better results than the other two models, with an F-score of 0.735 and an accuracy of 0.749.

Table 6. KNN Average Performance.

TP Rate	Precision	Recall	F-Measure	MCC	PRC Area	Classifier
0.730	0.742	0.730	0.730	0.660	0.793	KNN-3
0.735	0.744	0.735	0.733	0.666	0.811	KNN-5
0.735	0.749	0.735	0.735	0.668	0.812	KNN-7

Naive-Bayes (NB). NB correctly predicted 165 patients (82.5%) with a mean across-class accuracy of 0.836. It also revealed a MCC of 0.783 and an Area PRC of 0.861.

As shown in Table 7, the classes with the highest recall are, respectively, SLE (0.924), AIH (0.825) and DC (0.795). In contrast, MS and ALS have the worst recall results, with 0.697 and 0.429 in that order.

Random Forest (RF). The Random Forest correctly classified 158 cases (79%) with a weighted average accuracy of 0.793 and an equivalent recall of 0.790. As shown in Table 8, metrics vary according to each class, emphasizing SLE and ALS. There is a significant difference between these two classes' F-score (SLE = 0.943 and ALS 0.308) and the MCC (SLE = 0.906 and ALS 0.285).

Table 7. NB Detailed Performance by Class.

TP Rate	Precision	Recall	F-Measure	MCC	PRC Area	Class
0.924	0.986	0.924	0.954	0.927	0.984	SLE
0.429	0.600	0.429	0.500	0.492	0.529	ALS
0.795	0.660	0.795	0.722	0.638	0.794	DC
0.697	0.852	0.697	0.767	0.731	0.752	MS
0.838	0.756	0.838	0.795	0.747	0.836	AIH
0.825	0.836	0.825	0.827	0.783	0.861	Weighted Average

Table 8. RF Detailed Performance by Class.

TP Rate	Precision	Recall	F-Measure	MCC	PRC Area	Class
0.937	0.949	0.937	0.943	0.906	0.978	SLE
0.286	0.333	0.286	0.308	0.285	0.297	ALS
0.795	0.660	0.795	0.722	0.638	0.670	DC
0.545	0.750	0.545	0.632	0.582	0.722	MS
0.784	0.744	0.784	0.763	0.708	0.807	AIH
0.790	0.793	0.790	0.787	0.735	0.813	Weighted Average

Logistic Regression (LR). The Logistic Regression model had an accuracy of 84% (n = 168) with a weighted average precision of 0.852. According to Table 9, the general result of the classes of all metrics obtained values greater than 0.8.

The ALS class remains the one with the worst results among the others but showed an increase in recall (0.571), precision (0.667), F-score (0.615), MMC (0.604) and PRC Area (0.652) in comparison with the other classifiers.

Table 9. LR Detailed Performance by Class.

TP Rate	Precision	Recall	F-Measure	MCC	PRC Area	Class
0.924	1,000	0.924	0.961	0.938	0.984	SLE
0.571	0.667	0.571	0.615	0.604	0.652	ALS
0.864	0.704	0.864	0.776	0.710	0.836	DC
0.758	0.862	0.758	0.806	0.773	0.752	MS
0.757	0.737	0.757	0.747	0.688	0.831	AIH
0.840	0.852	0.840	0.843	0.803	0.873	Weighted Average

Multilayer Perceptron (MLP). The neural network correctly classified 156 instances (78%) with an average precision between classes equivalent to 0.778. In addition, an average F-score of 0.777 and an average PRC Area of 0.805 was obtained.

Furthermore, as with all other classifiers, SLE was the most accurately predicted class (0.944), while ALS was the lowest (0.333). The other quality metrics' specific values for each class are presented in Table 10.

Table 10. MLP Detailed Performance by Class.

TP Rate	Precision	Recall	F-Measure	MCC	PRC Area	Class
0.924	0.948	0.924	0.936	0.895	0.980	SLE
0.143	0.333	0.143	0.200	0.200	0.220	ALS
0.727	0.615	0.727	0.667	0.566	0.675	DC
0.636	0.677	0.636	0.656	0.591	0.660	MS
0.784	0.784	0.784	0.784	0.735	0.824	AIH
0.780	0.778	0.780	0.777	0.719	0.805	Weighted Average

4.4 Comparison Between Models

Training Set. Table 11 compares the algorithms used based on the weighted average of the quality and accuracy metrics achieved in each model. In this scenario, the KNN-7 was chosen as the model with the best performance for this classifier.

Visualizing that the results obtained between the classifiers are similar, not presenting significant divergences is possible. However, it is noted that Naive Bayes is the only model that gives results above 0.8 for all metrics, except for MCC, in addition to having the highest recall (0.817). On the other hand, the KNN-7 presents the worst results in general among the classifiers.

From the accuracy, sensitivity, and F-score perspective, NB, LR, and SVM performed best in that order. As for the MCC, no model achieved a value equal to or greater than 0.8. However, all except the SVM denoted a PRC Area greater than 0.8.

Test Set. Table 12 compares each of the algorithms used, with the choice of KNN-7 as the best model for this classifier. The values present in each metric refer to the weighted average obtained in each model with the test set.

From this, it is observed that there are no such significant discrepancies between each classifier. Still, among the six metrics, Logistic Regression performed the best in all of them. In contrast, the KNN-7 remains the worst model, with lower results in 5 metrics, except for the PRC Area.

From the precision, recall, and F-score perspective, only NB and LR obtained results above 0.8. As for the PRC Area, all models are above 0.8, excluding the SVM.

Table 11. Comparison between Training Models by Detailed Average Performance.

TP Rate	Precision	Recall	F-Measure	MCC	PRC Area	Classifier
0.774	0.778	0.774	0.765	0.705	0.838	KNN-7
0.783	0.790	0.783	0.783	0.727	0.728	SVM
0.809	0.817	0.817	0.809	0.761	0.878	NB
0.781	0.783	0.781	0.780	0.722	0.826	RF
0.796	0.800	0.796	0.797	0.744	0.860	LR
0.780	0.778	0.780	0.777	0.719	0.805	MLP

Table 12. Comparison between Test Models by Detailed Average Performance.

TP Rate	Precision	Recall	F-Measure	MCC	PRC Area	Classifier
0.735	0.749	0.735	0.735	0.668	0.812	KNN-7
0.790	0.811	0.790	0.795	0.743	0.737	SVM
0.825	0.836	0.825	0.827	0.783	0.861	NB
0.790	0.793	0.790	0.787	0.735	0.813	RF
0.840	0.852	0.840	0.843	0.803	0.873	LR
0.780	0.778	0.780	0.777	0.719	0.805	MLP

Confusion Matrix. Table 13 depicts an interesting result plotted in the classes of Autoimmune Hepatitis and Crohn's Disease. The Logistic Regression model's confusion matrix during the training phase shows these two classes.

Table 13. LR Test Model Confusion Matrix.

Class	Predict Class	
	AIH	DC
AIH	28	9
DC	6	38

Best Classifiers. Based on the results obtained, verifying the performance of each model against the quality metrics and comparing them between classes and generalized weighted averages, it was possible to find the best models, as shown in Table 14.

Table 14. Best Models by Detailed Average Performance.

TP Rate	Precision	Recall	F-Measure	MCC	PRC Area	Classifier
0.840	0.852	0.840	0.843	0.803	0.873	LR
0.825	0.836	0.825	0.827	0.783	0.861	NB
0.790	0.811	0.790	0.795	0.743	0.737	SVM

5 Discussion

Regarding the applied models, based on the general average results between the classes for each classifier, it appears that the three best algorithms are Logistic Regression, Naive Bayes, and Support Vector Machine, as shown in metric LR presents the best results on all six measures; NB has the second-best result among the metrics, and SVM has the third-best result among the five metrics, with the worst result for the PRC Area.

However, even though LR and NB are models with excellent metrics, with NB being the best machine learning method among the others, it is necessary to look at the confusion matrix and the quality parameters for each class. During the training phase, both algorithms had difficulty predicting the ALS class during the training phase, with sensitivity below 50%. Furthermore, an individual in the ALS class was normally classified as in the MS class. Thus, of the total ALS (n = 43), 35% were classified as MS for LR and 42% for NB.

In contrast, for the SLE class, both NB and LR scored 299 cases (93%) during training and 73 cases (92.4%) in the test phase. The main reason for this difference, which causes many classification errors for ALS and hits for SLE, is the class imbalance. Looking at the overall sample set (n = 1000), SLE represents 40% of the total, while MS only 5%. Such imbalance is present as it resembles the real world, where ALS is a rare disease with low prevalence, and SLE is rare in South America but with a much higher prevalence. By having a larger number of cases with SLE in the dataset, the algorithms can understand the attributes of this class. In contrast, for ALS, which has few samples, it becomes more complex to train and identify the possible factors that determine this disease.

In addition to dataset imbalance, which makes the prediction process more difficult, the similarity of symptoms between diseases is another factor. All diseases share at least three symptoms among them, and the more they have in common, the more complex it is for the algorithm to classify based on these attributes alone.

As seen in Table 13, it is noted that there is a tendency that when the classifier misses the DC class, it categorizes it as AIH and vice versa. These results denote a difficulty for the algorithm to distinguish these two classes depending on the scenario, in line with the literature, since AIH is a disease often associated concomitantly with other autoimmune diseases such as Crohn's Disease, Rheumatoid Arthritis, and thyroid diseases.

Finally, it is important to point out some caveats regarding this study. It is understood that a real database is a primordial point for the development of the work reliably and consistently with reality. Therefore, although the database has been created and supported by several references following logical criteria, it cannot fully portray the truth. Thus, it is recommended for future studies to search for a database of medical records of patients with rare diseases, such as the database from the Brazilian National Network of Rare Diseases (RARAS) [35], restricting it to autoimmune diseases and verifying which possible diagnoses are available.

Another relevant limiting factor concerns the sample set. As these are rare autoimmune diseases, this is a particular subset, making it difficult to find sufficient data for this type of analysis methodology. Furthermore, the lack of standardization among the data from medical records in such a small set is another point that hinders the use of certain attributes for the classification process. Even with these problems, there are articles described in the literature that address the standardization and structuring process for these types of data [36, 37].

Finally, this study uses only nine attributes (no class included), 3 of which are demographic data and 6 are symptoms. However, a patient's medical record has numerous attributes that can be considered during the analysis, not restricted to those of this study. Thus, it is interesting that these questions are raised in further studies, seeking the best possible methodology to work with these data types.

6   Conclusion

Given the results presented and the established validation metrics, it was possible to use machine learning models capable of predicting autoimmune disease, with Logistic Regression being the best general model and Naive Bayes the best machine learning classifier.

Both algorithms had satisfactory results regarding general quality parameters between classes. However, when considering the individual classes, none could accurately and reliably predict the ALS diagnosis. Furthermore, there is evidence of a slight but notable difficulty in distinguishing the DC and AIH classes. Therefore, the LR and NB models found can predict rare autoimmune diseases exclusively for this research. However, they are not maintained in real life due to the need for a feasible database, and they do not present any classification variant for diagnosing ALS.

Although a real dataset and medical records were not used, this work presents a promising approach to underpin the diagnosis of rare autoimmune diseases, supporting physicians in this challenging task. This methodology will be applied in future articles using real data from RARAS's database and exploring different kinds of predictions, like the prognosis of patients with rare diseases, for example.

Acknowledgments. This study is supported by the National Council for Scientific and Technological Development (CNPq) process no. 44303/2019-7.

References

1. Gattorno, M., Martini, A.: Immunology and rheumatic diseases. In: Textbook of Pediatric Rheumatology. 6th edn. W.B. Saunders (2011)
2. Smith, D.A., et al.: Introduction to immunology and autoimmunity. Environ. Health Perspect. **107**, 661–665 (1999)

3. Bioemfoco: Doenças Autoimunes: Porque são chamadas assim e os avanços nas pesquisas. https://bioemfoco.com.br/noticia/doencas-autoimunes-avancos-pesquisas/. Accessed 02 May 2022
4. Riedhammer, C., et al.: Antigen presentation, autoantigens, and immune regulation in multiple sclerosis and other autoimmune diseases. Frontiers Immunol. **6** (2015)
5. Autoimmune association: Diagnosis Tips. https://autoimmune.org/resource-center/diagnosis-tips/. Accessed 02 May 2022
6. Ministério da Saúde: Lúpus. https://www.gov.br/saude/pt-br/assuntos/saude-de-a-a-z/l/lupus. Accessed 02 May 2022
7. Araújo, A.D., et al.: Expressões e sentidos do lúpus eritematoso sistêmico (LES). Estudos de Psicologia (Natal). **12**, 119–127 (2007)
8. Borba, E.F., et al.: Consenso de lúpus eritematoso sistêmico. Rev. Bras. Reumatol. **48**, 196–207 (2008)
9. Santos, S.de.C.: Doença de Crohn: uma abordagem geral. Specialization dissertation, Universidade Federal do Paraná, Curitiba (2011)
10. Poli, D.D.: Impacto da raça e ancestralidade na apresentação e evolução da doença de Crohn no Brasil. Masters dissertation, Universidade de São Paulo, São Paulo (2007)
11. Stafford, I.S., et al.: A systematic review of the applications of artificial intelligence and machine learning in autoimmune diseases. Digit. Med. **3** (2020)
12. DeMarshall, C., et al.: Autoantibodies as diagnostic biomarkers for the detection and subtyping of multiple sclerosis. J. Neuroimmunol. **309**, 51–57 (2017)
13. Saavedra, Y.B.: Análisis de Autoreactividad de Anticuerpos Leucémicos Soportado por Estrategias de Inteligencia Artificial. Doctoral dissertation, Universidad de Talca, Talca (2021)
14. Forbes, J.D., et al.: A comparative study of the gut microbiota in immune-mediated inflammatory diseases—does a common dysbiosis exist? Microbiome **6** (2018)
15. Iwasawa, K., et al.: Dysbiosis of the salivary microbiota in pediatric-onset primary sclerosing cholangitis and its potential as a biomarker. Sci. Rep. **8** (2018)
16. Esclerose Múltipla (EM). https://www.einstein.br/doencas-sintomas/esclerose-multipla. Accessed 4 May 2022
17. Chiò, A., et al.: Global epidemiology of amyotrophic lateral sclerosis: a systematic review of the published literature. Neuroepidemiology **41**, 118–130 (2013)
18. Correia, L., et al.: Hepatite autoimune: os critérios simplificados são menos sensíveis? GE Jornal Português de Gastrenterologia. **20**, 145–152 (2013)
19. Pereira, A.B.C.N.G., et al.: Prevalence of multiple sclerosis in Brazil: a systematic review. Multiple Sclerosis Related Disord. **4**, 572–579 (2015)
20. Pearce, F., et al.: Can prediction models in primary care enable earlier diagnosis of rare rheumatic diseases? Rheumatology **57**, 2065–2066 (2018)
21. Sociedade Brasileira de Reumatologia: Lúpus Eritematoso Sistêmico (LES). https://www.reumatologia.org.br/doencas-reumaticas/lupus-eritematoso-sistemico-les. Accessed 10 Nov 2022
22. de Souza, M.M., et al.: Perfil epidemiológico dos pacientes portadores de doença inflamatória intestinal do estado de Mato Grosso. Revista Brasileira de Coloproctologia. **28**, 324–328 (2008)
23. Souza, M.H.L.P., et al.: Evolução da ocorrência (1980–1999) da doença de Crohn e da retocolite ulcerativa idiopática e análise das suas características clínicas em um hospital universitário do sudeste do Brasil. Arq. Gastroenterol. **39**, 98–105 (2002)
24. Tamega, A.de.A., et al.: Grupos sanguíneos e lúpus eritematoso crônico discoide. Anais Brasileiros de Dermatologia **84**, 477–481 (2009)
25. Chen, J., et al.: Systematic review with meta-analysis: clinical manifestations and management of autoimmune hepatitis in the elderly. Aliment. Pharmacol. Ther. **39**, 117–124 (2013)

26. Schaefer, J., et al.: The use of machine learning in rare diseases: a scoping review. Orphanet J. Rare Dis. **15** (2020)
27. Aslam, N., et al.: Multiple sclerosis diagnosis using machine learning and deep learning: challenges and opportunities. Sensors **22**, 7856 (2022)
28. Myszczynska, M.A., et al.: Applications of machine learning to diagnosis and treatment of neurodegenerative diseases. Nat. Rev. Neurol. **16**, 440–456 (2020)
29. Scikit Learn: Support Vector Machines. https://scikit-learn.org/stable/modules/svm.html. Accessed 14 May 2022
30. IBM: K-Nearest Neighbors Algorithm. https://www.ibm.com/topics/knn. Accessed 14 May 2022
31. Scikit Learn: Naive Bayes. https://scikit-learn.org/stable/modules/naive_bayes.html. Accessed 5 June 2022
32. IBM, What is random forest. https://www.ibm.com/topics/random-forest. Accessed 5 June 2022
33. Scikit Learn: Linear Models. https://scikit-learn.org/stable/modules/linear_model.html. Accessed 5 June 2022
34. Scikit Learn: Neural network models (supervised). https://scikit-learn.org/stable/modules/neural_networks_supervised.html. Accessed 5 June 2022
35. Rede Nacional de Doenças Raras. https://raras.org.br/. Accessed 10 Nov 2022
36. Alves, D., et al.: Mapping, infrastructure, and data analysis for the Brazilian network of rare diseases: protocol for the RARASnet observational cohort study. JMIR Res. Protoc. **10**(1), e24826 (2021)
37. Yamada, D.B., et al.: National network for rare diseases in Brazil: the computational infrastructure and preliminary results. In: Groen, D., de Mulatier, C. (eds.) Computational Science – ICCS 2022. ICCS 2022. LNCS, vol. 13352. Springer, Cham (2022). https://doi.org/10.1007/978-3-031-08757-8_4

Handwriting Analysis AI-Based System for Assisting People with Dysgraphia

Richa Gupta[1], Deepti Mehrotra[1], Redouane Bouhamoum[2],
Maroua Masmoudi[2], and Hajer Baazaoui[2]([✉])

[1] Amity University Uttar Pradesh, Noida, India
{rgupta6,dmeehrotra}@amity.edu
[2] ETIS UMR 8051, CY University, ENSEA, CNRS, Cergy, France
{redouane.bouhamoum,hajer.baazaoui}@ensea.fr,
maroua.kottimasmoudi@cy-tech.fr

Abstract. Dysgraphia is a learning disability of written expression, which affects the ability to write, mainly handwriting and coherence. Several studies have proposed approaches for assisting dysgraphic people based on AI algorithms. However, existing aids for dysgraphia take only one aspect of the problem faced by the patients into consideration. Indeed, while some provide writing assistance, others address spelling or grammatical problems.

In this paper, a novel system for helping people suffering from dysgraphia is proposed. Our system tackles several problems, such as spelling mistakes, grammatical mistakes and poor handwriting quality. Further, a text-to-speech functionality is added to improve the results.

The proposed system combines a plethora of solutions into a valuable approach for efficient handwriting correction: handwritten text recognition using a CNN-RNN-CTC model, a spelling correction model based on the SymSpell and Phoneme models, and a grammar correction using the GECToR model. Three machine learning models are proposed. The experimental results are compared based on the values of Character error rate, Word error rate, and the workflow of three handwritten text recognition models, and has led to an improvement of the quality of the results.

Keywords: Dysgraphia · handwritten text recognition · Artificial Intelligence · spelling correction · text correction

1 Introduction

In today's fast-paced world, many activities, such as school activities, bureaucratic procedures, etc., are difficult for people with learning disabilities such as dysgraphia. Dysgraphia is a learning disability that impairs a person's ability to write correctly. It can interfere with effective interaction between educators and students or colleagues, such as written communication via email, chat, or online courses. Some children cannot produce good writing, even with appropriate instructions and practice in writing. Tasks such as spelling, which rely primarily on hearing, can be difficult for people with dysgraphia. This is because a

© The Author(s), under exclusive license to Springer Nature Switzerland AG 2023
J. Mikyška et al. (Eds.): ICCS 2023, LNCS 14075, pp. 185–199, 2023.
https://doi.org/10.1007/978-3-031-36024-4_14

person with dysgraphia often makes phonemic errors in written text. Some students have problems with the sound of the word, while others may be bothered by the visual aspect. With respect to grammar, sentence structure and adherence to grammatical rules may be a problem. Difficulties may arise when trying to understand how words can be used together to form meaningful and complete sentences. In addition to poor handwriting, spelling and grammatical errors are common for people with dysgraphia.

There are mainly three ways to treat dysgraphia: the corrective approaches, the bypass approaches, and finally, the prevention methods [12]. The corrective approaches are appropriate when a physical remedy is needed. Mobile apps and graphic tablets are great tools for solving problems related to spelling and poor handwriting. Several works propose systems to help dysgraphic people based on artificial intelligence. However, the existing approaches only take into account one aspect of the problem encountered by the patients. Some of the existing approaches propose a writing aid, and others tackle spelling or grammar problems. The application of handwriting processing has been more widespread in the analysis of dysgraphic writing [2]. Some handy devices help with handwriting problems, but do not provide any help with spelling or grammar. Similarly, tools that provide spelling or grammar assistance do not focus on addressing poor handwriting. The need for a well-integrated system with multiple modules addressing a separate issue was observed.

In this article, we present an approach to help people with dysgraphia. Our motivation is to help dysgraphic people to improve their writing and communication skills, by building an automatic handwriting correction system. The originality of this work consists in defining an approach allowing to analyze and correct the poor handwriting, spelling mistakes and incorrect grammar of dysgraphic people. The proposed system would potentially help dysgraphic persons in their daily life, for example by improving the efficiency of communication with the teacher, using it to convert assignments and tests into legible and understandable text, taking notes in class and correcting them for better learning, communicating through letters and writing texts and messages. To achieve this, we have studied and implemented different solutions designed to work together for handwritten text recognition, spelling and grammar error correction, and text-to-speech conversion. Each solution is implemented on the basis of a well-known state-of-the-art solution. The novelty of our proposal lies in its ability to process the input handwritten text image to produce an output in a correct and usable format.

The remaining of this paper is organized as follows. In Sect. 2, we discuss and analyze the related works, along with our motivations and objectives. Section 3 describes the proposed system and details its components. Section 4 concentrates on the experimental evaluation. Section 5 concludes and proposes directions for future research works.

2   Literature Review

We discuss in the literature review different existing solutions for handwritten text recognition and text correction. By analyzing the existing solutions, we define the requirements that our system should cover. Then a deeper dive is taken into the existing literature on the approaches that can be implemented by our system for assisting people with dysgraphia.

Dysgraphia is generally related to Development Coordination Disorder (DCD), dyslexia, or attention deficit disorder [17]. All three conditions are Neurodevelopmental disorder, which is characterized as handwriting learning disability [10]. Handwriting generally involves a person's motor skills, perceptual and linguistics [16]. Usually, learning to handwrite proficiently requires 15 years, starting from the early age of 5 to 15 years. DCD is considered a lack of psychomotor development in children and adults. The inception of the disorder appears even before the child enters the school. Parents, friends, or teachers can realize the clinical expression and warning signs of DCD that something is not right in the child's activity. Children range 5 to 10 years show different means and have difficulties in performing simple talk like painting, eating with crockery, etc. DCD in adults may affect them differently, as their less mobility, lower visio-motor skill, and poor handwriting.

The detection of dysgraphia is crucial as early diagnosis allows children to perform well at home and school and allows parents to adapt to the environment according to the needs of the child [14]. The handwriting struggles can only be resolved through intervention and rehabilitation later [8]. The diagnosis of DCD is generally made using the Diagnosis and Statistical Manual (DSM-5) criteria. Smit-Eng et al. [23] presented various recommendations based on cutoff scores, which classify severity ranging from moderate to critical. The authors also advised diagnosis based on motor function, neurological disorder, intellectual deficit, and DSM score. Pediatricians and therapists use various tools and tests to assess. Initially, pediatricians carry out a differential diagnosis to rule out secondary development motor disorder then a therapist uses an assessment test to establish the effectiveness of the proposed care package. These tests generally use movement assessment batteries for children ranging from 3 to 10 years of age [7]. Finally, an intellectual deficit in a child is assessed using a psychometric test. The test and methods used by clinicians and therapists for neuro disorder mentioned establish a standard for diagnosis; however, like DCD, dysgraphia lacks any established tests and tools for assessment.

Reisman [21], in 1993, represented an evaluation scale for handwriting for children in European countries. The score achieved by children is used to evaluate legibility and speed; however, the test alone is not sufficient to establish the condition and sometimes requires clinical assessment. [5] proposed a detailed assessment of the speed of handwriting (DASH) for older kids. Rehabilitation of dysgraphia represents various difficulties due to the diverse origin of the disorder. In the new age, digitizers and pen tablets opt represent a promising tools to overcome the challenges of rehabilitation. As children suffering from writing disorder tend to avoid writing, tablets have the advantage of modifying the writer's willingness due to their liking of new technologies. Danna and Velay [9] presented the idea of providing auditory feedback to increase the sensory information of the written text. The existing test does not consider handwriting dynamics, tilt, and pressure on the pen while analyzing [19]. Thibault et al. [3] presented an automated diagnosis using a tablet that, compared to existing methods, is faster, cheaper, and does not suffer from human bias. Zolna et al. [27] presented the idea of improving the automated diagnosis of dysgraphia based on the recurrent Neural Network Model (RNN).

Various mobile applications provide interactive interfaces for improving motor skills. A mobile application designed by Mohd et al. [1] called Dysgraph-iCoach was built to help improve one's writing and motor skills. Khan et al. presented an augmented reality-based assistive solution called the AR-based dysgraphia assistance writing environment (AR-DAWE) model. It uses the Google Cloud Speech-to-text API. A popular solution for writing aid is a smart pen as a bypass approach. Boyle and Joyce [6] successfully described how smartpens can be used to support the note-taking skills of students with learning disabilities. Avishka et al. [4] proposed a mobile application, THE CURE, that attempts to detect the severity level of dysgraphia and dyslexia in its users. An interactive interface offers the user a place to practice reading and writing. Quenneville [20] mentioned various tools that help those with learning disabilities, like word predictors and auditory feedback. While the former deals with predicting the complete word as the user is typing each letter, the latter reinforces the writing process.

The approach presented by Scheidl et al. [11] used a popular machine learning tool in python known as TensorFlow to achieve accurate handwriting recognition. Another approach is using the Sequence-2-Sequence learning technique described by Sutskever et al. [24]. Sequence-2-Sequence Learning model uses neural networks and is a Deep Learning model. Toutanova et al. [25] described a spelling correction algorithm based on pronunciation. The proposed system uses the Brill and Moore noisy channel spelling correction model as an error model for letter strings. Another spelling correction model used nested RNN and pseudo-training data [13]. [11] used word beam search for decoding, enabling the model to constrain words in the dictionary.

Each presented approach considers only one aspect of the problems faced by dysgraphic persons. To the best of our knowledge, there does not exist such an approach that processes handwritten text images written by patients, to produce an output in a correct and usable format.

3   Handwriting Analysis AI-Based System for Dysgraphic Person's Aid

Our system aims to help persons suffering from dysgraphia by analyzing their handwritten text, correcting spelling and grammatical errors, and converting corrected text into audio. To this purpose, we have designed a system composed of four main processes, each of which serves an intrinsic purpose and is integrated one after another. A process is proposed to achieve the required functionality. Said processes are as follows:

i. Handwritten text recognition
ii. Spelling correction
iii. Grammar correction
iv. Text-to-speech conversion

The above steps consider the various problems observed in dysgraphic handwritten texts and make modifications to provide corrected information. The processed information is then displayed to the user and can be converted into an audio file using text-to-speech conversion. The flowchart given in Fig. 1 describes how the proposed system works.

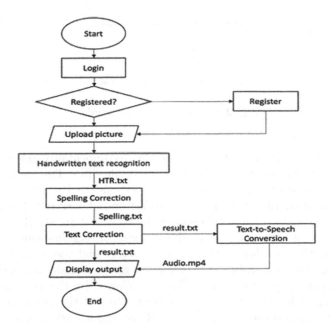

Fig. 1. The Flowchart of the Proposed System.

3.1 Handwritten Text Detection

This process comprises two steps, each for input image pre-processing and recognizing handwritten text. An input image typically will be a scanned copy of handwritten sentences that will be converted into a list of words and then given to the trained handwritten text recognition model. The basic functioning of this model is built on the works of Zhou et al. [26] is carried out by classifying each pixel in the input image as either an inner part/surrounding word, or a background one. Feature extractor used for this part of the project is ResNet18, and a U-shape architecture is used. The ResNet18 generates a feature map of size 14 × 14 by processing an image of size 448 × 448. Finally, the output of the neural network's size is 224 × 224. To establish relationships between two bounding boxes classified as inner parts, the concept of Jaccard distance is used. Subsequently a distance matrix that contains the Jaccard distance values for all pixel pairs is created. Using this matrix, clustering is performed using the DBSCAN algorithm. After the processing by neural networks and clustering, we get a list of coordinates of each word. Minimum value of 'x', maximum value of 'x', minimum value of 'y', and maximum value of 'y' are stored (cf. Fig. 2). Using these coordinate values, the words are cropped and arranged in the correct order as provided in the input image.

Fig. 2. Coordinates of the word.

The y-axis values are used to sort words by line appearance and x-axis values are used to sort words by their occurrence in the line. This is done by first calculating a range of pixel values that correspond to the lines in the input image. The list of coordinates for each word is iterated over and then sorted into the different segments or ranges present in the page that signify the line that the words are a part of. The maximum value of range of each line can be calculated by multiplying the calculated range of pixel values and the number of the line currently being assessed. A buffer is defined to accommodate words that extend from the default range of each line. This is followed by the application of the bubble sort algorithm on the x-axis coordinates to sort the words and correctly place them. Figure 3 presents an example of the segmentation of a handwritten text.

3.2 Handwritten Text Recognition Model

The neural network structure contains 5 layers of CNN, followed by 2 layers of RNN, and finally, a Connectionist Temporal Classification layer or CTC. The last layer, CTC, is used to compute the loss value once the processing is done by the layers of CNN and RNN. Figure 4 describes the flow of the model. The time stamp in LSTM is referred to as a feature considered for designing the model.

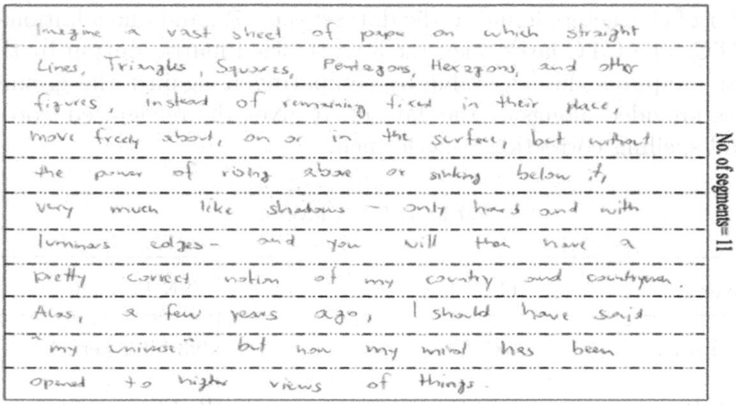

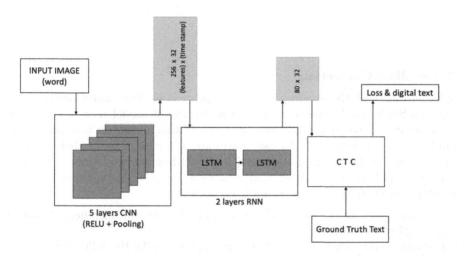

Fig. 3. Segments of Page.

Fig. 4. Handwritten text recognition model diagram.

The 5 layers of CNN in the model are each trained to extract the important features found in the input image. For the first two layers, during the convolutional operation, 5×5 sized kernel filter is applied, this is different from the 3×3 sized filter applied in the consequent 3 layers (Table 1). Next, application of the RELU function is done. RELU function can be described as:

$$R(Z) = \begin{cases} z, z > 0 \\ 0, z \leq 0 \end{cases} \tag{1}$$

Pooling layer downsizing in each layer is done by 2. Consequently, feature maps of size 32×256 are added. Long short-term memory or LSTM takes 32×256 feature map as input and produces a 32×80 sized matrix. The value 80 signifies the number of characters found in the dataset, i.e., 79, and one additional blank one for CTC. The CTC layer takes the 32×80 sized matrix output by the RNN layers and computes the loss value by comparing it against the ground truth text values provided alongside the dataset. It gives the recognized word that is stored, and spelling correction is performed.

Table 1. CNN layers.

CNN Layer 1	CNN Layer 2	CNN Layer 3
5x5 filter kernel	5x5 filter kernel	3x3 filter kernel
RELU function	RELU function	RELU function
Pooling layer (downsizing-2)	Pooling layer (downsizing-2)	Pooling layer (downsizing-2)

3.3 Spelling Correction

Implementation of the spelling correction process is done using the concepts from SymSpell and Phoneme models. The SymSpell model uses a Symmetric Delete spelling correction algorithm. The aim of this algorithm is to reduce the complexity of generation of edit candidates and the lookup in the dictionary for a given Damerau-Levenshtein distance [22]. The Phoneme model uses a Soundex algorithm which is an algorithm that indexes a name based on the sound of it. The steps for encoding a word using it are as follows:

- The first letter of the name is kept, and all the other occurrences of vowels, h, y and w are dropped
- The consonants after the first letter are replaced with the following
  - b, p, f, v with 1
  - g, k, j, c, s, q, z, x with 2
  - t, d with 3
  - l with 4
  - n, m with 5
  - r with 6
- If two or more adjacent letters have the same number, the first letter is retained. Letters with the same name that are separated by h, y or w, are coded as a single letter, and coded twice if separated by a vowel.
- If there are less than three numbers as the word is too short, 0s are appended until there are three numbers.

Here, we have initially subjected three to four letter words for correction. Later more complex scenarios like simple sentences such as "that is a bag", "this is a cat" are used. The dataset had many examples from above mentioned cases that were checked for analysis.

After the recognized handwritten text is spell corrected, it is processed to correct common grammatical mistakes.

3.4 Grammar Correction

For the process of correcting the grammar, the GEC tagging concept is implemented iteratively [18]. GEC sequence tagger made with a Transformer encoder that was pre-trained on a synthetic dataset and then fine-tuned. At each iteration the tags created by the GECToR model are processed to perform correction. This procedure is carried out until there are no further changes required. The base of this operation is set using a pre-trained encoder called RoBERTa.

3.5 Text-to-Speech Conversion

Text-to-speech conversion is performed to add the functionality of converting the processed text into audio format. This is done by using the Google Text to Speech API or TTS API. The extension of the output file is .mp3. This functionality provides the ability to allow the user of the system to choose the way he would prefer to receive the information; hence, text-to-speech offers learning benefits to all students, but especially to those with learning disabilities.

4 Experimental Evaluation

This section presents our experiments to show the effectiveness of our system to correct handwritten text including spelling and grammatical errors. We first present the datasets explored and used in the integrated system, then we present and discuss the results achieved by our proposal.

4.1 The Dataset

A vast database is required to develop a handwritten text recognition model to identify the basic features of handwritten text. In order to effectively evaluate our model, we have used the IAM dataset [15] and a self-created dataset built based on the observation of children with learning disabilities.

The IAM Handwriting Database was initially created for offline handwritten text recognition. It contains over 1500 pages of English text containing around 5600 sentences. The IAM Dataset was used to train the handwritten text recognition model to recognize the peculiar dysgraphic handwriting.

To collect handwritten images for children with learning disabilities and to build a real dataset for our experimental evaluations, we have approached the Shree Learning Centre, Sector 45, Noida, India. This center is a special learning school for children of all age groups who face difficulties due to learning disabilities and other disorders. Students here showed great enthusiasm for participating in the creation of the dataset. The age group of students varied from 5 to 12 years. The sample group size was approximately 40 students. Each student was given blank sheets and was asked to copy some sentences from their English textbooks. Required permissions from parents were collected for the usage of data for experimental purposes only.

The final collected dataset contains simple sentences using three to four letter words, usage of vowels and sight words. It has more than 100K words along with a text file containing the path to each image along with its ground truth text. The images of words are cropped precisely, and the contrast is significant.

In our work, the IAM dataset was used for the learning phase and the self-created dataset was used for the evaluation of the proposed model.

4.2   Results and Discussion

One of the first processes includes word segmentation which is performed on the input image. Figure 5 visually represents the words found in a handwritten document using bounding boxes during the process of word segmentation.

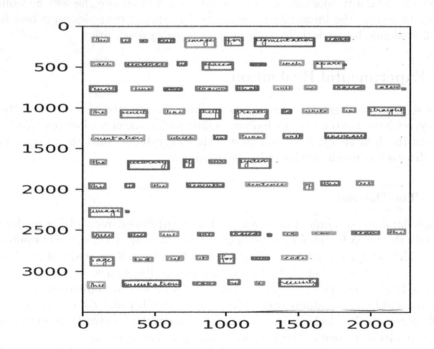

Fig. 5. Word Segmentation- Neural Networks.

To recognize the segmented words, the CNN-RNN-CTC model is trained on the merged IAM dataset and self-created dataset. Regarding the performance, the model is successfully trained as the process runs till 84 epochs, where the lowest character error rate is found at epoch 59. Lowest observed character error rate was 10.2% and the word error rate was noted to be 25% (cf. Fig. 6). The model classifies text whether it is written by a dyslexic pupil or not. Later grammar correction is done for the text written by dyslexic pupils using GECToR model. Figure 6 signifies the location of these metrics on the character error rate vs epoch plot and the word accuracy vs epoch plot. Similar plots were made for a CNN-RNN-CTC model that consisted of 7 layers of CNN instead of 5. Comparison of the two shows that the model used has a lower character error rate and word error rate than the one with 7 layers of CNN.

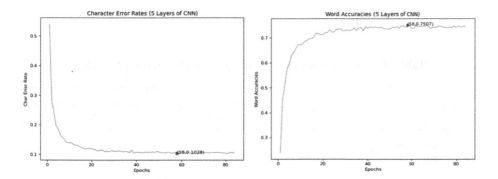

Fig. 6. Training Plots for 5 Layers of CNN.

Sequence-2-Sequence approach for handwritten text recognition was also explored and comparisons were drawn between the model used, 7 layers CNN-RNN-CTC model, and the Sequence-2-Sequence model (cf. Fig. 7). Table 2 represents the values of Character error rate, Word error rate, and the workflow of three handwritten text recognition models, Neural Networks, Sequence-2-Sequence and a 7 layers CNN. We can see that the character error rate for the Sequence-2-Sequence model is 12.62% and its word error rate is 26.65%. In comparison to this, the model used has a character error rate of 10% and word accuracy of 25%. The third model in the table is the altered version of the selected model.

Spelling correction and grammar correction modules were observed to mostly make the expected corrections. Grammar correction module yielded a confidence bias of 0.2 which had a significant improvement on the recognized text given out by the handwritten text recognition module. Table 3, shows scenarios where the input images are sentences with significantly poor handwriting, grammar mistakes, or both. The columns, Handwritten Text Recognition, Spelling Correction and Grammar Correction, in the table specify the different outputs given out by the three modules composing our system.

It can be observed that due to the presence of the spelling correction module, the anomalies like inverted letters are being corrected in the first scenario and

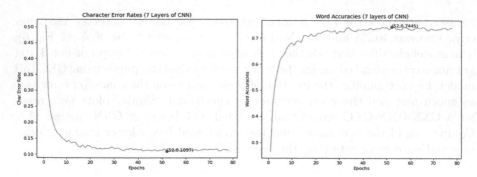

Fig. 7. Training Plots for 7 Layers of CNN.

Table 2. Comparison of Handwritten Text Recognition Models.

Model	Neural Networks	Sequence-2-Sequence with keras	Neural Networks (7 layers CNN)
Character error rate (CER)	10.2%	12.62%	10.9%
Word error rate (WER)	25%	26.65%	25.6%
Workflow	CNN (5 layers) RNN (2 layers of LSTM) CTC decode	RNN layer (as encoder) Another RNN layer (as decoder)	CNN (7 layers) RNN (2 layers of LSTM) CTC decode

Table 3. Comparison of Handwritten Text Recognition Models.

N°	Input Image	Expected output	Handwritten Text Recognition	Spelling Correction	Grammar Correction
1	The dottle is yelow	The bottle is yellow.	The . dottle . . . us a . yelow .	The . bottle . . . us a . yellow .	The bottle is yellow.
2	Thay have lernd a story	They have learned a story	Thay have leroad a story	They have lead a story	They have lead a story.
3	German pepl are hardworking.	German people are hardworking	German pepl awe handworking	German people awe handwriting	German people are hardworking.

the output is similar to the expected output. Output for the second is observed to be different from the expected output due to incorrect word recognition by the handwritten text recognition module, as consequence, the word 'lernd' is not converted to 'learned'. In the third scenario, the grammar correction module can be observed to be responsible for making the text resemble the expected output by correcting 'handwriting' to 'hardworking'.

5 Conclusion

Existing approaches for assisting dysgraphic persons include, generally, only one aspect of the problem faced by the patients into consideration, writing assistance, spelling or grammatical problems. In this work, a novel system for assisting persons suffering from dysgraphia, that addresses several aspects facing these persons is presented.

The proposal allows recognizing handwritten dysgraphia text, processes it and converts it into corrected digital text. The proposal consists of four main processes, namely, handwritten text recognition, spelling correction, grammar correction, and text-to-speech conversion. The integration of these processes provides a solution that accommodates more than one issue, like poor handwriting skills, poor spelling, and grammar skills. It can prove to be beneficial to educators, parents, colleagues, and guardians.

The experimental evaluations showed that our proposed system works well as an integrated system to correct poor handwritten text.

In our work in progress, we are considering the evolutions' monitoring of dysgraphic persons that are subjects to medical treatment or using our system. This would allow the evaluation of the improvement made by dysgraphic persons based on intelligent systems.

References

1. Ariffin, M., Othman, T., Aziz, N., Mehat, M., Arshad, N.: Dysgraphi coach: mobile application for dysgraphia children in Malaysia. Int. J. Eng. Technol. (UAE) **7**, 440–443 (2018). https://doi.org/10.14419/ijet.v7i4.36.23912
2. Asselborn, T., Chapatte, M., Dillenbourg, P.: Extending the spectrum of dysgraphia: a data driven strategy to estimate handwriting quality. Sci. Rep. **10**(1), 3140 (2020)
3. Asselborn, T., Gargot, T.: Automated human-level diagnosis of dysgraphia using a consumer tablet. npj Digit. Med. 1 (2018). https://doi.org/10.1038/s41746-018-0049-x
4. Avishka, I., Kumarawadu, K., Kudagama, A., Weerathunga, M., Thelijjagoda, S.: Mobile app to support people with dyslexia and dysgraphia. In: 2018 IEEE International Conference on Information and Automation for Sustainability (ICIAfS), pp. 1–6 (2018). https://doi.org/10.1109/ICIAFS.2018.8913335
5. Barnett, A., Henderson, S., Scheib, B., Schulz, J.: Handwriting difficulties and their assessment in young adults with DCD: extension of the dash for 17-to 25-year-olds. J. Adult Dev. **18**, 114–121 (2011). https://doi.org/10.1007/s10804-011-9121-3

6. Boyle, J., Joyce, R.: Using smartpens to support note-taking skills of students with learning disabilities. Interv. Sch. Clin. **55**, 105345121983764 (2019). https://doi.org/10.1177/1053451219837642
7. Brown, T., Lalor, A.: The movement assessment battery for children-second edition (MABC-2): a review and critique. Phys. Occup. Ther. Pediatr. **29**(1), 86–103 (2009)
8. Danna, J., Paz-Villagrán, V., Velay, J.L.: Signal-to-noise velocity peaks difference: a new method for evaluating the handwriting movement fluency in children with dysgraphia. Res. Dev. disabil. **34** (2013). https://doi.org/10.1016/j.ridd.2013.09.012
9. Danna, J., Velay, J.L.: Handwriting movement sonification: why and how? IEEE Trans. Hum. Mach. Syst. **47**, 299–303 (2017). https://doi.org/10.1109/THMS.2016.2641397
10. Deuel, R.K.: Developmental dysgraphia and motor skills disorders. J. Child Neurol. **10**(1_suppl), S6–S8 (1995)
11. Diem, M., Fiel, S., Garz, A., Keglevic, M., Kleber, F., Sablatnig, R.: ICDAR 2013 competition on handwritten digit recognition (HDRC 2013). In: 12th International Conference on Document Analysis and Recognition, ICDAR 2013, Washington, DC, USA, 25–28 August 2013, pp. 1422–1427. IEEE Computer Society (2013). https://doi.org/10.1109/ICDAR.2013.287
12. Khan, M., et al.: Augmented reality based spelling assistance to dysgraphia students. J. Basic Appl. Sci. **13**, 500–507 (2017). https://doi.org/10.6000/1927-5129.2017.13.82
13. Li, H., Wang, Y., Liu, X., Sheng, Z., Wei, S.: Spelling error correction using a nested RNN model and pseudo training data. CoRR abs/1811.00238 (2018). http://arxiv.org/abs/1811.00238
14. Magalhaes, L., Cardoso, A., Missiuna, C.: Activities and participation in children with developmental coordination disorder: a systematic review. Res. Dev. Disabil. **32**, 1309–16 (2011). https://doi.org/10.1016/j.ridd.2011.01.029
15. Marti, U.V., Bunke, H.: A full English sentence database for off-line handwriting recognition. In: Proceedings of the Fifth International Conference on Document Analysis and Recognition. ICDAR 1999 (Cat. No.PR00318), pp. 705–708 (1999). https://doi.org/10.1109/ICDAR.1999.791885
16. McCutchen, D.: From novice to expert: implications of language skills and writing-relevant knowledge for memory during the development of writing skill. J. Writ. Res. **3**(1), 51–68 (2011)
17. Mekyska, J., Faúndez-Zanuy, M., Mzourek, Z., Galaz, Z., Smékal, Z., Rosenblum, S.: Identification and rating of developmental dysgraphia by handwriting analysis. IEEE Trans. Hum. Mach. Syst. **47**(2), 235–248 (2017). https://doi.org/10.1109/THMS.2016.2586605
18. Omelianchuk, K., Atrasevych, V., Chernodub, A.N., Skurzhanskyi, O.: Gector - grammatical error correction: tag, not rewrite. In: Burstein, J., et al. (eds.) Proceedings of the Fifteenth Workshop on Innovative Use of NLP for Building Educational Applications, BEA@ACL 2020, Online, 10 July 2020, pp. 163–170. Association for Computational Linguistics (2020). https://doi.org/10.18653/v1/2020.bea-1.16
19. Pagliarini, E., et al.: Children's first handwriting productions show a rhythmic structure. Sci. Rep. **7** (2017). https://doi.org/10.1038/s41598-017-05105-6
20. Quenneville, J.: Tech tools for students with learning disabilities: infusion into inclusive classrooms. Preventing Sch. Fail. **5**, 167–170 (2001). https://doi.org/10.1080/10459880109603332

21. Reisman, J.E.: Development and reliability of the research version of the Minnesota handwriting test. Phys. Occup. Ther. Pediatr. **13**(2), 41–55 (1993)
22. Setiadi, I.: Damerau-Levenshtein algorithm and Bayes theorem for spell checker optimization (2013). https://doi.org/10.13140/2.1.2706.4008
23. Smits-Engelsman, B., Schoemaker, M., Delabastita, T., Hoskens, J., Geuze, R.: Diagnostic criteria for dcd: past and future. Hum. Mov. Sci. **42**, 293–306 (2015). https://doi.org/10.1016/j.humov.2015.03.010. https://www.sciencedirect. com/science/article/pii/S0167945715000524
24. Sutskever, I., Vinyals, O., Le, Q.V.: Sequence to sequence learning with neural networks. In: Ghahramani, Z., Welling, M., Cortes, C., Lawrence, N.D., Weinberger, K.Q. (eds.) Advances in Neural Information Processing Systems 27: Annual Conference on Neural Information Processing Systems 2014(December), pp. 8–13, Montreal, Quebec, Canada, pp. 3104–3112 (2014). https://proceedings.neurips.cc/ paper/2014/hash/a14ac55a4f27472c5d894ec1c3c743d2-Abstract.html
25. Toutanova, K., Moore, R.C.: Pronunciation modeling for improved spelling correction. In: Proceedings of the 40th Annual Meeting of the Association for Computational Linguistics, 6–12 July 2002, Philadelphia, PA, USA, pp. 144–151. ACL (2002). https://doi.org/10.3115/1073083.1073109. https://aclanthology.org/P02-1019/
26. Zhou, X., et al.: EAST: an efficient and accurate scene text detector. In: 2017 IEEE Conference on Computer Vision and Pattern Recognition, CVPR 2017, Honolulu, HI, USA, 21–26 July 2017, pp. 2642–2651. IEEE Computer Society (2017). https:// doi.org/10.1109/CVPR.2017.283
27. Zolna, K., et al.: The dynamics of handwriting improves the automated diagnosis of dysgraphia (2019)

Universal Machine-Learning Processing Pattern for Computing in the Video-Oculography

Albert Śledzianowski[✉], Jerzy P. Nowacki, Konrad Sitarz, and Andrzej W. Przybyszewski

The Faculty of Information Technology, Polish-Japanese Academy of Information Technology, 02-008 Warsaw, Poland
asledzianowski@gmail.com

Abstract. In this article, we present a processing pattern dedicated to video-oculography. It is a complete solution that allows for checking the external conditions accompanying the video oculography, conducting oculometric measurements based on different test models, estimating eye movement (EM) angles and detecting EM type in the stream of coordinates, and calculating its parameters. We based our architecture on neural networks (NN), machine-learning (ML) algorithms of different types, parallel/asynchronous computing, and compilations of the models, to achieve real-time processing in oculometric tests.

Oculometric tests provide significant insight into central neuro-motor states, but machine-learning methods are needed to estimate their meaning. A limitation of this cognitive-analytical trend was the reliance on dedicated measuring devices, such as eye-trackers, which are usually separate and expensive equipment. Presented approach goes beyond these limitations and was developed to use standard home computer equipment, with an internet connection and computer camera. Our set of dedicated algorithms embedded in the software compensates for hardware limitations.

We tested the results of the presented solution on reflexive saccades (RS) and a standard 30 frames per second (FPS) web-camera, and we were able to distinguish between young and old persons and between healthy and prodromal neurodegenerative (ND) subjects by analyzing RS parameters. Visual processes are connected to many brain structures and, by using ML methods, we are trying to dissect them. The development of solutions like the one presented in this article gives hope for the general availability of screening tests connected to ND risk and for collecting extensive data outside the laboratory.

We hope that this direction will contribute to the development of ND analytic means in computational health and consequently, to the faster development of new ND preventive measures.

Keywords: machine learning · deep learning · eye moves · eye tracking · eye move computations · video oculometry · computer vision · reflexive saccades · saccade detections · neurodegenerative diseases · Parkinson's disease

© The Author(s), under exclusive license to Springer Nature Switzerland AG 2023
J. Mikyška et al. (Eds.): ICCS 2023, LNCS 14075, pp. 200–212, 2023.
https://doi.org/10.1007/978-3-031-36024-4_15

1   Introduction

The video-oculography supported by computer vision algorithms introduces a non-invasive and economically efficient method of eye tracking based on standard cameras, rather than dedicated devices. It allows for measuring the same components of the EM, and thanks to the possibility of using standard devices such as computer webcams, it brings oculometry to widespread use.

Many researchers experiment with models and new approaches to estimate EM and gaze direction. It has already been confirmed that web-based eye tracking is suitable for studying fixation, pursuit eye movements, and free viewing with repeatability of well-known gazing patterns with slightly less accuracy and higher variance of the data compared to laboratory devices [1]. Furthermore, in the case of RS, such webcam systems are able to calculate parameters with sufficient results, allowing for clinical diagnosis and identification of neurological disorders such as Multiple Sclerosis (MS) [2]. Other researchers have proved that in terms of accuracy, webcam systems can be equal to infrared eye-trackers with 1000 Hz Hz sampling frequency [3].

The core elements of each video-oculographic system are algorithms associated with eye position estimation. There are many different approaches; Lin et al. experimented with estimation based on the appearance-features of eyes, Fourier Descriptors, and the Support Vector Machine, combining a position criterion, gaze direction detection, and lighting filtering [4]. Xu et al. experimented with grayscaling and histogram equalization, creating a 120D feature set for linear regression [5]. However, for the last period of time, for both computer vision and video-oculography, the golden standard has been Convolutional Neural Network (CNN) models. CNNs are based on convolution operations, which, in short, sample every possible pixel arrangement to find specific patterns. In case of video-oculography, this pattern will include the eye component, including the iris and the most important pupils. Akinyelu et al. based their estimation on the face component to extract the gaze features from the eyes, and a 39-point facial landmark component was used to encode the shape and location of the eyes into the network. Experiments confirmed better results of CNN in comparison to the Visual Geometry Group (VGG) NN [6]. Meng et al. experimented with CNN and webcams for an eye-tracking method based on detecting 6 eye features [7]. Gunawardena et al. explored this subject in terms of the efficiency of 4 lightweight CNN models for eye tracking: LeNet-5, AlexNet, MobileNet, and ShuffleNet. They indicate that MobileNet-V3 and ShuffleNet-V2 are the most suitable CNN models [8]. Harenstam-Nielsen experimented with adding Long-Short Term Memory (LSTM) cells to the convolutional layers and Recurrent Neural Network (RNN) cells to the feature layers to increase eyetracking performance [9]. Here, the importance of "memory" modeling in the context of the CNN should be emphasized. For example, the convolutional layer of a model trained for face detection might encode the information that "eyes" are present in the input image, but the receptive field will not have any information to explain whether the eyes are above the nose or not. This is because each convolutional kernel will not be large enough to process many features at one trial.

The cost of tracing wide-range dependencies within an image is ineffectiveness and weight of the model, which is most important for real-time processing. This is why several separate models are used for eyetracking based on the CNN architecture. Usually, this process starts from face detection, then facial landmarks are estimated to determine the location of the eyes. Finally, the image of the eyes is entered into the final network to determine gaze directions.

This issue seems to be already been resolved by the self-attention approach introduced first in transformer-based architectures, which are well known from neural language processing. In computer vision, this architecture takes a feature map as an input parameter and computes relations between each feature based on its weights, resulting in shared information about each object in the image. However, vision transformers in video-oculography are still a very fresh subject and will be elaborated in the next work, as CNN, despite its disadvantages, with the right approach allows for very fast and efficient video-oculographic processing. This is the subject of this text and general purpose was to propose a universal processing pattern based on our best knowledge, which would cover and resolve well-known issues and allow for easy embedding and conducting all standard oculometric tests. We hope that the proposed solution will contribute to the dissemination of oculometric tests also in the context of ND diagnosis.

2   Methods

We developed a pattern for general use in video-oculographic processing constructed of 3 base steps,

1. Analysis of information about sufficient/insufficient equipment or environmental conditions
2. Face detection and facial components estimations including estimation of gaze direction
3. EM estimation including signal smoothing and EM type detection and computations of its parameters.

The description of this processing pattern has been presented in subsequent sections and on Fig. 1.

2.1   Video Signal Quality Check

The environmental factors such as poor lighting or equipment, glasses on the subject's face, or insufficient concentration on the stimulus can bring an unacceptable quality of the signal and recordings of the EM test. This is why we decided to introduce several methods estimating quality components and established threshold values to validate the EM signal. Firstly, we set the frequency level threshold to be ~25 FPS, as our previous experiments showed that in lower frequencies it is impossible to calculate the parameter within the acceptable error range, due to an enormous temporal sample error. As explained in [3], in the case

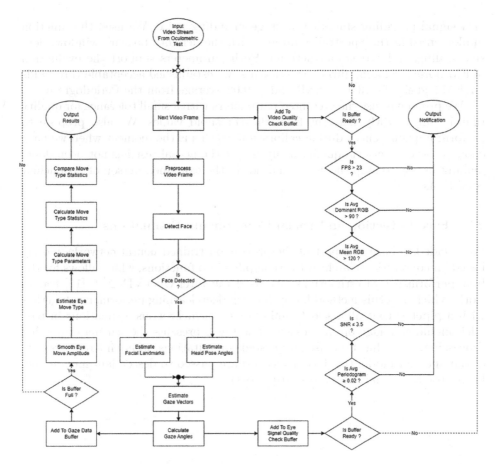

Fig. 1. The process pattern schema for video-oculometric test.

of web cameras, 30 FPS would be ideal, but many factors, such as insufficient light, can affect the video frequency.

This is why we determine lighting conditions from video signals by calculating the dominant and average colors in an image converted to grayscale. The dominant color is estimated with clustering by the k-means function implemented in the OpenCV library with 5 colors and selecting random initial centers in each attempt [10]. The mean color is just calculated as the mean value along the bitmap columns. These two parameters indicate acceptable lighting conditions in presented approach. We have experimentally determined that the average value obtained from the RGB of the dominant color should not be lower than 90 and the mean color should not be lower than 120. Similarly, for video signal estimation, we decided to use two parameters for noise level measurement of EM signals: Signal to Noise Ratio (SNR) as a simple relationship between the horizontal axis EM mean and its statistical deviation (SD). We also calculated the Periodogram of the horizontal axis EM, which estimates the spectral density of a signal providing statistical average including noise. We used the function implemented in the SpectralPy library with the default "boxcar" window, density scaling, and "constant" detrend. Both parameters support the evaluation of EM noise level and, during experiments, we established acceptable thresholds for EM signals of ≤ 3.5 for SNR and ≥ 0.02 average from the Periodogram.

We propose to use the presented parameters with a small tolerance for quality that is only slightly different from the threshold values. We also propose to perform a quality check during calibration, as this is the moment when we have enough data to perform quality analysis, yet the actual test has not been done, making it easy to stop the study and notify the examined person of insufficient conditions.

2.2 Face Detection and Facial Components Estimations

We decided to base this part of the process on trained neural networks mostly based on convolutional architecture compiled into functions, which allows for the best performance of estimations. We used the Intel OpenVINO™ [11] framework, which uses this method for optimizing deep learning performance together with a pipeline that allows for running many asynchronous estimations in parallel, including image pre-processing. The Fig. 2 presents the schema of parallel processing, which has been used for nested computations, where inference results of the ancestor model (i.e. face detection) are passed to descendant models (i.e. head pose and facial landmarks estimators).

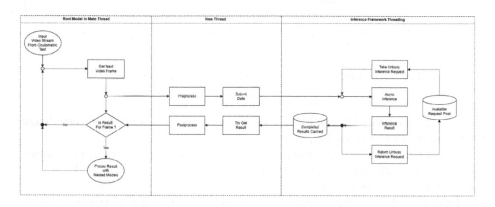

Fig. 2. The schema of parallel processing of nested models.

Additionally, we tried to select the lightest and quickest NN architectures we could find to minimize estimation latency. In order to estimate the eye angles, we connected four models working together in one sequence. The face detector finds the face in the main image from the camera. Next, we smooth the face bounding box coordinates to reduce jittering caused by the detector's uncertainty range with Weighted Moving Average (WMA) with a weight value set to 0.5.

Then, the cropped face image is sent in parallel to the next two models to estimate the position of the head relative to the camera lens and to estimate facial landmarks to find the eyes in the face image. Finally, gaze estimation is performed based on the images of both eyes and the head position angles. Pre-processing includes mostly resizing and rotating the image and reshaping the bitmap matrix into the model input [10,12], post-processing includes mostly reshaping the model outputs into a more convenient format and calculations, i.e. trigonometric conversion from gaze vectors into gaze angles.

For face detection, we used the lightweight "face-detection-0205" model, based on the MobileNetV2 scaffolding and Fully Convolutional One-stage Object Detection (FACOS) as a head [13]. The model was trained with indoor and outdoor scenes shot by a front-facing camera and outputs bounding boxes with 93.57% of average precision (AP) for faces larger than 64×64 pixels. For facial landmarks estimation, we used the "facial-landmarks-35-adas-0002" model [14]. The CNN was trained with a 1000-sample random subset of different facial expressions and outputting a row-vector of 70 floating point values (x, y) for normalized 35 landmarks coordinates with a Normed Error (NE) of 0.106 [13]. For head pose estimation, we used the "head-pose-estimation-adas-0001" model [15]. The CNN had ReLU activation function, batch normalization and angle regression layers as convolutions, fully connected with one output returning Tait-Bryan angles describing rotation in Euclidean space with an accuracy of: yaw 5.4 ± 4.4, pitch 5.5 ± 5.3 and roll 4.6 ± 5.6 [15]. For gaze estimation, we used the "gaze-estimation-adas-0002" model [16]. The Visual Geometry Group CNN model for gaze direction estimation outputs a 3-D vector in a Cartesian coordinate system (in which the z-axis is directed from the mid-point between eyes to the camera center, y-axis is vertical, and x-axis is perpendicular to both z and y) with a Mean Absolute Error (MAE) of 6.95 ± 3.58 of angle in degrees [16].

2.3   Eye Move Estimations

For the purposes of this study, we chose RS as one of the most common types of eye movement performed by humans thousands of times a day, which is also of great importance in oculometric research related to ND [17–25]. In presented approach, RS is stimulated by alternately displaying the fix-point and one of the left/right peripheral targets (markers) in random time intervals (in a step/no-delay model). All are green ellipses in the horizontal axis at a distance of about 8°, and the subjects sat at a distance of about 55–70 cm from the camera/monitor. The study included 20 RS tests performed by each subject, Fig. 3 presents the RS test pattern used in our experiment.

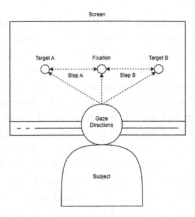

Fig. 3. The reflexive saccades test pattern schema.

For a more accurate estimation of the RS, the signal was smoothed. In presented approach, based on the EM fluctuation analysis, smoothing might be conducive to an accurate determination of the start and end of the amplitude peak. We smoothed the signal with a low-pass Butterworth 2nd order filter with a cutoff of 3.5, and for this purpose, we used the implementation of this function included in the SciPy library [26]. Smoothing was performed outside the RS estimated initially on the raw data, to prevent changing the shape of the RS amplitudes.

For RS detection, we implemented a rolling window function that estimates this type of EM in time series of gaze angles based on stabilization, deviation, and changes in the amplitude. In addition, we introduced a signal of markers to the algorithm for estimations of the RS latency and gain parameters. The algorithm first calculates fluctuation in the fixation state (as EM mean shift in the x-axis) in the small (300 ms) control window (fixation state window) preceding changes in the state of the marker. Then, the size of the next window is calculated with the starting point set between the moment when the peripheral target appears and below the possible minimal RS latency (40 ms) depending on the data frame rate. From the starting point, the window end is set to the maximum RS length (500 ms for ∼8°). Next, the window is iterated in search of longer amplitude change with minimal duration for longer RS (50 ms) towards the displayed target. For such amplitude change to be classified as start of the RS, the shifts between data points must be longer than the mean fluctuation in the fixation state window and must be ≥30% of the distance between fix-point and the peripheral target. With the assumption of the min RS duration (50 ms) algorithm looks for the data-point in which position relative to the previous point will be <30% which is considered as minimal inhibition of the RS and also its ending point. If no starting or ending point was found, a trail is classified as failed. For each detected RS, we calculated the following parameters: latency as the time difference between the start of the marker movement and the start of the eye movement, duration as the total movement time period, amplitude as the total EM angle, average and maximum velocity as the respectively average and maximum angular velocity, and gain as the ratio between the RS amplitude and fix-point-target distance.

2.4 Experimental Application

We evaluated the presented approach by creating a sample web application in which we implemented our processing pattern. Figure 4 presents its pipeline. In the first stage (Head Position Validation), we adjusted the head position using the camera view with landmarks markers applied to the subject's face. We estimated the distance between the head and the monitor by computing the distance between the subject's center of eyes, received from facial landmarks coordinates. We decided to accept an approximated range of 55–80 cm, allowing the subject to be able to use the mouse or keyboard, but not too far as the face image had to be large enough. We also estimated the perpendicular head position determined on the basis of the vertex positions of the face bounding box. We implemented this feature to obtain registration of subjects in the same head position. The next stage was a calibration. We decided on the simplest three-point schema mapping only the position of the fixation point and the peripheral targets in physical space. During calibration, we also computed the parameters of the video signal quality described in Subsect. 2.1. In all cases, the application reported poor or satisfactory parameter values, enabling corrections to be made. After successful completion of these stages, the main oculometric test for RS was performed, then RS were detected in the registered EM signal and their parameters were estimated as described in Sect. 2.3. Finally, the results were presented on the screen and recorded in a file system of a server.

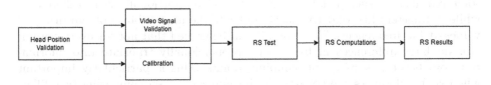

Fig. 4. The pipeline of the experimental application.

3 Results

We wanted to evaluate the presented processing pattern by using application described in previous section. We tested it with 8 subjects in range of 23–45 years old and different gender (158 RS samples). The group was small, but our goal was not to build a statistically large population of different individuals. Rather, we only wanted to collect data from sample equipment and examples of external conditions that may accompanying the study.

We also wanted to reference our results with laboratory measurements and laboratory-grade devices. We decided to compare our results with those published in the data sheet "Descriptive and Reproducibility Results of the Pro-Saccadic Task", where 8° horizontal RS were tested (28 subjects) [27]. All measurements in this research were performed with a professional infrared eye tracker, the Eyelink 1000 Plus, with a frequency of 1000 Hz Hz [27]. In our measurements we used a standard but good quality webcam, the Logitech C922 Pro Stream, at 30 FPS with Full HD mode.

For latency (ms), we obtained a mean of 192 ± 45, for gain (%), 1.16 ± 0.24, for average velocity we received means of 88 deg/sec $\pm$ 33.78, and for maximum velocity 145 (deg/s) $\pm$ 52. In the reference data, the latency mean was 176 ± 21 (a difference of 16 ms), the gain was 0.98 ± 0.05 (a difference of 0.18), and the maximum velocity was 342 deg/s $\pm$ 44 (a difference of approximately 200 ms) [27]. Table 1 presents comparison between experimental and reference statistics. All differences were statistically significant ($P < 0.05$), which is evident given the differences in frequency.

Table 1. Comparison between experimental and reference statistics.

Source	Latency (ms)	Gain (%)	Max Velocity (deg/s)
Webcam 30 Hz - experimental data	192 ± 45	1.16 ± 0.24	145 ± 52
Eye-tracker 1000 Hz - reference data	176 ± 21	0.98 ± 0.05	342 ± 44
Difference	16 ± 48.3	0.18 ± 0.25	197 ± 67.3

We can see that even with such a large difference in sampling rate (970 Hz), some parameters, such as latency, do not show proportionally large differences, while parameters like velocity clearly indicate the impossibility of capturing all values at low frequencies. Therefore, it can be argued that not for every parameter do large differences in frequencies proportionally translate into information loss between subsequent sampling points. This is particularly important when such differences are statistically insignificant when comparing two different groups of subjects, as indicated in the next section of this text.

However, in our system, the processing frequency depends only on the hardware capabilities. This is a fundamental limitation of an approach like ours that cannot be minimized by techniques such as synthetic oversampling, because no two oculomotor systems are the same and artificial data generated from statistical analysis will always be inadequate. In terms of eye tracking accuracy, with the very fast but not super accurate model we used, the results seem to be satisfactory. The difference in gain between our results (1.16 ± 0.24) and the results of the reference study (0.98 ± 0.05) does not seem to be great. Although gain expresses the differences between EM and marker amplitudes, which strongly depends on the subject, in the case of fundamental differences in accuracy, the difference in the resulting gains would certainly be much greater.

However, in terms of frequency, the results for a 100, 300, or 1000 FPS camera would be completely different due to temporal sampling error. The data are collected at some finite frequency, and if it is lower than 1000 Hz, changes in speed or direction of eye movement are not registered between two consecutive sampling points. As 1 ms is the maximal one-point temporal sampling error at a 1000 Hz [28] and the sampling error will be half of the sample time value $((1000/x)/2)$ we can calculate that 300 Hz this error would be 1.6, 100 Hz 5, and 30 Hz 16.6 ms, which corresponds to the difference between results from reference data we received for latency. It must be noted that the data frequency for the processing speed in the proposed pattern seems to be transparent, thanks to the use of parallel/asynchronous estimations and fast models with compiled form. This does not change the fact that in this approach, the frequency of data is independent from the processing, and in this context, the process may be responsible at most for notifications to the users of the system.

4   Discussions

The phenomenon of temporal sample error in lower frequencies, so clearly visible in the case of velocity results, is well described by Anderson et al. [28]. However, its effect has dissimilar influences on different parameters, as shown by the latency where, with a huge difference in sampling frequency, the nominal difference in parameter values is very small. When it comes to the classification of young and old people, or healthy people and ND patients, such an error margin may not affect the results. For example, in one of our previous studies on RS task with PD patients, specific disease stages (varying in progression) showed latency thresholds of 260.0 ms and 308.5 ms [29,30], while the healthy subjects for the RS show latency between 190–200 ms [31]. In this context, it is unlikely that any classifier will go wrong with such large differences. Similarly, for the classification of old and young, or other reasons for deviating from the average results of a regular person, high data rates may not be necessary, so standard computer hardware may be sufficient.

Therefore, it is important to be aware of the type of study and its requirements for sampling rate. Thus, one should be aware of the selection of the appropriate camera for an experiment or patient examination. For computational solutions such as ours, the sampling rate is transparent and depends on the equipment and lighting, which are also evaluated in presented approach by methods described in Sect. 2.

5   Conclusions

In this text, we presented a pattern for processing in modern video-oculometric applications, allowing us to get satisfactory results using consumer-grade computer equipment. We propose using this pattern also with standard web-cameras, where high frequency is not required and approximate results are enough to make assumptions. Because of the limitations of consumer-grade equipment, it will never compete in accuracy with laboratory equipment.

However, we have proved that it can bring reliable results, such as for latency in PD, which could be used for screening tests of ND. Due to the availability of computer devices, applications using this pattern may disseminate oculometric examinations, making them universally available.

For more accurate applications, high frame rate equipment is needed, but our solution is fully scalable and, in this context, parallel processing adapts to the capabilities of the video device.

Declaration of Competing Interest

Software frameworks and libraries: OpenVino, Tensorflow, OpenCV, SciPy, Sci-kit Learn, Numpy, Pandas, contributed to the development of the software methods presented in this article. The authors declare no conflict of interests.

References

1. Semmelmann, K., Weigelt, S.: Online webcam-based eye tracking in cognitive science: a first look. Behav. Res. Methods **50**(2), 451–465 (2017). https://doi.org/10.3758/s13428-017-0913-7
2. Aljaafreh, A., Alaqtash, M., Al-Oudat, N., Abukhait, J., Saleh, M.E.: A low-cost webcam-based eye tracker and saccade measurement system **14**, 04 (2020)
3. Naruniec, J., et al.: Webcam-based system for video-oculography. IET Comput. Vis. **11**(2), 173–180 (2017)
4. Lin, Y.-T., Lin, R.-Y., Lin, Y.-C., Lee, G.C.: Real-time eye-gaze estimation using a low-resolution webcam. Multimedia Tools Appl. **65**, 543–568 (2012)
5. Xu, P., Ehinger, K.A., Zhang, Y., Finkelstein, A., Kulkarni, S.R., Xiao, J.: TurkerGaze: crowdsourcing saliency with webcam based eye tracking (2015)
6. Akinyelu, A.A., Blignaut, P.: Convolutional neural network-based technique for gaze estimation on mobile devices. Frontiers Artif. Intell. **4** (2022)
7. Meng, C., Zhao, X.: Webcam-based eye movement analysis using CNN. IEEE Access **5**, 19581–19587 (2017)
8. Gunawardena, N., Ginige, J.A., Javadi, B., Lui, G.: Performance analysis of CNN models for mobile device eye tracking with edge computing. Procedia Comput. Sci. **207**, 2291–2300 (2022). Knowledge-Based and Intelligent Information And Engineering Systems: Proceedings of the 26th International Conference KES2022
9. Harenstam-Nielsen, L.: Deep convolutional networks with recurrence for eye-tracking [internet] [dissertation] (2018)
10. Bradski, G.: The OpenCV library. Dr. Dobb's J. Softw. Tools (2000)
11. Intel®. Intel® distribution of openvino™ toolkit (2022). https://docs.openvino.ai/
12. Harris, C.R., et al.: Array programming with NumPy. Nature **585**, 357–362 (2020)
13. Intel®. Openvino™ toolkit - open model zoo repository, Intel's pre-trained models (2023). https://docs.openvino.ai/latest/omz_models_model_face_detection_0205.html

14. Intel®. Openvino™ toolkit - open model zoo repository, Intel's pre-trained models (2022). https://docs.openvino.ai/latest/omz_models_model_facial_landmarks_35_adas_0002.html
15. Intel®. Openvino™ toolkit - open model zoo repository, Intel's pre-trained models (2022). https://docs.openvino.ai/2019_r1/_head_pose_estimation_adas_0001_description_head_pose_estimation_adas_0001.html
16. Intel®. Openvino™ toolkit - open model zoo repository, Intel's pre-trained models (2022) https://docs.openvino.ai/2019_r1/_gaze_estimation_adas_0002_description_gaze_estimation_adas_0002.html
17. Chambers, J.M., Prescott, T.J.: Response times for visually guided saccades in persons with Parkinson's disease: a meta-analytic review. Neuropsychologia 48(4), 887–899 (2010)
18. Turner, T., Renfroe, J., Delambo, A., Hinson, V.: Validation of a behavioral approach for measuring saccades in Parkinson's disease. J. Motor Behav. 49, 1–11 (2017)
19. Stuart, S., et al.: Pro-saccades predict cognitive decline in Parkinson's disease: ICICLE-PD. Mov. Disord. 34(11), 1690–1698 (2019)
20. Perneczky, R., Ghosh, B.C.P., Hughes, L., Carpenter, R.H.S., Barker, R.A., Rowe, J.B.: Saccadic latency in Parkinson's disease correlates with executive function and brain atrophy, but not motor severity. Neurobiol. Dis. 43(1), 79–85 (2011). Autophagy and Protein Degradation in Neurological Diseases
21. Antoniades, C., Zheyu, X., Carpenter, R., Barker, R.: The relationship between abnormalities of saccadic and manual response times in Parkinson's disease. J. Parkinsons Dis. 3, 10 (2013)
22. Yang, Q., Wang, T., Su, N., Xiao, S., Kapoula, Z.: Specific saccade deficits in patients with Alzheimer's disease at mid to moderate stage and in patients with amnestic cognitive impairment. Age (Dordrecht, Netherlands) 35 (2012)
23. Pereira, M.: Saccadic eye movements associated with executive function decline in mild cognitive impairment and Alzheimer's disease: Biomarkers (non-neuroimaging)/novel biomarkers. Alzheimer's Dement. 16 (2020)
24. Boxer, A.: Saccade abnormalities in autopsy-confirmed frontotemporal lobar degeneration and Alzheimer disease. Arch. Neurol. 69, 509–517 (2012)
25. Śledzianowski, A., Szymanski, A., Drabik, A., Szlufik, S., Koziorowski, D.M., Przybyszewski, A.W.: Combining results of different oculometric tests improved prediction of Parkinson's disease development. In: Nguyen, N.T., Jearanaitanakij, K., Selamat, A., Trawiński, B., Chittayasothorn, S. (eds.) Intelligent Information and Database Systems, pp. 517–526. Springer, Cham (2020). https://doi.org/10.1007/978-3-030-42058-1_43
26. Virtanen, P., et al.: SciPy 1.0: fundamental algorithms for scientific computing in Python. Nat. Methods 17, 261–272 (2020)
27. Bijvank, J.N., et al.: A standardized protocol for quantification of saccadic eye movements: demons. PLOS ONE 13, e0200695 (2018)
28. Andersson, R., Nyström, M., Holmqvist, K.: Sampling frequency and eye-tracking measures: how speed affects durations, latencies, and more. J. Eye Move. Res. 3, 1–12 (2010)
29. Szymanski, A., Szlufik, S., Koziorowski, D.M., Przybyszewski, A.W.: Building classifiers for Parkinson's disease using new eye tribe tracking method. In: ACIIDS (2017)

30. Przybyszewski, A., Kon, M., Szlufik, S., Szymański, A., Habela, P., Koziorowski, D.: Multimodal learning and intelligent prediction of symptom development in individual Parkinson's patients. Sensors **16**, 1498 (2016)
31. Ramsperger, E., Fischer, B.: Human express saccades: extremely short reaction times of goal directed eye movements. Exp. Brain Res. **57**(1), 191–5 (1984)

RuMedSpellchecker: Correcting Spelling Errors for Natural Russian Language in Electronic Health Records Using Machine Learning Techniques

Dmitrii Pogrebnoi[(✉)] [iD], Anastasia Funkner[iD], and Sergey Kovalchuk[iD]

ITMO University, Saint Petersburg, Russia
pogrebnoy.inc@gmail.com

Abstract. The incredible advances in machine learning have created a variety of predictive and decision-making medical models that greatly improve the efficacy of treatment and improve the quality of care. In healthcare, such models are often based on electronic health records (EHRs). The quality of this models depends on the quality of the EHRs, which are usually presented as plain unstructured text. Such records often contain spelling errors, which reduce the quality of intelligent systems based on them. In this paper we present a method and tool for correcting spelling errors in medical texts in Russian. By combining the Symmetrical Deletion algorithm and a finely tuned BERT model to correct spelling errors, the tool can improve the quality of original medical texts without significant cost. We have evaluated the correction precision and performance of the presented tool and compared it with other popular spelling error correction tools that support Russian language. Experiments have shown that the presented approach and tool are 7% superior to existing open-source tools for automatically correcting spelling errors in Russian medical texts. The proposed tool and its source code are available on GitHub[1] and pip[2] repositories([1]https://github.com/DmitryPogrebnoy/ MedSpellChecker [2]https://pypi.org/project/medspellchecker).

Keywords: spell checking · text correction · BERT · transformers · Russian natural language · electronic health records

1 Introduction

The integration of machine learning techniques in the field of healthcare has revolutionized the way treatments are administered and care is provided to patients. Predictive and decision-making models based on electronic health records help to take into account patient specifics more accurately and better adjust treatment for each patient. One important problem with EHRs, is that they are usually presented as unstructured text and contain spelling errors. According to

© The Author(s), under exclusive license to Springer Nature Switzerland AG 2023
J. Mikyška et al. (Eds.): ICCS 2023, LNCS 14075, pp. 213–227, 2023.
https://doi.org/10.1007/978-3-031-36024-4_16

a study by Toutanova et al. [17], spelling errors occur mainly for two reasons. The first reason is mainly related to the person himself and is that the writer may not know exactly how to spell a word correctly and therefore make mistakes. The second reason has to do with technology and is due to poor-quality typing devices, which can also lead to spelling errors. Such mistakes in EHRs can confuse machine-learning-based medical systems and reduce their quality and effectiveness of recommendations. Therefore to overcome this problem, we need a spelling error correction tool that will accurately correct errors and can improve model accuracy without additional cost.

There are many different open source tools for correcting spelling errors. However, most of them support English and a few other popular languages, while the number of such tools is much smaller for other, less common languages. In addition, medical texts differ significantly from general texts. They contain many medical terms that are almost never found in ordinary texts and require special processing. Correction of a valid word can change a sentence and lead to unpredictable results. Therefore special tools for medical texts aimed at correcting spelling errors in such texts are essential.

There are no known special open source tools for correcting Russian medical texts. And this is to be expected, since there are very few open datasets with medical texts in Russian. Nevertheless, there are several open source general purpose tools for correcting spelling errors in Russian texts.

In this paper, we present a method and a tool for correcting spelling errors specifically in Russian medical texts. The new tool is based on a combination of the Symmetric Deletion [6] algorithm for generating edit candidates and a finely tuned BERT-based [5] machine learning model for ranking candidates and selecting the most appropriate.

2   Related Works

There are a number of research papers devoted to different approaches to the processing of medical texts. However, most of them are mainly devoted to texts in English.

The study by Yifan Peng et al. [12] found that fine-tuning the BERT model on medical notes and PubMed articles outperformed most existing state-of-the-art models and demonstrating the effectiveness of fine-tuning language models for specific domains. The study by Jinhyuk Lee et al. [8] presents a BioBERT model that is fine-tuned to a biomedical corpus in English. Results of experiments have shown that the derived model outperforms the basic BERT model in various text mining tasks of medical texts. In another related paper by Emily Alsentzer et al. [2], the authors presents the ClinicalBERT model for English language. This model is finely tuned on clinical data and outperforms the basic BERT and BioBERT models in almost any type of task.

The processing of medical texts in Russian is much less well researched. This is mainly because there is not yet a significant amount of medical data collected for the Russian language that is comparable with the existing Russian-language medical data corpus. In spite of this, corpora are gradually forming and this field of processing Russian medical texts is actively developing.

The paper by Alexander Yalunin et al. [18] in Sberbank AI lab is one such study. The authors of the paper ideationally replicated Jinhyuk Lee et al. work with the BioBERT model. They fine-tuned several different BERT models on an open corpus of medical texts in Russian and compared the metrics of the new models on specific domain tasks. The evaluation results showed that the fine-tuned BERT models also outperformed the basic BERT models even though the fine-tuning corpus was smaller than in the BioBERT case.

The study by Ksenia Balabaeva et al. [3] is devoted specifically to correcting spelling errors in Russian medical texts. This work uses a combination of Damerau-Levenstein [4] distance to generate editing candidates and Word2vec [9] or FastText [7] embeddings to rank and select the best candidate. The evaluation showed quite strong results, but there is still room for improvement.

There have been impressive advances in the application of BERT models to various tasks in medical text processing. In this paper we present a new method and its implementation as a tool for correcting spelling errors in medical texts in Russian. We also compared the precision metrics and performance of the developed tool with other popular open source tools for spelling error correction in Russian.

3   Methods

3.1   Types of Spelling Errors

There are many different views on what errors should be corrected by a text correction tool. Alexey Sorokin et al. [15] in their study presented the results of the first competition on automatic text correction in Russian. Seven teams participated in the contest. Their task was to create tools that had to most accurately correct texts that contained predefined kinds of errors. In addition to syntactic errors, the texts contained grammatical and cognitive errors, as well as some other kinds of errors.

Medical texts are extremely sensitive to errors. An incorrectly corrected word, or worse, a corrected valid word, can greatly affect the meaning of a sentence and lead to unpredictable consequences. The more kinds of errors the tool tries to correct, the more cases when the tool can work false positive and change a valid word. To start gradually and reduce at first the number of false positive cases we consider only the kind of spelling errors. Developed spellchecker supports correction of six types of spelling errors. Four types of errors are related only to letters and two more types of errors related to spaces. Examples of supported spelling error types are shown in Fig. 1.

Type of mistake	Incorrect text	Correct text
Wrong characters	тубиркулез	туберкулез
Missing characters	туб☐ркулез	туберкулез
Extra characters	туберкпулез	туберкулез
Shuffled characters	тубрекулез	туберкулез
Missing word separator	острый|туберкулез	острый_туберкулез
Extra word separator	туб_еркулез	туберкулез

Fig. 1. Supported types of spelling errors by the spellchecking tool. As an example, the Russian words «tuberculosis» and «acute tuberculosis».

3.2 Datasets

Developing an accurate and efficient tool for correcting spelling errors in Russian-language medical texts is impossible without collecting enough data to train machine learning models and testing the tool.

The availability of open sources of Russian medical texts and data sets is extremely limited. This is mainly due to the complexity of collecting and processing sensitive data, since they can potentially contain personal data. In this paper, we use two open medical cases in Russian, as well as two closed datasets.

The summary of the datasets used is shown in Table 1.

Table 1. Summary information on the Russian medical datasets used.

Dataset Name	Number of Records	Avg Tokens in Record
RuMedNLI [11]	14717	8
RuMedPrimeData [16]	15250	31
Almazov Center	2355	42
Russian Academy of Sciences [14]	161	863

The first public datasets is RuMedNLI: A Russian Natural Language Inference Dataset For The Clinical Domain [11] by Pavel Blinov et al. This dataset is a manually translated from English MedNLI [13] dataset and contains 14717 medical records. Another public dataset is RuMedPrimeData [16] by Starovoytova Elena et al. This dataset contains 15250 medical anamneses of SSMU hospital visitors.

One of the private datasets is dataset with patients' medical anamneses which is provided by the Institute of Artificial Intelligence Problems of the Research Institute of the Russian Academy of Sciences (Russian Academy of Sciences) [14].

This dataset contains 161 large fragments of patients' anamneses. In addition, a closed corpus of anamneses provided by the Almazov National Medical Research Center (Almazov Center) was also used. This dataset contains 2355 patient anamneses for the period from 2010 to 2015.

Each of the datasets was pre-processed. All texts were converted to lower case and then lemmatised using the pymorphy2[1] tool. Such harsh pre-processing was done because of the extremely limited datasets. The lemmatisation helped to get rid of words in different forms and to use words in their initial form. This increased the number of example sentences for a particular word, which contributed to a better fine-tuning of the language models.

After preprocessing, all four datasets were combined into one. In total, all datasets together contain 30,737 medical records in Russian, which takes about 10.25 Mb.

3.3 Algorithm

The text correction algorithm takes raw, unprocessed text as input and returns a corrected text. The algorithm uses the Damerau-Levenstein [4] edit distance to generate candidates for correcting an invalid word. This algorithm generates candidates with an edit distance of one by default, but can potentially be extended to greater values. A special pre-calculated Symmetric Deletion index is used to optimize the computation of the edit distance and to speed up the generation of edit candidates. The diagram of the spellchecking process is shown in the Fig. 2.

The process is arranged as follows. First of all, the medical text is divided into tokens. Next, a couple of conditions are checked. Whether there are any non-Russian letters in the token and whether the word is a name. To check if a word is a name, we check if the word does not contain any capital letters, or if it is at the beginning of a sentence. If at least one condition is not fulfilled, the token is considered as not valid for correction and gets into the final result as it is. Otherwise, the token is reduced to lowercase letters, and the lemmatized form of the token and information about the form of the original word is retained in the internal representation. After that, it is checked whether the token or its lemmatised form is included in the dictionary of correct words. If it is included, then such a token gets into the final text as it is. Otherwise, a list of candidates is generated to replace the incorrect word. Then this list is ranked by a special language model and the most suitable candidate gets into the final text. The best candidate is transformed into the required form and case of the original word. This happens with every token. At the end, the corrected tokens are assembled into the final text.

The precision of the corrected algorithm depends mainly on the completeness and purity of the dictionary with correct words, and on how well the language model correctly ranks the edit candidates.

---

[1] https://github.com/pymorphy2/pymorphy2.

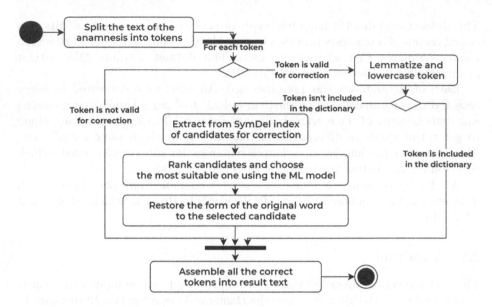

Fig. 2. Examples of supported types of spelling errors by the spellchecking tool.

3.4   Implementation

The new tool for correcting the spelling of medical texts is intended to work only with Russian text and is written in Python. The tool consists of seven components. The architecture of the tool is shown in the Fig. 3.

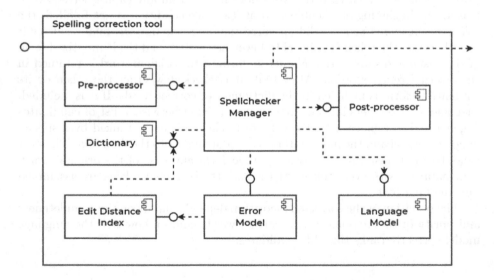

Fig. 3. Architecture of the new spelling correction tool.

The Pre-processor component is responsible for splitting the text into separate words, removing punctuation and capitalisation, and determining the lemmatised form and form information of the original word. In contrast, the Post-processor component restores the form of the original words to the corrected words and reassembles the entire text from the individual words.

The Dictionary component contains a dictionary of correct words and is used to quickly determine whether a word is correct or requires correction. This dictionary contains only lemmatised words from the Aspell Russian dictionary and several medical dictionaries. The final dictionary contains 214629 words in the primary form.

Error Model component is responsible for generating a list of candidates for correcting an incorrect word. The error model uses the Damerau-Levenstein edit distance to create a list of candidates. Generation of the candidate list is a computationally intensive operation. The Edit Distance Index component, which contains a special index, is used to significantly speed up the operation and improve the overall performance of the tool.

Language Model is responsible for ranking the editing candidates and choosing the most suitable one to replace the incorrect word. A finely tuned machine learning models based on the BERT architecture is used to rank candidates for Russian medical texts. The main advantage of this approach is that the model is able to take into account the context around the incorrect word when ranking candidates, which improves the quality of ranking and accuracy of correction.

3.5 Language Model

In this paper, three different BERT base models were fine-tuned on a collected dataset of Russian medical texts. Fine-tuning of the models took place using the transformers, datasets and accelerate libraries from the Hugging Faces platform. A common Fill Mask task was used to fine-tune the models. All hyperparameters for fine-tuning all models are the same and are shown in Table 2

Table 2. Hyperparameters of fine-tuning of all models.

Parameter	Value
Train epoch	5
Learning rate	0.00005
Weight decay	0.01
FP16 training	True
Gradient accumulation steps	256
Per device train batch size	1
Per device eval batch size	1
Gradient checkpointing	True

The first base model is pre-trained ruRoBERTa-large[2] model from Sberbank
AI. This model was chosen because it is one of the larger and more efficient models of the BERT architecture. This model is publicly available and pre-trained
for the Russian language, which makes it easy to use for various applications in
Russian. The fine-tuned model is published on Hugging Face as DmitryPogreb-
noy/MedRuRobertaLarge[3].

The second basic model was the distilbert-base-multilingual-cased model[4].
This model is slightly inferior in efficiency to the previous model, but has a
much smaller size and better performance. This model supports multiple languages including Russian. For the task of ranking editing candidates in Russian,
all other languages are superfluous. To reduce the size of the model, support for
unnecessary languages was discarded using the approach described in the paper
by Amine A. et al. [1]. But unlike the approach in the paper, all tokens not containing Russian letters except for special characters, numbers and punctuation
marks were discarded. This halves the size of the model. The converted model
is published on the Hugging Face service as DmitryPogrebnoy/distilbert-base-
russian-cased[5]. The fine-tuned model is published on Hugging Face as Dmitry-
Pogrebnoy/MedDistilBertBaseRuCased[6].

The third basic model was the cointegrated/rubert-tiny2[7] model. This model
only supports the Russian language and is based on the BERT architecture. The
main feature that distinguishes this model from the previous two is that it is
much smaller in size. It is half the size of the MedDistilBertBaseRuCased model.
The fine-tuned model is published on Hugging Face as DmitryPogrebnoy/Me-
dRuBertTiny2[8].

As a result, the Language Model component contains three different BERT-
based models fine-tuned to rank a edit candidate. With several models of different sizes, it is possible to use this spellchecker tool on a variety of technical
hardware, from a high-performance server with the MedRuRobertaLarge model
to an common personal computer with the MedRuBertTiny2 model.

The Language Model component can easily be extended by adding candidate
rankers based on other machine learning models. It is enough to implement the
necessary interface and everything will work out of the box. As an example, we
have added rankers based on the RuBioBERT and RuBioRoBERTa models from
the paper by Alexander Yalunin et al. [18].

3.6   Python Package

The developed spellchecker was assembled in a Python package and uploaded
to the official pip repository. The package with the new tool is called med-

---

[2] https://huggingface.co/sberbank-ai/ruRoberta-large.
[3] https://huggingface.co/DmitryPogrebnoy/MedRuRobertaLarge.
[4] https://huggingface.co/distilbert-base-multilingual-cased.
[5] https://huggingface.co/DmitryPogrebnoy/distilbert-base-russian-cased.
[6] https://huggingface.co/DmitryPogrebnoy/MedDistilBertBaseRuCased.
[7] https://huggingface.co/cointegrated/rubert-tiny2.
[8] https://huggingface.co/DmitryPogrebnoy/MedRuBertTiny2.

Listing 1.1. An example of using a new tool to correct an error in the sentence "The patient has been diagnosed with a heart attack"

```
1  from medspellchecker.tool.medspellchecker \
2      import MedSpellchecker
3  from medspellchecker.tool.distilbert_candidate_ranker \
4      import RuDistilBertCandidateRanker
5
6  candidate_ranker = RuDistilBertCandidateRanker()
7  spellchecker = MedSpellchecker(candidate_ranker)
8  fixed_text = spellchecker.fix_text(
9      "У больного диагностирован инфркт"
10 )
11
12 print(fixed_text)
```

spellchecker[9]. In addition to the source code and necessary classes, this package also contains a dictionary of correct words. Apart from the dictionary, the package does not contain any other additional datasets or models, which keeps the size of the package manageable.

The medspellchecker package does not contain any parts or compiled fragments of ranking models. They are all downloaded automatically when needed from the public repository on Hugging Faces. An internet connection is required to use the tool, at least for the first time. Once the model has been downloaded and cached, an internet connection is not required.

An example of the use of a package with the developed tool is shown in Listing 1.1.

The first two lines import the main class of the package. Lines 3 and 4 then import one of the three available classes to rank the edit candidates. Line 6 creates an instance of the candidate ranking class based on the fine-tuned DistilBert model. Line 7 creates an instance of the class to correct spelling errors. On line 8, the fix_text method is called, which takes the raw text and returns the corrected text. Finally line 12 prints the corrected result, which looks like «У больного диагностирован инфаркт». In this way, the package can be used to correct spelling errors in Russian medical texts in a few lines. The package also allows you to extend the ranking classes and add your own custom ones.

4 Experiments

There are several open source tools that can automatically correct spelling errors in words. Most of these tools focus mainly on English text, while the other languages fall by the wayside and cannot boast such a rich range of tools. Nevertheless, there are several tools that support the Russian language. However, none of them is intended for medical texts.

---

[9] https://pypi.org/project/medspellchecker.

We have evaluated the precision and performance of the tool, and compared it to other open source tools for correcting spelling errors in Russian texts.

The following parameters were identified for comparison.

– Error precision. This is the percentage of words with an error that the tool correctly corrected relative to all incorrect words.
– Lexical precision. This is the percentage of correct words without mistakes that the tool does not correct relative to all correct words.
– Overall precision. This is the average of error precision and lexical precision.
– Performance. Average number of words processed per second.

Precision metrics and tool performance were evaluated on two different datasets. The first dataset contains correct and incorrect words without their context. This set contains 100 entries for each of the first four error types, 900 entries for the fifth error type and 1000 for the sixth error type shown in Fig. 1. Thus this test dataset contains a total of 2300 records. Each record contains a pair of words. The first word contains a particular type of error and the second word is its correct version.

The second dataset also contains 100 entries for each of the first four error types, 900 entries for the fifth error type and 1000 for the sixth error type shown in Fig. 1. Thus this test dataset contains a total of 2300 records. This set does not contain single words, but words with context. Each record consists of three parts. The first part is a coherent passage of 10 words, one of which is written with an error of a certain type. The second part is the same 10 words, but fully correct. The third part is the number of the incorrect word, which is used to calculate the error and lexical precision.

All words and passages in the test datasets are collected from various Russian medical texts and anamneses.

Thus we have two test datasets with medical data. The first dataset allows us to evaluate how good the tool performs in correcting single words, and the second dataset allows us to calculate the quality of word correction in a coherent context.

In order to better compare the tools was also calculated overall precision, which is the average of lexical and error precision. This metric allows estimating the quality of spelling error correction in general. However, it should be noted that there are usually more correct words in a text than there are incorrect words, therefore lexical precision is more important than error precision.

In addition to the quality of correction of spelling errors, it is also necessary to take into account the time in which these errors are corrected. Of course, the faster the tool works, the better. The performance test was conducted on a computer on Ubuntu 20.04 with 24 GB RAM, Intel Core i5-10210U CPU @ 1.60GHz * 8 and NVIDIA Tesla V100.

5 Results and Discussion

Seven open source tools for correcting spelling errors in Russian texts were chosen for the experiments. For each tool, there are Python wrappers for easy handling from Python code. The selected tools and their Python wrappers are shown in the Table 3.

Table 3. Open source tools and their Python wrappers for correcting spelling errors in Russian chosen for experiments. Full links to Github repositories require «https:// www.github.com/» prefix.

Tool name	Wrapper name	Wrapper GitHub repository
Aspell	Aspell-python	WojciechMula/aspell-python
Hunspell	PyHunspell	blatinier/pyhunspell
Enchant	PyEnchant	pyenchant/pyenchant
LanguageTool	LanguageTool-python	jxmorris12/language_tool_python
Peter Norvig's spellchecker [10]	PySpellChecker	barrust/pyspellchecker
Symspell	SymspellPy	mammothb/symspellpy
Jumspell	Jumspell	bakwc/JamSpell

In addition to the existing tools, the experiment was also conducted with RuMedSpellchecker in ten different configurations, depending on the language model and the type of processor used. The first six configurations used models fine-tuned as part of this paper. The remaining four configurations used the RuBioBERT and RuBioRoBERTa models, which were obtained in the paper by Alexander Yalunin et al. [18] in Sberbank AI lab.

An example of correcting spelling errors with selected tools is shown in Table 4. The misspelled sentence is «У больногодиагностирован инфркт и туберкулз». The correct result of the misspelled sentence is «у больного диагностирован инфаркт и туберкулез». The case of letters is not taken into account. In this example, RuMedSpellchecker tool was run in CPU mode, as the quality and result of the fix is independent of which computing mode is used.

As you can see even such a relatively simple example causes problems with the correction. However, the Aspell-python and LanguageTool-python tools properly corrected the example sentence. The other existing tools made mistakes in endings, prepositions, or missing a space between words. The new tool corrected the example preposition with only two of the language models. The other three had incorrect results.

The results of the example sentence corrections cannot be used to evaluate the precision of the corrections and the performance of the tools. The following are the results of the single word corrections experiment and the word with context corrections experiment.

The result of the experiment with single word corrections is shown in Table 5. In addition to the tool names and four metric columns, the table contains a column with the CPU or GPU type of processor. GPUs can significantly improve the performance of spelling correction tools, but only those that support it. Unfortunately, none of the existing evaluated tools support GPU computing.

Table 4. The results of fixing the «У больногодиагностирован инфркт и туберкулз» sentence by the various tools selected.

Tool name	Result
Aspell-python	**У больного диагностирован инфаркт и туберкулез**
PyHunspell	Уф больного диагностирован инфаркт аи туберкул
PyEnchant	И больного диагностирован инфаркт аи туберкул
LanguageTool-python	**У больного диагностирован инфаркт и туберкулез**
PySpellChecker	У больногодиагностирован инфаркт и туберкулез
SymspellPy	и больногодиагностирован инфаркт и туберкулез
SymspellPy (compound mode)	у больного диагностирования инфаркт и туберкулез
Jumspell	У больногодиагностирован инфркт и туберкулз
RuMedSpellchecker (MedRuRobertaLarge)	У больного диагностирован инфект и туберкулз
RuMedSpellchecker (MedDistilBertBase)	**У больного диагностирован инфаркт и туберкулез**
RuMedSpellchecker (MedRuBertTiny2)	**У больного диагностирован инфаркт и туберкулез**
RuMedSpellchecker (RuBioBERT)	У больного диагностирован инфркт и туберкулз
RuMedSpellchecker (RuBioRoBERTa)	У больного диагностирован инфект и туберкулз

Table 5. Comparison of spelling correction tools in Russian medical texts and a new tool in the single word correction test.

Tool name	Processor type	Error precision	Lexical precision	Overall precision	Average words per second
Aspell-python	CPU	**0.86**	0.859	**0.859**	283.7
PyHunspell	CPU	0.812	0.539	0.675	9.4
PyEnchant	CPU	0.829	0.541	0.685	20
LanguageTool-python	CPU	0.762	0.904	0.833	25.1
PySpellChecker	CPU	0.354	0.86	0.607	3.4
SymspellPy	CPU	0.399	0.813	0.606	**9702.8**
SymspellPy (compound mode)	CPU	0.465	0.512	0.489	672
Jumspell	CPU	0.267	0.947	0.607	2552.1
RuMedSpellchecker (MedRuRobertaLarge)	CPU	0.715	**0.991**	0.853	2.1
	GPU				5.9
RuMedSpellchecker (MedDistilBertBaseRu-Cased)	CPU	0.701	**0.991**	0.846	12.7
	GPU				39.7
RuMedSpellchecker (MedRuBertTiny2)	CPU	0.681	**0.991**	0.836	24.2
	GPU				79.1
RuMedSpellchecker (RuBioRoBERTa)	CPU	0.695	**0.991**	0.843	2.2
	GPU				5.8
RuMedSpellchecker (RuBioBERT)	CPU	0.683	**0.991**	0.837	8.3
	GPU				20.1

The Aspell-python tool performed best in the single-word correction test in terms of error precision and overall precision. The new tool, on the other hand, was quite average in error precision. However, the tool was the best in lexical precision, which also influenced the overall precision and made it close to the best. This is probably due to the additionally extended vocabulary of valid words. In terms of performance, SymSpellPy showed the best result, while Jumspell also showed a high value. As expected, the performance of the new tool depends on the model and processor type used. The smaller the model, the faster the tool runs, but the precision is slightly reduced. Also the tool with our fine-tuned models shows slightly better precision metrics than with RuBioBERT and RuBioRoBERTa respectively. Overall, it can be said that the performance of the tool is average compared to competitors.

The results of the precision metrics of the new tool in the one-word correction test were not as high. However, the main feature of the language models used in the tool is to take into account the context around the word being corrected. Therefore, the new tool revealed itself in the test with word correction in context. The result of the experiment with word correction in context is shown in Table 6.

Table 6. Comparison of spelling correction tools in Russian medical texts and a new tool in the test of correcting a words with context.

Tool name	Processor type	Error precision	Lexical precision	Overall precision	Average words per second
Aspell-python	CPU	0.731	0.93	0.831	357.3
PyHunspell	CPU	0.706	0.719	0.713	11.8
PyEnchant	CPU	0.721	0.719	0.72	24.3
LanguageTool-python	CPU	0.727	0.942	0.835	43.6
PySpellChecker	CPU	0.304	0.868	0.586	6.7
SymspellPy	CPU	0.37	0.913	0.642	**26060.2**
SymspellPy (compound mode)	CPU	0.483	0.804	0.643	1604.2
Jumspell	CPU	0.307	0.969	0.638	4322.3
RuMedSpellchecker (MedRuRobertaLarge)	CPU	**0.792**	0.984	**0.888**	8.9
	GPU				29.1
RuMedSpellchecker (MedDistilBertBaseRu-Cased)	CPU	0.765	**0.99**	0.878	45.5
	GPU				153.8
RuMedSpellchecker (MedRuBertTiny2)	CPU	0.742	0.987	0.865	127.3
	GPU				356.2
RuMedSpellchecker (RuBioRoBERTa)	CPU	0.738	0.987	0.863	9.1
	GPU				31.3
RuMedSpellchecker (RuBioBERT)	CPU	0.715	0.988	0.852	30.7
	GPU				95.6

RuMedSpellchecker outperformed the competition in all precision metrics in the test of correcting words with context. In addition, in this test, the new tool showed significantly better performance than in the previous test. However, SymSpellPy was still the performance leader.

The new tool with the largest language model, MedRuRobertaLarge, scored higher in the error precision metric than the other models. However, lexical precision was higher with the MedDistilBertBaseRuCased model. This can probably be explained by the very limited dataset for fine-tuning large models. Also, due to the limited dataset, the generalization ability of the models may be biased. For example, for other types of medical texts the tool may perform worse. Despite this it is very likely that with more medical texts to train, the fine-tuning of models will be better and the precision metrics of the tool will also be even higher.

Nevertheless, the results show that the new approach for correcting medical texts in Russian is effective and outperforms the competition by 7% in overall precision. Moreover, this approach can be used not only to correct medical texts. This algorithm can be adapted to other specific domains and languages with limited available datasets.

6   Conclusion

We presented a new method for correcting medical texts in Russian and a tool that implements it. Experiments have shown that the new tool is slightly inferior to existing tools when correcting single words, but outperforms them by 7% in overall precision when correcting words with context. The achieved result can be improved by more medical data in Russian for better fine-tuning of language models.

The presented method can be used not only for correction of medical texts in Russian. The algorithm can be adapted to correct spelling errors in texts in other low-resource languages. In addition, the presented approach can also be applied not only to medical texts, but also to texts of other specific domains with a limited set of available data.

Acknowledgments. This work was supported by the Ministry of Science and Higher Education of Russian Federation, goszadanie no. 2019-1339.

References

1. Abdaoui, A., Pradel, C., Sigel, G.: Load what you need: Smaller versions of mutililingual BERT. In: Proceedings of SustaiNLP: Workshop on Simple and Efficient Natural Language Processing, pp. 119–123. Association for Computational Linguistics, Online (Nov 2020). https://doi.org/10.18653/v1/2020.sustainlp-1.16
2. Alsentzer, E., et al.: Publicly available clinical BERT embeddings. In: Proceedings of the 2nd Clinical Natural Language Processing Workshop, pp. 72–78. Association for Computational Linguistics, Minneapolis, Minnesota, USA (Jun 2019). https://doi.org/10.18653/v1/W19-1909

3. Balabaeva, K., Funkner, A., Kovalchuk, S.: Automated spelling correction for clinical text mining in russian. Stud. Health Technol. Inform. **270**, 43–47 (2020). https://doi.org/10.3233/SHTI200119

4. Damerau, F.J.: A technique for computer detection and correction of spelling errors. Commun. ACM **7**(3), 171–176 (1964). https://doi.org/10.1145/363958.363994

5. Devlin, J., Chang, M.W., Lee, K., Toutanova, K.: Bert: Pre-training of deep bidirectional transformers for language understanding (2018). https://arxiv.org/abs/1810.04805

6. Github repository of symspell tool (2018). https://github.com/wolfgarbe/SymSpell

7. Joulin, A., Grave, E., Bojanowski, P., Mikolov, T.: Bag of tricks for efficient text classification (2016). https://arxiv.org/abs/1607.01759

8. Lee, J., Yoon, W., Kim, S., Kim, D., Kim, S., So, C.H., Kang, J.: BioBERT: a pre-trained biomedical language representation model for biomedical text mining. Bioinformatics **36**(4), 1234–1240 (2019). https://doi.org/10.1093/bioinformatics/btz682

9. Mikolov, T., Chen, K., Corrado, G., Dean, J.: Efficient estimation of word representations in vector space (2013). https://arxiv.org/abs/1301.3781

10. Norvig, P.: Peter norvig's blog post about a simple spell checking algorithm (2007). https://norvig.com/spell-correct.html

11. Pavel, B., Aleksandr, N., Galina, Z., Arina, R., Vladimir, K., Chaitanya, S.: Rumednli: A russian natural language inference dataset for the clinical domain (2022). http://doi.org/10.13026/gxzd-cf80

12. Peng, Y., Yan, S., Lu, Z.: Transfer learning in biomedical natural language processing: An evaluation of bert and elmo on ten benchmarking datasets (2019). https://arxiv.org/abs/1906.05474

13. Romanov, A., Shivade, C.: Lessons from natural language inference in the clinical domain (2018). https://arxiv.org/abs/1808.06752

14. Shelmanov, A.O., Smirnov, I.V., Vishneva, E.A.: Information extraction from clinical texts in russian. In: Computational Linguistics and Intellectual Technologies: Papers from the Annual International Conference "Dialogue" , vol. 270, pp. 537–549 (2015)

15. Sorokin, A., Baytin, A., Galinskaya, I., Rykunova, E., Shavrina, T.: Spellrueval : the first competition on automatic spelling correction for russian (2016). https://www.dialog-21.ru/media/3427/sorokinaaetal.pdf

16. Starovoitova, E., et al.: RuMedPrimeData (2021). https://doi.org/10.5281/zenodo.5765873

17. Toutanova, K., Moore, R.C.: Pronunciation modeling for improved spelling correction. In: Proceedings of the 40th Annual Meeting on Association for Computational Linguistics, ACL 2002, pp. 144–151. Association for Computational Linguistics, USA (2002). https://doi.org/10.3115/1073083.1073109

18. Yalunin, A., Nesterov, A., Umerenkov, D.: Rubioroberta: a pre-trained biomedical language model for russian language biomedical text mining (2022). https://arxiv.org/abs/2204.03951

Named Entity Recognition for De-identifying Real-World Health Records in Spanish

Guillermo López-García[1]([✉]) [iD], Francisco J. Moreno-Barea[1] [iD], Héctor Mesa[1] [iD],
José M. Jerez[1] [iD], Nuria Ribelles[2] [iD], Emilio Alba[2] [iD],
and Francisco J. Veredas[1,3] [iD]

[1] Departamento de Lenguajes y Ciencias de la Computación,
Escuela Técnica Superior de Ingeniería Informática,
Universidad de Málaga, Málaga, Spain
`guilopgar@uma.es`
[2] Unidad de Gestión Clínica Intercentros de Oncología,
Instituto de Investigación Biomédica de Málaga (IBIMA),
Hospitales Universitarios Regional y Virgen de la Victoria, Málaga, Spain
[3] Research Institute of Multilingual Language Technologies, Universidad de Málaga,
Málaga, Spain

Abstract. A growing and renewed interest has emerged in Electronic
Health Records (EHRs) as a source of information for decision-making
in clinical practice. In this context, the automatic de-identification of
EHRs constitutes an essential task, since their dissociation from personal data is a mandatory first step before their distribution. However,
the majority of previous studies on this subject have been conducted
on English EHRs, due to the limited availability of annotated corpora
in other languages, such as Spanish. In this study, we addressed the
automatic de-identification of medical documents in Spanish. A private
corpus of 599 real-world clinical cases have been annotated with 8 different protected health information categories. We have tackled the predictive problem as a named entity recognition task, developing two different deep learning-based methodologies, namely a first strategy based
on recurrent neural networks (RNN) and an end-to-end approach based
on transformers. Additionally, we have developed a data augmentation
procedure to increase the number of texts used to train the models.
The results obtained show that transformers outperform RNN on the
de-identification of Spanish clinical data. In particular, the best performance was obtained by the XLM-RoBERTa large transformer, with a
strict-match micro-averaged value of 0.946 for precision, 0.954 for recall
and 0.95 for F1-score, when trained on the augmented version of the corpus. The performance achieved by transformers in this study proves the
viability of applying these state-of-the-art models in real-world clinical
scenarios.

Keywords: Named Entity Recognition · Natural Language
Processing · Electronic Health Records · De-Identification · Spanish

© The Author(s), under exclusive license to Springer Nature Switzerland AG 2023
J. Mikyška et al. (Eds.): ICCS 2023, LNCS 14075, pp. 228–242, 2023.
https://doi.org/10.1007/978-3-031-36024-4_17

1   Introduction

The adoption of electronic health records (EHR) [12] is a key component for
health systems and medical professionals, as well as representing an important source of information to advance medical research and improve healthcare-
related services. However, for their widespread use in medical research, it is
necessary to remove identifiable information to protect patients' data privacy.
EHRs store information in a wide variety of formats that contain information
related to clinical diagnoses, treatments, procedures, and especially, the privacy
of patients and medical professionals. However, the unstructured nature of the
textual fields makes the task of automatically extracting the relevant concepts
from them especially difficult, but the manual extraction of concepts is non-
reusable, time-consuming and costly [7].

The automatic extraction and masking of the concepts related to individually identifiable data thus becomes the primary task to treat the information
contained in the EHR in other medical-analytical processes. This task, called
de-identification, is not only an ethical prerequisite, but also a legal requirement imposed by data privacy legislation. In the United States (US), the Health
Insurance Portability and Accountability Act (HIPAA) requires the deletion of
18 categories of protected health information (PHI) [29]. Similarly, the General
Data Protection Regulation (GDPR) of the European Union (EU) [4], and the
Ley Orgánica Española de Protección de Datos Personales y Garantía de Derechos Digitales (LOPD-GDD) of Spain [3] in particular, prohibit the processing of
personal data unless identifiable information is masked. In this paper, we focus
on the de-identification of Spanish EHRs for compliance with the LOPD-GDD.

De-identifying clinical texts is a named entity recognition (NER) task from
the standpoint of natural language processing (NLP). NER is the process of
identifying sections of text that reference rigid designators belonging to predefined semantic types, such as person, organisation, location, etc. The term
Named Entity (NE) was first used at the 6th Message Understanding Conference (MUC6) [8], a scientific event designed to promote and evaluate research in
information extraction. In the case of the de-identification process in EHR, the
PHI categories are treated as NEs.

In the last decades, text de-identification has been addressed following three
different approaches: rule-based methods, machine learning (ML) systems and
deep learning (DL) models. Early de-identification systems were generally rule-
based. However, considering that these methods are not reproducible for different domains, researchers began designing ML algorithms, especially motivated
by the organisation of various NLP de-identification challenges. ML algorithms
used for this task include decision tree, hidden Markov model, support vector
machines and conditional random field (CRF) [14]. The three main systems in
the 2014 i2b2 de-identification challenge [27] were based on CRF, since it was
the state-of-the-art (SOTA) method at the time the shared task was held.

In recent years, deep neural networks, which have the ability to automatically learn effective features from large-scale datasets, have been extensively
applied in different NPL tasks. Architectures based on feedforward neural net-

works and Recurrent Neural Networks (RNN), as well as other modified and combined deep neural networks, show impressive results in the NER task. Especially successful are the Long Short-Term Memory networks (LSTM) [10] and its variations, e.g., bidirectional LSTM (BiLSTM) and BiLSTM-CRF. Currently, large pretrained language models based on the multi-head self-attention mechanism [30], specifically the Bidirectional Encoder Representations from Transformers (BERT) model [6], outperform other ML systems for the task of NER, particularly in the biomedical domain [16].

In this paper, we have addressed the problem of automatic detection of personally identifiable information in Real-World Data (RWD) from EHR written in Spanish. For this purpose, we have used several models based on RNN (BiLSTM, 2-BiLSTM and BiLSTM-CRF) and the Transformer architecture (XLM-RoBERTa [2] and RoBERTa-BNE [9]). The models studied herein represent the SOTA in various NER tasks [5]. However, to the best of our knowledge, this is the first study that analyses the application of these models to the problem of identifying PHI using real-world medical texts in Spanish. Through this study, the transformers analysed in this work achieve a higher performance in this task with respect to the RNNs applied, demonstrating why they currently represent the SOTA in many NER tasks.

2 Related Works

With the organisation of different NER challenges and NLP shared tasks [27], and the increased adoption of EHRs in healthcare systems worldwide, text de-identification studies proliferated. In particular, great progress has been made in the development of de-identification systems based on the CRF model [32]. Recently, the rise of DL has increased the number of proposed architectures based on RNNs. LSTM networks [10] and their combinations with CRF (LSTM-CRF) [15], showed better performance on the 2016 CEGS N-GRID de-identification task than CRF models. On the other hand, in [5], the authors developed a BiLSTM model to tackle the de-identification problem, which achieved SOTA in the 2014 i2b2/UTHealth dataset. This model implements a layer of BiLSTM units at a character level input to obtain character embeddings that are concatenated with pretrained token embeddings. The enhanced embeddings are returned to the BiLSTM units and the sequence of probabilities is adjusted with the CRF sequence optimiser to produce the system output. In this way, with the superior performance demonstrated by the LSTM-CRF architectures, most of the previous works addressing the de-identification of medical texts have implemented these models [13].

Although the majority of developed techniques for de-identifying clinical cases have focused on English, national laws around the world vary, thus language-specific methodologies are needed. Consequently, automatic de-identification strategies have been proposed for documents in other languages [11,26]. Being the second most spoken language in the world in terms of the number of native speakers [31], there is a pressing need to develop medical NLP methodologies focused on Spanish. In fact, some competitions and projects have been organised to exploit the content of unstructured medical records in Spanish, although due to the limited availability of annotated corpus with clinical-entity information, the task remains a challenge. One example of this type of initiatives is Cantemist (Cancer Text Mining SharedTask), a shared task focused on the recognition of NEs of a critical type of concept related to cancer and tumor morphology in Spanish medical records [19]. Another example is MED-DOCAN (Medical Document Anonymization), a shared task organised in 2019 that focused on the de-identification of clinical EHRs in Spanish [21]. As in the most recent NER and de-identification competitions worldwide, the NLP models that showed the best performance in MEDDOCAN used methodologies based on DL [22].

Even though MEDDOCAN represents the first shared task specifically devoted to the de-identification of medical texts in Spanish, the corpus derived from the competition does not closely resemble the documents found in clinical practice. In this way, the organisers created a synthetic collection of curated and well-structured clinical texts enriched with PHI expressions [21], in contrast with the unstructured and complex nature of the real-world clinical cases. In this study, we have addressed the de-identification of EHRs in Spanish written by physicians during clinical practice. For this purpose, we have systematically analysed the performance of RNN and transformers based models as automatic systems to detect PHI contained in medical texts. Our results show that transformer models outperform RNNs in de-identifying Spanish medical cases, demonstrating the ability of these models to be successfully applied not only in general and biomedical domains in Spanish [19,22,23], but also in real-world clinical scenarios.

Finally, for reproducibility purposes, all the code needed to replicate our work is publicly available at https://github.com/guilopgar/DeIdentSpanishEHR.

3   Materials

3.1   Galén Texts Annotation

In this study, we have used a private collection of 599 clinical cases retrieved from the Galén Oncology Information System [25,28], which collects information on more than 62,250 cancer patients from the *Hospital Regional Universitario* and the *Hospital Universitario Virgen de la Victoria* in Málaga, Spain. In total, Galén stores 600,000 documents corresponding to clinical episodes as well as a significant number of structured fields (which are completed both in real time, during clinical care activity, and later by specific personnel in charge of this supervised task).

Once the medical documents were obtained, we proceeded, as it is mandatory according to the LOPD-GDD [3], with the de-identification of the medical records to guarantee their correct anonymization and dissociation of the personal information of the individuals involved in the health system. In order to label each of the 599 available records for the subsequent training and evaluation of automatic de-identification algorithms, a manual supervised annotation was performed exclusively by authorised clinical personnel of the Intercentre Clinical Management Unit of Oncology (UGCOI). Thus, the collection of clinical records was processed and labelled by our authorised staff, also replacing sensitive information with standardised labels in a non-reversible way.

Finally, with the aim of evaluating the quality of the annotations, we measured the inter-annotator agreement (IAA) on a subset of 100 documents—which corresponds to the test subset (see Sect. 3.2). For this purpose, we computed a widely used metric to quantify the IAA in NER tasks [17], namely the F1-score, comparing the labelling produced by two different authorised annotators. The IAA value was a micro-averaged F1-score of 0.9633, which indicates a significant agreement between the labelling performed by both annotators.

3.2 Named Entities

In this section, we provide a detailed description of the distinct PHI categories considered to perform the annotation process. We examined the presence of the PHI categories defined by the HIPAA from the US. A reliable interpretation of the HIPAA guidelines was made, adapting some PHI entities to fit the reality of health records in Spain. The Spanish legal system does not provide specific guidance on what information must be removed to de-identify medical texts, but the annotation guidelines made by the Spanish National Plan for the Advancement of Language Technology (Plan TL) for the MEDDOCAN task [21] were taken into account. The task was also carried out from a position of "risk aversion", due to the great variability of users who would later view and interpret the information based on the de-identification.

A sub-selection of 8 PHI categories was made after manually reviewing the data, with the aim of adapting the guidelines mentioned above to the specificities of our real-world clinical corpus. The chosen NEs were:

- CENTRO (healthcare centre): includes any reference to names of clinical centres, general hospitals, institutions or health centres.
- CONTACTO (contact): includes any form of contact with a patient, doctor, nurse or health centre, such as telephone numbers and email addresses.
- DIRECCION (address): includes the appearance of a physical address, such as streets, avenues or buildings.
- HISTORIA (EHR number): includes the identifiers used to control hospital medical records (NHC) and Andalusian health history (NUHSA).
- IDENT (identifier): includes personal identifiers such as the national identity document (DNI), the social security number (NSS), identifiers associated with insurers, the personal numerical code in the Andalusian Health Service (CNP), or any other type of unique identifier.
- PERSONA (person): includes names and surnames of people, as well as initials.
- UBICACION (location): includes references to the location of a person or centre, without this representing a specific physical address, but rather locations relative to the name of a city/town, region or country.

- REFERENCIA (reference): includes identifiers related to medical tests performed, such as tests, biopsies, scans or x-rays.

Table 1 shows the distribution of NEs in the selected documents from the Galén corpus. The collection of 599 clinical cases was randomly split into 3 subsets: the training set (399 documents), the validation set (100 documents) and the test set (100 documents). In Table 1, for the different training, validation and test sets (including DA sets, see Sect. 3.3), the columns show the absolute number (abs) of NEs for each of the 8 classes considered as well as their relative frequency (%). The majority of NEs in the Galén corpus are related to Centre, Person and Location, with approximately a presence greater than 20% in all the considered sets. Meanwhile, the entities related to the Address are almost non-existent in non-DA sets (2 annotations in train, 1 in val and 1 in test sets).

Table 1. Description of the number of annotations per corpus subset: the training and validation sets with and without DA, and the test set.

NE	Train		Train + DA		Val		Val + DA		Test	
	abs	%	abs	%	abs	%	abs	%	abs	%
CENTRO	468	.3145	5148	.3153	114	.2953	1254	.2960	91	.2600
CONTACTO	41	.0276	451	.0276	14	.0363	154	.0364	17	.0486
DIRECCION	2	.0013	22	.0013	1	.0026	11	.0026	1	.0029
HISTORIA	18	.0121	198	.0121	14	.0363	154	.0364	15	.0429
IDENT	63	.0423	682	.0418	9	.0233	99	.0234	8	.0229
PERSONA	439	.2950	4828	.2957	147	.3808	1617	.3817	130	.3714
REFERENCIA	115	.0773	1235	.0756	13	.0337	133	.0314	22	.0629
UBICACION	342	.2298	3762	.2304	74	.1917	814	.1922	66	.1886
Total	1488		16326		386		4236		350	

3.3 Data Augmentation

The nature of the NER task assumes the need of recognising words with a high probability of being out of vocabulary. In order to mitigate the lack of context from a small corpus like Galén, a method for augmenting the amount of documents has been applied. For each document in the training and validation dataset, 10 different synthetic documents are generated by text surrogation of entities susceptible to natural replacement. Entities which present a numerical pattern (e.g., telephone numbers, numerical identifiers) are replaced by perturbing digits randomly. Alpha-numeric entities (e.g., DNI, NSS) are replaced by perturbing digits and characters. The surrogation of the entities made primarily of proper nouns—such as people, locations, countries and centres—is a dictionary-based replacement using data from the Spanish National Statistics Institute (INE)[1].

---

[1] https://www.ine.es/inebmenu/indiceAZ.htm.

Table 1 shows the distribution of NEs in documents created by this augmentation process from the Galén corpus. Although the number of NEs present in the training and validation sets increases, the proportion with respect to the rest of the entities remains essentially the same.

4 Methods

This section presents the two distinct NLP methodologies developed in this study to tackle the de-identification of real-world EHRs in Spanish. The first methodology addresses the problem using an approach based on RNN models, while the second methodology applies transformer-based models. In both cases, the de-identification problem is tackled as a sequence labelling NER task, using the IOB2 tagging scheme [24].

4.1 Recurrent Neural Models

Features. Due to the small order of magnitude of the training corpus, several decisions were made in the tokenization stage to reduce the size of the vocabulary and maximise the inference of contextual relationships. The tokenization was not case sensitive and the easily detectable expressions with a low degree of error based on alpha-numerical sequences (e.g., DNI, NSS, dates, telephone numbers) were replaced by special tokens. To keep information at the character level, a series of descriptors are added to indicate the presence of initial case, uppercase, lowercase, and digits. In addition, a flag is added to tokens that have been detected as a new line, a numeric expression, or if the token is a separator character. Finally, a fastText skip-gram model [1] was used as embedding model trained with the Gálen corpus (training documents and stored clinical documents not used for de-identification).

Thus, we have three types of entry descriptors: indicator descriptors character ranges in the word, descriptors of detected expressions and vector from word embedding. These descriptors are joined in an embedding layer and constitute the input fed to RNN models, as seen in Fig. 1.

Recurrent Neural Networks. RNNs are a generalization of feed-forward neural networks specially designed to deal with temporal problems. In particular, the recurrent SOTA models are based on the LSTM network [10]. An LSTM network has three different gates: the input gate infers the values used to modify the block memory, the forget gate infers features to be discarded and the output gate determines the output using both input and block memory.

BiLSTM consists of two LSTM networks which learns each token in the sequence based on its future and past context [15]. The sequence is processed from left-to-right by one LSTM to learn the past context (current step and previous unit state), while the other LSTM network processes it from right-to-left to learn the future context (current step and subsequent unit state). The designed model includes two time distributed layers, in order to process each

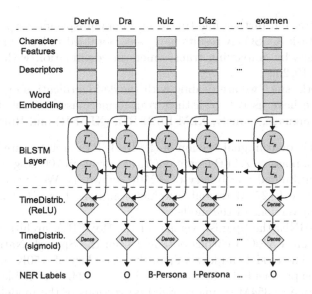

Fig. 1. The BiLSTM model structure for the NER system.

of the time steps and infer the label for each input sequence. This RNN model is shown in Fig. 1. Additionally, 2 BiLSTM layers can be linked to produce a more context-aware algorithm. In this case, the hidden unit forward layer of the second BiLSTM layer receives the output of the previous state and the output of the hidden unit forward layer of the first BiLSTM layer. In contrast, the hidden unit backward layer is computed based on the hidden unit backward layer of the first BiLSTM and the future hidden state.

The BiLSTM-CRF is a type of RNN model that has been used to improve NER performance [5]. A Bidirectional LSTM and a CRF [14] are stacked together for sequence learning. CRF is an undirected discriminative probabilistic graph model, composed of a set of random variables that are used to represent probabilities about structured outputs based on a particular input sequence. The result processed by a BiLSTM layer and a time distributed layer without activation function is fed to a CRF model, which obtains the entity assigned to each word by the system.

4.2 Transformers

The second methodology developed herein to address the de-identification problem is based on transformers. The Transformer model [30] uses the self-attention mechanism to create a contextual numeric representation of each input word, as well as to increase computing efficiency through the parallelization of its network architecture. For the past three years, transformers have become one of the most successful models in multiple areas of NLP [6,22]. One of the reasons that explain their enormous popularity nowadays is that, by following a transfer learning (TL) approach, these attention models can be pretrained on general domain corpora and further fine-tuned on a domain-specific corpus to tackle a certain NLP task [6]. SOTA results have been obtained in both biomedical and clinical domains by employing transformers in combination with different TL strategies [16,19,20].

In this work, since we are dealing with the de-identification of medical texts in Spanish, we have used two distinct transformer-based models that support the Spanish language, namely, XLM-RoBERTa (XLM-R) and RoBERTa-BNE:

- XLM-R: this multilingual version of the RoBERTa architecture [18] was pre-trained on a massive 2.4TB CommonCrawl Corpus in 100 languages [2], using a large multilingual vocabulary of ∼250K subwords. We experimented with both the Base (∼277M trainable weights) and the Large (∼559M trainable parameters) versions of the model.
- RoBERTa-BNE: the Spanish version of the RoBERTa architecture [18] was pretrained on a 570GB corpus obtained from the Spanish National Library (BNE) [9]. The model employs a Spanish vocabulary of ∼50K subtokens and, again, we experimented with both the Base (∼124M trainable parameters) and the Large (∼354M trainable weights) versions of the model.

We have developed an end-to-end approach to address the de-identification problem using transformers, by fine-tuning the models on the clinical corpus obtained from Galén. Figure 2 shows a visual description of the developed methodology. In this way, on the one hand, since the transformers further segment words into a sequence of subwords, each sequence of words from the medical documents was further tokenized into a sequence of subwords. Therefore, each sequence of subwords constitutes the input to the models, without using any additional feature as input data. On the other hand, at inference time, since transformer-based models produce predictions at subword-level, the outputted labels had to be converted to word-level (see Fig. 2). For this purpose, we used the maximum probability criterion proposed in [19], which consists in, for each word, selecting the label predicted with the maximum probability across all subwords obtained from the same word. Finally, the sequence of word-level tags could be further compared with the gold standard (GS) annotations with the aim of evaluating the performance of the models.

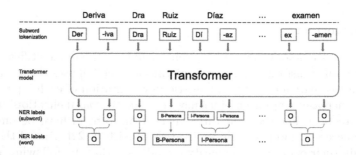

Fig. 2. Illustration of the transformer-based methodology applied to tackle the de-identification problem.

5 Experiments and Results

To experiment with both LSTM-based and transformer models, the validation set was used as an evaluation set. Hyper-parameter fine-tuning was performed by training the models and early stopping the networks using the validation set. The hyper-parameters of the networks were therefore chosen based on obtaining a higher macro averaged F1-score for the validation set. Finally, the performance evaluation of the different architectures was based on the predictions made by the models on the test set.

The standard method for performance evaluation of a NER system is to compare the GS annotations with the tagged output obtained. In this experimentation overall evaluation was considered, namely strict-match and exact-match. For each NE, the type of entity identified and its spans—start and end characters—are obtained. In the strict-match evaluation, both the entity type and spans must match with a GS annotation for a prediction to be considered as correct, while the exact-match evaluation only requires the spans to match.

In addition, different error categories introduced at the MUC6 conference [8] are also considered in this work. For every NE type, the categories are based on the comparison between the GS annotations and the NER system output. Therefore, a predicted entity is considered as correct (cor) if it matches a GS annotation; incorrect (inc) if it does not match the annotation exactly; missed (mis) if the GS annotation is not captured; and spurious (spu) if the prediction does not match any GS annotation.

5.1 NER systems comparison

The experimentation process described above was followed, and Table 2 shows the results obtained by the two methodologies developed in this study to tackle the de-identification problem. Two main conclusion can be drawn from the results described in Table 2. On the one hand, according to the strict-match F1-score—which represents the principal metric used to evaluate automatic de-identification systems [17,21]—, all predictive models applied in this work benefit from the DA procedure. Thus, every system obtains a higher F1-score when trained on the augmented Galén de-identification corpus than when trained solely on the Galén dataset.

On the other hand, considering both strict and exact match evaluations, transformer-based models outperform RNN-based systems for the de-identification of Spanish clinical documents. Among the RNNs, the best performance is achieved by the 2-BiLSTM model when trained on the augmented Galén corpus, obtaining a strict and exact match F1-score of 0.9060 and 0.9201, respectively. In the case of transformers, the XLM-R large model achieves the highest performance obtained in this study, with a strict and exact match F1-score of 0.9502 and 0.9531, respectively, when following the DA procedure. In fact, although the RoBERTa-BNE model was pretrained on a corpus exclusively containing Spanish texts, the XLM-R model surpasses RoBERTa-BNE on the de-identification of Spanish medical cases. In this way, the base versions of XLM-R

238   G. López-García et al.

Table 2. Micro-averaged metrics computed on Galen's test set. We report the performance of each model when trained on both the Galén de-identification corpus and its augmented version. Finally, for each evaluation strategy (strict and exact match), precision (P), recall (R) and F1-score (F1) metrics are computed.

Model	Corpora	NER (strict)			Spans (exact)		
		P	R	F1	P	R	F1
BiLSTM	Gálen	.7625	.8257	.7929	.7731	.8371	.8038
	Gálen + DA	.7855	.8686	.8250	.7933	.8771	.8331
2-BiLSTM	Gálen	.8189	.8657	.8417	.8378	.8857	.8611
	Gálen + DA	.8898	.9229	.9060	.9036	.9371	.9201
BiLSTM-CRF	Gálen	.9062	.8829	.8944	.9238	.9000	.9117
	Gálen + DA	.8840	.9143	.8989	.8895	.9200	.9045
XLM-R	Gálen	.8937	.8886	.8911	.9080	.9029	.9054
(Base)	Gálen + DA	.9218	.9429	.9322	.9302	.9514	.9407
XLM-R	Gálen	.9224	.9514	.9367	.9280	**.9571**	.9423
(Large)	Gálen + DA	**.9462**	**.9543**	**.9502**	**.9490**	**.9571**	**.9531**
RoBERTa-BNE	Gálen	.8620	.8743	.8681	.8732	.8857	.8794
(Base)	Gálen + DA	.8825	.9229	.9022	.8852	.9257	.9050
RoBERTa-BNE	Gálen	.9202	.9229	.9215	.9259	.9286	.9272
(Large)	Gálen + DA	.9405	.9486	.9445	.9405	.9486	.9445

and RoBERTa-BNE obtain a strict-match F1-score of 0.9322 and 0.9022, respectively, while the large versions of the models achieve a strict F1-score of 0.9502 and 0.9445, respectively. Hence, for this particular task, the large-scale multilingual pretraining followed by the XLM-R model has proved to be more effective than the Spanish-specific pretraining followed by the RoBERTa-BNE model.

5.2   Metrics for Each NE

With the aim of conducting a thorough analysis of the performance of the XLM-R large transformer—the system achieving the highest de-identification results (see Table 2)—, in Table 3, we show the results obtained by the model for each NE separately. Apart from the F1-score, following the "risk aversion" principle, recall is often considered as the reference metric to evaluate the performance of de-identification systems, since false negative errors have the potential to threaten the privacy of patients and medical professionals [17]. An automatic system showing a recall value over 0.95 is generally considered as reliable for de-identifying a clinical corpus [17,27]. As we can see from Table 3, the XLM-R transformer achieves a recall value over 0.95 in 6 of the 8 NEs, proving the viability of the model as a medical de-identification system.

The two NEs for which the model does not reach a recall value of 0.95 are CONTACTO (a recall score of 0.9412) and UBICACION (a recall value of 0.8788). Most of the errors the model makes for the UBICACION label are caused by the difficulty the model has in differentiating between centres and locations. In many cases, when the name of a centre is not sufficient to unambiguously identify it, its location is labelled as part of its name. For instance, in the text "Complexo Hospitalario Universitario de Lugo", the name of the Spanish city "Lugo" is labelled as part of the centre's name. However, in the text "Hospital General De Almansa de Lugo", "Lugo" is not tagged as part of the centre's name, but instead is labelled as a location. This is not only a challenge for the automatic models, but also for the human annotators.

Table 3. NER (strict-match) metrics for each NE obtained by the XLM-R (large) system on the test set when trained using Gálen and the augmented documents. Additionally, we describe the number of NEs inferred correctly (cor) and incorrectly (inc), missed (mis) or spurious (spu), and the total number of NEs in the GS set and inferred by the NER system (pred).

NE	P	R	F1	COR	INC	MIS	SPU	GS	PRED
CENTRO	.9167	.9670	.9412	88	3	0	4	91	96
CONTACTO	1.000	.9412	.9697	16	0	1	0	17	16
DIRECCION	1.000	1.000	1.000	1	0	0	0	1	1
HISTORIA	.8824	1.000	.9375	15	0	0	2	15	17
IDENT	.8889	1.000	.9412	8	0	0	1	8	9
PERSONA	.9618	.9692	.9655	126	3	1	2	130	131
REFERENCIA	.9565	1.000	.9778	22	0	0	1	22	23
UBICACION	.9667	.8788	.9206	58	5	3	1	66	60
macro-avg	.9466	.9695	.9567						
micro-avg	.9462	.9543	.9502	334	12	4	20	350	353

Finally, for illustration purposes, Fig. 3 shows the predictions made by the XLM-R large model on a sample medical document in Spanish. We elaborated a text with a similar structure to the real-world clinical cases contained in the Gálen corpus. As we can see from the image, for this particular clinical document, the transformer-based model was able to correctly identify all PHI entities contained in it.

240   G. López-García et al.

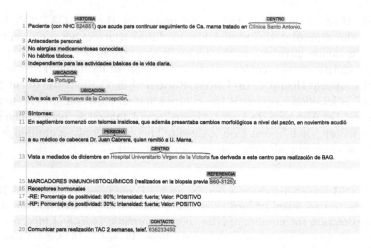

Fig. 3. De-identification performed by the XLM-R (Large) system on a sample clinical text in Spanish.

6 Conclusions

In this work, we have addressed the problem of automatic de-identification of real-world clinical documents in Spanish. For this purpose, we have produced a corpus of 599 de-identified medical cases obtained from Galén [25]. We have systematically analysed the performance of RNNs and transformer-based models when applied to this NER task. Additionally, we have developed a DA strategy to obtain a ×10 augmentation of the number of documents used to train the models. The obtained results show that, on the one hand, all the models applied herein benefit from the DA procedure, since their predictive performance increases when using the augmented version of the corpus. On the other hand, transformers outperform RNNs for the de-identification of medical texts in Spanish. Among the RNN-based systems, the best performance is obtained by the 2-BiLSTM model, with strict-match micro-averaged precision, recall and F1-score of 0.8898, 0.9229 and 0.9060, respectively. For its part, the multilingual XLM-R large transformer achieves the highest performance obtained in this study, with strict-match micro-averaged precision, recall and F1-score of 0.9462, 0.9543 and 0.9502, respectively.

In future works, given the observed superiority of transformer-based models analysed in this study, we will explore how domain-specific models perform on the de-identification problem, since, as it has been shown in the literature, transformers adapted to the specificities of the medical documents obtain SOTA performance in many distinct tasks in the clinical NLP domain [16,19,20]. Finally, given the promising results obtained in this work, we will try to validate our developed methodology on external real-world corpora from other medical centres in Spain.

Acknowledgements. The authors acknowledge the support from the Ministerio de Economía y Empresa (MINECO) through grant TIN2017-88728-C2-1-R, from the Ministerio de Ciencia e Innovación (MICINN) under project PID2020-116898RB-I00, from the Universidad de Málaga and Junta de Andalucía through grant UMA20-FEDERJA-045, from the Malaga-Pfizer consortium for AI research in Cancer - MAPIC, and from the Instituto de Investigación Biomédica de Málaga - IBIMA (all including FEDER funds).

References

1. Bojanowski, P., Grave, E., Joulin, A., Mikolov, T.: Enriching word vectors with subword information. Trans. Associat. Comput. Linguist. **5**, 135–146 (2017)
2. Conneau, A., et al.: Unsupervised cross-lingual representation learning at scale. In: Proceedings of the 58th Annual Meeting of the Association for Computational Linguistics, pp. 8440–8451, Online (Jul 2020)
3. Cortes Generales de España: Ley Orgánica 3/2018, de 5 de diciembre. de Protección de Datos Personales y garantía de los derechos digitales, Boletìn Oficial del Estado (2018)
4. Council of the European Union: Regulation (EU) 2016/679 of the European Parliament and of the Council of 27 April 2016 on the protection of natural persons with regard to the processing of personal data and on the free movement of such data, and repealing Directive 95/46/EC (General Data Protection Regulation). Off. J. Eur. Union **119**, 1–88 (2016)
5. Dernoncourt, F., Lee, J.Y., Uzuner, O., Szolovits, P.: De-identification of patient notes with recurrent neural networks. J. Am. Med. Inform. Assoc. **24**(3), 596–606 (2017). https://doi.org/10.1093/jamia/ocw156
6. Devlin, J., Chang, M.W., Lee, K., Toutanova, K.: BERT: Pre-training of deep bidirectional transformers for language understanding. In: Proceedings of the 2019 Conference of the North American Chapter of the Association for Computational Linguistics: Human Language Technologies, vol. 1, pp. 4171–4186 (2019)
7. Dorr, D.A., Phillips, W., Phansalkar, S., Sims, S.A., Hurdle, J.F.: Assessing the difficulty and time cost of de-identification in clinical narratives. Methods Inf. Med. **45**(03), 246–252 (2006). https://doi.org/10.1055/s-0038-1634080
8. Grishman, R., Sundheim, B.M.: Message Understanding Conference-6: A brief history. In: COLING 1996 Volume 1: The 16th International Conference on Computational Linguistics (1996)
9. Gutiérrez-Fandiño, A., et al.: MarIA: Spanish Language Models. Procesamiento del Lenguaje Natural 68(0), 39–60 (2022). https://doi.org/10.26342/2022-68-3
10. Hochreiter, S., Schmidhuber, J.: Long short-term memory. Neural Comput. **9**(8), 1735–1780 (1997). https://doi.org/10.1162/neco.1997.9.8.1735
11. Jan, T., Trienschnigg, D., Seifert, C., Hiemstra, D.: Comparing rule-based, feature-based and deep neural methods for de-identification of dutch medical records. In: ACM Health Search and Data Mining Workshop, HSDM 2020 (2020)
12. Jha, A., et al.: Use of electronic health records in US hospitals. N. Engl. J. Med. **360**(16), 1628–1638 (2009)
13. Jiang, Z., Zhao, C., He, B., Guan, Y., Jiang, J.: De-identification of medical records using conditional random fields and long short-term memory networks. J. Biomed. Inform. **75**, S43–S53 (2017)
14. Lafferty, J.D., McCallum, A., Pereira, F.C.: Conditional Random Fields: Probabilistic Models for Segmenting and Labeling Sequence Data. In: Proceedings of the Eighteenth International Conference on Machine Learning, pp. 282–289 (2001)

15. Lample, G., Ballesteros, M., Subramanian, S., Kawakami, K., Dyer, C.: Neural architectures for named entity recognition. arXiv preprint arXiv:1603.01360 (2016)
16. Lee, J., et al.: Biobert: a pre-trained biomedical language representation model for biomedical text mining. Bioinformatics 36(4), 1234–1240 (2020)
17. Liu, L., Perez-Concha, O., Nguyen, A., Bennett, V., Jorm, L.: De-identifying Australian hospital discharge summaries: An end-to-end framework using ensemble of deep learning models. J. Biomed. Inform. 135, 104215 (2022)
18. Liu, Y., et al.: RoBERTa: A robustly optimized BERT pretraining approach. arXiv [cs.CL] (2019)
19. López-García, G., Jerez, J.M., Ribelles, N., Alba, E., Veredas, F.J.: Detection of tumor morphology mentions in clinical reports in spanish using transformers. In: Advances in Computational Intelligence, pp. 24–35. Springer International Publishing, Cham (2021). https://doi.org/10.1007/978-3-030-85030-2_3
20. López-Garcìa, G., Jerez, J.M., Ribelles, N., Alba, E., Veredas, F.J.: Transformers for Clinical Coding in Spanish. IEEE Access 9, 72387–72397 (2021)
21. Marimon, M., et al.: Automatic de-identification of medical texts in spanish: the meddocan track, corpus, guidelines, methods and evaluation of results. In: IberLEF@ SEPLN, pp. 618–638 (2019)
22. Perez, N., García-Sardiña, L., Serras, M., Del Pozo, A.: Vicomtech at MEDDO-CAN: Medical Document Anonymization. In: IberLEF@ SEPLN, pp. 696–703 (2019)
23. Pérez-Díez, I., Pérez-Moraga, R., López-Cerdán, A., Salinas-Serrano, J.M., la Iglesia-Vayá, M.d.: De-identifying Spanish medical texts-named entity recognition applied to radiology reports. J. Biomed. Semant. 12(1), 1–13 (2021)
24. Ramshaw, L.A., Marcus, M.P.: Text chunking using Transformation-Based learning. In: Natural Language Processing Using Very Large Corpora, pp. 157–176. Springer, Netherlands, Dordrecht (1999). https://doi.org/10.1007/978-94-017-2390-9_10
25. Ribelles, N., et al.: Galén: Sistema de información para la gestión y coordinación de procesos en un servicio de oncología. RevistaeSalud 6(21), 1–12 (2010)
26. Richter-Pechanski, P., Amr, A., Katus, H.A., Dieterich, C.: Deep learning approaches outperform conventional strategies in de-identification of german medical reports. In: GMDS, pp. 101–109 (2019). https://doi.org/10.3233/SHTI190813
27. Stubbs, A., Kotfila, C.: Özlem Uzuner: Automated systems for the de-identification of longitudinal clinical narratives: Overview of 2014 i2b2/UTHealth shared task Track 1. J. Biomed. Inform. 58, S11–S19 (2015)
28. Urda, D., Ribelles, N., Subirats, J.L., Franco, L., Alba, E., Jerez, J.M.: Addressing critical issues in the development of an oncology information system. Int. J. Med. Informatics 82(5), 398–407 (2013)
29. U.S. Dept. of Health & Human Services: Guidance Regarding Methods for De-identification of Protected Health Information in Accordance with the Health Insurance Portability and Accountability Act (HIPAA) Privacy Rule. Office for Civil Rights (OCR) (2012)
30. Vaswani, A., et al.: Attention is all you need. In: Advances in Neural Information Processing Systems 30 (2017)
31. Vítores, D.F.: El español: una lengua viva. Instituto Cervantes (2019). https://www.cervantes.es/imagenes/File/espanol_lengua_viva_2019.pdf
32. Yang, H., Garibaldi, J.M.: Automatic detection of protected health information from clinic narratives. J. Biomed. Inform. 58, S30–S38 (2015)

Discovering Process Models from Patient Notes

Rolf B. Bänziger(✉) [iD], Artie Basukoski [iD], and Thierry Chaussalet [iD]

University of Westminster, London, UK
{r.banziger,a.basukoski,chausst}@westminster.ac.uk

Abstract. Process Mining typically requires event logs where each event is labelled with a process activity. That's not always the case, as many process-aware information systems store process-related information in the form of text notes. An example are patient information systems (PIS), which store much information in the form of free-text patient notes. Labelling text-based events with their activity is not trivial, because of the amount of data involved, but also because the activity represented by a text note can be ambiguous. Depending on the requirements of a process analyst, we might need to label events with more or fewer unique activities: two similar events could represent the same activity (e.g. screen referral) or two different activities (e.g. screen adult ADHD referral and screen depression referral). We can therefore view activities as ontologies with an arbitrary number of entries.

This paper proposes a method that produces an ontology for the activities of a process by analysing a text-based event log. We implemented an interactive tool that generates process models based on this ontology and the text-based event log. We demonstrate the proposed method's usefulness by discovering a mental health referral process model from real-world data.

Keywords: Process Mining · Text Mining · Healthcare · Mental Healthcare

1 Introduction

Process Mining [1] is a set of techniques and tools to extract knowledge from event logs – traces left by process-aware IT systems describing who executed what activity and when. Event logs typically contain a list of events associated with a case. Each event is described by an activity (the process step it represents) and often includes additional data, such as date/time, the user involved and activity-specific information.

While the extracted knowledge can take many forms, Process Mining is often used to discover a process model as a flowchart that can be used to visualise and improve workflows. Many mature open and commercial Process Discovery algorithms emerged [2–6]. However, they all need event logs where each event is labelled with the activity. While this is usually the case in more structured systems and processes, such as a purchase-to-pay process managed in an ERP system, more flexible systems often do not include activity labels in their event logs.

© The Author(s), under exclusive license to Springer Nature Switzerland AG 2023
J. Mikyška et al. (Eds.): ICCS 2023, LNCS 14075, pp. 243–249, 2023.
https://doi.org/10.1007/978-3-031-36024-4_18

Patient Information Systems (PIS) and their electronic patient records are an example of such flexible systems. Events are often documented using patient notes, i.e., free-text notes with no prescribed structure. Each event has to be labelled with its associated activity to discover a process model from such an event log. This is not a trivial task because the activity is not always obvious, especially without an apriori known list of activities. Allard et al. [7] present a multi-step manual workflow involving multiple to label a relatively small text-based event dataset. Clearly, the amount of work required to label an unlabeled event log becomes quickly prohibitive.

Furthermore, it is often not apparent whether two events represent the same activity or two different but related ones. Indeed, this may depend on the requirements of the process analyst using the process model. For instance, a process analyst analysing mental health data may be explicitly interested in patient journeys involving adult Attention Deficit Hyperactivity Disorder (ADHD). In this case, the process model needs to separate events related to adult ADHD from other events. However, if the analyst is interested in the high-level process flow, such a differentiation is not only unnecessary, it is detrimental to the comprehensibility of the resulting process models (process models with fewer activities are usually more accessible). Therefore, we must view activity labels not as a flat list but as a hierarchical ontology where the activity an event represents can be expressed using different (but related) concepts, depending on the required level of granularity of the process model.

We propose a method that discovers process models from text notes, leveraging an automatically created ontology describing activities. We implemented the method as an interactive tool and evaluated it with real-world data from a community mental health hub.

2 Related Work

While text data has been identified as an event log source, most of the research in Process Mining concentrates on using labelled event logs; far less research concerns itself with text-based event logs.

One of the first attempts at using free-text data as a basis for automated process discovery was described by [8]. They extracted email messages from Microsoft Outlook. Each email message had to be tagged manually by the Outlook user with the associated activity. They demonstrated the usefulness of applying Process Mining to text data but having users reliably tag emails (or other text notes) is often unrealistic or not feasible.

Jlailaty and Grigori [9] described a framework to extract business process activities from email logs using hierarchical clustering and compare the clustering quality using Latent Semantic Analysis (LSA) and Word2Vec. They did not suggest a method to label clusters automatically but instead relied on manual labelling.

We [10] proposed a framework to automatically extract process models from text notes stored in Customer Relationship Management (CRM) systems. We suggested using Latent Dirichlet Allocation (LDA) [11] to group notes representing the same activity and generate keywords for each group.

Chambers et al. [12] proposed a pipeline to discover business processes from emails. Their approach requires a ground truth dataset, as they use supervised machine learning and a ground truth excerpt of the data set to label emails.

de Medeiros et al. [13] gave an early overview of the opportunities of semantic Process Mining using ontologies. The same authors presented how using ontologies together with Process Mining can answer queries that regular Process Mining cannot. They provided a concrete implementation of their method.

To the best of our knowledge, there is no published approach that combines Process Mining from unlabelled text-based event logs with the semantic concepts of ontologies.

3   The Mining Process

Figure 1 gives an overview of our method. We start by extracting an event log containing free-text patient notes. The text data and the positional information of each event are used to create two distance matrices. These matrices are combined and used as a basis for hierarchical clustering. We use the output of the hierarchical clustering, the dendrogram, as the ontology describing the process activities. We use feature selection techniques to label each node of the ontology. Finally, an interactive process miner shows the process model at various levels of detail selectable by the user, which can be exported for use in other Process Mining tools.

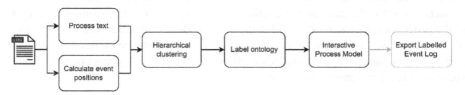

Fig. 1. Overview of the components of the mining process

3.1   Unlabelled Event Log Requirements

In the first step, we extract an unlabelled event log from the source system (in our case, a patient information system). Each event has three fields: a case id, a timestamp and a free-text field containing the patient note. Timestamps allow us to determine the order in which events occurred. We are deliberately only looking at the minimum of information necessary to mine processes to prove the general viability of our approach. In practice, a process-aware information system might contain more information, e.g., the note's author, date, and other process-related information, which could be used to improve results.

3.2   Text Processing

We are treating each event (or free-text note) as a document; all notes together make up our corpus. Each note is turned into a bag-of-n-grams [14]. We found that using n-grams of length (1,2) provide a good compromise between quality of the resulting distance matrix and computing performance.

We create a document-term-matrix (*dtm*) containing term-frequency/inverse-document-frequency (*tf-idf*) values for each document/term combination. *tf-idf* values favour terms (or n-grams) occurring often in few documents and penalise terms occurring in many documents. This improves the eventual clustering by emphasising more important keywords while neglecting common words.

Finally, we use the cosine distance to calculate a distance matrix, yielding values between 0 and 1 for each pair of notes.

3.3   Calculating Event Positions

While text clustering works very well to identify certain activities, in other cases very similar notes can describe different activities. In our case, we have two activities that share similar vocabulary: a screening activity, where a clinician summarises the medical problem and an assessment, where another clinician writes about the medical situation of the patient in more detail. Naturally, both activities will use similar words. Using text data alone will put events from both activities in the same group. However, the activities are, in reality two different process steps that occur at different times.

As we are importing events in the order they occur, we know whether a note is at the start, the end, or anywhere in a process and can use this information to improve clustering results. We calculate two values for each event: distance from the beginning of the process and distance to the end of the process. Both values are scaled to fit between 0 and 1; then we use the Euclidean distance to create a distance matrix between each pair of events. Since we scaled values to (0,1), the distances use the same scales as the values in the matrix created from the text data.

3.4   Hierarchical Clustering

In this step, we calculate the weighted average between both distance matrices. The weight is configurable in the interactive tool, however, giving both matrices the same weight seems to produce reasonable results. We expect that the more unique the vocabulary of each activity is, the less weighting should be given to the positional data.

We conduct hierarchical agglomerative clustering to generate a dendrogram. This dendrogram indicates which activities represent the same activity at different levels of detail. At the lowest level of detail, each event represents a unique activity, at the highest level of detail, each event represents the same activity. Since neither of these extreme levels of detail is useful, in practice, the user of the process mining tool will need to select a suitable level of detail.

3.5   Labelling the Ontology

To turn the dendrogram of activities into an ontology, we need to label each node. We create a node-term-matrix, which contains the summed term-frequencies of all events/notes belonging to the respective node. We then calculate mutual information for each node and term and select the six highest weighted terms of each node as the node label. Selecting six terms creates concise labels which allow users to infer the activity quickly.

3.6 Interactive Process Miner

Events following each other have a *directly follows* relationship. As our events are unique, each *directly follows* relationship is unique and therefore, all relationships have the same weight. When the user selects the level of detail of the process model, these relationships are aggregated through the activity ontology. The hierarchical ontology is "cut" at the specified level and the interactive process miner shows the process graph. It indicates frequencies by using bolder connecting arrows for common activities and transitions. It also provides interactive filters to hide infrequent activities and transitions.

This simple process discovery algorithm has some issues, the biggest being that it cannot detect parallelism. As there are many mature process discovery tools that might produce better result, our tool allows exporting the process as a regular, labelled event log.

4 Use Case

We are evaluating our process discovery tool using data from a London NHS trust providing mental health care. Part of this service are community adult mental health hubs. These hubs receive patient referrals from several sources, mostly general practitioners (GPs), but also social care, police, etc. Each referral must be assessed and referred to the appropriate secondary healthcare service.

This process is supposedly straightforward with only four activities: First, referrals are screened, and then admin staff books a tele-triage appointment with the patient. A clinician will assess the patient in the tele-triage appointment and finally refer and discharge the patient. After each of these activities, staff create a note in their PIS (see Fig. 2a).

Referrals are supposed to be completed within two weeks. However, there are various problems with the process and sometimes referrals are abandoned due to patients not showing up, referrals not being appropriate, staff being overworked, etc. Since the process is mostly documented with patient notes, there is little visibility of the process. The hubs do not know how many processes deviate from the straightforward process, or when deviations occur.

We exported all patient notes that were created during a timespan of several months from one of the hubs. Each note is associated with an anonymised patient id. We also checked that notes do not contain patient names. After preparing the event log, we load it into our tool and choose appropriate parameters and filters. Figure 2b shows one of the discovered process models, clearly showing the most frequent process pathways ("process highways") following the supposed process, but also showing infrequent deviations from the supposed process. We showed the process model to subject matter experts, who were able to recognise their process flow and its activities from the automatically selected keywords, thus demonstrating that the technique is viable.

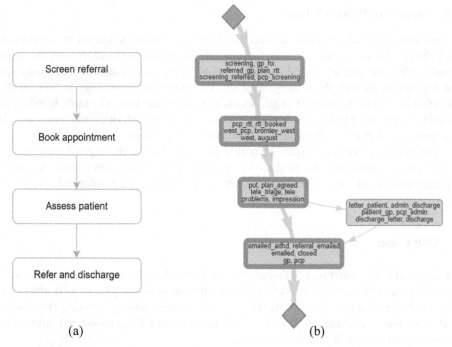

(a)   (b)

Fig. 2. (a) the supposed process (b) one of the discovered process models

5   Conclusion and Future Works

We introduced a novel approach to mine process models from text notes by using the concept of an ontology. Furthermore, we showed that using positional information can improve the clustering of events into activities. We demonstrated the viability of this approach using real-world clinical data.

In future research, we plan to use other available structured data, e.g., the note's author to further improve clustering. We also plan to evaluate text summarisation techniques to label the ontology. We will also evaluate other, more complex clustering techniques, such as hierarchical topic models ([15]) or word2vec ([16]).

References

1. van der Aalst, W.M.P.: Process mining: discovery, conformance and enhancement of business processes. Springer, Berlin (2011). https://doi.org/10.1007/978-3-642-19345-3
2. Alves de Medeiros, A.K., Van Dongen, B.F., van der Aalst, W.M.P., Weijters, A.J.M.M.: Process mining: Extending the α-algorithm to mine short loops. Univ. Technol. **113**, 145–180 (2004). https://doi.org/10.1016/0076-6879(95)52025-2
3. Weijters, A.J.M.M., van der Aalst, W.M.P., Medeiros, A.K.A.D.: Process Mining with the Heuristics Miner Algorithm. Technische Universiteit Eindhoven, Tech. Rep. WP. 166, 1–34 (2006)

4. Günther, C.W., van der Aalst, W.M.P.: Fuzzy mining – adaptive process simplification based on multi-perspective metrics. In: Alonso, G., Dadam, P., Rosemann, M. (eds.): BPM 2007. LNCS, vol. 4714. Springer, Heidelberg (2007). https://doi.org/10.1007/978-3-540-75183-0

5. Leemans, S.J.J., Fahland, D., Van Aalst, W.M.P.: Process and deviation exploration with inductive visual miner. In: CEUR Workshop Proceedings, vol. 1295, pp. 46–50 (2014)

6. Engert, M., Chu, Y., Hein, A., Krcmar, H.: Managing the Interpretive Flexibility of Technology: A Case Study of Celonis and its Partner Ecosystem. (2021)

7. Allard, T., Alvino, P., Shing, L., Wollaber, A., Yuen, J.: A dataset to facilitate automated workflow analysis. PLoS ONE **14**, e0211486 (2019). https://doi.org/10.1371/journal.pone.0211486

8. van der Aalst, W.M.P., Nikolov, A.: EMailAnalyzer: An E-Mail Mining Plug-in for the ProM Framework, (2007). 10.1.1.143.2975

9. Jlailaty, D., Grigori, D.: Mining business process activities from email logs. In: 2017 IEEE International Conference on Cognitive Computing (ICCC). IEEE (2017). https://doi.org/10.1109/IEEE.ICCC.2017.28

10. Banziger, R.B., Basukoski, A., Chaussalet, T.J.: Discovering business processes in crm systems by leveraging unstructured text data. Presented at the The 4th IEEE International Conference on Data Science and Systems (DSS-2018), Exeter, UK January 24 (2019). https://doi.org/10.1109/HPCC/SmartCity/DSS.2018.00257

11. Blei, D.M., Ng, A.Y., Jordan, M.I.: Latent dirichlet allocation. J. Mach. Learn. Res. **2**(3), 993–1022 (2003). https://doi.org/10.1162/jmlr.2003.3.4-5.993

12. Chambers, A.J., et al.: Automated business process discovery from unstructured natural-language documents. LNBIP, pp. 232–243. Springer Science and Business Media Deutschland GmbH (2020). https://doi.org/10.1007/978-3-030-66498-5_18

13. de Medeiros, A.K.A., et al.: An outlook on semantic business process mining and monitoring. In: Meersman, R., Tari, Z., Herrero, P. (eds.) OTM 2007. LNCS, vol. 4806, pp. 1244–1255. Springer, Heidelberg (2007). https://doi.org/10.1007/978-3-540-76890-6_52

14. Manning, C.D., Raghavan, P., Schütze, H.: Introduction to Information Retrieval. Cambridge University Press, New York (2008)

15. Blei, D.M., Griffiths, T.L., Jordan, M.I., Tenenbaum, J.B.: Hierarchical topic models and the nested Chinese restaurant process. In: Advances in Neural Information Processing Systems (2004)

16. Mikolov, T., Chen, K., Corrado, G., Dean, J.: Efficient Estimation of Word Representations in Vector Space (2013). http://arxiv.org/abs/1301.3781,

Knowledge Hypergraph-Based Multidimensional Analysis for Natural Language Queries: Application to Medical Data

Sana Ben Abdallah Ben Lamine[1] , Marouane Radaoui[1] ,
and Hajer Baazaoui Zghal[2]([⊠])

[1] Riadi Laboratory, ENSI, University of Manouba, Manouba, Tunisia
`sana.benabdallah@riadi.rnu.tn`
[2] ETIS UMR8051, ENSEA, CY University, CNRS, Cergy-Pontoise, France
`hajer.baazaoui@ensea.fr`

Abstract. In recent years, data is continuously evolving not only in volume but also in types and sources, which makes the multidimensional analysis and decision making using traditional approaches a complex and difficult task. In this paper, we propose a three-layer-based architecture to perform multidimensional analysis of natural language queries on health data: 1/ Treatment layer aiming at xR2RML mappings generation and knowledge hypergraph building; 2/ Storage layer allowing mainly to store the RDF triples returned by the query of NoSQL databases, and 3/ Semantic layer, based on a domain ontology which constitutes the knowledge base for the generation of the mappings and the building of the knowledge hypergraph. The originality of our proposal lies in the knowledge hypergraph and its capacity to support multidimensional queries. A prototype is developed and the experiments have shown the relevance of the returned multidimensional query results as well as an improvement over traditional approaches.

Keywords: Knowledge hypergraph · Multidimensional analysis ·
Natural language queries · NoSQL databases · Health decision support

1 Introduction

One of the main needs of decision-makers is going through Data Warehouses (*DWs*), building OnLine Analytical Processing (*OLAP*) cubes and performing MultiDimensional Analysis (*MDA*)[1]. The transition to *NoSQL* systems in *DWs* gives decision-makers the possibility of storing and querying unstructured data [2] in large amounts. *DWs* allow *MDA* but remain an expensive solution. Thus, some works proposed to achieve *MDA* on *NoSQL DBs* without going through *DWs* [4,5]. Nonetheless, for decision-makers, and health experts it seems difficult to formulate a MultiDimensional Query (*MDQ*).

© The Author(s), under exclusive license to Springer Nature Switzerland AG 2023
J. Mikyška et al. (Eds.): ICCS 2023, LNCS 14075, pp. 250–257, 2023.
https://doi.org/10.1007/978-3-031-36024-4_19

In this paper, we propose an Knowledge HyperGraph(*KHG*)-based approach to perform *MDA* of natural language queries over multi-source health *NoSQL* data aiming at improving the decision-making process. A prototype is developed and the experiments have shown the results' relevance. In the remainder of this paper, a brief overview of related works is given in Sect. 2. Then, Sects. 3 and 4 detail our proposal and experimental results. Finally, we conclude and present our future works in Sect. 5.

2   Related Works

To address *MDA*, *DWs* use *OLAP* to query data and analyze it from multiple perspectives. Nonetheless, relational models, usually used to implement *DWs* don't permit managing massive data, in addition to data non-freshness and cost of *DWs*. To overcome these issues, researchers have proposed alternative solutions for *MDA*. The use of *NoSQL DWs* is increasingly envisaged [2]. In [4,5] *MDA* is done via direct access to a document-oriented *NoSQL DBs*.

On the other hand, Knowledge Graphs (*KGs*) and *KHGs* allow solving interoperability problems in order to interrogate efficiently massive and heterogeneous data [6]. *KGs* have been also attractive to researchers for they support analytics and decision-making [7]. Some works have used *KGs* to address *MDA*. In [11], *OLAP* is adapted to perform analysis on *KGs*. In [12], a graph-based *DW* is proposed. Our motivation is to exploit the advantages of *KHGs* in *MDA*.

3   Proposed Approach

Figure 1 presents the three-layers architecture of our *MDA* approach.

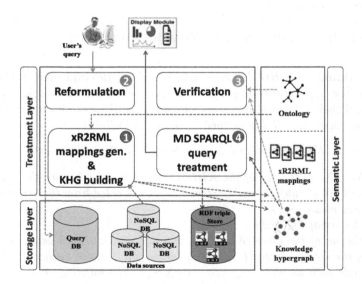

Fig. 1. Knowledge hypergraph-based multidimensional analysis architecture

1. **Treatment layer**: its four modules are detailed in the following subsections.
2. **Storage layer**: it consists of:
   - **NoSQL DBs**: are the data sources targeted by the user's query
   - **Query DB**: stores the queries and their respective reformulations.
   - **RDF triples store**: stores the *RDF* triples returned by the query.
3. **Semantic layer**: it consists of:
   - **Ontology**: it is the domain ontology developed in [9] and constituting the knowledge base for the generation of the mappings and the building of the *KHG*. It is also used in the verification module.
   - **KHG**: is a data integration framework allowing for unified querying.
   - **xR2RML mappings**: are *RDF* documents representing logical sources extracted from the input databases.

3.1 *xR2RML* mappings generation and *KHG* building module

xR2RML **Mappings Generation** Is done in the following steps:

1. For each collection c_i in the *NoSQL* database DB_J, a logical source (*xR2RML* semantic view) is extracted using the property **xrr:logicalSource**.
2. For each document d_k of c_i, a triples map (tp) is created. For each tp, a subject map is generated, which represents the unique identifier used in all the *RDF* triples generated from it.
3. For each tp, a **predicateObjectmap** is generated. The predicate is extracted from the input data or the ontology. The object corresponds to the document's value field according to its type:
   - If it is simple, it is mapped to a predicate object and a data property using **xrr:reference**
   - Else, if it is complex, it is mapped to another triples map and an object property using the **rr:ParentTriplesMap** property.

KHG Building. A *KHG* describes real world entities and their interrelations organized in a hypergraph. It permits the representation of complex structures (classes and their relationships) into a hypernode. Hypernodes are interrelated, using hyperedges. A *KHG* is defined formally in Definition 1.

Definition 1. *KHG: is defined as a tuple* $< N, V, A, S_M, \zeta >$ *with :*

- $N = N_s \cup N_o$ *is the set of the KHG's nodes;* N_s *and* N_o *are the sets respectively of triples' subjects and triples' objects extracted from the set of xR2RML mappings views* (S_M)
- V: *is the set of hyperedges*
- A: *is the set of the directed arcs; an arc is a pair* $< u, v >$ *where* $u, v \in N$
- S_M: *is the set of xR2RML mapping views* (m_i), *where each* $m_i \in S_M$ *is an hypernode such as:* $m_i = S_n \cup S_a$, *with* $S_n \subset N$ *and* $S_a \subset A$
- ζ *is the set of the concepts of the used domain ontology*

The construction of the *KHG* is done via the definition of its:

1. **Entities**: each semantic view is a hypernode (a directed graph which nodes and arcs are respectively the ontology's concepts and properties).
2. **Relations**: relations between the semantic views are the hyperedges of the *KHG*. Two types of hyperedges are constructed:
   - *DBRef* fields from data sources are transformed into hyperedges.
   - Domain ontology's object properties are transformed into hyperedges.

3.2   Reformulation Module

The reformulation of a user's query (*U-Q*) into a *MDQ* (Def. 2) is detailed below.

Definition 2. *MDQ is a tuple* $Q = (G, S, M, \psi)$, *with:*

- *G: is the non-empty set of GROUP BY attributes of the request,*
- *S: is the selection predicate (facultative)*
- *M: is the set of measure attributes which are numeric.*
- *ψ: is the aggregation operator (average, sum, ...).*

Preprocessing. Consists of decomposing *U-Q* into words (or set of words) according to their grammatical function in the query phrase using a grammatical resource. A vector of pairs is obtained $\overrightarrow{V} = <p_1, ..., p_n>$, with $p_i = (S_{wi}, f_i)$ and $S_{wi} = \{w_{i1}, ..., w_{ik}\}$ is a set of k words of the decomposed query, where $k >= 1$ and f_i is its grammatical function.

BGPQ Schema Extraction Is done in two steps:

- **Triples's extraction**: Transforms *U-Q* into a set of triple patterns (S_{tp}). Having $\overrightarrow{V}$ as input, it parses *U-Q* into sub-sentences using the lexical resource. Each sub-sentence turns into a *tp (s, p, o)*, where *s*, *p* and *o* are respectively the subject, verb, and object of the sub-sentence.
- **Triples's aggregation**: Transforms S_{tp} into a *BGPQ* Schema (Algorithm 1). To determine S_{class} (set of classes) and $S_{predicat}$ (set of predicates), for each pair $p_k = <t_i, t_j>$ in S_{tp}, where $k \in [1, |S_{tp}|^2]$, $i \in [1, |S_{tr}| - 1]$ and $j \in [2, |S_{tp}|]$, I_p is the set of common classes between t_i and t_j. If $|I_p| > 0$, non common classes are added to S_{class}, and the respective predicates to $S_{predicate}$. The *BGPQ* Schema is the union of S_{class} and $S_{predicate}$ (line 10).

Algorithm 1. Triples Agregation

Input: S_{tp} : Set of triples
Output: $BGPQSchema$
Begin

1: $S_{class} \leftarrow \{\}$
2: $S_{predicate} \leftarrow \{\}$
3: **for each** $p_k = < t_{pi}, t_{pj} > \in S_{tp}$ **do**
4:     $I_p \leftarrow \{t_{pi}.subject, t_{pi}.object\} \cap \{t_{pj}.subject, t_{pj}.object\}$
5:     **if** $|I_p| > 0$ **then**
6:         $S_{class} \leftarrow S_{class} \cup \{t_{pi}.subject, t_{pi}.object, t_{pj}.subject, t_{pj}.object\} - I_p$
7:         $S_{predicate} \leftarrow S_{predicate} \cup \{t_{pi}.predicate, t_{pj}.predicate\}$
8:     **end if**
9: **end for**
10: $BGPQSchema \leftarrow S_{class} \cup S_{predicate}$
11: **return** $BGPQSchema$

End

Algorithm 2. MDQComponentsExtraction

Input: V: Vector of pairs
$BoolList$: list of boolean operators $(>, <, >=, <=, =, in, \ldots)$
$ListOp$: list of aggregation operators (sum, average, percentage...)
$BGPQSchema$
Output: Ψ, M, S, G
Begin

1: $max \leftarrow 0$
2: **for each** $p_i \in \overrightarrow{V}$ **do**
3:     **for each** $w \in p_i.S_{wi}$ **do**
4:         $m \leftarrow MaxSimAg(w, ListOp, index)$
5:         **if** $m > max$ **then**
6:             $max \leftarrow m$
7:             $j \leftarrow i$
8:             $indexMax \leftarrow index$
9:         **end if**
10:     **end for**
11: **end for**
12: $\Psi \leftarrow ListOp[indexMax]$
13: $M \leftarrow SearchNumAtt(\overrightarrow{V}[j].S_{wi}, ListNumAtt(BGPQSchema))$
14: $S \leftarrow SearchPredAtt(\overrightarrow{V}, BoolList, ListAtt(BGPQSchema))$
15: $G \leftarrow SearchGroupByAtt(BGPQSchema, \overrightarrow{V})$

End

BGPQ Formulation. Algorithm 2 extracts the *MDQ* 's components (Def. 2):

- **Aggregation operator** ψ: for each p_i of $\overrightarrow{V}$, *SimAg()* seeks for the word w of $p_i.S_{wi}$ and the operator of *ListOp*, which are the most similar (Jaccard similarity coefficient). *indexMax* is the index of ψ in *ListOp*.
- **Measure attribute** M: j is the index of S_{wi} containing ψ (line 7). Thus, *SearchNumAtt()* assigns to M the most similar among the numeric attributes of the *BGPQ* schema to the words of S_{wi}.
- **Selection predicate** S **(optional)**: *SearchPredAtt()* searches if a set of words S_w of $\overrightarrow{V}$ contains an operator from *BoolList* and finds the most similar among *ListAtt(BGPQ Schema)*.
- **GROUP BY attributes** G: *SearchGroupByAtt()* returns a set of atomic attributes G, excluding M and S.

3.3   Verification Module

In a *MDQ* the measure attributes M and the GROUP BY attributes G must not be on the same dimension hierarchy. The verification is double:

- λ: based on the *KHG*, allows to check the correctness of the request based on the graph of functional dependencies of the grammatical resource (Eq. 1). The query is correct if $\lambda <> 0$ (a Roll-up if $\lambda > 0$ else a Drill-down).

$$\lambda\left(Att\left(m_i\right), Att\left(g_k\right)\right) = 1 - \frac{|Root\left(m_i\right)|}{|Root\left(g_k\right)|} \tag{1}$$

- λ_s: checks the validity of the query against the domain ontology. In Eq. 2, *Cpt (a)* returns the ontological concept with the attribute a. The query is correct if $\lambda_s <> O$.

$$\lambda_S\left(Cpt\left(m_i\right), Cpt\left(g_k\right)\right) = 1 - \frac{SemanticDepth\left(Cpt\left(m_i\right)\right)}{SemanticDepth\left(Cpt\left(g_k\right)\right)} \tag{2}$$

3.4   Multidimensional SPARQL Query Treatment Module

MD SPARQL Query Generation is done in three steps:

1. ***SELECT*** **clause**: followed by the attributes of *BGPQ* excluding G, then ψ followed by M.
2. ***WHERE*** **clause**: followed by the list of *tp* of the query and if $S <> \emptyset$, the selection predicates are added between parentheses after *FILTER*.
3. ***GROUP BY*** **clause**: all the attributes of G are added. It should be mentioned that all the attributes of the request are preceded by '*?*'.

Display and Storage of Results. The obtained *MD SPARQL* query is syntactically checked and then executed on the *KHG*. The obtained triples are sent to the display module and the *RDF* store for further use in similar queries.

4   Evaluation and Discussion

To implement the prototype we used: xR2RML[1], Jena API, OWL (Ontology Web Language), OWL API to check syntactically and execute the *MD SPARQL*[2] query, BabelNET API as lexical resource, Stanford dictionary API[3]: as lexical and grammatical resource, and Allegrograph[4]as *RDF* store.

The data collection used is *Patient_ survey (Data.gov* site[5]) with more than 700,000 records. *Json* Generator tool[6] is used to produce large-scale data with a data schema presented in [8]. These files are loaded in a *MongoDB* database.

4.1   Evaluation of the KHG's Completeness

The completeness of information in a *KHG* influences the relevance of data query results. Three completeness metrics are calculated using Sieve[7] and KBQ[8]:

- **Schema Completeness (*SC*)**: the rate of ontology's classes and properties in the *KHG*. *SC*=0.97, hence the *KHG* represents a large range of knowledge.
- **Interlinking Completeness (*IC*)**: the ratio of interlinked triples. *IC*=0.873, hence the richness of the *KHG*'s properties.
- **Currency Completeness (*CC*)**: the ratio of unique triples. *CC*=0.819, so no redundancy.

4.2   Evaluation of the KHG-Based MDA

Precision, recall and *F-measure* are used to evaluate the relevance of a set of queries. The average values obtained are respectively 0.82, 0.53 and 0.63. Table 1 reports average *precision* and *recall* for two traditional approaches [9] and ours for which the relevance is improved. In [6], it is reported that after 80% of integrated data sources, these values tend towards 1, when using *KHG*.

Table 1. Comparison of relevance results

Approaches	Average precision	Average recall
Without domain ontology [9]	0.59	0.36
Domain ontology + *NoSQL DB* [9]	0.62	0.52
Our *KHG*-based *MDA* approach	0.82	0.53

---

[1] https://github.com/frmichel/morph-xr2rml.
[2] https://www.w3.org/TR/sparql11-query/.
[3] https://nlp.stanford.edu/software/lex-parser.shtml.
[4] https://allegrograph.com/.
[5] http://healthdata.gov/dataset/patient-survey-hcahps-hospital/.
[6] https://json-generator.com/.
[7] http://sieve.wbsg.de/.
[8] https://github.com/KBQ/KBQ.

5   Conclusions and Future Work

In this paper, a *KHG*-based *MDA* approach is proposed. The idea is to help health experts expressing *MDQ* on multi-source data to improve decision-making. The relevance of the results is improved. In our future work, we intend to study the performance of the approach with real-time treatment and scaling up data size. For the reformulation of queries, deep learning will be used based on our previous works [10].

References

1. Selmi, I., Kabachi, N., Ben Abdalah Ben Lamine, S., Baazaoui Zghal, H.: adaptive agent-based architecture for health data integration. In: Yangui, S., et al. (eds.) ICSOC 2019. LNCS, vol. 12019, pp. 224–235. Springer, Cham (2020). https://doi.org/10.1007/978-3-030-45989-5_18
2. Dehdouh, K.: Building OLAP cubes from columnar NoSQL data warehouses. In: Bellatreche, L., Pastor, Ó., Almendros Jiménez, J.M., Aït-Ameur, Y. (eds.) MEDI 2016. LNCS, vol. 9893, pp. 166–179. Springer, Cham (2016). https://doi.org/10.1007/978-3-319-45547-1_14
3. Chevalier, M., El Malki, M., Kopliku, A., Teste, A., Tournier, R.: document-oriented models for data warehouses - nosql document-oriented for data warehouses. In: Proceedings of the 18th International Conference on Enterprise Information Systems, pp. 142–149. SciTePress, Rome, Italy (2016)
4. Chouder, M.L., Rizzi, S., Chalal, R.: EXODuS: exploratory OLAP over document stores. Inf. Syst. **79**, 44–57 (2019)
5. Gallinucci, E., Golfarelli, M., Rizzi, S.: Approximate OLAP of document-oriented databases: A variety-aware approach. Inf. Syst. **85**, 114–130 (2019)
6. Masmoudi, M., Ben Abdallah Ben Lamine, S., Baazaoui Zghal, H., Archimède; B., Karray, M.: Knowledge hypergraph-based approach for data integration and querying: Application to Earth Observation. Future Gener. Comput. Syst. (2021)
7. Gomez-Perez, J.M., Pan, J.Z., Vetere, G., Wu, H.: Enterprise knowledge graph: an introduction. In: Exploiting Linked Data and Knowledge Graphs in Large Organisations, pp. 1–14. Springer, Cham (2017). https://doi.org/10.1007/978-3-319-45654-6_1
8. Ait Brahim, A., Tighilt Ferhat, R., Zurfluh, G.: Model driven extraction of NoSQL databases schema: case of MongoDB. In: International Conference on Knowledge Discovery and Information Retrieval, pp. 145–154. ScitePress, Vienna, Austria (2019)
9. Radaoui, M., Ben Abdallah Ben Lamine, S., Baazaoui Zghal, H., Ghedira, C., Kabachi, N.: Knowledge guided knowledge guided integration of structured and unstructured data in health decision process. In: Information Systems Development: Information Systems Beyond 2020, ISD 2019 Proceedings, Association for Information Systems, Toulon, France (2019)
10. Ben Abdallah Ben Lamine, S., Dachraoui, M., Baazaoui-Zghal, H.: Deep learning-based extraction of concepts: a comparative study and application on medical data. J. Inform. Knowl. Manag. (2022)
11. Schuetz, C.G., Bozzato, L., Neumayr, B., Schrefl, M., Serafini, L.: Knowledge Graph OLAP. Semantic Web **12**(4), 649–683 (2021)
12. Friedrichs, M.: BioDWH2: an automated graph-based data warehouse and mapping tool. J. Integr. Bioinform. **18**(2), 167–176 (2021)

Coupling Between a Finite Element Model of Coronary Artery Mechanics and a Microscale Agent-Based Model of Smooth Muscle Cells Through Trilinear Interpolation

Aleksei Fotin$^{(\boxtimes)}$ and Pavel Zun

ITMO University, Saint Petersburg, Russia
alexseixiv@gmail.com

Abstract. Finite element (FE) simulation is an established approach to mechanical simulation of angioplasty and stent deployment. Agent-based (AB) models are an alternative approach to biological tissue modeling in which individual cells can be represented as agents and are suitable for studying biological responses. By combining these two approaches, it is possible to leverage the strengths of each to improve the in silico description of angioplasty or stenting and the following healing response.

Here we propose a couping between FE and AB vascular tissue models using trilinear interpolation, where the stresses (and strains) in the AB model arise directly from the forces of interaction between individual agents. The stress values for FE and AB models are calculated and compared.

Keywords: arterial tissue · mechanical model · agent-based model · finite-element model

1 Introduction

Finite element simulation is now often used to study stent implantation in coronary arteries *in silico*. Using finite element method (FEM) reproduces the mechanical behavior on the continuous scale of the stent, balloon, and artery. On the other hand, an approach that allows to explore the scale of cellular interaction is offered by agent-based models (ABM) [1]. Although the FEM [2–4] and ABM [5–7] approaches could be used independently to study both the mechanical processes during stenting and the biological response after stent deployment, nevertheless, combining the two methods will provide a more accurate and reliable approach to analysis both from the mechanical point of view (through FEM) and from the biological point of view (through ABM).

The strength of FE modelling lies in the ability to accurately simulate large-scale mechanical behaviour. Cell-resolved AB models, on the other hand, can naturally include cell-scale biological behaviour [6]. Most of the AB models proposed so far are center-based (CB) models, meaning that each cell is approximated as a sphere, and the forces act between cell centers, e.g. [6,8–11], which

---

The original version of chapter 20 was revised: the authors' names are displayed correctly. The correction to this chapter is available at
https://doi.org/10.1007/978-3-031-36024-4_54

© The Author(s), under exclusive license to Springer Nature Switzerland AG 2023, corrected publication 2023
J. Mikyška et al. (Eds.): ICCS 2023, LNCS 14075, pp. 258–269, 2023.
https://doi.org/10.1007/978-3-031-36024-4_20

allow for simulations on the order of several millions of cells, but simplify the microscopic properties. Deformable cell (DC) models present an alternative approach. They produce a much more detailed microscopic behaviour, but also have much higher computational costs per cell, e.g. [12,13].

Here, a method is proposed to combine the two approaches to analyze stent deployment in coronary arteries. The purpose of this paper is to propose a method for integrating the results of finite element modeling of stent deployment into an agent-based model, which, in turn, will be used to study tissue growth and vessel restenosis. Thus, we want to obtain an equivalent mechanical response for the agent-based model based on the results of the finite element simulation, which is important for obtaining reliable predictions of the development of restenosis.

One-way couplings of FE and AB models for vascular walls have been proposed before, e.g. in [14–16], as well as a two-way coupling approach [17]. Note that in all these papers the agents are uniformly seeded on the deformed postdeployment vessel, do not interact mechanically, and the stress value is passed to them as an external parameter. Here we propose an alternative approach where the stress (and strain) in the AB model are computed directly from the interaction forces between individual agents.

2 Methods

This section introduces the AB and FE models used in this study, the coupling method, and outlines the computational experiments performed.

2.1 Finite Element Model

The geometric model of an idealized artery represented as a straight cylindrical tube was discretized with linear hexahedral elements with reduced integration (C3D8R). For simplicity, the wall of the vessel consists only of the tunica media, the middle layer of the artery. The reason is that tunica media is responsible for a large part of the vessel's mechanical behaviour, and also its mechanical properties are relatively less varied between individuals. A material model with hyperelastic behavior was calibrated to replicate the experimental stress-strain relationships reported in [18]. In particular, the model parameters were selected to reproduce the circumferential behavior, since it is the prevalent strain direction in the vessel during the pressurization and stenting procedure. Simulation of vessel pressurization was performed by imposing an increase in pressure on the lumen surface until the desired value ($100 \ mmHg$) was reached.

All finite element analyses were performed using Abaqus/Explicit (SIMULIA Corp., USA), and each was checked to work in the quasi-static regime.

2.2 Agent-Based Model

The model presented here uses a design similar to the models presented in [5,6, 19]. It is a center-based AB model, where each cell is represented as an elastic sphere, which can interact with its neighbours via elastic repulsion and pairwise bond forces. The elastic repulsion force pushes overlapping agents away, while the bonds are established for all initially neighbouring cells and are used to simulate tissue's tensile properties, which *in vivo* are provided by cell-cell adhesion as well as by extracellular molecules such as collagen and elastin [20]. Both the elastic and the bond forces in the model act between the cell centers.

For elastic repulsion, a Neo-Hookean extension of Hertz elastic sphere contact is used as described in [5,19]:

$$F_{hertz} = \frac{8a^3 B(16a^2 - 36\pi aR + 27\pi^2 R^2)}{3R(4a - 3\pi R)^2} \tag{1}$$

where R is the effective radius calculated as

$$R = \frac{R_1 R_2}{R_1 + R_2}$$

and a is the contact area and is approximated as

$$a = \sqrt{R \cdot (R_1 + R_2 - d)}$$

Here, R_1 and R_2 are the agents' radii, d is the distance between their centres, B is the elastic constant. For all experiments, the elastic constant was set to $B = 0.2\ MPa$ based on [5].

The attractive bond force is based on several different mechanisms (cell adhesion and extracellular fibers) and its purpose in the model is to provide a realistic macroscopic behaviour. We opt to use a polynomial attractive force which is fitted to mimic the macroscopic behaviour of the FE model and the experimental data. The force between neighbouring cells is calculated by the following formula:

$$F_{bond} = (R_1 + R_2)^2 \cdot \sum_{k=1}^{N} c_k \sigma^k, \quad \sigma > 0 \tag{2}$$

where the bond strain σ equals

$$\sigma = \frac{d - R_1 - R_2}{R_1 + R_2}$$

All c_k coefficients were restricted to non-negative values, to ensure that the energy minimum for each bond is located at zero strain. Following the FE model, we use 6th order polynomials for the bond forces ($N = 6$). Separate sets of coefficients $c_1..c_6$ are used for intima, media and adventitia in the three-layer wall model considered here.

The cells are placed randomly while maintaining a minimal distance between each pair of cells. For this, we use a three-dimensional Poisson disc sampling generated by Bridson's algorithm [21]. This sampling allows us to produce isotropic tissue with an almost constant density. A sample of the generated tissue is shown in Fig. 1. Each cell was assigned the same radius $r = 0.015$ mm.

Fig. 1. Isotropic tissue sample generated by a 3D Poisson Disc sampling.

The AB model is implemented as a part of the ISR3D model[1].

2.3   Coupling AB and FE Simulations

For inflation and stenting simulations, AB vessels were generated based on FE geometries. For each finite element in the arterial wall and in the stent, agents were placed inside using Poisson Disc (PD) sampling. For each agent, bonds were added to agents within a cutoff distance. The tissue was then equilibrated to improve the structure at the interfaces between different finite elements, reducing gaps and overlaps of agents.

To verify the equivalence between the finite element model and the agent-based model, the displacement history recorded with the finite element simulation was imposed on the AB model, and the lumen inner diameter-pressure relation was observed. The correspondence of this relationship obtained with the finite element method and with the agent-based model allows to affirm that the two models are equivalent.

The displacement is imposed by recording the trajectories for all nodes in the FE artery. Then, for each agent, its trajectory is calculated using trilinear interpolation. This method is an extension of linear and bilinear interpolation for the 3D case. In essence, the method is a sevenfold application of linear interpolation according to the formula:

$$c = c_0 + \frac{dist(\tilde{c}, \tilde{c}_0)}{dist(\tilde{c}_1, \tilde{c}_0)}(c_1 - c_0) \tag{3}$$

where $\tilde{c}_i$ is the coordinate of the point i, c_i is the known offset at the point i, c is the result of the interpolation.

---

[1] https://github.com/ISR3D/ISR3D.

Thus, for each agent, it is necessary to determine the nodes and the offset of the finite element in which the agent is located, and sequentially calculate the coefficients c_{00}, c_{01}, c_{10}, c_{11}, c_0, c_1, c (Fig. 2). The last coefficient is the required trajectory for agent.

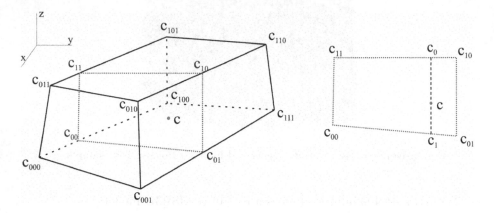

Fig. 2. Trilinear interpolation for point c in cuboid.

During the displacement, the agents in the vessel wall interact via the forces described above, providing stress-strain dynamics for the deployment. After the agents reach the end of their trajectories, the lumen surface is fixed in place, and the rest of the agents are equilibrated.

2.4   Pressurization Tests

Pressurization tests were performed in a simple cylindrical vessel (length 4.5 mm, inner radius 1.26 mm, outer radius 2.32 mm). The FE vessel was generated first (Fig. 3), and then the AB vessel (containing 1767269 agents, Fig. 4) was generated. Variant of this vessel was considered where the cylinder is made entirely of the media tissue. Pressurization to 100 mmHg (13.332 kPa) was simulated with the FE model, followed by the AB model, coupled as described before. Pressurization was considered: one, where no specific boundary conditions were imposed on the nodes at the ends of the vessel; the other, where the nodes (FE) or agents (AB) on both ends of the vessel were blocked from moving along the longitudinal direction.

2.5   Assessing the Results

The values of inner radius change and intramural stress were used for comparison between AB and FE. The stress was considered in three directions relative to the vessel axis: circumferential, radial, and axial. The intramural stress for the FE model was calculated from the per-element stress matrix. In AB model, the

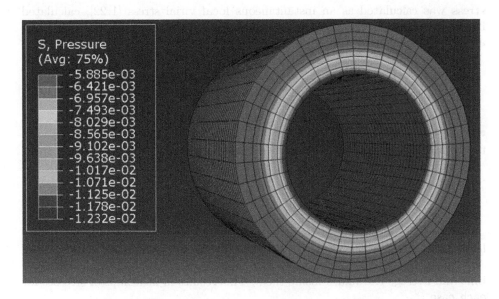

Fig. 3. Results of the finite element model of the vessel.

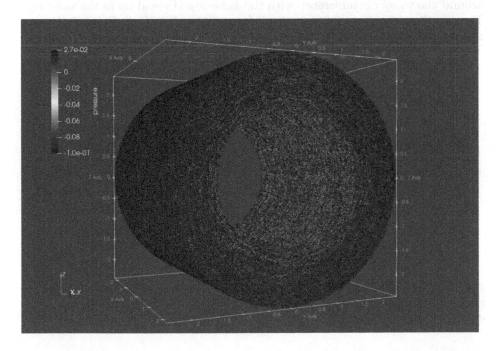

Fig. 4. The agent-based vessel generated from the finite element model after pressurization displacement.

stress was calculated as an instantaneous local virial stress [1,22], calculated from pairwise interaction forces and agent size. The virial stress for i-th agent along m-th coordinate is calculated as:

$$\tau_m^i = \frac{3}{4\pi R_i^3} \sum_{j\in neighbours} f_{ij}^m \cdot (R_i + \frac{d_{ij} - R_i - R_j}{2}), \ m \in \{x,y,z\} \tag{4}$$

where j is an index that goes over all neighbouring agents interacting with the i-th agent, and f_{ij}^m is the force exerted by j-th on i-th agent along m-th coordinate. The AB pressure then is calculated as

$$p_{virial} = -\frac{\tau_x^i + \tau_y^i + \tau_z^i}{3} \tag{5}$$

To enable comparison of stresses on the same scale, and to smooth out the local fluctuations in the virial stress that arise from the inhomogeneities in the AB tissue, we allocate the stress values from both FE and AB simulations to bins and average the values inside them. This results in heatmaps of stresses for each case.

Since the vessels are cylindrically symmetric, we choose to allocate the bins based on the longitudinal coordinate x and the radial distance r; all the points around the vessel circumference with the same x and r end up in the same bin. The bin dimensions are 0.32 mm both for x and r directions. Figure 5 shows a schematic illustration of this approach.

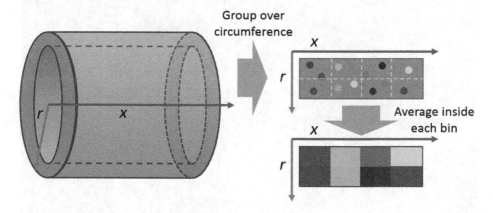

Fig. 5. A schematic depiction of the averaging using bins in the rx plane. First, the points are grouped over the circumferential direction. Then, the value of each bin is calculated as the average of all points in it.

The results are analyzed with NumPy[2] and Pandas[3], and the plots are generated with Matplotlib[4] and Seaborn[5]. 3D results from the AB model are plotted with Paraview[6].

3   Results

To compare FE and AB models, we performed pressurization tests at a pressure of 100 $mmHg$. For the case of a segment of media with free ends, a cross-sectional distribution of intramural pressure for the FE and AB vessels is shown in the Fig. 6 and in the Table 2 . The case of a segment of media with fixed ends is shown in the Fig. 7 and in the Table 1.

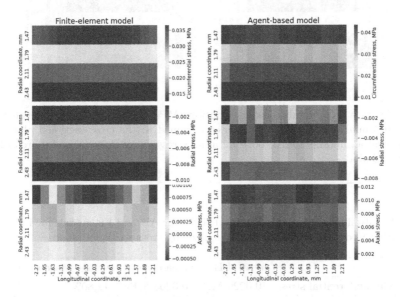

Fig. 6. Average intramural stress distribution (MPa) for AB and FE models for inflation at 100 mmHg. Media tissue, unconstrained ends. Top: circumferential, middle: radial, bottom: axial stress. Note that the scales are different for each subplot.

The inner radius of the AB model vessel has been increased from 1.26 mm to 1.47 mm, the same as the FE vessel. As for the outer radius, for the agent-based model it changed from 2.32 mm to 2.45 mm, and in the finite element simulation from 2.32 mm to 2.43 mm.

---

[2] https://numpy.org.
[3] https://pandas.pydata.org.
[4] https://matplotlib.org.
[5] https://seaborn.pydata.org.
[6] https://www.paraview.org.

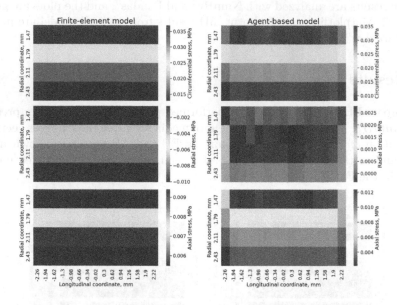

Fig. 7. Average intramural stress distribution (MPa) for AB and FE models for inflation at 100 mmHg. Media tissue, ends constrained to the cross-sectional plane. Top: circumferential, middle: radial, bottom: axial stress. Note that the scales are different for each subplot.

Table 1. Intramural stress distribution (MPa) for AB and FE models for inflation at 100 mmHg. Media tissue, ends constrained to the cross-sectional plane.

	mean, FE	mean, AB	average inaccuracy
Circumferential stress	0,0229 MPa	0,0203 MPa	12,80%
Radial stress	–0,0049 MPa	0,0006 MPa	116,10%
Axial stress	0,0069 MPa	0,0065 MPa	23,00%

Table 2. Intramural stress distribution (MPa) for AB and FE models for inflation at 100 mmHg. Media tissue, unconstrained ends.

	mean, FE	mean, AB	average inaccuracy
Circumferential stress	0,0225 MPa	0,0206 MPa	25,40%
Radial stress	–0,0049 MPa	-0,0043 MPa	100,30%
Axial stress	0,0001 MPa	0,0044 MPa	6764,60%

4   Discussion

The inflation tests for a cylindrical segment of the vessel demonstrate a good agreement between AB and FE for the final inner and outer radii of the vessel. The inner radius in AB is exactly the same as FE, and the outer radius is slightly higher. The likely cause is that incompressibility is not enforced in any way in the AB model, unlike FE. Even if individual agents stay at the same equilibrium distance from each other, the enveloping volume may differ. There are ways to enforce a constant volume in particle-based methods: for example, [23,24] describe such a method for smooth particle hydrodynamics (SPH). However, testing these methods is outside the scope of this article.

The agreement between the two considered models is good in terms of circumferential stress (the average deviation is no more than 25%), and somewhat weaker for the radial and axial stress components. It should be noted that the circumferential stress is the largest in the pressurization scenario consivered here and also in clinically relevant stenting scenarios, and is considered the most important for the biological response. A possible reason for the discrepancy in the axial and the circumferential components, in addition to the compressibility of AB tissue, is one limitation intrinsic to center-based AB models: since the cells cannot change their shape, virial stress calculations are known to be inaccurate, especially for compressive behaviour [1]. The only way to avoid this is by using deformable cell models [12], which are a lot more expensive computationally.

5   Conclusions

The pressurization tests show that the presented AB model is in a good match with the geometry of the FE model. This allows the AB model to capture deformation-based cues, important for the cells' biological response.

The averaged stress in the tissue is also close between the two models. However, there are noticeable fluctuations on the scale of individual cells.

This means that the coupling mechanism considered in this article can be used to directly impose mechanical strains and stresses from finite element models onto agent-based biological models, to inform the biological response to mechanical cues in tissue simulations. However, stresses should be used more cautiously than strains, since the difference between FE and AB is larger, although it is comparable with biological variability between individual vessels [18].

Acknowledgements. The authors acknowledge funding from the Russian Science Foundation under agreement #20-71-10108.

This work has been carried out using computing resources of the federal collective usage center Complex for Simulation and Data Processing for Mega-science Facilities at NRC "Kurchatov Institute", http://ckp.nrcki.ru/

References

1. Van Liedekerke, P., Palm, M.M., Jagiella, N., Drasdo, D.: Simulating tissue mechanics with agent-based models: concepts, perspectives and some novel results. Comput. Particle Mech. **2**(4), 401–444 (2015). https://doi.org/10.1007/s40571-015-0082-3
2. Morlacchi, S., Pennati, G., Petrini, L., et al.: Influence of plaque calcifications on coronary stent fracture: A numerical fatigue life analysis including cardiac wall movement. J. Biomech. **47**, 899–907 (2014). https://doi.org/10.1016/j.jbiomech.2014.01.007
3. Javier, E., Martínez, M.A., McGinty, S.P.E.: Mathematical modelling of the restenosis process after stent implantation. J. R. Soc. Interface. **16** (2019). https://doi.org/10.1098/rsif.2019.0313
4. Gierig, M., Wriggers, P., Marino, M.: Computational model of damage-induced growth in soft biological tissues considering the mechanobiology of healing. Biomech. Model. Mechanobiol. **20**(4), 1297–1315 (2021). https://doi.org/10.1007/s10237-021-01445-5
5. Melnikova, N., Svitenkov, A., Hose, D., Hoekstra, A.: A cell-based mechanical model of coronary artery tunica media. J. R. Soc. Interface **14** (2017). https://doi.org/10.1098/rsif.2017.0028
6. Zun, P.S., Narracott, A.J., Chiastra, C., Gunn, J., Hoekstra, A.G.: Location-specific comparison between a 3D In-stent restenosis model and Micro-CT and histology data from Porcine *In Vivo* Experiments. Cardiovasc. Eng. Technol. **10**(4), 568–582 (2019). https://doi.org/10.1007/s13239-019-00431-4
7. Corti, A., Chiastra, C., Colombo, M., Garbey, M., Migliavacca, F., Casarin, S.: A fully coupled computational fluid dynamics - agent-based model of atherosclerotic plaque development: Multiscale modeling framework and parameter sensitivity analysis. Comput. Biol. Med. **118** (2020). https://doi.org/10.1016/j.compbiomed.2020.103623
8. Walker, D., Georgopoulos, N., Southgate, J.: Anti-social cells: Predicting the influence of E-cadherin loss on the growth of epithelial cell populations. J. Theor. Biol. **262**, 425–440 (2010). https://doi.org/10.1016/j.jtbi.2009.10.002
9. Schlüter, D., Ramis-Conde, I., Chaplain, M.: Multi-scale modelling of the dynamics of cell colonies: insights into cell-adhesion forces and cancer invasion from in silico simulations. J. Royal Soc. Interf. Royal Soc. **12**, 20141080- (2015). https://doi.org/10.1098/rsif.2014.1080
10. Ingham-Dempster, T., Walker, D., Corfe, B.: An agent-based model of anoikis in the colon crypt displays novel emergent behaviour consistent with biological observations. Royal Soc. Open Sci. **4** (2017). https://doi.org/10.1098/rsos.160858
11. Tahir, H., et al.: Multi-scale simulations of the dynamics of in-stent restenosis: impact of stent deployment and design. Interface Focus **1**, 365–373 (2011). https://doi.org/10.1098/rsfs.2010.0024
12. Van Liedekerke, P., et al.: A quantitative high-resolution computational mechanics cell model for growing and regenerating tissues. Biomech. Model. Mechanobiol. **19**(1), 189–220 (2019). https://doi.org/10.1007/s10237-019-01204-7
13. Dabagh, M., Jalali, P., Butler, P., Randles, A., Tarbell, J.: Mechanotransmission in endothelial cells subjected to oscillatory and multi-directional shear flow. J. Royal Soc. Interface. **14** (2017). https://doi.org/10.1098/rsif.2017.0185
14. Keshavarzian, M., Meyer, C.A., Hayenga, H.N.: Mechanobiological model of arterial growth and remodeling. Biomech. Model. Mechanobiol. **17**(1), 87–101 (2017). https://doi.org/10.1007/s10237-017-0946-y

15. Nolan, D., Lally, C.: An investigation of damage mechanisms in mechanobiological models of in-stent restenosis. J. Comput. Sci. **24**, 132–142 (2018). https://doi.org/10.1016/j.jocs.2017.04.009
16. Boyle, C., Lennon, A., Prendergast, P.: Application of a mechanobiological simulation technique to stents used clinically. J. Biomech. **46**, 918–924 (2013). https://doi.org/10.1016/j.jbiomech.2012.12.014
17. Li, S., Lei, L., Hu, Y., Zhang, Y., Zhao, S., Zhang, J.: A fully coupled framework for in silico investigation of in-stent restenosis. Comput. Methods Biomech. Biomed. Eng. **22**, 217–228 (2019). https://doi.org/10.1080/10255842.2018.1545017
18. Holzapfel, G., Sommer, G., Gasser, C., Regitnig, P.: Determination of layer-specific mechanical properties of human coronary arteries with nonatherosclerotic intimal thickening and related constitutive modeling. Am. J. Physi.-Heart Circulatory Physiol. **289**, 2048–2058 (2005). https://doi.org/10.1152/ajpheart.00934.2004
19. Zun, P., Anikina, T., Svitenkov, A., Hoekstra, A.: A comparison of fully-coupled 3D in-stent restenosis simulations to In-vivo Data. Front. Physiol. **8**, 284 (2017). https://doi.org/10.3389/fphys.2017.00284
20. Ratz, P.: Mechanics of Vascular Smooth Muscle. In: Terjung, R., (ed.) Comprehensive Physiology, vol. 2, pp. 111–168 (2015). https://doi.org/10.1002/cphy.c140072
21. Bridson, R.: Fast Poisson disk sampling in arbitrary dimensions. In: ACM SIGGRAPH 2007 sketches (SIGGRAPH 2007), p. 22-es. Association for Computing Machinery, New York (2007). https://doi.org/10.1145/1278780.1278807
22. Subramaniyan, A., Sun, C.: Continuum interpretation of virial stress in molecular simulations. Int. J. Solids Struct. **45**, 4340–4346 (2008). https://doi.org/10.1016/j.ijsolstr.2008.03.016
23. Bender, J., Koschier, D.: Divergence-free smoothed particle hydrodynamics. In: Proceedings of the 14th ACM SIGGRAPH/Eurographics Symposium on Computer Animation, pp. 147–155 (2015). https://doi.org/10.1145/2786784.2786796
24. Bender, J., Koschier, D.: Divergence-Free SPH for Incompressible and viscous fluids. IEEE Trans. Visualiz. Comput. Graph. **23**, 1193–1206 (2017). https://doi.org/10.1109/TVCG.2016.2578335

Does Complex Mean Accurate: Comparing COVID-19 Propagation Models with Different Structural Complexity

Israel Huaman$^{(\boxtimes)}$ (iD) and Vasiliy Leonenko$^{(\boxtimes)}$ (iD)

ITMO University, 49 Kronverksky Pr., St. Petersburg, Russia 197101
israel.huaman@ucsp.edu.pe, vnleonenko@yandex.ru

Abstract. During the last years, a wide variety of epidemic models was employed to analyze the spread of COVID-19. Finding the most suitable model according to the available epidemic data is an important task to consider. In this project, we perform a comparison of several models of COVID-19 dynamics and analyze the dependence of their accuracy on their structural complexity, using COVID-19 incidence data for St. Petersburg. The assessment is based on Akaike information criterion (AIC). The results of the study contribute to understanding how to properly choose the complexity of an explanatory model for a given epidemic dataset.

Keywords: mathematical modeling · epidemiology · COVID-19 · structural complexity · logistic regression · compartmental models · Akaike information criterion

1 Introduction

During the last years, a wide variety of epidemic models was employed to analyze the spread of COVID-19. As a rule, the modeling aim is to assess the impact of the epidemics and the efficiency of control measures to reduce the social and economic toll. In this regard, finding the most suitable model according to the available epidemic data is an important task to consider. The most common methodologies used for COVID-19 modeling include classical compartmental SEIR models based on difference and differential equations [1–4], metapopulation models which simulate population migration between countries and cities [5,6], and multi-agent models [7] which are beneficial in simulating outbreaks in high detail. In the majority of the investigations, the choice of an optimal model structure suitable for the task is not discussed and there is no opportunity to compare the accuracy of different models calibrated to a fixed dataset. At the same time, it is known that the selected model type affects the modeling outcome, and, consequentially, the choice of the most effective control measures [17]. Particularly, due to remaining blind spots

---

This research was supported by The Russian Science Foundation, Agreement #22-71-10067.

© The Author(s), under exclusive license to Springer Nature Switzerland AG 2023
J. Mikyška et al. (Eds.): ICCS 2023, LNCS 14075, pp. 270–277, 2023.
https://doi.org/10.1007/978-3-031-36024-4_21

regarding COVID-19 incidence dynamics, simpler models may have an advantage over more complicated ones, because the output of the latter might demonstrate higher levels of uncertainty. In this study, following the logic of our earlier research [13], we perform a retrospective analysis of COVID-19 dynamics using a family of models based on distinct modeling approaches. Several types of logistic models and SEIR compartmental models are calibrated to COVID-19 data, with the disease incidence in St. Petersburg in 2020–2022 used as a test dataset, and their accuracy is compared using the modification of the indicator based on Akaike information criterion [16], namely, AICc [17]. We demonstrate the resulting goodness of fit of the models and discuss how to find a best compromise between the model complexity and model calibration accuracy for a given dataset. Since COVID-19 modeling research mostly inherit the methods from earlier works in mathematical epidemiology, particularly, dedicated to influenza modeling (e.g., [9,12]), the described methodology can be easily generalized to be used for any epidemic ARIs.

2   Methods

The data for model calibration was taken from the repository [18] which contains daily dynamics of COVID-19 in St. Petersburg, Russia. At the time of our study, the sample contained disease data from 02-03-2020 to 15-12-2022. The following datasets were formed for analysis:

- Daily incidence, i.e. number of active cases — taken directly from the source data;
- Total cumulative number of registered cases — calculated from the source data.

The whole dataset ('Multiwave') was also split into six separate outbreaks of COVID-19 ('Waves') according to the information available from open sources. The following model types were compared:

- Models based on the logistic equations: a single equation for separate COVID waves and a sum of logistic equations for the multiwave epidemic;
- Compartmental differential SEIR models similar to those used by the authors for influenza prediction [8], in two modifications: for single COVID waves (constant intensity of infection) and the multiwave epidemic (variable intensity of infection).

The description of the models follows.

2.1   Logistic Models

The single-wave model calibrated to the total number of registered cases is based on the logistic model [14] and has the following form:

$$\frac{dC}{dt} = rC\left(1 - \frac{C}{K}\right), \; C(0) = C_0 \geq 0,$$

where C is the total number of registered cases of COVID-19, t is the current time, r is the infection propagation intensity and K is the maximum load, which is equal to the maximal possible number of infection cases. We use the *optimize.fmin* function from Python *scipy* library to find the optimal values for r and K. Figure 1a demonstrates the calibration result for the second wave of COVID-19. With the help of some simple arithmetic operations, this model can also produce output in a form of daily incidence, which makes it possible to use daily incidence data for model calibration (Fig. 1b).

To describe a whole period of available COVID-19 epidemic data, corresponding to a multiwave epidemic, a multiwave logistic model was build according to the approach proposed in [15]. This model is comprised of a sum of several logistic equations, each of them reflecting the introduction of a new COVID wave. The fitted model trajectory is shown in Fig. 1c. Similarly to the single-wave case, the modeling output can be recalculated to obtain simulated daily incidence. The resulting modeling curve calibrated to data is shown in Fig. 1d.

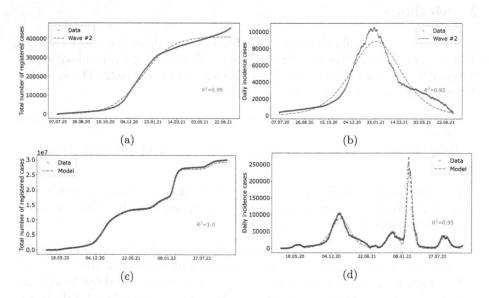

(a)   (b)

(c)   (d)

Fig. 1. Fitted logistic models: (a) single-wave, total number of registered cases; (b) single-wave, daily incidence; (c) multiwave, total number of registered cases; (d) multiwave, daily incidence data.

It is worth mentioning that the multiwave model in the described form generates a 'smooth' (i.e. differentiable) curve which worsens the fit quality of late COVID waves. In Fig. 1c it can be seen that the model trajectory starting approximately from April, 2022 is lower than the data. To fix this problem and thus enhance the fit quality, we present an additional 'adjusted' multiwave logistic model. In this model, the initial position of each simulated wave is artificially matched to the corresponding point in the dataset point of each single wave (thus, simulated incidence in the first day of each COVID wave is equal to the real incidence). The resulting simulated function is non-differentiable in the points of change of COVID waves.

2.2   SEIR Models

The compartmental SEIR model can be expressed as a system of ordinary differential equations in the following form:

$$\frac{dS(t)}{dt} = -\beta S(t)I(t) + \epsilon R(t);$$

$$\frac{dE(t)}{dt} = \beta S(t)I(t) - \gamma E(t);$$

$$\frac{dI(t)}{dt} = \gamma E(t) - \delta I(t);$$

$$\frac{dR(t)}{dt} = \delta I(t) - \epsilon R(t),$$

$$S(t_0) = S^{(0)} \geq 0, E(t_0) = E^{(0)} \geq 0, I(t_0) = I^{(0)} \geq 0;$$

$$S^{(0)} + E^{(0)} + I^{(0)} = \alpha \in [0;1], R(t_0) = 1 - \alpha,$$

where $S(t)$ is the proportion of susceptible individuals in the population at time t, $E(t)$ is the proportion of individuals in incubation period at time t, $I(t)$ is the proportion of infected individuals at a given time t, $R(t)$ is the proportion of recovered individuals at the time t, α is the initial proportion of susceptible individuals in the population, δ is the recovery rate of infected individuals, γ is the intensity of transition to the stage of infected individuals, ϵ is the percentage of recovered population that will become susceptible again and β is the coefficient of intensity of effective contacts of individuals (i.e. contacts with subsequent infection).

For a single wave case (Fig. 2a), we consider the intensity β a constant. In a multiwave case, the intensity is a function of time, i.e. $\beta = \beta(t)$. This modification reflects the influence of introduced control measures on intensity of contacts, as well as the influence of change in circulating SARS-CoV-2 virus strains on disease infectivity. In the current research, we used a piece-wise constant $\beta(t)$. The time moments corresponding to the days when the intensity of contacts in St. Petersburg might have changed were selected from the portal of the Government of St. Petersburg where reports on initiated control measures were published [20]. Initially, 38 dates were selected. Several events corresponding to close dates were combined, which resulted in 35 potential moments of changes in the intensity of contacts. Each moment was assigned a characteristic that assumes the direction of changes in intensity (the number of contacts decreases with increasing restrictions on disease control and increases with their weakening), as well as a subjective categorical assessment of the possible strength of the impact of the corresponding event on the epidemic dynamics. Using this information, we defined the moments of change of $\beta(t)$, splitted the incidence dataset into subsets corresponding to each of the values of intensity β_i (see Fig. 2b) and found those values by sequentially optimizing the model on each of the incidence subsets separately.

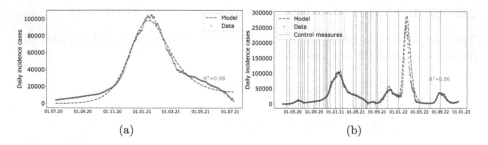

(a)   (b)

Fig. 2. SEIR models calibrated to daily incidence data: (a) single-wave case; (b) multiwave case.

2.3   Accuracy Indicators

Our main indicator for model comparison is Akaike information criterion (AIC) for models calibrated with the least square method, in its corrected version $AICc$ which is more suitable for small samples ($k > n/40$):

$$AICc = n\ln(RSS) + 2k + \frac{2k^2 + 2k}{n - k - 1},$$

where RSS is the residual sum of squares, k is the number of free parameters and n is the data sample size. As an auxiliary indicator, which is not dependent on model complexity (i.e. the number of free parameters k), we employed the coefficient of determination R^2:

$$R^2 = 1 - \frac{RSS}{TSS},$$

where TSS is total sum of squares (the explained sum of squares plus the residual sum of squares).

3   Results

The calibration speed for different models is shown in Table 1. It is noticeable that calibration algorithms for logistic models are approximately twice faster than those for SEIR models, which could be important for experiments with numerous calibration runs, such as calibrating a multitude of datasets or sensitivity analysis. The results of comparison of accuracy are demonstrated in Tables 2–4. As Table 2 demonstrates, the logistic model fitted to incidence data presented the lower AIC value for multiwave case[1]. In case of model fitting to separate waves, the SEIR model almost always demonstrated the lowest AIC value, except for wave 5. At the same time, according to Table 3, logistic models fitted to total cases have the highest R^2 values, including a maximum score of 1.0 obtained on multiwave data. However, single-wave SEIR models present higher R^2 values in most of the cases compared to logistic models fitted to the same data format (daily incidence). Table 4 demonstrates that an adjusted logistic model is slightly better by its AIC than the original logistic multiwave model which generates differentiable simulated curve.

Table 1. Execution time for different models

Model	Multiwave	Wave 1	Wave 2	Wave 3	Wave 4	Wave 5	Wave 6
Logistic, total cases	2.11	0.70	0.87	0.71	0.68	0.74	0.74
Logistic, incidence	2.33	0.70	0.80	0.69	0.65	0.74	0.71
SEIR	5.72	1.48	2.63	0.63	1.55	0.73	1.84

Table 2. AIC for incidence models (best calibration in bold)

Model	Multiwave	Wave 1	Wave 2	Wave 3	Wave 4	Wave 5	Wave 6
Logistic	**1.75 · 10⁷**	0.47·10⁷	1.55·10⁷	0.65·10⁷	1.28·10⁷	**1.68 · 10⁷**	1.17·10⁷
SEIR	5.24·10⁷	**0.35 · 10⁷**	**1.38 · 10⁷**	**0.61 · 10⁷**	**0.91 · 10⁷**	1.91·10⁷	**1.05 · 10⁷**

Table 3. R^2 for all models (best calibration in bold)

Model	Multiwave	Wave 1	Wave 2	Wave 3	Wave 4	Wave 5	Wave 6
Logistic, total cases	1.0	**0.99**	**0.99**	**0.99**	1.0	**0.99**	**0.90**
Logistic, incidence	0.95	**0.99**	0.93	0.46	0.79	0.95	**0.90**
SEIR	0.86	0.99	0.98	0.58	0.98	0.8	0.95

Table 4. Calibration accuracy of multiwave logistic models

Indicator	AIC	R^2
Original	2.96 · 10⁷	1.0
Adjusted	**2.64 · 10⁷**	1.0

4 Discussion

In this study, we showed how accuracy of the model fit could be compared accounting for their structural complexity. The selection of the winning model could depend on data representation (daily incidence vs total registered cases),

---

[1] Since AIC can be compared only for the models fitted to the same data, in Table 2 we do not demonstrate results for logistic model calibrated to total cases.

the period of regarded data (one wave vs multiwave) and, last but not least, the desired output indicators. It is demonstrated that comparing models solely by the quality of fit (R^2 in our case) may be misleading, which becomes clear by looking at Tables 3 and 4. According to the values of AIC, SEIR models should be preferred in case of analysing single COVID-19 waves. At the same time, while regarding multiwave disease dynamics, a multiwave SEIR model loses to logistic models because of its increased number of parameters. Thus, we can conclude that in some cases complex indeed does not mean accurate.

It is important to mention that the comparison of models solely based on AIC cannot be regarded as exhaustive, because many aspects might be missing in this case. First of all, unlike R^2, AIC is calculated in absolute values and thus is dependent on the data representation. Particularly, the direct comparison of AIC for the logistic model calibrated to total cases number and the same model calibrated to incidence data is impossible, whereas we can compare them via R^2. Also, some important epidemic indicators which could be provided by one model type are unavailable in another model type. For instance, logistic models are unable to deliver the value of basic reproduction number R_0 [19] because the recovery process in such models is entirely omitted. This aspect can be crucial for epidemiologists, since this parameter is informative when a particular outbreak is studied. Regarding the assessment of the cost-effectiveness of control measures, multiwave models subjectively seem to be more efficient because they allow to replicate the whole process instead of separating it into parts like in single-wave models. When talking about incidence prediction, which is another meaningful task for the models, the application of the mentioned models might differ in efficiency. The authors believe that logistic models will give a smoother trend thus reducing variation in error sample, although they potentially give a biased solution. In their turn, SEIR models due to bigger number of parameters might demonstrate larger confidence intervals for the predicted incidence. In another words, fitting SEIR multiwave models to incomplete incidence data could be challenging due to the greater variability of the output. In the forthcoming studies, we plan to explicitly quantify our hypotheses related to applicability of the models for assessing epidemic indicators (including incidence forecasting with uncertainty assessment). Also, an important research aim is to generalize the proposed methods and algorithms to make them suitable for the other model types, including multiagent models [10,11].

References

1. Read, J., et al.: Novel coronavirus 2019-nCoV (COVID-19): early estimation of epidemiological parameters and epidemic size estimates. Philos. Trans. R. Soc. B. **376**, 20200265 (2021)
2. Maier, B., Brockmann, D.: Effective containment explains subexponential growth in recent confirmed COVID-19 cases in China. Science **368**, 742–746 (2020)
3. Iboi, E., et al.: Mathematical modeling and analysis of COVID-19 pandemic in Nigeria. Math. Biosci. Eng. **17**, 7192–7220 (2020)

4. López, L., Rodo, X.: A modified SEIR model to predict the COVID-19 outbreak in Spain and Italy: simulating control scenarios and multi-scale epidemics. Results Phys. **21**, 103746 (2021)

5. Chinazzi, M., et al.: The effect of travel restrictions on the spread of the 2019 novel coronavirus (COVID-19) outbreak. Science **368**, 395–400 (2020)

6. Bajardi, P., et al.: Human mobility networks, travel restrictions, and the global spread of 2009 H1N1 pandemic. PloS One. **6**, e16591 (2011)

7. Hoertel, N., et al.: Facing the COVID-19 epidemic in NYC: a stochastic agent-based model of various intervention strategies. MedRxiv (2020)

8. Leonenko, V.N., Ivanov, S.V.: Prediction of influenza peaks in Russian cities: comparing the accuracy of two SEIR models. Math. Biosci. Eng. **15**(1), 209 (2018)

9. Leonenko, V., Bobashev, G.: Analyzing influenza outbreaks in Russia using an age-structured dynamic transmission model. Epidemics **29**, 100358 (2019)

10. Leonenko, V., Lobachev, A., Bobashev, G.: Spatial Modeling of Influenza Outbreaks in Saint Petersburg Using Synthetic Populations. In: Rodrigues, J.M.F. (ed.) ICCS 2019. LNCS, vol. 11536, pp. 492–505. Springer, Cham (2019). https://doi.org/10.1007/978-3-030-22734-0_36

11. Leonenko, V., Arzamastsev, S., Bobashev, G.: Contact patterns and influenza outbreaks in Russian cities: A proof-of-concept study via agent-based modeling. J. Comput. Sci. **44**, 101156 (2020)

12. Leonenko, V.: Herd immunity levels and multi-strain influenza epidemics in Russia: a modelling study. Russ. J. Numer. Anal. Math. Model. **36**, 279–291 (2021)

13. Huaman, I., Plesovskaya, E., Leonenko, V.: Matching model complexity with data detail: influenza propagation modeling as a case study. In: 2022 IEEE International Multi-Conference on Engineering, Computer and Information Sciences (SIBIRCON), pp. 650–654 (2022)

14. Yang, W., et al.: Rational evaluation of various epidemic models based on the COVID-19 data of China. Epidemics **37**, 100501 (2021)

15. Kurkina, E., Koltsova, E.: Mathematical modeling of the propagation of Covid-19 pandemic waves in the World. Comput. Math. Model. **32**, 147–170 (2021)

16. Roda, W., Varughese, M., Han, D., Li, M.: Why is it difficult to accurately predict the COVID-19 epidemic? Infect. Dis. Model. **5**, 271–281 (2020)

17. Mathieu, J. et al.: Tactical robust decision-making methodology: Effect of disease spread model fidelity on option awareness. ISCRAM (2010)

18. Kouprianov, A.: Monitoring COVID-19 epidemic in St. Petersburg, Russia: Data and scripts (2021). https://github.com/alexei-kouprianov/COVID-19.SPb.monitoring

19. Harko, T., Lobo, F., Mak, M.: Exact analytical solutions of the Susceptible-Infected-Recovered (SIR) epidemic model and of the SIR model with equal death and birth rates. Appl. Math. Comput. **236**, 184–194 (2014)

20. St Petersburg against the coronavirus. Official Information (In Russian). https://www.gov.spb.ru/covid-19/dokument/

Multi-granular Computing Can Predict Prodromal Alzheimer's Disease Indications in Normal Subjects

Andrzej W. Przybyszewski[1,2]([✉]) [iD]

[1] Polish-Japanese Academy of Information Technology, 02-008 Warsaw, Poland
przy@pjwstk.edu.pl, Andrzej.Przybyszewski@umassmed.edu
[2] Dept. Neurology, University of Massachusetts Medical School, Worcester, MA 01655, USA

Abstract. The processes of neurodegeneration related to Alzheimer's disease (AD) begin several decades before the first symptoms. We have used multi-granular computing (MGC) to classify cognitive data from BIOCARD study that have been started over 20 years ago with 354 normal subjects. Patients were evaluated every year by a team of neuropsychologists and neurologists and classified as normal, with MCI (mild cognitive impairments), or with dementia. As the decision attribute, we have used CDRSUM (Clinical Dementia Rating Sum of Boxes) as a more quantitative measure than the above classification. Based on 150 stable subjects with different stages of AD, and on the group of 40 AD, we have found sets of different granules that classify cognitive attributes with *CDRSUM* as the disease stage. By applying these rules to normal (*CDRSUM = 0*) 21 subjects we have predicted that one subject might get mild dementia (*CDRSUM > 4.5*), one very mild dementia (*CDRSUM > 2.25*), four might get very mild dementia or questionable impairment and one other might get questionable impairment (*CDRSUM > 0.75*). AI methods can find, invisible for neuropsychologists, patterns in cognitive attributes of normal subjects that might indicate their pre-dementia stage, also in longitudinal testing.

Keywords: Granular Computing · Rough Set · Rules · Cognition · Genotype

1 Introduction

The prevalence of Alzheimer's Disease (AD) related dementia is fast increasing due to our aging population, and it may reach 139 million in 2050. There is no cure for AD, as during the first clinical symptoms and neurological diagnosis many parts of the

And the BIOCARD Study Team—Data used in preparation of this article were derived from BIOCARD study, support[1]ed by grant U19 - AG033655 from the National Institute on Aging. The BIOCARD study team did not participate in the analysis or writing of this report, however, they contributed to the design and implementation of the study. A listing of BIOCARD investigators can be found on the BIOCARD website (on the 'BIOCARD Data Access Procedures' page, 'Acknowledgement Agreement' document).

© The Author(s), under exclusive license to Springer Nature Switzerland AG 2023
J. Mikyška et al. (Eds.): ICCS 2023, LNCS 14075, pp. 278–285, 2023.
https://doi.org/10.1007/978-3-031-36024-4_22

brain are already affected without the possibility to recover. As the neurodegenerations begin two to three decades before observed symptoms, the best chance to fight AD is to estimate the beginning period of the AD-related brain changes. The BIOCARD* study was initiated in 1995 by NIH with 354 normal individuals interrupted in 2005 and continued from 2009 as Johns Hopkins (JHU) study. At JHU, patients have yearly cognitive and clinical visits that measured a total of over 500 attributes with 96 cognitive parameters [1, 2]. Albert et al. [2] have successfully predicted conversion from normal to MCI (Mild Cognitive Impairment) due to AD, 5 years after baseline, for 224 subjects by using the following parameters: CSF: beta-amyloid and p-tau, MRI hippocampal and entorhinal cortex volumes, cognitive tests scores, and APOE genotype. But our approach is different, as we have performed classification by using granular computing (GC) [3] connecting cognitive test results with genetic data (related to the apolipoprotein E ApoE genotype) and AD-related clinical symptoms in the group of different subjects from normal to AD. We took this group as our Model for the supervised training of different granules that are related to various stages of the disease, from normal subjects to MCI (Mild Cognitive Impairment) and to subjects with AD-related dementia.

In the next step, we applied these granules to individual, normal subjects to predict in every single tested subject, the possibility of the beginning of the neurodegeneration (AD-related symptoms). To **validate our method**, we have also applied it to the early stages in patients that were diagnosed with AD if we can predict their future AD stage (the preclinical classification of all potential patients).

Our GC method implemented with a rough set (RS) gave better classifications than such ML methods as Random Forest, Decision Tables, Bayes classifier, and Tree ensembles for Parkinson's disease patients [4], see review for more comparisons [5].

2 Methods

It is a continuation of our previous study [6] therefore methods are similar, but in this part, we are using MGC (Multi GC) in addition to analysis of the longitudinal changes in our subjects. We have analyzed predominantly cognitive and in addition to genotype (*APEO*) attributes of several different groups of subjects. The first group consists of 150 subjects with 40 normal subjects, 70 MCI (Mild Cognitive Impairment), and 40 subjects with dementias (AD). It was chosen this way as in the whole population of 354 normal subjects followed from 1995, only 40 subjects became demented. Therefore, we have added 40 normal subjects and 70 MCI as they are in between AD and normal subjects. The second group was 40 AD subjects, and the last group was 21 subjects, clinically classified as normal. We have estimated stages of the disease based on *CDRSUM* values (see abstract) as (0.0) – normal; (0.5–4.0) – questionable cognitive impairment; (0.5–2.5) – questionable impairment; (3.0–4.0) – very mild dementia; (4.5–9.0) – mild dementia [7]. We have used the same attributes as before [6]: Logical Memory Immediate (*LOGMEM1A*), Logical Memory Delayed (*LOGMEM2A*), Trail Making, Part A (*TrailA*) and B (*TrailB*), Digit Symbol Substitution Test (*DSST*), Verbal Fluency Test (*FCORR*), Rey Figure Recall (*REYRECAL*), Paired Associate Immediate (*PAIRED1*), Paired Associate Delayed (*PAIRED2*), Boston Naming Test (*BOSTON*), and new California Verbal Learning Test (*CVLT*). In addition, we have registered *APOE* genotype;

individuals who are *ApoE4* carriers vs. non-carriers (digitized as 1 vs. 0). Based on our classification, we have estimated Clinical Dementia Rating Sum of Boxes (*CDRSUM*), compared with *CDRSUM* obtained by neurologists, and determined the predicted stage of an individual patient.

2.1 Rough Set Implementation of GC

Our data mining granular computing (GC) analysis was implemented by rough set theory (RST) discovered by Zdzislaw Pawlak [8], whose solutions of the vague concept of boundaries were approximated by sharp sets of the upper and lower approximations (Pawlak 1991). More details in our previous paper [9]

We have used Rough Set Exploration System RSES 2.2 as a toolset for analyzing data with rough set methods [10, 11].

3 Results

3.1 Statistics

We have performed statistical analysis for all 15 attributes, and we found that 7 attributes had stat. Sig. Difference of means: FCORR, REYRECAL, PAIRED1, PAIRED2, BOSTON, CVLT, CDRSUM. We analyzed different groups of subjects: normal (GroupN), a mixture of normal MCI, AD (Group1), and AD (Group2).

3.2 Rules from the General Model (Group1)

In this study, by using MGC, we reduced the number of attributes from 14 that were used before [6] to the following five: *APOE, FCORR, DDST, TrailB,* and the decision attribute was *CDRSUM.* The APOE genotype; individuals who have *ApoE4* is an important genetic factor, which influences the probability of AD. One of the early predictors of AD is poor language performance, quantified by the *FCORR* test. Another early indication is difficulties in reasoning that may be estimated by the *DDST* test. Slowing processing speed is also observed as an early AD indicator that can be quantified by *TrailB* tests.

We put all data in the decision table as in [6], and with RSES help, after discretization, we found that because large data set and a small number of parameters, the decision attribute has 7 ranges (related to a very precise classification): "(-Inf,0.25)", "(0.25,0.75)", "(0.75,1.25)", "(1.25,2.25)", "(2.25,3.25)", "(3.25,4.25)", "(4.25,Inf)". After generalization, there were 82 rules, below are some examples:

$$(APOE = 0)\&(DSST =" (66.5, Inf)") => (CDRSUM =" (-Inf, 0.25)"[4]) 4 \quad (1)$$

$$(APOE = 1)\&(TRAILB =" (128.5, Inf)")\&(FCORR =" (6.5, 10.5)") =>$$
$$(CDSUM =" (4.25, Inf)" [2]) 2 \quad (2)$$

One significant attribute in the genetic *APOE* genotype, and in these approximate rules lack of *ApoE4* carriers ($APOE = 0$) related to health (Eq. 1), where ($APOE = 1$) increases the probability of AD (Eq. 2). The *DSST* - digit symbol substitution test is related to associative learning, and higher numbers are better (Eq. 1). In contrast to *TrailB* higher value means slow execution that is bad (Eq. 2) and language fluency problems (low value of *FCORR*) are the main factors that such patients have indications of mild dementia (*CDRSUM* is larger than 4.25) (Eq. 2).

By applying all 82 rules to the healthy patients (GroupN) with clinically confirmed $CDRSUM = 0$, we found one patient with *CDRSUM* significantly larger than 0 in the following classification:

$$(Pat = 164087)\&(APOE = 1))\&(DSST ='' (46.5, 49.5)'')\&(FCORR ='' (10.5, 13.5)'')\&$$
$$(TRAILB ='' (72.5, 128.5)'' => (CDRSUM ='' (2.25, 3.25)''$$

(3)

The Eq. 3 indicates based on 4 condition attributes that patient *164087* might have. *CDRSUM* = "$(2.25,3.25)$" suggests very mild dementia [7].

In the next step of the MGC method, we have increased the number of attributes to seven: *APOE, BOSTON, FCORR, DDST, TrailB,* and *REYRECAL,* with *CDRSUM* as the decision attribute. We added the Boston naming test (*BOSTON*) forgetting the names of objects, and problems related to a visual memory of the complex figure (*REYRECAL*). After discretization with RSES help, the decision attribute has 3 ranges: "(-Inf,0.75)", "(0.75,1.25)", and "(1.25, Inf)". We have obtained 104 rules from Group1 patients and applied them to GroupN normal subjects, and got the following classifications, e.g.:

$$(Pat = 164087)\&(APOE = 1))\&(FCORR ='' (10.5, 13.5)''))\&(REYRCAL ='' (15.75, 25.25)''$$
$$)'')\&(TRAILB ='' (75.0, 114.5)'')\&(DSST ='' (47.5, 53.5)'')\&(BOSTON ='' (25.5, 26.5)'') =>$$
$$(CDRSUM ='' (2.0, 3.25)$$

(4)

In this example the same patient *Pat* = *164087* with more condition attributes in Eq. 4 in comparison to Eq. 3 gives almost identical results for *CDRSUM*, as the main factors are related to bad speech fluency (*FCORR*) and the *APOE* genome. Comparing with 14 attributes from our previous work [6] actual results are at least partly overlapping:

$$(Pat = 164087)\&(LOGMEM\,1A = ''(-Inf, 15.5)''))\&(LOGMEM\,2A = ''(-Inf, 16.5)'')\&$$
$$(TRAILA = ''(35.5, Inf)'')\&(TRAILB = ''(74.5, 153.0)'')\&(FCORR ='' (-Inf, 16.5)'')\&$$
$$(REYRECAL = ''(15.75, 25.25)'')\&(PAIRD2 ='' (-Inf, 6.5)'')\&(age = ''(-Inf, 76.5)'')\&$$
$$(APOE = 1) => (CDRSUM = ''(2.25, Inf)''$$

(5)

3.3 Granular Computing for Reference of Group2 Patients

In this part with our model is based on AD patients (Group2), we have reduced the number of attributes from 14 [6] to five: *APOE, DDST, FCORR, TrailB*, and the decision attribute was *CDRSUM*. After discretization (RSES) we obtained e.g.

$$(APOE = 1)\&(FCORR ='' (-Inf, 15.5)'') => (CDRSUM ='' (3.25, Inf)''[6]) 6 \tag{6}$$

We applied the above rules from Group2 to predict the *CDRSUM* of GroupN:

$$(Pat = 164087)\&(APOE = 1))\&(DSST ='' (45.0, 56.0)'')\&(FCORR ='' (-Inf, 15.5)'')\&$$
$$(TRAILB ='' (-Inf, 128.5)'' => (CDRSUM ='' (3.25, Inf)'' \tag{7}$$

Equation 7 confirms the previous classifications (Eq. 5) but now is based on AD patients.

If we add to all our original [6] classifications, an attribute related to the verbal learning and memory test *CVLT* (California Verbal Learning Test), we obtained the following classification:

$$Pat = 164087)\&(TRAILA = ''(-Inf, 73.5)'')\&(TRAILB ='' (52.5, Inf)''))\&(FCOR =$$
$$''(-Inf, 17.0)'')\&(PAIRD1 ='' (14.5, Inf)'')\&(PAIRD2 ='' (-Inf, 5.5)''))\&(BOSTON =$$
$$''(-Inf, 27.5)'')\&(CVLT = ''(33.5, Inf)'')\&(APOE = 1) => (CDRSUM ='' (5.75, 6.5)'' \tag{8}$$

In our previous work [6] patient *Pat = 164087* was classified with 9 attributes (after reduction of non-significant ones) that resulted in an estimation of his/her *CDRSUM* between 4.5 and 6, which means that this patient might have mild dementia [6]. We have repeated the same subject classification using an additional attribute *CVLT* as a more universal test of verbal learning and memory (now 10 attributes). The result is in Eq. 8 that not only confirm our previous results [6], but also gives a narrower *CDRSUM* range between 5.75 and 6.5 which means mild dementia [6] for a clinically normal patient. The doctor's estimation of *CDRSUM* was 0.

3.4 GC Classification for Longitudinal Reference of Early Stages in AD Patients

We have applied GC to the psychophysical data to estimate *CDRSUM* in subjects in their normal, Impaired Not MCI, MCI, and dementia stages as determined by the diagnostic data (neurological diagnosis).

There are four different patients tested clinically every year (time in months from the beginning of their participation). The only times when changes in their symptoms have occurred are in Table 1. **These results validate our method.** Our predictions have higher values than clinical, fluctuate as clinical, and predict dementia.

In summary, we have demonstrated that our method gives similar results to neurological diagnostic and functional evaluation tests. In many cases, as shown above, is more sensitive and it gave predictive values in some cases. These findings are very important for our future clinical applications.

Table 1. Clinical and GC patients state estimations from normal to dementia

Pat#	Time (month)	CDRSUM Clinical	CDRSUM Predicted
653735	146	0 -ImpNotMCI	1.25–3.75
	157	0.5 - MCI	1.25–3.75
	169	1 - MCI	1.25–3.75
	182	4 - dementia	> 5.75
921569	143	0.5 - MCI	1.25–3.75
	217	5 - dementia	> 5.75
411007	149	1 – MCI	1.25–3.75
	213	1-ImpNotMCI	3.75–5.75
	224	1- dementia	3.75–5.75
703257	96	1 - MCI	1.25–3.75
	108	2 - MCI	3.75–5.75
	120	1.5 - MCI	1.25–3.75
	131	5 - dementia	3.75–5.75

4 Discussion

As Alzheimer's disease has a long (20–30 years) prodromal phase, during which individual compensatory processes may develop differently between subjects. Therefore, our aim was **to detect the beginning of compensatory changes reflective of underlying neurodegeneration,** as it might give a chance to prevent dementia. We have developed a novel tool to monitor ongoing progression in normal subjects more easily and accurately by looking into patterns of cognitive attributes' values and comparing them with our two groups of patients (Group1: general and Group2: AD).

We have studied these patterns with a multi-granular computing (MGC) method and by comparing different sets of attributes (granules) to find possible patterns in normal subjects (n = 21) that might have similarities to granules observed in AD patients.

In this study, we have changed our classifications by removing/adding attributes. As in the previous study [6] we have always used 14 attributes, in this part we have changed from 5 to 7, or even to 15 attributes, and compared classification results. The other new and important part was the interpretability of obtained rules.

Also, rules can be created with different granularity and algorithms that might give different classifications. Therefore, we were looking for classifications that are complete e.g., they give similar results with different sets of rules.

Group11 has given us subtle rules that determine the beginning of possible symptoms. These new granules gave us rules supporting our previous classifications like for 5 attributes or gave some new but consistent rules for 7 or 14 attributes.

In the next step, we used a more advanced model – Group2 that gave rules based on AD patients. Thanks to the AD group, we could get higher values of the *CRDSUM* that gave us classifications of the possible subjects with very mild or mild dementia.

Using only 5 attributes, we obtained a new classification confirming our previous findings for *Pat = 164087* (compare eqs. 4 and 6). Both equations suggest that the patient might have at least very might dementia or maybe even mild dementia. Medium dementia was earlier confirmed for *Pat = 164087* in 14 attributes classifications [6]. To confirm our previous [6] result, we performed classifications with 15 attributes (Eq. 8) that gave us confirmation of our actual (eq. 7) and previous [6] results and better precision in the estimation of patient's mild dementia. An advantage of the multi-granular computations is spectrum of new rules, and classifications with smaller numbers of attributes that were easier to understand and interpret. This approach might be important for clinicians if they want to estimate a patient's state in a simple and fast but approximate way (small number of tests). Depending on obtained results doctors might perform the following tests to get more precise classifications. This MGC approach follows the functioning of the visual brain [12] where object recognition starts from light spot classification (retina, LGN), through edge orientations (V1) to faces in higher cortical areas (IT). As we learn to recognize new objects, we expand our models (here Group1 and 2) that give advantages of better and faster classifications (descending brain pathways) [12]. We have used previously this approach to discover the concept of Parkinson's disease [13].

As it is the first, to our knowledge, work that estimates singular complex patterns of the individual patient's symptoms, our rules are taken from one population and applied to different subjects, so they are not certain. Therefore, the next step is to find different methods (e.g., tests in the clinic) for their confirmation.

References

1. Albert, M., Soldan, A., Gottesman, R., et al.: The BIOCARD Research Team, Cognitive changes preceding clinical symptom onset of mild cognitive impairment and relationship to *ApoE* genotype. Curr Alzheimer Res. **11**(8), 773–784 (2014)
2. Albert, M., Zhu,Y., Moghekar, et al.: Predicting progression from normal cognition to mild cognitive impairment for individuals at 5years Brain. **141**(3), 877–887 (2018)
3. Pedrycz, W. (ed.): Granular Computing: an emerging paradigm. Physica Verlag, Heidelberg, New York (2001)
4. Przybyszewski, A.W., Kon, M., Szlufik, S., Szymanski, A., Koziorowski, D.M.: Multimodal learning and intelligent prediction of symptom development in individual parkinson's patients. Sensors **16**(9), 1498 (2016). https://doi.org/10.3390/s16091498
5. Przybyszewski, A.W., Śledzianowski, A., Chudzik, A., Szlufik, S., Koziorowski, D.: Machine learning and eye movements give insights into neurodegenerative disease mechanisms. Sensors **23**(4), 2145 (2023). https://doi.org/10.3390/s23042145
6. Przybyszewski, A.W.: AI classifications applied to neuropsychological trials in normal individuals that predict progression to cognitive decline. In: Groen, D., et al. (eds.): ICCS 2022. LNCS, vol. 13352, pp. 150–156 (2022). https://doi.org/10.1007/978-3-031-08757-8_14
7. O'Bryant, S.E., Waring, S.C., Cullum, C.M., et al.: Staging dementia using clinical dementia rating scale sum of boxes scores: a Texas Alzheimer's research consortium study. Arch Neurol. **65**(8), 1091–1095 (2008)

8. Pawlak, Z.: Rough sets: Theoretical aspects of reasoning about data. Kluwer, Dordrecht (1991)
9. Przybyszewski, A.W.: Theory of mind helps to predict neurodegenerative processes in Parkinson's Disease. In: Paszynski, M., Kranzlmüller, D., Krzhizhanovskaya, V.V., Dongarra, J.J., Sloot, P.M.A. (eds.) ICCS 2021. LNCS, vol. 12744, pp. 542–555. Springer, Cham (2021). https://doi.org/10.1007/978-3-030-77967-2_45
10. Bazan, J.G., Szczuka, M.: The rough set exploration system. In: Peters, J.F., Skowron, A. (eds.) Transactions on Rough Sets III. LNCS, vol. 3400, pp. 37–56. Springer, Heidelberg (2005). https://doi.org/10.1007/11427834_2
11. Bazan, J.G., Szczuka, M.: RSES and RSESlib - a collection of tools for rough set computations. In: Ziarko, W., Yao, Y. (eds.) RSCTC 2000. LNCS (LNAI), vol. 2005, pp. 106–113. Springer, Heidelberg (2001). https://doi.org/10.1007/3-540-45554-X_12
12. Przybyszewski, A.W.: The neurophysiological bases of cognitive computation using rough set theory. In: Peters, J.F., Skowron, A., Rybiński, H. (eds.) Transactions on Rough Sets IX. LNCS, vol. 5390, pp. 287–317. Springer, Heidelberg (2008). https://doi.org/10.1007/978-3-540-89876-4_16
13. Przybyszewski, A.W., Nowacki, J.P., Drabik, A., Szlufik, S., Koziorowski, D.M.: Concept of parkinson leading to understanding mechanisms of the disease. In: Nguyen, N.T., Iliadis, L., Maglogiannis, I., Trawiński, B. (eds.) ICCCI 2021. LNCS (LNAI), vol. 12876, pp. 456–466. Springer, Cham (2021). https://doi.org/10.1007/978-3-030-88081-1_34

Accounting for Data Uncertainty in Modeling Acute Respiratory Infections: Influenza in Saint Petersburg as a Case Study

Kseniya Sahatova, Aleksandr Kharlunin, Israel Huaman(iD),
and Vasiliy Leonenko[(✉)](iD)

ITMO University, 49 Kronverksky Pr., St. Petersburg, Russia 197101
aleksandrharlunin03@gmail.com , vnleonenko@yandex.ru

Abstract. Epidemics of acute respiratory infections, such as influenza and COVID-19, pose a serious threat to public health. To control the spread of infections, statistical methods and mathematical models are often used to estimate and forecast epidemic indicators. The assessment of values of these indicators might be impacted by the uncertainty in the initial data used for model calibration. The dependence of modeling results on accuracy of the data can be huge, and the lack of its consideration, which is typical for most works on modeling the spread of epidemic ARIs, makes it difficult to correctly predict the effectiveness of anti-epidemic measures. In this research, we present methods and algorithms for uncertainty estimation in retrospective analysis and forecasting of the incidence of epidemic ARIs. The uncertainty assessment is performed by replicating simulated incidence curves with assumed random errors which distribution is defined by the variance of the original incidence dataset. The application of the methods is demonstrated on an example, with influenza outbreak in St. Petersburg, Russia, as a case study.

Keywords: mathematical modeling · epidemiology · influenza · data uncertainty

1 Introduction

Epidemics of acute respiratory viral infections, with influenza and COVID-19 being the leading ones, pose a serious threat to public health and the economic development of the affected countries [13]. To contain the epidemic spread, health authorities use comprehensive measures, such as quarantine and vaccination. To provide adequate justification to the planned interventions, health organs often rely on statistical and mechanistic models calibrated to disease incidence data. One of the most popular approaches is connected with so called compartmental

This research was supported by The Russian Science Foundation, Agreement #22-71-10067.

© The Author(s), under exclusive license to Springer Nature Switzerland AG 2023
J. Mikyška et al. (Eds.): ICCS 2023, LNCS 14075, pp. 286–299, 2023.
https://doi.org/10.1007/978-3-031-36024-4_23

models. Compartmental models have been a powerful tool in mathematical modeling of epidemiological processes for many years. During the COVID-19 they were actively used to predict the evolution of the pandemic and to estimate the effect of health interventions. Comparing to the prediction and decision support systems based on machine learning methods, such as deep neural networks, the output of the compartmental models is much easier to interpret. Using interpretable models is beneficial for decision makers compared to 'black box' approaches, which contributes to the popularity of classical mechanistic models in analysing epidemic dynamics. A comparison of the strengths and weaknesses of the mentioned approaches can be found in [21].

Modeling of epidemiological processes is often a challenging task, considering the fact that the epidemiological data is prone to errors. As a consequence, the modeling output might not properly represent the real situation, which leads to subpar policies in epidemic containment. It is crucial to employ the systems that consider and unite all potential sources of uncertainty to increase the trust to decisions made on their basis. Main difficulties that can cause a misleading prognosis by the models are described in [27]. Since one of the main components of the modeling framework is input data, its quality and quantity influences the uncertainty in the output of the model.

The aim of this work is to characterize uncertainty in ARI incidence data, estimate the confidence intervals for model parameters and assess the corresponding variation in disease dynamics forecasts. For this purpose, we use an influenza epidemic model employed in our earlier research [8]. This model is described in detail in Sect. 2. Section 3 provides a description of experiments for parameter uncertainty estimation using the computational approach presented by Chowell et al. [3]. The results are discussed in Sect. 4.

2   Methods

2.1   Incidence Data

The dataset used in the study was based on the data provided by the Research Institute of Influenza. By combining ARI incidence data in 2010–2011 for Saint Petersburg, Russia, and the results of virological testing performed during the same season, we found the numbers for weekly clinically registered cases of ARI attributed to different influenza strains [15]. Although in this research only one season was used in the experiments, the described algorithm is versatile enough to be applied to any other epidemic season in a given Russian city which has sufficient volume of reported incidence and virological testing.

2.2   Modeling Framework

The epidemic model used in this study is a SEIR–type discrete compartmental model represented by a system of difference equations. This model has a flexible structure which allows to distinguish infections caused by different influenza strains [15] and track them separately in different age groups [14]. In this paper, we use the simplest version of the model structure, in which we do not regard age groups and specific strains. The following system of equations is used in the model:

$$x_{t+1}^{(h)} = \max\left\{0, \left(1 - a\frac{\beta}{\rho}\overline{y}_t\right)x_t^{(h)}\right\}, h \in \overline{1,2},$$

$$y_{t+1} = \frac{\beta}{\rho}\overline{y}_t \sum_{h=1}^{n_s+1} ax_t^{(h)},$$

$$\overline{y}_t = \sum_{\tau=0}^{T} y_{t-\tau}g_\tau,$$

$$x_0^{(h)} = \alpha^{(h)}((1-\mu)\rho - y_0) \geq 0, h \in \overline{1,2},$$

$$y_0 = \varphi_0 \geq 0.$$

where $x_{t+1}^{(h)}$ corresponds to the fraction of susceptible individuals at the time moment $t+1$ with exposure history $h \in \overline{1,2}$, y_{t+1} corresponds to the amount of newly infected individuals at the time moment $t+1$, $\overline{y}_t$ corresponds to the total amount of infected by the time t. A group $x_t^{(1)}$ of susceptible individuals with exposure history state $h = 1$ is composed of those individuals who were subjected to infection in the previous epidemic season, whereas a group $x_t^{(2)}$ with exposure history state $h = 2$ is regarded as naive to the infection. The parameter $\alpha^{(h)}$ is a fraction of population exposed to the infection in the preceding epidemic season, $h \in \overline{1,2}$. The variable $\mu \in [0; 1)$ reflects the fraction of population which do not participate in infection transmission. The piece-wise constant function g_τ reflects the change in individual infectiousness over time from the moment of infection. The parameter β is the intensity of effective contacts, ρ is the population size. Due to immunity waning, the individuals with the history of exposure to influenza virus in the preceding season might lose immunity in the following epidemic season. We assume that the fraction a of those individuals, $a \in (0; 1)$, becomes susceptible, whereas $1 - a$ individuals retain their immunity by the moment of the modeled epidemic. We also assume that $\alpha^{(1)} = 1 - \alpha^{(2)} = \alpha$, with the proportion α of people exposed to our generic influenza virus strain being a free parameter. More details about the model can be found in [8].

2.3   Calibration Algorithm

Model calibration procedure consists in finding parameter values which deliver the optimum for the optimization function (1). In this function, weighted sum of squared residuals is minimized. The residuals are calculated as the difference between the incidence data ($Z^{(\mathrm{dat})}$) and the simulated incidence values ($Z^{(\mathrm{mod})}$) for each week. In the previous studies [16], we estimated a plausible range for the epidemiological parameters, which is given in Table 1.

$$F\left(Z^{(\mathrm{mod})}, Z^{(\mathrm{dat})}\right) = \sum_{i=0}^{t_1} w_i \cdot \left(z_i^{(\mathrm{mod})} - z_i^{(\mathrm{dat})}\right)^2 \longrightarrow \min \qquad (1)$$

Table 1. Estimated epidemiological parameters for model fitting

Definition	Description	Value range
α	Individuals with the history of exposure in the previous season	(0.005, 0.9)
β	Intensity of infectious contacts	(0.03, 0.24)
a	Fraction of previously exposed individuals who will lose immunity in the following epidemic season	(0.0, 1.0)
μ	Fraction of population which do not participate in the infection transmission	0.9

The weights w_i included in the formula (1) are used to reflect the 'importance' of fitting the model curve to the particular data points. In case of seasonal influenza, the data closer to the peak have higher plausibility, because the laboratory sample sizes tend to become bigger during a full-fledged outbreak and thus the proportion of different virus strains in the overall ARI incidence is calculated more precisely. Thus, for weight estimation we use the formula $w_i = \sigma^{-d}$, where d is the distance of the current data point from the peak in time units (weeks), σ is a parameter related to decreasing rate with growing distance from the peak. More details on model calibration could be found in [14].

The time step of the model is one day, whereas the incidence data has a time step of one week, so to be able to compare the incidences we derive simulated weekly incidence by summing and interpolating the obtained daily values.

To align the simulation timeline $t = 0, 1, \ldots$ with the dates of the epidemic dataset, we assume that the maximum weekly incidence should be attributed to the same time moment in real and simulated data, which gives us a reference point for matching in time the model with data.

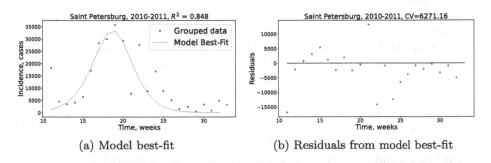

(a) Model best-fit   (b) Residuals from model best-fit

Fig. 1. (a) The model calibration to incidence data of the single epidemic season of ARI in St. Petersburg, Russia. Green circles represent the reported weekly case incidence, red line shows the model trajectory, R^2 stands for the coefficient of determination. (b) The plot of residuals, CV stands for the coefficient of variation. (Color figure online)

Figure 1 displays the obtained model best fit. The pattern of the residuals justifies that our model offers a fairly adequate match for the epidemic data.

2.4 Forecasting

In addition to the epidemic indicators that are assessed by retrospective analysis of the full incidence dataset, we aim to assess the forecasting uncertainty. The prediction algorithm is based on model calibration on incomplete data which was implemented by the authors in earlier research [17]. In the present study, we modified the prediction algorithm to make it suitable for the current model, and added a new method of aligning in time the simulated curve with the incidence dataset.

In overall, the calibration algorithm on incomplete data which produces a prediction curve does not differ much from the procedure of calibration on complete data. The only difference is in the fact that we do not know a day of the maximum incidence which we used to align the model time frame with real time. To solve this issue, two types of prediction algorithms were proposed with distinct alignment procedures:

- **Primary algorithm**, in which we loop through possible time moments of maximum incidence t_n to t_{n+k}, where t_n is the last known incidence from the incomplete dataset and $k = const$ is a calibration parameter. During each iteration, we assume that the regarded point t_{n+i} is the time moment of maximum incidence and we perform the model calibration based on that assumption. As a result, instead of one calibration run in case of the complete dataset we have to perform k calibrations. In the end, we compare the obtained k values of function (1) and select the parameter values which deliver the optimum for (1).
- **Auxiliary algorithm**, in which we assume that the time moment $t = 0$ corresponds to the first date of the incidence dataset, thus we replace the alignment by peak with the alignment by the first day of the epidemic. In general case, due to the uncertainty in detection of the influenza epidemic onset in the ARI dataset [17], it is hardly possible to establish the starting date in an incidence dataset with certainty, so the auxiliary algorithm cannot be employed. At the same time, this algorithm is suitable in case of COVID-19 outbreaks when the starting moment of each wave is more or less defined, and, as it will be discussed later, for making predictions on bootstrapped data. Since for any incidence dataset the first day may be clearly defined, there is no need in looping through possible time alignment alternatives (i.e. only one calibration run is needed), which makes the auxiliary algorithm much faster compared to the primary algorithm. The described algorithm might be also used without changes for the model calibration on complete data (i.e for the retrospective analysis).

2.5   Uncertainty Quantification

Sources of Uncertainty. Uncertainty in epidemiological data is an important factor to consider since it plays one of the key roles [9] in accuracy of the modeling output along with the structural complexity of models employed [8]. Error measurement in epidemic studies stays one of the main problems of mathematical epidemiology [23]. In [27], the authors highlight the major sources of uncertainty such as a lack of knowledge, biased data, measurement errors, stochasticity of the process, parameter uncertainty, redundant complexity and erroneous structure of the utilized model. Among the sources of uncertainty related in particular to input data, the following ones could be distinguished:

- Noise in incidence data;
- Bias caused by missed historic records during national holidays, scarce data;
- Under-reported cases at the time of weekends;
- Implausible data, which is not reflecting the trend of the epidemic;
- Confounding factors that can or cannot be measured.

The complexity of combination of these factors makes it difficult to estimate the resulting error structure. Thus synthetic error structure models are often used based on known probabilistic distributions.

Relevant Research. There are various ways to perceive and interpret the uncertainty of the model parameters related to the input data. Valero and Morillas [20] used the bootstrap method based on resampling of incidence data, relying on the fact that they have the greatest influence on the uncertainty of modeling process. Another method is parametric bootstrapping approach from the best-fit model solution proposed by Chowell [24]. This method was successfully used in a handful of studies [4,19] to characterize the uncertainty of model parameters. The paper [1] identifies the problem of characterization of uncertainty to improve decision making procedures in risk assessments. There is a plethora of research such as [1,27,28] that discuss the necessity to consider the corruption of data or the potential bias in it during the assessment of an epidemic indicators. However, not all of these studies provide quantification of this impact on the modeling results. Lloyd [18] and Samsuzzoha et al. [26] performed uncertainty analysis of the basic reproduction number with various mathematical and statistical methods. Chowell [3] discusses methods of parameter uncertainty quantification in dynamic models, as well as evaluation of confidence intervals. In paper [11] Krivorot'ko et al. formulated an inverse problem of refining the model parameters and studied the reliability of forecasts by the epidemic models based on the COVID-19 incidence. For the purpose of our research, we selected a method of uncertainty quantification from [3], which is used further in the study.

Modeling Uncertainty. Our main objective in modeling uncertainty is to select and test particular error structures, which are compatible with incidence data and allow us to encapsulate the sources of uncertainty mentioned earlier.

To model the uncertainty of the input data, the Poisson error structure may be used, as it is one of the most popular error models. The modeling framework described by Chowell [2,3] introduces a bias via generating a random value with Poisson distribution with the mean equal to the number of new registered infection cases in a fixed time step. In our case, we applied the given distribution to weekly incidence dataset. Figure 2 illustrates the variance of simulated residuals acquired after $n = 5000$ resamples on the initial dataset. The empirical distribution demonstrates that the simulated uncertainty is higher on the epidemic peak, while sample values attributed to low incidence (the beginning and the end of the epidemic wave) do not show much variability.

Another approach, which is negative binomial error structure [12], can be used to model higher variance levels in the data as in this case the mean and variance could be set independently, based on the variability of residuals. It can be useful for disease modeling in situations, where distributions with overdispersion provide better representation of the actual data [25]. Taking the epidemic data overdispersion into account in simulation is often crucial, considering that in practical situations, especially in sparse time series, variance can be much larger than the mean and this fact should not be ignored [10]. For cases of excessive data variability we used error structure based on zero-truncated negative binomial distribution [7] to exclude zeros that were not eliminated in the data correction process.

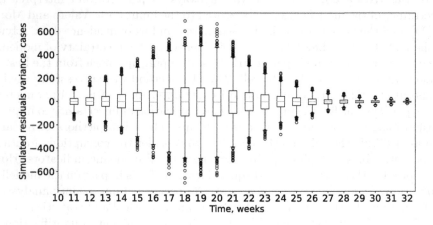

Fig. 2. The empirical distribution of residuals obtained using Poisson error structure

3   Experiments and Results

3.1   Interval Estimates of Model Parameters

To actually perform the estimation of the parameter uncertainty, we use the parametric bootstrap method [6] based on best-fit model curve resampling [5]. Such choice allows us to take into account different approaches to treating parameter uncertainty. The implemented algorithm has the following steps:

- Obtain the best-fit parameter values $\hat{\theta} = (\hat{\alpha}, \hat{\beta}, \hat{a})$ for simulated incidence $Z_1^{(\mathrm{mod})}(\hat{\theta})$ and corresponding residuals $\rho_i = z_i^{(\mathrm{mod})}(\hat{\theta}) - z_i^{(\mathrm{dat})}$ by calibrating the model to the original incidence data incidence.
- Generate $m = 200$ datasets $Z_1^{'(\mathrm{dat})}, ..., Z_{200}^{'(\mathrm{dat})}$ based on $Z^{(\mathrm{mod})}(\hat{\theta})$ incidence curve with the addition of generated residuals: $z_i^{'(\mathrm{dat})} = z_i^{(\mathrm{mod})}(\hat{\theta}) + \delta_i$ In the first set of experiments, to determine the residual value δ_i at time t_i we used Poisson distribution function $\delta_i \sim Pois(\lambda(t_i))$ in which $\lambda(t_i) = z_i^{(\mathrm{mod})}(\hat{\theta})$ [2]. In the second set of experiments, the assumed error structure was based on zero-truncated negative binomial distribution. In the latter case, the residual values δ_i have a distribution $\delta_i \sim NB(r, p)$ with parameters r and p defined so that the expressions $\mu = z_i^{(\mathrm{mod})}(\hat{\theta})$ and $\sigma^2 = \nu$ hold true. The value ν was derived experimentally due to the fact that there is no generally accepted form for calculating the variance for the simulated error structure, given that it must cover the potential error spread in the incidence data. As a result of empirical observations, the variance of the error is considered equal to the variance of residual sample ρ, i.e. $\nu = \sigma_\rho^2$.
- Re-estimate the model parameters by calibrating the model on the generated data $Z_1^{'(\mathrm{dat})}, ..., Z_{200}^{'(\mathrm{dat})}$. At this step we assume that the initial day of an epidemic is clearly defined (it was essentially established during the previous steps when the model was calibrated to original data), thus, the calibration procedure might be performed by means of an auxiliary algorithm (matching by initial day of the epidemic).
- Characterise the distribution of model parameters. $\theta_1^*, ..., \theta_m^*$ The percentile method [22] was used to obtain confidence intervals.

The described approach allows flexibility in choosing the error structure, which is its advantage. The computational experiments revealed the major drawback of this method, which is its computational complexity [3]. However, the usefulness of this technique exceeds its limitations, which justified its recurrent usage in various studies [4, 19]

Figure 3 demonstrates a set of curves compared to the initial model in bootstrap procedure, while Fig. 4 and Table 2 show results of the uncertainty estimation for epidemiological parameters, with total number of the infected chosen as an example.

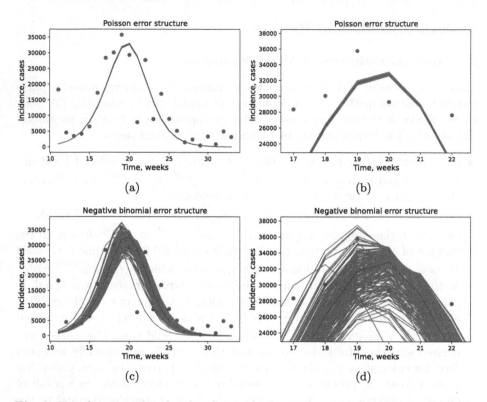

Fig. 3. Visualization of re-fitted curves after bootstrap procedure for the epidemic season of ARI at 2010–2011 in St. Petersburg, Russia, for Poisson (a,b) and negative binomial (c,d) error structures

4   Model Forecast

In this section, we demonstrate how the uncertainty quantification approach can be applied for interval forecasting. In this case, the bootstrapped datasets are generated in the same way as it was described earlier, i.e. based on the simulated modeling curve. The simulated incidence values for each week were used as mean values of the bias with Poisson and negative binomial distributions. Lower incidence values produce narrower distribution of the synthetic data points around the mean, and vice versa. As it is demonstrated in Fig. 5, the sample distribution of 200 bootstrapped incidence values with Poisson error for 15th week is wider than those distributions for 25th and 10th weeks, because it is closer in time to the peak of the outbreak.

During the prediction procedure, we use the calibration algorithms described in Sect. 2.4. The first step includes the calibration based on iterative alignment of the epidemiological peak (primary algorithm). During the resampling procedure, an auxiliary algorithm is used to calibrate the model to the bootstrapped incidence curves. Figure 6 shows a result of the forecasting procedure. The pre-

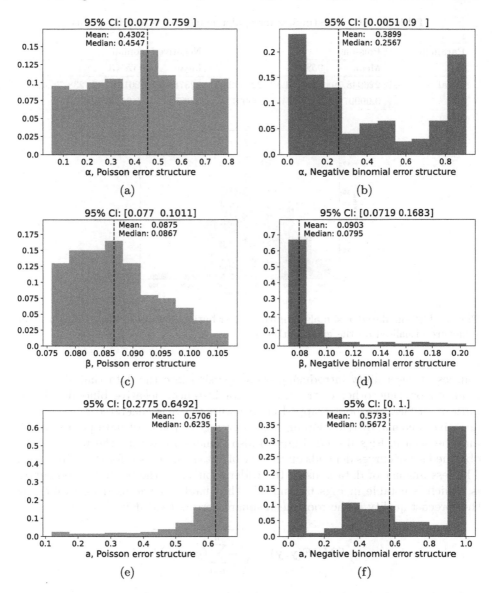

Fig. 4. The empirical distributions for parameter estimations after re-fitting of parameters to the single epidemic season with different error structures. Dashed vertical line represents the median of given parameter.

diction was performed on the data samples comprised of first 5 and 7 data points after the outbreak inception. The grey curves show the variability of the forecast. A bunch of curves which correspond to simulated datasets with Poisson error distribution (Fig. 6a, 6c) is quite narrow which does not correspond to the distribution of incidence data. Thus, it could be assumed that there are some

Table 2. Interval estimates for epidemic indicators and fit quality

Parameter	Poisson		Negative binomial	
	Mean	95% CI	Mean	95% CI
Total recovered	226948.79	(226091.13, 227849.43)	223786.87	(201687.25, 238776.42)
R^2	0.99997	(0.99994, 0.99999)	0.9576	(0.9116, 0.9842)

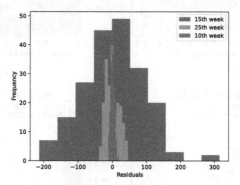

Fig. 5. The simulated residuals empirical distribution obtained using Poisson error structure visualised on the given outbreak

sources of uncertainty introducing overdispersion into the data that should be modeled by some other distribution — for instance, negative binomial distribution. In fact, the bootstrapped curves with the bias generated according to negative binomial distribution provided coverage of the most data points, which may be seen in Figs. 6b, 6d. Figure 6 also demonstrates that the width of the obtained set of curves depends on the size of data sample used for the calibration. The less amount of data is used, the wider a bunch of the bootstrapped curves is, which is notable in Figs. 6b and 6d. The quality metric used for evaluating the forecast quality is the root mean squared errors (RMSE):

$$RMSE(\mathbf{y}, \hat{\mathbf{y}}) = \sqrt{\frac{1}{n}\sum_{i=1}^{n}(y_i - \hat{y}_i)^2}.$$

This value is calculated for every prediction curve, thus forming a sample. The means and CI for the samples of RMSE values are provided in Table 3 for the forecasts made with 5, 6, and 7 initial data points and two error distributions (Poisson and negative binomial).

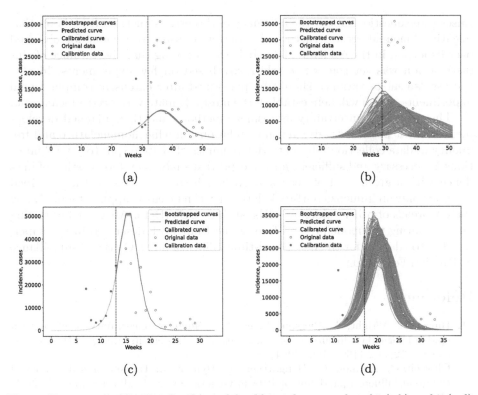

Fig. 6. Forecast visualization for the model calibrated on samples of 5 (a,b) and 7 (c,d) data points; gray curves represent the uncertainty in the forecast with Poisson (a,c) and negative binomial (b,d) error structures, cyan curve is the model best fit to the data sample, the red line is a prediction. (Color figure online)

Table 3. The root mean squared errors of the forecasts

Sample size	RMSE, Poisson		RMSE, Negative binomial	
	Mean	95% CI	Mean	95% CI
5	11823.2168	(11819.3652, 11827.0683)	13229.16249	(13012.460, 13445.8649)
6	9001.0666	(8996.8640, 9005.2692)	8529.351404	(8369.3499, 8689.35282)
7	9775.0136	(9767.2357, 9782.7916)	7564.4317	(7459.4304, 7669.4331)

5 Conclusion

In this paper, we proposed an algorithm which is able to estimate parameter and forecasting uncertainty caused by the variance in the incidence data. For that purpose, by exploiting a framework for uncertainty estimation in epidemiological setting proposed by Chowell in [3], we have performed uncertainty quantification procedures and the interval prediction of disease incidence on the limited data points. In our case study, the confidence intervals obtained by assuming Poisson error structure are rather narrow if compared with variance in initial data, which raises some questions about the adequacy of this error structure. It is shown that negative binomial error structure is a better solution to account for observed overdispersion in the epidemic data. It is worth noting that the variance of this distribution was set somewhat arbitrarily based on few experiments. Ideally, the assessment of variance should be performed after extensive computational experiments, which will help establish the most adequate variance-to-mean ratio.

The presented uncertainty assessment methods and their planned development make it possible to give recommendations on what nomenclature and frequency of data collection related to the dynamics of acute respiratory viral infections is necessary and sufficient for retrospective analysis and forecasting of incidence with a given level of accuracy. Although so far we tested the described methods only on influenza outbreak data, they can be easily applied for analysing the outbreaks of other epidemic ARIs, such as COVID-19. After further testing and validating the modeling and forecasting framework described in the paper, we plan to adapt it for the use in real-time influenza and COVID-19 surveillance in Russian Federation.

References

1. Burns, C., Wright, J., Pierson, J., et al.: Evaluating uncertainty to strengthen epidemiologic data for use in human health risk assessments. Environ. Health Perspect. **122**(11), 1160–1165 (2014)
2. Chowell, G., Ammon, C., Hengartner, N., Hyman, J.: Transmission dynamics of the great influenza pandemic of 1918 in Geneva, Switzerland: assessing the effects of hypothetical interventions. J. Theor. Biol. **241**, 193–204 (2006)
3. Chowell, G.: Fitting dynamic models to epidemic outbreaks with quantified uncertainty: a primer for parameter uncertainty, identifiability, and forecasts. Infect. Dis. Model. **2**(3), 379–398 (2017)
4. Chowell, G., Luo, R.: Ensemble bootstrap methodology for forecasting dynamic growth processes using differential equations: application to epidemic outbreaks. BMC Med. Res. Methodol. **21**(1), 1–18 (2021)
5. Davison, A.C., Hinkley, D.V., Young, G.A.: Recent developments in bootstrap methodology. Stat. Sci. **18**, 141–157 (2003)
6. Efron, B., Tibshirani, R.: Bootstrap methods for standard errors, confidence intervals, and other measures of statistical accuracy. Stat. Sci. **1**, 54–75 (1986)
7. Hilbe, J.M.: Negative Binomial Regression Second Edition. Cambridge University Press, Cambridge (2011)
8. Huaman, I., Plesovskaya, E.P., Leonenko, V.N.: Matching model complexity with data detail: influenza propagation modeling as a case study. In: 2022 IEEE International Multi-Conference on Engineering, Computer and Information Sciences (SIBIRCON), pp. 650–654 (2022)
9. Jurek, A., Maldonado, G., Greenland, S., et al.: Exposure-measurement error is frequently ignored when interpreting epidemiologic study results. Eur. J. Epidemiol. **21**, 871–876 (2006)
10. Kimab, D.R., Hwang, S.Y.: Forecasting evaluation via parametric bootstrap for threshold-INARCH models. Commun. Stat. Appl. Methods **27**, 177–187 (2020)
11. Krivorot'ko, O., Kabanikhin, S., Zyat'kov, N., et al.: Mathematical modeling and forecasting of COVID-19 in Moscow and Novosibirsk region. Numer. Anal. Appl. **13**, 332–348 (2020)

12. Kulesa, A., Krzywinski, M., Blainey, P., Altman, N.: Sampling distributions and the bootstrap. Nature Methods **12**, 477–478 (2015)
13. Lee, V.J., Tok, M.Y., Chow, V.T., Phua, K.H., et al.: Economic analysis of pandemic influenza vaccination strategies in Singapore. PLOS One **4**(9), 1–8 (2009). https://doi.org/10.1371/journal.pone.0007108
14. Leonenko, V., Bobashev, G.: Analyzing influenza outbreaks in Russia using an age-structured dynamic transmission model. Epidemics **29**, 100358 (2019)
15. Leonenko, V.N.: Herd immunity levels and multi-strain influenza epidemics in Russia: a modelling study. Rus. J. Numer. Anal. Math. Model. **36**(5), 279–293 (2021)
16. Leonenko, V.N., Ivanov, S.V.: Fitting the SEIR model of seasonal influenza outbreak to the incidence data for Russian cities. Rus. J. Numer. Anal. Math. Model. **26**, 267–279 (2016)
17. Leonenko, V.N., Ivanov, S.V.: Prediction of influenza peaks in Russian cities: comparing the accuracy of two SEIR models. Math. Biosci. Eng. **15**(1), 209 (2018)
18. Lloyd, A.L.: Sensitivity of model-based epidemiological parameter estimation to model assumptions. In: Mathematical and Statistical Estimation Approaches in Epidemiology, pp. 123–141. Springer, Dordrecht (2009). https://doi.org/10.1007/978-90-481-2313-1_6
19. López, M., Peinado, A., Ortiz, A.: Characterizing two outbreak waves of COVID-19 in Spain using phenomenological epidemic modelling. PLOS One **16**, e0253004 (2021)
20. Morillas, F., Valero, J.: Random resampling numerical simulations applied to a SEIR compartmental model. Eur. Phys. J. Plus **136**(10), 1–24 (2021). https://doi.org/10.1140/epjp/s13360-021-02003-9
21. Ning, X., Jia, L., et al.: Epi-DNNs: epidemiological priors informed deep neural networks for modeling COVID-19 dynamics. Comput. Biol. Med. **158**, 106693 (2023)
22. Puth, M., Neuhäuser, M., Ruxton, G.: On the variety of methods for calculating confidence intervals by bootstrapping. J. Anim. Ecol. **84**, 892–897 (2015)
23. Ranker, L., Petersen, J., Fox, M.: Awareness of and potential for dependent error in the observational epidemiologic literature: a review. Ann. Epidemiol. **36**, 15–19 (2019)
24. Roosa, K., Chowell, G.: Assessing parameter identifiability in compartmental dynamic models using a computational approach: application to infectious disease transmission models. Theor. Biol. Med. Model. **16**, 1 (2019)
25. Roosa, K., Luo, R., Chowell, G.: Comparative assessment of parameter estimation methods in the presence of overdispersion: a simulation study. Math. Biosci. Eng. **16**, 4299–4313 (2019)
26. Samsuzzoha, M., Singh, M., Lucy, D.: Uncertainty and sensitivity analysis of the basic reproduction number of a vaccinated epidemic model of influenza. Appl. Math. Model. **37**(3), 903–915 (2013)
27. Swallow, B., Birrell, P., Blake, J., et al.: Challenges in estimation, uncertainty quantification and elicitation for pandemic modelling. Epidemics **38**, 100547 (2022)
28. Zelner, J., Riou, J., Etzioni, R., Gelman, A.: Accounting for uncertainty during a pandemic. Patterns **2**(8), 100310 (2021)

A Web Portal for Real-Time Data Quality Analysis on the Brazilian Tuberculosis Research Network: A Case Study

Victor Cassão[1]([⊠]) [ID], Filipe Andrade Bernardi[1] [ID], Vinícius Costa Lima[2] [ID],
Giovane Thomazini Soares[2] [ID], Newton Shydeo Brandão Miyoshi[2] [ID],
Ana Clara de Andrade Mioto[1] [ID], Afrânio Kritski[3] [ID], and Domingos Alves[2] [ID]

[1] São Carlos School of Engineering, University of São Paulo, São Carlos, SP, Brazil
{victorcassao,filipepaulista12,anaclara.mioto}@usp.br
[2] Ribeirão Preto Medical School, University of São Paulo, Ribeirão Preto, SP, Brazil
viniciuslima@alumni.usp.br, giovane.soares@usp.br, quiron@fmrp.usp.br
[3] School of Medicine, Federal University of Rio de Janeiro, Rio de Janeiro,
RJ, Brazil

Abstract. Research projects with Tuberculosis clinical data generate large volumes of complex data, requiring sophisticated tools to create processing pipelines to extract meaningful insights. However, creating this type of tool is a complex and costly task, especially for researchers who need to gain experience with technology or statistical analysis. In this work, we present a web portal that can connect to any database, providing easy access to statistical analysis of the clinical data in real-time using charts, tables, or any other data visualization technique. The tool is user-friendly and customizable, reaching the project's needs according to its particularities. The developed portal in this work was used as a use case for the research project developed by the Federal University of Rio de Janeiro (UFRJ) for the validation and cost of performance of the Line Probe Assay 1 and 2 (LPA) as a method of diagnosing resistant Tuberculosis in Brazilian's reference centers. In the use case, the tool proved to be a valuable resource for researchers, bringing efficiency and effectiveness in analyzing results for quick and correct clinical data interpretation.

Keywords: Web Portal · Data Processing · Clinical Data · Statistical Analyzes

1 Introduction

A web portal is a web-based repository of data, information, facts, results of analyzes, and also knowledge [1,2]. Portals can enable data search and filtering [3], facilitating access to information of interest. In health, information is usually gathered from multiple sources and organized in a user-friendly way [4].

© The Author(s), under exclusive license to Springer Nature Switzerland AG 2023
J. Mikyška et al. (Eds.): ICCS 2023, LNCS 14075, pp. 300–312, 2023.
https://doi.org/10.1007/978-3-031-36024-4_24

Web portals provide a unified interface for applications and databases that, without this approach, would be isolated elements [3]. Also, they are an efficient and effective form of communication and dissemination, allowing the production of knowledge and intelligence [2,5]. Given the relevance of these products for planning processes [6], web portals in the health area [1] can assist in the planning of public health programs [7]. In this context, it is understandable that the content of web portals must be valid, trustworthy, coherent, representative, sensitive, comprehensive, and ethical [5,8].

However, health web portals stand out for the lack of interoperability [2,5], accessibility, and quality [9] that data in this area can present. Also, in the health domain, many web portals suffer from usability problems [10], undermining their effectiveness [11]. When we discuss health portals, we need to understand in which context and disease it is present. In Brazil, one of the diseases that is still a challenge, both in terms of control and cure, is Tuberculosis.

Tuberculosis (TB) is an infectious disease caused by the bacterium Mycobacterium tuberculosis, which can affect different organs, mainly the lungs. In most cases, it is a curable disease if the treatment is carried out correctly. In 2021, it was estimated that 10.6 million people globally were sick with TB, with 1.6 million deaths. The Brazilian scenario is also very challenging, with around 88099 cases reported in 2021, aggravated by the Covid-19 pandemic [12].

Several studies point to a higher disease prevalence in countries with low socioeconomic status, especially among vulnerable groups such as homeless people and people deprived of liberty [13]. The current cycle between poverty and illness is fed back between individuals and their social environment, influencing the outcome of TB treatment, which is the primary way to cure and reduce transmission of the disease.

It evidences problems like the low effectiveness and adherence to treatment, which is around 70%, the national average [14]. Between its causes, we can mention treatment abandonment (people that did not attend medical follow-ups and stopped taking all medications), incorrect use (people that use only prescribed medications), and irregular use of drugs (people that use in wrong periods). It is also worth highlighting that engagement problems can lead to treatment failure, drug resistance, and TB relapse [15].

It is still a worrying disease across the national and global scene, so WHO has developed the End TB Strategy. The main objectives of this mission are to zero the number of deaths and people living with the disease through 3 pillars: the first - integrated, patient-centered care and prevention; the second - bold policies and supportive systems and the third - intensified research and innovation [12].

2 Objectives

This paper aims to present the development, evaluation, and usability validation of a new web portal to integrate tuberculosis databases to a data processing pipeline for clinical data analysis using initially statistical methods and support for future use of machine learning based approaches. The results of this study will provide insights into the potential of web-based tools for clinical data analysis on the REDE-TB project and contribute to the development of more efficient and effective data analysis methods for improving patient outcomes.

3   Methods

In this section, the context in which this research was developed will be detailed, highlighting the dataset selection as a use case for this work. The data pre-processing phase and the tools and technologies will also be presented. Figure 1 shows the detailed workflow of how the application works, from the data acquisition to the pre-processing and the tools to make information available to the user. These steps will be detailed next.

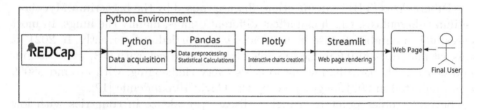

Fig. 1. Software workflow.

3.1   Scenario: Research Data on TB in Brazil

Created in 2001 by an interdisciplinary group of Brazilian researchers, the Brazilian Tuberculosis Research Network (REDE-TB) aims to promote interaction between government, academia, health service providers, civil society, and industry in developing and implementing new technologies and strategies to improve tuberculosis control across the country. REDE-TB is organized into ten research areas, comprising around 320 members from 65 institutions representing 16 of the 27 Brazilian states [16].

Data collection and management of studies carried out by REDE-TB are done through the Research Electronic Data Capture (REDCap) platform. Developed by Vanderbilt University and available through an open-source license, REDCap is a web application that manages case report forms (CRF), surveys, and research databases online.

REDCap also provides tools for exporting data in different formats. The standard used in developing this project was the Comma Separated Values (CSV), which is easy to handle through other third-party software.

Currently, the REDE-TB REDCap instance has 21 projects in progress and 261 registered users, which has a mechanism for sharing and tracking data captured through the software. [17].

3.2   Use Case: Line Probe Assay (LPA) Project

The LPA project started in 2019 and aims to evaluate the performance and cost of the Line Probe Assay 1 and 2 techniques in reference centers in Brazil to diagnose drug-resistant tuberculosis compared to the Xpert and MGIT 960 methods. It is an analytical, prospective, and multicenter study that aims to incorporate the method in the diagnostic routine. Reference centers with sputum samples and MTB cultures from patients with TB processed in 9 (nine) local reference laboratories for mycobacteria in the following Brazilian states participate in the study: Rio de Janeiro, São Paulo, Amazonas, and Bahia.

The study population is adults (18 years or older) with symptoms of pulmonary TB whose clinical specimens are positive on Xpert MTB/RIF and who have not received treatment for sensitive, drug-resistant, or multidrug-resistant TB for more than seven days in the past 30 d. Thus, it is expected to estimate, compare and analyze the diagnostic accuracy of the LPA compared to the gold standard adopted in health services. Secondary results aim to identify barriers and facilitators for implementing a quality system in participating laboratories and analyze the time elapsed between screening and a) detection of resistant TB; b) initiation of patient treatment.

The study variables distributed in 14 CRFs were defined and divided into primary and secondary data. The primary data are gathered directly from health centers, where health agents collect data through interviews with eligible patients. Demographic, socioeconomic, recruitment, and patient eligibility data are collected in these cases. Secondary data are collected through the results obtained in tests performed by specialized laboratories participating in the study.

The research evaluates the accuracy of the test by using the Positive Predictive Value (PPV) and Negative Predictive Value (NPV). PPV represents the proportion of patients who tested positive, and NPV for the negatives, in a given exam result, e.g., rifampicin sensitive, isoniazid sensitive. Those metrics show how many patients were correctly diagnosed for both cases, with positives and negatives. Through this metric, we can compare the results of different exams and evaluate their accuracy to extract information about their performance and viability. Furthermore, the clinical impact between the time of sample collection and resistant TB detection is an important indicator of the research, especially, to reach the objective of screening time reduction and appropriate treatment initiation. This also impacts the reduction of mortality, treatment abandonment, and TB transmission in the community.

Regarding data quality, managing data at source and applying Findable, Accessible, Interoperable, and Reusable (FAIR) Guiding Principles for its transparency use are recognized as fundamental strategies in interdisciplinary scientific collaboration.A standard protocol designed for this study has been used based on these principles and in applied routinely algorithm data-driven monitoring. This algorithm is responsible for data verification according to the metrics mentioned before.

3.3   Knowledge Discovery in Databases Methods

This work used the main processes and methods of Knowledge Discovery in Databases (KDD). KDD is a well-known knowledge extraction process from large databases that aims to improve data quality. It is divided into well-organized steps that aim to reach that finality. Figure 2 exemplifies the steps of the organization on KDD.

The first stage, data selection, was carried out with the support of a technical team that computerized the entire data acquisition process through electronic forms deployed on REDCap. Health agents are responsible for entering study data directly into the system. To increase the quality of this collected data, actions and procedures are adopted before and after data insertion in REDCap. Before the data entry phase, manuals and training are provided for users. The subsequent approach to data insertion is their validation, aiming to identify potentially false or inconsistent data with other information about the patients. Such validation is carried out by software developed by the team responsible for maintaining the REDCap platform and manually by the team members. After these procedures, data can be analyzed with more reliability.

The LPA data selected for analysis is divided into three main groups: socioeconomic data, exam results, and key dates. Socioeconomic data includes descriptive patient information, such as age, height, weight, BMI, skin color, gender, etc. These data are essential in profiling patients and identifying correlations with TB outcomes. Laboratory examination results provide information about the type of TB detected in the collected sample and for which medication was detected resistance. The selected variables present information on the results of gold-standard testing examinations such as Xpert, MGIT 960, LPA1, and LPA2. Auxiliary variables that discriminate the sample's resistance to certain medications, such as Rifampicin and Isoniazid, were also selected. The data belonging to key dates represent the beginning and end of two specific stages of the study. The purpose is to analyze the elapsed time between such dates to validate the time interval and its performance. The main selected dates are the date of collection, the date of release of examination results, the date of DNA hybridization, and other relevant dates.

The second stage of KDD, data preprocessing, has as its main purpose the improvement of data quality, making it possible to use it in machine learning algorithms, data mining, and statistical analysis [3]. It is a crucial stage in extracting knowledge from a clinical database because this type of data is stored in its raw format, often without any treatment, containing inconsistencies, noise, and missing data [4]. Another important process in this stage is data cleaning. The cleaning process aims to remove errors and inconsistencies. It is responsible for reducing noise, correcting or removing outliers, treating missing data, duplicate data, or any other routine that aims to remove anomalies in the database.

This stage focuses on correctly cleaning, treating, and standardizing invalid data and any noise or inconsistency. In the case of LPA use, the lack or inconsistency of data originates from various factors, such as errors in filling out the data on the form, delay in delivering the test results by the laboratories, or some other unknown factor. Identifying these missing or inconsistent data cases is considered one of the software's functionalities so that the centers can request correct filling, avoiding problems for the patient and the study itself.

In the case of unavailable data, REDCap has, by default, specific data types that handle its absence and reason. Table 1 provides these different data types and their meaning.

Table 1. Codes for Missing Data.

Code	Description
NI	No information
INV	Invalid
UNK	Unknown
NASK	Not asked
ASKU	Asked but unknown
NA	Not applicable
NAVU	Not available

In addition to the types of invalid data provided by REDCap, there is also the possibility of data not existing. In these cases, this step is also responsible for transforming the data into the types provided by the platform. This process makes it possible to extract which patients have missing records in a standardized way, allowing for correct completion by the responsible center. Inconsistencies found in variables that store dates are also treated in this step. Errors in date insertion can cause inconsistencies in the elapsed time calculation(in cases where the end date is sooner than the start date). In these cases, the inconsistencies are not included in the treated database, avoiding anomalies in the analysis and requesting a correction.

The third step is data transformation. This step transforms the cleaned and pre-processed data into the correct format for data mining algorithms. The main need met by the data transformation step was the standardization of test results, making it possible to compare different tests. Some variables in laboratory test results are stored in different formats, making it impossible to compare them directly. An example of this transformation is the manipulation of data from the TSA and LPA1 test results to detect resistance to the drug Isoniazid. While the first is stored in a categorical data format, where each row represents the situation of the test result, the second is stored in boolean form, stored in several columns that indicate one of the possible test results as the column name, having true or false values in the rows to represent whether the condition is valid or not. The applied operation transforms this boolean data into categorical data, facilitating comparison with other tests and ensuring better analysis of their values through the available charts.

With all KDD steps performed, algorithms are applied to extract patterns and generate reports and dashboards. The results section provides more information on all analysis generated after data processing.

3.4   Tools and Technologies

Python programming developed the dedicated web portal for the LPA project. Due to its versatility, all the particularities of the software were developed solely and exclusively using Python and auxiliary libraries.

The Pandas library was used for data manipulation, preprocessing, statistical calculations, and other operations on descriptive data analysis. It is currently one of the most famous libraries for data manipulation developed in Python.

Plotly is a library for creating different charts, with native integration with Pandas, making its use more straightforward in the project. Plotly can create interactive charts where the user can filter data by legend and periods, apply zoom in/out, download the chart as an image, and many other functionalities. These filtering options run in real-time, improving the user experience.

Streamlit is a Python library for web page creation, with no need for in-depth knowledge of front-end technologies. Due to its native rendering of tables, graphs, filters, and other features, it is an excellent tool for sharing data-oriented applications. It has a native integration with Plotly, rendering the graphics generated by it and all its functionalities directly on the web page.

4   Results

This section brings detailed information about the results of the analyzes available on the web portal and the insights extracted by the researchers who used the tool.

The web portal developed to fit the needs of the LPA use case can compare 11 different laboratory exam results with charts and tables to make knowledge extraction easier from it. It also can compare five key dates for elapsed time checking between crucial steps from research. More than 20 variables related to the patient's socioeconomic data are also available in the web portal for review.

The analysis of laboratory test results is interactive, where the user can choose between two different options, filtering by center or making a general comparison between them. The results are calculated according to the filter and dynamically rendered as a bar chart. The primary purpose is an accuracy check between the studied test (LPA) and all the other gold-standard TB tests. An example, a comparison between Xpert and LPA1 to detect resistance against Rifampicin is shown in Fig. 2.

Besides comparing test results, mismatching and missing data were seen and removed from the cleaned dataset. From that, contacting the research centers and asking for data correction is possible. Figure 3 shows a bar chart with the relationship between the test results and missing values.

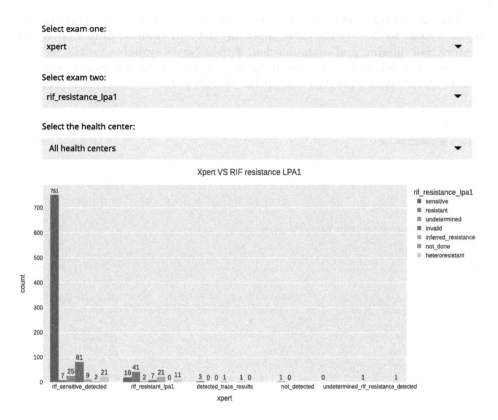

Fig. 2. Comparison between Xpert and LPA1 to detect resistance against Rifampicin.

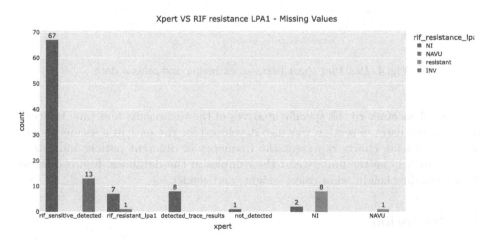

Fig. 3. Comparison between Xpert and LPA1 for missing values.

The elapsed time between key dates is generated in a Box Plot chart type to make it easier to identify the mean times and data distribution from all over the centers. For filtering and manipulation, a slide bar is available for the user to configure a threshold to set a minimum value in days between those two dates. Figure 4 presents this functionality in the Box Plot chart.

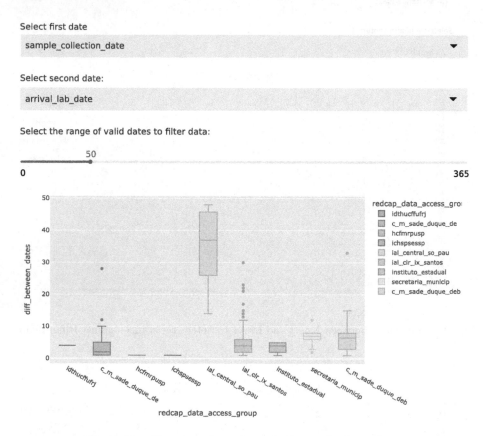

Fig. 4. Box Plot chart between gathering and release date.

Complementary to the specific analyzes of the test results and time intervals between key dates, a section was also developed for the patient's socioeconomic data. In general, charts represent the frequency of different patient indicators, which can help better understand the samples in the database. Figure 5 shows data regarding height, skin color, weight, and gender.

5 Discussion

During the initial stage, the tool demonstrated highly satisfactory results in meeting all the expected requirements for the LPA project use case. One of

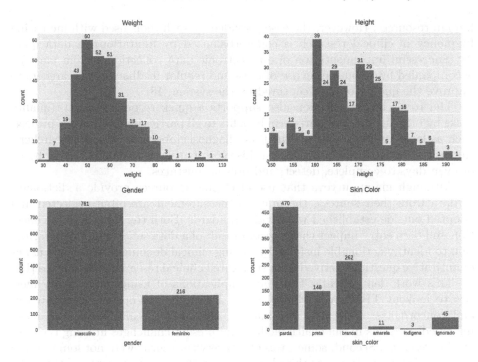

Fig. 5. Bar Charts with weight, height, gender and skin color histograms.

the major challenges faced during the development process was to create an analysis of indicators and metrics to assess the accuracy of the tests being evaluated against the gold standard. The web portal provides up-to-date data and easy access to the values required for the statistical analysis and indicators as described on Sect. 3.2.

Before the adoption of the web portal, this process was completely manual, expending a lot of human resources on a repetitive task, resulting in overdue reports delivery. This tool automated this entire process, greatly reducing the report creation time. In a few seconds, the researchers have all the analysis available, always updated with the most recent data from the dataset, with several auxiliary results to improve the study database maintenance and guarantee data quality.

The main contribution of this work is to demonstrate, through a web portal, a pipeline for processing and analyzing clinical data on tuberculosis using statistical methods. The results raise important questions about the need to provide researchers with a tool capable of extracting information from the data collected by their research, presenting them briefly, clearly, and objectively.

The main positive point was the accuracy of developing a tool capable of meeting the needs of the traditional clinical trial use cases. Transitioning from paper records to electronic records and reports is a complex task that requires an advanced information technology infrastructure and a significant investment in human resources. However, the ease provided by such tools used with increasing frequency in clinical research is counterbalanced by limitations in data quality. Successful implementation of an electronic data collection system must be accompanied by a routine review of data and regular feedback to researchers to improve the quality of data captured by the system [18].

The tool's implementation also supports a quick response to data quality tasks by bringing several short-term benefits to all parties involved in the project, such as local and national coordinators, diagnosis laboratories, and researchers. In many studies, researchers perform this type of analysis manually, taking hours or even days to complete, detect, and prevent mistakes.

Although many surveys that use electronic resources provide a dictionary of data from their sources, the units of measurement are often neglected and adopted outside established worldwide standards. Both the form of data collection and data entry impact the expected result of a data set. Extracting information to identify actionable insights by mining clinical documents can help answer quantitative questions derived from structured clinical research data sources [19].

This work demonstrates two main implications of using data quality metrics to lookout TB data in clinical trials. First, the sub-optimal completeness and concordance of the data can make it difficult for health services to make informed decisions in the health facility and present challenges in using the data for TB research. Second, some cases of TB service records were not found in the different screening forms analyzed, suggesting that some reports needed to be transcribed into some computerized systems or were transcribed imprecisely. It may be overcome through an institutional inventory study to understand patient follow-up better.

Although using quantitative research methods is more frequent to assess data accuracy, consistency, completeness, and availability, subjective evaluation can help design effective strategies to improve data quality. Critical and non-critical variables and multiple audits (before, during, and after) with quality improvement feedback should be included in the quality monitoring activities of survey data. This combination is considered a cost-effective solution to visualizing project issues and ensuring data quality.

6   Conclusion

The development of the web portal for statistical analysis delivery on Tuberculosis research has provided a powerful and user-friendly tool for researchers to analyze and interpret clinical data, even in the early stage of development. The portal has been developed to collect data from third-party databases, apply extract-load-transform pipelines, and provide a range of visualizations (such as box plots, bar plots, tables, and data frames) to help researchers to visualize their data and extract meaningful insights.

Going forward, we plan to continue to develop and refine our web portal to attempt the evolving needs of clinical research in the REDE-TB project. Specifically, we aim to enhance the researcher's experience by introducing new functionalities and improving data visualization capabilities. We also plan to create a machine learning section to make available prediction models to identify bad outcomes, resistant TB, or any other future need of the project. Overall, we believe that our web portal has the potential to make a significant impact on the Tuberculosis clinical research community and contribute to the advancement of medical science.

Acknowledgments. The study was partially supported by the Brazilian National Council for Scientific and Technological Development (CNPq) - grant numbers 440758/2018-1, 22411-2019, coordinated by author A.K., and by the São Paulo Research Foundation (FAPESP) - grant number 2020/ 01975-9, coordinated by author D.A.

References

1. Rodrigues, R.J., Gattini, C.H.: Chapter 2 - National Health Information Systems and Health Observatories. In: Global Health Informatics: How Information Technology Can Change Our Lives In A Globalized World, pp. 14–49. Elsevier (2016)
2. Yoshiura, V.T.: Desenvolvimento de um modelo de observatório de saúde baseado na web semântica: o caso da rede de atenção psicossocial. Ph.D. thesis, University of São Paulo, Brazil (2020)
3. World Health Organization - Guide for the establishment of health observatories. https://apps.who.int/iris/handle/10665/246123. Accessed 03 Mar 2023
4. Xiao, L., Dasgupta, S.: Chapter 11 - User satisfaction with web portals: An empirical study. In: Web systems design and online consumer behavior, pp. 192–204. IGI Global (2005)
5. Bernardi, F.A., et al.: A proposal for a set of attributes relevant for Web portal data quality: The Brazilian Rare Disease Network case. Procedia Comput. Sci. (in press)
6. Oliveira, D.P.R.: Planejamento estratégico: conceitos, metodologia e práticas, 23rd edn. Editora Atlas, São Paulo (2007)
7. Alves, D., et al.: Mapping, infrastructure, and data analysis for the Brazilian network of rare diseases: protocol for the RARASnet observational cohort study. JMIR Res. Protoc. **10**(1), e24826 (2021)
8. World Health Organization. Regional Office for Africa. Guide for the establishment of health observatories. World Health Organization. Regional Office for Africa. https://apps.who.int/iris/handle/10665/246123. Accessed 03 Mar 2023
9. Pereira, B.S., Tomasi, E.: Instrumento de apoio à gestão regional de saúde para monitoramento de indicadores de saúde. Epidemiol. Serv. Saúde **25**(2), 411–418 (2016)
10. Nahm, E.S., Preece, J., Resnick, B., Mills, M.E.: Usability of health Web sites for older adults: a preliminary study. CIN: Comput. Inf. Nurs. **22**(6), 326–334 (2004)
11. Saeed, M., Ullah, S.: Usability Evaluation of a Health Web Portal. Master's thesis, Blekinge Institute of Technology, MSC-2009:16 Ronneby (2009)
12. World Health Organization (WHO). Global Tuberculosis Report 2018. Geneva: WHO. 2018. https://apps.who.int/iris/handle/10665/274453. Accessed 03 Mar 2023

13. Macedo, L.R., Maciel, E.L.N., Struchiner C.J.: Populações vulneráveis e o desfecho dos casos de tuberculose no Brasil. Ciênc saúde coletiva [Internet]. (Ciênc. saúde coletiva) (2021). https://doi.org/10.1590/1413-812320212610.24132020
14. Rabahi, M.F., et all. Tuberculosis treatment. Jornal Brasileiro de Pneumologia (2017). https://doi.org/10.1590/S1806-37562016000000388
15. Sales, O.M.M., Bentes, P.V.: Tecnologias digitais de informação para a saúde: revisando os padrões de metadados com foco na interoperabilidade. Rev Eletron Comun Inf Inov Saúde. https://www.reciis.icict.fiocruz.br/index.php/reciis/article/view/1469. Accessed 03 Mar 2023
16. Kritski, A. et al.: The role of the Brazilian Tuberculosis Research Network in national and international efforts to eliminate tuberculosis. Jornal Brasileiro Pneumologia **44**, 77–81. https://doi.org/10.1590/s1806-37562017000000435
17. Bernardi, F., et al.: Blockchain based network for tuberculosis: a data sharing initiative in Brazil. Stud. Health Technol. Inform. **262**, 264–267 (2019). https://doi.org/10.3233/SHTI190069
18. Sharma, A., Ndisha, M., Ngari, F., et al.: A review of data quality of an electronic tuberculosis surveillance system for case-based reporting in Kenya. Eur. J. Publ. Health **25**, 1095–1097 (2015). https://doi.org/10.1093/eurpub/ckv092
19. Malmasi, S., Hosomura, N., Chang, L.-S., et al.: Extracting healthcare quality information from unstructured data. AMIA Annu. Symp. Proc. **2017**, 1243–1252 (2017)

Use of Decentralized-Learning Methods Applied to Healthcare: A Bibliometric Analysis

Carolina Ameijeiras-Rodriguez[1] , Rita Rb-Silva[1] , Jose Miguel Diniz[1,3] ,
Julio Souza[1,2] , and Alberto Freitas[1,2(✉)]

[1] MEDCIDS - Department of Community Medicine, Information and Health Decision Science,
Faculty of Medicine, University of Porto, Porto, Portugal
alberto@med.up.pt
[2] CINTESIS@RISE, Faculty of Medicine, University of Porto, Porto, Portugal
[3] PhD Program in Health Data Science Faculty of Medicine of University of Porto, Porto,
Portugal

Abstract. The use of health data in research is fundamental to improve health care, health systems and public health policies. However, due to its intrinsic data sensitivity, there are privacy and security concerns that must be addressed to comply with best practices recommendations and the legal framework. Decentralized-learning methods allow the training of algorithms across multiple locations without data sharing. The application of those methods to the medical field holds great potential due to the guarantee of data privacy compliance. In this study, we performed a bibliometric analysis to explore the intellectual structure of this new research field and to assess its publication and collaboration patterns. A total of 3023 unique documents published between 2013 and 2023 were retrieved from Scopus, from which 488 were included in this review. The most frequent publication source was the IEEE Journal of Biomedical and Health Informatics (n = 27). China was the country with the highest number of publications, followed by the USA. The top three authors were Dekker A (n = 14), Wang X (n = 13), Li X (n = 12). The most frequent keywords were "Federated learning" (n = 218), "Deep learning" (n = 62) and "Machine learning" (n = 52). This study provides an overall picture of the research literature regarding the application of decentralized-learning in healthcare, possibly setting ground for future collaborations.

Keywords: Federated learning · Decentralized-learning · Privacy-preserving protocols; Machine learning · Smart Healthcare · Medicine · Bibliometric analysis

1 Introduction

In recent years, data production has increased exponentially in most industries. In the health industry, data is crucial for the continuous improvement of healthcare (HC), health systems and public health policies. On one hand, data generated by this field represents a significant percentage of the global data volume [1]. On the other hand,

© The Author(s), under exclusive license to Springer Nature Switzerland AG 2023
J. Mikyška et al. (Eds.): ICCS 2023, LNCS 14075, pp. 313–320, 2023.
https://doi.org/10.1007/978-3-031-36024-4_25

this type of data is extremely sensitive, so the preservation of the patient's privacy becomes the main objective and the emerging concerns regarding privacy protection led to the creation of demanding regulations and increasing data governance and privacy barriers. Thus, decentralized-learning solutions have gained increased attention by allowing to extract the maximum possible value from health data while preserving data privacy. Decentralized-learning methods include collaborative methods, such as split- and federated-learning (FL). FL is probably the most attractive decentralized distributed machine learning (ML) paradigm, with its first publications related to health published in 2013 [2, 3]. In 2017, FL gained great exposure after being presented by Google [4]. Published applications of FL to HC include patient similarity learning [5], cardiology [6], oncology [7, 8], population health [9], among others.

This is the first bibliometric overview of the literature regarding the use of decentralized analysis in HC, fostering new ideas for investigation and potential collaborations.

2 Material and Methods

2.1 Bibliographic Database

Scopus was chosen as a broad and generalized scientific database of peer-reviewed research, comprising all research fields. It allows data extraction and aggregation in several formats, providing detailed information for the analyses.

2.2 Study Design

A bibliometric analysis was performed to evaluate the related research on decentralized-learning in HC, published from January 2013 to the date of the search, January 25, 2023. The search was restricted to documents originally written in English.

2.2.1 Eligibility Criteria

For the purpose of this review, decentralized-learning was defined as a ML approach to use data available from multiple parties, without sharing them with a single entity.

Inclusion criteria: Articles had to be peer-reviewed original research or review; and describe an application of decentralized learning techniques to human health.

Exclusion criteria: All the results comprising books, book chapters, conference proceedings, editorials, perspectives, study protocols, commentaries, pre-prints and other unpublished studies; studies with quality issues; publications made before 2013; and full text publications not written in English were excluded. This was intended to reduce the potential bias caused by different publications from the same study (duplication) and restrict the analysis to high-quality original studies.

2.2.2 Search Strategy

A composite search strategy was adopted to include terms related to "decentralized-learning" and "healthcare". The final search expression was accomplished by two queries, using TITLE-ABS-KEY filter as presented in the Supplementary Material (Sup Mat).

2.3 Software and Data Analysis

After a review by two independent reviewers and a third reviewer for disagreements, we selected the articles that met the inclusion criteria. To facilitate this selection, we used the Rayyan tool [10]. Bibliometrix R package and the Biblioshiny platform [11] were used to perform the analysis of the selected articles. The Keyword Plus engine from the Web of Science was used to analyze the most used keywords.

2.4 Ethical Considerations

This study has been granted exemption from requiring ethics approval as it did not involve animal or human participants and only publicly available data was used.

3 Results

3.1 General Information

A total of 3050 documents were retrieved from Scopus, corresponding to 3023 unique documents published between January 1, 2013, to the day of the search, January 25, 2023. After the eliminating duplicated records, title and abstract screening, and full-text review, 488 documents were included (16.1% of total retrieved documents) - https://bit.ly/41rCjDL. The list of included articles is available in the Sup Mat.

The average annual growth rate was 16.49%. However, from 2013 to 2019, the number of published articles was relatively stable, with a mean of 7.4 articles published per year. Since 2020, there has been exponential growth, with 41 articles published in 2020, 101 articles in 2021, 271 articles in 2022, and 23 in the first 25 days of 2023. The average of citations per document was 12.3.

3.2 Countries

A total of 45 countries presented a corresponding author with at least one publication. Nevertheless, 7 countries alone concentrated approximately half of the world's production: China (71 publications), the USA (68), Germany (23), South Korea (22), Netherlands (22), India (21), and Canada (15). International co-authorship comprised 49.8% of the total publications. In terms of total citations, publications from the USA stood out in the first position (1895 citations), followed by the Netherlands (702), Germany (432), Canada (373) and China (338). However, analyzing the average number of citations per article, the Netherlands took first place (31.9 citations/article), followed by the USA (27.9), Canada (24.9), Italy (23.2) and Germany (18.8).

A country-collaboration bibliometric network was constructed and is presented in Fig. 1. Researchers from 50 countries reported international collaborations. Some collaboration clusters can be seen, namely one dominated by the USA (red cluster), another dominated by China (blue cluster) and another one mostly constituted by European countries, with a prominent position for Germany and the UK (green cluster).

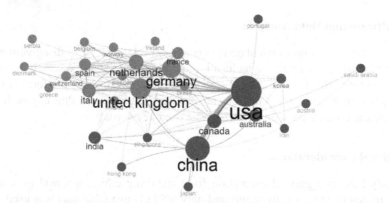

Fig. 1. Countries collaboration network regarding research in decentralized analysis in HC. The size of each circle is proportional to the number of collaborations with the respective country.

3.3 Affiliation and Institutions

There were 1250 different institutions worldwide having published at least one article in this area, averaging 2.6 institutions per article Apart from Harvard Medical School, no institutions that have concentrated production in the field of decentralized learning in HC. Harvard Medical School (133 publications), Technical University of Munich (45), University of Michigan (42), University of Pennsylvania (39) and University of California (38) ranked in the top-5 most productive institutions.

3.4 Sources and Publications

The 488 articles were published in 245 different international scientific journals with an impact factor. No single source concentrated a significant percentage of the overall production. The most relevant sources, with at least 10 published articles were IEEE Journal of Biomedical and Health Informatics (27), IEEE Access (17), IEEE Transactions On Medical Imaging (13), IEEE Internet Of Things Journal (12), Journal Of Biomedical Informatics (12), Journal Of The American Medical Informatics Association (12), IEEE Transactions On Industrial Informatics (10), JMIR Medical Informatics (10). Table 1 presents the most cited articles and their respective Normalized Total Citations (NTCs). Brismiti et al. (2018) presents the highest number of global citations, whereas Xu et al. (2021) presents the highest NTC, despite being the fourth most cited article.

Table 1. Top-5 most globally cited articles regarding research in decentralized analysis in HC.

Paper	DOI	Citations	NTC
Brimisi TS et al., 2018, Int J Med Inform	https://doi.org/10.1016/j.ijmedinf.2018.01.007	297	4.18
Sheller MJ et al., 2020, Sci Rep	https://doi.org/10.1038/s41598-020-69250-1	216	7.47
Chen Y et al., 2020, IEEE Intell Syst	https://doi.org/10.1109/MIS.2020.2988604	201	6.95
Xu J et al., 2021, J Healthc Inform Res	https://doi.org/10.1007/s41666-020-00082-4	175	9.81
Chang K et al., 2018, J Am Med Infor As	https://doi.org/10.1093/jamia/ocy017	158	2.23

3.5 Authors

A total of 4038 researchers participated in the 488 articles, corresponding to a mean of 8.3 authors per article. Only 6 publications (1.2% of the total) were single-authored articles. Considering only the 488 included articles, the top 3 most productive authors in this field, based on the number of publications, are Andre Dekker (n = 14 publications), Philippe Lambin (n = 11), and Johan van Soest (n = 10). Figure 2 further presents the production of these authors over time, where the size of the circle is proportional to the number of published articles and the intensity of the color is proportional to the number of citations.

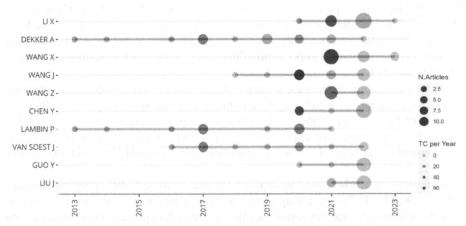

Fig. 2. Production of the top-10 most relevant authors over time.

3.6 Keywords and Concepts

In the last 3 years, the most common keywords, considering the Keyword Plus engine from Web of Science, were "federated learning" (217 counts), "deep learning" (62), "machine learning" (52), "distributed learning" (29), "medical imaging" (9), "electronic health records" (8), "big data" (7), "pharmacoepidemiology" (7) and "distributed computing" (5). Other common keywords reveal some commonly targeted research fields, namely "COVID-19", "hospital care", "medical imaging", "electronic health records" and "internet of things" (IoT).

4 Discussion

This is the first bibliometric review focused on the application of decentralized-learning to the HC field. This research domain has emerged in recent years as a response to the privacy and security barriers to the use of health data from multiple sources. The number of included articles is relatively small due to the novelty of this topic.

The exponential production after 2019 may be explained by several factors. First, the use of FL by Google, a company with worldwide projection, may have attracted research groups attention to these methods. Second, the General Data Protection Regulation (GDPR) became enforceable beginning 25 May 2018 [12], which forced collaborative research network to find solutions to overcome some privacy barriers. In addition, the SARS-COV-2 worldwide pandemic promoted an intense development and dissemination of privacy-preserving technologies in the HC field.

China and the USA stood out for their research volume, considering the corresponding author's affiliation, with the latter being by far the country with the highest number of citations, presenting a dominance of the intellectual structure in this field. Moreover, four of the top-5 most productive institutions are based in the USA, highlighting the USA' investments in this field. In terms of country collaboration, China and the USA present a substantial network with countries geographically widespread, whereas there are robust collaboration networks within European countries, who must jointly comply with the GRDP [12].

The most relevant sources are journals on health informatics, rather than computer science or engineering, which might be explained by the specificity of the HC field. In terms of methods, FL is the most popular one, followed by blockchain, which has introduced a wide range of applications in HC, such as protection of HC data and management of genomics databases, electronic health records (EHRs) and drug supply chain [13].

Common keywords related to application areas include EHRs, imaging analysis, COVID-19 and IoT. The increasing adoption of EHRs facilitates multi-institutional collaboration, although concerns regarding infrastructure, data privacy and standardization introduces several constraints. In this sense, decentralized analysis, in particular FL, allows multiple medical and research sites to jointly train a global predictive model without data sharing, fostering collaborative medical research [14]. Moreover, a decentralized approach is useful for avoiding the issue of having a single point of failure, which is common in current EHR systems, which are usually centralized [15]. Imaging analysis, paired with advanced techniques such as deep learning, has become an effective diagnostic method, e.g., detection of several types of cancer [16]. For COVID-19, imaging analysis and decentralized learning have been employed for processing chest X-ray images to predict clinical outcomes [17]. This disease brought unprecedented challenges for health systems worldwide, including mandatory testing and geographic tracking and mapping of outbreaks, with FL being mainly proposed for COVID detection using large amounts of multisite data [18]. Emerging technologies such as IoT devices forced Industry and Academia to join efforts for building medical applications [19]. Research on decentralized analysis in this field addresses the several security threats resulting from the use of these types of devices. In particular, the use of blockchain and its decentralized storage system facilitates secure data storing and sharing through networks composed of distributed nodes, as noted with IoT-based systems [20].

We recognize several limitations of this study, including: 1) the use of a multiple bibliographic databases would be preferable, however adding results from searches in other databases would not be feasible due to time restrictions; 2) the existence of errors in the Scopus database, e.g. duplicated publications; 3) considering the continuous update of Scopus and the interval between the day of search and the report of the review results, there is a lag between these results and the actual research progress, and 4) limitations of the bibliometric methodology itself [21]. However, this study provides a useful and comprehensive overview of the intellectual structure in this research domain.

5   Conclusion

This bibliometric review analyzed 488 published articles to identify the intellectual structure and trends of research on decentralized-learning applied to the HC field. This review may be a useful resource for gaining insights into this emergent research domain and fostering novel ideas for investigation.

Acknowledgments. This research work was developed under the project "Secur-e-Health" (ITEA 20050), co-financed by the North Regional Operational Program (NORTE 2020) under the Portugal 2020 and European Regional Development Fund (ERDF), with the reference NORTE-01–0247-FEDER-181418.

Supplementary Material. The Sup Mat can be found at: https://bit.ly/41rCjDL.

References

1. Callaway, A.: The healthcare data explosion. https://www.rbccm.com/en/gib/healthcare/epi sode/the_healthcare_data_explosion
2. Doiron, D., Burton, P., Marcon, Y., Gaye, A., Wolffenbuttel, B.H., Perola, M., et al.: Data harmonization and federated analysis of population-based studies: the BioSHaRE project. Emerg. Themes Epidemiol. **10**(1), 1–8 (2013)
3. El Emam, K., Samet, S., Arbuckle, L., Tamblyn, R., Earle, C., Kantarcioglu, M.: A secure distributed logistic regression protocol for the detection of rare adverse drug events. J. Am. Med. Inform. Assoc. **20**(3), 453–461 (2013)

4. McMahan, B., Moore, E., Ramage, D., Hampson, S., y Arcas, B.A. (eds.), Communication-efficient learning of deep networks from decentralized data. In: Artificial intelligence and statistics. PMLR (2017)
5. Lee, J., Sun, J., Wang, F., Wang, S., Jun, C.-H., Jiang, X.: Privacy-preserving patient similarity learning in a federated environment: development and analysis. JMIR Med. Inform. **6**(2), e7744 (2018)
6. Brisimi, T.S., Chen, R., Mela, T., Olshevsky, A., Paschalidis, I.C., Shi, W.: Federated learning of predictive models from federated electronic health records. Int. J. Med. Informatics **112**, 59–67 (2018)
7. Naeem, A., Anees, T., Naqvi, R.A., Loh, W.-K.: A comprehensive analysis of recent deep and federated-learning-based methodologies for brain tumor diagnosis. J. Personalized Med. **12**(2), 275 (2022)
8. Pati, S., Baid, U., Edwards, B., Sheller, M., Wang, S.-H., Reina, G.A., et al.: Federated learning enables big data for rare cancer boundary detection. Nat. Commun. **13**(1), 7346 (2022)
9. Huang, L., Shea, A.L., Qian, H., Masurkar, A., Deng, H., Liu, D.: Patient clustering improves efficiency of federated machine learning to predict mortality and hospital stay time using distributed electronic medical records. J. Biomed. Inform. **99**, 103291 (2019)
10. Ouzzani, M., Hammady, H., Fedorowicz, Z., Elmagarmid, A.: Rayyan—a web and mobile app for systematic reviews. Syst. Rev. **5**, 1–10 (2016)
11. Ahmi, A.: Bibliometric Analysis using R for Non-Coders: A practical handbook in conducting bibliometric analysis studies using Biblioshiny for Bibliometrix R package (2022)
12. General Data Protection Regulation (GDPR) Regulation (EU) 2016/679 (2016)
13. Haleem, A., Javaid, M., Singh, R.P., Suman, R., Rab, S.: Blockchain technology applications in healthcare: An overview. Int. J. Intell. Netw. **2**, 130–139 (2021)
14. Dang, T.K., Lan, X., Weng, J., Feng, M.: Federated learning for electronic health records. ACM Trans. Intell. Syst. Technol. (TIST) **13**(5), 1–17 (2022)
15. Kimovski, D., Ristov, S., Prodan, R.: Decentralized Machine Learning for Intelligent Health Care Systems on the Computing Continuum. arXiv preprint arXiv:220714584 (2022)
16. Subramanian, M., Rajasekar, V., Sathishkumar, V.E., Shanmugavadivel, K., Nandhini, P.: Effectiveness of decentralized federated learning algorithms in healthcare: a case study on cancer classification. Electronics **11**(24), 4117 (2022)
17. Dayan, I., Roth, H.R., Zhong, A., Harouni, A., Gentili, A., Abidin, A.Z., et al.: Federated learning for predicting clinical outcomes in patients with COVID-19. Nat. Med. **27**(10), 1735–1743 (2021)
18. Naz, S., Phan, K.T., Chen, Y.P.P.: A comprehensive review of federated learning for COVID-19 detection. Int. J. Intell. Syst. **37**(3), 2371–2392 (2022)
19. Al-Kahtani, M.S., Khan, F., Taekeun, W.: Application of Internet of Things and sensors in healthcare. Sensors **22**(15), 5738 (2022)
20. Sivasankari, B., Varalakshmi, P.: Blockchain and IoT technology in healthcare: a review. Challenges Trustable AI Added-Value Health: Proc. MIE **2022**(294), 277 (2022)
21. Wallin, J.A.: Bibliometric methods: pitfalls and possibilities. Basic Clin. Pharmacol. Toxicol. **97**(5), 261–275 (2005)

Computational Modelling of Cellular Mechanics

Simulating Initial Steps of Platelet Aggregate Formation in a Cellular Blood Flow Environment

Christian J. Spieker[✉][iD], Konstantinos Asteriou, and Gabor Zavodszky[iD]

Computational Science Lab, Informatics Institute, Faculty of Science,
University of Amsterdam, Amsterdam, The Netherlands
c.j.spieker@uva.nl

Abstract. The mechano-chemical process of clot formation is relevant in both hemostasis and thrombosis. The initial phase of thrombus formation in arterial thrombosis can be described by the mechanical process of platelet adhesion and aggregation via hemodynamic interactions with von Willebrand factor molecules. Understanding the formation and composition of this initial blood clot is crucial to evaluate differentiating factors between hemostasis and thrombosis. In this work a cell-based platelet adhesion and aggregation model is presented to study the initial steps of aggregate formation. Its implementation upon the pre-existing cellular blood flow model HemoCell is explained in detail and the model is tested in a simple case study of initial aggregate formation under arterial flow conditions. The model is based on a simplified constraint-dependent platelet binding process that coarse-grains the most influential processes into a reduced number of probabilistic thresholds. In contrast to existing computational platelet binding models, the present method places the focus on the mechanical environment that enables the formation of the initial aggregate. Recent studies highlighted the importance of elongational flows on von Willebrand factor-mediated platelet adhesion and aggregation. The cell-resolved scale used for this model allows to account for important hemodynamic phenomena such as the formation of a red blood cell free layer and platelet margination. This work focuses on the implementation details of the model and presents its characteristic behavior at various coarse-grained threshold values.

Keywords: Platelet aggregation · Platelet adhesion · Cellular blood flow · Thrombosis · Hemostasis

1 Introduction

Clot formation in thrombosis and hemostasis shares similar chemical pathways during the adherence and aggregation of platelets as well as the formation of fibrin reinforcements [34,38]. While the chemical pathways are shared, under hemostasis and arterial thrombus formation, the latter has a different mechanical trigger [6,22]. To improve treatment of pathological thrombosis, in the form of antithrombotic agents, without inhibiting bleeding cessation functionality of

© The Author(s), under exclusive license to Springer Nature Switzerland AG 2023
J. Mikyška et al. (Eds.): ICCS 2023, LNCS 14075, pp. 323–336, 2023.
https://doi.org/10.1007/978-3-031-36024-4_26

hemostasis, differentiating factors between the processes have to be evaluated. Both processes, hemostasis and thrombosis, incorporate mechano-driven and chemical processes, where the former are mostly based on hemodynamic interactions between platelets and the plasma molecule von Willebrand factor (VWF), while the chemical processes involve the activation and cleavage of clotting factors enabling the catalyzation of reactions of the coagulation cascade, which ultimately leads to the reinforcement of the clot through fibrin depositions.

In hemostasis the balance between mechanical and chemical processes is mirrored in their onset sequence, from primary to secondary hemostasis, respectively [15]. In a recent work, Rhee et al. [28] showed that the onset and balance of these mechanisms significantly impact the structure of the forming thrombus, where tightly adherent platelets make up the outer most layer of a wound closing thrombus. In relation to this, work by Yakusheva et al. [46] proposed that hemostatic environments can be sites of high shear levels, which are traditionally associated with arterial thrombosis. Another recent work by Kim et al. [21] has shown that arterial thrombosis follows the same initial order of events as the healthy wound healing process in an arterial injury. In this work, a clot forming in a stenosed vessel geometry is shown to consist almost exclusively of VWF at the vessel wall, followed by an adjacent high concentration of platelets.

This initial adhesion and aggregation of platelets is mediated by VWF, revealing its A1 domain (VWF-A1) binding sites while unfolding from a globular to an extended form under shear flow. In the case of platelet adhesion to a coagulation surface, these platelet to VWF (platelet glycoprotein Ib (GPIb)α to VWF-A1) bonds can form under low shear flow with VWF adhered to adhesive glycoproteins such as collagen or fibrin. Platelet aggregation mediated by plasma suspended VWF requires a high shear environment [4,22]. An exception to this are sites of high elongational flow which are caused by shear gradients and defined by exerting tensile forces. These elongational flow fields, which can be conceptualized as shear flow without the rotational flow proportion, are found to unfold plasma VWF at comparatively low shear rates [37,39,40]. Additionally, Abidin et al. recently proposed that platelets can mechanosense the rate of change in elongational flow leading to their activation [1].

The mechanical nature of initial platelet aggregate formation establishes why the hemodynamic flow conditions as well as cell interactions and distributions are crucial for its understanding. While experimental in vivo and in vitro analyses are limited in their temporal and spatial scale, computational models are increasingly employed to complement or even replace their experimental counterparts. To study the cellular effects and accurately evaluate clot composition in regards to porosity and permeability, platelet binding models are of special interest. In the past, Xu et al. presented a multi-scale cellular Potts model covering the continuum scale of thrombosis [45] and Fogelson et al. proposed to use spring and hinge forces to model platelet binding, while considering the effects of the coagualation cascade via added advection-diffusion equations [13]. Other aggregate models have used the dissipative particle method [11,41,42], the moving particle semi-implicit method [18,19] and the Monte Carlo method [12].

More recently, Yazdani et al. presented a spectral element method model, which considers different activation states of platelets [47]. Liu et al. used a shear-induced platelet aggregation model including immobilized and unfolded VWF as well as platelets modeled as rigid spheres [22, 23].

While existing platelet binding models cover a range of scales and blood clotting processes, they oftentimes reduce cellular complexity to save computational cost, by neglecting the presence of red blood cells or simplifying the ellipsoidal platelet shape. This in turn reduces the accuracy of the fluid mechanical predictions that are of pivotal importance to investigate mechanosensitive processes. Several cellular mechanisms contribute to the mechanical environments. The formation of a red blood cell free layer is associated with the margination of platelets, which is crucial for their deployment in aggregate formation [27, 49]. Additionally, the effects of elongational flows on binding mechanics are not considered by previous models.

In this work, a simplified platelet binding model is presented with focus on initial aggregate formation. The implementation is build on the existing cell-resolved blood flow model HemoCell [48], in the form of constraint-dependent platelet immobilization. The theoretical concept and early implementation was previously discussed by van Rooij [29]. In the following chapters the methodology is explained in detail and the model is tested in proof-of-concept case study simulations. While the case study focuses on initial aggregate growth in arterial thrombosis, the structure of the model allows for the simulation of different scenarios, including hemostatic platelet adhesion and aggregation, via adjustment of the individual constraint parameters. This work focuses on the implementation details and methodology of the model and its characteristic behavior is presented at various coarse-grained threshold values.

2   Initial Platelet Aggregate Formation Model

The cell-based platelet adhesion and aggregation model presented in this work is built upon the cell-resolved blood flow model HemoCell. Its fundamental concept is based on previous work by van Rooij [29], and its implementation are described in the following paragraphs. While the constraint-dependent nature of the model allows for the adjustment of threshold parameters to simulate different environments for aggregate formation, the presented work focuses on initial clot formation in arterial thrombosis.

2.1   Conceptual Model Considerations

HemoCell. The aggregate formation model is added upon the open-source framework HemoCell. The cellular blood flow model couples a discrete element method membrane solver to the lattice Boltzmann method-based fluid solver via the immersed boundary method. In the current state the model includes membrane models for platelets and red blood cells, which can be initialized at different concentrations in a predefined geometry. The single cell and bulk flow behavior of the model were previously validated by Závodszky et al. [48]. Furthermore, HemoCell has found a wide field of application such as modelling the flow of diabetic and healthy human as well as mouse blood in a variety of geometries and differing in their complexity [8–10,30,31,39,49].

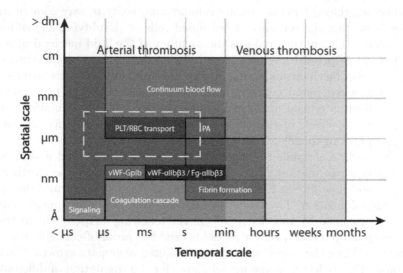

Fig. 1. HemoCell scale separation map. The dashed line represents the range of operation for HemoCell. Therefore, the processes occurring below this scale are represented in a coarse-grained fashion. 'PA' denotes platelet aggregation. Adapted with permission from van Rooij [29].

The temporal and spatial scale covered by HemoCell are visualized in the scale separation map in Fig. 1. The temporal resolution up to the order of seconds and spatial resolution up to 500 µm are in the necessary ranges to cover initial platelet aggregate formation in arterial thrombosis.

Model Concept. In line with the scale of operation of HemoCell, the presented model focuses on the initial phase of platelet aggregate formation, excluding the subsequent mechanisms of platelet activation, the coagulation cascade and fibrin formation. Furthermore, the adherence of red blood cells can be omitted since in arterial thrombosis, which typically leads to a platelet-rich thrombus, red blood cells assume a passive function in aggregate formation [5]. The initial phase of this aggregate formation can be described as a mechanically driven process, regulated by hemodynamic thresholds in the flow environment. The process consists of platelets binding to a thrombogenic surface, known as platelet adhesion and to each other, in the form of platelet aggregation. Initial platelet adhesion and aggregation in the high shear environment of arterial thrombosis is mediated by bonds between VWF-A1 to the platelet receptor GPIbα of the GPIb-IX-V complex [32,35,43].

In the proposed model, initial aggregate formation relies on the constraint-dependent immobilization of platelets, coarse-graining the process of VWF-mediated platelet adhesion and aggregation into a probabilistic threshold dependent on the local fluid mechanics. The model includes three threshold constraints - shear rate, rate of elongation and distance - which will induce the immobilization of individual platelets when reached. The numerical implementation of platelet immobilization is a process denoted as 'solidification' during which the location of the immobilized platelet is turned into a solid boundary. Additional to three constraints, a solidification probability is added as a further calibration and scaling parameter.

2.2   Model Design and Implementation

Binding Sites. Platelet adhesion to a thrombogenic surface, such as collagen, fibronectin or laminin is a necessity for subsequent platelet aggregation [32]. In the model, the thrombogenic surface is defined as a list of boundary nodes of the simulated domain wall, labeled as *BindingSites*. For the immobilization of platelets in the solidification process, the vicinity of each binding site node is searched for platelets that reach the threshold constraints. The platelets that are solidified become binding sites for other platelets and their nodes are therefore added to the *BindingSites* list (see Fig. 2(b) and (c)).

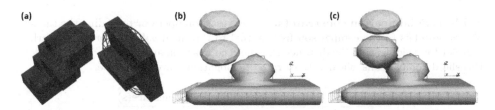

Fig. 2. Platelet solidification process. (a): Two platelets (in red) with the inner fluid nodes determined using the even-odd rule (in blue). Adapted with permission from van Rooij [29]. (b) & (c): Solidification process, with one and two solidified platelets, respectively. The boundary contours of the binding site (in yellow) are surrounded by a wireframe view of boundary nodes. (Color figure online)

Constraints. The included constraints are shear rate, rate of elongation and distance to the closest binding site, which are evaluated for platelets in vicinity to binding sites. The distance at which a platelet solidifies includes the platelet-VWF bond length, since the mediating VWF is not explicitly modeled. While the bond length between VWF and GPIbα is around 100 nm, VWF molecules unfolded under shear flow are shown to extend to a contour length of up to 250 μm [25, 40]. In this model a distance threshold of 1.5 μm is chosen, which is sufficiently resolved by the immersed boundary method that couples the particles with the fluid in HemoCell, since it is three times the length of one lattice node. Each platelet that has a ≤ 1.5 µm large distance from at least one of its membrane nodes to the closest binding site is evaluated for the local shear rate and rate of elongation.

For the shear rate, a threshold of 1000 s^{-1} is defined, which is in line with values where immobilized VWF begins to unfold [14,33]. Although, shear-induced platelet aggregation, mediated via plasma suspended VWF, occurs at much higher shear rates around 5000 s^{-1}, the lower value is chosen, since platelet adhesion to a thrombogenic surface is the initial step of aggregate formation and the model does not differentiate between binding sites of the initial thrombogenic surface and binding sites of solidified platelets [36]. Additionally, the rate of elongation threshold is set to 300 s^{-1}, which corresponds to the critical value for VWF unfolding, determined by Sing et al. [37]. The shear rate threshold is evaluated based on the strain rate of the fluid nodes overlapping with the respective platelets. The largest difference in principal strain rates, calculated as the eigenvalues of the strain rate tensor, is used as the shear rate of the specific node and is compared to the defined threshold. The rate of elongation is calculated as the magnitude of the diagonal elements of the rate of strain tensor of the inner platelet nodes. If the platelet in close distance to a binding site experiences shear rates ≥ 1000 s^{-1} or elongational flows $\geq 300^{-1}$ it becomes a candidate for solidification. The constraint parameters are summarized in Table 1.

Table 1. Solidification constraints. The comparison operator preceding the threshold value refers to the comparison from actual to threshold value. In order to mark a platelet for solidification the distance threshold comparison and at least one of the flow threshold comparisons, shear rate or rate of elongation, have to be true.

Constraint	Threshold value	Reference(s)
Distance	≤ 1.5 µm	[25,40]
Shear rate	≥ 1000 s^{-1}	[14,33]
Rate of elongation	≥ 300 s^{-1}	[37,39]

Solidification Probability. The candidate platelets that reach the threshold constraints are tagged for solidification with a predefined probability, labeled *solidificationProbability*, between 0 and 1. This parameter serves as a calibration and scaling parameter and represents the binding affinity of the modeled platelet adhesion and aggregation process [2]. To implement this binding affinity in HemoCell, platelets that reach the thresholds of the set constraints (see Table 1) are only tagged for solidification if a generated (pseudo-)random number is below the defined *solidificationProbability* value of the range of generated numbers (between 0 and *rand_max*). Otherwise, the respective platelet is not considered for solidification in the current iteration. The plot in Fig. 3 visualizes the effect of the *solidificationProbability* value on the time of solidification. For a probability of 1 the solidification delay Δt is 0 s, since a platelet eligible for solid-ification will be tagged and subsequently solidified immediately. With decreasing probability, the mean delay and its standard deviation increase steeply. The res-olution of the solidification delay is limited by the time step of 1^{-1} s defined in HemoCell. For initial assessment of the model, the *solidificationProbability* parameter is set to 0.5.

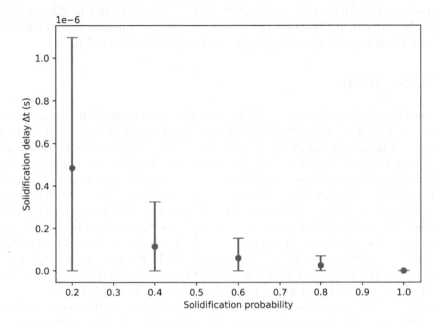

Fig. 3. Solidification probability vs. solidification delay Δt. The plot is based on single platelet simulations. To isolate the effects of the *solidificationProbability*, the platelet reaches the thresholds of the set constraints (see Table 1) at every iteration. Each measured value is based on 20 simulations.

Solidification. Platelet solidification is the numerical implementation of platelet immobilization as a consequence of binding during adhesion and aggrega-tion. The platelets tagged for solidification are converted to solid wall boundary nodes and added to the *bindingSites* list in the subsequent iteration. This 'con-version' is performed by establishing which fluid nodes the platelet intersects with and changing them to solid boundary nodes. These nodes at the position of the platelet to be solidified (see Fig. 2(a)) are determined via ray casting by applying the even-odd rule. This method was previously used by De Haan et al. to identify the internal fluid nodes of a red blood cell in order to change their viscosity to match the increased viscosity of hemoglobin [16]. The intersections between a ray and a polygon, in this case a triangle on the surface of a platelet, is determined by the Möller-Trumbore ray-triangle intersection algorithm [26].

The ray is cast from an arbitrary distant point outside the respective shape, in this case the cell, towards the point to be tested for being within the shape or outside. To improve performance of this operation, the ray casting is fixed in its direction and an octree data structure is used to create a spatial subdivision of the triangle vertices [24]. The sum of intersections are counted for each node. An even sum places the point outside the platelet and if the number is odd; it is located inside the platelet. The identified internal nodes are converted to boundary nodes and added to the *bindingSites* list, as displayed in Fig. 2(b) and (c). From the subsequent iteration they are considered as binding sites themselves and their surface is made available for further platelets to bind to.

3   Case Study

Simulation Setup. To assess the functionality and evaluation process of the presented model, a 'case study' simulation is set up. The simulated domain, shown in Fig. 4, is based on the parallel plate flow chamber design commonly used in microfluidics to study shear-dependent platelet adhesion and aggregation [17,20]. The setup consists of a 100 µm long channel with a width and height of 50 and 62.5 µm, respectively. In order to mimic the center of a parallel plate flow chamber, the domain utilizes periodic boundary conditions along the width (y-axis). The central 90 µm of the top channel wall are initialized as binding sites (see yellow section of channel wall in Fig. 4). Additionally, a $25 \times 50 \times 62.5$ µm^{-1} large periodic pre-inlet supplies the inlet of the domain with a constant inflow of cells, preventing a decrease in platelet concentration due to solidification of platelets in aggregate formation [3,39]. Furthermore, with regular periodic boundary conditions in the flow direction, a growing aggregate could affect the surrounding cellular flow conditions and therefore possibly distort the aggregate formation results.

The pre-inlet is initialized with 162 red blood cells and 157 platelets, corresponding to a discharge hematocrit of 25%. In order to study the effects of shear rate magnitude on the initial aggregate formation, two cases with differing flow velocities are defined. The simulated flow is set up to yield a Reynolds number of 0.75 and 0.6, leading to initial wall shear rates of around 1350 and 1000 s^{-1}, respectively. The final simulations are executed on 2 nodes (consisting of 128 cores each) on the Snellius supercomputer (SURF, Amsterdam, Netherlands; https://www.surf.nl/en/dutch-national-supercomputer-snellius) with an average performance of 0.049 ± 0.001 s/iteration. Before simulating cellular flow and platelet binding in the entire domain, flow is simulated in the pre-inlet exclusively, to converge from the random initial cell packing towards a marginated state. The shortened domain for this 'warm-up' simulation excludes the binding sites and has the added benefit of reduced computational cost.

Simulation Results. Figure 5 displays the boundary contours of the solidified platelets for the two simulated cases. The difference in flow conditions causes different initial aggregate formations. The Re = 0.75 simulation creates a 35%

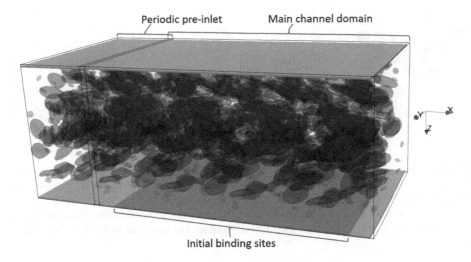

Fig. 4. Case study setup. Simulation domain with periodic pre-inlet and binding site location (in yellow). Defined flow direction is the positive x-direction. The red blood cells and platelets displayed in the channel domain are visualized at 30% opacity. (Color figure online)

longer (in x-direction) and 35% higher (in z-direction) aggregate, than the Re = 0.6 simulation. While the aggregate in the Re = 0.75 simulation formed close to the domain inlet, the Re = 0.6 simulation aggregate formed at the center of the channel. As the color scale in Fig. 5 showcases, this difference also affects the shear rate applied to the surface of the aggregates. The Re = 0.75 simulation aggregate experiences higher shear rates on a larger area, due to protruding deeper into the channel. Both cases exhibit aggregate formations that extend towards the oncoming flow direction. Furthermore, both aggregates are connected to only a single platelet that solidified upon close enough contact to the initial binding site area and all additional solidifications of the aggregates occurred within binding distance to previously solidified platelets. In addition to the aggregates shown in Fig. 5, the Re = 0.6 simulation exhibits one additional solidified platelet close to the channel outlet.

4   Discussion

The presented work explains the methodology of a cellular platelet adhesion and aggregation model. The model is build upon the pre-existing cell-resolved blood flow framework HemoCell and focuses on simulating the mechanical process of initial platelet aggregate formation, realized via constraint-dependent platelet immobilization. The model considers red blood cells and their effect on platelet distribution, as well as the role of elongational flows in VWF-mediated platelet adhesion and aggregation, distinguishing it from previous models.

The case study simulations act as a proof-of-concept of the model and its implemented functionalities. Furthermore, the simulations show that flow

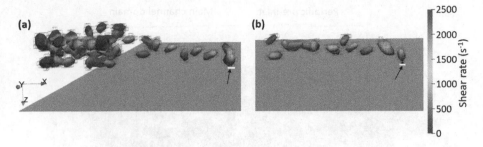

Fig. 5. Case study evaluation. (a) & (b): Boundary contour of solidified platelets, for the Re = 0.75 and Re = 0.6 simulations, respectively. The color scale represents the shear rate applied to the surface of the aggregates. Both results are captured after 0.15 s of simulated flow. The arrows point towards the area of connection (in white) of the first solidified platelet to the initial binding site area (in yellow). (Color figure online)

conditions impact the aggregate shape in the model. The implementation in HemoCell allows for extensive evaluation of cellular distributions, flow conditions and aggregate properties, such as forces acting on the solidified platelets, aggregate volume, aggregate growth over time, porosity and permeability, among others. Additionally, the model allows for adjustments of the clotting environment in regards to channel geometry, flow conditions and platelet immobilization constraints. These clotting environments can cover sites of arterial thrombosis, like a stenosed vessel section, as well as hemostatic environments, such as an injured vessel.

The larger aggregate size observed in the higher flow velocity case study simulation (see Fig. 5(a)) could be explained by an increased likelihood of platelets reaching the shear threshold as well as more platelets passing by the binding sites, as both arguments are dependent on flow velocity. Though in order to verify these results and apply the model to investigate initial platelet aggregate formation in different clotting environments, its validation is required. Complete validation is a currently ongoing effort that is performed by comparing simulated results of the model to experimental aggregate formations. Additionally, decoupling the initial binding sites, acting as the thrombogenic surface, from binding sites in the form of solidified platelets, could increase accuracy of the model. Platelet solidifications occurring in the vicinity of the initial binding sites represent platelet adhesion, whereas platelets solidifying close to previously solidified platelets depict platelet aggregation. While both processes can be mediated by VWF, they require different flow environments [32]. By decoupling their coarse-grained representation in the model, the mechanical thresholds for VWF elongation, namely distance, shear rate and rate of elongation, could be defined separately. The current simplified implementation could explain why the aggregates of both case study simulations are structured upon a single platelet - thrombogenic surface bond and their subsequent shape extends into the oncoming flow direction. An additional simplification of the model is the solid and constant modeling of a platelet bond. In reality, platelet aggregates are slightly deformable and active structures that exhibit contraction over time. Furthermore, depending on the strength of the bond, platelets can detach from an aggregate, individually or in bulk [44]. Explicitly modeling the platelet bonds, for example with a bead-spring model, could further increase accuracy of the model.

The original model concept idea by van Rooij included the usage of so called CEPAC (Combined Effect of Pro- and Anticoagulants) fields [29]. Implemented as advection-diffusion fields to reduce the amount of free parameters in the model, the CEPAC fields are set to model the secretion of agonists by activated platelets (CEPAC1), as well as pro- and anticoagulants of the coagulation cascade (CEPAC2). Combined with different activation states of platelets, these additions to the model could widen its focus from an initial aggregate model to a more complete thrombosis and hemostasis model. To cover the large temporal scale from initial platelet adhesion to coagulation and fibrin formation (see Fig. 1), increasing the reaction rates and the total platelet count could accelerate aggregate growth. Ultimately, scale bridging techniques, such as time splitting or amplification could be applied in a multiscale model approach [7].

5   Conclusion

The introduced cellular platelet adhesion and aggregation model offers a platform to investigate initial platelet aggregate formation in individually defined aggregation environments. Its constraint-dependent structure is an adjustable and modular addition to the cellular blood flow framework HemoCell. While the presented case study simulations showcase the functionality of the model and already hint towards the importance of hemodynamic flow conditions in initial aggregate formation, further calibration and validation of the model is required before significant conclusions can be made.

Acknowledgements. C.J.S., K.A. and G.Z. acknowledge financial support by the European Union Horizon 2020 research and innovation programme under Grant Agreement No. 675451, the CompBioMed2 Project. C.J.S., K.A. and G.Z. are funded by CompBioMed2. The use of supercomputer facilities in this work was sponsored by NWO Exacte Wetenschappen (Physical Sciences).

References

1. Abidin, N.A.Z., et al.: A microfluidic method to investigate platelet mechanotransduction under extensional strain. Res. Pract. Thromb. Haemost. **7**, 100037 (2023). https://doi.org/10.1016/j.rpth.2023.100037
2. Auton, M., Zhu, C., Cruz, M.A.: The mechanism of VWF-mediated platelet GPIbα binding. Biophys. J . **99**(4), 1192–1201 (2010). https://doi.org/10.1016/j.bpj.2010.06.002
3. Azizi Tarksalooyeh, V.W., Závodszky, G., van Rooij, B.J., Hoekstra, A.G.: Inflow and outflow boundary conditions for 2d suspension simulations with the immersed boundary lattice Boltzmann method. Comput. Fluids **172**, 312–317 (2018)

4. Bergmeier, W., Hynes, R.O.: Extracellular matrix proteins in hemostasis and thrombosis. Cold Spring Harb. Perspect. Biol. **4**(2), a005132–a005132 (2011). https://doi.org/10.1101/cshperspect.a005132
5. Byrnes, J.R., Wolberg, A.S.: Red blood cells in thrombosis. Blood **130**(16), 1795–1799 (2017). https://doi.org/10.1182/blood-2017-03-745349
6. Casa, L.D., Ku, D.N.: Thrombus formation at high shear rates. Ann. Rev. Biomed. Eng. **19**(1), 415–433 (2017). https://doi.org/10.1146/annurev-bioeng-071516-044539
7. Chopard, B., Falcone, J.-L., Kunzli, P., Veen, L., Hoekstra, A.: Multiscale modeling: recent progress and open questions. Multiscale Multidisc. Model. Exp. Des. **1**(1), 57–68 (2018). https://doi.org/10.1007/s41939-017-0006-4
8. Czaja, B., Závodszky, G., Tarksalooyeh, V.A., Hoekstra, A.G.: Cell-resolved blood flow simulations of saccular aneurysms: effects of pulsatility and aspect ratio. J. Roy. Soc. Interface **15**(146), 20180485 (2018). https://doi.org/10.1098/rsif.2018.0485
9. Czaja, B., et al.: The effect of stiffened diabetic red blood cells on wall shear stress in a reconstructed 3d microaneurysm. Comput. Meth. Biomech. Biomed. Eng. **25**(15), 1691–1709 (2022). https://doi.org/10.1080/10255842.2022.2034794
10. Czaja, B., Gutierrez, M., Závodszky, G., de Kanter, D., Hoekstra, A., Eniola-Adefeso, O.: The influence of red blood cell deformability on hematocrit profiles and platelet margination. PLOS Comput. Biol. **16**(3), e1007716 (2020). https://doi.org/10.1371/journal.pcbi.1007716. https://dx.plos.org/10.1371/journal.pcbi.1007716
11. Filipovic, N., Kojic, M., Tsuda, A.: Modelling thrombosis using dissipative particle dynamics method. Philos. Trans. R. Soc. A Math. Phys. Eng. Sci. **366**(1879), 3265–3279 (2008). https://doi.org/10.1098/rsta.2008.0097
12. Flamm, M.H., Sinno, T., Diamond, S.L.: Simulation of aggregating particles in complex flows by the lattice kinetic Monte Carlo method. J. Chem. Phys. **134**(3), 034905 (2011). https://doi.org/10.1063/1.3521395
13. Fogelson, A.L., Guy, R.D.: Immersed-boundary-type models of intravascular platelet aggregation. Comput. Meth. Appl. Mech. Eng. **197**(25–28), 2087–2104 (2008). https://doi.org/10.1016/j.cma.2007.06.030
14. Fu, H., Jiang, Y., Yang, D., Scheiflinger, F., Wong, W.P., Springer, T.A.: Flow-induced elongation of von Willebrand factor precedes tension-dependent activation. Nat. Commun. **8**(1), 324 (2017). https://doi.org/10.1038/s41467-017-00230-2
15. Gale, A.J.: Continuing education course #2: current understanding of hemostasis. Toxicol. Pathol. **39**(1), 273–280 (2010). https://doi.org/10.1177/0192623310389474
16. de Haan, M., Zavodszky, G., Azizi, V., Hoekstra, A.: Numerical investigation of the effects of red blood cell cytoplasmic viscosity contrasts on single cell and bulk transport behaviour. Appl. Sci. **8**(9), 1616 (2018). https://doi.org/10.3390/app8091616
17. Hao, Y., Závodszky, G., Tersteeg, C., Barzegari, M., Hoekstra, A.G.: Image-based flow simulation of platelet aggregates under different shear rates (2023). https://doi.org/10.1101/2023.02.22.529480
18. Kamada, H., Imai, Y., Nakamura, M., Ishikawa, T., Yamaguchi, T.: Computational study on thrombus formation regulated by platelet glycoprotein and blood flow shear. Microvasc. Res. **89**, 95–106 (2013). https://doi.org/10.1016/j.mvr.2013.05.006

19. Kamada, H., ichi Tsubota, K., Nakamura, M., Wada, S., Ishikawa, T., Yamaguchi, T.: A three-dimensional particle simulation of the formation and collapse of a primary thrombus. International Journal for Numerical Methods in Biomedical Engineering **26**(3–4), 488–500 (2010). https://doi.org/10.1002/cnm.1367

20. Kent, N.J., et al.: Microfluidic device to study arterial shear-mediated platelet-surface interactions in whole blood: reduced sample volumes and well-characterised protein surfaces. Biomed. Microdevices **12**(6), 987–1000 (2010). https://doi.org/10.1007/s10544-010-9453-y

21. Kim, D.A., Ku, D.N.: Structure of shear-induced platelet aggregated clot formed in an in vitro arterial thrombosis model. Blood Adv. **6**(9), 2872–2883 (2022). https://doi.org/10.1182/bloodadvances.2021006248

22. Liu, Z.L., Ku, D.N., Aidun, C.K.: Mechanobiology of shear-induced platelet aggregation leading to occlusive arterial thrombosis: a multiscale in silico analysis. J. Biomech. **120**, 110349 (2021). https://doi.org/10.1016/j.jbiomech.2021.110349

23. Liu, Z.L., Bresette, C., Aidun, C.K., Ku, D.N.: SIPA in 10 milliseconds: VWF tentacles agglomerate and capture platelets under high shear. Blood Adv. **6**(8), 2453–2465 (2022). https://doi.org/10.1182/bloodadvances.2021005692

24. Meagher, D.: Geometric modeling using Octree Encoding. Comput. Graph. Image Process. **19**(2), 129–147 (1982). https://doi.org/10.1016/0146-664x(82)90104-6

25. Mody, N.A., King, M.R.: Platelet adhesive dynamics. part II: high shear-induced transient aggregation via GPIbα-vWF-GPIbα bridging. Biophys. J. **95**(5), 2556–2574 (2008). https://doi.org/10.1529/biophysj.107.128520

26. Möller, T., Trumbore, B.: Fast, minimum storage ray/triangle intersection. In: ACM SIGGRAPH 2005 Courses on - SIGGRAPH 2005. ACM Press (2005). https://doi.org/10.1145/1198555.1198746

27. Qi, Q.M., Shaqfeh, E.S.G.: Theory to predict particle migration and margination in the pressure-driven channel flow of blood. Phys. Rev. Fluids **2**(9), 093102 (2017). https://doi.org/10.1103/physrevfluids.2.093102

28. Rhee, S.W., et al.: Venous puncture wound hemostasis results in a vaulted thrombus structured by locally nucleated platelet aggregates. Commun. Biol. **4**(1), 1090 (2021). https://doi.org/10.1038/s42003-021-02615-y

29. van Rooij, B.J.M.: Platelet adhesion and aggregation in high shear blood flow: an insilico and in vitro study. Ph.D. thesis, Universiteit van Amsterdam (2020)

30. van Rooij, B.J.M., Závodszky, G., Azizi Tarksalooyeh, V.W., Hoekstra, A.G.: Identifying the start of a platelet aggregate by the shear rate and the cell-depleted layer. J. R. Soc. Interface **16**(159), 20190148 (2019). https://doi.org/10.1098/rsif.2019.0148, https://royalsocietypublishing.org/doi/10.1098/rsif.2019.0148

31. van Rooij, B.J.M., Závodszky, G., Hoekstra, A.G., Ku, D.N.: Haemodynamic flow conditions at the initiation of high-shear platelet aggregation: a combined in vitro and cellular in silico study. Interface Focus **11**(1), 20190126 (2021). https://doi.org/10.1098/rsfs.2019.0126, https://royalsocietypublishing.org/doi/10.1098/rsfs.2019.0126

32. Ruggeri, Z.M., Mendolicchio, G.L.: Adhesion mechanisms in platelet function. Circ. Res. **100**(12), 1673–1685 (2007). https://doi.org/10.1161/01.res.0000267878.97021.ab

33. Ruggeri, Z.M., Orje, J.N., Habermann, R., Federici, A.B., Reininger, A.J.: Activation-independent platelet adhesion and aggregation under elevated shear stress. Blood **108**(6), 1903–1910 (2006). https://doi.org/10.1182/blood-2006-04-011551. https://ashpublications.org/blood/article/108/6/1903/22637/Activationindependent-platelet-adhesion-and

34. Sang, Y., Roest, M., de Laat, B., de Groot, P.G., Huskens, D.: Interplay between platelets and coagulation. Blood Reviews **46**, 100733 (2021). https://doi.org/10.1016/j.blre.2020.100733
35. Savage, B., Saldívar, E., Ruggeri, Z.M.: Initiation of platelet adhesion by arrest onto fibrinogen or translocation on von Willebrand factor. Cell **84**(2), 289–297 (1996). https://doi.org/10.1016/s0092-8674(00)80983-6
36. Schneider, S.W., et al.: Shear-induced unfolding triggers adhesion of von Willebrand factor fibers. Proc. Natl. Acad. Sci. **104**(19), 7899–7903 (2007). https://doi.org/10.1073/pnas.0608422104
37. Sing, C.E., Alexander-Katz, A.: Elongational flow induces the unfolding of von Willebrand factor at physiological flow rates. Biophys. J. **98**(9), L35–L37 (2010). https://doi.org/10.1016/j.bpj.2010.01.032. https://linkinghub.elsevier.com/retrieve/pii/S0006349510001979
38. Smith, S.A., Travers, R.J., Morrissey, J.H.: How it all starts: initiation of the clotting cascade. Crit. Rev. Biochem. Mol. Biol. **50**(4), 326–336 (2015). https://doi.org/10.3109/10409238.2015.1050550
39. Spieker, C.J., et al.: The effects of micro-vessel curvature induced elongational flows on platelet adhesion. Ann. Biomed. Eng. **49**(12), 3609–3620 (2021). https://doi.org/10.1007/s10439-021-02870-4
40. Springer, T.A.: von willebrand factor, Jedi knight of the bloodstream. Blood **124**(9), 1412–1425 (2014). https://doi.org/10.1182/blood-2014-05-378638
41. Tosenberger, A., Ataullakhanov, F., Bessonov, N., Panteleev, M., Tokarev, A., Volpert, V.: Modelling of thrombus growth in flow with a DPD-PDE method. J. Theoret. Biol. **337**, 30–41 (2013). https://doi.org/10.1016/j.jtbi.2013.07.023
42. Tosenberger, A., Ataullakhanov, F., Bessonov, N., Panteleev, M., Tokarev, A., Volpert, V.: Modelling of platelet–fibrin clot formation in flow with a DPD–PDE method. J. Math. Biol. **72**(3), 649–681 (2015). https://doi.org/10.1007/s00285-015-0891-2
43. Ulrichts, H., et al.: Shielding of the a1 domain by the d′d3 domains of von Willebrand factor modulates its interaction with platelet glycoprotein Ib-IX-V. J. Biol. Chem. **281**(8), 4699–4707 (2006). https://doi.org/10.1074/jbc.m513314200
44. Xu, S., Xu, Z., Kim, O.V., Litvinov, R.I., Weisel, J.W., Alber, M.: Model predictions of deformation, embolization and permeability of partially obstructive blood clots under variable shear flow. J. R. Soc. Interface **14**(136), 20170441 (2017). https://doi.org/10.1098/rsif.2017.0441
45. Xu, Z., Chen, N., Kamocka, M.M., Rosen, E.D., Alber, M.: A multiscale model of thrombus development. J. R. Soc. Interface **5**(24), 705–722 (2007). https://doi.org/10.1098/rsif.2007.1202
46. Yakusheva, A.A., et al.: Traumatic vessel injuries initiating hemostasis generate high shear conditions. Blood Adv. **6**(16), 4834–4846 (2022). https://doi.org/10.1182/bloodadvances.2022007550
47. Yazdani, A., Li, H., Humphrey, J.D., Karniadakis, G.E.: A general shear-dependent model for thrombus formation. PLoS Comput. Biol. **13**(1), e1005291 (2017). https://doi.org/10.1371/journal.pcbi.1005291
48. Závodszky, G., van Rooij, B., Azizi, V., Hoekstra, A.: Cellular level in-silico modeling of blood rheology with an improved material model for red blood cells. Front. Physiol. **8**, 563 (2017). https://doi.org/10.3389/fphys.2017.00563
49. Závodszky, G., Van Rooij, B., Czaja, B., Azizi, V., De Kanter, D., Hoekstra, A.G.: Red blood cell and platelet diffusivity and margination in the presence of cross-stream gradients in blood flows. Phys. Fluids **31**(3), 031903 (2019). https://doi.org/10.1063/1.5085881

Estimating Parameters of 3D Cell Model Using a Bayesian Recursive Global Optimizer (BaRGO)

Pietro Miotti$^{(\boxtimes)}$ iD, Edoardo Filippi-Mazzola iD, Ernst C. Wit iD, and Igor V. Pivkin iD

Institute of Computing, Università della Svizzera Italiana, Lugano, Switzerland
pietro.miotti@usi.ch
https://www.ci.inf.usi.ch/

Abstract. In the field of Evolutionary Strategy, parameter estimation for functions with multiple minima is a difficult task when interdependencies between parameters have to be investigated. Most of the current routines that are used to estimate such parameters leverage state-of-the-art machine learning approaches to identify the global minimum, ignoring the relevance of the potential local minima. In this paper, we present a novel Evolutionary Strategy routine that uses sampling tools deriving from the Bayesian field to find the best parameters according to a certain loss function. The Bayesian Recursive Global Optimizer (BaRGO) presented in this work explores the parameter space identifying both local and global minima. Applications of BaRGO to 2D minimization problems and to parameter estimation of Red Blood Cell model are reported.

Keywords: Bayesian inference · Markov-Chain-Monte-Carlo · Cell model

1 Introduction

Estimating parameters of computational cell models from experimental measurements is often a difficult task that may involve handling large number of degrees of freedom, high computational costs, and scarcity of data. The estimation is typically performed by manually changing the values of the parameters, which can become laborious and time consuming.

With the developments in the machine learning field, researchers have adapted optimization routines to provide fast and automatic approaches for parameter estimation [1,5]. In this context, the idea of finding the global minimum of some objective function through a Bayesian-like approach that relies on Markov-Chain-Monte-Carlo (MCMC) method was introduced in reference [2]. Building on similar assumptions about the probability distribution of the population behavior, we propose a Bayesian Recursive Global Optimizer (BaRGO), a novel Bayesian evolutionary strategy for performing parameter estimation. BaRGO combines the reliability of Markov-Chain-Monte-Carlo

© The Author(s), under exclusive license to Springer Nature Switzerland AG 2023
J. Mikyška et al. (Eds.): ICCS 2023, LNCS 14075, pp. 337–344, 2023.
https://doi.org/10.1007/978-3-031-36024-4_27

(MCMC) approaches to estimate the posterior parameters together with probabilistic cluster analysis to find multiple minima by assuming a mixture of probability distributions that are estimated in the routine by an Expectation-Maximization algorithm [4]. In this sense, BaRGO explores all the minima it encounters in the domain, to find the best set of parameters for the objective function that are considered to match the experimental data [13] which can provide deeper insight and understanding of dependencies between parameters.

Following a brief introduction to Evolutionary Strategy using a statistical Bayesian perspective in Sect. 2, we present the algorithmic procedure of BaRGO in Sect. 3. In Sect. 4, we evaluate performance of BaRGO on 2D minimization problems and parameter estimation of Red Blood Cell (RBC) model.

2   Bayesian Inference on Population Parameters in Evolutionary Optimization

Evolutionary Strategies (ES) are a subfamily of stochastic optimization algorithms that investigate a potential solution space using biologically inspired mechanisms such as selection, mutation, and gene crossover. As a result, in ES, iterations are referred to as generations, and the set of values evaluated at each generation is referred to as the population [6].

The main steps in the optimization process are as follows. Given the loss function f and the solution space S, the goal of the ES algorithm is to find $x^* \in S$ such that $x^* = \arg\min_x f(x)$. Assuming that λ different elements are sampled from the solution space (population), only k are saved (selection). These are considered to be the best elements that minimize the loss function f. Such elements are then recombined (via crossover and mutation) to sample additional λ elements that will represent the new population (the next generation).

From a probabilistic perspective, the operations of crossover and mutation can be performed by sequentially updating the parameters of the distributions from which the population is sampled. From the Bayesian perspective of continuously updating prior beliefs, it is possible to design a MCMC approach that infers the parameter's population through sampling using the previous generation parameters as priors (see [9] for an overview on Bayesian inference).

In this regard, let's assume that a population $\mathbf{x}_g$ of size λ, at generation g, for $g = 1, \ldots, G$, is multivariate Normally distributed in d dimensions with mean $\boldsymbol{\beta}_g$ and covariance matrix Σ_g, where

$$
\mathbf{x}_g = \begin{pmatrix} x_{1,g} \\ x_{2,g} \\ \vdots \\ x_{\lambda,g} \end{pmatrix} \quad
\boldsymbol{\beta}_g = \begin{pmatrix} \beta_{1,g} \\ \beta_{2,g} \\ \vdots \\ \beta_{d,g} \end{pmatrix} \quad
\Sigma_g = \begin{pmatrix} \sigma_1^2 & \sigma_{1,2} & \ldots & \sigma_{1,d} \\ \sigma_{1,2} & \sigma_2^2 & \ldots & \sigma_{2,d} \\ \vdots & \vdots & & \vdots \\ \sigma_{1,d} & \sigma_{2,d} & \ldots & \sigma_d^2 \end{pmatrix}.
$$

From an ES perspective, we can consider $\boldsymbol{\beta}_g$ to be the candidate value which minimizes f (i.e. $\boldsymbol{\beta}_g \approx x^*$) whereas Σ_g quantifies the uncertainty we have in proposing that candidate.

Using the previous generation parameters as prior beliefs, it is possible to update such parameters for the current generation $\boldsymbol{\beta}_g$ and Σ_g in a Bayesian fashion through a Gibbs sampler routine that sequentially samples from the full conditional distributions. Indeed, this involves defining semi-conjugate priors for the mean and for the covariance matrix. A convenient one for $\boldsymbol{\beta}_g$ is the multivariate normal distribution $p(\boldsymbol{\beta}_g) \sim \mathcal{N}_d(\boldsymbol{\mu}_{g-1}, \Lambda_{g-1})$. By selecting the $k < \lambda$ population elements which minimize the loss function value within all the generations, the full conditional distribution for the mean will then be a multivariate Normal distribution with the following updated parameters:

$$p(\boldsymbol{\beta}_g \mid x_1, \ldots, x_k, \Sigma_g) \sim \mathcal{N}_d(\boldsymbol{\mu}_g, \Lambda_g),$$
$$\boldsymbol{\mu}_g = (\Lambda_{g-1}^{-1} + n\Sigma_g^{-1})^{-1}(\Lambda_0^{-1}\boldsymbol{\mu}_g + n\Sigma_g^{-1}\bar{x}),$$
$$\Lambda_g = (\Lambda_{g-1}^{-1} + n\Sigma_g^{-1})^{-1},$$

where $\bar{x}$ is the sample average of the current generation. Similarly, an appropriate semi-conjugate prior for Σ_g is the inverse-Wishart distribution $p(\Sigma) \sim$ inv-Wis(v_{g-1}, S_{g-1}^{-1}). Thus, the full conditional distribution is

$$p(\Sigma_g \mid x_1, \ldots, x_k) \sim \text{inv-Wis}(v_g, S_g^{-1}),$$
$$v_g = v_{g-1} + \lambda,$$
$$S_g = S_{g-1} + S_\beta,$$

where S_β is the residual sum of squares from the population mean $\boldsymbol{\beta}_g$.

From the full conditional distributions, it is possible to construct a Gibbs sampler routine that generates posterior samples for $\boldsymbol{\beta}_g$ and Σ_g. Given a set of starting conditions, namely $\{\boldsymbol{\mu}_{g-1}, \Lambda_{g-1}, v_{g-1}, S_g\}$, the Gibbs sampler generates at iteration t, for $t = 1, \ldots, T$, $\{\boldsymbol{\beta}_g^{(t+1)}, \Sigma_g^{(t+1)}\}$ from $\{\boldsymbol{\beta}_g^{(t)}, \Sigma_g^{(t)}\}$ according to the following two steps:

1. sample $\boldsymbol{\beta}_g^{(t+1)}$ from its full conditional distribution:
   - update $\boldsymbol{\mu}_g$ and Λ_g through $x_1, \ldots, x_k$ and $\Sigma_g^{(t)}$;
   - sample $\boldsymbol{\beta}_g^{(t+1)} \sim \mathcal{N}_d(\boldsymbol{\mu}_g, \Lambda_g)$;
2. sample $\Sigma_g^{(t+1)}$ from its full conditional distribution:
   - update S_g through $x_1, \ldots, x_k$ and $\boldsymbol{\beta}_g^{(t+1)}$;
   - sample $\Sigma_g^{(t+1)} \sim \text{inv-Wis}(v_g, S_g^{-1})$.

Therefore, after a sufficient amount of iterations to ensure the convergence of the chain, it is possible to have estimates for the parameters of the posterior by computing the empirical average across the sampled values:

$$\hat{\boldsymbol{\beta}}_g = \mathbb{E}[\boldsymbol{\beta}_g^{t=1\ldots T}],$$
$$\hat{\Sigma}_g = \mathbb{E}[\Sigma_g^{t=1\ldots T}].$$

Thus, having updated the new parameters, it is possible to generate a new population $\boldsymbol{x}_{g+1}$ by sampling from a multivariate Normal with the posterior estimates $\hat{\boldsymbol{\beta}}_g$ and $\hat{\Sigma}_g$.

3   Bayesian Recursive Global Optimizer (BaRGO)

We propose a Bayesian Recursive Global Optimizer (BaRGO), a routine that leverages on Bayesian MCMC described above, to iteratively converge at x^* that minimizes f. BaRGO provides posterior estimates for each generation g, using previous generation information as prior beliefs.

At every generation, g, β_g and Σ_g are updated considering the k-best population elements encountered following the Bayesian inference approach introduced in Sect. 2. For functions with a unique global minimum, after every generation we can expect a decay in the uncertainty $\Sigma_g \to 0$ as β_g converges to x^*. However, when functions exhibit multiple minima $\tilde{x}_1^*, \ldots, \tilde{x}_m^*$, the best values selected for computing the posteriors may belong to different minima, which may result in the update of β_g as the weighted average of the priors and the minima captured by the k-best elements selected. As a consequence, since there are multiple minima, Σ_g values will never converge to 0. To solve this issue, BaRGO splits the non-convex problem into small local convex problems and solves them separately. Specifically, once the KullBack Leiber divergence [10] between the multivariate Normal densities of two subsequent generations is smaller than a predefined threshold, the algorithm realizes that there are multiple minima $\tilde{x}_1^*, \ldots, \tilde{x}_m^*$ which are not allowing BaRGO to fully converge. As a consequence, we can consider this case as the result of population data being generated by a mixture of multivariate Gaussian distributions. The k-best population elements can then be re-grouped into m distinct clusters using an Expectation-Maximization (EM) approach [4,15]. Once the clustering is deployed, BaRGO is recursively applied to each cluster, producing a local result for each of them $\beta_1, \ldots, \beta_m$. Local results are then compared, while the best one is returned as global minimum.

The upside of using a model-based approach to cluster data is the ability to use information criteria such as the Bayesian Information Criterion (BIC) to perform model selection on the most suited number of clusters given the current population. Indeed, the EM step is applied multiple times on a grid of a potential number of clusters. On each of these, the BIC is evaluated and the one with the lowest score on average is selected as the most suited one.

From implementation perspective, ES are in general highly parallelizable, which makes them perfectly suitable for the parameter estimation of computationally intensive models. Specifically, the evaluation of the population elements can be performed in parallel, hence the computational time of a single generation can be reduced to the computational time of single population element. Each population element can in turn be evaluated in parallel, exploiting HPC solvers.

4   Applications of BaRGO

4.1   Estimating Function Parameters in 2D Minimization Problems

The current state-of-the-art ES method, the Covariance Matrix Adaptation Evolutionary Strategy (CMA-ES) [8], exploits a similar idea as it samples directly from a multivariate Normal distribution [7]. In preliminary studies, we compare performance of BaRGO and CMA-ES in a 2D minimization problem setup on three traditional functions: Sphere (also known as the cone), Schwefel and Rastring functions. The peculiarities of these functions are entailed in their number of minima: the Sphere function exhibits a single minimum, whereas Rastring and Schwefel have multiple minima. Obtained results are reported in Fig. 1. The evolution of the mean and the standard deviation of the sampling distributions $\mathcal{N}_2(\boldsymbol{X}_g, \Sigma_g)$ with $\boldsymbol{X} = [X_1, X_2]$ and Σ_g with standard deviations σ_1 and σ_2, are reported to compare the speed of convergence of both algorithms. In order to perform proper comparisons, since CMA-ES only search for the global optima, we apply the recursive call of BaRGO only to the cluster which exhibits the best candidate value, which means that we do not explore all the possible minima. In all three cases BaRGO converges with smaller number of function evaluations than CMA-ES, which translates to faster convergence properties, as seen from the decay-rate of standard deviations in Fig. 1. These preliminary studies demonstrate interesting properties of BaRGO that deserve further investigation and analysis.

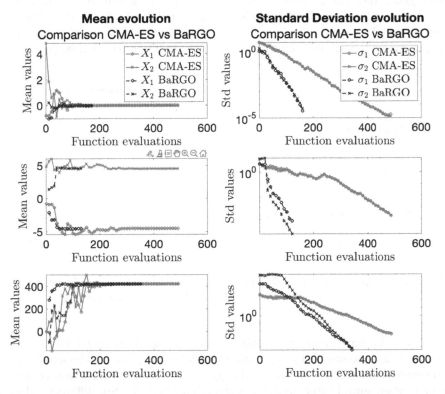

Fig. 1. Performance comparison between CMA-ES and BaRGO on Sphere (upper row), Rastring (middle row) and Schwefel (lower row) functions.

4.2   Estimating the RBC Model Parameters Using Data from Optical Tweezers Experiment

Optical tweezers is an experimental technique based on optical or laser traps that is used to stretch cells in one or more directions by trapping beads that are strategically attached to the cell surface [3]. The different deformations in the axial and transverse diameter of the cell (Fig. 2), resulting from the stretching force applied, provide information regarding the elastic properties of the cell. For this application, using experimental data from [14], we tested BaRGO by estimating the parameters of the RBC model from [16]. In this coarse-grained model, which we implemented in LAMMPS [12], the cell membrane is modeled using surface triangulation with $N_v = 5000$ vertices connected with Worm Like Chain (WLC) links. The parameters of the model that we estimate are the WLC persistence length, p, the membrane bending coefficient, k_{bend}, as well as the area and volume conservation coefficients, k_a and k_v. These parameters play a key role in defining the elastic properties of the RBC model. For detailed description of the model and its parameters we refer to reference [16]. In order to speed up convergence the search space of each coefficient is constrained. Specifically, we limit the search for the persistence length between 0 and 0.512×10^{-7} (the later value is equal to maximum length of the links in the WLC model, l_{max} [16]), k_{bend} between 0 and 1000, and both k_v and k_a between 0 and 1000000.

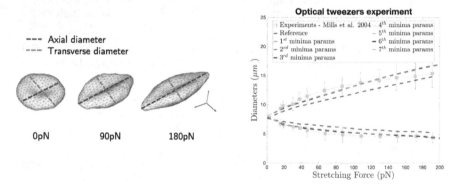

Fig. 2. (Left) Optical tweezers experiment: RBC shape evolution at different stretching forces (0, 90, and 180 pN). (Right) Estimation RBC model's parameters

Similar to experiment, we apply in simulations stretching force and measure the cell deformation by computing the deformation along the axial (x) and transverse diameter (y). With x_F and y_F, we refer to the axial and transverse diameters measured after applying a stretching force F. Six different values of stretching force are considered to perform the fitting of parameters, with corresponding values for axial and transverse diameters taken from experimental data from reference [14]. Thus, the loss function to minimize is defined as follows:

$$Loss = \sum_{F \in [16,38,88,108,150,192]} |\tilde{x}_F - x_F| + |\tilde{y}_F - y_F|.$$

where $\tilde{x}_F$ and $\tilde{y}_F$ are the deformations in the axial and transverse diameter computed in simulation.

In the optimization process, BaRGO managed to find multiple minima, reported in Table 1. The values of the loss function in μm are also reported for each minima in the table. The overall cell stiffness is mainly defined by the WLC potential used in the RBC model. Specifically, the values of the persistence length, p are very similar in all minima, while the values of bending coefficient k_{bend} and the coefficients for the constant area and volume, k_a and k_v, can vary significantly without affecting the simulation results (Fig. 2). We note that the surface area and volume are well conserved in all cases with the average fluctuations of 0.5% and 0.1%, respectively. The non uniqueness of the set of parameters that minimizes the loss function either might be due to the experiment itself or to the selected measurements, axial and transverse diameters, which may not be not sufficient to capture the detailed behaviour of the RBC membrane. Related discussion can be found in [11], where additional data are proposed to be collected in optical tweezers experiments.

Table 1. Parameters resulted from of the optimization process

Parameters	1^{st}minima	2^{nd}minima	3^{rd}minima	4^{th}minima	5^{th}minima	6^{th}minima	7^{th}minima	Params from [16]
p	4.14×10^{-9}	4.15×10^{-9}	4.17×10^{-9}	4.11×10^{-9}	4.23×10^{-9}	4.13×10^{-9}	4.11×10^{-9}	3.43×10^{-9}
k_{bend}	362	438	486	674	602	372	625	200
k_a	752794	597984	525380	663420	722226	513549	743761	6000
k_v	262911	698463	471066	186937	108131	328961	83505	6000
$Loss$	2.85	2.89	2.9	2.88	2.83	2.89	2.87	

5   Conclusions

We propose BaRGO, a novel Evolutionary Strategy algorithm that follows a Bayesian approach and by recursively calling itself is able to find multiple minima of loss function f. Convergence performance of BaRGO was shown to be comparable to the current state-of-the-art CMA-ES model in preliminary studies. Finally, we have applied BaRGO for estimating RBC model parameters using data from the optical tweezers experiment.

One of the drawbacks of BaRGO is its computational intensity, requiring an MCMC cycle at each iteration. Despite this, it can still be a useful tool in cases where complex models require parameter estimation and multiple minima need to be explored to better understand the model and dependencies among parameters.

Acknowledgements. The authors acknowledge support from the Swiss National Science Foundation grants 200021L_204817 and 192549. Simulations were carried out at the Swiss National Supercomputer Center under project u4.

References

1. Balogh, P., Gounley, J., Roychowdhury, S., Randles, A.: A data-driven approach to modeling cancer cell mechanics during microcirculatory transport. Sci. Rep. **11**(1), 15232 (2021)
2. Benhamou, E., Saltiel, D., Vérel, S., Teytaud, F.: BCMA-ES: a Bayesian approach to CMA-ES. CoRR abs/1904.01401 (2019)
3. Dao, M., Lim, C.T., Suresh, S.: Mechanics of the human red blood cell deformed by optical tweezers. J. Mech. Phys. Solids **51**(11–12), 2259–2280 (2003)
4. Dempster, A.P., Laird, N.M., Rubin, D.B.: Maximum likelihood from incomplete data via the EM algorithm. J. R. Stat. Soc. Ser. B (Methodol.) **39**(1), 1–38 (1977). http://www.jstor.org/stable/2984875
5. Economides, A., et al.: Hierarchical Bayesian uncertainty quantification for a model of the red blood cell. Phys. Rev. Appl. **15**, 034062 (2021)
6. Goldberg, D.E.: Genetic Algorithms in Search, Optimization and Machine Learning. Addison-Wesley Longman Publishing Co., Inc., USA (1989)
7. Hansen, N.: The CMA evolution strategy: a tutorial. arXiv preprint arXiv:1604.00772 (2016)
8. Hansen, N., Müller, S.D., Koumoutsakos, P.: Reducing the time complexity of the derandomized evolution strategy with covariance matrix adaptation (CMA-ES). Evol. Comput. **11**(1), 1–18 (2003)
9. Hoff, P.: A First Course in Bayesian Statistical Methods. Springer Texts in Statistics, Springer, New York (2009). https://doi.org/10.1007/978-0-387-92407-6
10. Kullback, S., Leibler, R.A.: On information and sufficiency. Ann. Math. Stat. **22**, 79–86 (1951)
11. Sigüenza, J., Mendez, S., Nicoud, F.: How should the optical tweezers experiment be used to characterize the red blood cell membrane mechanics? Biomech. Model. Mechanobiol. **16**(5), 1645–1657 (2017). https://doi.org/10.1007/s10237-017-0910-x
12. LAMMPS: Lammps (2015). http://lammps.sandia.gov/bench.html
13. Lim, C., Zhou, E., Quek, S.: Mechanical models for living cells - a review. J. Biomech. **39**(2), 195–216 (2006)
14. Mills, J.P., Qie, L., Dao, M., Lim, C.T., Suresh, S.: Nonlinear elastic and viscoelastic deformation of the human red blood cell with optical tweezers. Mech. Chem. Biosyst. **1**(3), 169–80 (2004)
15. Pedregosa, F., et al.: Scikit-learn: machine learning in Python. J. Mach. Learn. Res. **12**, 2825–2830 (2011)
16. Pivkin, I.V., Karniadakis, G.E.: Accurate coarse-grained modeling of red blood cells. Phys. Rev. Lett. **101**(11), 118105 (2008)

A Novel High-Throughput Framework to Quantify Spatio-Temporal Tumor Clonal Dynamics

Selami Baglamis[1,2,3,4], Joyaditya Saha[1,2,3,4], Maartje van der Heijden[1,2,3,4], Daniël M. Miedema[1,2,3,4], Démi van Gent[1,2,3,4], Przemek M. Krawczyk[2,5], Louis Vermeulen[1,2,3,4], and Vivek M Sheraton[1,2,3,4(✉)]

[1] Laboratory for Experimental Oncology and Radiobiology, Center for Experimental and Molecular Medicine, Amsterdam University Medical Centers (location AMC), Amsterdam, The Netherlands
v.s.muniraj@amsterdamumc.nl
[2] Cancer Center Amsterdam, Amsterdam, The Netherlands
[3] Oncode Institute, Amsterdam, The Netherlands
[4] Amsterdam Gastroenterology Endocrinology Metabolism, Amsterdam, The Netherlands
[5] Department of Medical Biology, Amsterdam University Medical Centers (location AMC), Amsterdam, The Netherlands

Abstract. Clonal proliferation dynamics within a tumor channels the course of tumor growth, drug response and activity. A high-throughput image screening technique is required to analyze and quantify the spatiotemporal variations in cell proliferation and influence of chemotherapy on clonal colonies. We present two protocols for generating spatial, Lentiviral Gene Ontology (LeGO) fluorescent tag based, mono- and co-culture systems with provisions for spatio-temporal tracking of clonal growth at the nucleus- and cytoplasm-level. The cultured cells are subjected to either drug treatment or co-cultured with fibroblasts and analyzed with a novel image processing framework. This framework enables alignment of cell positions based on motion capture techniques, tracking through time and investigation of drug actions or co-culturing on individual cell colonies. Finally, utilizing this framework, we develop agent-based models to simulate and predict the effects of the microenvironment and clonal density on cell proliferation. The model and experimental findings suggest growth stimulating effects of local clonal density irrespective of overall cell confluency.

Keywords: LeGO · cell tracking · diffusion dynamics · agent-based model

1 Introduction

Multicellular organisms exhibit complex biological phenomena. In order to delineate the biological processes associated with the observed phenomena, it is imperative to study them at the cellular level. In recent years, increasing emphasis is

© The Author(s), under exclusive license to Springer Nature Switzerland AG 2023
J. Mikyška et al. (Eds.): ICCS 2023, LNCS 14075, pp. 345–359, 2023.
https://doi.org/10.1007/978-3-031-36024-4_28

being placed on tracking specific populations of cells found to be enriched in a range of physiological and pathological conditions [4,12]. Temporal tracking of cells constituting different clones enables a greater understanding of associated clonal dynamics [15].

A range of factors influences clonal dynamics. The abundance of nutrients such as amino acids, nucleotides, lipids, and glucose influences the metabolic profiles of cells constituting clones which in turn influences their phenotype and dynamics [20]. While the role of nutrients on clonal dynamics has been studied extensively, in recent times, increasing attention is being directed to the role of cellular communication within and between clones. It is observed that cell-cell interactions often induce the activation of pathways favoring the establishment of unique cellular phenotypes that promote survival [10]. For instance, it is observed that cancer stem cells communicate with each other via Connexin46 (Cx46) gap junction proteins to synchronize biological processes such as cellular growth and proliferation. Disruption of the Cx46 gap protein specific cell-cell communication induces apoptosis in cancer stem cells [9].

In addition to cell-cell interactions taking place within the same population of cells, interactions can also occur between distinct populations of cells. Fibroblasts, in particular, are implicated to play a crucial role in the development of multiple pathologies such as cancer and rheumatoid arthritis (RA) [3,16,20]. The mechanisms via which fibroblasts promote the development and progression of tumors has been the focus of intense research and as such have been well documented. Fibroblast-cancer cell interactions are bi-directional [16]. Cancer-associated fibroblasts (CAFs), can either be tumor promoting or suppressive based on their interactions with cancer cells. In a recent study, it was observed that pancreatic cancer cells with a gain-of-function (GOF) mutant p53 interacted with CAFs in a paracrine manner via NF-κB signaling [16]. This paracrine interaction mediated by NF-κB signaling results in increased secretion of heparin sulphate proteoglycan 2 (HSPG2) which in turn improves the metastatic capacity of cancer cells [16]. Another pioneering study showed that fibroblast activating protein (FAP) expression specifically increased in CAFs upon being cultured in conditioned medium derived from colorectal cancer (CRC) cells [3].

Many techniques to track clonal dynamics have been developed over the years. Most are often limited in scope and require sufficient technical expertise to execute successfully [12]. Currently used techniques to track clonal dynamics include fluorescence activated cell sorting (FACS) or sequencing of barcoded DNA or RNA from the same source at multiple time intervals [12]. In addition, manual cell tracking via time lapse microscopy also continues to be used to track clonal populations in a spatio-temporal manner [4]. Given the inherent complexity of these techniques and the degree of subjectivity involved in the interpretation of the results obtained thereof, there is a need to develop an automated and standardized framework to reliably study clonal dynamics in a wide range of conditions [6].

Lentivirus Gene Ontology (LeGO) vectors have proven to be strong candidates that can reliably and stably transduce a wide range of primary cell cultures and established cell lines upon being packaged with the appropriate coat proteins [17]. These vectors consist of a selection of fluorescent and drug-selectable markers [17]. Cells can be transduced simultaneously with multiple LeGO vectors [8] to give rise to distinct cell lineages characterized by different colors [17]. The fluorescent proteins can further be co-expressed with a nucleus-localizing signal (NLS) to ensure fluorescence from a cell is restricted to its nucleus.

Given the ease with which cells can be fluorescently tagged using LeGO vectors, in this paper, we develop a novel, high-throughput image-processing framework to analyze clonal dynamics of CRC cells. We further develop a simple two function agent-based cell proliferation model to explore the effects of local density on colony growth.

2   Materials and Methods

2.1   Cell Culture

HT55 (Sanger Institute) cells and primary human embryonic fibroblasts (HEFs), established previously by our lab [7], were cultured in DMEM/F12 (Life Technologies) supplemented with 1% Penicillin-Streptomycin (Gibco) and 10% fetal calf serum (FCS;Gibco). For the co-cultures, cells were cultured in DMEM/F12 (Life Technologies) supplemented with 1% Penicillin-Streptomycin (Gibco) and 0.1% fetal calf serum (Gibco). Cell lines were incubated at $37\,°C$ and $5.0\,\%$ CO_2.

2.2   Multicolor Marking

To introduce LeGO markers into the HT55 cells, we followed the techniques previously utilised by Weber et al. [18] and Heijen et al. [2]. To improve data acquisition, we inserted a nuclear localization signal of the human c-Myc proto-oncogene (3' GGACGACGCTTCTCCCAGTTTAACCTG 5') [1] into the LeGO-C2 (27339), LeGO-V2 (27340), and LeGO-Cer2 (27338) (Addgene) vectors, using standard DNA cloning protocols. This enabled nuclear visualization. Briefly, 50,000 cells were seeded in a single well of a 12-well plate in 1 mL culture medium and incubated at $37\,°C$ and 5.0% CO_2 for 24 h, at 70 % confluency. The medium was refreshed with 1 mL culture medium containing 8 µg mL^{-1} polybrene (Sigma-Aldrich) and 50 µL of either LeGO or LeGO-NLS lentivirus, was added into the medium. Cells were incubated overnight at $37\,°C$ and $5.0\,\%$ CO_2 and the medium was refreshed the next day. Transduced cells were selected by a Sony SH800 cell sorter.

2.3   Treatment Assay

2000 HT55 LeGO cells were seeded in the wells of 96 wells-plate three days prior to treatment. After 3 days, cells were either treated with 2.5 µM oxaliplatin (Sigma) or 100 µM 5-fluorouracil (Sigma). Drug concentrations were based on the IC_{50} found in Genomics of Drug Sensitivity in Cancer (GDSC) database [19] and our preliminary drug response experiments. Cell lines were treated with different chemotherapy drugs for 3 days, followed by a 4 day wash out cycle and then imaged at these days.

2.4   The Effect of Fibroblasts on Cell Growth

To investigate the effect of fibroblasts on cell growth, we co-cultured 1000 HT55 cells with 7000 human fibroblasts per well in a 96-well plate. As a control, 1000 HT55 cells were cultured without fibroblasts per well in the same 96-well plate. Three independent experiments were performed. Image acquisition was started 3 h post-seeding and continued daily for at least 5 time points to monitor the growth of HT55 cells in the presence or absence of fibroblasts.

2.5   Imaging

HT55 cells were imaged using an EVOS FL Cell Imaging System (Thermo Fisher Scientific) with a 4X objective, and the following LED light cubes were utilized: Texas Red (excitation 445/45 nm and emission 510/42 nm), YFP (excitation 500/24 nm and emission 524/27 nm), and CFP (excitation 585/29 nm and emission 624/40 nm). High resolution images were captured using a Leica Thunder Wide Field Fluorescence Microscope at 10x magnification, with the following settings: Quad Filter Block; DFT51010 or CYR70010 both of which employ an excitation wavelength of 375–412 nm, 483–501 nm, and 562–588 nm, and an emission wavelength of 441–471 nm, 512–548 nm, and 600–660 nm for mCerulean, mVenus, and mCherry respectively.

2.6   Image Analysis

An image processing pipeline "LeGO cell track" was developed to analyze the experimentally obtained microscopy images. First, high resolution microscopy images were exported to PNG image format. The raw PNG images were processed using ImageJ v1.53k to remove background fluorescence. An image correction method to reduce temporal background noise and shading [11] was implemented via the BaSiC plugin in ImageJ. The background corrected images were then aligned based on their temporal sequence. To enable temporal tracking, we used OpenCV's motion tracking protocol "MOTION AFFINE". The aligned images were segmented for individual bright field (black and white) segments. These segments may contain multiple colonies overlapping on each other. The identified segments were separated based on the colors of cells into individual colonies. To accomplish this color-based colony sorting, the pixel colors within a segment was clustered into eight different channels in the Hue, Saturation and Value (HSV) color map. Geometric and other statistical metrics such as the center of mass of a colony, area occupied and color channel were extracted from the color sorted colonies. These metrics were used to track individual colonies through images from different consequent time points.

2.7   Model Development

We developed a spatiotemporal agent-based model to quantify the clonal growth dynamics. Each agent was assumed to represent a colony. The agent attributes

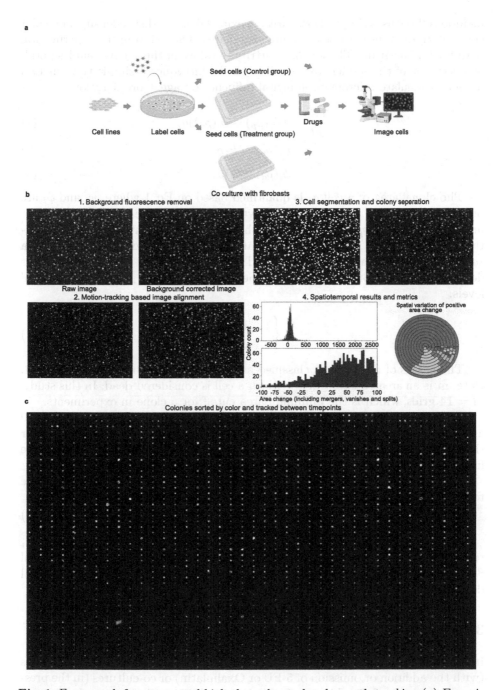

Fig. 1. Framework for automated high-throughput clonal growth tracking (a) Experimental setup illustration created with BioRender.com (b) Steps involved in processing the microscopy image data and (c) Color sorted and time-tracked colonies placed in a (time point 1, time point 2) manner.

include cell status, grid positions, area, proximal density, distal density, and color channel. In this study, these attribute values are extracted from the experimental data for initialization. The localized attribute values of the agents enables parallel execution of the model [13] at cell level. We introduce a simple two-function model to simulate microenvironmental influence on agent proliferation.

$$\frac{da_i}{dt} = (\psi_i(\rho_p) + \phi_i(\rho_d))a_i \tag{1}$$

$$\psi_i(\rho_p) = k_1\rho_p \tag{2}$$

$$\phi_i(\rho_d) = k_2\rho_d \tag{3}$$

The clonal area of a cell a_i is quantified based on Eq. 1 where, ψ_i and ϕ_i are proximal and distal density-based growth contributions respectively from the microenvironment, ρ_p and ρ_d indicate the proximal and distal densities respectively. The densities are calculated as ratio of area occupied by clones within a zone to the total area of the zone. k_1 and k_2 from Eqs. 2 and 3 are rate constants denoting contribution to area-growth per unit density at proximal and distal levels.

$$S_i = \begin{cases} 1, & \text{if } a_i > m \\ 0, & \text{otherwise} \end{cases} \tag{4}$$

The status of an agent is classified as viable (1) or extinct (0) based Eq. 4. Here, m is an area constant below which a cell is considered dead. In this study, m = 14 grids, which is the minimum area cutoff for a clone in experiments.

For model parametric fitting, we extracted a smaller (nx * ny) 1000×1000 area from EXP67 microscopy images to limit compute resource requirements for the fitting process. The radius for proximal and distal density measurements were fixed at approximately 1/3 of image length and full image length respectively.

A loss function indicated in Eq. 5 was minimized to carry out the parametric fitting,

$$\chi = M_{SE}(\sum a_i, \overline{\rho_p}, \overline{\rho_d}, med(\rho_p), med(\rho_d)) \tag{5}$$

In Eq. 5, χ indicates the loss, M_{SE} indicates the mean square error, $\sum a_i$ represents the sum of colony areas, $\overline{\rho_p}$, $\overline{\rho_d}$ are the mean proximal and distal densities respectively and $med(\rho_p)$, $med(\rho_d)$ are the median proximal and distal densities respectively.

3   Results and Discussions

In order to analyze the clonal dynamics of HT55 cells cultured as mono-cultures (with the addition or omission of 5-FU or Oxaliplatin) or co-cultures (in the presence or absence of fibroblasts), LeGO labeled HT55 cells were seeded in well plates corresponding to different experimental conditions (Fig. 1a). Cells cultured as mono-cultures were imaged at varying time points with or without cytotoxic drugs. For the co-cultures, seeded cells within the same plate were cultured with or without fibroblasts and imaged at multiple time points up to eight days post-seeding. Figure 1b highlights the framework through which clones across different conditions were tracked over time. The background of the raw images obtained at different time intervals were first corrected to remove excess fluorescence arising from additional constituents of the microenvironment such as fibroblasts or added cytotoxic agents. The background corrected images from two time points were then aligned on the basis of colony motion-tracking (see Sect. 2.6). Irrespective of the distinct and varied fluorescent colors expressed by colonies, the motion-tracking technique employed was able to reliably segregate individual colonies (Fig. 1c). Following colony motion-tracking, clones were separated and color sorted on the basis of their fluorescence (Fig. 1b). The developed high-throughput framework was first benchmarked against synthetic microscopy images with known cell count and area increase. A seed image with dimensions similar to the experimental images (5439×3238 pixels) was constructed with a random distribution of 100 colonies of equal area of disc shape. To mimic colony growth, two additional synthetic images were created. These images use the same seed positions of colonies from the seed image but with equal incremental area coverage of the individual colonies as shown in Fig. 2a (t1, t2 and t3). The colony sizes were varied in a linear fashion (1x, 2x and 3x) with respect to the disc radius in the synthetic images generated at the three time points (t1, t2 and t3). There was no overlap of colonies in the first two synthetic images (interval 1). In the second and third synthetic images (interval 2) two colonies overlapped over each other. Analytically estimated ratio of area increase between the intervals was 1.67. These three synthetic images were analyzed with the presented framework. As expected, there is no variation in colony number with time and total colony count remains at 100 as evidenced in Fig. 2b and Fig. 2d. Figure 2c provides a comparison of area increase between different intervals. The framework overestimated the ratio of area increase by approximately 11% compared to the analytical solution. This could potentially be due to image compression artifacts when creating the synthetic images. Overall, the benchmarking of the developed framework against synthetic images shows that it segments and tracks with sufficient accuracy in an ideal setting. Additionally, the framework is capable of handling overlap of colonies by separating and color sorting them into individual colonies.

Throughout the study we used three major clonal growth indicators, (i) area increase per area - calculated as ratio of area increase between time points (interval) to the area occupied by the colonies at the start of the interval, (ii) area increase per colony - calculated as ratio of area increase between time points (interval) to the number of colonies at the start of the interval and (iii) area increase - indicating total area change between a period of time. Negative values indicate decrease in colony area. In the control conditions of the mono-culture drug experiments, a consistent reduction in area increase per area was observed upon increase of total area and normalized neighbor area of clonal colonies (Figs. 2e, 3g, 3i, 3k, 3m, 3o, 3q) despite experiment 5 behaving as an outlier. The same phenomenon was observed when comparing area increase per area to the total number of clonal colonies (Figs. 2f, 3h, 3j, 3l, 3n, 3p, 3r). These results show that overall area increase would be lesser in densely packed or confluent systems. However, if we consider local den-

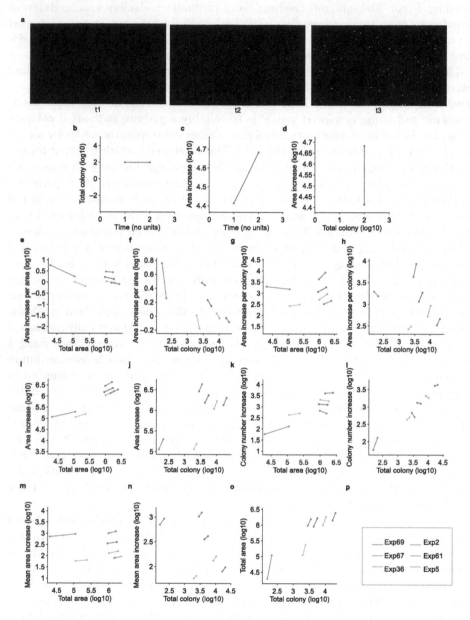

Fig. 2. Clonal proliferation profiles resulting from mono-culture experiments (a) Synthetic images for different timepoints (b) (c), (d) are comparisons of various growth parameters for synthetic images, (e)–(o) plots of growth indicators for the control conditions of the mono-culture drug experiments and (p) figure legend for plots (e)–(o).

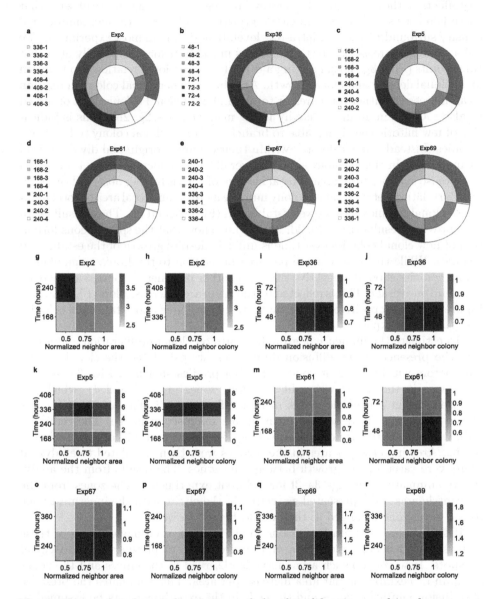

Fig. 3. Temporal and spatial variations of clonal proliferation resulting from mono-culture experiments, (a)–(f) Concentric area profiles for area increase per area for different time points and control experiments, the gradients indicate the magnitude of area change per area and the legends indicate the 'time point - the relative concentric area position' in the experiments and (g)–(r) heatmaps showing the variations in area increase per area at various normalized density zones (neighbor area and neighbor colony count) for different time points and control experiments.

sity effects on the clonal growth, the trend reverses. In Figs. 3g–3r, we observe that there is a consistent increase in clonal growth in colonies with maximum clonal density surrounding them at intra-day level. In addition, in most experiments, we see a gradual increase of growth indicators in the heat maps especially at earlier time points (Figs. 3g–3r) suggesting a directly proportional relationship between local clonal density and clonal growth. Area increase per clonal colony in relation to an increase in either total area or total number of clonal colonies was observed to also increase in an unexpectedly linear manner (Figs. 2g, 2h). This is indicative of few mitotic cells being able to branch off from a clonal colony to form new colonies. Instead, these cells stay within their colony of origin and divide to lead to an increase in clonal colony area. An overall trend of increasing area compared to increase in either total area or total number of cells is further observed (Figs. 2i, 2j). Correlation between clonal colony number increase and total area or total number of clonal colonies was not well established (Figs. 2k and 2l). These results support previous results (Figs. 2b, 2d, 2g, 2h) and show that the spontaneous formation of new clonal colonies over time is unlikely despite growth of the established colonies. While the area increase per area in relation to total area or number of clonal colonies showed a negative trend, the other parameters showed an increasing trend on average. This is reflected by the mean area increase plots (Fig. 2m, 2n). Figure 2o shows a positive correlation between total area and total clonal colonies over time, as is expected.

To estimate the influence of growth solutes on clonal proliferation and ascertain the presence of any diffusion limitations, we calculated the clonal growth parameters within different concentric area profiles shown in Figs. 3a–3f. The total area in the image was split into four zones (1, 2, 3 and 4), which are enclosed within concentric circles of four different radii. The maximum radius of the circle was fixed to the width of the microscopy image. The reminder radii were calculated as normalized values of 0.25, 0.5 and 0.75 of the maximum radius. Area increase per area in these zones were then calculated to analyze if there were variations in growth between the zones. The results from the analysis are summarized in Figs. 3a–3f for different experiments. The zones from the concentric area are represented as sectors in the circle, with the innermost zone to the outer zone arranged as they would be within wells.

If there were diffusion limitations in the system, then one or more of these sectors should occupy most area of the circle due to presence of local clonal hotspots. However, no such asymmetry is observed in the circles and the zones are found to exhibit similar area increase per area. This eliminates the presence of diffusion limitation zones and the cells in the experiment may be assumed to grow in the presence of excess nutrients. We analyzed the effects of two drugs 5-FU and Oxaliplatin on the proliferation of clonal colonies. Negative clonal growth indicators were observed in the drug treated clonal growth analysis. Figures 4e–4h and 4m–4n summarize the changes in clonal growth. Other than the one outlier in Fig. 4h (area increase) all other plots show consistent decrease in clonal growth both in terms of clonal area and colony count. Density dependent growth contribution was not observed from the heatmaps 4a-4d and 4i-4l. Overall, drug treated systems behave vastly dissimilar to the control population in

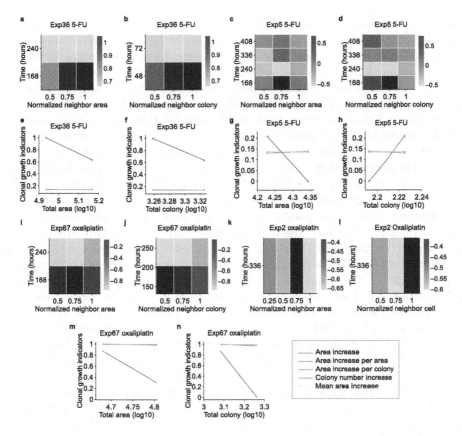

Fig. 4. Effect of spatial density on drug treated clonal growth (a)–(d) Heatmaps showing the variations in area increase per area at various normalized density zones for different time points and 5-FU experiments, (e)–(h) clonal growth indicators at different time points for 5-FU experiments, (i)–(l) heatmaps showing the variations in area increase per area at various normalized density zones for different time points and oxaliplatin experiments, (m)–(n) clonal growth indicators at different time points for oxaliplatin experiment. The clonal growth indicators are included in the figure legend.

terms of clonal growth and local and global density-driven effects. Clonal colony dynamics were also studied in co-cultures to ascertain the effect of fibroblasts on HT55 colonies in vitro (Fig. 5). Area increase was observed to be reduced in the mono-culture condition while the opposite trend was observed in the case of the co-culture condition (Fig. 5a). Similar to the control results of the cytotoxic drug experiment presented in Fig. 2, a reduction in area increase per area was observed over time in the mono- and co-culture condition, albeit to a lesser extent in the case of the co-culture condition (Fig. 5b). Area increase per clonal colony was also observed in both conditions (Fig. 5c), in line with expectations based on Fig. 2. While the clonal colony number increase remained similar over time for the mono-culture condition, a marked decrease was observed in the co-culture condition (Fig. 5d). This observation might be due to the mono-culture condition being less confluent therefore, supporting the spontaneous formation of new clonal colonies for a longer interval of time. Furthermore, the mean area increase over time was reduced in both conditions (Fig. 5e) and was lower in the co-culture condition suggesting confluency is attained faster in the presence of fibroblasts. Taken together, these results suggest that similar to the control HT55 cultures in the cytotoxic drug experiments, confluent co-culture systems also exhibit overall reduced area increase over time. Additionally, the results support the notion that fibroblasts produce factors to create a pro-clonal colony environment which supports faster clonal outgrowth. The agent-based model was numerically simulated using the experimental findings as input for model initialization. Three time points 168, 240 and 360 h from Exp67 were used for parametric quantification. Loss functions were minimized for two different time intervals I (168–240) and II (240–336). To include intermediate density-driven dynamics between the large intervals, one hour was chosen as the minimum time step Δt for the simulations. Loss minimization landscape is summarized in Fig. 5f and 5g. The z-axis of the plots is set in log scale to capture the vast range of loss values. In both intervals I and II multiple local minima can be observed. The values of k1 and k2 are positive and negative respectively indicating the contributions to total area change from the proximal and distal clonal densities. The values of k1 and k2 at different time intervals I and II are k1 (I) = 0.08, k1 (II) 0.021, k2 (I) = −0.061 and k2(II) = −0.013. Values of k1 decreases with time while k2 increases. Thus, in the simple two-function clonal growth model, k1 may be considered as the growth contribution factor, while k2 as a growth retardation parameter. The numerical range of model parameters, k1 and k2, do not exhibit drastic change between the different intervals of growth analyzed. This is an expected behavior since the density effects cannot drive overall clonal proliferation higher than its threshold growth capacity and nutrient availability. The contribution to growth retardation by parameter k2 increases with time indicating the increase in energy requirement for overall clonal growth. Therefore, k1 and k2, may be treated as energy requirement parameters varying based on the cell type. Both k1 and k2, are seen to tend towards the value of 0 with approximately equal rate of change. This is reflected in the heatmaps of Figs. 3g, k and o, where, at later time points, clonal density does not play a significant role in clonal growth as growth saturates. The developed nutrient-independent model could enable simulation of tumor proliferation through elimination of computationally intensive mass-transfer kinetics modelling [5,14]. Even though the simple two-function model can capture the growth contributions from proximal and distal densities, the current model is restricted to systems with presence of excess nutrients. In presence of a limiting solute, clonal growth would be restricted by diffusion dynamics. One other limitation of the current agent-based model is that area overlap between clones is not considered. This could mean that loss function minimization may not result in a global minimum and produce underestimated parametric values.

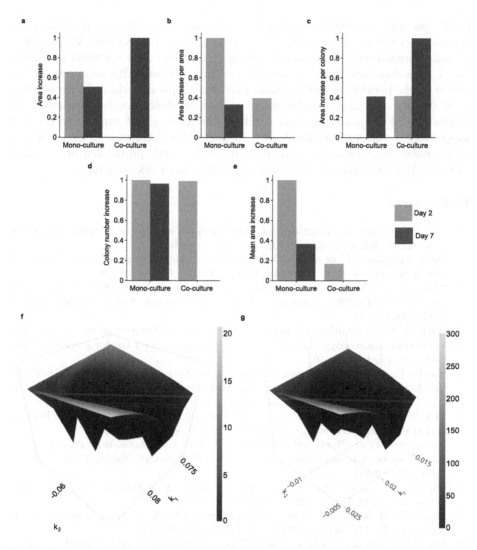

Fig. 5. Comparison of clonal growth indicators between mono- and co-culture systems (a)–(e) at different time points, the clonal growth indicators are normalized between 0–1 at different time points and (f) and (g) loss function minimization landscape for k1 and k2 parameters in the computational model, the color bar indicates the loss calculated from the χ function.

4 Conclusions

We developed a high throughput framework for clonal tracking and growth estimation. From the framework outputs, we discovered that local density of clones could affect the growth rate of the colonies. To test this hypothesis, a simple two function agent-based model of clonal proliferation was developed. The model parameters were estimated through loss minimization principles, using inputs from the clonal growth experiments. Parameters k1 and k2 provide a straight-forward way of simulating clonal growth based on proximal and distal clonal area densities. Both k1 and k2 tend towards zero with progress of time, in the model parametric estimations, indicating a possible diminishing effect of local density in confluent older colonies. The developed model currently lacks mechanistic descriptions to simulate drug interactions and capture presence of supporting factors such as fibroblasts. As part of the future work, the model can be improved with provisions for including cell growth promoting or retarding factor(s) that are derived or inferred from the framework outputs.

References

1. Dang, C.V., Lee, W.: Identification of the human c-myc protein nuclear transloca-tion signal. Mol. Cell. Biol. **8**(10), 4048–4054 (1988)
2. van der Heijden, M., et al.: Spatiotemporal regulation of clonogenicity in colorectal cancer xenografts. Proc. Natl. Acad. Sci. **116**(13), 6140–6145 (2019)
3. Henriksson, M.L., et al.: Colorectal cancer cells activate adjacent fibroblasts result-ing in FGF1/FGFR3 signaling and increased invasion. Am. J. Pathol. **178**(3), 1387–1394 (2011)
4. Huth, J., et al.: Significantly improved precision of cell migration analysis in time-lapse video microscopy through use of a fully automated tracking system. BMC Cell Biol. **11**, 1–12 (2010)
5. Kaura, P., Mishra, T., Verma, N., Dalal, I.S., Sheraton, V.: Effects of combined chemotherapeutic drugs on the growth and survival of cancerous tumours-an in-silico study. J. Computat. Sci. **54**, 101421 (2021)
6. Kok, R.N., et al.: OrganoidTracker: efficient cell tracking using machine learning and manual error correction. PLoS ONE **15**(10), e0240802 (2020)
7. Lenos, K.J., et al.: Stem cell functionality is microenvironmentally defined during tumour expansion and therapy response in colon cancer. Nat. Cell Biol. **20**(10), 1193–1202 (2018)
8. Mohme, M., et al.: Optical barcoding for single-clone tracking to study tumor heterogeneity. Mol. Ther. **25**(3), 621–633 (2017)
9. Mulkearns-Hubert, E.E., et al.: Development of a Cx46 targeting strategy for can-cer stem cells. Cell reports **27**(4), 1062–1072. e5 (2019)
10. Noonan, J., et al.: A novel triple-cell two-dimensional model to study immune-vascular interplay in atherosclerosis. Front. Immunol. **10**, 849 (2019)
11. Peng, T., et al.: A basic tool for background and shading correction of optical microscopy images. Nat. Commun. **8**(1), 14836 (2017)
12. Pogorelyy, M.V., et al.: Precise tracking of vaccine-responding t cell clones reveals convergent and personalized response in identical twins. Proc. Natl. Acad. Sci. **115**(50), 12704–12709 (2018)
13. Sheraton, M.V., Sloot, P.M.A.: Parallel performance analysis of bacterial biofilm simulation models. In: Shi, Y., et al. (eds.) ICCS 2018. LNCS, vol. 10860, pp. 496–505. Springer, Cham (2018). https://doi.org/10.1007/978-3-319-93698-7_38
14. Sheraton, V.M., Ma, S.: Exploring ductal carcinoma in-situ to invasive ductal carci-noma transitions using energy minimization principles. In: Computational Science-ICCS 2022: Proceedings of the 22nd International Conference, London, UK, 21–23 June 2022, Part I, pp. 375–388. Springer, Heidelberg (2022). https://doi.org/10.1007/978-3-031-08751-6_27

15. Tang, R., et al.: A versatile system to record cell-cell interactions. eLlife **9**, e61080 (2020)
16. Vennin, C., et al.: CAF hierarchy driven by pancreatic cancer cell p53-status creates a pro-metastatic and chemoresistant environment via perlecan. Nat. Commun. **10**(1), 3637 (2019)
17. Weber, K., Bartsch, U., Stocking, C., Fehse, B.: A multicolor panel of novel lentiviral "gene ontology" (LeGO) vectors for functional gene analysis. Mol. Ther. **16**(4), 698–706 (2008)
18. Weber, K., Thomaschewski, M., Benten, D., Fehse, B.: RGB marking with lentiviral vectors for multicolor clonal cell tracking. Nat. Protoc. **7**(5), 839–849 (2012)
19. Yang, W., et al.: Genomics of drug sensitivity in cancer (GDSC): a resource for therapeutic biomarker discovery in cancer cells. Nucleic Acids Res. **41**(D1), D955–D961 (2012)
20. Zhu, J., Thompson, C.B.: Metabolic regulation of cell growth and proliferation. Nat. Rev. Mol. Cell Biol. **20**(7), 436–450 (2019)

Computational Optimization, Modelling
and Simulation

Expedited Metaheuristic-Based Antenna Optimization Using EM Model Resolution Management

Anna Pietrenko-Dabrowska[1] , Slawomir Koziel[1,2]([✉]) , and Leifur Leifsson[3]

[1] Faculty of Electronics Telecommunications and Informatics, Gdansk University of
Technology, Narutowicza 11/12, 80-233 Gdansk, Poland
anna.dabrowska@pg.edu.pl
[2] Engineering Optimization and Modeling Center, Department of Engineering, Reykjavík
University, Menntavegur 1, 102 Reykjavík, Iceland
koziel@ru.is
[3] School of Aeronautics and Astronautics, Purdue University, West Lafayette, IN 47907, USA
leifur@purdue.edu

Abstract. Design of modern antenna systems heavily relies on numerical optimization methods. Their primary purpose is performance improvement by tuning of geometry and material parameters of the antenna under study. For reliability, the process has to be conducted using full-wave electromagnetic (EM) simulation models, which are associated with sizable computational expenditures. The problem is aggravated in the case of global optimization, typically carried out using nature-inspired algorithms. To reduce the CPU cost, population-based routines are often combined with surrogate modeling techniques, frequently in the form of machine learning procedures. While offering certain advantages, their efficiency is worsened by the curse of dimensionality and antenna response nonlinearity. In this article, we investigate computational advantages of combining population-based optimization with variable-resolution EM models. Consequently, a model management scheme is developed, which adjusts the discretization level of the antenna under optimization within the continuous spectrum of acceptable fidelities. Starting from the lowest practically useful fidelity, the resolution converges to the highest assumed level when the search process is close to conclusion. Several adjustment profiles are considered to investigate the speedup-reliability trade-offs. Numerical results have been obtained for two microstrip antennas and particle swarm optimizer as a widely-used nature-inspired algorithm. Consistent acceleration of up to eighty percent has been obtained in comparison to the single-resolution version with minor deterioration of the design quality. Another attractive feature of our methodology is versatility and easy implementation and handling.

Keywords: Antenna design · global optimization · variable resolution models · EM-driven design · nature-inspired optimization

© The Author(s), under exclusive license to Springer Nature Switzerland AG 2023
J. Mikyška et al. (Eds.): ICCS 2023, LNCS 14075, pp. 363–377, 2023.
https://doi.org/10.1007/978-3-031-36024-4_29

1 Introduction

Contemporary antenna systems are developed to satisfy stringent performance requirements imposed by existing and emerging applications (internet of things (IoT) [1], body area networks [2], 5G technology [3], implantable devices [4], etc.), enable a range of functionalities (multi-band [5] and MIMO operation [6], reconfigurability [7], beam scanning [8]), and, in many cases, feature compact physical dimensions [9]. Fulfilling such performance demands leads to topologically intricate structures, whose parameters necessitate meticulous tuning. At the same time, they can be reliably evaluated solely using full-wave electromagnetic (EM) analysis. As a matter of fact, EM simulation tools are indispensable at all design stages, starting from geometry evolution, through parametric studies, to final tuning of antenna parameters.

Given the complexity of modern antennas but also the need for handling multiple objectives and constraints, performance-oriented parameter adjustment has to be carried out using rigorous numerical optimization methods [10]. The most problematic issue thereof is high computational cost, which may be troublesome even for local tuning. Global optimization entails incomparably higher expenses, yet it is recommended in a growing number of situations, e.g., design of frequency-selective surfaces [11], array pattern synthesis [12], EM-driven miniaturization [13], re-design of antennas over broad ranges of operating frequencies.

Nowadays, global optimization is primarily conducted using nature-inspired methods [14, 15]. Some of the popular techniques include evolutionary algorithms [16], particle swarm optimizers (PSO) [17], differential evolution (DE) [18], or firefly algorithm [19]. New methods are reported on almost daily basis (e.g., [20–22]), yet the differences between them are often cosmetic. The global search capability is arguably a result of exchanging information between candidate solutions processed by the algorithm [23], using exploratory/exploitative operators, as well as mimicking social or biological phenomena [24]. Popularity of nature-inspired methods stems from their simplicity, both in terms of implementation and handling. The downside is remarkably poor computational efficiency. Typical running costs measured in thousands of objective function evaluations are prohibitive from the perspective of EM-driven design. A possible workaround is the incorporation of surrogate modeling [25–27]. Shifting the computational burden into a fast metamodel enables acceleration. In practice, iterative procedures, often referred to as machine learning [28], are utilized, where the surrogate serves as predictor which undergoes refinement using the accumulated EM simulation data. The strategies for generating the infill points may be based on parameter space exploration (identifying the most promising regions), exploitation (pursuing the optimum) or combination of both [29]. In the context of global optimization, the employment of metamodels is impeded by the curse of dimensionality, broad ranges of geometry parameters and frequency, as well as antenna response nonlinearity. These can be alleviated by domain confinement [30, 31], variable-fidelity approaches [32], or feature-based methodologies [33, 34].

The mitigation methods mentioned above address some of the problems pertinent to global EM-based antenna design but are not free for the issues on their own. These include, among others, limited versatility and implementation complexity. From this perspective, the employment of variable-resolution models seems to be the simplest yet offering sizable computational benefits. In most cases, it utilizes two levels of fidelity (equivalent networks vs. EM analysis [35]) or resolution (coarse- and fine-discretization EM simulations [36]). Using a continuous range of model resolutions might be a more flexible option. In the realm of nature-inspired procedures, this idea has been pursued in [37]; however, it was demonstrated mainly using analytical functions.

This article investigates potential merits of incorporating variable-resolution EM analysis into nature-inspired optimization of antenna systems. A model management scheme is developed, which establishes the model fidelity from a continuous spectrum of resolutions. The latter is controlled by a discretization density of the computational model of the antenna under design. The search process starts from the minimum usable resolution and gradually increases it as the algorithm reaches convergence. The speedup-reliability trade-offs can be worked out by adjusting the model selection profile. Numerical experiments have been conducted using two microstrip antennas and a particle swarm optimizer (PSO) as a representative nature-inspired optimization routine. The results demonstrate that the search process can be considerably expedited with cost savings of up to eighty percent as compared to the single-fidelity PSO version. At the same time, design quality degradation is practically negligible. The proposed approach is straightforward to implement and handle, and can be incorporated into any population-based metaheuristic.

2   Antenna Optimization. Variable-Resolution Models

In this section, formulation of the antenna optimization problem and introduction of variable-resolution computational models are recalled. The latter are illustrated using a microstrip antenna example.

2.1   EM-driven Design. Problem Formulation

In this work, we use the following formulation of the simulation-based antenna optimization task. Given the parameter vector x, the aim is to minimize a scalar objective function U quantifying the design quality. The optimum parameter vector x^* is found as

$$\mathbf{x}^* = \arg \min_{\mathbf{x}} U(\mathbf{x}) \tag{1}$$

Representative examples of optimization scenarios and the associated objective function can be found in Fig. 1. Therein, f stands for the frequency, whereas $|S_{11}(x,f)|$, $G(x,f)$, and $AR(x,f)$ are the reflection coefficient, gain, and axial ratio at design x and frequency f; $A(x)$ is the antenna size (e.g., the footprint area). Note that in all cases, the primary objective is directly optimized, whereas secondary objectives are cast into constraints handled using the penalty function approach.

2.2   Variable-Resolution EM Models

In the design of antennas and microwave components, variable-resolution EM simulations have been already employed for expediting simulation-driven design optimization procedures [35, 39]. Yet, in majority of frameworks, only two levels of resolution are utilized, i.e., coarse (low-fidelity) and fine (high-fidelity). The performance of any variable-fidelity procedure strongly depends on the evaluation cost and accuracy of the low-fidelity model, whose appropriate selection is a challenging task of fundamental importance [40].

Design scenario: verbal description	Objective function U
Design for best in-band matching within the frequency range F	$U(x) = S(x) = \max\{f \in F : |S_{11}(x,f)|\}$
Design for maximum average in-band gain (in frequency range F); Ensuring that in-band matching does not exceed −10 dB in F	$U(x) = \dfrac{1}{F}\int_F G(x,f)df + \beta_1 c_1(x)^2$ where $c_1(x) = \left[\dfrac{\max(S(x)+10,0)}{10}\right]^2$
Design for size reduction of a circularly polarized antenna; Ensuring that in-band matching (in frequency range F) does not exceed −10 dB, and axial ratio does not exceed 3 dB, both over the frequency range F	$U_P(x) = \max\{f \in F : A(x,f)\} + \beta_1 c_1(x)^2 + \beta_2 c_2(x)^2$ where $c_1(x) = \left[\dfrac{\max(S(x)+10,0)}{10}\right]^2$ and $c_2(x) = \left[\dfrac{\max(\max\{f \in F : AR(x,f)\}-3,0)}{3}\right]^2$

Fig. 1. Representative antenna design optimization scenarios.

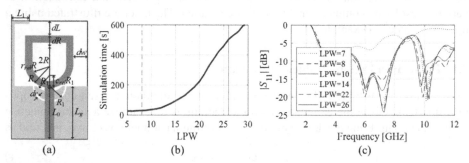

Fig. 2. Multi-resolution EM models: (a) exemplary broadband monopole antenna, (b) average simulation time versus LPW (the vertical lines indicate the values of LPW for the high-fidelity model (—) and the lowest practically useful low-fidelity model (- - -)); (c) reflection responses corresponding to various discretization densities.

Here, the low-fidelity EM models are realized using coarse-discretization EM analysis, which a usual approach in the case of antenna structures. Other possibilities, e.g., equivalent networks or analytical models, are not readily obtainable and difficult to parameterize. In our approach, a major mechanism employed to accelerate the simulation process is a reduction of the model discretization level. Our numerical experiments are carried out using CST Microwave Studio [41], one of the most popular commercial EM solvers. Therein, the discretization density is controlled with the use of a single parameter, LPW (lines per wavelength).

Figure 2 presents an exemplary antenna: its geometry and reflection response $|S_{11}|$ evaluated for various values of the LPW parameter. Both evaluation accuracy and the simulation cost increase with an increase of the LPW value. The acceptable range of model resolutions should be cautiously selected, because below a certain LPW value, the model is no longer usable due to largely inaccurate rendition of antenna characteristics as shown in Fig. 2. In practice, a visual inspection of family of antenna responses suffices to assess the admissible range of LPW: from L_{min}, being the lowest value suitable for carrying out antenna optimization, up to L_{max}, i.e., the highest value representing the model of the maximum fidelity. The former is normally estimated as the value for which the respective model accounts for all meaningful features of the antenna response, such as the resonances. Whereas the latter corresponds to the accuracy level deemed adequate by the designer. It is most often estimated as the LPW value increasing of which does not bring in any further changes to antenna responses.

3 Population-Based Optimization with Variable-Fidelity EM Models

The aim of this section is to outline the metaheuristic-based antenna optimization technique considered in this work. Section 3.1 delineates a generic structure of nature-inspired optimization procedures, whereas its integration with variable-resolution model management is provided in Sect. 3.2. Demonstration experiments are delineated in Sect. 4.

3.1 Nature-Inspired Algorithms. Generic Structure

Let us first define the main entities, which are conventionally utilized in virtually any given nature-inspired algorithm. In the kth iteration, we have the population $\boldsymbol{P}^{(k)} = [P_1^{(k)} \dots P_N^{(k)}]$ of size N, which, depending on algorithm type, may also be referred to as a swarm or pack. The assumed number of iterations k_{max} defines the computational budget, which, in turn, decides upon algorithm termination. The aim is to minimize the cost function $E(P)$, which quantifies the solution quality. Here, we will use a shortened symbol $E_{k,j}$ instead of a full notation $E(P_j^{(k)})$. In particle swarm optimization algorithm [17], utilized in this work as a base search engine, the best particle found so far is passed over throughout the subsequent iterations (this feature is referred to as elitism).

The particular ways of creating a new population $\boldsymbol{P}^{(k+1)}$ from the previous one vary between different nature-inspired algorithms. For example, in PSO [17] (but also in DE [18], and firefly algorithm [19]), the replacement of the individuals is, in general, not performed. Instead, the individuals are repositioned in the search space according to the assumed rules, usually, by random modifications biased in the direction of the best local and global solutions identified in the previous iterations. In PSO, a velocity vector governing the transfer to a new location is assigned to each particle. The said vector is modified with the use of a linear combination of three factors: (i) a random factor, (ii) a vector in the direction of the best location of a given particle, and (iii) a vector pointing into the direction of the global optimum.

3.2 Variable-Resolution Model Management

We aim at accelerating metaheuristic-based optimization procedure delineated in Sect. 3.1 by exploiting variable-resolution EM models of Sect. 2.2. In our algorithm, the model resolution L is to be continuously modified (from the minimum value L_{min} to the highest one L_{max}) based on the iteration count $k \leq k_{max}$, where k_{max} is the maximum number of iterations. The following adjustment scheme is employed [38].

---

1. Set the iteration index $k = 0$;

2. Set the model resolution $L(k) = L_{min}$;

3. Initialize population $P^{(k)}$;

4. Evaluate population $P^{(k)}$ at the resolution level $L(k)$ to find
    $E_{k,j}, j = 1, ..., N$;

5. Find the best individual $[P_{best}, E_{best}]$ in $P^{(k)}$, where $E_{best} = \min\{j = 1, ..., N :$
    $E_{k,j}\}$, and P_{best} is the individual associated with E_{best};

6. **while** $k < k_{max}$ **do**

7.   Set $k = k + 1$;

8.   Generate a new population $P^{(k)}$ from $P^{(k-1)}$ using the algorithm-
        specific rules;

9.   Update model resolution $L(k)$ according to (2);

10.   Evaluate population $P^{(k)}$ at the resolution level $L(k)$ to find
        $E_{k,j}, j = 1, ..., N$;

11.   Evaluate P_{best} at the resolution level $L(k)$ to find updated E_{best};

12.   Find the best individual $[P_{best.tmp}, E_{best.tmp}]$ in $P^{(k)}$;

13.   **if** $E_{best.tmp} < E_{best}$ **then**

14.   Update global best: $P_{best} = P_{best.tmp}$ and $E_{best} = E_{best.tmp}$;

15. **return** P_{best} and E_{best}.

Fig. 3. Accelerated nature-inspired algorithm incorporating multi-resolution EM models: pseudocode. The steps specific to the adopted acceleration mechanism: Step 2 (initialization of the model resolution level), Step 9 (adjustment of the current model resolution level), and Step 11 (re-evaluation of the best individual identified so far using the new resolution level). The remaining steps are common for both the basic and accelerated version of the algorithm.

$$L(k) = L_{min} + (L_{max} - L_{min}) \left[\frac{k}{k_{max}} \right]^p \tag{2}$$

where p denotes power parameter. The resolution adjustment scheme (2) is sufficiently flexible: (i) for $p > 1$, $L \approx L_{min}$ throughout majority of optimization course, when close to convergence L quickly increases towards L_{max}; (ii) for $p < 1$, L_{min} is used only at the beginning of the optimization process, with $L \approx L_{max}$ utilized throughout the rest of the optimization course.

A pseudocode of an expedited population-based algorithm employing variable-resolution EM models is shown in Fig. 3. At the onset of the optimization process, we set $L = L_{\min}$ (Step 2). Next, the individuals are evaluated at the current resolution level $L(k)$ adjusted according to (2) (Step 9). At the end of the current iteration, the best solution P_{best} is re-evaluated using an updated resolution level, and it is subsequently compared (Step 13) to the best solution from the current population (to ensure that the comparison is carried out at the same resolution level). This is because the individual being the best at previous fidelity level is not necessarily the best at new resolution.

Clearly, it is to be anticipated that the higher the p, the higher are potential savings. As a matter of fact, these saving may be assessed a priori. The computational expenditures T_I of the basic single-fidelity algorithm using the fine EM model of resolution $L_{\max}$ may be expressed as

$$T_I = N \cdot k_{\max} \cdot T(L_{\max}) \tag{3}$$

where $T(L_{\max})$ refers to the evaluation time of the antenna structure under design at the highest resolution level $L_{\max}$ (for any given model fidelity L, we denote the corresponding evaluation time as $T(L)$). Whereas the cost of the proposed multi-fidelity optimization procedure equals to

$$T_{II} = N \cdot T(L_{\min}) + (N + 1)T(L(1)) + (N + 2)T(L(2)) + \ldots + (N + 1)T(L(k_{\max})) \tag{4}$$

which may be approximated as

$$T_{II} \approx (N + 1) \cdot \sum_{k=0}^{k_{\max}} T(L(k)) \tag{5}$$

In (5), the multiplier $N + 1$ stems from the necessity of re-evaluating the best individual using current L in Step 11.

Let us analyse the algorithm running times predicted using (5) for the antenna of Fig. 2 (see Table 1). The control parameters of the core PSO algorithm are: the population size $N = 10$, with the maximum iteration number $k_{\max} = 100$. Observe that both N and $k_{\max}$ are kept low (for a typical population-based algorithm). The reason is the necessity to curb the optimization cost as the responses of the antenna structures are evaluated using expensive full-wave EM simulations. Despite the fact that the antenna shown in Fig. 2 is relatively simple, the computational expenses provided in Table 1 are still high (over five days). Such cost level is, however, unavoidable in simulation-driven antenna design. Potential savings due to the proposed accelerated algorithm with respect to basic procedure depend on the power factor p. Even for the lowest value of $p = 1$, the anticipated savings reach 50% Whereas for $p = 3$ savings of over 70% may be obtained. Decreasing the computational cost to such degree is highly desirable. Yet, the reliability of the proposed multi-fidelity procedure remains to be verified. It is especially of interest, whether and to what extent the computational speedup might be detrimental to the design quality. This is going to be verified in Sect. 4.

Table 1. Estimated cost of a generic metaheuristic-based algorithm for antennas of Fig. 2

EM model setup		Computational cost of the optimization process ($N = 10$, $k_{max} = 100$)	
		Execution time	Savings w.r.t. single-fidelity-based algorithm
High-fidelity ($L = L_{max}$)		132.1 h	-
Variable resolution	$p = 0.5$	82.2 h	37.7 %
	$p = 1.0$	57.0 h	56.8 %
	$p = 2.0$	37.7 h	71.5 %
	$p = 3.0$	29.6 h	77.6 %

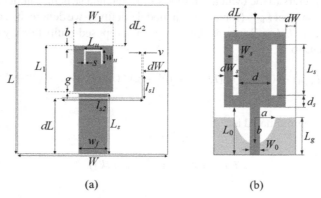

(a)   (b)

Fig. 4. Antennas used for verification of the introduced porcedure: (a) Antenna I [42], and (b) Antenna II [43] (ground-plane metallization is marked using light grey color).

4   Demonstration Experiments

The introduced multi-resolution metaheuristic-based antenna optimization algorithm delineated in Sect. 3 is validated using a triple-band antenna (Antenna I), as well as a wideband monopole antenna (Antenna II). The core optimization procedure is the particle swarm optimizer (PSO) [41], which has been selected as a widely utilized population-based technique.

4.1   Test Cases

The test antenna structures utilized for numerical validation of our approach are shown in Fig. 4, whereas the relevant details on their parameters, design space (delimited by the vectors *l* and *b*, i.e., the lower and upper bounds for antenna dimensions, respectively) and objectives, as well as the setup of variable-resolution EM models are provided in Table 2. Antenna I, presented in Fig. 4(a), is a triple band U-slotted patch using L-slot

Table 2. Verification antenna structures

	Case study	
	Antenna I	Antenna II
Substrate	$\varepsilon_r = 3.2$ $h = 3.1$ mm	$\varepsilon_r = 4.3$ $h = 1.55$ mm
Design parameters	$x = [L_1\ L_s\ L_u\ W\ W_1\ dL\ dW\ g\ l_{s1}\ l_{s2}\ w_u]^T$	$x = [L_g\ L_0\ L_s\ W_s\ d\ dL\ d_s\ dW_s\ dW\ a\ b]^T$
Other parameters	$b = 1,\ w_f = 7.4,$ $s = 0.5,\ w = 0.5,$ $dL_2 = L_1, L = L_s + g + L_1 + dL_2$	$W_0 = 3.0$
Operating bands	80 MHz bandwidth centered at frequencies 3.5 GHz, 5.8 GHz, and 7.5 GHz	UWB frequency band from 3.1 GHz to 10.6 GHz
Parameter space	$l = [10\ 17\ 5\ 45\ 8\ 15\ 9\ 0.2\ 4\ 20\ 2]^T$ $u = [16\ 25\ 8\ 55\ 12\ 20\ 12\ 0.4\ 6\ 24\ 3]^T$	$l = [5\ 5\ 5\ 0.2\ 0.2\ 5\ 0.3\ 0.5\ 1.0\ 0.1\ 0.2]^T$ $u = [15\ 15\ 15\ 1.2\ 8\ 15\ 1.5\ 2.5\ 5\ 0.5\ 0.5]^T$
	Low-fidelity model	
L_{min}	8	6
Simulation time [s]	32	33
	High-fidelity model	
L_{max}	25	25
Simulation time [s]	114	378

defected ground structure (DGS) [42]. Whereas Antenna II of Fig. 4(b) is a compact ultra-wideband (UWB) monopole antenna with radiator slots [43]. For both structures, the design goal is to minimize the maximum in-band reflection levels; the formulation of the design problems follows that of the second row of Fig. 1.

Antenna characteristics are evaluated using the transient solver of CST Microwave Studio. The model of Antenna II includes the SMA connector [44]. Table 2 presents the setup of variable-resolution EM models for both antennas: the lowest applicable resolution (L_{min}) and the highest one (L_{max}), along with the corresponding simulation times. The relationships between the model fidelity and the simulation time for both antennas are given in Fig. 5. The time evaluation ratios (fine/coarse model) equals 3.5 for Antenna I and 15 for Antenna II. This implies higher possible speedup to be obtained for the latter structure.

4.2 Setup and Numerical Results

Both verification structures have been optimized with the use of the PSO algorithm (swarm size $N = 10$, $k_{max} = 100$, the standard values of the control parameters, $\chi = 0.73$, $c_1 = c_2 = 2.05$, cf. [41]). Benchmarking included four versions of the proposed

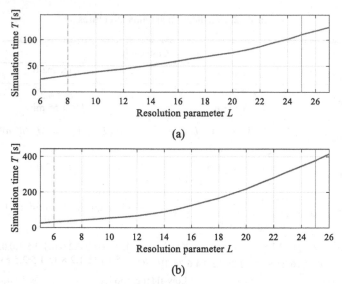

(a)

(b)

Fig. 5. Relationship of simulation time on EM model fidelity for antennas of Fig. 4: (a) Antenna I, and (b) Antenna II. The resolutions $L_{\min}$ (the minimum usable resolution) and $L_{\max}$ (the maximum resolution of high-fidelity model) are marked using (- - -) and (—) values, respectively.

multi-fidelity algorithm (for the power factor $p = 1, 2, 3$, and 4), as well as single-fidelity basic PSO algorithm.

The results are provided in Table 3. Each algorithm has been run fifteen times independently. The algorithm performance is assessed using the following indicators: the average value of the merit function (i.e., the maximum in-band reflection), and its standard deviation, which serves to quantify solution repeatability. Algorithm cost-efficacy is assessed in terms of the overall execution time, as well as the savings w.r.t single-fidelity PSO algorithm. Figures 6 and 7 present the representative optimized antenna responses.

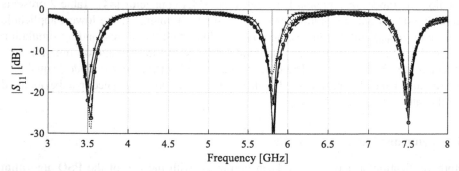

Fig. 6. Optimized designs of Antenna I for the representative runs of the single-fidelity procedure (—), as well as the proposed variable-fidelity algorithm: $p = 1$ (- - -), $p = 2$ (····), $p = 3$ (- o -), $p = 4$ (- x -). Target operating frequencies are indicated using vertical lines.

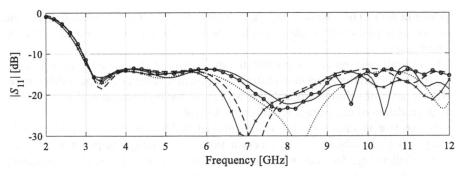

Fig. 7. Optimized designs of Antenna II for the representative runs of the single-fidelity procedure (—), as well as the proposed variable-fidelity algorithm: $p = 1$ (- - -), $p = 2$ (⋯), $p = 3$ (- o -), $p = 4$ (- x -). Target operating frequencies are indicated using vertical lines.

Table 3. Optimization results

Algorithm setup		Antenna I				Antenna II			
		Execution time [hours]	Saving s[1]	U [dB][2]	std(U) [dB]	Execution time [hours]	Saving s[1]	U [dB][2]	std(U) [dB]
High-fidelity ($L = L_{max}$)		31.7	-	-15.7	2.5	105.1	-	-13.1	1.6
Variable resolution	$p = 1.0$	19.6	38.2 %	-18.4	2.1	45.6	56.6 %	-13.2	1.5
	$p = 2.0$	16.5	47.9 %	-17.9	1.5	32.1	69.5 %	-13.0	1.6
	$p = 3.0$	14.5	54.3 %	-15.8	2.2	26.3	75.0 %	-12.9	1.6
	$p = 4.0$	13.4	57.7 %	-14.0	3.0	22.5	78.6 %	-12.5	1.7

[1]Percentage savings w.r.t. single-fidelity PSO algorithm
[2]Objective function value averaged over 15 algorithm runs

To verify the sample normality, we performed a Kolmogorov-Smirnov test for the merit function values rendered by consecutive algorithm runs: the null hypothesis that the results follow a normal distribution of the mean and standard deviation provided in Table 3 has not been rejected at the 5% significance level. Moreover, the typical p-values vary from 0.4 to 0.9. This corroborates that the (normalized) distribution of the merit function values is close to normal. This also indicates that the mean and standard deviation allow for reliable assessment of the algorithm performance.

4.3 Discussion

The following summary of the numerical results of Table 3 may be formulated:

- Considerable computational speedup w.r.t the single-fidelity procedure has been achieved through the employment of variable-resolution EM models into PSO optimization procedure. The actual level of cost-efficacy depends on the value of the power factor p. The lowest savings have been obtained for $p = 1$ (around 48 percent on average), whereas the highest ones have been reached for $p = 4$ (around 68 percent on average).

- The optimization process is reliable for the values of the power factor up 3. For higher p, the standard deviation of the merit function value increases, which implies deterioration of solution repeatability. Moreover, for both antennas, the $p > 3$ leads to degradation of the average merit function value.
- Observe that in this work, the computational budget for PSO algorithm is relatively low (1,000 objective function evaluations), even though the presented tasks are quite challenging. The reason for such a low budget is the necessity of maintaining practically acceptable CPU cost of the optimization procedure.

Overall, the employment of variable-resolution models leads to a considerable reduction in computational cost of the metaheuristic-based search without degrading design quality. Thus, the introduced algorithm may constitute an attractive alternative to direct metaheuristic-based optimization of antenna structures. The main advantages of our approach are simplicity of implementation and practically acceptable cost. At the same time, it should be emphasized that the procedure is generic because the arrangement and handling of variable-resolution models is straightforward, here, realized using a single parameter of the EM solver selected for antenna evaluation.

5 Conclusion

This article investigated a possibility of reducing computational costs of nature-inspired antenna optimization by incorporating variable-resolution EM models. A simple model management scheme has been developed to adjust the fidelity of the antenna analysis. The search process starts with the lowest practically useful fidelity and gradually converges towards the high-fidelity representation upon the conclusion of the algorithm run. Different adjustment profiles have been tested. The proposed approach has been validated using several microstrip antennas. Extensive numerical experiments demonstrate up to eighty percent reduction of the computational costs with regard to the single-resolution algorithm without degrading the design quality.

Acknowledgement. The authors would like to thank Dassault Systemes, France, for making CST Microwave Studio available. This work is partially supported by the Icelandic Centre for Research (RANNIS) Grant 206606 and by Gdańsk University of Technology Grant DEC-41/2020/IDUB/I.3.3 under the Argentum Triggering Research Grants program - 'Excellence Initiative - Research University'.

References

1. Wang, Y., Zhang, J., Peng, F., Wu, S.: A glasses frame antenna for the applications in internet of things. IEEE Internet Things J. **6**(5), 8911–8918 (2019)
2. Le, T.T., Yun, T.-Y.: Miniaturization of a dual-band wearable antenna for WBAN applications. IEEE Ant. Wirel. Propag. Lett. **19**(8), 1452–1456 (2020)

3. Yuan, X.-T., Chen, Z., Gu, T., Yuan, T.: A wideband PIFA-pair-based MIMO antenna for 5G smartphones. IEEE Ant. Wirel. Propag. Lett. **20**(3), 371–375 (2021)
4. Xu, L., Xu, J., Chu, Z., Liu, S., Zhu, X.: Circularly polarized implantable antenna with improved impedance matching. IEEE Ant. Wirel. Propag. Lett. **19**(5), 876–880 (2020)
5. Ameen, M., Thummaluru, S.R., Chaudhary, R.K.: A compact multilayer triple-band circularly polarized antenna using anisotropic polarization converter. IEEE Ant. Wirel. Propag. Lett. **20**(2), 145–149 (2021)
6. Wong, K., Chang, H., Chen, J., Wang, K.: Three wideband monopolar patch antennas in a Y-shape structure for 5G multi-input–multi-output access points. IEEE Ant. Wirel. Propag. Lett. **19**(3), 393–397 (2020)
7. Shirazi, M., Li, T., Huang, J., Gong, X.: A reconfigurable dual-polarization slot-ring antenna element with wide bandwidth for array applications. IEEE Trans. Ant. Prop. **66**(11), 5943–5954 (2018)
8. Karmokar, D.K., Esselle, K.P., Bird, T.S.: Wideband microstrip leaky-wave antennas with two symmetrical side beams for simultaneous dual-beam scanning. IEEE Trans. Ant. Prop. **64**(4), 1262–1269 (2016)
9. Sambandam, P., Kanagasabai, M., Natarajan, R., Alsath, M.G.N., Palaniswamy, S.: Miniaturized button-like WBAN antenna for off-body communication. IEEE Trans. Ant. Prop. **68**(7), 5228–5235 (2020)
10. Kovaleva, M., Bulger, D., Esselle, K.P.: Comparative study of optimization algorithms on the design of broadband antennas. IEEE J. Multiscale Multiphys. Comp. Techn. **5**, 89–98 (2020)
11. Genovesi, S., Mittra, R., Monorchio, A., Manara, G.: Particle swarm optimization for the design of frequency selective surfaces. IEEE Ant. Wirel. Propag. Lett. **5**, 277–279 (2006)
12. Liang, S., Fang, Z., Sun, G., Liu, Y., Qu, G., Zhang, Y.: Sidelobe reductions of antenna arrays via an improved chicken swarm optimization approach. IEEE Access **8**, 37664–37683 (2020)
13. Kim, S., Nam, S.: Compact ultrawideband antenna on folded ground plane. IEEE Trans. Ant. Prop. **68**(10), 7179–7183 (2020)
14. Li, W., Zhang, Y., Shi, X.: Advanced fruit fly optimization algorithm and its application to irregular subarray phased array antenna synthesis. IEEE Access **7**, 165583–165596 (2019)
15. Jia, X., Lu, G.: A hybrid Taguchi binary particle swarm optimization for antenna designs. IEEE Ant. Wirel. Propag. Lett. **18**(8), 1581–1585 (2019)
16. Michalewicz, Z.: Genetic Algorithms + Data Structures = Evolution Programs. Springer, Heidelberg (1996). https://doi.org/10.1007/978-3-662-03315-9
17. Wang, D., Tan, D., Liu, L.: Particle swarm optimization algorithm: an overview. Soft. Comput. **22**(2), 387–408 (2017). https://doi.org/10.1007/s00500-016-2474-6
18. Jiang, Z.J., Zhao, S., Chen, Y., Cui, T.J.: Beamforming optimization for time-modulated circular-aperture grid array with DE algorithm. IEEE Ant. Wirel. Propag. Lett. **17**(12), 2434–2438 (2018)
19. Baumgartner, P., et al.: Multi-objective optimization of Yagi-Uda antenna applying enhanced firefly algorithm with adaptive cost function. IEEE Trans. Magn. **54**(3), 1–4 (2018). Article no. 8000504
20. Yang, S.H., Kiang, J.F.: Optimization of sparse linear arrays using harmony search algorithms. IEEE Trans. Ant. Prop. **63**(11), 4732–4738 (2015)
21. Li, X., Luk, K.M.: The grey wolf optimizer and its applications in electromagnetics. IEEE Trans. Ant. Prop. **68**(3), 2186–2197 (2020)
22. Darvish, A., Ebrahimzadeh, A.: Improved fruit-fly optimization algorithm and its applications in antenna arrays synthesis. IEEE Trans. Antennas Propag. **66**(4), 1756–1766 (2018)
23. Bora, T.C., Lebensztajn, L., Coelho, L.D.S.: Non-dominated sorting genetic algorithm based on reinforcement learning to optimization of broad-band reflector antennas satellite. IEEE Trans. Magn. **48**(2), 767–770 (2012)

24. Cui, C., Jiao, Y., Zhang, L.: Synthesis of some low sidelobe linear arrays using hybrid differential eution algorithm integrated with convex programming. IEEE Ant. Wirel. Propag. Lett. **16**, 2444–2448 (2017)
25. Queipo, N.V., Haftka, R.T., Shyy, W., Goel, T., Vaidynathan, R., Tucker, P.K.: Surrogate-based analysis and optimization. Prog. Aerosp. Sci. **41**(1), 1–28 (2005)
26. Easum, J.A., Nagar, J., Werner, P.L., Werner, D.H.: Efficient multi-objective antenna optimization with tolerance analysis through the use of surrogate models. IEEE Trans. Ant. Prop. **66**(12), 6706–6715 (2018)
27. Liu, B., Aliakbarian, H., Ma, Z., Vandenbosch, G.A.E., Gielen, G., Excell, P.: An efficient method for antenna design optimization based on evolutionary computation and machine learning techniques. IEEE Trans. Ant. Propag. **62**(1), 7–18 (2014)
28. Alzahed, A.M., Mikki, S.M., Antar, Y.M.M.: Nonlinear mutual coupling compensation operator design using a novel electromagnetic machine learning paradigm. IEEE Ant. Wirel. Prop. Lett. **18**(5), 861–865 (2019)
29. Couckuyt, I., Declercq, F., Dhaene, T., Rogier, H., Knockaert, L.: Surrogate-based infill optimization applied to electromagnetic problems. Int. J. RF Microw. Computt. Aided Eng. **20**(5), 492–501 (2010)
30. Koziel, S., Pietrenko-Dabrowska, A.: Performance-based nested surrogate modeling of antenna input characteristics. IEEE Trans. Ant. Prop. **67**(5), 2904–2912 (2019)
31. Koziel, S., Pietrenko-Dabrowska, A.: Performance-driven surrogate modeling of high-frequency structures. Springer, New York (2020). https://doi.org/10.1007/978-3-030-389 26-0
32. Pietrenko-Dabrowska, A., Koziel, S.: Antenna modeling using variable-fidelity EM simulations and constrained co-kriging. IEEE Access **8**(1), 91048–91056 (2020)
33. Koziel, S., Pietrenko-Dabrowska, A.: Expedited feature-based quasi-global optimization of multi-band antennas with Jacobian variability tracking. IEEE Access **8**, 83907–83915 (2020)
34. Koziel, S.: Fast simulation-driven antenna design using response-feature surrogates. Int. J. RF Micr. CAE **25**(5), 394–402 (2015)
35. Koziel, S., Bandler, J.W.: Reliable microwave modeling by means of variable-fidelity response features. IEEE Trans. Microwave Theor. Tech. **63**(12), 4247–4254 (2015)
36. Rayas-Sanchez, J.E.: Power in simplicity with ASM: tracing the aggressive space mapping algorithm over two decades of development and engineering applications. IEEE Microwave Mag. **17**(4), 64–76 (2016)
37. Koziel, S., Unnsteinsson, S.D.: Expedited design closure of antennas by means of trust-region-based adaptive response scaling. IEEE Antennas Wirel. Prop. Lett. **17**(6), 1099–1103 (2018)
38. Li, H., Huang, Z., Liu, X., Zeng, C., Zou, P.: Multi-fidelity meta-optimization for nature inspired optimization algorithms. Appl. Soft. Comp. **96**, 106619 (2020)
39. Tomasson, J.A., Pietrenko-Dabrowska, A., Koziel, S.: Expedited globalized antenna optimization by principal components and variable-fidelity EM simulations application to microstrip antenna design. Electronics **9**(4), 673 (2020)
40. Koziel, S., Ogurtsov, S.: Model management for cost-efficient surrogate-based optimization of antennas using variable-fidelity electromagnetic simulations. IET Microwaves Ant. Prop. **6**(15), 1643–1650 (2012)
41. Kennedy, J., Eberhart, R.C.: Swarm Intelligence. Morgan Kaufmann, San Francisco (2001)
42. Consul, P.: Triple band gap coupled microstrip U-slotted patch antenna using L-slot DGS for wireless applications. In: Communication, Control and Intelligent Systems (CCIS), Mathura, India, pp. 31–34 (2015)

43. Haq, M.A., Koziel, S.: Simulation-based optimization for rigorous assessment of ground plane modifications in compact UWB antenna design. Int. J. RF Microwave CAE **28**(4), e21204 (2018)
44. SMA PCB connector, 32K101-400L5, Rosenberger Hochfrequenztechnik GmbH & C. KG (2021)

Dynamic Core Binding for Load Balancing of Applications Parallelized with MPI/OpenMP

Masatoshi Kawai[1]([✉]), Akihiro Ida[2], Toshihiro Hanawa[3],
and Kengo Nakajima[3,4]

[1] Nagoya University, Nagoya, Japan
kawai@cc.nagoya-u.ac.jp
[2] Research Institute for Value-Added-Information Generation (VAiG),
Yokosuka, Japan
[3] The University of Tokyo, Tokyo, Japan
[4] Riken, Wakō, Japan

Abstract. Load imbalance is a critical problem that degrades the performance of parallelized applications in massively parallel processing. Although an MPI/OpenMP implementation is widely used for parallelization, users must maintain load balancing at the process level and thread (core) level for effective parallelization. In this paper, we propose dynamic core binding (DCB) to processes for reducing the computation time and energy consumption of applications. Using the DCB approach, an unequal number of cores is bound to each process, and load imbalance among processes is mitigated at the core level. This approach is not only improving parallel performance but also reducing power consumption by reducing the number of using cores without increasing the computational time. Although load balancing among nodes cannot be handled by DCB, we also examine how to solve this problem by mapping processes to nodes. In our numerical evaluations, we implemented a DCB library and applied it to the lattice $\mathcal{H}$-matrixes. Based on the numerical evaluations, we achieved a 58% performance improvement and 77% energy consumption reduction for the applications using the lattice $\mathcal{H}$-matrix.

Keywords: Dynamic Core Binding · Dynamic Load Balancing · Power-aware · Hybrid Parallelization · Simulated Annealing · Lattice $\mathcal{H}$-matrix

1 Introduction

Load balancing is one of the most important issues in parallelizing applications for massively parallel processing. Load imbalance results in poor application performance and power wastage due to the long waiting for barrier synchronizations. Nevertheless, there are many algorithms for which load balancing is difficult, such as the particle-in-cell (PIC) method, lattice Boltzmann method,

© The Author(s), under exclusive license to Springer Nature Switzerland AG 2023
J. Mikyška et al. (Eds.): ICCS 2023, LNCS 14075, pp. 378–394, 2023.
https://doi.org/10.1007/978-3-031-36024-4_30

and hierarchical matrix ($\mathcal{H}$-matrix) method. In the case of the PIC and lattice Boltzmann methods, the amount of computation per unit area exhibits drastic variations with respect to time. In the H-matrix method, it is difficult to estimate the load bound to each process before creating the matrix. Moreover, hybrid programming by MPI (process) and OpenMP (thread) is widely used for parallelization, and the load must be equal at both the process and thread levels. Considering load balancing on multiple levels might be very complex work.

Dynamic load balancing is a well-known concept for reducing such load imbalance. Task parallelism emphasizes a parallelized nature in the applications, but it may cause poor performance due to the large overhead. To achieve good performance, a specialized approach for each application is used. The disadvantage of the specialized approach is required cumbersome jobs for optimization because of low portability. Therefore, we need an effective load balancing approach with low overhead, low implementation costs, and good portability.

In this study, we propose a dynamic core binding (DCB) approach that supports effective and simple load balancing. Whereas in the standard MPI/OpenMP hybrid parallelization environment, an identical number of cores is bound to all processes, in the DCB environment, the number of cores bound to each process is changed to absorb load imbalance among processes inside of a node. DCB can achieve improving the parallel performance of reducing power consumption by the different core binding policies. To improve the performance of the applications, it binds all cores to processes so that the loads per core are similar. Depending on the process with the maximum load, it reduces the cores mapped to the other processes; as a result, the energy consumption can be reduced without changing the computation time. This approach is suitable for many-core architectures, which have recently become more general. Load balancing by DCB has low overhead and high versatility as it only involves changing the number of bound cores. For supporting the DCB environment, an appropriate system call(not required administrator permission) is required. Then we implement the DCB library to black box this system call.

In addition, to maintain the effectiveness of the DCB approach, we consider the load imbalance among nodes. The DCB library only supports load balancing inside a node. Therefore, loads among nodes must be balanced using another method. The load balancing problem among nodes is similar to the job scheduling problem [12], which is a type of combinatorial optimization problem (COP). Job scheduling problems are categorized as nondeterministic polynomial-time complete (NP-complete) [12], and it is difficult to find the optimal solution to these problems with many jobs and machines. Studies on job scheduling problems discussed the use of simulated annealing (SA) to find appropriate approximate solutions [1,6] and have reported good results. Therefore, we also use the SA to absorb the load imbalance among nodes. In order to use the results obtained from the SA, the DCB library has a mechanism to generate a new MPI communicator, which is balanced loads among nodes.

The originality of this study is that it proposes a single approach (changing the cores bound to each process) to reduce the computation time and/or energy consumption of applications based on load balancing. In addition, we consider load imbalance among nodes using a metaheuristic approach to improve the effectiveness of DCB. Two existing studies [2,4] proposed ideas similar to ours. By changing the number of cores, one study [2] achieved performance improvement of hybrid parallelized applications, while the other study [4] reduced the energy consumption of applications parallelized using OpenMP. Our proposed approach integrates these studies [2,4] and also considers load balancing among nodes. Various studies achieved a reduction in the computation time of hybrid parallelized applications by dynamic load balancing. In one study [14], the imbalance among threads caused by communication hiding was solved by the task parallelization of OpenMP. Another study [5,11] proposed load balancing with task parallelization by original libraries in MPI/OpenMP or MPI-parallelized applications. Our proposed approach fundamentally differs from the aforementioned studies, and performance improvement can be achieved using a simple interface. Studies have also been conducted on power awareness [3,13,15] by throttling core and memory clocks to reduce energy consumption. In particular, one study [13] focused on applications parallelized with MPI/OpenMP. The difference between our proposed approach and these studies is that we achieve a reduction in energy consumption by changing the cores bound to each process; furthermore, our approach supports performance improvement and reduction in energy consumption simultaneously.

We discuss the effectiveness and usability of the DCB library based on a sample implementation and numerical evaluations with a practical application using a lattice $\mathcal{H}$-matrix [9,10]. The lattice $\mathcal{H}$-matrix method is utilized to approximate naive dense matrices and focuses on reducing communication overhead during parallel computation. However, this improvement increases load imbalance among the processes. Therefore, the lattice $\mathcal{H}$-matrix is a suitable target for evaluating the effectiveness of the DCB library.

2   Dynamic Core Binding (DCB)

In this section, we introduce a method for improving parallel performance and/or reducing energy consumption using DCB. In addition, we indicate an overview of the interfaces of the DCB library that provide the DCB approach and describe its use.

The basic concept of the DCB approach is to equalize loads among cores by changing the number of cores bound to each process. The target of the DCB is parallelized applications based on OpenMP/MPI hybrid. In the general hybrid parallelization environment, the number of cores bound to each process is identical. Balancing the loads at the core level leads to reducing computational time and/or energy consumption. Then, as shown in Fig. 1(a), if "Process 1" has a three times larger amount of computation than "Process 2", the allocated amount of computation to cores 1–4 is also three times larger than cores 5–8. By the basic concept of DCB, we expect to reduce the load imbalance among the cores.

In OpenMP, there is a function "*omp_set_num_threads*" for changing the number of threads. However, this function does not change the number of usable cores

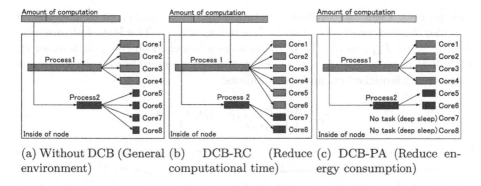

(a) Without DCB (General (b)   DCB-RC   (Reduce (c) DCB-PA (Reduce en-
environment)   computational time)   ergy consumption)

Fig. 1. Policies of dynamic core binding(DCB) library for core binding

for each process. When users increase the number of threads such that they exceed the number of cores, some threads are bound to the same core. When the number of threads is less than the number of cores, the threads can be moved among the cores by controlling the operating system. Changing the number of cores bound to each process requires a system call such as "$sched\_set\_affinity$" or alternative functions with unique arguments. The DCB library internally calls these system calls, allowing users to easily benefit from an environment with an unequal number of cores bound to each process. The permission of the system calls inside of the DCB library is not limited to the super-user, and the DCB library can be used with normal permissions.

The DCB library determines the number of bound cores based on a parameter received from the user. The DCB library implements two core-binding policies to reduce computational time and/or energy consumption. In Sect. 2.2, we introduce these two policies. The DCB library focuses on load balancing inside each node, and load imbalance among nodes must be addressed using a different approach. In Sect. 2.3, we describe the use of SA to solve load imbalance among nodes.

2.1   Concept of DCB

2.2   Binding Policy

In the DCB library, we implement two policies for improving parallel performance or reducing the energy consumption of applications. The policies of the DCB library change how the cores are bound to each process.

One policy of the DCB library focuses on reducing the computation time of the application (referred to as the RC policy). In the RC policy, all cores are bound to a process based on the load. Here we explain how to determine the number of cores bound to each process using the RC policy. The bound cores are determined by an argument l_p, which denotes the load executed by processes p, passed to the DCB library by the user. The optimal l_p depends on the application and is assumed to include the amount of computation and memory transfer for each process. Then, we consider a system whose nodes have c cores, and the IDs of nodes constructed the system is denoted as set N_a. The IDs of all processes denotes as set P_a, and the IDs of the processes launched on the node n as set P_n. The DCB library with the RC policy determines the number of cores c_p bound to process p launched on node n as follows:

$$c_p = \left\lfloor \frac{l_p}{\Sigma_{\tilde{p} \in P_n} l_{\tilde{p}}} \left(c - |P_n| \right) \right\rfloor + 1 \tag{1}$$

In this equation, the term $c - |P_n| + 1$ guarantees binding more than one core to all processes. If there are remaining cores ($c - \sum_{p \in P_n} c_p \neq 0$), the DCB library bounds one more core ($c_p = c_p + 1$) in order to the value $\beta = (l_p / \Sigma_{\tilde{p} \in P_n} l_{\tilde{p}}) c - c_p$ until there are no more cores remaining. When we apply the RC policy in the state Fig. 1(a), six cores are bound to "Process 1" and two cores are bound to "Process 2" as shown in Fig. 1(b). In this situation, the values of (1) are $l_1 = 3$, $l_2 = 1$, $c = 8$, $P_1 = 2$, respectively. The expected minimum computation time by using DCB is as follows:

$$t_{dcb} = t_{cmp} \left(\frac{l_m}{c_m} \right) / \left(\frac{\max L_n}{c} \right) + t_{oth}, \quad L_n = \left\{ \sum_{p \in P_n} l_p : n \in N_a \right\} \tag{2}$$

Here, t_{cmp} denotes the time of computations that are parallelized with OpenMP in the applications, and t_{oth} denotes the time of the other part of the application such as the communication and sequential computational part. m denotes the ID of the process that has the largest load in the node n which has the largest sum of load. In (2), t_{dcb} is calculated as the ratio of the maximum load allocated to the core with and without DCB on the node which has the largest sum of the loads. (2) also shows that the performance improvement of the DCB is capped by the load imbalance among the node. If the loads among the nodes are equalized, $maxL_n$ is minimized, and the effectiveness of DCB is maximized.

The second policy of the DCB library focuses on reducing energy consumption. In the PA policy, the DCB library reduces the number of cores bound to a process based on the largest load of the process. Figure 1(c) illustrates the result of applying the PA policy to the state Fig. 1(a). In this example, two cores are not mapped to any process, and their power status is changed to "deep sleep". The energy consumption of the deep sleeping cores is drastically reduced relative to cores in the running state. The number of bound cores with the PA policy is determined as follows:

$$c_p = \left\lceil \frac{l_p}{\max L_a} c \right\rceil, \quad L_a = \{ l_p : p \in P_a \} \tag{3}$$

The bound cores c_p by PA policy are simply decided as the ratio of l_p and maximum load in all processes. The expected reduction ratio ρ of the energy consumption is calculated as follows:

$$\rho = (J_g - J_{idle}) \frac{\sum_{p \in P_a} c_p}{c \times |N_a|} + J_{idle} \tag{4}$$

Here, J_g and J_{idle} denote the energy consumption of the application without DCB and the idle CPU state, respectively. The ratio ρ is expressed as the ratio of actually bound cores to all cores that construct the target system. In contrast to the RC policy, the effectiveness of DCB with the PA policy is not capped by the load imbalance among nodes.

2.3   Load Balancing Among Nodes

DCB only supports load balancing within a node. Load balancing among nodes must be maintained by another approach. Load balancing among nodes can be considered a massive COP; this COP is similar to a job scheduling problem and is categorized as NP-complete. Solving the COP with the existing libraries for classical computers takes a long time and is impractical. Then we consider solving the COP by using SA, which can be easy to solve. SA has a lot of experience in solving complex COPs, and various companies provide services that enable the use of SA. In order to solve the COP using SA, it is necessary to convert the COP into a quadratic unconstrained binary optimization (QUBO) and submit it to SA.

Then, a requirement and constraints of COPs for load balancing among the nodes are as follows:

Item.1 The loads among nodes are equalized
Item.2 Every process is mapped to one of the node
Item.3 Every node has more than one process

These requirements are expressed as the following Hamiltonian of the QUBO model:

$$H = \underbrace{\sum_{n=1}^{|N_a|} \left(\frac{\sum_{p=1}^{|P_a|} l_p}{|N_a|} - \sum_{p=1}^{|P_a|} l_p x_{p,n} \right)^2}_{Item.1} + \lambda \left\{ \underbrace{\sum_{p=1}^{|P_a|} \left(1 - \sum_{n=1}^{|N_a|} x_{p,n} \right)^2}_{Item.2} \right.$$

$$\left. + \underbrace{\sum_{n=1}^{|N_a|} \left(\sum_{s=0}^{\lceil \log_2 m \rceil} 2^s y_s - \sum_{p=1}^{|P_a|} x_{p,n} \right)^2}_{Item.3} \right\} \quad (5)$$

In this Hamiltonian, the first, second, and third terms express the Item.1, Item.2, and Item.3 of the requirements, respectively. Here, $x_{p,n}$ denotes the binary values appearing in the mapping. If $x_{p,n} = 1$, process p is mapped to node n. Also, binary values y_s denote sticky bits for expressing inequality of Item.3. λ denotes the penalty term for expressing the constraints. The objective of SA is looking for a combination of $x_{p,n}$ when Hamiltonian H becomes zero. (5) is constructed such that H is zero if all conditions are satisfied. We use pyQUBO [16], a Python module, for handling, and Fujitsu Digital Annealer (FDA) [8] to solve the QUBO model.

3   DCB Library

In this section, we introduce the implementation of and how to use the DCB library.

3.1   Interface

Important functions of the DCB libraries are two, *"DCB_init"* and *"DCB_balance"*. Users must call the *"DCB_init"* function before calling *"DCB_balance"*. *"DCB_balance"* is the main function, and the number of cores bound to each process is changed by referencing l_p that is passed as an argument of the function. After calling the *"DCB_balance"* function, users can use the DCB environment inside *"#pragma omp parallel"* regions. If the load of each process is expected to vary significantly from one OMP parallel region to another, calling *"DCB_balance"* can maintain a uniform amount of computation for each process in a node.

- DCB_MODE_POLICY = ("compute" or "power"): change the binding policy of the DCB
- DCB_DISABLE = (0 or 1): disable the DCB
- DCB_PROCESS_ON_DIFFERENT_SOCKETS = (0 or 1): map a process on a different socket

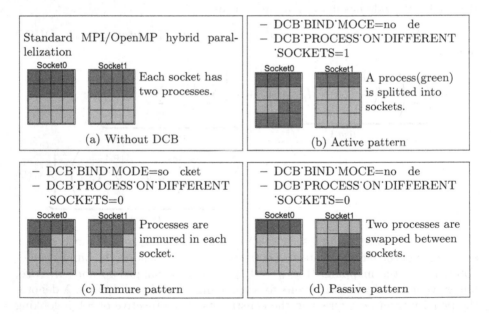

Fig. 2. Mapping pattern of cores on node controlled by numerical environment. Two sockets are on the node and four processes are mapped to the node. The loads l_p (passed parameters from the user) of each process are as follows: Process0:20, Process1:30, Process2:50, Process3:60

– DCB_BIND_MODE = ("node" or "socket"): Bounding processes to the whole node or a socket domain

DCB_MODE_POLICY is used to change the policies of the DCB library. If users set DCB_MODE_POLICY = compute or DCB_MODE_POLICY = power, the DCB library uses the RC or PA policy, respectively. DCB_DISABLE = 1 disables the DCB library. DCB_PROCESS_ON_DIFFERENT_SOCKETS and DCB_BIND_MODE are used to control the patterns of the core bounds on systems with the NUMA domain. Figure 2 presents the relationship between the numerical environments and mapping patterns on the NUMA domain. In this figure, there are two sockets and four processes mapped to a node. The four processes have 20, 30, 50, and 60 loads, respectively. Figure 2(b) presents the mapping pattern with the numerical environments DCB_BIND_MODE = node and DCB_PROCESS_ON_DIFFERENT_SOCKET = 1. This is the most active pattern for obtaining balanced loads among the cores. One process is mapped to different sockets depending on the process loads, as illustrated in Fig. 2(b). If users set BIND MODE = socket and DCB_PROCESS_ON_DIFFERENT_SOCKET = 0, the process is mapped as displayed in Fig. 2(c). With this pattern, the core bound to the process is immured to the inside of each socket. Figure 2(d) presents the mapping pattern with the numerical environments DCB_BIND_MODE = node and DCB_PROCESS_ON_DIFFERENT_SOCKET = 0. Although this pattern is similar to the most immure pattern (Fig. 2(c)), it permits moving the process among sockets.

3.2 Creation of Load-Balanced Communicator

In this section, we describe applying the approximate solution obtained by SA to the execution of the application. A common approach for changing the mapping of processes to nodes is to spawn processes and move them by internode communication. This approach requires many application-side implementation tasks and a large overhead. Here, we focus on applications where the operations allocated to each process are determined by the rank ID. In these applications, we achieve load balancing among nodes by creating a new MPI communicator that has a new rank ID derived from the solution of the COP. Hereafter, this communicator is referred to as a load-balanced communicator. In the DCB library, load balancing among nodes is achieved by process spawning using the usual MPI functions and the result of the COP. Based on the result of the COP, MPI_Comm_spawn_multiple is called inside the DCB library to spawn a different number of additional processes on each node. There are unused processes created when launching the program; thus, these processes wait at a barrier synchronization until the end of the program. In addition, one core is allocated as if it were a process not used for the operation, and all processes are allocated to that core. The core bound to unused processes is not mapped to the used processes for computation. Within the DCB library, a new communicator containing all the processes used for computation is generated and returned to the user.

4   Numerical Evaluations

In this section, we demonstrate the effectiveness of DCB using numerical evaluations.

4.1   Target Application

In this section, we introduce the target application and its load imbalance.

We consider a lattice $\mathcal{H}$-matrix-vector multiplication (LHMVM) using the MPI/OpenMP hybrid programming model for distributed memory systems. Its efficient communication patterns ensure that communication costs remain constant even as the number of MPI processes increases. The parallel scalability of the LHMVM should be significantly improved if the DCB resolves the load imbalance among MPI processes.

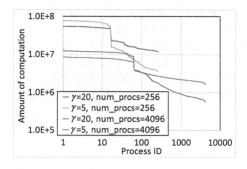
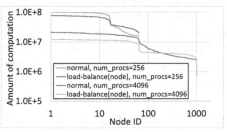

Fig. 3. load imbalance among processes (problem = 1188kp25)

Fig. 4. Result of applying obtained result of simulated annealing(SA) (problem = 1188kp25, parameter $\gamma = 5$)

The load imbalance of the lattice $\mathcal{H}$-matrix depends on the number of lattice blocks into which the target dense matrix is divided. Basically, the large number of lattice blocks leads the better load balancing. However, the reduction of memory usage and amount of computation due to approximation becomes smaller. Figure 3 illustrates the amount of computation bound to each process in 256 and 4,096 processes parallelization with a particular model(1188kp25), which is approximated by the lattice $\mathcal{H}$-matrix. A parameter γ in Fig. 3 is used to control the number of lattice blocks. The amount of computation with $\gamma = 5$ is smaller than $\gamma = 20$, but the load imbalance of the $\gamma = 5$ among the node is larger than $\gamma = 20$. This trend is observed regardless of the number of processes. Therefore, the advantage of the lattice $\mathcal{H}$-matrix can be maximized if the load imbalance is absorbed by approximating with smaller γ.

The evaluation was performed by 50 LHMVMs. To use the DCB library in LHMVM, we use the amount of computation executed by each process as the argument l_p. Target problems utilizing the lattice $\mathcal{H}$-matrix are "1188kp25" and "human-1x100" (models for the surface charge method of an electromagnetic simulation)

4.2 Environments and Conditions

For the numerical evaluations, we used the Oakbridge-CX (OBCX) and Wisteria/BDEC-01 Odyssey (WO) systems at the Information Technology Center at the University of Tokyo. Tables 1 and 2 present the system specifications and compiler information, respectively. On the OBCX, the turbo boost technology is enabled.

In the numerical evaluations, we set four processes per node to bring out the effectiveness of DCB. This condition is the same in the evaluation without the DCB library. In advance evaluations of the original LHMVM implementation, we have confirmed that the performance of four processes per node is better than that of one process per node. We have confirmed that the performance of FlatMPI is slightly better than that of four processes per node, but we do not use flatMPI. This is because LHMVM has a sequential execution part and it requires a large amount of memory when handling large problems.

To measure the energy consumption of the application, we used the "power-stat" command on OBCX and the PowerAPI [7] library on WO. The result of energy consumption shows a sum of CPUs and memory in the nodes in OBCX,

Table 1. System specifications

Specifications		Oakbridge-CX	Wisteria/BDEC-01 Odyssey
CPU	Model	Xeon Platinum 8280 (Cascade Lake)	A64FX
	Number of cores	56 (2 Sockets)	48
	Clock	2.7 GHz	2.2 GHz
	L2-cache	1 kB/core	8 MB/CMG
Memory	Technology	DDR4	HBM2
	Size	192 GB	32 GB
	Bandwidth	281.6 GB/s	1,024 B/sec
Network	Interconnect	Omni-Path	Tofu Interconnect D
	Topology	Full-bisection Fat Tree	6D mesh/Torus
	Bandwidth	100 Gbps	56 Gbps

Table 2. Compiler and options on each system

Oakbridge-CX	mpiifort	2021.5.0	- xHost -O3 -ip -qopt-zmm-usage = high
Wisteria/BDEC-01 Odyssey	mpifrtpx	4.8.0 tcsds-1.2.35	- O3 -Kfast, lto, openmp, zfill, A64FX - KARMV8_A, ocl, noalias = s

and the sum of all modules in the nodes in WO. In the WO system, before measuring the energy consumption and computation time, we enabled the retention feature of every core. The retention feature provides a type of deep sleep in an A64FX CPU.

4.3   Effectiveness of the Load-Balancing Among Nodes Using SA

We evaluated the effect of load balancing among nodes by applying the approximate solution of SA. Figure 4 illustrates the effectiveness of load balancing from the approximate solution in the 1188kp25 problem with 256 and 4,096 processes and parameter $\gamma = 5$. By applying the approximate solution of SA, the maximum load was reduced for each condition. We observed similar effectiveness for the other conditions and problems.

When applying the approximate solution of SA, the method of generating the load-balanced communicators described in Sect. 3.2 was used. Then, we executed LHMVM with one process per node initially, and spawned the required processes when creating the load-balanced communicator. In the WO system, the Fujitsu MPI does not support a non-uniform number of processes spawning. Thus, we spawned the same number of processes on all nodes corresponding to the node requiring the most number of processes.

4.4   Performance Improvement

Figures 5 and 6 display the computation time of 50 LHMVMs with and without the DCB library, including the results using the load-balanced communicator (denoted as "RC (node-balanced)" in the figures). The core-binding pattern used for execution with the load-balanced communicator was the active pattern (Fig. 2(b)). The figures also display the estimated computation time with the DCB library (2) and the best cases (there is no load imbalances among nodes). As the computation node of OBCX is a two-socket NUMA domain, we evaluated all mapping patterns of the cores on the node to the processes in Fig. 2. For WO, we only evaluated the active patterns. For all problems and parameters on all systems, the DCB library improved the computation performance. The performance improvement of DCB tended to be larger at smaller-parallelism and smaller at larger-parallelism. This is because the performance in the high-parallel condition was degraded by the load imbalance among nodes. When using the load-balanced communicator, further performance improvement was obtained on OBCX. The effectiveness of the load-balanced communicator was maximum at 64 processes, and the computation time with the load-balanced communicator was close to the estimated time on OBCX. By using the DCB library without the load-balanced communicator, we achieved a performance improvement of 31%–52% for 16 processes and 9%–31% for 1,024 processes. With the load-balanced communicator, we achieved a performance improvement of 39%–58% compared with the result without DCB. Compared with using DCB (active), we achieved a 33%–50% performance improvement for 256 processes. However, on WO, we could not achieve performance improvement using the load-balanced communicator with high-parallel conditions. This was due to the increase in communication overhead, as illustrated in Fig. 7. The communication pattern among the processes became complex when applying the load-balanced communicator. The communication pattern among processes was not considered when we constructed the COP of load balancing among nodes. In particular, the network topology of WO is a six-dimensional torus, and the communication overhead is large in complex communication patterns.

We also examined the effect of the DCB library on the parameter γ of the lattice $\mathcal{H}$-matrix. Without the DCB library, the computation time with $\gamma = 5$ was longer than that with $\gamma = 20$ in spite of lower memory usage and amount of computation. This was due to load imbalance, as displayed in Fig. 3. In contrast, using the DCB library, the computation time with $\gamma = 5$ was shorter than or similar to that with $\gamma = 20$. This indicates that the use of the DCB library improved the benefits of the lattice $\mathcal{H}$-matrix.

The computation time of the PA policy was expected to be similar to that in conditions without the DCB library. In OBCX, the computation time of the PA policy was slightly lower than that without DCB. This was because the number of cores used for each node was reduced. This created a margin for memory access, and resources for accessing memory by processes that had a large load were obtained. In WO, the computation time of the PA policy was similar to that without DCB, as we expected.

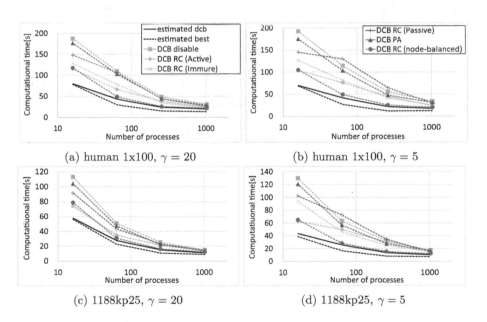

(a) human 1x100, $\gamma = 20$ (b) human 1x100, $\gamma = 5$

(c) 1188kp25, $\gamma = 20$ (d) 1188kp25, $\gamma = 5$

Fig. 5. Computational time of LHMVM with DCB on Oakbridge-CX

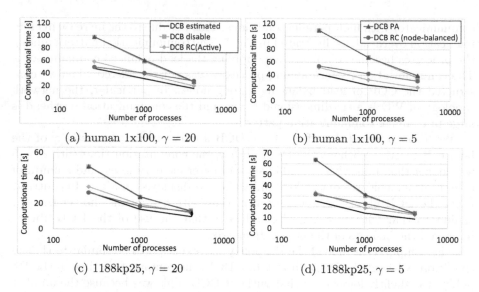

(a) human 1x100, $\gamma = 20$ (b) human 1x100, $\gamma = 5$

(c) 1188kp25, $\gamma = 20$ (d) 1188kp25, $\gamma = 5$

Fig. 6. Computation time of LHMVM with DCB on Wisteria/BDEC-01 Odyssey

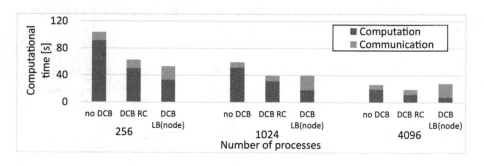

Fig. 7. Breakdown of execution time in WO (1188kp25, $\gamma = 20$)

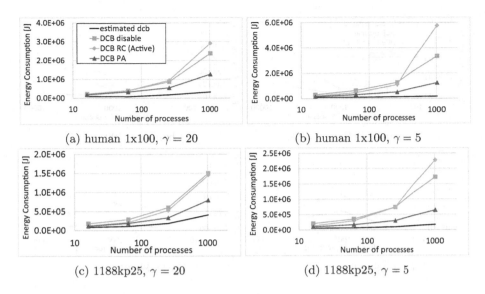

(a) human 1x100, $\gamma = 20$

(b) human 1x100, $\gamma = 5$

(c) 1188kp25, $\gamma = 20$

(d) 1188kp25, $\gamma = 5$

Fig. 8. Energy consumption of LHMVM with DCB on Oakbridge-CX

4.5   Reducing Energy Consumption

This section evaluates the effectiveness of the PA policy in reducing energy consumption. Figures 8 and 9 display the energy consumption for the number of processes in each system. A reduction in energy consumption using the PA policy of the DCB library was observed in all conditions. In OBCX, the energy consumption tended to decrease with high parallelism. This was because the load imbalance among processes increased with high parallelism, and many cores were not used for computation due to the DCB library. In OBCX, we achieved a 47%–63% reduction in energy consumption with 1,024 processes. In WO, we achieved a 61%–77% reduction in energy consumption with 256 processes and a 46%–56% reduction in energy consumption with 4,096 processes.

The energy consumption of the RC policy of the DCB library was higher than that without the DCB even though the computation time was shorter in OBCX. We are assumed to be due to the turbo boost. That is also why the energy consumption with the PA policy was higher than estimated. In WO, the energy consumption was higher than estimated because the energy consumption of the network modules and assistant cores was not considered.

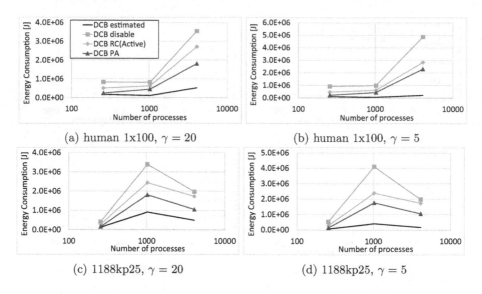

(a) human 1x100, $\gamma = 20$ (b) human 1x100, $\gamma = 5$

(c) 1188kp25, $\gamma = 20$ (d) 1188kp25, $\gamma = 5$

Fig. 9. Energy consumption of LHMVM with DCB on Wisteria/BDEC-01 Odyssey

5 Conclusion

In this paper, we propose the DCB approach to reduce load imbalance among processes. By this approach, we can expect to reduce the computation time and/or energy consumption for applications for which load balancing is difficult. In the DCB environment, the load imbalance among processes is rectified by changing the number of cores bound to each process to equalize the loads of the cores. We also use SA to consider load balancing among nodes, which cannot be achieved using DCB. The results of applying DCB to the lattice $\mathcal{H}$-matrix demonstrate a reduction in computation time of more than 50% and a reduction in energy consumption of more than 70% for OBCX and WO.

In this study, core binding to processes was determined based on the parameters received from the user in the DCB library. In addition, the number of using cores is based on the ideal condition that the bottleneck is only load balancing. In practice, it is necessary to consider the computer architecture and NUMA. However, it is difficult to create a realistic model that takes into account the complexity of the computer architecture and the characteristics of the applications. Therefore, in the future, we will study an algorithm for automatically determining core binding based on a CPU performance counter and other information. The load of each process with the lattice $\mathcal{H}$-matrix was determined when the matrix was generated. There are many applications in which the process loads are changed dynamically in runtime, and we will consider applying DCB to these applications. Then, the small overhead of changing the core binding is important to minimize; therefore, we will attempt to reduce this overhead by examining system calls and other aspects.

For load balancing among nodes, we did not consider the communication patterns among processes. Therefore, we could not achieve performance improvement using the load-balanced communicator. In future work, we will consider the communication patterns among processes and the network topology of systems to improve the effectiveness of DCB. We will also investigate an approach for automatically changing the process mapping to nodes to support dynamic load balancing. For this investigation, we will refer to research on fault tolerance.

Acknowledgment. This work was supported by JSPS KAKENHI Grant Number 18K18059, 21H03447, and 19H05662. This work is also supported by "Joint Usage/Research Center for Interdisciplinary Large-scale Information Infrastructures (JHPCN)" in Japan (Project ID: jh230058).

References

1. Attiya, I., et al.: Job scheduling in cloud computing using a modified Harris hawks optimization and simulated annealing algorithm. Comput. Intell. Neurosci. **2020** (2020)
2. Corbalan, J., et al.: Dynamic load balancing of MPI+OpenMP applications. In: 2004 International Conference on Parallel Processing, ICPP 2004, vol. 1, pp. 195–202 (2004)
3. Curtis-Maury, M., et al.: Online power-performance adaptation of multithreaded programs using hardware event-based prediction. In: Proceedings of the 20th Annual International Conference on Supercomputing, pp. 157–166 (2006)
4. Curtis-Maury, M., et al.: Prediction-based power-performance adaptation of multithreaded scientific codes. IEEE Trans. Parallel Distrib. Syst. **19**(10), 1396–1410 (2008)
5. Garcia, M., Corbalan, J., Badia, R.M., Labarta, J.: A dynamic load balancing approach with SMPSuperscalar and MPI. In: Keller, R., Kramer, D., Weiss, J.-P. (eds.) Facing the Multicore - Challenge II. LNCS, vol. 7174, pp. 10–23. Springer, Heidelberg (2012). https://doi.org/10.1007/978-3-642-30397-5_2
6. Garza-Santisteban, F., et al.: A simulated annealing hyper-heuristic for job shop scheduling problems. In: 2019 IEEE Congress on Evolutionary Computation (CEC), pp. 57–64 (2019)
7. Grant, R.E., et al.: Standardizing power monitoring and control at exascale. Computer **49**(10), 38–46 (2016)
8. Hiroshi, N., et al.: Third generation digital annealer technology (2021). https://www.fujitsu.com/jp/documents/digitalannealer/researcharticles/DA_WP_EN_20210922.pdf
9. Ida, A.: Lattice $\mathcal{H}$-matrices on distributed-memory systems. In: IEEE International Parallel and Distributed Processing Symposium (IPDPS), pp. 389–398 (2018)
10. Iwashita, T., et al.: Software framework for parallel BEM analyses with H-matrices using MPI and OpenMP. Procedia Comput. Sci. **108**, 2200–2209 (2017)
11. Klinkenberg, J., et al.: CHAMELEON: reactive load balancing for hybrid MPI+OpenMP task-parallel applications. J. Parallel Distrib. Comput. **138**, 55–64 (2020)
12. Korte, B.H., et al.: J. Comb. Optim. **1** (2011)
13. Li, D., et al.: Strategies for energy-efficient resource management of hybrid programming models. IEEE Trans. Parallel Distrib. Syst. **24**(1), 144–157 (2012)

14. Nakajima, K., et al.: Communication-computation overlapping with dynamic loop scheduling for preconditioned parallel iterative solvers on multicore and manycore clusters. In: 2017 46th International Conference on Parallel Processing Workshops (ICPPW), pp. 210–219 (2017)
15. Suleman, M.A., et al.: Feedback-driven threading: power-efficient and high-performance execution of multi-threaded workloads on CMPs. ACM SIGPLAN Not. **43**(3), 277–286 (2008)
16. Zaman, M., et al.: PyQUBO: Python library for mapping combinatorial optimization problems to QUBO form. IEEE Trans. Comput. **71**(4), 838–850 (2022)

Surrogate-Assisted Ship Route Optimisation

Roman Dębski[✉][ID] and Rafał Dreżewski[ID]

Institute of Computer Science, AGH University of Science and Technology,
Al. Mickiewicza 30, 30-059 Kraków, Poland
{rdebski,drezew}@agh.edu.pl

Abstract. A new surrogate-assisted, pruned dynamic programming-based optimal path search algorithm—studied in the context of *ship weather routing*—is shown to be both effective and (energy) efficient. The key elements in achieving this—the fast and accurate physics-based surrogate model, the pruned simulation, and the OpenCL-based SPMD-parallelisation of the algorithm—are presented in detail. The included results show the high accuracy of the surrogate model (relative approximation error medians smaller than 0.2%), its efficacy in terms of computing time reduction resulting from pruning (from 43 to 60 times), and the notable speedup of the parallel algorithm (up to 9.4). Combining these effects gives up to 565 times faster execution. The proposed approach can also be applied to other domains. It can be considered as a dynamic programming based, optimal path planning framework parameterised by a problem specific (potentially variable-fidelity) cost-function evaluator.

Keywords: simulation-based optimisation · surrogate model · ship weather routing · optimal path planning · heterogeneous computing

1 Introduction

In international trade, approximately 80% of goods are transported by sea and this number is likely to increase [25]. This is why, in 2018, the International Maritime Organisation approved an agreement to reduce carbon emissions (from fuel consumption) per transport unit by 40% by 2030 as compared with 2008. As a result, fuel consumption reduction—one of the objectives in most ship route optimisation tasks—and optimal navigation itself have become more important than ever [7,19,23].

A model of fuel consumption usually assumes its dependence on the ship propulsion system and hull characteristics, the sea state, and the ship speed and its heading angle against the waves [8,13,30]. Any application of such a complex model in an optimisation task in most cases leads to a simulation-based algorithm, which in many instances can be computationally expensive. This issue is of particular importance in the context of algorithms based on dynamic programming, mostly due to search space size.

© The Author(s), under exclusive license to Springer Nature Switzerland AG 2023
J. Mikyška et al. (Eds.): ICCS 2023, LNCS 14075, pp. 395–409, 2023.
https://doi.org/10.1007/978-3-031-36024-4_31

One way to address this is to parallelise the algorithm. In many scenarios, however, it is at most a partial solution to the problem. This can be due to target computer system constraints, the intrinsic strong sequential component of the algorithm, and/or the cost of a single simulation, which is often by far the most computationally expensive part of the optimisation process.

Another option is to simplify the model (often significantly) so that the computational cost of a single simulation is acceptable, or even no simulation is required at all. Two examples of such simplifications are the assumption of a constant ship speed along a single segment of a route [19, 26] and the use of the ship speed as the control input/signal [27]. But to ignore completely the ship acceleration may lead to inaccurate results, especially from the fuel consumption perspective. This approach also does not generalise to more complex, often multi-objective ship route optimisation tasks, which require simulators based on ship dynamics.

A surrogate-assisted algorithm could potentially address all the previously mentioned issues [11, 12]. But, to the best of our knowledge, surrogate-assisted ship route optimisation based on dynamic programming has not been studied yet. Our aim is to present such an algorithm.

The main contributions of the paper are:

1. effective, physics-based surrogate model of ship motion (with fuel consumption included) defined in the spatial-domain rather than the time domain, in which the ODE-based (hi-fidelity) model is set (Sect. 4.3),
2. admissible heuristic function used to accelerate (by search space pruning) both the surrogate-based estimation and hi-fidelity model based simulation (Sect. 4.2),
3. surrogate-assisted, SPMD-parallel, dynamic programming based ship route optimisation algorithm, which incorporates a refinement of the search space (Sect. 4.4),
4. results which demonstrate three important aspects of the algorithm: the accuracy of the surrogate-model, the pruning-related speedup, and the parallelisation capabilities (Sect. 5).

The remainder of this paper is organised as follows. The next section presents related research. Following that, the optimisation problem under consideration is defined, and the proposed algorithm is described. After that, experimental results are presented and discussed. The last section contains the conclusion of the study.

2 Related Research

Ship safety navigation and energy efficiency are crucial factors for enhancing the competitiveness and sustainability of ship operations [7]. Consequently, an increasing number of researchers work on ship route optimisation based on factors such as weather forecasts and sea conditions, which will significantly impact the ship velocity, primary engine output, fuel oil consumption (FOC), and emissions [6, 7, 19, 23]. Typically, the ship route optimisation problem is reduced to a sub-path search problem in two dimensions, taking into account only the position and assuming that either the ship velocity or primary engine output remains constant [7,27]. The methods applied to solve such simplified ship weather routing problem include [7]: the modified isochrone method [9], the isopone method for the minimum-time problem [20], dynamic programming for ship route optimisation with minimising FOC [2], and Dijkstra's algorithm taking into account the weather forecast data [18]. Also, evolutionary algorithms were applied to find the optimal ship route [22]. Nonetheless, in reality the long routes and severe fluctuations in weather and sea conditions make it impossible to maintain a constant sailing velocity and primary engine output for ships crossing the ocean [7].

The quest for a more realistic simulation of fuel consumption has led to the proposition of a number of three-dimensional optimisation algorithms. In [16] a real-coded genetic algorithm was used for weather routing optimisation taking into account the weighted criteria of fuel efficiency and ship safety. An alternative strategy is to integrate the time factor into the existing two-dimensional path search algorithm and solve the three-dimensional path search optimisation problem. This can be accomplished, for example, by adapting algorithms like the three-dimensional isochrones method with weighting factors [14].

Three-dimensional dynamic programming algorithms taking into account meteorological factors were also proposed by researchers. A novel three-dimensional dynamic programming based method for ship weather routing, minimising ship fuel consumption during a voyage was presented in [19]. In contrast to the approaches mentioned above, which modify only the ship heading while keeping the engine output or propeller rotation speed constant, the method proposed in [19] took into account both factors: the engine power and heading settings. The ship heading and speed planning taking into account the weather conditions (and some additional constraints and objectives) was realised using an improved three-dimensional dynamic programming algorithm for ship route optimisation [7]. The model of ship route estimation based on weather conditions, and additional constraints like the main engine rated power and ship navigation safety was proposed in [6]. The model could analyse multiple routes taking into account the indicators such as travel speed, fuel consumption, estimated time of arrival (ETA), and carbon emissions.

The selected machine learning models were also applied to the ship weather routing problem. The proposed approaches included training neural networks and machine learning models using available ship navigation data for different weather conditions. Such trained models were then used to predict FOC and optimise the ship route, taking into account fuel consumption and time of arrival [31]. Statistical models taking into account wave height, wave period, wind speed, and main engine's RPM data [17] were used to plan ship routes while maintaining safety, on-time arrival and reducing fuel consumption.

The approaches based on multi-objective optimisation to solving the ship route optimisation problem taking into account weather conditions were also proposed. A new algorithm for solving the two-objective problem (minimising the fuel consumption and the total risk) formulated as a nonlinear integer programming problem was introduced in [26]. The authors considered time-dependent point-to-point shortest path problem with constraints including constant nominal ship speed and the total travel time.

Multi-objective evolutionary algorithms were also applied in several works to solve the ship weather routing problem, taking into account several objectives like estimated time of arrival, fuel consumption and safety [10,21,22]. The preference-based multi-objective evolutionary algorithm with weight intervals for ship weather routing optimisation was introduced in [23]. In the problem considered, the route consisted of control points. Each of them stored the information about its location, estimated time of arrival (based on the previous part of the route, weather forecast in the immediate vicinity, and speed of the ship), engine settings, and ship heading.

Computer simulations are now extensively employed to validate engineering designs, to refine the parameters of created systems and in the optimal search problems. Unfortunately, accurate (high-fidelity) simulation models for such tasks tend to be computationally expensive, making them often difficult or even unfeasible to apply in practice. When dealing with such situations, methods that employ surrogates, simplified (low-fidelity) simulation models, are utilised. Although these models are less complex than the original simulation model of a system or process, they accurately represent it and are significantly more efficient in terms of computation [11,12].

Broadly speaking, surrogate simulation models can be classified into two main categories. The first type, called *approximation-based models*, involves constructing a function approximation using data obtained from high-fidelity simulation models that are precise. The second type, known as *physics-based models*, involves building surrogates based on simplified physical models of systems or processes [12].

The approximation-based models are usually developed with the use of radial basis functions, Kriging, polynomial response surfaces, artificial neural networks, support vector regression, Gaussian process regression, or multidimensional rational approximation [12,28]. Moreover, recently deep-learning based surrogate models were proposed, for example, for characterisation of buried objects using Ground Penetrating Radar [29], design optimisation procedure of a Frequency Selective Surface based filtering antenna [15], and modelling of microwave transistors [3].

Physics-based surrogates typically incorporate simplified knowledge of the system or processes, as they are based on low-fidelity models. As a result, they usually only need a few high-fidelity simulations to be properly configured and provide reliable results [12]. Physics-based surrogates possess inherent knowledge about the simulated system or process, which gives them strong generalisation abilities. Consequently, they are capable of producing high-quality predictions of the accurate simulation model, even for system designs or configurations that were not used during the training phase [12].

It appears that the surrogate-based approach can help solve the problems that arise when using high-accuracy simulation models in the problem of optimising the ship's route when weather conditions are taken into account. These problems have so far limited the ability to use accurate simulation models, and to the best of the authors' knowledge, the surrogate-based approach has not been used so far in the case of the ship weather routing problem.

Contrary to the above-mentioned approaches to the ship weather routing problem, in this paper we propose the physics-based surrogate model of ship motion, with fuel consumption included, defined in the spatial-domain. Furthermore, the admissible heuristic function for search space pruning (used to accelerate both the surrogate-based and hi-fidelity simulation model) is proposed. Finally, surrogate-assisted, SPMD-parallel, dynamic programming based ship weather routing algorithm, which incorporates a refinement of the search space, is introduced.

3   Problem Formulation

Consider a ship sailing in the given area—S_A, from point A to B, along the path/route:

$$\widetilde{AB}(t) = \left(x_1^{(AB)}(t), x_2^{(AB)}(t) \right)^T \tag{1}$$

resulting from a specific control input:

$$\mathbf{u}(t) = \mathbf{u}^{(AB)}(t) = (c(t), n(t))^T, \tag{2}$$

where: c denotes the ship course and n - the propeller speed of rotation (RPM). Given the sea state[1] at time t_k (see Fig. 1), we can express the ship dynamics in the following way:

$$\dot{\mathbf{x}}(t) = \mathbf{a}\left(\mathbf{x}(t), \mathbf{u}(t), t_k\right) = \mathbf{a}^{(k)}\left(\mathbf{x}(t), \mathbf{u}(t)\right) \tag{3}$$

with $\mathbf{x}(t) = (x_1(t), x_2(t), \dot{x}_1(t), \dot{x}_2(t))^T$ denoting the (ship) state vector.

The performance of the ship is evaluated in the following way:

$$J_s = C_1\, t_f + C_2 \int_0^{t_f} f_{cr}\left(\mathbf{x}(t), \mathbf{u}(t)\right) dt \tag{4}$$

where: t_f is the sailing duration (the final time), f_{cr} - the ship fuel consumption rate function, and C_1, C_2 are user-defined constants.

Problem Statement. The route optimisation problem under consideration can be defined as follows[2]: *find an admissible control* $\mathbf{u}^*$ *which causes the ship to follow an admissible trajectory* $\mathbf{x}^*$ *that minimises the performance measure* J_s.

Remark 1. We assume that the values of J_s can be found only through simulation and only on-board, off-line computers can be used.

---

[1] either explicitly or derived (computed) from the wind vector field.
[2] $\mathbf{u}^*$ denotes an optimal control and $\mathbf{x}^*$ an optimal trajectory (they may not be *unique*).

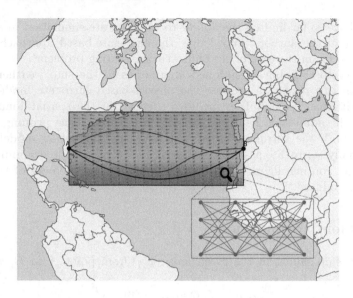

Fig. 1. Conceptual diagram of the ship route optimisation problem under consideration (source of the North Atlantic Ocean map: [24]).

4   Proposed Solution

The approach we propose in this paper is based on the following two main steps:

1. transformation of the continuous optimisation problem into a (discrete) search problem over a specially constructed graph;
2. application of surrogate-assisted, pruned dynamic programming to find the approximation of the optimal control input sequence:

$$(\mathbf{u}_k)_{k=1}^N = \left([c_1, n_1]^T, [c_2, n_2]^T, \ldots, [c_N, n_N]^T\right). \tag{5}$$

Remark 2. The sequence $(c_k)_{k=1}^N$ represents a *continuous piecewise-linear* sailing path[3], whilst $(n_k)_{k=1}^N$ is the sequence of the propeller RPMs, which forms a *piecewise-constant* function; in both cases $t \in [0, t_f]$.

Remark 3. The above two steps repeated several times form an adaptive version of the algorithm in which subsequent search spaces are generated through mesh refinement making use of the best solution found so far.

The key elements of the proposed algorithm, i.e.:

1. *3D-graph* based solution space and the SPMD-parallel computational topology it generates,
2. pruning-accelerated simulation and estimation,
3. fast and accurate estimator of the performance measure (surrogate model)

are discussed in the following sub-sections.

---

[3] formed by the sequence of the graph edges that correspond to (c_k).

4.1   The Solution Space Representation

Discretisation of the original problem domain may be seen as a two-phase process. In the first phase, we construct a multi-stage graph (G_2) that creates the *'route space'* (i.e., c_k-space/spatial-dimension of $(\mathbf{u}_k)$, see Remark 2). In most instances, this graph is regular with equidistant nodes grouped in rows and columns: n_c columns and $(n_r - 2)$ regular rows, plus two special (single-node) rows—one with point A and the other with point B (see Fig. 2).

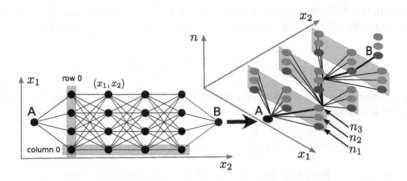

Fig. 2. Solution space representation.

In the second phase, each node of G_2 (apart from the one corresponding to point A) is replicated n_p times to add the n_k-dimension to the solution space (of control inputs), which can be seen as an addition of the 'third dimension' to G_2. This new, *'3D-graph'* (G_3):

- has $n_p \left[n_c(n_r - 2) + 1 \right] + 1$ nodes,
- has $n_c n_p \left[(n_r - 3)n_c n_p + n_p + 1 \right]$ edges,
- represents $n_p(n_c n_p)^{n_r - 2}$ different control input sequences.

Remark 4. The piecewise-linear approximation of the ship route simplifies the problem significantly. Instead of one complex, continuous two-dimensional problem, we have a series of simple one-dimensional sub-problems—each corresponding to a single route segment only.

4.2   Pruning in Simulation and Estimation

The evaluation of a single route segment, s, can be aborted as soon as we know that the computation cannot lead to a better solution than the best one found so far. This process can be based on any effective *admissible heuristic*[4].

Such a heuristic can be derived in the following way (see Eq. 4):

$$J_s \leq J_s^{(adm)} = C_1 \frac{|s|}{v_{max}} + C_2 \, f_{cr}^{(min)} \frac{|s|}{v_{max}} = \left(C_1 + C_2 \, f_{cr}^{(min)} \right) \frac{|s|}{v_{max}} \qquad (6)$$

---

[4] an *admissible heuristic* never overestimates the actual value of performance measure.

where: $f_{cr}^{(min)}$ denotes the *minimum fuel consumption rate* and v_{max} is the *maximum possible speed* of the ship.

Remark 5. The values of $f_{cr}^{(min)}$ and v_{max} can be either taken from the ship characteristics or approximated during the segment estimation phase.

4.3 Surrogate-Based Performance Measure Estimator

The proposed estimator of the performance measure/cost of a single, straight route segment AB (corresponding to one edge of G_2) is an extended version of the one introduced in [4,5], and based on the *work-energy principle*:

$$\int_A^B \boldsymbol{F} \cdot d\boldsymbol{r} = \frac{1}{2}\, m(v_B^2 - v_A^2),$$

This formula can be used to transform the original problem from *time domain* to *spatial domain*, i.e.,

$$m\frac{dv}{dt} = mv\frac{dv}{ds} = F \rightarrow mvdv = Fds = dW.$$

From this, we can find the distribution of the velocity along the sailing line:

$$\begin{cases} s_0 = s_A, \\ v_0 = v_A, \\ v_i^2 = v_{i-1}^2 + \frac{2}{m}F(s_{i-1}, v_{i-1})\Delta s_i \end{cases} \tag{7}$$

where: $i = 1, 2, \ldots, i_B$, $\Delta s_i = s_i - s_{i-1}$, and $s_{i_B} = s_B$. Knowing $(v_i)_{i=1}^{i_B}$ and assuming a constant value of F in each sub-interval, we can find the sailing duration and fuel consumption (see Appendix A and [4,5]).

Remark 6. Operating in the *spatial domain* is the key property of this estimator because it takes a *predetermined number of steps*, N_s, stemming from the spatial discretisation of a route segment. In the original, hi-fidelity model—given in the time domain—the number of time-steps to be taken by the simulator (i.e., the ODE solver) to reach the final point is unknown upfront. In some cases, it can be several orders of magnitude larger than N_s.

4.4 The Algorithm

Graph G_3, representing the solution space, is directed, acyclic (DAG) and has a layered structure[5]. Since, at the beginning of the search process, the cost of each route segment is unknown, it has to be obtained from simulation. The cost of reaching each node of G_3 can be computed using the *Principle of Optimality* [1]. It can be expressed for an example path A-$N_{r,c,n}$ in the following way:

$$J_s^*(A \rightarrow N_{r,c,n}) = \min_{c_j, n_k} \left\{ J^*(A \rightarrow N_{r-1,c_j,n_k}) + J(N_{r-1,c_j,n_k} \rightarrow N_{r,c,n}) \right\} \tag{8}$$

---

[5] therefore, it is a multistage graph.

where: $c_j = 0, \ldots, (n_c - 1)$, $n_k = 0, \ldots, (n_p - 1)$, $J_s (N_s \rightarrow N_e)$ is the cost corresponding to the path $N_s \rightarrow N_e$ (N_s—the start node, N_e—the end node), J_s^* represents the optimal value of J_s, and $N_{r,c,s}$ is the node of G_3 with 'graph coordinates' $(row, column, RPM) = (r, c, n)$.

The structure of the computation (i.e., simulation flow) is reflected in the SPMD-structure of Algorithm 1 (see annotation **@parallel**). The computation begins from point A in layer 1, taking into account the corresponding initial conditions, and is continued (layer by layer) for the nodes in subsequent rows. On the completion of the simulations for the last layer (i.e., reaching the end node B), we get the optimal route and its cost.

Algorithm 1: Surrogate-assisted optimal route search

Input:
- $G_3^{(AB)}$: initial search space with start point A and target point B,
- S_{st}: sea state,
- HFM: ship simulator (high-fidelity model),
- LFM: surrogate (low-fidelity model).

Output: the optimal route and RPMs

1 **foreach** *refinement r_G of $G_3^{(AB)}$* **do**
2   **foreach** *layer l_{r_G} in r_G* **do**
3   **@parallel foreach** *entry point e_p of l_{r_G}* **do**
4           $(s_k)_{k=1}^d \leftarrow$ all segments ending in e_p // d - the in-degree of e_p
            `/* 1. pruned estimation using LFM, `$d_1 \leq d$`, Sect. 4.3      */`
5           $(s_k, \mathrm{cost}_k^{est})_{k=1}^{d_1} \leftarrow$ pruned estimation of $(s_k)_{k=1}^d$
            `/* 2. pruned simulation using HFM, see Appendix A       */`
6           $(s_k, \mathrm{cost}_k^{sim})_{k=1}^{d_1} \leftarrow$ pruned simulation of $(s_k, \mathrm{cost}_k^{est})_{k=1}^{d_1}$
            `/* 3. saving the optimal `e_p`-entry info, `$J_s^*(A \rightarrow e_p)$`    */`
7           $(s_{e_p}^{min}, \mathrm{cost}_{e_p}^{min}) \leftarrow \min_{\text{by cost}} (s_k, \mathrm{cost}_k^{sim})_{k=1}^{d_1}$
8   **save** $(e_p, (s_{e_p}^{min}, \mathrm{cost}_{e_p}^{min}))$

Complexity Analysis. Algorithm 1 average-case time complexity is determined by the number of solution space refinements, n_i, the average number of reduced force evaluations[6] for a single path/route segment, $\bar{n}_F$, and the number of such segments, $n_c n_p [(n_r - 3) n_c n_p + n_p + 1]$ (see Sect. 4.1). For the sequential version of the algorithm it can be expressed as:

$$T_s = \Theta \left(n_i \, n_r \, n_c^2 \, n_p^2 \, \bar{n}_F \right). \qquad (9)$$

---

[6] values of F are used both in HFM and LFM; Runge-Kutta-Fehlberg 4(5) method, used in the simulator, requires at each step six evaluations of F.

In the SPMD-parallel version of the algorithm, the evaluations for all nodes in a given row can be performed in parallel (using p processing units), thus:

$$T_p = \Theta \left(n_i \; n_r \; n_c \; n_p \left\lceil \frac{n_c \; n_p}{p} \right\rceil \bar{n}_F \right). \tag{10}$$

The Algorithm 1 *space complexity* formula, $\Theta \left(n_r \; n_c \; n_p \right)$, arises from the solution space representation.

Remark 7. Significant reduction of the average-case of computational cost of the algorithm is important for at least two reasons. Firstly, we often need to know the solution as soon as possible (sometimes for safety). Secondly, since we only use on-board computers, energy efficiency is critical while at sea.

5   Results and Discussion

To demonstrate the effectiveness of the algorithm, a series of experiments was carried out using a MacBook Pro[7] with macOS 12.6.3 and OpenCL 1.2. This system has one operational OpenCL-capable device: Intel Iris Graphics 6100, 1536 MB (the integrated GPU). The aim of the experiments was to investigate three important aspects of the algorithm: the accuracy of the surrogate-model, the pruning-related computational time cost reduction, and the SPMD-parallelisation efficiency. The results are presented in the subsequent paragraphs.

The Accuracy of the Surrogate-Model. This element has a direct and significant impact on the computational cost reduction since it is strictly related to pruning efficiency. Indeed, the more accurate the estimator is, the more simulations can be aborted (or even completely omitted). The results are given in the form of a violin plot in Fig. 3. The plot shows the distributions of relative estimation errors, i.e.:

$$\mathrm{err}(J_s|_{sim}^{est}) = \left| \frac{J_s^{(est)} - J_s^{(sim)}}{J_s^{(sim)}} \right| \tag{11}$$

for route segments from the solution spaces corresponding to $n_c = 16, 32, 64,$ and 128.

Remark 8. In all cases, the medians of relative estimation errors were smaller than 0.002 (i.e., 0.2%), which confirms the very high accuracy of the proposed surrogate-based segment cost estimator.

Efficacy of the Surrogate-Based Pruning. Having verified the accuracy of the surrogate-model, we can test the computation speedup resulting from surrogate-based pruning applied both in the simulation and estimation phases. The corresponding results are given in Table 1 and Fig. 4.

---

[7] Retina, 13-inch, Early 2015, with 16GB of DDR3 1867 MHz MHz RAM.

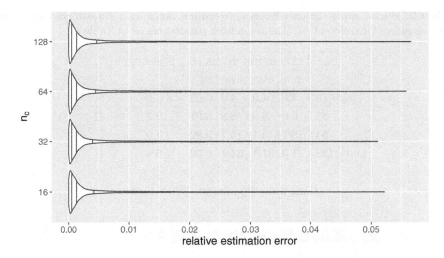

Fig. 3. Distributions of relative errors of the performance measure/cost estimate (see Eq. 11) for solution spaces with different n_c, with outliers ($e > \bar{e}+3\,\sigma_e$, if any) excluded.

Table 1. Efficacy of surrogate-based pruning: execution times (in seconds), and the pruning-related speedup for different n_c. The solution space (one refinement) with $n_r = 32$, $n_p = 8$. Statistics from ten runs.

n_c	EXECUTION TIMES (t_{sim})								SPEEDUP
	BASE MODEL				SURROGATE-ASSISTED				
	min	max	avg	sd	min	max	avg	sd	
16	271.1	272.6	271.8	0.52	6.2	6.3	6.2	0.05	43.5
32	1077.3	1080.7	1078.9	1.25	18.5	18.7	18.6	0.08	58.0
64	4399.9	4424.3	4410.8	7.24	74.5	75.5	74.7	0.22	59.1
128	17494.2	17609.7	17559.3	39.62	292.4	294.3	292.7	0.59	60.0

SPMD-*Parallelisation Efficiency.* Parallelisation is another way of lowering the total computation time. Contemporary mobile/on-board computers are usually equipped with more than one type of processor, typically one CPU and at least one GPU. OpenCL makes it possible to use these heterogeneous platforms effectively, since the same code can be executed on any OpenCL-capable processor. The execution times for different sizes of the solution space and the corresponding parallel-speedups are presented in Table 2 and Fig. 4. Its maximum recorded value was 9.4 (see Table 2 and Fig. 4). With reference point set as the sequential search based on the full simulation, it gives in total 565.4 times faster execution.

Table 2. SPMD-parallelisation efficiency: execution times, t_{sim} (in seconds) and (parallel) speedup for different n_c. The remaining parameters as in Table 1.

n_c	EXECUTION TIMES (t_{sim})				SPEEDUP
	min	max	avg	sd	
16	4.0	4.0	4.0	0.02	1.6
32	6.9	6.9	6.9	0.01	2.7
64	15.1	15.1	15.1	0.02	4.9
128	31.0	31.1	31.1	0.02	9.4

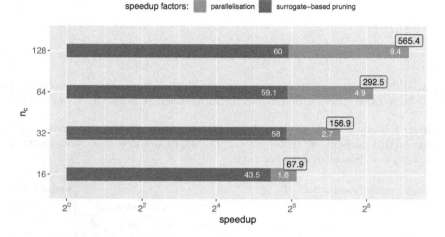

Fig. 4. Total speedup (boxed numbers at the end of each bar) and its factors: surrogate-based pruning (dark-grey part) and SPMD-parallelisation (light-grey part) for different n_c. The remaining parameters as in Table 1. (Color figure online)

6   Conclusions

It has been shown that the surrogate-assisted, pruned dynamic programming based ship route optimisation algorithm can be both effective and (energy) efficient. The key elements in achieving this have been the fast and accurate physics-based surrogate model, the pruned simulation, and the OpenCL-based SPMD-parallelisation of the algorithm.

The results show the high accuracy of the surrogate model (the medians of relative approximation errors were smaller than 0.2%, see Fig. 3), its efficacy in terms of the reduction of computing time resulting from pruning (from 43 to 60 times, see Table 1 and Fig. 4), and the notable speedup of the parallel algorithm (its maximum observed value was 9.4, see Fig. 4). Combining these effects has given up to 565 times faster execution time (see Fig. 4).

The proposed approach can also be applied to other scenarios. In fact, it can be considered as a dynamic programming-based, optimal path planning framework parameterised by a problem specific (potentially variable-fidelity) cost-function evaluator.

Future research work could concentrate on verification of the proposed algorithm with a more accurate ship simulator and different types of surrogate models (e.g., data-driven deep learning based).

Acknowledgement. The research presented in this paper was partially supported by the funds of Polish Ministry of Education and Science assigned to AGH University of Science and Technology.

A   The Ship Simulation Model

1. The *ship equation of motion* along a straight route segment:

$$(M + M_A)\frac{dv_s}{dt} = T_p(t) - (R_T(t) + R_{AW}(t)) \tag{12}$$

where: M denotes the ship mass, M_A - the 'added' mass, v_s - speed of ship, T_p - trust force of the propeller, R_T - frictional resistance, and R_{AW} - wave-making resistance; in addition: $M_R = M + M_A = 43\,341\,239$,

2. *Trust force:*

$$T_p(t) = 425 \left(1 - 4.45\frac{v_s(t)}{n}\right) n^2 \tag{13}$$

3. *Frictional resistance:*

$$R_T(t) = 11000 \left(v_s(t)\right)^2 \tag{14}$$

4. *Wave-making resistance:*

$$R_{AW}(t) = 562.5 \left[\frac{2\pi}{0.75v_w(t)} \left(1 - \frac{2\pi}{0.75v_w(t)}v_s(t)\cos\mu\right) v_s(t)\right]^2 \tag{15}$$

where: v_w is wave speed, v_s - ship speed, and μ is the 'encountering angle', i.e., the angle between the direction of wave travel and the direction of ship heading, which is measured in a clockwise manner from the direction of wave travel,

5. *Fuel consumption* (for a single route segment):

$$FOC_s = \int_0^{t_f} \max\left\{2\,\frac{R_T(t) + R_{AW}(t)}{M_R}\,v_s(t), 0.0625\right\} dt \tag{16}$$

6. *Weights of the objectives* (see Eq. 4):

$$C_1 = C_2 = 0.5 \tag{17}$$

References

1. Bellman, R., Dreyfus, S.: Applied Dynamic Programming. Princeton University Press, Princeton (1962)
2. Bijlsma, S.J.: On the applications of optimal control theory and dynamic programming in ship routing. Navigation **49**(2), 71–80 (2002). https://doi.org/10.1002/j.2161-4296.2002.tb00256.x
3. Calik, N., Güneş, F., Koziel, S., Pietrenko-Dabrowska, A., Belen, M.A., Mahouti, P.: Deep-learning-based precise characterization of microwave transistors using fully-automated regression surrogates. Sci. Rep. **13**(1), 1445 (2023). https://doi.org/10.1038/s41598-023-28639-4
4. Dębski, R., Dreżewski, R.: Adaptive surrogate-assisted optimal sailboat path search using onboard computers. In: Groen, D., de Mulatier, C., Paszynski, M., Krzhizhanovskaya, V.V., Dongarra, J.J., Sloot, P.M.A. (eds.) Computational Science – ICCS 2022. pp. 355–368. Springer International Publishing, Cham (2022). DOI: https://doi.org/10.1007/978-3-031-08757-8_30
5. Dębski, R., Sniezynski, B.: Pruned simulation-based optimal sailboat path search using micro HPC systems. In: Paszynski, M., Kranzlmüller, D., Krzhizhanovskaya, V.V., Dongarra, J.J., Sloot, P.M.A. (eds.) ICCS 2021. LNCS, vol. 12745, pp. 158–172. Springer, Cham (2021). https://doi.org/10.1007/978-3-030-77970-2_13
6. Du, W., Li, Y., Zhang, G., Wang, C., Chen, P., Qiao, J.: Estimation of ship routes considering weather and constraints. Ocean Eng. **228**, 108695 (2021). https://doi.org/10.1016/j.oceaneng.2021.108695
7. Du, W., Li, Y., Zhang, G., Wang, C., Zhu, B., Qiao, J.: Energy saving method for ship weather routing optimization. Ocean Eng. **258**, 111771 (2022). https://doi.org/10.1016/j.oceaneng.2022.111771
8. Ghaemi, M.H., Zeraatgar, H.: Analysis of hull, propeller and engine interactions in regular waves by a combination of experiment and simulation. J. Marine Sci. Technol. **26**(1), 257–272 (2020). https://doi.org/10.1007/s00773-020-00734-5
9. Hagiwara, H., Spaans, J.A.: Practical weather routing of sail-assisted motor vessels. J. Navigat. **40**(1), 96–119 (1987). https://doi.org/10.1017/S0373463300000333
10. Hinnenthal, J., Clauss, G.: Robust pareto-optimum routing of ships utilising deterministic and ensemble weather forecasts. Ships Offshore Struct. **5**(2), 105–114 (2010). https://doi.org/10.1080/17445300903210988
11. Koziel, S., Leifsson, L. (eds.): Surrogate-Based Modeling and Optimization. Springer, New York (2013). https://doi.org/10.1007/978-1-4614-7551-4
12. Koziel, S., Ogurtsov, S.: Antenna design by simulation-driven optimization. SO, Springer, Cham (2014). https://doi.org/10.1007/978-3-319-04367-8
13. Li, J., Jia, Y.: Calculation method of marine ship fuel consumption. IOP Conf. Ser.: Earth Environ. Sci. **571**(1), 012078 (2020). https://doi.org/10.1088/1755-1315/571/1/012078
14. Lin, Y.H., Fang, M.C., Yeung, R.W.: The optimization of ship weather-routing algorithm based on the composite influence of multi-dynamic elements. Appl. Ocean Res. **43**, 184–194 (2013). https://doi.org/10.1016/j.apor.2013.07.010
15. Mahouti, P., Belen, A., Tari, O., Belen, M.A., Karahan, S., Koziel, S.: Data-driven surrogate-assisted optimization of metamaterial-based filtenna using deep learning. Electronics **12**(7), 1584 (2023). https://doi.org/10.3390/electronics12071584
16. Maki, A., et al.: A new weather-routing system that accounts for ship stability based on a real-coded genetic algorithm. J. Marine Sci. Technol. **16**(3), 311–322 (2011). https://doi.org/10.1007/s00773-011-0128-z

17. Mao, W., Rychlik, I., Wallin, J., Storhaug, G.: Statistical models for the speed prediction of a container ship. Ocean Eng. **126**, 152–162 (2016). https://doi.org/10.1016/j.oceaneng.2016.08.033
18. Padhy, C., Sen, D., Bhaskaran, P.: Application of wave model for weather routing of ships in the north Indian ocean. Nat. Hazards **44**, 373–385 (2008). https://doi.org/10.1007/s11069-007-9126-1
19. Shao, W., Zhou, P., Thong, S.K.: Development of a novel forward dynamic programming method for weather routing. J. Marine Sci. Technol. **17**(2), 239–251 (2012). https://doi.org/10.1007/s00773-011-0152-z
20. Spaans, J.: New developments in ship weather routing. Navigation, pp. 95–106 (1995)
21. Szlapczynska, J.: Multiobjective approach to weather routing. TransNav Int. J. Marine Navigat. Safety Sea Transp. **1**(3) (2007)
22. Szlapczynska, J., Smierzchalski, R.: Multicriteria optimisation in weather routing. TransNav Int. J. Marine Navigat. Safety Sea Transp. **3**(4), 393–400 (2009). https://doi.org/10.1201/9780203869345.ch74
23. Szlapczynska, J., Szlapczynski, R.: Preference-based evolutionary multi-objective optimization in ship weather routing. Appl. Soft Comput. **84**, 105742 (2019). https://doi.org/10.1016/j.asoc.2019.105742
24. The North Atlantic Ocean map. https://commons.wikimedia.org/wiki/File:North_Atlantic_Ocean_laea_location_map.svg (2012). licensed under the https://creativecommons.org/licenses/by-sa/3.0/deed.en
25. United Nations Conference on Trade and Development (UNCTAD): Review of Maritime Transport (2022). https://unctad.org/topic/transport-and-trade-logistics/review-of-maritime-transport
26. Vencti, A., Makrygiorgos, A., Konstantopoulos, C., Pantziou, G., Vetsikas, I.A.: Minimizing the fuel consumption and the risk in maritime transportation: a bi-objective weather routing approach. Comput. Oper. Res. **88**, 220–236 (2017). https://doi.org/10.1016/j.cor.2017.07.010
27. Wang, H., Mao, W., Eriksson, L.: A three-dimensional Dijkstra's algorithm for multi-objective ship voyage optimization. Ocean Eng. **186**, 106131 (2019). https://doi.org/10.1016/j.oceaneng.2019.106131
28. Wang, X., Song, X., Sun, W.: Surrogate based trajectory planning method for an unmanned electric shovel. Mechan. Mach. Theory **158**, 104230 (2021). https://doi.org/10.1016/j.mechmachtheory.2020.104230
29. Yurt, R., Torpi, H., Mahouti, P., Kizilay, A., Koziel, S.: Buried object characterization using ground penetrating radar assisted by data-driven surrogate-models. IEEE Access **11**(1), 13309–13323 (2023). https://doi.org/10.1109/ACCESS.2023.3243132
30. Zeraatgar, H., Ghaemi, H.: The analysis of overall ship fuel consumption in acceleration manoeuvre using Hull-Propeller-Engine interaction principles and governor features. Polish Maritime Res. **26**, 162–173 (2019). https://doi.org/10.2478/pomr-2019-0018
31. Zis, T.P., Psaraftis, H.N., Ding, L.: Ship weather routing: a taxonomy and survey. Ocean Eng. **213**, 107697 (2020). https://doi.org/10.1016/j.oceaneng.2020.107697

Optimization of Asynchronous Logging Kernels for a GPU Accelerated CFD Solver

Paul Zehner[✉][iD] and Atsushi Hashimoto[iD]

JAXA, 7-44-1, Jindiaiji Higashi-machi, Choufu 182-0012, Tokyo, Japan
{zehner.paul,hashimoto.atsushi}@jaxa.jp

Abstract. Thanks to their large number of threads, GPUs allow massive parallelization, hence good performance for numerical simulations, but also make asynchronous execution more common. Kernels that do not actively take part in a computation can be executed asynchronously in the background, in the aim to saturate the GPU threads. We optimized this asynchronous execution by using mixed precision for such kernels. Implemented on the FaSTAR solver and tested on the NASA CRM case, asynchronous execution gave a speedup of 15% to 27% for a maximum memory overhead of 4.5% to 9%.

Keywords: asynchronous · CFD · GPU · mixed precision · OpenACC

1 Introduction

General-Purpose computing on GPU (GPGPU) allows massive parallelization thanks to its large number of threads, and is considered a promising path to exascale computing ($1 \cdot 10^{18}$ FLOP s^{-1}) [18,22]. Its use in Computational Fluid Dynamics (CFD) allows to perform more complex, more precise, and more detailed simulations, but requires either to adapt existing codes or to create new ones [3], with the help of specific techniques such as CUDA, OpenCL or OpenACC [17]. CFD codes can not only benefit from the computational power of Graphical Processing Units (GPUs), but also from their asynchronous execution capabilities.

The asynchronous execution of kernels stems from the capacity of the Streaming Multiprocessor (SM) to execute different instructions on its different warps [8,15]. The programmer specifies an asynchronous execution by using *streams* for CUDA (with a fourth argument to the triple chevron <<<>>> syntax), or by using the `async` clause for OpenACC. Uncommon among traditional High Performance Computing (HPC) paradigms, asynchronous execution offers interesting uses.

Its immediate benefit is to allow kernels execution and memory transfers to overlap. Multi-GPU performance is improved by this approach, as it allows to hide network transfers. Micikevicius [13], on a CUDA-accelerated three-dimensional structured finite difference code, used asynchronous execution along with non-blocking Message Passing Interface (MPI) calls. He proposed a two-step algorithm: for each timestep, the first step consists in computing only the

© The Author(s), under exclusive license to Springer Nature Switzerland AG 2023
J. Mikyška et al. (Eds.): ICCS 2023, LNCS 14075, pp. 410–424, 2023.
https://doi.org/10.1007/978-3-031-36024-4_32

boundary cells; the second step consists in computing all the other cells and, at the same time, transferring the boundary cells. McCall [12] accelerated an artificial compressibility Navier–Stokes structured solver using OpenACC. The same approach was used, but reversely: the first step consists in calculating the artificial compressibility and numerical damping terms on the interior cells and, at the same time, transferring the boundary cells with MPI; the second step consists in calculating the remaining cells. McCall considered that this overlapping technique improved weak scaling significantly. Shi et al. [20], accelerating an incompressible Navier–Stokes solver with CUDA, also used streams to overlap computations and non-blocking MPI communications. Choi et al. [2] used a more aggressive approach with an overdecomposition paradigm which uses Charm++ instead of MPI. Charm++ [11] is a C++ library which allows to decompose a problem in several tasks, named *chares*, that are executed asynchronously by processing elements and communicate with messages. In their paper, Choi et al. applied this technique to a test Jacobi solver, and obtained a weak scaling performance gain of 61% to 65% on 512 nodes, compared to a traditional MPI approach, and a strong scaling gain of 40% on 512 nodes.

Another benefit resides in the concurrent execution of independent kernels; where each kernel operates on different arrays at the same time. Hart et al. [5], for the OpenACC acceleration of the Himeno benchmark on the Cray XK6 supercomputer, overlapped data transfers and computation, and executed computation and MPI data packing concurrently. Compared to synchronous GPU code, performances were increased by 5% to 10%. The authors predicted that a higher gain could be obtained for larger codes, as they considered Himeno simple. Searles et al. [19] accelerated the wavefront-based test program Minisweep with OpenACC and MPI, in order to improve the nuclear reactor modeling code Denovo. They proposed to parallelize the Koch–Baker–Alcouffe algorithm, which consists in different sweeps. They used asynchronous kernel execution to realize the 8 sweeps simultaneously, in order to saturate the GPU threads. The authors reported good performance.

Asynchronous execution also allows to run background tasks with low resources, while a main task is running synchronously, in the objective to saturate the GPU threads. The unstructured CFD solver FaSTAR, developed at Japan Aerospace eXploration Agency (JAXA) [6,7,10] and accelerated with OpenACC [28,29], uses this feature in order to run logging kernels asynchronously. These kernels are executed with a certain number of OpenACC gangs, and require memory duplication to avoid race conditions. This process allowed a speedup of 23% to 35%, but increased memory occupancy by up to 18%. In this paper, we propose to optimize this process by reducing the memory overhead with lower precision storage of floating point arrays.

This paper is organized as follows. We introduce the FaSTAR solver, its current state of acceleration, and its asynchronous execution of logging kernels in Sect. 2. The use of mixed precision to reduce the memory overhead of the asynchronous execution is presented in Sect. 3. Then, we validate and check the performance of this improvements on the NASA Common Research Model (NASA CRM) test case in Sect. 4. To finish, we conclude in Sect. 5.

2 Description of the FaSTAR Solver

2.1 Solver Introduction

The FaSTAR code solves the compressible Reynolds-Averaged Navier-Stokes (RANS) equations on unstructured meshes. It consists in 80,000 lines of Fortran code, has an Array of Structures memory layout, is parallelized with MPI, and uses Metis for domain decomposition. The Cuthill–McKee algorithm is used to reorder cells. The solver was designed to run a simulation of 10 million cells on 1000 Central Processing Unit (CPU) cores within 2min [6].

2.2 Acceleration of the Solver

The solver was partially accelerated with OpenACC in a previous work [29]. OpenACC was selected in favor of CUDA, as we wanted to keep CPU compatibility without having to duplicate the source code.

Parts of the code that were accelerated so far are the Harten-Laxvan Leer-Einfeldt (HLLEW) scheme [16], the Lower-Upper Symmetric Gauss- Seidel (LU-SGS) [27] and the Data-Parallel Lower-Upper Relaxation (DP-LUR) [26] implicit time integration methods, the GLSQ algorithm [21] (a hybrid method of Green-Gauss and Least-Square), the Hishida slope limiter (a Venkatakrishnan-like limiter that is complementary with difference of neighboring cell size) and the Monotonic Upstream-centered Scheme for Conservation Laws (MUSCL) reconstruction method. During execution, all of the computation takes place on the GPU, and data movements between the CPU and the GPU memories are minimized.

2.3 Asynchronous Execution of Logging Kernels

Some kernels only used for logging purpose can take a significant computation time, while their result is not used anywhere in the computation. Such kernels are run asynchronously in the background [28], and use just enough resources to saturate the GPU threads. In the case of FaSTAR, the kernel to log the right-hand side (RHS) values of the Navier–Stokes equation and the kernel to compute the L^2 norm of residuals are concerned by this optimization. These two kernels are called at different times during the iteration, respectively after the computation of the RHS term, and after the time integration.

The subroutine hosting such a kernel is separated in two parts: one computing the values, offloaded on the GPU, and another one writing the values (on disk or on screen), executed by the CPU. A module stores the value between the two subroutines:

```
1  module store
2    real :: value(n_value)
3  end module store
4
5  module compute
6  contains
```

```
7     subroutine compute_value(array)
8       use store, only: value
9       ...
10
11      !$acc kernels async(STREAM) &
12      !$acc           copyout(value) &
13      !$acc           copyin(...)
14      ! compute value from array
15      !$acc end kernels
16    end subroutine compute_value
17
18    subroutine write_value
19      use store, only: value
20
21      !$acc wait(STREAM)
22      print "('value(1) = ', g0)", value(1)
23      ...
24    end subroutine write_value
25  end module compute
```

As seen on line 11, the kernel in the compute part is executed asynchronously
with the OpenACC async clause in a specific STREAM, while the write part is
called as late as possible and, on line 21, waits for the asynchronous kernel to
end with the wait directive. At the same time, other kernels are executed synchronously. The write part (for each logging subroutine in the case of FaSTAR)
is called at the end of the iteration.

In order to avoid read race conditions as the main synchronous kernels may
modify the memory, arrays used by the logging kernel must be duplicated. The
following modifications are added to the code:

```
1   module store
2     ...
3     real, allocatable :: array_d(:)
4   end module store
5
6   module compute
7   contains
8     subroutine compute_value(array)
9       use store, only: array_d
10      ...
11
12      !$acc kernels
13      array_d(:) = array(:)
14      !$acc end kernels
15
16      ...
17    end subroutine compute_value
18    ...
19  end module compute
```

In subroutine `compute_value`, the memory duplication takes place before the asynchronous kernel. This is however a costly operation in term of memory. The complete process workflow for FaSTAR is sketched in Fig. 1. The two asynchronous kernels use the same stream to not be concurrent to each other.

Main sync. kernels
Sync. kernels for copying data
Async. kernels for compute part
Subroutines for write part

Fig. 1. Workflow of asynchronous execution of the two logging kernels.

Finding the optimal number of threads for the logging kernels is an optimization problem, as too few dedicated threads makes its execution time too long (hence increasing the simulation time), and too much slows down the synchronous kernels (hence increasing the simulation time again). The number of cells of the simulation as well as the hardware itself has an influence on this number of threads. Some assumptions are used to simplify the problem. We consider that the compute kernel represents enough operations to fill at least the threads of one SM, which translates in one gang in the OpenACC terminology of *gang* (SM), *worker* (Streaming Processor (SP)) and *vector* (thread). We consequently investigate the number of gangs, instead of the number of threads, which can be set with the OpenACC clause `num_gangs`. We also consider that each iteration is similar and takes the same amount of time to compute (with the same amount of resources). The number of gangs can either be set by the user, or can be automatically estimated with what we name a sloped descend gradient method:

$$n_g^{(n+1)} = n_g^{(n)} - \text{sign} \frac{\Delta t^{(n)} - \Delta t^{(n-1)}}{n_g^{(n)} - n_g^{(n-1)} + \epsilon}, \tag{1}$$

where $n_g^{(n+1)}$, respectively $n_g^{(n)}$ and $n_g^{(n-1)}$, is the number of gangs for the next iteration (or group of iterations), respectively for the current iteration and the previous iteration, $\Delta t^{(n)}$, respectively $\Delta t^{(n-1)}$, is the duration of the current iteration, respectively of the previous iteration, and ϵ is a small value. We consider, as for many Fortran implementations, that sign0 = 1. This formulation is robust and increases the number of gangs by ±1 for each evaluation, but has two drawbacks: the optimal value may be slow to reach, depending on the initial value to start the evaluation from, and the number of gangs can only fluctuate close to this optimum. Using the iteration time to compute the number of gangs makes this algorithm vulnerable to external factors, such as network flutters in case of MPI parallelization, or input/output latency when writing intermediate results to disk. This problem is mitigated by taking a group of iterations in eq. (1), by default 100, instead of one, and by disabling the evaluation of the number of gangs for iterations when results are outputted to a file.

3 Implementation of Mixed Precision Kernels

The asynchronous execution process presented in the previous section has the disadvantage to trade GPU memory for execution speed, while memory is limited on this hardware. FaSTAR uses double precision floating point arrays, which means by the Institute of Electrical and Electronics Engineers (IEEE) 754 norm [9] that each element in an array is stored on 64 bits (8 bytes). Such a precision is the standard in CFD, but is not required for logging purpose, where the order of magnitude is usually enough as logs are read by a human. Consequently, we propose to use a lower precision for the logging kernels, such as single precision (32 bits per element), and even half precision (16 bits per element). This can reduce the memory cost of the asynchronous execution process by respectively two and four. Even if mixed precision is commonly used in a context of performance gain [24], we only aim in this paper to reduce memory occupancy.

3.1 Single Precision

We first implement the storage of duplicated arrays and the computation of the value using single precision. As the selection of precision has to be decided at compile time, we use preprocessor commands, which are only executed by the compiler before converting the source code in binary instructions. Since the Fortran standard does not specify a preprocessor, we use the traditional C preprocessor syntax, which is commonly available in compilers.

We use a set of two preprocessor macros, that are substituted by the compiler in the rest of the source code:

```
1  #if LOGGING_REDUCED_PRECISION == 4
2  #   define TYPE 4
3  #   define STORE(value) (real(value, TYPE))
4  #else
5  #   define TYPE 8
6  #   define STORE(value) (value)
7  #endif
```

The macro TYPE qualifies the kind of the floating point variables, and is used in the declaration of the duplicated arrays and in the compute and write subroutines, using the kind argument of type real. Line 2 defines it to single precision (for most environments), and line 5 to double precision. STORE explicitly converts variables to the desired precision using the real Fortran intrinsic function; it is used when duplicating the arrays. At line 3, it converts values to single precision, and at line 6 it is an invariance. Using mixed precision can be disabled completely by the LOGGING_REDUCED_PRECISION macro at line 1.

3.2 Half Precision

Half precision has several limitations that must be taken into consideration first. The Fortran standard does not explicitly define which precision is available for the programmer, and this choice is left to the compiler maker. Single and double precision are usually always implemented, but half precision is less common, hence not supported by all compilers. In this study, we chose to use the NVIDIA HPC Software Development Kit (NVIDIA HPC SDK) compiler, as it is the most common and mature OpenACC compiler available when writing this paper, and as it has support for half precision. Also, half precision has an exponent range limited to $\pm 1 \cdot 10^4$, with any other value replaced by $\pm\infty$. This limitation may be unacceptable if the values to manipulate fall outside of this range. Another representation considered was *bfloat16* [25], which has a larger exponent range of $\pm 1 \cdot 10^{38}$ while using the same amount of memory, but is not covered by the Fortran standard [1]. Another shortcoming of half precision is that not all operations may be implemented for this precision. Consequently, when using half precision, only the duplicated arrays are stored using this precision, whereas work on these arrays is done using single precision.

The set of preprocessor macros is extended with two other macros, TYPE_DUPLICATED and UNSTORE:

```
1  #if LOGGING_REDUCED_PRECISION == 2
2  #   define TYPE 4
3  #   define TYPE_DUPLICATED 2
4  #   define STORE(value) (real(value, TYPE_DUPLICATED))
5  #   define UNSTORE(value) (real(value, TYPE))
6  #if LOGGING_REDUCED_PRECISION == 4
7  #   define TYPE 4
8  #   define TYPE_DUPLICATED 4
9  #   define STORE(value) (real(value, TYPE))
10 #   define UNSTORE(value) (value)
11 #else
12 #   define TYPE 8
13 #   define TYPE_DUPLICATED 8
14 #   define STORE(value) (value)
15 #   define UNSTORE(value) (value)
16 #endif
```

The macro TYPE is now only used for the kind of arrays in the compute and write subroutines. At line 2 it is defined for single precision. TYPE_DUPLICATED is only used for the kind of the duplicated arrays. At line 3 it is defined for half precision, at lines 8 and 13 as the current TYPE. UNSTORE explicitly converts back duplicated variables to the desired precision, and is used in the compute part. At line 5, it converts values from half precision to single precision, at lines 10 and 15 it is an invariance. The LOGGING_REDUCED_PRECISION macro is used to select the desired precision.

4   Validation and Performance

4.1   Description of the Cases

Now mixed precision is implemented, we test the validity and the performance of the modified solver on the NASA CRM test case [23]. This geometry aims to propose a research model of a commercial airplane, it has a wing and a cabin. We simulated a cruise flight at Mach number $M = 0.85$, angle of attack $\alpha = 2.5°$, and Reynolds number $Re = 5 \cdot 10^6$. The upstream temperature was $T_\infty = 100°\text{F} \approx 310.93K$ and the air specific heat ratio was $\gamma = 1.4$.

Table 1. Number of cells for the different meshes.

Mesh	Tetrahedron	Pyramid	Prism	Hexahedron	Total
Coarse	146,062	587,161	0	2,946,009	3,679,232
Medium	258,724	1,040,044	27,424	5,260,349	6,586,541
Fine	467,168	1,897,302	77,256	9,491,406	11,933,132
Extra-fine	1,633,430	6,554,906	147,004	38,410,538	46,745,878

We used a set of four meshes of increasing number of cells, as described in Table 1, created with HexaGrid. Minimum cell size was 5in, 3.5in, 2in and 1in $(1.27 \cdot 10^{-1}\text{m}, 8.89 \cdot 10^{-2}\text{m}, 5.08 \cdot 10^{-2}\text{m}, 2.54 \cdot 10^{-2}\text{m})$ for the coarse, medium, fine and extra-fine mesh respectively. Maximum cell size was 5in $(1.27 \cdot 10^{-1}\text{m})$, and the dimensionless wall distance was $y^+ = 1.00$.

Methods cited in Sect. 2.1 were used; the DP-LUR implicit time integration method was used with 6 sub-iterations. We performed steady state simulations of 10,000 iterations, using local time stepping and setting a Courant-Friedrichs-Lewy (CFL) number $CFL = 50$. The solver was compiled using NVIDIA HPC SDK version 22.7, simulations were executed on JAXA Supercomputer System generation 3 (JSS3) [4] using NVIDIA V100 GPUs [14]. Cases were executed on single GPU and multi-GPU using 2 and 4 domains. The extra-fine case memory occupancy was too high for some executions: for single GPU and multi-GPU on 2 domains with double precision.

Compared to reference computations using synchronous logging kernels, we aim to analyze the speedup and the memory overhead of the different implementations. Two batches of computations were executed: one for which the number of gangs was automatically estimated by the algorithm presented in Sect. 2.3, another one with a fixed number of gangs based on the outcome of the first batch.

4.2 Accuracy Analysis

Before analyzing performance, we analyze the accuracy of the results for computations on the fine mesh on single GPU. We first made sure that solution files were similar with an absolute tolerance of $1 \cdot 10^{-5}$ for the different simulations. Letting the algorithm estimate the number of gangs or setting it manually did not change the results.

The evolution of the L^2 norm of the residual of density $\|\rho\|_2$ is displayed in Fig. 2. Residual of computations using asynchronous execution with double,

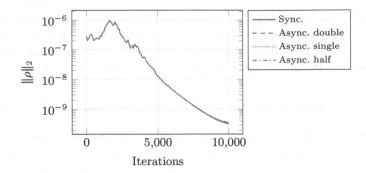

Fig. 2. Evolution of the L^2 norm of residual.

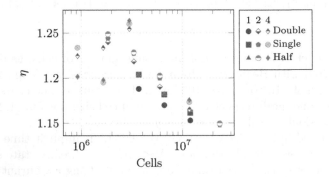

Fig. 3. Evolution of the speedup compared with synchronous execution.

single, or half precision are compared with a reference synchronous computation. Results are indistinguishable. A closer analysis reveals that residual for single and half precision are a bit lower than reference results, which can be explained as they are both computed with single precision.

The evolution of the sum of the RHS terms, not displayed in this paper, was also indistinguishable from the results of the synchronous execution. However, some values were replaced by $+\infty$ in results of the asynchronous execution using half precision.

We conclude that using reduced precision when computing residual did not affect the quality of the solution. The quality of the logging files was almost identical when using single precision; on the other hand, half precision gave almost identical results in most cases, but some values could not be represented.

4.3 Performance Analysis

We analyze now the performance of the new implementations, both in term of speedup and in term of memory overhead.

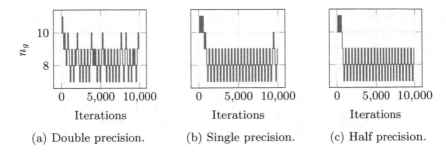

(a) Double precision.   (b) Single precision.   (c) Half precision.

Fig. 4. Evolution of the number of gangs for asynchronous execution on the fine mesh on 1 domain.

We define the speedup η as the ratio of wall clock times spent in the time integration loop, which excludes the initial time for loading the mesh and the final time for writing the solution:

$$\eta = \frac{t_{\text{sync}}}{t_{\text{async}}} \tag{2}$$

where t_{sync} is related to the reference synchronous execution, and t_{async} is related to the asynchronous execution of the different implementations. The speedup is displayed in Fig. 3 as a function of the number of cells, when the number of gangs is automatically estimated. We represented speedups for 1 domain (single GPU), 2 and 4 domains (multi-GPU) on the same graph. For any precision, the overall tendency is that the speedup decreases as the number of cells increases. As the number of cells increases, there are less idle threads to saturate, and the asynchronous execution is hence less efficient. The speedup ranged from 1.15 to 1.21 for 1 domain, from 1.15 to 1.25 for 2 domains, and from 1.17 to 1.27 for 4 domains. Computations on 1 and 2 domains exhibited a similarly decreasing speedup, whereas computations on 4 domains exhibited a more chaotic trend with high variability, which can be explained by other factors contributing more to the speedup, such as network latency. Executions with single precision had a higher speedup compared with double precision by on average 1% for 1 domain and 0.6% for 2 domains. Half precision had a higher speedup by on average 1.1% for 1 domain and 0.9% for 2 domains. This can either be due to less time spent in duplicating the memory, or by more resources available for the main kernels. Using reduced precision had a limited influence on speedup.

When automatically estimated by the algorithm, the number of gangs stabilized close to a single value. Figure 4 displays the evolution of the number of gangs as a function of the number of iteration for the computations on the fine mesh on single GPU only. Despite being specific to this mesh, the evolution of the number of gangs is representative of the other simulations. For all cases, the initial number of gangs was set to 10. The optimum value was 8 for the different precision values, meaning that the precision of the logging kernels did not have an impact on the estimation of the number of gangs. Compared with the maximum number of 84 gangs available on a V100 GPU, the number of gangs dedicated to the logging kernels remained reasonable: 9%. For multi-GPU execution, the number of gangs was usually the same on the different devices. However, for the simulation of the coarse case on 4 domains using half precision, the number of gangs on one device converged to a different value.

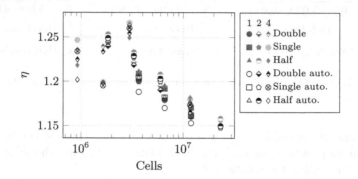

Fig. 5. Evolution of the speedup compared with synchronous execution for a fixed number of gangs.

The speedup for a fixed number of gangs is displayed in Fig. 5. Speedup for automatic estimation of the number of gangs is also represented for reference with black empty markers. Using a fixed number of gangs improved speedup by 0.6% on average (-1.1% to 1.3%). The gain is very moderate, the automated estimation gave satisfactory performance without any input from the user.

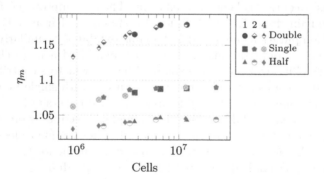

Fig. 6. Evolution of the memory overhead compared with synchronous execution.

We analyze now the performance in term of memory overhead. We define the memory overhead η_m as the ratio of the GPU average memory used during execution and reported by the job submission system on JSS3:

$$\eta_m = \frac{m_{\text{async}}}{m_{\text{sync}}} \qquad (3)$$

where m_{sync} is the memory of the synchronous execution, and m_{async} is the memory of the asynchronous execution of the different implementations. Evolution of the memory overhead is displayed in Fig. 6 as a function of the number of cells. Independently from the precision, the memory overhead of the the asynchronous execution increases with the number cells up to $1 \cdot 10^7$ cells. Simulations on different number of domains coincide well. Asynchronous execution using double precision has a maximum memory overhead of 18%, while single precision has a maximum overhead of 9% (i.e. a half of double precision), and half precision has an overhead of 4.5% (i.e. a fourth of double precision). This is in agreement with theoretical expectations. The overhead on the CPU memory was also analyzed, and was close to 1 for all simulations.

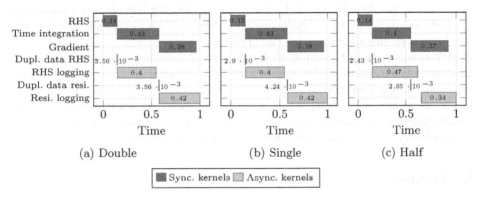

(a) Double   (b) Single   (c) Half

Sync. kernels   Async. kernels

Fig. 7. Timeline execution of the different parts of the solver for asynchronous execution on the fine mesh on 1 domain.

We finalize this performance analysis by profiling the execution of the code for a fixed number of 8 gangs for one iterations with NVIDIA Nsight Systems. A synthetic timeline of the execution of the kernels is displayed in Fig. 7 for the asynchronous execution on the fine mesh on 1 domain, for the three precision values. In the figure, the execution time is expressed as a ratio of the iteration time. The asynchronous logging kernels (named *RHS logging* and *resi. logging*) were executed concurrently with the other kernels, and the end of the second logging kernel coincides with the end of the last group of main kernels (named *gradient*), but occurs always a bit later. The timings for each kernels is similar for any precision requested. The time spent in arrays duplication for asynchronous execution (named *dupl. data RHS* and *dupl. data resi.*) was always negligible (less than 0.5% for any precision value). In the first iteration only, not shown in the graph, a consequent time is spent in creating the pinned memory buffer when preparing to transfer the computed logging values back to the host memory.

5   Conclusion

In this study, we improved the GPU execution of asynchronous logging kernels used in the CFD solver FaSTAR that was initiated in a previous work. We implemented mixed precision in the execution of the logging kernels in order to reduce the memory overhead, by reducing the floating point precision of the arrays they work on. The optimization was tested on four different meshes of the NASA CRM case, ranging from 3.7 to 46.7 million cells, on single GPU and multi-GPU (2 and 4 domains). Simulation results were identical, but a few logged values were incorrectly rendered as $+\infty$ using half precision. Asynchronous execution gave a speedup of 15% to 27%, which decreased as the number of cells increased, as there are less remaining threads to saturate. Simulations on 4 domains had more variation in speedup. Using double precision for logging kernels had a maximum memory overhead of 18%, single precision 9%, and half precision 4.5%; it did not have a significant impact on the speedup.

We conclude from this study that background asynchronous execution of logging kernels can improve the speedup of computations for an acceptable memory cost. Single precision is enough to have good fidelity, while half precision is a more aggressive approach that may not be suited for all computations. We believe that the optimization technique presented in this paper can be beneficial to other numerical solvers and improve their performance.

References

1. Bleikamp, R.: BFLOAT16 (Feb 2020), https://j3-fortran.org/doc/year/20/20-118.txt
2. Choi, J., Richards, D.F., Kale, L.V.: Improving scalability with GPU-aware asynchronous tasks. In: 2022 IEEE International Parallel and Distributed Processing Symposium Workshops (IPDPSW), pp. 569–578. IEEE, Lyon (2022). https://doi.org/10.1109/IPDPSW55747.2022.00097
3. Duffy, A.C., Hammond, D.P., Nielsen, E.J.: Production level CFD code acceleration for hybrid many-core architectures. Tech. Rep. NASA/TM-2012-217770, L-20136, NF1676L-14575, NASA (2012). https://ntrs.nasa.gov/citations/20120014581
4. Fujita, N.: JSS3/TOKI overview and large-scale challenge breaking report. In: Proceedings of the 53rd Fluid Dynamics Conference/the 39th Aerospace Numerical Simulation Symposium, vol. JAXA-SP-21-008, pp. 95–100. JAXA, Online (2022). https://id.nii.ac.jp/1696/00048362/
5. Hart, A., Ansaloni, R., Gray, A.: Porting and scaling OpenACC applications on massively-parallel, GPU-accelerated supercomputers. Eur. Phys. J. Spec. Top. **210**(1), 5–16 (2012). https://doi.org/10.1140/epjst/e2012-01634-y
6. Hashimoto, A., Ishida, T., Aoyama, T., Hayashi, K., Takekawa, K.: Fast parallel computing with unstructured grid flow solver. In: 28th International Conference on Parallel Computational Fluid Dynamics, Parallel CFD'2016 (2016)
7. Hashimoto, A., Ishida, T., Aoyama, T., Takekawa, K., Hayashi, K.: Results of three-dimensional turbulent flow with FaSTAR. In: 54th AIAA Aerospace Sciences Meeting. American Institute of Aeronautics and Astronautics, San Diego (2016). https://doi.org/10.2514/6.2016-1358

8. Hennessy, J.L., Patterson, D.A.: Chapter four: Data-level parallelism in vector, SIMD, and GPU architectures. In: Computer Architecture: A Quantitative Approach, 6th edn, pp. 281–365. Morgan Kaufmann Publishers, Cambridge (2019)

9. IEEE: IEEE Standard for Floating-Point Arithmetic. IEEE Std 754-2019 (Revision of IEEE 754-2008) pp. 1–84 (2019). https://doi.org/10.1109/IEEESTD.2019.8766229

10. Ito, Y., et al.: TAS code, FaSTAR, and Cflow results for the sixth drag prediction workshop. J. Aircraft **55**(4), 1433–1457 (2018). https://doi.org/10.2514/1.C034421

11. Kale, L.V., Krishnan, S.: CHARM++: A portable concurrent object oriented system based on C++. In: Proceedings of the Eighth Annual Conference on Object-Oriented Programming Systems, Languages, and Applications - OOPSLA '93. pp. 91–108. ACM Press, Washington (1993). https://doi.org/10.1145/165854.165874

12. McCall, A.J.: Multi-level parallelism with MPI and OpenACC for CFD applications. Master of Science thesis, Virginia Tech (2017). https://vtechworks.lib.vt.edu/handle/10919/78203

13. Micikevicius, P.: 3D finite difference computation on GPUs using CUDA. In: Proceedings of 2nd Workshop on General Purpose Processing on Graphics Processing Units (GPGPU-2), pp. 79–84. ACM Press, Washington (2009). https://doi.org/10.1145/1513895.1513905

14. NVIDIA: NVIDIA V100 Datasheet. Tech. rep. (2020). https://images.nvidia.com/content/technologies/volta/pdf/volta-v100-datasheet-update-us-1165301-r5.pdf

15. NVIDIA: CUDA Toolkit Documentation (2022). https://docs.nvidia.com/cuda/index.html

16. Obayashi, S., Guruswamy, G.P.: Convergence acceleration of a Navier-Stokes solver for efficient static aeroelastic computations. AIAA J. **33**(6), 1134–1141 (1995). https://doi.org/10.2514/3.12533

17. OpenACC: OpenACC API 2.7 Reference Guide (2018). https://www.openacc.org/sites/default/files/inline-files/API%20Guide%202.7.pdf

18. Riley, D.: Intel 4th Gen Xeon series offers a leap in data center performance and efficiency (2023). https://siliconangle.com/2023/01/10/intel-4th-gen-xeon-series-offers-leap-data-center-performance-efficiency/

19. Searles, R., Chandrasekaran, S., Joubert, W., Hernandez, O.: MPI + OpenACC: Accelerating radiation transport mini-application, minisweep, on heterogeneous systems. Comput. Phys. Commun. **236**, 176–187 (2019). https://doi.org/10.1016/j.cpc.2018.10.007

20. Shi, X., Agrawal, T., Lin, C.A., Hwang, F.N., Chiu, T.H.: A parallel nonlinear multigrid solver for unsteady incompressible flow simulation on multi-GPU cluster. J. Comput. Phys. **414**, 109447 (2020). https://doi.org/10.1016/j.jcp.2020.109447

21. Shima, E., Kitamura, K., Haga, T.: Green–Gauss/weighted-least-squares hybrid gradient reconstruction for arbitrary polyhedra unstructured grids. AIAA J. **51**(11), 2740–2747 (2013). https://doi.org/10.2514/1.J052095

22. Trader, T.: How argonne is preparing for Exascale in 2022. HPCwire (2021). https://www.hpcwire.com/2021/09/08/how-argonne-is-preparing-for-exascale-in-2022/

23. Vassberg, J., Dehaan, M., Rivers, M., Wahls, R.: Development of a common research model for applied CFD validation studies. In: 26th AIAA Applied Aerodynamics Conference. American Institute of Aeronautics and Astronautics, Honolulu (2008). https://doi.org/10.2514/6.2008-6919

24. Walden, A., Nielsen, E., Diskin, B., Zubair, M.: A mixed precision multicolor point-implicit solver for unstructured grids on GPUs. In: 2019 IEEE/ACM 9th Workshop

on Irregular Applications: Architectures and Algorithms (IA3), pp. 23–30. IEEE, Denver (2019). https://doi.org/10.1109/IA349570.2019.00010

25. Wang, S., Kanwar, P.: BFloat16: The secret to high performance on Cloud TPUs (2019).   https://cloud.google.com/blog/products/ai-machine-learning/bfloat16-the-secret-to-high-performance-on-cloud-tpus

26. Wright, M.J., Candler, G.V., Prampolini, M.: Data-parallel lower-upper relaxation method for the Navier-Stokes equations. AIAA J. **34**(7), 1371–1377 (1996). https://doi.org/10.2514/3.13242

27. Yoon, S., Jameson, A.: An LU-SSOR scheme for the Euler and Navier-Stokes equations. In: 25th AIAA Aerospace Sciences Meeting. American Institute of Aeronautics and Astronautics, Reno (1987). https://doi.org/10.2514/6.1987-600

28. Zehner, P., Hashimoto, A.: Asynchronous execution of logging Kernels in a GPU accelerated CFD solver. In: Proceedings of the 54th Fluid Dynamics Conference/the 40th Aerospace Numerical Simulation Symposium, vol. JAXA-SP-22-007, pp. 331–339. Japan Aerospace Exploration Agency (JAXA), Morioka (2022). https://id.nii.ac.jp/1696/00049141/

29. Zehner, P., Hashimoto, A.: Acceleration of the data-parallel lower-upper relaxation time-integration method on GPU for an unstructured CFD solver. Comput. Fluids 105842 (2023). https://doi.org/10.1016/j.compfluid.2023.105842

Constrained Aerodynamic Shape Optimization Using Neural Networks and Sequential Sampling

Pavankumar Koratikere[1], Leifur Leifsson[1](✉)(iD), Slawomir Koziel[2,3](iD),
and Anna Pietrenko-Dabrowska[3](iD)

[1] School of Aeronautics and Astronautics, Purdue University,
West Lafayette, IN 47907, USA
{pkoratik,leifur}@purdue.edu
[2] Engineering Optimization & Modeling Center, Department of Engineering,
Reykjavík University, Menntavegur 1, 102 Reykjavík, Iceland
koziel@ru.is
[3] Faculty of Electronics Telecommunications and Informatics,
Gdansk University of Technology, Narutowicza 11/12, 80-233 Gdansk, Poland
anna.dabrowska@pg.edu.pl

Abstract. Aerodynamic shape optimization (ASO) involves computational fluid dynamics (CFD)-based search for an optimal aerodynamic shape such as airfoils and wings. Gradient-based optimization (GBO) with adjoints can be used efficiently to solve ASO problems with many design variables, but problems with many constraints can still be challenging. The recently created efficient global optimization algorithm with neural network (NN)-based prediction and uncertainty (EGONN) partially alleviates this challenge. A unique feature of EGONN is its ability to sequentially sample the design space and continuously update the NN prediction using an uncertainty model based on NNs. This work proposes a novel extension to EGONN that enables efficient handling of nonlinear constraints and a continuous update of the prediction and prediction uncertainty data sets. The proposed algorithm is demonstrated on constrained airfoil shape optimization in transonic flow and compared against state-of-the-art GBO with adjoints. The results show that the proposed constrained EGONN algorithm yields comparable optimal designs as GBO at a similar computational cost.

Keywords: Aerodynamic shape optimization · global surrogate modeling · neural networks · sequential sampling

1 Introduction

The goal of aerodynamic design is to find a shape (or adjusting an existing one) in airflow (such as airfoils, wings, helicopter rotor blades, wind turbine blades, and the external shape of aircraft) that improves a given quantity of interest

© The Author(s), under exclusive license to Springer Nature Switzerland AG 2023
J. Mikyška et al. (Eds.): ICCS 2023, LNCS 14075, pp. 425–438, 2023.
https://doi.org/10.1007/978-3-031-36024-4_33

(QoI) (such as the drag force), while adhering to appropriate constraints (such as a specified lift force) [10, 21, 25]. Computational aerodynamic design involves the use of computer simulations of the airflow past the shape, using computational fluid dynamics (CFD), to numerically evaluate the QoI and the associated constraints [22]. Aerodynamic shape optimization (ASO) is the automation of computational aerodynamic design by embedding the computer simulations within an optimization framework to search for the constrained optimal shape [8]. Key challenges of ASO include (1) time-consuming simulations, (2) a large number of design variables, (3) a large number of constraints, and (4) many model evaluations.

The state-of-the-art ASO is gradient-based optimization (GBO) with adjoints [7]. The main advantage of the adjoint approach is that the cost of a gradient calculation can be made nearly independent of the number of design variables. In the context of solving ASO problems with GBO and adjoints, this means that for each design evaluation the objective function and constraints need to be computed, which involves one primal computer simulation and one adjoint simulation for the objective and each constraint. For example, if there is one objective and two constraints, GBO with adjoints needs one primal computer solution and three adjoint solutions. Typically, the time per one primal CFD solution is roughly the same as one adjoint CFD solution [23]. This means, for the given example, that each design evaluation involves four simulations which yields the objective function value, the constraint function values, and the gradients of the objective and constraints. This is independent of the number of design variables, which renders the approach scalable to high-dimensional problems. It should be noted, however, that the computational cost grows quickly with the number of constraints.

Another way to solve ASO problems is to use surrogate-based optimization (SBO) where a surrogate replaces the time-consuming simulations in the computation of the objective and constraint functions (as well as its gradients, if needed) [26]. SBO has been around for a long time. A widely used approach is the efficient global optimization (EGO) algorithm [9]. In EGO, the design space is sampled initially using design of experiments [26], such as Latin hypercube sampling, and an initial surrogate model is constructed using kriging [2]. The kriging surrogate is then iteratively improved by sequentially sampling the design space using both the prediction and prediction variance. The key advantage of using kriging is that it can improve the surrogate accuracy for a given number samples since the samples can be assigned to regions where the surrogate shows poor accuracy (exploration) or where a local minimum is found (exploitation). A key disadvantage of kriging is that the computational cost grows quickly with the number of samples [15].

Neural network (NN) regression modeling [5], on the other hand, scales more efficiently for large data sets [20, 29]. A major limitation, however, is that uncertainty estimates are not readily available for a single prediction [20], and it is necessary to make use of an ensemble of NNs with a range of predictions [4, 18, 32] or use dropout to represent model uncertainty [3]. These algorithms are, however, computationally very intensive.

A recently created EGO algorithm with neural network (NN)-based prediction and uncertainty (called EGONN) partially alleviates these challenges [14]. In EGONN, a NN model is utilized to approximate nonlinear high-dimensional objective functions. The unique feature of EGONN is its ability to sequentially sample the design space and continuously update the NN-based prediction with a prediction uncertainty which is modeled by a second NN.

In this paper, a novel extension to EGONN is proposed that enables efficient handling of nonlinear constraints. Furthermore, the utilization of data for the prediction and prediction uncertainty is made more efficient by a continuously updating all data sets. The EGONN algorithm only sequentially updated the prediction data set. The proposed constrained EGONN (cEGONN) algorithm is demonstrated on an airfoil shape optimization problem in transonic flow involving one objective, two constraints, and thirteen design variables. The proposed algorithm is compared against state-of-the-art GBO with adjoints.

The next section describes ASO using state-of-the-art GBO and the proposed cEGONN algorithm. The following section presents the numerical results of the constrained airfoil shape optimization problem using those algorithms. The last section provides concluding remarks and possible next steps in this work.

2   Aerodynamic Shape Optimization

This section states the ASO problem formulation, and then describes its solution with GBO and adjoints and with the proposed cEGONN algorithm.

2.1   Problem Formulation

ASO involves minimization of the drag coefficient (C_d) of a baseline airfoil at a fixed free-stream Mach number1 (M_∞) and Reynolds number (Re_∞) with respect to design variables ($\mathbf{x}$) controlling the shape, subject to inequality constraints on the lift coefficient (C_l), and the airfoil cross-sectional area (A). Specifically, the constrained nonlinear minimization problem is formulated as:

$$\min_{\mathbf{x}} C_d(\mathbf{x})$$

subject to

$$C_l(\mathbf{x}) \geq C_{l_{ref}}$$
$$A(\mathbf{x}) \geq A_{ref} \tag{1}$$

and

$$\mathbf{x}_l \leq \mathbf{x} \leq \mathbf{x}_u$$

where $\mathbf{x}_l$ and $\mathbf{x}_u$ are the lower and upper bounds, respectively, $C_{l_{ref}}$ is a reference lift coefficient, and A_{ref} is the cross-sectional area of the baseline airfoil nondimensionalized with the square of the chord length.

The design variable vector usually consists of the shape parameterization variables and the angle of attack of the free-stream to the airfoil chordline. The shape of the airfoil can be parameterized using various methods such as free form deformation (FFD) [27], class-shape transformation (CST) [16], PARSEC [30], and hicks-henne bump function [6]. In this work, the CST parametrization method is used with a total of twelve variables, six for the upper surface and six for the lower surface.

The next two subsections describe the algorithms to solve the ASO problem, specifically, the GBO algorithm with adjoints, and proposed cEGONN algorithm.

2.2 Constrained GBO Algorithm with Adjoints

The extended design structure matrix (XDSM) [17] diagram shown in Fig. 1 outlines the GBO with adjoints algorithm implemented in this work. There are five major modules in the process which are arranged in the diagonal of the matrix. The input to the process (topmost row) are airfoil coordinates, volume mesh, and starting point, and output (leftmost column) is the optimized design. The coordinate file consists of points describing shape of the baseline airfoil in selig format. A structured surface mesh is created using the coordinate file and pyHyp [28] is used for creating an o-mesh grid by extruding the generated surface mesh outwards using hyperbolic mesh marching method.

The first module in the process is an optimizer which drives the entire process. The starting point for optimizer is essentially CST coefficients for baseline airfoil and angle of attack at which lift constraint is satisfied. In this work, sequential quadratic programming is used as optimizer which is implemented in pyOptSparse [33].

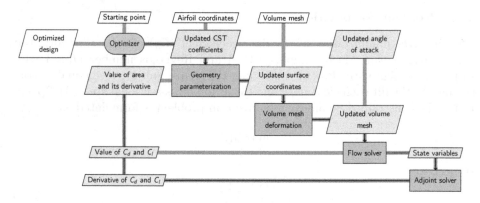

Fig. 1. Extended design structure matrix for gradient-based optimization with adjoints.

The second module consists of pyGeo [11] which provides CST parameterization and is initialized using the airfoil coordinate file. The third module performs mesh deformation using IDWarp [28] and uses volume mesh generated earlier for initialization. The flow solver module consists of ADflow [23] which is a finite-volume structured multiblock mesh solver. In this work, Approximate Newton

Krylov method [34] is used to start the sovler and reduce the residual norm to a value of 10^{-6}, relative to initial value, and then full Newton Krylov method is employed to further reduce the residual norm to 10^{-14}, relative to initial norm. Last module consists of a jacobian free discrete adjoint (computed using algorithmic differentiation) method which is implemented within ADflow [12]. The generalized minimal residual method is used for solving the adjoint equations with termination criteria set to a value of 10^{-11}, relative to initial norm.

The process starts with updated design variables from pyOptSparse. pyGeo receives updated CST coefficients and updates the surface coordinates, it also returns the cross-sectional area of updated airfoil shape and its derivative with respect to the variables. The updated surface coordinates are then passed to IDWarp which deforms volume mesh and sends it to flow solver. With the updated angle of attack and mesh, ADflow computes various field variables like pressure, velocity, density, etc. It also computes integral quantities like C_d, C_l, C_m. The converged results are then passed on to adjoint solver which calculates the derivative of objectives and constraints with respect to the design variables. This process is continued until one of the convergence criteria is met within optimizer. Each iteration involves one primal solution (which gives C_d, C_l, C_m) and multiple adjoint solutions (which give derivatives of C_d, C_l, C_m with respect to the design variables) depending on the number of objectives and constraints.

2.3   Constrained EGONN Algorithm

Algorithm 1 describes the proposed cEGONN algorithm for aerodynamic shape optimization. Two different data sets are generated using CFD which contain the objective function values $\mathbf{Y}$ and the constraint function values $\mathbf{G}$ for the sampling plan $\mathbf{X}$. In the first step, a neural network (NN_y) learns the mapping between $\mathbf{X}$ and $\mathbf{Y}$ in the first data set. Then, NN_y is used to get prediction $\widehat{\mathbf{Y}}$ and $\widehat{\mathbf{Y}}_u$ at $\mathbf{X}$ and $\mathbf{X_u}$, respectively. Following that the squared prediction error for both the predictions are computed as

$$\mathbf{S} = \sqrt{(\mathbf{Y} - \widehat{\mathbf{Y}})^2}, \tag{2}$$

and

$$\mathbf{S}_u = \sqrt{(\mathbf{Y}_u - \widehat{\mathbf{Y}}_u)^2}. \tag{3}$$

The samples, and the values of the prediction errors and constraints are combined to get a larger data set. A second neural network (NN_u) learns the mapping between the combined input $\widetilde{\mathbf{X}}$ and the prediction error $\widetilde{\mathbf{S}}$. A third neural network NN_g learns the mapping between the combined input $\widetilde{\mathbf{X}}$ and the combined constraint values $\widetilde{\mathbf{G}}$. Once the NN models are trained, the expected improvement, computed by

$$EI(\mathbf{x}) = \begin{cases} [y(\mathbf{x}^*) - \hat{y}(\mathbf{x})]\Phi(Z) + s(\mathbf{x})\phi(Z) & \text{if } s(\mathbf{x}) > 0 \\ 0 & \text{if } s(\mathbf{x}) = 0, \end{cases} \tag{4}$$

Algorithm 1. Constrained EGONN

Require: initial data sets $(\mathbf{X}, \mathbf{Y}, \mathbf{G})$ and $(\mathbf{X}, \mathbf{Y}, \mathbf{G})_u$

  repeat

    fit NN_y to data $(\mathbf{X}, \mathbf{Y})$

    use NN_y to get $\widehat{\mathbf{Y}}$ at $\mathbf{X}$ and $\widehat{\mathbf{Y}}_u$ at $\mathbf{X}_u$

    compute prediction error: $\mathbf{S} \leftarrow \sqrt{(\mathbf{Y} - \widehat{\mathbf{Y}})^2}$ and $\mathbf{S}_u \leftarrow \sqrt{(\mathbf{Y}_u - \widehat{\mathbf{Y}}_u)^2}$

    combine data: $\widetilde{\mathbf{X}} \leftarrow \mathbf{X} \cup \mathbf{X_u}$, $\widetilde{\mathbf{S}} \leftarrow \mathbf{S} \cup \mathbf{S_u}$ and $\widetilde{\mathbf{G}} \leftarrow \mathbf{G} \cup \mathbf{G_u}$

    fit NN_u to data $(\widetilde{\mathbf{X}}, \widetilde{\mathbf{S}})$

    fit NN_g to data $(\widetilde{\mathbf{X}}, \widetilde{\mathbf{G}})$

    $\mathbf{P} \leftarrow \arg\max EI(\boldsymbol{x})$ such that $\widehat{\mathbf{G}}(x) \leq 0$

    $\mathbf{X} \leftarrow \mathbf{X} \cup \mathbf{P}$

    get observations Y and G

    $\mathbf{Y} \leftarrow \mathbf{Y} \cup Y$ and $\mathbf{G} \leftarrow \mathbf{G} \cup G$

  until convergence

  $y^* \leftarrow \min(\mathbf{Y})$ such that constraints are satisfied

  $\mathbf{x}^* \leftarrow \arg\min(\mathbf{Y})$

  return $(\mathbf{x}^*, y^*)$

---

is maximized to get a new infill point $\mathbf{P}$, where $\hat{y}(\mathbf{x})$ is the NN_y prediction and $s(\mathbf{x})$ is NN_u prediction. The Z is a standard normal variable, and Φ and ϕ are cumulative distribution function and probability density function of standard normal distribution, respectively. CFD analysis is performed at the new point $\mathbf{P}$ to obtain the objective function Y and constraint functions G, which are then appended to the first data set. This process is continued until the convergence criteria is met. Unlike GBO, there is no need for adjoint solutions. Compared to the original EGONN, two data sets are utilized in a more efficient manner in the process. Moreover, the algorithm is adapted to handle constrained optimization problems instead of an unconstrained one. The neural networks in cEGONN are implemented within Tensorflow [1].

The next section gives the results of applying GBO algorithm with adjoints and the proposed cEGONN algorithm to the ASO of an airfoil in transonic flow.

3 Results

The general ASO problem formulation is given in (1). In this work, the baseline airfoil is the RAE 2822 (shown in Fig. 2) with the nondimensional reference cross-sectional area $A_{ref} = 0.777$. The free-stream Mach number is fixed at 0.734 and the Reynolds number at 6.5×10^6. The reference lift coefficient is set to $C_{l_{ref}} = 0.824$.

The airfoil shape is parameterized using CST [16] with a total of twelve coefficients, six for the upper surface and six for the lower surface. Thus, the design variable vector $\mathbf{x}$ consists of 13 elements in total, the angle of attack and twelve shape variables. The upper and lower bounds for shape variables are set to +30% and -30% perturbations, respectively, of the fitted CST coefficients for

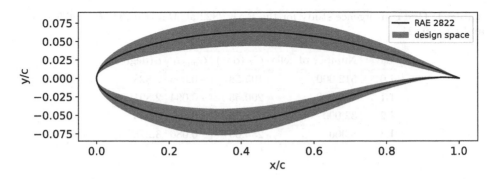

Fig. 2. The baseline RAE 2822 airfoil shape and the design space.

the baseline RAE 2822 airfoil, and the angle of attack is bounded between 1.5°
and 3.5°. Figure 2 shows the design space obtained with these bounds on the
shape variables.

An O-mesh grid around the airfoil is created using pyHyp. Table 1 shows
the result of grid convergence study for the mesh. The y^+ plus value for all the
levels is less than 1 and the mesh is extruded until $100c$. In this work, all the
computations are performed using the L1 mesh. Figure 3 shows the generated
far-field and surface L1 mesh.

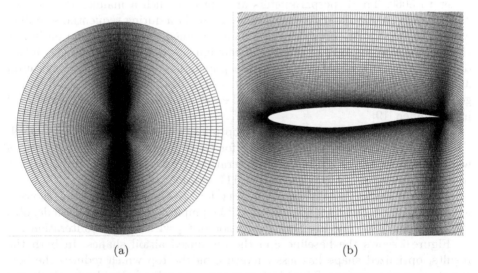

(a)   (b)

Fig. 3. Computational mesh for the airfoil flow simulation: (a) the far-field, and (b) a
zoom in near the airfoil surface

Table 1. Grid convergence study for the RAE 2822 at $M_\infty = 0.734$, $Re = 6.5 \times 10^6$, and $C_l = 0.824$.

Level	Number of cells	C_d (d.c.)	$C_{m,c/4}$	α(degree)
L0	512,000	195.58	−0.096	2.828
L1	128,000	200.55	−0.094	2.891
L2	32,000	213.26	−0.091	3.043
L3	8,000	235.05	−0.086	3.278

The initial data sets required for the cEGONN algorithm are generated using a Latin hypercube sampling plan [24] with the bounds described earlier. The first data set contains 50 samples and second data set contains 25 samples, hence, a total of 75 CFD simulations are performed initially. The convergence criteria is set to a maximum of 100 iterations. The NN_y consists of two hidden layers having eight and six neurons, respectively. The NN_u contains two hidden layers, each having eight neurons. The total number of neurons in NN_u is slightly higher than NN_y to avoid underfitting since it trains on a larger data set. The NN_g also contains two hidden layers but with four and three neurons, respectively. The area constraint is computed based on the airfoil shape using numerical integration, while NN_g provides values for the C_l constraint. In all the NNs, the hyperbolic tan is used as activation function and the number of epochs is set to 5000. The hyperparameters are tuned in such a manner that all the NNs slightly overfit the data since adding more data during sequential sampling will make the NN fit good. All the NNs are trained using the Adam optimizer [13] with a learning rate of 0.001. For maximizing the expected improvement, differential evolution [31] is used with a population size of 130. The mutation and recombination is set to 0.8 and 0.9, respectively, with a maximum of 100 generations. The constraints are handled explicitly using the strategy described in Lampinen [19].

Table 2 summarizes the result of optimization for GBO and cEGONN. Figure 4 shows the convergence history for GBO where a total of 37 iterations were needed for convergence. At each iteration, 1 primal solution and 2 adjoint solutions are computed which totals to 111 solutions.

Figure 5 shows the convergence history for cEGONN. A green dot shows a feasible sample, a red dot shows an infeasible sample, and the gray region denotes the initial samples. The line shows variation of C_d with respect to iterations.

Figure 6 shows the baseline and the optimized airfoil shapes. In both the results, optimized shape has less curvature on the top which reduces the flow speed on upper surface. This decreases the strength of the shock which can be clearly noted in the pressure plot shown in Fig. 7. Due to the decrease in curvature, the lift generated also decreases which is compensated by an increase in the aft curvature of the lower surface.

Table 2. Characteristics of the baseline and optimized airfoil shapes and the number of CFD simulations for each algorithm.

	RAE 2822	GBO	cEGONN
Primal	–	37	175
Adjoint	–	74	0
C_d (d.c.)	200.55	110	114
C_l	0.824	0.824	0.830
α	2.89°	2.83°	2.80°
A	0.777	0.777	0.779

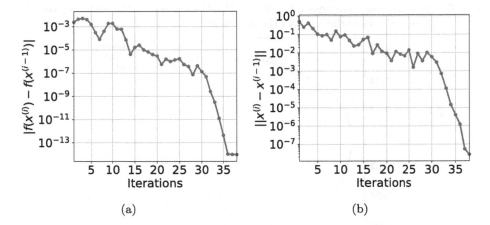

(a)   (b)

Fig. 4. Convergence history for gradient based optimization showing the change between iterations of: (a) the objective function, and (b) the design variable vector.

In Fig. 5 it is observed that in the early stages of sequential sampling, many infeasible samples are added, but as the iteration progresses the number of feasible samples increase. This is attributed to the fact that the number of samples in the initial data set is low which leads to inaccurate constraint predictions. As more samples are added, the constraint fitting improves and the number of feasible infills increase. Figure 8 shows the Mach contours for the baseline and the optimized shapes at the given free-stream conditions. In the baseline contour Fig. 8a, it can seen that there is a strong shock on the upper surface, whereas in the contour plots of optimized shapes, Figs. 8b and 8c, shock is not present. This shows the capability of the proposed cEGONN for ASO.

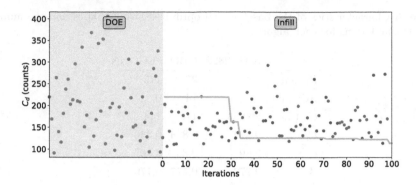

Fig. 5. Optimization convergence history for cEGONN showing the sampling and the drag coefficient values variation with the iterations.

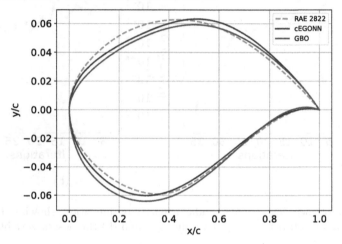

Fig. 6. The baseline and optimized airfoil shapes.

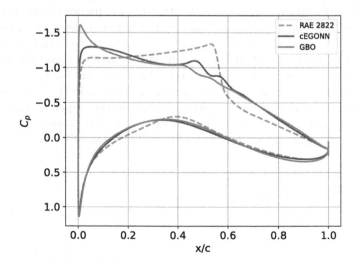

Fig. 7. Coefficient of pressure for the baseline and optimized airfoils.

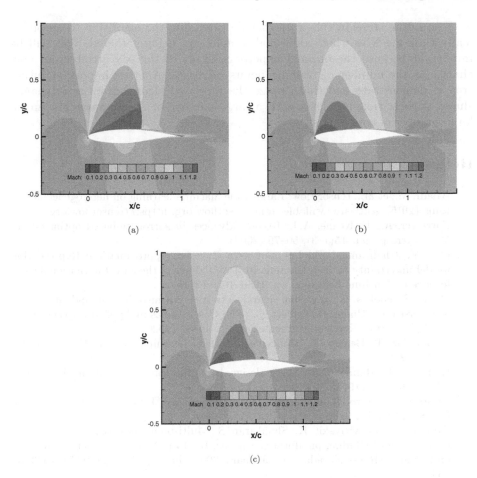

Fig. 8. Flow field Mach contours for (a) the baseline, (b) the GBO optimum, and (c) the cEGONN optimum.

4   Conclusion

In this work, a novel extension to the efficient global optimization with neural network (NN)-based prediction and uncertainty (EGONN) algorithm is proposed which enables the handling of nonlinear constraints. A unique feature of the proposed constrained EGONN algorithm is its ability to perform sequential sampling of the design space and updating the NN predictions of nonlinear objective and constraint functions.

A demonstration example involving airfoil shape optimization in transonic flow with one objective, two constraints, and twelve design variables shows that the proposed algorithm can obtain comparable optimal designs as gradient-based optimization with adjoints with similar computational cost.

The future steps in this work include extending the algorithm to automatically tune the NNs when additional samples are being gathered. This will be important because the optimal hyperparameters of the NN architecture may change as the number of training points increases. Another important future step is demonstrate and characterize the proposed algorithm on aerodynamic shape optimization problems with a large number of design variables, and a large number of constraints.

References

1. Abadi, M., et al.: TensorFlow: Large-scale machine learning on heterogeneous systems (2015), software available from tensorflow.org. https://tensorflow.org/
2. Forrester, A.I.J., Keane, A.J.: Recent advances in surrogate-based optimization. Prog. Aerosp. Sci. **45**(1–3), 50–79 (2009)
3. Gal, Y., Ghahramani, Z.: Dropout as a Bayesian approximation: Representing model uncertainty in deep learning. In: Proceedings of the 33rd International Conference on Machine Learning, pp. 1050–1059 (2016)
4. Goan, E., Fookes, C.: Bayesian neural networks: An introduction and survey. In: Mengersen, K., Pudlo, P., Robert, C. (eds) Case Studies in Applied Bayesian Data Science. Lecture Notes in Mathematics, pp. 45–87 (2020)
5. Goodfellow, I., Bengio, Y., Courville, A.: Deep Learning. The MIT Press, Cambridge (2016)
6. Hicks, R.M., Henne, P.A.: Wing design by numerical optimization. J. Aircraft **15**(7), 407–412 (1978)
7. Jameson, A.: Aerodynamic design via control theory. J. Sci. Comput. **3**, 233–260 (1988). 10.1.1.419.9280
8. Jameson, A., Leoviriyakit, K., Shankaran, S.: Multi-point aero-structural optimization of wings including planform variations. In: 45th Aerospace Sciences Meeting and Exhibit (Reno, Nevada 8–11, January, 2007). https://doi.org/10.2514/6.2007-764
9. Jones, D.R., Schonlau, M., Welch, W.J.: Efficient global optimization of expensive black-box functions. J. Glob. Optimiz. **13**(4), 455–492 (1998)
10. Kenway, G., Martins, J.R.R.A.: Aerostructural shape optimization of wind turbine blades considering site-specific winds. In: 12th AIAA/ISSMO Multidisciplinary Analysis and Optimization Conference (Victoria, British Columbia, 10–12 September, 2008). https://doi.org/10.2514/6.2008-6025
11. Kenway, G., Kennedy, G., Martins, J.R.: A CAD-free approach to high-fidelity aerostructural optimization. In: 13th AIAA/ISSMO Multidisciplinary Analysis and Optimization Conference (2010)
12. Kenway, G.K., Mader, C.A., He, P., Martins, J.R.: Effective adjoint approaches for computational fluid dynamics. Prog. Aerosp. Sci. **110**, 100542 (2019)
13. Kingma, D.P., Ba, J.: Adam: A method for stochastic optimization. arXiv preprint arXiv:1412.6980 (2014)
14. Koratikere, P., Leifsson, L.T., Barnet, L., Bryden, K.: Efficient global optimization algorithm using neural network-based prediction and uncertainty. In: AIAA SCITECH 2023 Forum, p. 2683 (2023)

15. Koziel, S., Echeverria-Ciaurri, D., Leifsson, L.: Surrogate-based methods. In: Koziel, S., Yang, X.S. (eds.) Computational Optimization, Methods and Algorithms, Series: Studies in Computational Intelligence. Springer, Heidelberg (2011). https://doi.org/10.1007/978-3-642-20859-1_3
16. Kulfan, B.M.: Universal parametric geometry representation method. J. Aircraft **45**(1), 142–158 (2008)
17. Lambe, A.B., Martins, J.R.R.A.: Extensions to the design structure matrix for the description of multidisciplinary design, analysis, and optimization processes. Struct. Multidiscip. Optimiz. **46**, 273–284 (2012)
18. Lampinen, J., Vehtari, A.: Bayesian approach for neural networks - review and case studies. Neural Netw. **14**(3), 257–274 (2001)
19. Lampinen, J.: A constraint handling approach for the differential evolution algorithm. In: Proceedings of the 2002 Congress on Evolutionary Computation. CEC'02 (Cat. No. 02TH8600), vol. 2, pp. 1468–1473. IEEE (2002)
20. Lim, Y.F., Ng, C.K., Vaitesswar, U.S., Hippalgaonkar, K.: Extrapolative Bayesian optimization with Gaussain process and neural network ensemble surrogate models. Adv. Intell. Syst. **3**, 2100101 (2021)
21. Lyu, Z., Kenway, G.K.W., Martins, J.R.R.A.: Aerodynamic shape optimization investigations of the common research model wing benchmark. AIAA J. **53**(4), 968–984 (2015). https://doi.org/10.2514/1.J053318
22. Mader, C.A., Martins, J.R.R.A.: Derivatives for time-spectral computational fluid dynamics using an automatic differentiation adjoint. AIAA J. **50**(12), 2809–2819 (2012). https://doi.org/10.2514/1.J051658
23. Mader, C.A., Kenway, G.K., Yildirim, A., Martins, J.R.: ADflow: An open-source computational fluid dynamics solver for aerodynamic and multidisciplinary optimization. J. Aerosp. Inf. Syst. **17**(9), 508–527 (2020)
24. McKay, M.D., Beckman, R.J., Conover, W.J.: A comparison of three methods for selecting values of input variables in the analysis of output from a computer code. Technometrics **21**(2), 239–245 (1979)
25. Mousavi, A., Nadarajah, S.: Heat transfer optimization of gas turbine blades using an adjoint approach. In: 13th AIAA/ISSMO Multidisciplinary Analysis and Optimizaiton Conference (Fort Worth, Texas, 13–15 September, 2010). https://doi.org/10.2514/6.2010-9048
26. Queipo, N.V., Haftka, R.T., Shyy, W., Goel, T., Vaidyanathan, R., Tucker, P.K.: Surrogate-based analysis and optimization. Prog. Aerosp. Sci. **21**(1), 1–28 (2005)
27. Samareh, J.: Aerodynamic shape optimization based on free-form deformation. In: 10th AIAA/ISSMO Multidisciplinary Analysis and Optimization Conference, p. 4630 (2004)
28. Secco, N.R., Kenway, G.K., He, P., Mader, C., Martins, J.R.: Efficient mesh generation and deformation for aerodynamic shape optimization. AIAA J. **59**(4), 1151–1168 (2021)
29. Snoek, J., et al.: Scalable bayesian optimization using deep neural networks. In: Proceedings of the 32nd International Conference on Machine Learning, pp. 2171–2180 (2015)
30. Sobieczky, H.: Parametric airfoils and wings. In: Recent Development of Aerodynamic Design Methodologies: Inverse Design and Optimization, pp. 71–87 (1999)
31. Storn, R., Price, K.: Differential evolution - A simple and efficient heuristic for global optimization over continuous spaces. J. Glob. Optimiz. **11**, 341–359 (1997)

32. Titterington, D.M.: Bayesian methods for neural networks and related models. Statist. Sci. **19**(1), 128–139 (2004)
33. Wu, N., Kenway, G., Mader, C.A., Jasa, J., Martins, J.R.: pyoptsparse: A python framework for large-scale constrained nonlinear optimization of sparse systems. J. Open Source Softw. **5**(54), 2564 (2020)
34. Yildirim, A., Kenway, G.K., Mader, C.A., Martins, J.R.: A Jacobian-free approximate Newton-Krylov startup strategy for rans simulations. J. Comput. Phys. **397**, 108741 (2019)

Optimal Knots Selection in Fitting Degenerate Reduced Data

Ryszard Kozera[1,2(✉)] and Lyle Noakes[2]

[1] Institute of Information Technology, Warsaw University of Life Sciences - SGGW,
ul. Nowoursynowska 159, 02-776 Warsaw, Poland
ryszard_kozera@sggw.edu.pl, ryszard.kozera@uwa.edu.au,
ryszard.kozera@gmail.com
[2] School of Physics, Mathematics and Computing, The University of Western
Australia, 35 Stirling Highway, Crawley 6009, Perth, Australia
lyle.noakes@uwa.edu.au

Abstract. The problem of fitting a given ordered sample of data points in arbitrary Euclidean space is addressed. The corresponding interpolation knots remain unknown and as such must be first somehow found. The latter leads to a highly non-linear multivariate optimization task, equally non-trivial for theoretical analysis and for derivation of a computationally efficient numerical scheme. The non-degenerate case of at least four data points can be handled by Leap-Frog algorithm merging generic and non-generic univariate overlapping optimizations. Sufficient conditions guaranteeing the unimodality for both cases of Leap-Frog optimization tasks are already established in the previous research. This work complements the latter by analyzing the degenerate situation i.e. the case of fitting three data points, for which Leap-Frog cannot be used. It is proved here that the related univariate cost function is always unimodal yielding a global minimum assigned to the missing internal-point knot (with no loss both terminal knots can be assumed to be fixed). Illustrative examples supplement the analysis in question.

Keywords: Optimization · Curve Modelling · Data Fitting

1 Introduction

The problem of fitting data is a classical problem for which numerous interpolation techniques can be applied (see e.g. [1,11,16]). In classical setting, most of such schemes admit a sequence of $n + 1$ input points $\mathcal{M}_n = \{x_n\}_{i=0}^n$ in arbitrary Euclidean space $\mathbb{E}^m$ accompanied by the sequence of the corresponding interpolation knots $\mathcal{T} = \{t_i\}_{i=0}^n$. The problem of data fitting and modeling gets complicated for the so-called *reduced data* for which only the points $\mathcal{M}_n$ are given. Here, for a given fitting scheme, different choices of ordered interpolation knots $\{\hat{t}_i\}_{i=0}^n$ render different curves. An early work on this topic can be found in [12] subsequently investigated e.g. in [10,11]. In particular, various

© The Author(s), under exclusive license to Springer Nature Switzerland AG 2023
J. Mikyška et al. (Eds.): ICCS 2023, LNCS 14075, pp. 439–453, 2023.
https://doi.org/10.1007/978-3-031-36024-4_34

quantitative criteria (often for special $m = 2, 3$) are introduced to measure the suitability of a special choice of $\{\hat{t}_i\}_{i=0}^n$ - e.g. the convergence rate of the interpolant to the unknown curve once $\mathcal{M}_n$ is getting denser. A more recent work in which different parameterization of the unknown knots are discussed (including the so-called cumulative chord or its extension called the exponential parameterization) can be found e.g. in [2,8,11]. One of the approaches (see [3]) to substitute the unknown knots $\mathcal{T}_{n-1} = (t_1, t_2, \ldots, t_{n-1})$ (here one can set $t_0 = 0$ and $t_n = T$, e.g. with T as cumulative chord) is to minimize $\int_{t_0}^{t_n} \|\ddot{\gamma}^{NS}(t)\|^2 dt$ subject to $0 < t_1 < t_2 < \ldots < t_{n-1} < T$, where $\gamma^{NS} : [0, T] \to \mathbb{E}^m$ defines a natural spline (see [1]) based on $\mathcal{M}_n$ and $\mathcal{T}_{n-1}$. Such constrained optimization task can be transformed into minimizing (see e.g. [3]) the following multivariate cost function:

$$\mathcal{J}(t_1, t_2, \ldots, t_{n-1}) = 4 \sum_{i=0}^{n-1} \left(\frac{-1}{(\Delta t_i)^3} (-3\|x_{i+1} - x_i\|^2 + 3\langle v_i + v_{i+1} | x_{i+1} - x_i \rangle \Delta t_i \right.$$
$$\left. -(\|v_i\|^2 + \|v_{i+1}\|^2 + \langle v_i | v_{i+1} \rangle)(\Delta t_i)^2 \right), \quad (1)$$

where the set $\{v_i\}_{i=0}^n$ represents the respective velocities at $\mathcal{M}_n$ which are expressible in terms of $\mathcal{M}_n$ and parameters $\mathcal{T}_{n-1}$ (see [1]). The latter constitutes a highly non-linear multivariate optimization task difficult to analyze and to solve numerically (see [4]). For technical reason it is also assumed that $x_i \neq x_{i+1}$, for $i = 0, 1, 2, \ldots, n-1$. Leap-Frog algorithm is a possible remedy here ($n \geq 3$ yields a non-degenerate case of (1) yielding at least four interpolation points) which with the aid of iterative overlapping univariate optimizations (generic and non-generic one) computes a critical point $(t_1^{opt}, t_2^{opt}, \ldots, t_{n-1}^{opt})$ to (1). More information on Leap-Frog in the context of minimizing (1) together with the analysis on establishing sufficient conditions enforcing the unimodality of the respective univariate cost functions can be found in [6,7].

This work[1] extends theoretical analysis on minimizing (1) via Leap-Frog algorithm (performed for $n \geq 3$) to the remaining degenerate case of $n = 2$. Here neither generic nor non-generic case of Leap-Frog is applicable. Indeed for the univariate generic Leap-Frog optimization (see [6]) both velocities in the k-iteration process i.e. v_i^k and v_{i+2}^k (at points x_i and x_{i+2}, respectively) are assumed to be temporarily fixed and a local complete spline $\gamma_i^C : [t_i^k, t_{i+2}^k] \to \mathbb{E}^m$ (see [1]) is used to recompute the knot t_{i+1}^{k-1} to t_{i+1}^k and consequently the velocity v_{i+1}^{k-1} to v_{i+1}^k (at point x_{i+1}) upon minimizing $\mathcal{E}_i(t_{i+1}) = \int_{t_i^k}^{t_{i+2}^k} \|\ddot{\gamma}_i^C(t)\| dt$. Similarly univariate non-generic Leap-Frog optimization (see [7]) relies on $a_0 = \mathbf{0}$ and v_2^k or on v_{n-2}^k and $a_n = \mathbf{0}$ given (with velocities v_2^k and v_{n-2}^k fed by generic Leap-Frog iterations), where a_0 and a_n represent the corresponding accelerations fixed at x_0 and x_n, respectively. Here knots t_1^{k-1} (and t_{n-1}^{k-1}) are recomputed into t_1^k (and t_{n-1}^k) with the corresponding velocities v_1^{k-1} (and v_{n-1}^{k-1}) updated to v_1^k (and v_{n-1}^k). For the exact local energy formulation and the related analysis see [7].

───────────────

[1] This work is a part of Polish National Centre of Research and Development Project POIR.01.02.00-00-0160/20.

Noticeably for $n = 2$ (i.e. degenerate case of reduced data with three interpolation points) accompanied by $a_0 = \mathbf{0}$ and $a_2 = \mathbf{0}$ minimizing (1) reformulates into:

$$\mathcal{E}_{deg}(t_1) = \mathcal{J}(t_1) = 4\sum_{i=0}^{1} \left(\frac{-1}{(\Delta t_i)^3}(-3\|x_{i+1} - x_i\|^2 + 3\langle v_i + v_{i+1}|x_{i+1} - x_i\rangle \Delta t_i \right.$$

$$\left. - (\|v_i\|^2 + \|v_{i+1}\|^2 + \langle v_i|v_{i+1}\rangle)(\Delta t_i)^2 \right), \quad (2)$$

which does not fall into one of the Leap-Frog so-far derived schemes (i.e. generic and non-generic case handling $\mathcal{M}_n$ with $n \geq 3$).

In this paper we prove the unimodality of (2) and supplement illustrative examples. Consequently, computing the optimal knot t_1 is not susceptible to the initial guess and forms a global minimum of (2). Fitting reduced data $\mathcal{M}_n$ via optimization (1) (or with other schemes replacing the unknown knots $\mathcal{T}$) applies in modeling [16] (e.g. computer graphics and computer vision), in approximation and interpolation [5,11] (e.g. trajectory planning, length estimation, image segmentation or data compression) as well as in many other engineering and physics problems [13] (e.g. robotics or particle trajectory estimation).

2   Degenerate Case: First and Last Accelerations Given

Assume that for three data points $x_0, x_1, x_2 \in \mathcal{M}_n$ (for $n = 2$) the interpolation knots t_0 and t_2 are somehow known together with respective first and terminal accelerations $a_0, a_2 \in \mathbb{R}^m$. In fact, one can safely assume here (upon coordinate shift and rescaling) that $t_0 = 0$ and $t_2 = T_{cc} = \|x_1 - x_0\| + \|x_2 - x_1\|$ representing a cumulative chord parameterization. Let a C^2 piecewise cubic (depending on varying $t_1 \in (t_0, t_2)$) denoted by $\gamma_{deg}^c : [t_0, t_2] \to \mathbb{E}^m$ (i.e. a cubic on each $[t_0, t_1]$ and $[t_1, t_2]$) satisfy:

$$\gamma_{deg}^c(t_j) = x_j, \quad j = 0, 1, 2; \qquad \ddot{\gamma}_{deg}^c(t_j) = a_k, \quad k = 0, 2,$$

and be C^2 class over $[t_0, t_2]$. Upon introducing the mapping $\phi_{deg} : [t_0, t_2] \to [0, 1]$

$$\phi_{deg}(t) = \frac{t - t_0}{t_2 - t_0} \qquad (3)$$

the reparameterized curve $\tilde{\gamma}_{deg}^c : [0, 1] \to \mathbb{E}^m$ defined as $\tilde{\gamma}_{deg}^c = \gamma_{deg}^c \circ \phi_{deg}^{-1}$ satisfies, for $0 < s_1 < 1$ (where $s_1 = \phi_{deg}(t_1)$):

$$\tilde{\gamma}_{deg}^c(0) = x_0, \quad \tilde{\gamma}_{deg}^c(s_1) = x_1, \quad \tilde{\gamma}_{deg}^c(1) = x_2,$$

with the adjusted first and terminal accelerations $\tilde{a}_0, \tilde{a}_2 \in \mathbb{R}^m$

$$\tilde{a}_0 = \tilde{\gamma}_{deg}^{c''}(0) = (t_2 - t_0)^2 a_0, \quad \tilde{a}_2 = \tilde{\gamma}_{deg}^{c''}(1) = (t_2 - t_0)^2 a_2. \qquad (4)$$

Remark 1. An easy inspection shows (for each $s_1 = \phi_{deg}(t_1)$) that $\tilde{\mathcal{E}}_{deg}(s_1)$

$$= \int_0^1 \|\tilde{\gamma}_{deg}^{c''}(s)\| ds = (t_2 - t_0)^3 \int_{t_0}^{t_2} \|\ddot{\gamma}_{deg}^c(t)\|^2 dt = (t_2 - t_0)^3 \mathcal{E}_{deg}(t_1). \qquad (5)$$

Hence all critical points s_1^{crit} of $\tilde{\mathcal{E}}_{deg}$ are mapped (one-to-one) onto critical points $t_1^{crit} = \phi_{deg}^{-1}(s_1^{crit}) = s_1^{crit}(t_2 - t_0) + t_0$ of $\mathcal{E}_{deg}$. Consequently all optimal points of $\mathcal{E}_{deg}$ and $\tilde{\mathcal{E}}_{deg}$ are conjugated with $t_1^{opt} = \phi_{deg}^{-1}(s_1^{opt})$. $\qquad\square$

We determine now the explicit formula for $\tilde{\mathcal{E}}_{deg}$. In doing so, for $\tilde{\gamma}_{deg}^c$ (with $s_1 \in (0,1)$ as additional parameter) define now (here $\tilde{\gamma}_{deg}^{lc}(s_1) = \tilde{\gamma}_{deg}^{lc}(s_1)$):

$$\tilde{\gamma}_{deg}^c(s) = \begin{cases} \tilde{\gamma}_{deg}^{lc}(s) \text{ , for } s \in [0, s_1] \\ \tilde{\gamma}_{deg}^{rc}(s) \text{ , for } s \in [s_1, 1] \end{cases}$$

where, with $c_j^{deg}, d_j^{deg} \in \mathbb{R}^m$ (for $j = 0, 1, 2, 3$)

$$\tilde{\gamma}_{deg}^{lc}(s) = c_0^{deg} + c_1^{deg}(s - s_1) + c_2^{deg}(s - s_1)^2 + c_3^{deg}(s - s_1)^3 \text{ ,}$$
$$\tilde{\gamma}_{deg}^{rc}(s) = d_0^{deg} + d_1^{deg}(s - s_1) + d_2^{deg}(s - s_1)^2 + d_3^{deg}(s - s_1)^3 \text{ ,}$$

the following must hold:

$$\tilde{\gamma}_{deg}^{lc}(0) = x_0 \text{ ,} \quad \tilde{\gamma}_{deg}^{lc}(s_1) = \tilde{\gamma}_{deg}^{rc}(s_1) = x_1 \text{ ,} \quad \tilde{\gamma}_{deg}^{rc}(1) = x_2 \text{ ,} \tag{6}$$

and also

$$\tilde{\gamma}_{deg}^{lc''}(0) = \tilde{a}_0 \text{ ,} \qquad \tilde{\gamma}_{deg}^{rc''}(1) = \tilde{a}_2 \text{ ,} \tag{7}$$

together with the C^1 and C^2 class constraints at $s = s_1$:

$$\tilde{\gamma}_{deg}^{lc'}(s_1) = \tilde{\gamma}_{deg}^{rc'}(s_1) \text{ ,} \qquad \tilde{\gamma}_{deg}^{lc''}(s_1) = \tilde{\gamma}_{deg}^{rc''}(s_1) \text{ .} \tag{8}$$

Without loss, we may assume that

$$\tilde{x}_0 = x_0 - x_1 \text{ ,} \quad \tilde{x}_1 = \mathbf{0} \text{ ,} \quad \tilde{x}_2 = x_2 - x_1 \text{ ,} \tag{9}$$

and hence by (6) we have

$$\tilde{\gamma}_{deg}^{lc}(0) = \tilde{x}_0 \text{ ,} \quad \tilde{\gamma}_{deg}^{lc}(s_1) = \tilde{\gamma}_{deg}^{rc}(s_1) = \mathbf{0} \text{ ,} \quad \tilde{\gamma}_{deg}^{rc}(1) = \tilde{x}_2 \text{ .} \tag{10}$$

Therefore combining now (8) with $\tilde{x}_1 = \mathbf{0}$ we obtain

$$\tilde{\gamma}_{deg}^{lc}(s) = c_1^{deg}(s - s_1) + c_2^{deg}(s - s_1)^2 + c_3^{deg}(s - s_1)^3 \text{ ,}$$
$$\tilde{\gamma}_{deg}^{rc}(s) = c_1^{deg}(s - s_1) + c_2^{deg}(s - s_1)^2 + d_3^{deg}(s - s_1)^3 \text{ ,} \tag{11}$$

with $c_0^{deg} = d_0^{deg} = \mathbf{0}$. The unknown vectors $c_1^{deg}, c_2^{deg}, c_3^{deg}, d_3^{deg}$ appearing in (11) are uniquely determined by solving the following system of four linear vector equations obtained from and (7) and (10):

$$\tilde{x}_0 = -c_1^{deg} s_1 + c_2^{deg} s_1^2 - c_3^{deg} s_1^3 \text{ ,}$$
$$\tilde{x}_2 = c_1^{deg}(1 - s_1) + c_2^{deg}(1 - s_1)^2 + d_3^{deg}(1 - s_1)^3 \text{ ,}$$
$$\tilde{a}_0 = 2c_2^{deg} - 6c_3^{deg} s_1 \text{ ,}$$
$$\tilde{a}_2 = 2c_2^{deg} + 6d_3^{deg}(1 - s_1) \text{ .} \tag{12}$$

An inspection reveals that vector coefficients:

$$c_1^{deg} = -\frac{\tilde{a}_0 s_1^2 - \tilde{a}_2 s_1^2 - 2\tilde{a}_0 s_1^3 + 2\tilde{a}_2 s_1^3 + \tilde{a}_0 s_1^4 - \tilde{a}_2 s_1^4 - 6\tilde{x}_0 + 12 s_1 \tilde{x}_0 - 6 s_1^2 \tilde{x}_0}{6(s_1 - 1)s_1}$$
$$\quad - \frac{6 s_1^2 \tilde{x}_2}{6(s_1 - 1)s_1} \, ,$$

$$c_2^{deg} = -\frac{-\tilde{a}_2 s_1 - \tilde{a}_0 s_1^2 + 2\tilde{a}_2 \, s_1^2 + \tilde{a}_0 s_1^3 - \tilde{a}_2 s_1^3 + 6\tilde{x}_0 - 6 s_1 \tilde{x}_0 + 6 s_1 \tilde{x}_2}{4(s_1 - 1)s_1} \, ,$$

$$c_3^{deg} = -\frac{-2\tilde{a}_0 s_1 - \tilde{a}_2 s_1 + \tilde{a}_0 s_1^2 + 2\tilde{a}_2 s_1^2 + \tilde{a}_0 s_1^3 - \tilde{a}_2 s_1^3 + 6\tilde{x}_0 - 6 s_1 \tilde{x}_0 + 6 s_1 \tilde{x}_2}{12(s_1 - 1)s_1^2} \, ,$$

$$d_3^{deg} = -\frac{-3\tilde{a}_2 s_1 - \tilde{a}_0 s_1^2 + 4\tilde{a}_2 s_1^2 + \tilde{a}_0 s_1^3 - \tilde{a}_2 s_1^3 + 6\tilde{x}_0 - 6 s_1 \tilde{x}_0 + 6 s_1 \tilde{x}_2}{12(s_1 - 1)^2 s_1} \, , \quad (13)$$

solve (as functions in s_1) the system (12). Alternatively one may e.g. use *Mathematica* function *Solve* to find explicit formulas for (13) solving (12). In our special degenerate case of $n = 2$ we have $a_0 = a_2 = 0$ (with $\tilde{a}_0 = \tilde{a}_2 = \mathbf{0}$ - see (4)) and thus (13) reads as:

$$c_1^{dego} = -\frac{-6\tilde{x}_0 + 12 s_1 \tilde{x}_0 - 6 s_1^2 \tilde{x}_0 + 6 s_1^2 \tilde{x}_2}{6(s_1 - 1)s_1} \, ,$$

$$c_2^{dego} = -\frac{6\tilde{x}_0 - 6 s_1 \tilde{x}_0 + 6 s_1 \tilde{x}_2}{4(s_1 - 1)s_1} \, ,$$

$$c_3^{dego} = -\frac{6\tilde{x}_0 - 6 s_1 \tilde{x}_0 + 6 s_1 \tilde{x}_2}{12(s_1 - 1)s_1^2} \, ,$$

$$d_3^{dego} = -\frac{6\tilde{x}_0 - 6 s_1 \tilde{x}_0 + 6 s_1 \tilde{x}_2}{12(s_1 - 1)^2 s_1} \, . \quad (14)$$

Note that the formula for the energy $\tilde{\mathcal{E}}_{deg}$ reads as:

$$\tilde{\mathcal{E}}_{deg}(s_1) = \int_0^{s_1} \|\tilde{\gamma}_{deg}^{lc''}(s)\|^2 ds + \int_{s_1}^1 \|\tilde{\gamma}_{deg}^{rc''}(s)\|^2 ds \, . \quad (15)$$

Combining (15) with (11) and (14) yields (upon e.g. using *Mathematica* functions *Integrate* and *FullSimplify*):

$$\tilde{\mathcal{E}}_{deg}(s_1) = \frac{3(\|\tilde{x}_0\|^2 (s_1 - 1)^2 + s_1(\|\tilde{x}_2\|^2 s_1 + (2 - 2s_1)\langle \tilde{x}_0 | \tilde{x}_2 \rangle))}{(s_1 - 1)^2 s_1^2}$$

$$= 3 \left\| \frac{\tilde{x}_0(s_1 - 1) - \tilde{x}_2 s_1}{(s_1 - 1)s_1} \right\|^2 \, , \quad (16)$$

for arbitrary m. To justify the latter one proves first that (16) holds for $m = 1$. Then the same vector version is derived by using m times one dimensional formula with the observation that one dimensional sections of homogeneous boundary conditions $a_0 = a_2 = \mathbf{0}$ are additive. Coupling $\tilde{x}_0 = x_0 - x_1$ and $\tilde{x}_2 = x_2 - x_1$ with (16) yields:

$$\tilde{\mathcal{E}}_{deg}(s_1) = 3 \left\| \frac{x_1 - x_0 - s_1(x_2 - x_0)}{(s_1 - 1)s_1} \right\|^2. \tag{17}$$

By (16) or (17) we have

$$\lim_{s_1 \to 0^+} \tilde{\mathcal{E}}_{deg}(s_1) = \lim_{s_1 \to 1^-} \tilde{\mathcal{E}}_{deg}(s_1) = +\infty. \tag{18}$$

Thus as $\tilde{\mathcal{E}}_{deg} \geq 0$, $\tilde{\mathcal{E}}_{deg} \in C^1((0,1))$ there exists a global minimum $s_1^{opt} \in (0,1)$ of $\tilde{\mathcal{E}}_{deg}$ for which $\tilde{\mathcal{E}}'_{deg}$ vanishes. We use next the assumption that $x_i \neq x_{i+1}$ (for $i = 0, 1$) as then $\tilde{x}_0 \neq 0$ and $\tilde{x}_2 \neq 0$. Similarly, (18) holds for $x_0 = x_2$.

An easy inspection shows that (use alternatively symbolic differentiation in *Mathematica* and *FullSimplify*):

$$\tilde{\mathcal{E}}'_{deg}(s_1) = -\frac{6(\|\tilde{x}_0\|^2(s_1-1)^3 + \|\tilde{x}_2\|^2 s_1^3 - (s_1-1)s_1(2s_1-1)\langle\tilde{x}_0|\tilde{x}_2\rangle)}{(s_1-1)^3 s_1^3}. \tag{19}$$

The numerator of (19) is a polynomial of degree 3 ($\tilde{\mathcal{E}}'_{deg}(s_1) = \frac{-1}{(s_1-1)^3 s_1^3} N_{deg}(s_1)$)

$$N_{deg}(s_1) = b_0^{deg} + b_1^{deg} s_1 + b_2^{deg} s_1^2 + b_3^{deg} s_1^3 \tag{20}$$

with vector coefficients $b_j^{deg} \in \mathbb{R}^m$ (for $j = 0, 1, 2, 3$) equal to (apply e.g. *Mathematica* functions *Factor* and *FullSimplify*):

$$\frac{b_0^{deg}}{6} = -\|\tilde{x}_0\|^2 = -\|x_1\|^2 - \|x_0\|^2 + 2\langle x_0|x_1\rangle,$$

$$\frac{b_1^{deg}}{6} = 3\|\tilde{x}_0\|^2 - \langle\tilde{x}_0|\tilde{x}_2\rangle = 3\|x_0\|^2 + 2\|x_1\|^2 - 5\langle x_0|x_1\rangle + \langle x_1|x_2\rangle - \langle x_0|x_2\rangle,$$

$$\frac{b_2^{deg}}{6} = 3(\langle\tilde{x}_0|\tilde{x}_2\rangle - \|\tilde{x}_0\|^2) = 3(-\langle x_1|x_2\rangle + \langle x_1|x_0\rangle + \langle x_0|x_2\rangle - \|x_0\|^2),$$

$$\frac{b_3^{deg}}{6} = \|\tilde{x}_2 - \tilde{x}_0\|^2 = \|x_2\|^2 + \|x_0\|^2 - 2\langle x_0|x_2\rangle. \tag{21}$$

Note that by (16) the sufficient condition for $\mathcal{E}_{deg}$ to vanish is the *colinearity* of shifted data $x_1 - x_0$ and $x_2 - x_1$ and the existence of $s_1^{opt} \in (0,1)$ satisfying $x_1 - x_0 = s_1^{opt}(x_2 - x_0)$. Such s_1^{opt} yields the global minimum of $\mathcal{E}_{deg}$.

In a search for critical points and a global optimum of $\tilde{\mathcal{E}}_{deg}$, one can invoke *Mathematica Package Solve* which can find all roots (real and complex) for a given low order polynomial. Upon computing the roots of (20) we select only these which are real and belong to $(0,1)$. Next we evaluate $\tilde{\mathcal{E}}_{deg}$ on each critical point $s_1^{crit} \in (0,1)$ and choose s_1^{crit} with minimal energy $\tilde{\mathcal{E}}_{deg}$ as global optimizer over $(0,1)$. We shall perform more exhaustive analysis for the existence of critical points of $\tilde{\mathcal{E}}_{deg}$ over $(0,1)$ in Sect. 3 (for arbitrary m).

3   Critical Points for Degenerate Case

We investigate now the character of critical points for $\tilde{\mathcal{E}}_{deg}$ over the interval $(0, 1)$ (for degenerate case of reduced data $\mathcal{M}_n$ with $n = 2$). In Sect. 2 the existence of a global minimum for $\tilde{\mathcal{E}}_{deg}$ over $(0, 1)$ is justified.

Example 1. Consider first a special case i.e. of *co-linear data* $\mathcal{M}_n$ for which $\tilde{x}_0 = k\tilde{x}_2$. Recall that cumulative chord assigns to $\mathcal{M}_n$ the knots $\hat{t}_0 = 0$, $\hat{t}_1 = \|x_1 - x_0\| = \|\tilde{x}_0\|$ and $\hat{t}_2 = \hat{t}_1 + \|x_2 - x_1\| = \|\tilde{x}_0\| + \|\tilde{x}_2\|$. Thus the normalized cumulative chord reads as $s_0^{cc} = 0$, $s_1^{cc} = \|\tilde{x}_0\|/(\|\tilde{x}_0\| + \|\tilde{x}_2\|)$ and $s_2^{cc} = 1$.

For $k < 0$ i.e. *co-linearly ordered data* the cumulative chord $\hat{s}_1^{cc}$ is the global optimizer s_g nullifying $\tilde{\mathcal{E}}_{deg}(s_1^{cc}) = 0$. Indeed by (16) we see that $\tilde{\mathcal{E}}_{deg}$ vanishes if and only if $\tilde{x}_0(1 - s_1) + s_1\tilde{x}_2 = \mathbf{0}$. Since $\tilde{x}_0 = k\tilde{x}_2$ we have $(k(1 - s_1) + s_1)\tilde{x}_2 = \mathbf{0}$, which as $\tilde{x}_2 \neq \mathbf{0}$ yields $k(1 - s_1) + s_1 = 0$ and therefore as $k < 0$ the global optimizer $s_1 = -k/(1 - k) = |k|/(1 + |k|) \in (0, 1)$. The latter coincides with cumulative chord $\hat{s}_1^{cc}$ for $\mathcal{M}_n$. This fact can be independently inferred by defining $x(s) = \tilde{x}_0(1 - s) + \tilde{x}_2 s$ which satisfies $x(0) = \tilde{x}_0$, $x(1) = \tilde{x}_2$, $x''(0) = x''(1) = \mathbf{0}$. As $\tilde{x}_0 = k\tilde{x}_2$ with $k < 0$ we also have $x(\hat{s}_1^{cc}) = \tilde{x}_1 = \mathbf{0}$, where $\hat{s}_1^{cc} = |k|/(1 + |k|)$ is a cumulative chord. As also $x(s)$ is a piecewise cubic and is of class C^2 at $\hat{s}_1^{cc}$ since $x''(s) \equiv \mathbf{0}$ the curve $x(s)$ minimizes $\tilde{\mathcal{E}}_{deg}$ with $\hat{s}_1^{cc}$.

The case of $k = 0$ in $\tilde{x}_0 = k\tilde{x}_2$ is impossible as otherwise $\tilde{x}_0 = \mathbf{0}$ which leads to $x_0 = x_1$ contradictory to $x_i \neq x_{i+1}$ (for $i = 0, 1$).

For $0 < k < 1$ (*co-linearly unordered data*) in $\tilde{x}_0 = k\tilde{x}_2$ the energy at global minimum $\tilde{\mathcal{E}}_{deg}(s_g) > 0$, as otherwise $s_g = k/(k - 1) < 0$.

Similarly for $k > 1$ in $\tilde{x}_0 = k\tilde{x}_2$ the energy at global minimum $\tilde{\mathcal{E}}_{deg}(s_g) > 0$ as otherwise $s_g = k/(k - 1) > 1$.

Lastly, for $k = 1$ in $\tilde{x}_0 = k\tilde{x}_2$ the energy $\tilde{\mathcal{E}}_{deg}(s_g) > 0$ satisfies $s_g = s_1^{cc} = 1/2$. Indeed, by (16) here $\tilde{\mathcal{E}}_{deg}(s_1) = \|x_0\|^2/((s_1 - 1)^2 s_1^2) > 0$ (as $x_0 \neq x_1$) and yields a single critical point $s_g = 1/2$ which coincides with $s_1^{cc} = 1/2$ as $\|\tilde{x}_0\| = \|x_2\|$. □

In a search for *all critical points* of $\tilde{\mathcal{E}}_{deg}$ recall (20) and (21) which yield (modulo factor $1/6$) $N_{deg}(0) = -\|\tilde{x}_0\|^2 < 0$ and $N_{deg}(1) = \|\tilde{x}_2\|^2 > 0$ - note that the remaining factor in $\tilde{\mathcal{E}}'_{deg}$ is always positive for $s_1 \in (0, 1)$ (see (19)). To guarantee the existence of a single critical point it suffices by Intermediate Value Theorem to show that either

$$N'_{deg}(s_1) = c_0^{deg} + c_1^{deg} s_1 + c_2^{deg} s_1^2 > 0$$

over $(0, 1)$ (yielding N_{deg} as strictly increasing with exactly one root $\hat{s}_1 \in (0, 1)$) or that $N'_{deg} = 0$ has exactly one root $\hat{u}_1 \in (0, 1)$ (i.e. N_{deg} has exactly one max/min/saddle at some $\hat{s}_1$). The latter combined with $N_{deg}(0) \cdot N_{deg}(1) < 0$ results in $N_{deg}(s_1) = 0$ having exactly root $\hat{s}_1 \in (0, 1)$ (a critical point of $\tilde{\mathcal{E}}_{deg}$). Note that if $\hat{s}_1 = \hat{u}_1$ then $\hat{u}_1$ is a saddle point of N_{deg}. Here the quadratic $N'_{deg}(s_1)$ has the following coefficients (see (21)):

$$c_0^{deg} = 3\|\tilde{x}_0\|^2 - \langle\tilde{x}_0|\tilde{x}_2\rangle\,, c_1^{deg} = 6(\langle\tilde{x}_0|\tilde{x}_2\rangle - \|\tilde{x}_0\|^2)\,, c_2^{deg} = 3\|\tilde{x}_2 - \tilde{x}_0\|^2 > 0. \quad (22)$$

We introduce *two auxiliary parameters* $(\lambda, \mu) \in \Omega = (\mathbb{R}_+ \times [-1, 1]) \setminus \{(1, 1)\}$:

$$\|\tilde{x}_0\| = \lambda \|\tilde{x}_2\|, \quad \langle \tilde{x}_0 | \tilde{x}_2 \rangle = \mu \|\tilde{x}_0\| \|\tilde{x}_2\| . \qquad (23)$$

Note that geometrically μ stands for $\cos(\alpha)$, where α is the angle between vectors $\tilde{x}_0$ and $\tilde{x}_2$ - hence $\mu = \lambda = 1$ results in $\tilde{x}_0 = \tilde{x}_2$. Such case is analyzed in Ex. 1 for $k = 1$ rendering $(\mu, \lambda) = (1, 1)$ as admissible.

Upon substituting two parameters $(\mu, \lambda) \in \Omega$ (see (23)) into $N'_{deg}(s_1) > 0$ (see also (22)) we arrive at genuine quadratic inequality in $u \in (0, 1)$ (for $u = s_1$):

$$W_{\lambda\mu}(u) = 3\lambda^2 - \mu\lambda + u(6\mu\lambda - 6\lambda^2) + u^2(3\lambda^2 - 6\mu\lambda + 3) > 0 , \qquad (24)$$

with positive coefficient (standing with u^2) as $3\lambda^2 - 6\mu\lambda + 3 = 3((\lambda - \mu)^2 + 1 - \mu^2) > 0$ over Ω.

We examine now various constraints on $(\mu, \lambda) \in \Omega$ ensuring existence of either no roots of the quadratic $N'_{deg}(s_1) = 0$ (as then since $N_{deg}(0) < 0$ and $N_{deg}(1) > 0$ a cubic N_{deg} is an increasing function) or exactly one root of the quadratic $N'_{deg}(s_1) = 0$ over $(0, 1)$. As already pointed out the satisfaction of both these cases yield exactly one critical point of $\tilde{\mathcal{E}}_{deg}$). The discriminant $\tilde{\Delta}_{crit}$ for $W_{\lambda\mu}(u) = 0$ (see (24)) reads as:

$$\tilde{\Delta}_{deg}(\mu, \lambda) = 12\lambda(-3\lambda + \mu + \lambda^2\mu + \lambda\mu^2) .$$

We consider now 3 cases in a search for admissible zones to enforce unimodality of $\tilde{\mathcal{E}}_{deg}$.

1. $\tilde{\Delta} < 0$. Here N'_{deg} has no real roots over $(0, 1)$. Since $c_2^{deg} > 0$, clearly $N'_{deg} > 0$ over $(0, 1)$ - note here $N_{deg}(0) < 0$ and $N_{deg}(1) > 0$. The latter

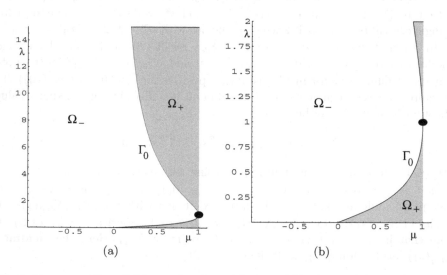

Fig. 1. Decomposition of Ω into subregions: (a) over which $\Delta_{deg} > 0$ (i.e. Ω_+), $\Delta_{deg} = 0$ (i.e. Γ_0) or $\Delta_{deg} < 0$ (i.e. Ω_-), (b) only for λ small.

inequality amounts to (with $\Delta_{deg} = (\tilde{\Delta}_{deg}/12\lambda$ and $\lambda > 0)$

$$\Delta_{deg} = -3\lambda + \mu + \lambda^2\mu + \lambda\mu^2 < 0 . \tag{25}$$

In order to decompose Ω into subregions Ω_- (with $\Delta_{deg} < 0$), Ω_+ (with $\Delta_{deg} > 0$) and Γ_0 (with $\Delta_{deg} \equiv 0$) we resort to *Mathematica* functions *InequalityPlot*, *ImplicitPlot* and *Solve*. Figure 1 a) shows the resulting decomposition and Fig. 1 b) shows its magnification for λ small. The intersection points of Γ_0 and boundary $\partial\Omega$ (found by *Solve*) read: for $\mu = 1$ it is a point $(1,1)$ (already excluded though analyzed - see dotted point in Fig. 1), for $\mu = -1$ it is a point $(-1,0)$ (excluded as $\lambda > 0$) and for $\lambda = 0$ it is a point $(0,0) \notin \Omega$ (also excluded as $\lambda > 0$).

The admissible subset $\Omega_{ok} \subset \Omega$ of parameters (μ, λ) (for which there is one local minimum of $\tilde{\mathcal{E}}_{deg}$ and thus a unique global one) satisfies $\Omega_- \subset \Omega_{ok}$. The complementary set to $\Omega\setminus\Omega_-$ forms a potential *exclusion zone* i.e. $\Omega_{ex} \subset \Omega\setminus\Omega_-$. Next we limit furthermore an exclusion zone $\Omega_{ex} \subset \Omega$ (currently not bigger than shaded region in Fig. 1).

2. $\tilde{\Delta}_{deg} = 0$. There is only one root $\hat{u}_1^0 \in \mathbb{R}$ of $N'_{deg}(s_1) = 0$. As already explained, irrespectively whether $\hat{u}_1^0 \in (0,1)$ or $\hat{u}_1^0 \notin (0,1)$ this results in exactly one root $\hat{s}_1 \in (0,1)$ of $N_{deg}(s_1) = 0$, which in turn yields exactly one local minimum for $\tilde{\mathcal{E}}_{deg}$ (turning out to be a global one). Hence $\Omega_- \cup \Gamma_0 \subset \Omega_{ok}$.

3. $\tilde{\Delta}_{deg} > 0$. There are two different roots $\hat{u}_1^\pm \in \mathbb{R}$ of $N'_{deg}(s_1) = 0$. They are either (in all cases we use Vieta's formulas):

(a) of *opposite signs*: i.e. $(c_0^{deg}/c_2^{deg}) < 0$ or

(b) *non-positive*: i.e. $(c_0^{deg}/c_2^{deg}) \geq 0$ and $(-c_1^{deg}/c_2^{deg}) < 0$ (here $\hat{u}_1^- < \hat{u}_1^+$ as $c_2^{deg} > 0$) or

(c) *non-negative*: i.e. $(c_0^{deg}/c_2^{deg}) \geq 0$ and $(-c_1^{deg}/c_2^{deg}) > 0$ - split into:

    (c1) $\hat{u}_1^+ \geq 1$: i.e.

    (c2) $0 < \hat{u}_1^+ < 1$ (recall $\hat{u}_1^- < \hat{u}_1^+$).

Evidently cases $a)$, $b)$ and $c1)$ yield *up to one root* $\hat{u}_1 \in (0,1)$ of $N_i^{deg'}(s_1) = 0$. Hence as already explained there is only one root $\hat{s}_1 \in (0,1)$ of $N_i^{deg}(s_1) = 0$, which is the unique critical point of $\tilde{\mathcal{E}}_{deg}^c$ over $(0,1)$ (in fact a global minimum).

In the next step we show that *the inequalities* from $a)$ or $b)$ or $c1)$ *extend* (contract) the admissible (exclusion) zone to $\Omega_{ok} = \Omega$ (to $\Omega_{ex} = \emptyset$). Indeed:

a) the constraint $(c_0^{deg}/c_2^{deg}) < 0$ upon using (24) reads (as $\lambda > 0$ and $c_2^{deg} > 0$):

$$3\lambda^2 - \mu\lambda < 0 \quad \equiv \lambda < \frac{\mu}{3} . \tag{26}$$

Fig. 2 a) shows Ω_1 (over which (26) holds) cut out from the exclusion zone Ω_{ex} of parameters $(\mu, \lambda) \in \Omega$ (again *Mathematica InequalityPlot* is used here). Thus $\Omega_- \cup \Gamma_0 \cup \Omega_1 \subset \Omega_{ok}$. The intersection $\Gamma_1 \cap \partial\Omega = \{(0,0), (1,1/3)\}$ (here $\Gamma_1 = \{(\mu, \lambda) \in \Omega : 3\lambda - \mu = 0\}$) and $\Gamma_0 \cap \Gamma_1 = \{(0,0)\}$ (we invoke here *Solve* function).

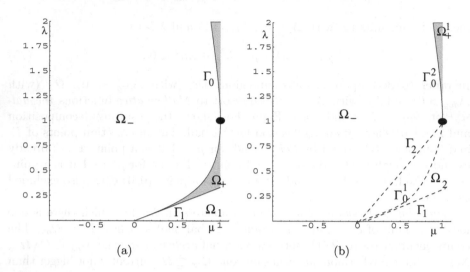

Fig. 2. Extension of admissible zone Ω_{ok} by cutting from exclusion zone Ω_{ex}: (a) Ω_1, (b) Ω_2.

b) the constraints $(c_0^{deg}/c_2^{deg}) \geq 0$ and $(-c_1^{deg}/c_2^{deg}) < 0$ combined with (24) yield:

$$\lambda \geq \frac{\mu}{3} \quad \text{and} \quad ((6\mu\lambda - 6\lambda^2) > 0 \quad \equiv \lambda < \mu). \tag{27}$$

Again with the aid of *ImplicitPlot* and *InequalityPlot* we find $\Omega_2 \cup \Gamma_1$ (to be cut out of Ω_{ok}) as the intersection of three sets determined by (27) and $\Delta > 0$ (for Ω_2 see Fig. 2 b) - a set bounded by Γ_0^1, Γ_1). Thus the admissible set Ω_{ok} satisfies $\Omega_- \cup \Gamma_0 \cup \Omega_1 \cup \Omega_2 \cup \Gamma_1 \subset \Omega_{ok}$ (see Fig. 2 b)). Note that for $\Gamma_2 = \{(\mu, \lambda) \in \Omega : \mu - \lambda = 0\}$ the intersection of curves $\Gamma_0 \cap \Gamma_2 = \{(0, 0), (1, 1)\}$ and $\Gamma_1 \cap \Gamma_2 = \{(0, 0)\}$, and of Γ_2 with the boundary $\partial\Omega$ $\{(0, 0), (1, 1)\}\}$ (use e.g. *Solve*). The exclusion zone Ω_{ex} must now satisfy $\Omega_{ex} \subset \Omega_+^1$ (see Fig. 2 b)).

c1) the constraints $(c_0^{deg}/c_2^{deg}) \geq 0$, $(-c_1^{deg}/c_2^{deg}) > 0$ and $u_1^+ \geq 1$ combined with (24) render (as $c_2^{deg} > 0$):

$$\lambda \geq \frac{\mu}{3}, \quad \lambda > \mu, \quad \sqrt{\tilde{\Delta}_{deg}} \geq 6(1 - \lambda\mu). \tag{28}$$

As we are only interested in reducing Ω_+^1 one intersects first the corresponding two sets (determined by the first two inequalities in (28)) with Ω_+^1 which clearly yields Ω_+^1. To complete solving (28) it suffices to find the intersection of Ω_+^1 with the set determined by the third inequality of (28). Note that for $(\mu, \lambda) \in \Omega_+^1$ we have $\mu > 0$. Indeed, by inspection or by applying the *Mathematica* function *Solve* to $\tilde{\Delta}_{deg}(\mu, \lambda) = 0$ (with $\mu(\lambda)$ treated as variable and λ as parameter) by (25) we have

$$\mu_\pm(\lambda) = \frac{-1 - \lambda^2 \pm \sqrt{1 + 14\lambda^2 + \lambda^4}}{2\lambda}.$$

A simple verification shows that $\mu_+ > 0$ as $12\lambda^2 > 0$ (since $\lambda > 0$ over Ω_+^1). Similarly $\mu_- < -1$ as $-\sqrt{1 + 14\lambda^2 + \lambda^4} < (1 - \lambda)^2$ holds. Thus for a fixed pair $(\mu, \lambda) \in \Omega_+^1$ (where $\tilde{\Delta}_{deg}(\mu, \lambda) > 0$) we must have $\mu(\lambda) \in (-\infty, \mu_-) \cup (\mu_+, +\infty)$. The latter intersected with $\mu \in [-1, 1]$ yields that $0 < \mu$ for all pairs $(\mu, \lambda) \in \Omega_+^1$. We show now that the third inequality in (28) results in $\Omega_{ex} = \emptyset$. Indeed note that $\sqrt{\tilde{\Delta}_{deg}} \geq 6(1 - \lambda\mu)$ is satisfied if $1 - \lambda\mu < 0$ which is equivalent (as $\mu > 0$ - shown to hold over Ω_+^1) to $\lambda > 1/\mu$ (for $\mu \in (0, 1]$). The case when $1 - \lambda\mu \geq 0$ yields $\lambda \leq 1/\mu$ which gives the set not intersecting with Ω_+^1. *Mathematica* function *InequalityPlot* yields the region $\Omega_5 = \Omega_3 \cup \Omega_4 \cap \Gamma_0^2$ (bounded by the curve $\Gamma_3 = \{(\mu, \lambda) \in \Omega : 1 - \mu\lambda = 0, \ \mu > 0\}$ and the boundary $\partial\Omega$ - see Fig. 3). Clearly as $\Omega_4 = \Omega_+^1$ we can cut out from Ω_{ex} the set Ω_+^1 and thus $\Omega_{ex} = \emptyset$ (or $\Omega_{ok} = \Omega$). Hence there is only one critical point of $\tilde{\mathcal{E}}_0^{deg}$ over $(0, 1)$.

The last geometric argument exploits the observation that Γ_0^2 is positioned above Γ_3. To show the latter algebraically one solves $\tilde{\Delta}_{deg} = 0$ in $\lambda(\mu)$ (use e.g. *NSolve*) by treating λ as variable and μ as a parameter. We obtain then (see (25))

$$\lambda_\pm(\mu) = \frac{3 - \mu^2 \pm \sqrt{9 - 10\mu^2 + \mu^4}}{2\mu}.$$

Note that the expression inside the square root is always non-negative. It suffices to show now that $\lambda_\pm(\mu) > \lambda_1(\mu) = 1/\mu$ for all $(\mu, \lambda) \in (0, 1] \times (1, +\infty)$. The inequality $\lambda_+(\mu) > \lambda_1(\mu)$ holds as for $\mu > 0$ it amounts to

$$\sqrt{9 - 10\mu^2 + \mu^4} > \mu^2 - 1.$$

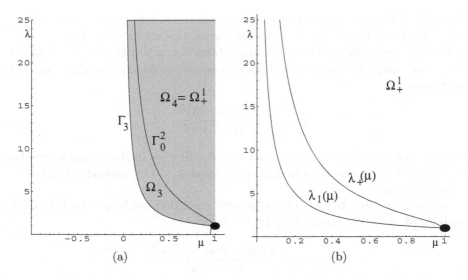

Fig. 3. (a) Cutting from Ω_{ex} the set $\Omega_5 = \Omega_3 \cup \Omega_4 \cup \Gamma_0^2$, (b) The graphs of two functions $\lambda_1(\mu)$ and $\lambda_+(\mu)$.

The latter holds as $\mu^2 - 1 < 0$ for $\mu \neq \pm 1$ - $\mu = -1$ excluded and $\mu = 1$ results in $\lambda_+(1) = 1$ which also is excluded as $(1,1) \notin \Omega$. The second function $\lambda_-(\mu)$ cannot be bigger that $\lambda_1(\mu)$ and thus is excluded. Indeed for $\mu \in [0,1)$ the latter would amount to

$$-\sqrt{9 - 10\mu^2 + \mu^4} \geq \mu^2 - 1 .$$

As both sides are non-positive we have

$$(9 - 10\mu^2 + \mu^4) \leq (\mu^2 - 1)^2$$

which is false as $8(1 - \mu^2) \leq 0$ does not hold for $\mu \in [0,1)$. Recall that $\mu = 1$ would result here in $\lambda_-(\mu) = 1$, which gives already excluded pair $(1,1)$.

Thus for the degenerate case we proved that there is always *exactly one critical point* of $\tilde{\mathcal{E}}_{deg}$ over $(0,1)$. However if we wish to have global minimum of $\tilde{\mathcal{E}}_{deg}$ to be close to cumulative chord $\hat{s}_1^{cc}$ (which optimizes special case of co-linear data) then the perturbation analysis (similar to the generic and non-generic case of Leap-Frog algorithm - see [6,7]) can be invoked. Note also if Decartes's sign rule is applied to $N_{deg}(s_1) = 0$ (see (20)) since $b_0^{deg} > 0$ and $b_3^{deg} < 0$ a sufficient condtion for one positive rule is that $b_1^{deg} b_2^{deg} \geq 0$. A simple verification shows that this condtion does not yield unique critical point of $\tilde{\mathcal{E}}'_{deg}$ for arbitrary data $\{\tilde{x}_0, \tilde{x}_1, \tilde{x}_2\}$.

4   Experiments

We illustrate now the theoretical results from Sect. 3 in the following example.

Example 2. Consider first three ordered co-linear points $\tilde{x}_2 = (-1,-3)$, $\tilde{x}_0 = k\tilde{x}_2 = (3,9)$ (with $k = -3$) $\tilde{x}_1 = \mathbf{0}$. The cumulative chord $\hat{s}_1^{cc} = |k|/(1 + |k|) = 3/4$ is expected to be the global minimum (also the only critical point) of $\tilde{\mathcal{E}}_{deg}$ at which the energy vanishes. Indeed the corresponding formula for $\tilde{\mathcal{E}}_{deg}$ (see (16)) reads here as:

$$\tilde{\mathcal{E}}_{deg}(s) = \frac{30(4s - 3)^2}{(s-1)^2 s^2} .$$

The *Mathematica Plot* function (used here for all tests in this example) renders the graph of $\tilde{\mathcal{E}}_{deg}$ with global minimum attained at $\hat{s} = \hat{s}_1^{cc}$ satisfying $\tilde{\mathcal{E}}_{deg}(\hat{s}_1^{cc}) = 0$ (see Fig. 4 a)).

A slight perturbation of the co-linearity conditions with $\tilde{x}_0 = (3,10)$ and $\tilde{x}_2$ unchanged yields the energy $\tilde{\mathcal{E}}_{deg}$ (see (16))

$$\tilde{\mathcal{E}}_{deg}(s) = \frac{3(109 - 284s + 185s^2)}{(s-1)^2 s^2} .$$

Again *Mathematica Plot* function renders the graph of $\tilde{\mathcal{E}}_{deg}$ with one global minimum at $\hat{s}_1 \approx 0.76748$ (see Fig. 4 b) and c)). Cumulative chord, which reads here $\hat{s}_1^{cc} \approx 0.767524$, together with $\hat{s}_1$ satisfy $0.509341 \approx \tilde{\mathcal{E}}_{deg}(\hat{s}_1) < \tilde{\mathcal{E}}_{deg}(\hat{s}_1^{cc}) \approx 0.509375$.

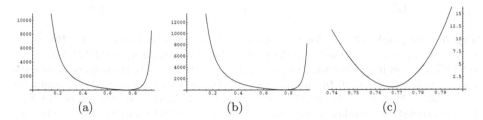

(a)   (b)   (c)

Fig. 4. The graphs of $\tilde{\mathcal{E}}_{deg}$ for (a) co-linear data $\tilde{x}_0 = (3,9)$, $\tilde{x}_1 = (0,0)$ and $\tilde{x}_2 = (-1,-3)$ vanishing at a global minimum $\hat{s}_1^{cc} = 3/4$, (b) slightly non-co-linear data $\tilde{x}_0 = (3,10)$, $\tilde{x}_1 = (0,0)$ and $\tilde{x}_2 = (-1,-3)$ non-vanishing at a global minimum at $\hat{s}_1 = 0.76748 \neq \hat{s}_1^{cc} = 0.767524$, (c) as in (b) but with visible $\hat{\mathcal{E}}_{deg}(\hat{s}_1) = 0.509341 > 0$.

On the other hand adding a large perturbation to the co-linearity by taking e.g. $\tilde{x}_0 = (4,-15)$ and $\tilde{x}_2$ unchanged yields the corresponding energy $\tilde{\mathcal{E}}_{deg}$ (see (16))

$$\tilde{\mathcal{E}}_{deg}(s) = \frac{3(241 - 400s + 169s^2)}{(s-1)^2 s^2}.$$

Again *Mathematica Plot* function renders the graph of $\tilde{\mathcal{E}}_{deg}$ with one global minimum at $\hat{s}_1 \approx 0.695985$ (see Fig. 5 a) and b)). Cumulative chord which reads here $\hat{s}_1^{cc} \approx 0.830772$, together with $\hat{s}_1$ satisfy $2979.8 \approx \tilde{\mathcal{E}}_{deg}(\hat{s}_1) < \tilde{\mathcal{E}}_{deg}(\hat{s}_1^{cc}) \approx 3844.87$. The value $\tilde{\mathcal{E}}_{deg}(\hat{s}_1) = 2979.8 >> 0$ (see Fig. 5 a) and b)).

Finally, for $\tilde{x}_0 = (0.05, -1)$ and $\tilde{x}_2$ unchanged the corresponding energy $\tilde{\mathcal{E}}_{deg}$ (see (16)) reads

$$\tilde{\mathcal{E}}_{deg}(s) = \frac{15.3075(0.196472 + 0.763351\, ss + s^2)}{s^2(s-1)^2}.$$

Again *Mathematica Plot* function renders the graph of $\tilde{\mathcal{E}}_{deg}$ with one global minimum at $\hat{s}_1 \approx 0.357839$ (see Fig. 5 c)). Cumulative chord which reads here $\hat{s}_1^{cc} \approx 0.240481$, together with $\hat{s}_1$ satisfy $173.264 \approx \tilde{\mathcal{E}}_{deg}(\hat{s}_1) < \tilde{\mathcal{E}}_{deg}(\hat{s}_1^{cc}) \approx 200.916$. Again, the value $\tilde{\mathcal{E}}_{deg}(\hat{s}_1) = 173.264 > 0$ (see Fig. 5 c)). $\qquad\square$

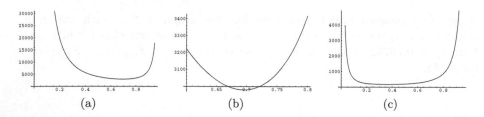

(a)   (b)   (c)

Fig. 5. The graphs of $\tilde{\mathcal{E}}_{deg}$ for (a) strongly non-co-linear data $\tilde{x}_0 = (4, -15)$, $\tilde{x}_1 = (0,0)$ and $\tilde{x}_2 = (-1, -3)$ non-vanishing at a global minimum $\hat{s}_1 \approx 0.695985 \neq \hat{s}_1^{cc} \approx 0.830772$, (b) as in (a) but with visible $\tilde{\mathcal{E}}_{deg}(\hat{s}_1) = 2979.8 >> 0$, (c) non-co-linear data $\tilde{x}_0 = (0.05, -1)$, $\tilde{x}_1 = (0,0)$ and $\tilde{x}_2 = (-1, -3)$ non-vanishing at a global minimum $\hat{s}_1 \approx 0.357839 \neq \hat{s}_1^{cc} \approx 0.240481$, here $\tilde{\mathcal{E}}_{deg}(\hat{s}_1) = 173.264 > 0$. The graphs of $\tilde{\mathcal{E}}_{deg}$ for (a) strongly non-co-linear data $\tilde{x}_0 = (4, -15)$, $\tilde{x}_1 = (0,0)$ and $\tilde{x}_2 = (-1, -3)$ non-vanishing at a global minimum $\hat{s}_1 \approx 0.695985 \neq \hat{s}_1^{cc} \approx 0.830772$, (b) as in (a) but with visible $\tilde{\mathcal{E}}_{deg}(\hat{s}_1) = 2979.8 >> 0$, (c) non-co-linear data $\tilde{x}_0 = (0.05, -1)$, $\tilde{x}_1 = (0,0)$ and $\tilde{x}_2 = (-1, -3)$ non-vanishing at a global minimum $\hat{s}_1 \approx 0.357839 \neq \hat{s}_1^{cc} \approx 0.240481$, here $\tilde{\mathcal{E}}_{deg}(\hat{s}_1) = 173.264 > 0$.

5   Conclusions

In this work *the unimodality of optimization task* (2) to fit reduced data $\mathcal{M}_n$ is proved for $n = 2$. The latter constitutes a degenerate variant of (1) investigated earlier for $n \geq 3$ in the context of using Leap-Frog algorithm. This scheme forms an iterative numerical tool to compute the substitutes of the unknown interpolation knots for $n \geq 3$ while minimizing (1) - see [3,4,6,7,9]. In contrast to the degenerate case of reduced data i.e. to $\mathcal{M}_2$, local iterative univariate functions of Leap-Frog are not in general unimodal though some specific sufficient conditions enforcing the latter are established in [6,7]. Minimizing the univariate and unimodal function (2) makes any numerical scheme to compute the unique global minimum (i.e. an optimal knot t_1) insensitive to the choice of initial guess which e.g. can be taken e.g. as cumulative chord. More information on Leap-Frog in the context of other applications can be found among all in [14,15].

References

1. de Boor, C.: A Practical Guide to Splines, 2nd edn. Springer, New York (2001). https://link.springer.com/book/9780387953663
2. Farouki, R.T.: Optimal parameterizations. Comput. Aided Geom. Des. 14(2), 153–168 (1997). https://doi.org/10.1016/S0167-8396(96)00026-X
3. Kozera, R., Noakes, L.: Optimal knots selection for sparse reduced data. In: Huang, F., Sugimoto, A. (eds.) PSIVT 2015. LNCS, vol. 9555, pp. 3–14. Springer, Cham (2016). https://doi.org/10.1007/978-3-319-30285-0_1
4. Kozera, R., Noakes, L.: Non-linearity and non-convexity in optimal knots selection for sparse reduced data. In: Gerdt, V.P., Koepf, W., Seiler, W.M., Vorozhtsov, E.V. (eds.) CASC 2017. LNCS, vol. 10490, pp. 257–271. Springer, Cham (2017). https://doi.org/10.1007/978-3-319-66320-3_19

5. Kozera, R., Noakes, L., Wilkołazka, M.: Parameterizations and Lagrange cubics for fitting multidimensional data. In: Krzhizhanovskaya, V.V., Závodszky, G., Lees, M.H., Dongarra, J.J., Sloot, P.M.A., Brissos, S., Teixeira, J. (eds.) ICCS 2020. LNCS, vol. 12138, pp. 124–140. Springer, Cham (2020). https://doi.org/10.1007/978-3-030-50417-5_10
6. Kozera, R., Noakes, L., Wiliński, A.: Generic case of Leap-Frog Algorithm for optimal knots selection in fitting reduced data. In: Paszynski, M., Kranzlmüller, D., Krzhizhanovskaya, V.V., Dongarra, J.J., Sloot, P.M.A. (eds.) ICCS 2021. LNCS, vol. 12745, pp. 337–350. Springer, Cham (2021). https://doi.org/10.1007/978-3-030-77970-2_26
7. Kozera, R., Noakes L.: Non-generic case of Leap-Frog Algorithm for optimal knots selection in fitting reduced data. In: Groen, D., et al. (eds.) ICCS 2022, pp. 341–354. LNCS vol. 13352, Part III, Springer Cham (2022). https://doi.org/10.1007/978-3-031-08757-8_29
8. Kozera, R., Noakes, L., Wilkołazka, M.: Exponential parameterization to fit reduced data. Appl. Math. Comput. **391**, 125645 (2021). https://doi.org/10.1016/j.amc.2020.125645
9. Kozera, R., Wiliński, A.: Fitting dense and sparse reduced data. In: Pejaś, J., El Fray, I., Hyla, T., Kacprzyk, J. (eds.) ACS 2018. AISC, vol. 889, pp. 3–17. Springer, Cham (2019). https://doi.org/10.1007/978-3-030-03314-9_1
10. Kuznetsov, E.B., Yakimovich A.Y.: The best parameterization for parametric interpolation. J. Comput. Appl. Math. **191**(2), 239–245 (2006). https://core.ac.uk/download/pdf/81959885.pdf
11. Kvasov, B.I.: Methods of Shape-Preserving Spline Approximation. World Scientific Publication, Singapore (2000). https://doi.org/10.1142/4172
12. Lee, E.T.Y.: Choosing nodes in parametric curve interpolation. Comput. Aided Des. **21**(6), 363–370 (1989). https://doi.org/10.1016/0010-4485(89)90003-1
13. Matebese, B., Withey, D., Banda, M.K.: Modified Newton's method in the Leapfrog method for mobile robot path planning. In: Dash, S.S., Naidu, P.C.B., Bayindir, R., Das, S. (eds.) Artificial Intelligence and Evolutionary Computations in Engineering Systems. AISC, vol. 668, pp. 71–78. Springer, Singapore (2018). https://doi.org/10.1007/978-981-10-7868-2_7
14. Noakes, L.: A global algorithm for geodesics. J. Aust. Math. Soc. Ser. A **65**(1), 37–50 (1998). https://doi.org/10.1017/S1446788700039380
15. Noakes, L., Kozera, R.: Nonlinearities and noise reduction in 3-source photometric stereo. J. Math. Imaging Vision **18**(2), 119–127 (2003). https://doi.org/10.1023/A:1022104332058
16. Piegl, L., Tiller, W.: The NURBS Book. Springer-Verlag, Berlin Heidelberg (1997). https://doi.org/10.1007/978-3-642-97385-7

A Case Study of the Profit-Maximizing Multi-Vehicle Pickup and Delivery Selection Problem for the Road Networks with the Integratable Nodes

Aolong Zha[1(✉)] [iD], Qiong Chang[2], Naoto Imura[1], and Katsuhiro Nishinari[1]

[1] The University of Tokyo, Tokyo, Japan
{a-zha,nimura}@g.ecc.u-tokyo.ac.jp, tknishi@mail.ecc.u-tokyo.ac.jp
[2] Tokyo Institute of Technology, Tokyo, Japan
q.chang@c.titech.ac.jp

Abstract. This paper is a study of an application-based model in profit-maximizing multi-vehicle pickup and delivery selection problem (PPDSP). The graph-theoretic model proposed by existing studies of PPDSP is based on transport requests to define the corresponding nodes (i.e., each request corresponds to a pickup node and a delivery node). In practice, however, there are probably multiple requests coming from or going to an identical location. Considering the road networks with the integratable nodes as above, we define a new model based on the integrated nodes for the corresponding PPDSP and propose a novel mixed-integer formulation. In comparative experiments with the existing formulation, as the number of integratable nodes increases, our method has a clear advantage in terms of the number of variables as well as the number of constraints in the generated instances, and the accuracy of the optimized solution obtained within a given time.

Keywords: Combinatorial optimization · Mixed-integer programming · Logistics

1 Introduction

With the increasing expansion of the supply chain, logistics has become the backbone of the modern economy. Adequate transportation capacity is a necessary condition for commerce. However, during the peak period of logistics order trading, the shortage of transportation resources still occurs from time to time. As an operator (i.e., transportation company), it is a challenge to provide limited transportation resources to some of the requests in a competitive market in order to maximize profit.

Recently, the profit-maximizing multi-vehicle pickup and delivery selection problem (PPDSP) has gained a lot of attention in the field of practical transportation and logistics, such as *sensor transportation* [8], *dial-a-ride servise* [11], *green pickup-delivery* [3], and *customized bus network design* [7]. This problem

© The Author(s), under exclusive license to Springer Nature Switzerland AG 2023
J. Mikyška et al. (Eds.): ICCS 2023, LNCS 14075, pp. 454–468, 2023.
https://doi.org/10.1007/978-3-031-36024-4_35

was first proposed in [10], which involves three classical problem models: *routing optimization, pickup and delivery*, and *selective pickup* (a.k.a. *knapsack*). Solving this problem quickly and optimally can both help improve the operational efficiency of the carriers and contributes to more eco-friendly transportation.

For solving PPDSP, previous studies have presented a *branch-and-price algorithm* [1], an exact method of graph search [4], and several metaheuristic methods including *tabu search, genetic algorithm, scatter search* [12], *simulated annealing algorithm* [2], and *variable neighborhood search* [6]. To the best of our knowledge, in the graph-theoretic models constructed in the existing studies, the definitions of nodes are based on the pickup and delivery location from the requests. This means that one request needs to correspond to two nodes. However, in application scenarios, there are often plural requests coming from or to arrive at the same location. The number of variables to be required in the existing mixed-integer formulation heavily depends on the number of nodes in the model, and since PPDSP is an $\mathcal{NP}$-hard problem, its computational complexity is exponential with respect to the number of variables.

The motivation of this study is to provide a more reasonable and effective mathematical model for practical logistics and transportation problems. Considering the road networks with the integratable nodes as above, we proposes a new concise model in which the definition of nodes is based on the locations rather than the requests, and give a tailored mixed-integer programming formulation that reduces the number of required variables. It is necessary to claim that the solution set of our proposed location-based method is a subset of the solution set of the request-based method. This is because the integrated node can only be passed at most once under the *Hamiltonian cycle* constraint, where "at most" is due to the objective function of profit-maximizing (i.e., it is possible that any node will not be traversed).

In Fig. 1, we take the delivery of single truck as an example, and assume that the truck is transported according to the route given in Fig. 1a as a feasible solution in this scenario, where the pickup coordinates of request 1 (abbr. req. 1 pickup coord.) and req. 2 pickup coord. are the identical locations. We can integrate them as location 1 as shown in Fig. 1b, and after integrating all the same coordinates into one location respectively, our model becomes more concise. This integration changes the nodes of the model from repeatable coordinates based on the request to the unique locations, which results in each location can only be visited once. Therefore, the location-based model can also correspond to the feasible solutions in Fig. 1a, but not to some of the feasible solutions corresponded to the request-based model, such as the solution given in Fig. 1c. Whereas in real long-distance logistics application scenarios, making multiple retraces to the same location is rare (e.g., insufficient vehicle capacity, goods cannot be mixed, etc.), and also non-efficient. For such locations with a large number of requests or high loading volume, it is a more common strategy to assign multiple vehicles to them. Therefore, we argue that the location-based model can improve the optimization efficiency although it reduces the range of feasible solutions.

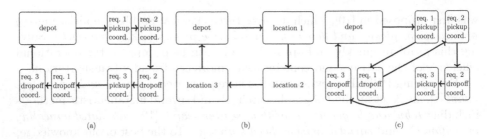

Fig. 1. A simple example for explaining the relationship between the request-based model (i.e., Fig. 1a and 1c) and the location-based model (i.e., Fig. 1b), where {req. 1 pickup coord.} and {req. 2 pickup coord.}, {req. 3 pickup coord.} and {req. 2 dropoff coord.}, and {req. 3 dropoff coord.} and {req. 1 dropoff coord.} can be integrated as {location 1}, {location 2} and {location 3}, respectively. The route shown in Fig. 1c can be regarded as a feasible solution of the request-based model, but cannot be corresponded to in the location-based model.

It should be emphasized that although the location-based model has been widely used on the vehicle routing problem, there is no existing study that employs the location-based model on the pickup and delivery probem because it is relatively difficult to formulate the location-based capacity constraint, especially for PPDSP. We propose a novel formulation for the location-based model on PPDSP and conduct comparative experiments with the existing formulation through simulations, which is the main contribution of our study.

2   Preliminaries

Given a directed graph $G = (V, E)$, where $V = \{0, 1, 2, \ldots, |V|\}$ is the set of nodes representing the location points (0 is depot) and E is the set of arc, denoted as $\widehat{od}$, and given a set of trucks $T = \{1, 2, \ldots, |T|\}$, we define a type of Boolean variables x^t_{od} that is equal to 1 if the truck t passes through the arc $\widehat{od}$ and 0 otherwise. We denote the load capacity of truck t as c^t, and the cost of the truck t traversing the arc $\widehat{od}$ as l^t_{od}, where $t \in T$ and $o, d \in V$.

Let $R = \{1, 2, \ldots, |R|\}$ be a set of requests. Each request r $(r \in R)$ is considered as a tuple $r = \langle w_r, q_r, f(r), g(r) \rangle$, where

- w_r is the payment that can be received for completing the shipping of request r;
- q_r is the volume of request r;
- $f(r)$ is the loading point of request r;
- $g(r)$ is the unloading point of request r,

and $f : R \to V \setminus \{0\}$ (resp. $g : R \to V \setminus \{0\}$) is a function for mapping the loading (resp. unloading) point of requests r. We also define another type of Boolean variables y^t_r that is equal to 1 if request r is allocated to truck t and 0 otherwise, where $r \in R$ and $t \in T$. Besides, we denote the number of location points visited by truck t before it reaches v $(v \in V \setminus \{0\})$ as u^t_v.

Definition 1 (Delivery of Truck t). *Delivery of truck t is denoted by $D_t = \bigcup_{r \in R} \{r \mid y_r^t = 1\}$, where $t \in T$.*

Definition 2 (Route of Truck t). *Route of truck t is denoted by $S_t = \bigcup_{o \in V} \bigcup_{d \in V} \{\widehat{od} \mid x_{od}^t = 1\}$, where $t \in T$. S_t satisfies the following conditions if $D_t \neq \emptyset$:*

Hamiltonian Cycle
- *Denote $P_t = \{0\} \cup \bigcup_{r \in D_t} \{f(r), g(r)\}$ as the set of location points visited and departed exactly once by truck t, where the predecessor and successor are the same node is not counted (i.e., even if $\widehat{vv} \mid x_{vv}^t = 1$, neither this time can be included in the number of visits or departures of truck t to/from location v);*
- *Ensure that no subtour exists in S_t.[1]*

Loading Before Unloading $\forall i \in D_t, u_{f(i)}^t < u_{g(i)}^t$.

Capacity Limitation *At any time, the total volume of cargo carried by truck t cannot exceed its capacity c^t.*

Definition 3 (Delivery Routing Solution). *Delivery routing solution $DS = \bigcup_{t \in T} \{(D_t, S_t)\}$ that is a partition of R into disjoint and contained D_t with the corresponding S_t:*

$$\forall i, j \, (i \neq j), \ D_i \cap D_j = \emptyset, \ \bigcup_{D_i \in DS} D_i \subseteq R, \ \bigcup_{S_i \in DS} S_i \subseteq E.$$

Denote the set of all possible delivery routing solutions as $\Pi(R, T)$.

Definition 4 (Profit-Cost Function). *A profit-cost function assigns a real-valued profit to every D_t: $w : D_t \to \mathbb{R}$ and a cost to every S_t: $l : S_t \to \mathbb{R}$, where $w(D_t) = \sum_{r \in D_t} w_r$ and $l(S_t) = \sum_{\widehat{od} \in S_t} l_{od}^t$. For any delivery routing solution $DS \in \Pi(R, T)$, the value of DS is calculated by*

$$\xi(DS) = \sum_{D_t \in DS} w(D_t) - \sum_{S_t \in DS} l(S_t).$$

In general, a delivery routing solution DS that considers only maximizing profits or minimizing costs is not necessarily an optimal DS. Therefore, we have to find the optimal delivery routing solution that maximizes the sum of the values of profit-cost functions. We define a delivery routing problem in profit-cost function as follows.

Definition 5 (PPDSP). *For a set of requests and trucks (R, T), a profit-maximizing multi-vehicle pickup and delivery selection problem (PPDSP) is to find the optimal delivery routing solution DS^* such that*

$$DS^* \in \arg \max_{DS \in \Pi(R, T)} \xi(DS).$$

---

[1] The time window constraint included in the existing studies can be regarded as a special MTZ-based formulation [9]. In this study, we omit the time window constraint and instead use the most basic MTZ-formulation to eliminate subtour.

Here we show an example of PPDSP.

Example 1. Assume that two trucks $T = \{t_1, t_2\}$ are responsible for three requests $R = \{r_1, r_2, r_3\}$ with the following conditions.

– The information about the requests:

Request	w_r	q_r	$f(r)$	$g(r)$
r_1	13	4	a	c
r_2	7	2	a	b
r_3	4	1	b	c

– The information about the trucks:
  • The capacities of the trucks $c^{t_1} = 6$ and $c^{t_2} = 3$.
  • The cost matrices of each truck through each arc (δ is depot).

$l^{t_1}_{od}$	δ	a	b	c
δ	0	2	2	2
a	2	0	4	7
b	2	4	0	2
c	2	7	2	0

$l^{t_2}_{od}$	δ	a	b	c
δ	0	1	1	1
a	1	0	3	5
b	1	3	0	1
c	1	5	1	0

For all $DS \in \Pi(R, T)$, we have their respective profit-cost function as follows.

$\xi(\{(D_{t_1} = \emptyset, S_{t_1} = \emptyset), (D_{t_2} = \emptyset, S_{t_2} = \emptyset)\}) = 0,$
$\xi(\{(D_{t_1} = \emptyset, S_{t_1} = \emptyset), (D_{t_2} = \{r_2\}, S_{t_2} = \{\widehat{\delta a}, \widehat{ab}, \widehat{b\delta}\})\}) = 2,$
$\xi(\{(D_{t_1} = \emptyset, S_{t_1} = \emptyset), (D_{t_2} = \{r_3\}, S_{t_2} = \{\widehat{\delta b}, \widehat{bc}, \widehat{c\delta}\})\}) = 1,$
$\xi(\{(D_{t_1} = \emptyset, S_{t_1} = \emptyset), (D_{t_2} = \{r_2, r_3\}, S_{t_2} = \{\widehat{\delta a}, \widehat{ab}, \widehat{bc}, \widehat{c\delta}\})\}) = 5,$
$\xi(\{(D_{t_1} = \{r_1\}, S_{t_1} = \{\widehat{\delta a}, \widehat{ac}, \widehat{c\delta}\}), (D_{t_2} = \emptyset, S_{t_2} = \emptyset)\}) = 2,$
$\xi(\{(D_{t_1} = \{r_1\}, S_{t_1} = \{\widehat{\delta a}, \widehat{ac}, \widehat{c\delta}\}), (D_{t_2} = \{r_2\}, S_{t_2} = \{\widehat{\delta a}, \widehat{ab}, \widehat{b\delta}\})\}) = 4,$
$\xi(\{(D_{t_1} = \{r_1\}, S_{t_1} = \{\widehat{\delta a}, \widehat{ac}, \widehat{c\delta}\}), (D_{t_2} = \{r_3\}, S_{t_2} = \{\widehat{\delta b}, \widehat{bc}, \widehat{c\delta}\})\}) = 3,$
$\xi(\{(D_{t_1} = \{r_1\}, S_{t_1} = \{\widehat{\delta a}, \widehat{ac}, \widehat{c\delta}\}), (D_{t_2} = \{r_2, r_3\}, S_{t_2} = \{\widehat{\delta a}, \widehat{ab}, \widehat{bc}, \widehat{c\delta}\})\}) = 7,$
$\xi(\{(D_{t_1} = \{r_2\}, S_{t_1} = \{\widehat{\delta a}, \widehat{ab}, \widehat{b\delta}\}), (D_{t_2} = \emptyset, S_{t_2} = \emptyset)\}) = -1,$
$\xi(\{(D_{t_1} = \{r_2\}, S_{t_1} = \{\widehat{\delta a}, \widehat{ab}, \widehat{b\delta}\}), (D_{t_2} = \{r_3\}, S_{t_2} = \{\widehat{\delta b}, \widehat{bc}, \widehat{c\delta}\})\}) = 0,$
$\xi(\{(D_{t_1} = \{r_3\}, S_{t_1} = \{\widehat{\delta b}, \widehat{bc}, \widehat{c\delta}\}), (D_{t_2} = \emptyset, S_{t_2} = \emptyset)\}) = -2,$
$\xi(\{(D_{t_1} = \{r_3\}, S_{t_1} = \{\widehat{\delta b}, \widehat{bc}, \widehat{c\delta}\}), (D_{t_2} = \{r_2\}, S_{t_2} = \{\widehat{\delta a}, \widehat{ab}, \widehat{b\delta}\})\}) = 0,$
$\xi(\{(D_{t_1} = \{r_1, r_2\}, S_{t_1} = \{\widehat{\delta a}, \widehat{ab}, \widehat{bc}, \widehat{c\delta}\}), (D_{t_2} = \emptyset, S_{t_2} = \emptyset)\}) = 10,$
$\xi(\{(D_{t_1} = \{r_1, r_2\}, S_{t_1} = \{\widehat{\delta a}, \widehat{ac}, \widehat{cb}, \widehat{b\delta}\}), (D_{t_2} = \emptyset, S_{t_2} = \emptyset)\}) = 7,$
$\xi(\{(D_{t_1} = \{r_1, r_2\}, S_{t_1} = \{\widehat{\delta a}, \widehat{ab}, \widehat{bc}, \widehat{c\delta}\}), (D_{t_2} = \{r_3\}, S_{t_2} = \{\widehat{\delta b}, \widehat{bc}, \widehat{c\delta}\})\}) = 11,$
$\xi(\{(D_{t_1} = \{r_1, r_2\}, S_{t_1} = \{\widehat{\delta a}, \widehat{ac}, \widehat{cb}, \widehat{b\delta}\}), (D_{t_2} = \{r_3\}, S_{t_2} = \{\widehat{\delta b}, \widehat{bc}, \widehat{c\delta}\})\}) = 8,$
$\xi(\{(D_{t_1} = \{r_1, r_3\}, S_{t_1} = \{\widehat{\delta a}, \widehat{ab}, \widehat{bc}, \widehat{c\delta}\}), (D_{t_2} = \emptyset, S_{t_2} = \emptyset)\}) = 7,$
$\xi(\{(D_{t_1} = \{r_1, r_3\}, S_{t_1} = \{\widehat{\delta b}, \widehat{ba}, \widehat{ac}, \widehat{c\delta}\}), (D_{t_2} = \emptyset, S_{t_2} = \emptyset)\}) = 2,$

$\xi(\{(D_{t_1} = \{r_1, r_3\}, S_{t_1} = \{\widehat{\delta a}, \widehat{ab}, \widehat{bc}, \widehat{c\delta}\}), (D_{t_2} = \{r_2\}, S_{t_2} = \{\widehat{\delta a}, \widehat{ab}, \widehat{b\delta}\}))\}) = 9,$
$\xi(\{(D_{t_1} = \{r_1, r_3\}, S_{t_1} = \{\widehat{\delta b}, \widehat{ba}, \widehat{ac}, \widehat{c\delta}\}), (D_{t_2} = \{r_2\}, S_{t_2} = \{\widehat{\delta a}, \widehat{ab}, \widehat{b\delta}\}))\}) = 4,$
$\xi(\{(D_{t_1} = \{r_2, r_3\}, S_{t_1} = \{\widehat{\delta a}, \widehat{ab}, \widehat{bc}, \widehat{c\delta}\}), (D_{t_2} = \emptyset, S_{t_2} = \emptyset))\}) = 1.$

In this example, the optimal delivery routing solution DS^* for the given PPDSP is $\{(D_{t_1} = \{r_1, r_2\}, S_{t_1} = \{\widehat{\delta a}, \widehat{ab}, \widehat{bc}, \widehat{c\delta}\}), (D_{t_2} = \{r_3\}, S_{t_2} = \{\widehat{\delta b}, \widehat{bc}, \widehat{c\delta}\})\}$, and its value is 11.

3   Problem Formulation

In this section, we present the following MIP formulation of the location-based model for PPDSP and prove its correctness as well as the space complexity of the generated problem.

$$\max \sum_{r \in R} \sum_{t \in T} (w_r \cdot y_r^t) - \sum_{t \in T} \sum_{o \in V} \sum_{d \in V} (l_{od}^t \cdot x_{od}^t), \tag{1}$$

$$\text{s.t. } x_{od}^t, y_r^t \in \{0, 1\}, \qquad \forall (t, o, d, r) : t \in T, o, d \in V, r \in R, \tag{2}$$

$$\sum_{t \in T} y_r^t \leq 1, \qquad \forall r : r \in R, \tag{3}$$

$$y_r^t \leq \sum_{\substack{o \in V \\ o \neq f(r)}} x_{of(r)}^t, \qquad \forall (t, r) : t \in T, r \in R, \tag{4}$$

$$y_r^t \leq \sum_{\substack{o \in V \\ o \neq g(r)}} x_{og(r)}^t, \qquad \forall (t, r) : t \in T, r \in R, \tag{5}$$

$$\sum_{d \in V} x_{od}^t - \sum_{d \in V} x_{do}^t = 0, \qquad \forall (t, o) : t \in T, o \in V, \tag{6}$$

$$\sum_{\substack{d \in V \\ d \neq o}} x_{od}^t \leq 1, \qquad \forall (t, o) : t \in T, o \in V, \tag{7}$$

$$u_d^t - u_o^t \geq 1 - |V|(1 - x_{od}^t),$$
$$\qquad\qquad \forall (t, o, d) : t \in T, o, d \in V \setminus \{0\}, o \neq d, \tag{8}$$

$$u_{f(r)}^t - u_{g(r)}^t < |V|(1 - y_r^t), \qquad \forall (t, r) : t \in T, r \in R, \tag{9}$$

$$\begin{cases} \Gamma - M \cdot (1 - x_{od}^t) \leq h_d^t - h_o^t \leq \Gamma + M \cdot (1 - x_{od}^t), \\ \Gamma \triangleq \sum_{r \in f^{-1}(d)} (q_r \cdot y_r^t) - \sum_{r \in g^{-1}(d)} (q_r \cdot y_r^t), \\ M \triangleq c^t + \sum_{r \in R} q_r, \end{cases}$$
$$\qquad\qquad \forall (t, o, d) : t \in T, o, d \in V \setminus \{0\}, o \neq d, \tag{10}$$

$$0 \leq h_v^t \leq c^t, \qquad \forall (t, v) : t \in T, v \in V \setminus \{0\}, \tag{11}$$

$$0 \leq u_v^t \leq |V| - 2, \qquad \forall (t, v) : t \in T, v \in V \setminus \{0\}. \tag{12}$$

3.1   Correctness

The objective function in Eq. (1) maximizes the profit-cost function for all possible delivery routing solutions. Equation (2) introduces two types of Boolean variables x_{od}^t and y_r^t. Equation (3) guarantees that, each request can be assigned to at most one truck.

Theorem 1. *For PPDSP, the Hamiltonian cycle constraints can be guaranteed by the simultaneous Eqs. (4)–(8) and Eq. (12).*

Proof. According to Eq. (4) (resp. Equation (5)), if request r is assigned to truck t, then truck t reaches the pickup (resp. dropoff) point of r at least once via a point that is not the pickup (resp. dropoff) point of r. Equation (6) ensures that the number of visits and the number of departures of a truck at any location must be equal. Equation (7) restricts any truck departing from a location to at most one other location. It is clear from the above that, all nodes of $\bigcup_{r \in D_t} \{f(r), g(r)\}$ are ensured to be visited and departed by truck t exactly once.

Equation (8), a canonical MTZ subtour elimination constraint [9], resrticts the integer variables u_v^t, where $v \in V \setminus \{0\}$, such that they form an ascending series to represent the order that truck t arrives at each location v. Equation (12) gives the domain of such variables u_v^t. Furthermore, Eq. (8) associated with Eqs. (4)–(7) also enforces that depot 0 must be visited and departed by truck t exactly once if $D_t \neq \emptyset$.   □

Equation (9) ensures that, for any request r, the arrival of its pickup point must precede the arrival of its dropoff point by truck t if $y_r^t = 1$.

Theorem 2. *For PPDSP, the capacity constraint can be guaranteed by the Eqs. (10) and (11).*

Proof. We introduce another integer variables h_v^t, where $v \in V \setminus \{0\}$, in Eq. (10), whose domains are between 0 and c^t (i.e., the capacity of truck t) as given in Eq. (11). Such a variable h_v^t can be considered as the loading volume of truck t when it departs at location v. We define the amount of change in the loading of truck t at location v as Γ, which equals the loaded amount at v (i.e., $\sum_{r \in f^{-1}(v)} (q_r \cdot y_r^t)$) minus the unloaded amount at v (i.e., $\sum_{r \in g^{-1}(v)} (q_r \cdot y_r^t)$). The main part of Eq. (10) guarantees that $h_d^t - h_o^t$ is exactly equal to Γ when $x_{od}^t = 1$, which is written as a big-M linear inequality, where we specify M as $c^t + \sum_{r \in R} q_r$.   □

3.2   Space Complexity

Consider that the number of trucks is m (i.e., $|T| = m$) and the number of requests is n (i.e., $|R| = n$). Assume that the average repetition rate of the same locations is k, where $k \geq 1$, then we can estimate the number of unique locations as $2nk^{-1}$, which is also the number of integrated nodes (viz., $|V \setminus \{0\}| = 2nk^{-1}$).

Theorem 3. *The number of linear (in)equations corresponding to Eqs. (3)–(10) is always of a pseudo-polynomial size, and both this number and the number of required variables are bounded by $\Theta(mn^2 k^{-2})$.*

Proof. The number of Boolean variables x_{od}^t (resp. y_r^t) required in proposed formulation is $m(1 + 2nk^{-1})^2$ (resp. mn); while both the number of required integer variables u_d^t and h_d^t are $m(1 + 2nk^{-1})$. In Table 1, we list the bounded numbers of linear (in)equations that correspond to Eqs. (3)–(10) involved in the proposed PPDSP formulation. Therefore, the number of required variable as well as the number of corresponded linear (in)equations are of $\Theta(mn^2k^{-2})$. □

Table 1. The bounded numbers of linear (in)equations corresponding to the constraints that are formulated in Eqs. (3)–(10).

Constraint	#(In)equations	Constraint	#(In)equations
Eq. (3)	n	Eq. (7)	$m(1 + 2nk^{-1})$
Eq. (4)	mn	Eq. (8)	$m\binom{2nk^{-1}}{2}$
Eq. (5)	mn	Eq. (9)	mn
Eq. (6)	$m(1 + 2nk^{-1})$	Eq. (10)	$2m\binom{2nk^{-1}}{2}$

4 Experiment

We compare the performance of a MIP optimizer in solving randomly generated PPDSP instances based on the proposed formulation (i.e., location-based model), and based on the existing formulation (i.e., request-based model), respectively. Due to page limitations of the submission, please refer to the appendix section in the preprint version of this manuscript [13] for the specific description of the existing formulation used for the comparative experiments.

4.1 Simulation Instances Generation

The directed graph informations for generating instances are set based on the samples of TSPLIB benchmark with displaying data as the coordinates of nodes.[2] We denote the number of nodes contained in the selected sample as $|V|$, and set the first of these nodes to be the depot node. The pickup and dropoff points of each request are chosen from the non-depot nodes. Therefore, given an average repetition rate k for the same locations, we need to repeatably select n pairs of non-depot nodes (i.e., a total of $2n$ repeatable non-depot nodes) to correspond to n requests, where $n = \text{round}(\frac{k(|V|-1)}{2})$.

In Algorithm 1, we construct the *repeaTimeList* of length $|V| - 1$ to mark the number of times of each non-depot node being selected. In order to ensure that each non-depot node is selected at least once, each value on such list is initialized to 1. We continuously generate a random index of *repeaTimeList* and add one to the value corresponding to the generated index (i.e., the number of times of the non-depot node being selected increases by one) until the sum of these numbers reaches $2n$.

---
[2] http://comopt.ifi.uni-heidelberg.de/software/TSPLIB95/tsp/.

Algorithm 1: Randomly specify the number of repetitions of each non-depot node

---

Input: $G = (V, E)$, k

Init. : $n \leftarrow \text{round}(\frac{k(|V|-1)}{2})$, $repeaTimeList \leftarrow [|V| - 1 \text{ of } 1]$

1 **while** $\sum_i repeaTimeList[i] < 2n$ **do**

2     $i \leftarrow \text{round}(\text{RANDOMUNIFORM}(0, |V| - 2))$

3     $repeaTimeList[i] \leftarrow repeaTimeList[i] + 1$

4 **return** $repeaTimeList$

The pickup and dropoff nodes for each request are randomly paired up in Algorithm 2. We first rewrite *repeaTimeList* as a list of length $2n$, *shuffList*, which consists of all selected non-depot nodes, where the number of repetitions indicates the number of times they are selected. For example, we have *shuffList* = $[0, 0, 0, 1, 2, 2]$ for *repeaTimeList* = $[3, 1, 2]$. Next, we shuffle *shuffList* and clear *pairList*, which is used to store pairs of nodes indicating the pickup and dropoff points of all requests. We then divide *shuffList* into n pairs in order. If two nodes of any pair are identical (i.e., the pickup and dropoff are the same point), or the pair with considering order is already contained in *pairList*, then we are back to Line 4. Otherwise, we append the eligible pair of nodes into *pairList* until the last pair is added.

Algorithm 2: Randomly pair up the pickup and dropoff nodes for each request

---

Input: $G = (V, E)$, $n = \text{round}(\frac{k(|V|-1)}{2})$, $repeaTimeList$

Init. : $shuffList \leftarrow []$, $pairList \leftarrow []$, $reshuffle \leftarrow \text{true}$

1 **for** $i \leftarrow 0$ **to** $|V| - 2$ **do**

2     $shuffList.\text{EXTEND}([repeaTimeList[i] \text{ of } i])$

3 **while** $reshuffle$ **do**

4     $\text{RANDOMSHUFFLE}(shuffList)$

5     $pairList \leftarrow []$

6   **for** $i \leftarrow 0$ **to** $n - 1$ **do**

7   **if** $shuffList[2i] = shuffList[1 + 2i]$ **or** $[shuffList[2i], shuffList[1 + 2i]]$ **in** $pairList$ **then**

8   **break**

9   **else**

10           $pairList.\text{APPEND}([shuffList[2i], shuffList[1 + 2i]])$

11   **if** $i = n - 1$ **then** $reshuffle \leftarrow \text{false}$

12 **return** $pairList$

Since we intend to generate instances corresponding to different k for each selected sample of TSPLIB benchmark, and the number of requests n gets smaller as k decreases, we need to produce a decrementable list of node pairs such that we can remove some of the node pairs to correspond to smaller k, but the remaining part of the node pairs still contains all locations in the sample other than that as depot (i.e., they still need to be selected at least once). Therefore, in Algorithm 3, we assume that a sorted list of node pairs pops the end elements out of it one by one, yet it is always guaranteed that every non-depot node is selected at least once. We insert the node pair in which both nodes are currently selected only once into the frontmost of *head* in Line 7, and then append the node pair in which one of them is selected only once into *head* in Line 11. Then we keep inserting the node pair in which the sum of the repetitions of the two nodes is currently maximum into the frontmost of *tail* in Line 22. Note that such insertions and appends require popping the corresponding node pair out of *pairList* and updating $\mathcal{L}$. We merge *head* and *tail* to obtain a sorted list of node pairs *sortedPairs* at the end of Algorithm 3.

Algorithm 3: Sort the node pairs by the sum of the repeat times of each node in the node pair

---

Input: *repeaTimeList, pairList*
Init. : $\mathcal{L} \leftarrow$ *repeaTimeList*, *head* $\leftarrow []$, *tail* $\leftarrow []$, *max* $\leftarrow 0$, *maxIndex* $\leftarrow -1$, *sortedPairs* $\leftarrow []$

1 **while** $|pairList| > 0$ **do**
2      $i \leftarrow 0$
3   **while** $i < |pairList|$ **do**
4   **if** $\mathcal{L}[pairList[i][0]] = 1$ **and** $\mathcal{L}[pairList[i][1]] = 1$ **then**
5            $\mathcal{L}[pairList[i][0]] \leftarrow \mathcal{L}[pairList[i][0]] - 1$
6            $\mathcal{L}[pairList[i][1]] \leftarrow \mathcal{L}[pairList[i][1]] - 1$
7            $head.\text{INSERT}(0, pairList.\text{POP}(i))$
8   **else if** $\mathcal{L}[pairList[i][0]] = 1$ **or** $\mathcal{L}[pairList[i][1]] = 1$ **then**
9            $\mathcal{L}[pairList[i][0]] \leftarrow \mathcal{L}[pairList[i][0]] - 1$
10           $\mathcal{L}[pairList[i][1]] \leftarrow \mathcal{L}[pairList[i][1]] - 1$
11           $head.\text{APPEND}(pairList.\text{POP}(i))$
12   **else** $i \leftarrow i + 1$
13     $max \leftarrow 0$
14     $maxIndex \leftarrow -1$
15   **for** $j \leftarrow 0$ **to** $|pairList| - 1$ **do**
16   **if** $\mathcal{L}[pairList[j][0]] + \mathcal{L}[pairList[j][1]] > max$ **then**
17           $max \leftarrow \mathcal{L}[pairList[j][0]] + \mathcal{L}[pairList[j][1]]$
18           $maxIndex \leftarrow j$
19   **if** $maxIndex \neq -1$ **then**
20        $\mathcal{L}[pairList[maxIndex][0]] \leftarrow \mathcal{L}[pairList[maxIndex][0]] - 1$
21        $\mathcal{L}[pairList[maxIndex][1]] \leftarrow \mathcal{L}[pairList[maxIndex][0]] - 1$
22        $tail.\text{INSERT}(0, pairList.\text{POP}(maxIndex))$

23 *sortedPairs* $\leftarrow head.\text{EXTEND}(tail)$
24 **return** *sortedPairs*

Parameter Settings Finally, we can generate a list containing n requests for each selected sample of TSPLIB benchmark, as shown in Algorithm 4. In addition, we generate data for the three types of trucks recursively in the order of maximum load capacity of 25, 20, and 15 until the number of generated trucks reaches m. And these three types of trucks correspond to their respective cost coefficients for each traversing arc of 1.2, 1, and 0.8. For example, for a truck t of load capacity 15, l_{od}^t, i.e., the cost of it passing through the arc $\widehat{od}$ is $0.8 \times \text{DISTANCE}(\widehat{od})$.[3] According to the average volume of 5 for each request set in Algorithm 4, it is expected that each truck can accommodate four requests at the same time.

Algorithm 4: Randomly generate the list of requests

Input: $G = (V, E)$, $n = \text{round}(\frac{k(|V|-1)}{2})$, sortedPairs

Init. : $avgDistance \leftarrow \frac{1}{|V| \cdot |V-1|} \sum_{o \in V} \sum_{\substack{d \in V \\ d \neq o}} \text{DISTANCE}(\widehat{od})$, $avgVolume \leftarrow 5$,
$\qquad$ $requestList \leftarrow [\,]$

1 **for** $r \leftarrow 0$ **to** $n - 1$ **do**
2 $\quad$ $q_r \leftarrow \text{round}(\text{RANDOMUNIFORM}(1, 2 \times avgVolume - 1))$
3 $\quad$ $w_r \leftarrow \text{round}(2 \times avgDistance \times q_r \div avgVolume)$
4 $\quad$ $f(r) \leftarrow sortedPairs[r][0]$
5 $\quad$ $g(r) \leftarrow sortedPairs[r][1]$
6 $\quad$ $requestList.\text{APPEND}(\langle w_r, q_r, f(r), g(r)\rangle)$

7 **return** $requestList$

We generate a total of $3 \times 5 \times 5 = 75$ PPDSP instances for our comparative experiments based on the proposed formulation and on the existing formulation, respectively, according to the following parameter settings.

- The selected TSPLIB samples are $\{burma14, ulysses16, ulysses22\}$;
- The average repetition rate of the same locations, $k \in \{1, 1.5, 2, 2.5, 3\}$;
- The number of trucks $m \in \{2, 4, 6, 8, 10\}$.

4.2 Experimental Setup

All experiments are performed on an Apple M1 Pro chip, using the Ubuntu 18.04.6 LTS operating system via the Podman virtual machine with 19 GB of allocated memory. A single CPU core is used for each experiment. We implemented the instance generators for both the proposed formulation and the existing formulation by using Python 3. Each generated problem instance is solved by the MIP optimizer–CPLEX of version 20.1.0.0 [5] within 3,600 CPU seconds time limit. Source code for our experiments is available at https://github.com/ReprodSuplem/PPDSP.

---

[3] $\text{DISTANCE}(\widehat{od})$ is the *Euclidean distance* between the coordinates of node o and the coordinates of node d.

Table 2. Comparison of the existing formulation-based method with the proposed formulation-based method for TSPLIB sample *burma14* in terms of the number of variables and constraints generated, as well as the optimal values achieved.

m	item	$k=1$ ($n=7$)		$k=1.5$ ($n=10$)		$k=2$ ($n=13$)		$k=2.5$ ($n=16$)		$k=3$ ($n=20$)	
2	#Var	576	**458**	1056	**464**	1680	**470**	2448	**476**	3696	**484**
	#Con	**1027**	1041	1942	**1062**	3145	**1083**	4636	**1104**	7072	**1132**
	Opt	30	30	**42**	40	**56**	53	65	**75**	100	**107**
4	#Var	1152	**916**	2112	**928**	3360	**940**	4896	**952**	7392	**968**
	#Con	**2047**	2075	3874	**2114**	6277	**2153**	9256	**2192**	14124	**2244**
	Opt	**33**	30	**49**	44	58	**61**	53	**77**	68	**103**
6	#Var	1728	**1374**	3168	**1392**	5040	**1410**	7344	**1428**	11088	**1452**
	#Con	**3067**	3109	5806	**3166**	9409	**3223**	13876	**3280**	21176	**3356**
	Opt	**33**	31	**46**	42	57	**58**	43	**81**	46	**103**
8	#Var	2304	**1832**	4224	**1856**	6720	**1880**	9792	**1904**	14784	**1936**
	#Con	**4087**	4143	7738	**4218**	12541	**4293**	18496	**4368**	28228	**4468**
	Opt	**33**	31	**46**	44	56	56	4	**79**	81	**104**
10	#Var	2880	**2290**	5280	**2320**	8400	**2350**	12240	**2380**	18480	**2420**
	#Con	**5107**	5177	9670	**5270**	15673	**5363**	23116	**5456**	35280	**5580**
	Opt	**33**	31	44	44	57	**58**	24	**80**	49	**96**

4.3 Results

Tables 2, 3 and 4 show the performance of the existing formulation-based method and our proposed formulation-based method in terms of the number of generated variables (#Var.), the number of generated constraints (#Con.), and the optimal values (Opt.) for the various average repetition rates of the same locations (k) and the different numbers of trucks (m), corresponding to TSPLIB samples *burma14*, *ulysses16* and *ulysses22*, respectively. Each cell recording the left and right values corresponds to a comparison item, where the left values refers to the performance of the existing formulation-based method, while the right values corresponds to the performance of the proposed formulation-based method. We compare the performance of these two methods in such cells and put the values of the dominant side in bold.

We can see that either the number of variables or the number of constraints generated by our proposed method is proportional to m, while neither the number of variables nor the number of constraints generated by our proposed method increases significantly as k becomes larger. Such a result is consistent with Theorem 3. In contrast, although the number of variables and the number of constraints produced by the existing method is also proportional to m, with k becoming larger, exponential increases in both of them are observed. Furthermore, it is interesting to note that even when $k = 1$, the proposed method generates fewer variables than the existing method.

As k increases, the optimal value obtained by the proposed method increases within the time limit; while there is no such trend in the optimal values obtained

Table 3. Comparison of the existing formulation-based method with the proposed formulation-based method for TSPLIB sample *ulysses16* in terms of the number of variables and constraints generated, as well as the optimal values achieved.

m	item	$k = 1$ $(n = 8)$		$k = 1.5$ $(n = 11)$		$k = 2$ $(n = 15)$		$k = 2.5$ $(n = 19)$		$k = 3$ $(n = 23)$	
2	#Var	720	**588**	1248	**594**	2176	**602**	3360	**610**	4800	**618**
	#Con	**1300**	1380	2311	**1401**	4107	**1429**	6415	**1457**	9235	**1485**
	Opt	94	94	**96**	94	109	**112**	157	**187**	168	**230**
4	#Var	1440	**1176**	2496	**1188**	4352	**1204**	6720	**1220**	9600	**1236**
	#Con	**2592**	2752	4611	**2791**	8199	**2843**	12811	**2895**	18447	**2947**
	Opt	**99**	98	**104**	101	104	**124**	140	**198**	29	**247**
6	#Var	2160	**1764**	3744	**1782**	6528	**1806**	10080	**1830**	14400	**1854**
	#Con	**3884**	4124	6911	**4181**	12291	**4257**	19207	**4333**	27659	**4409**
	Opt	**99**	98	**105**	101	104	**139**	92	**205**	0	**226**
8	#Var	2880	**2352**	4992	**2376**	8704	**2408**	13440	**2440**	19200	**2472**
	#Con	**5176**	5496	9211	**5571**	16383	**5671**	25603	**5771**	36871	**5871**
	Opt	**99**	98	91	**98**	89	**138**	75	**176**	96	**241**
10	#Var	3600	**2940**	6240	**2970**	10880	**3010**	16800	**3050**	24000	**3090**
	#Con	**6468**	6868	11511	**6961**	20475	**7085**	31999	**7209**	46083	**7333**
	Opt	**99**	98	96	**101**	111	**130**	82	**186**	27	**221**

by the existing method. In addition, as m grows up, in theory, the upper bound of the optimal value cannot be smaller. However, the optimal values obtained by the proposed method and the existing method do not maintain the trend of increasing, instead, they both have inflection points. This is due to the fact that the search space of the problem becomes huge, which makes it inefficient for the solver to update the optimized solution. There are even cases on the existing method where the initial solution is not obtained until the end of time.

Last but not least, we can clearly see that for those problems with larger k (i.e., more nodes that can be integrated), the method based on our proposed formulation generates fewer variables and constraints, as well as achieves larger optimal values, than the method based on the existing formulation.

4.4   Discussion

In a real-world scenario, different logistics orders may be sent from the same place to various places, or from different places to the same place. With a certain range of shipping services, as the number of orders increases, the average repetition rate (k) of pickup locations as well as delivery locations becomes higher. From the experimental results, we can see that, although the proposed method is more efficient than the existing methods, PPDSP is still difficult to be solved exactly, even for problem instances generated based on small-scale road networks in the limited time of 3,600 CPU seconds to find the optimal

Table 4. Comparison of the existing formulation-based method with the proposed formulation-based method for TSPLIB sample *ulysses22* in terms of the number of variables and constraints generated, as well as the optimal values achieved.

m	item	$k=1$ ($n=11$)		$k=1.5$ ($n=16$)		$k=2$ ($n=21$)		$k=2.5$ ($n=26$)		$k=3$ ($n=32$)	
2	#Var	1248	**1074**	2448	**1084**	4048	**1094**	6048	**1104**	8976	**1116**
	#Con	**2311**	2685	4636	**2720**	7761	**2755**	11686	**2790**	17452	**2832**
	Opt	116	**120**	161	**206**	182	**225**	85	**297**	79	**336**
4	#Var	2496	**2148**	4896	**2168**	8096	**2188**	12096	**2208**	17952	**2232**
	#Con	**4611**	5359	9256	**5424**	15501	**5489**	23346	**5554**	34872	**5632**
	Opt	**134**	120	141	**185**	53	**216**	34	**277**	15	**327**
6	#Var	3744	**3222**	7344	**3252**	12144	**3282**	18144	**3312**	26928	**3348**
	#Con	**6911**	8033	13876	**8128**	23241	**8223**	35006	**8318**	52292	**8432**
	Opt	115	**125**	136	**192**	0	**227**	45	**242**	0	**343**
8	#Var	4992	**4296**	9792	**4336**	16192	**4376**	24192	**4416**	35904	**4464**
	#Con	**9211**	10707	18496	**10832**	30981	**10957**	46666	**11082**	69712	**11232**
	Opt	**124**	117	102	**206**	25	**224**	58	**228**	41	**276**
10	#Var	6240	**5370**	12240	**5420**	20240	**5470**	30240	**5520**	44880	**5580**
	#Con	**11511**	13381	23116	**13536**	38721	**13691**	58326	**13846**	87132	**14032**
	Opt	**124**	123	31	**173**	44	**204**	42	**243**	44	**319**

solution. We do not consider the time window constraint in this paper because adding more constraints would make the solution of the problem more difficult. Nonetheless, for solving larger scale PPDSP within a reasonable computation time, we still need to tailor heuristic algorithms for the location-based model in our future work. Compared to the request-based model, we propose a surrogate optimization model for PPDSP, whose solution set is a subset of the former. This may result in a lower upper bound on the optimized value of exact solutions. However, our approach remains competitive in obtaining approximate solutions.

5 Conclusion

In this paper, we revisit PPDSP on road networks with the integratable nodes. For such application scenarios, we define a location-based graph-theoretic model and give the corresponding MIP formulation. We prove the correctness of this formulation as well as analyze its space complexity. We compare the proposed method with the existing formulation-based method. The experimental results show that, for the instances with more integratable nodes, our method has a significant advantage over the existing method in terms of the generated problem size, and the optimized values.

Acknowledgements. The authors are grateful for the support sponsors from *the Progressive Logistic Science Laboratory*, the Research Center for Advanced Science and Technology, the University of Tokyo.

References

1. Ahmadi-Javid, A., Amiri, E., Meskar, M.: A profit-maximization location-routing-pricing problem: A branch-and-price algorithm. Eur. J. Oper. Res. **271**(3), 866–881 (2018). https://doi.org/10.1016/j.ejor.2018.02.020
2. Al-Chami, Z., Flity, H.E., Manier, H., Manier, M.: A new metaheuristic to solve a selective pickup and delivery problem. In: Boukachour, J., Sbihi, A., Alaoui, A.E.H., Benadada, Y. (eds.) 4th International Conference on Logistics Operations Management, GOL 2018, Le Havre, 10–12 April, 2018. pp. 1–5. IEEE (2018). https://doi.org/10.1109/GOL.2018.8378089
3. Asghari, M., Mirzapour Al-e-hashem, S.M.J.: A green delivery-pickup problem for home hemodialysis machines; sharing economy in distributing scarce resources. Transp. Res. Part E: Logist. Transp. Rev. **134**, 101815 (2020). https://doi.org/10.1016/j.tre.2019.11.009
4. Bruni, M.E., Toan, D.Q., Nam, L.H.: The multi-vehicle profitable pick up and delivery routing problem with uncertain travel times. Transp. Res. Procedia **52**, 509–516 (2021). https://doi.org/10.1016/j.trpro.2021.01.060
5. CPLEX, I.I.: IBM ILOG CPLEX Optimizer (2020). https://www.ibm.com/products/ilog-cplex-optimization-studio/cplex-optimizer
6. Gansterer, M., Küçüktepe, M., Hartl, R.F.: The multi-vehicle profitable pickup and delivery problem. OR Spectrum **39**(1), 303–319 (2016). https://doi.org/10.1007/s00291-016-0454-y
7. Huang, D., Gu, Y., Wang, S., Liu, Z., Zhang, W.: A two-phase optimization model for the demand-responsive customized bus network design. Transp. Res. Part C: Emerg. Technol. **111**, 1–21 (2020). https://doi.org/10.1016/j.trc.2019.12.004
8. Liu, C., Aleman, D.M., Beck, J.C.: Modelling and solving the senior transportation problem. In: van Hoeve, W.-J. (ed.) CPAIOR 2018. LNCS, vol. 10848, pp. 412–428. Springer, Cham (2018). https://doi.org/10.1007/978-3-319-93031-2_30
9. Miller, C.E., Tucker, A.W., Zemlin, R.A.: Integer programming formulation of traveling salesman problems. J. ACM **7**(4), 326–329 (1960). https://doi.org/10.1145/321043.321046
10. Qiu, X., Feuerriegel, S., Neumann, D.: Making the most of fleets: A profit-maximizing multi-vehicle pickup and delivery selection problem. Eur. J. Oper. Res. **259**(1), 155–168 (2017). https://doi.org/10.1016/j.ejor.2016.10.010
11. Riedler, M., Raidl, G.R.: Solving a selective dial-a-ride problem with logic-based benders decomposition. Comput. Oper. Res. **96**, 30–54 (2018). https://doi.org/10.1016/j.cor.2018.03.008
12. Ting, C., Liao, X., Huang, Y., Liaw, R.: Multi-vehicle selective pickup and delivery using metaheuristic algorithms. Inf. Sci. **406**, 146–169 (2017). https://doi.org/10.1016/j.ins.2017.04.001
13. Zha, A., Chang, Q., Imura, N., Nishinari, K.: A case study of the profit-maximizing multi-vehicle pickup and delivery selection problem for the road networks with the integratable nodes (2022). https://doi.org/10.48550/arXiv.2208.14866

Symbolic-Numeric Computation in Modeling the Dynamics of the Many-Body System TRAPPIST

Alexander Chichurin[1] [iD], Alexander Prokopenya[2(✉)] [iD],
Mukhtar Minglibayev[3,4] [iD], and Aiken Kosherbayeva[4] [iD]

[1] The John Paul II Catholic University of Lublin, ul. Konstantynow 1H,
20-708 Lublin, Poland
achichurin@kul.lublin.pl
[2] Warsaw University of Life Sciences–SGGW, Nowoursynowska 159, 02-776 Warsaw,
Poland
alexander_prokopenya@sggw.edu.pl
[3] Fesenkov Astrophysical Institute, Observatoriya 23, 050020 Almaty, Kazakhstan
[4] Al-Farabi Kazakh National University, Al-Farabi av. 71, 050040 Almaty,
Kazakhstan

Abstract. Modeling the dynamics of the exoplanetary system TRAP-
PIST with seven bodies of variable mass moving around a central parent
star along quasi-elliptic orbits is discussed. The bodies are assumed to
be spherically symmetric and attract each other according to Newton's
law of gravitation. In this case, the leading factor of dynamic evolution
of the system is the variability of the masses of all bodies. The problem
is analyzed in the framework of Hamiltonian's formalism and the differential equations of motion of the bodies are derived in terms of the
osculating elements of aperiodic motion on quasi-conic sections. These
equations can be solved numerically but their right-hand sides contain
many oscillating terms and so it is very difficult to obtain their solutions
over long time intervals with necessary precision. To simplify calculations
and to analyze the behavior of orbital parameters over long time intervals
we replace the perturbing functions by their secular parts and obtain a
system of the evolutionary equations composed by 28 non-autonomous
linear differential equations of the first order. Choosing some realistic
laws of mass variations and physics parameters corresponding to the
exoplanetary system TRAPPIST, we found numerical solutions of the
evolutionary equations. All the relevant symbolic and numeric calculations are performed with the aid of the computer algebra system Wolfram
Mathematica.

Keywords: Non-stationary many-body problem · Isotropic change of
mass · Secular perturbations · Evolution equations · Poincaré variables

This research is partly funded by the Committee of Science of the Ministry of Science
and Higher Education of the Republic of Kazakhstan (Grant No. AP14869472).

© The Author(s), under exclusive license to Springer Nature Switzerland AG 2023
J. Mikyška et al. (Eds.): ICCS 2023, LNCS 14075, pp. 469–482, 2023.
https://doi.org/10.1007/978-3-031-36024-4_36

1   Introduction

In the wake of discoveries of exoplanetary systems [1], study of dynamic evolution of such planetary systems has become highly relevant. Observational data show that celestial bodies in such systems are non-stationary, their characteristics such as mass, size, and shape may vary with time [2–6]. At the same time, it is very difficult to take into account non-stationarity of the bodies because the corresponding mathematical models become very complicated. Even in the case of classical two-body problem, a general solution of which is well-known, dependence of masses on time makes the problem non-integrable; only in some special cases its exact solution can be found in symbolic form (see [7]). However, the masses of the bodies influence essentially on their interaction and motion and so it is especially interesting to investigate the dynamics of the many-body system of variable masses. One of the first works in this direction were done by T.B. Omarov [8] and J.D. Hadjidemetriou [9] (see also [10]) who started investigation of the effects of mass variability on the dynamic evolution of non-stationary gravitating systems. Later these investigations were continued in a series of works [11–15], where the systems of three interacting bodies with variable masses were considered. It was shown also that application of the computer algebra systems is very fruitful and enables to get new interesting results because very cumbersome symbolic computations are involved (see [16–18]).

It should be noted that most of the works on the dynamics of planetary systems are devoted to the study of evolution of multi-planet systems of many point bodies with constant masses. As the many-body problem is not integrable the perturbation theory based on the exact solution of the two-body problem is usually used (see [19]). This approach turned out to be very successful and many interesting results were obtained in the investigation of the motion of planet or satellite in the star-planets or double star system (see [20,21]). Paper [22] describes the problem of constructing a theory of four planets' motion around the central star, while the bodies masses are constant. The Hamiltonian functions are expanded into Poisson series in the osculating elements of the second Poincaré system up to the third power of the small parameter. Evolution of the planetary systems Sun - Jupiter - Saturn - Uranus - Neptune is studied in [23]. The averaged equations of motion are constructed analytically up to the third order in a small parameter for a four-planetary system. Paper [24] studies the orbital evolution of three-planet exosystem HD 39194 and the four-planet exosystems HD 141399 and HD 160691 (μ Ara). As a result, the authors have developed an averaged semi-analytical theory of motion of the second order in terms of exoplanet masses.

In the present paper, we investigate a classical problem of 8 bodies of variable masses which may be considered as model of the exoplanetary system TRAPPIST with a central star and 7 planets orbiting the star (see [1,25,26]). The work is aimed at calculating secular perturbations of planetary systems on non-stationary stage of its evolution when mass variability is the leading factor of evolution. Equations of motion of the system are obtained in a general form in the relative coordinate system with the star located at the origin. The masses of the bodies are variable and change isotropically which means that the reactive forces do not arise. We describe main computational problems occurring when the perturbing functions are written in terms of the second Poincaré system and the evolutionary equations are obtained. For this paper, all symbolic computations were performed with the aid of the computer algebra system Wolfram Mathematica [30] which has a convenient interface and allows one to combine various kinds of computations.

The paper is organized as follows. In Sect. 2 we formulate the physical problem and describe the model. Then in Sect. 3 we derive the equations of motion in the osculating elements which are convenient for applying the perturbation theory. Section 4 is devoted to computing the perturbing functions in terms of the second Poincaré system. As a result, we obtain the evolutionary equations in Sect. 5 and write out them in terms of dimensionless variables. In Sect. 6 we describe numerical solution of the evolutionary equations. At last, we summarize the results in Conclusion.

2   Statement of the Problem and Differential Equations of Motion

Let us consider the motion of a planetary system consisting of $n + 1$ spherical bodies with isotropically changing masses mutually attracting each other according to Newton's law. Let us introduce the following notation: S is a parent star of the planetary system of mass $m_0 = m_0(t)$, P_i are planets of masses $m_i = m_i(t), (i = 1, 2, ..., n)$. We will study the motion in a relative coordinate system with the origin at the center of the parent star S the axes of which are parallel to the corresponding axes of the absolute coordinate system.

The positions of the planets are such that P_i is an inner planet relative to the P_{i+1} planets, but at the same time it is an outer planet relative to P_{i-1}. We assume that this position of the planets is preserved during the evolution.

Let the rate of mass change be different

$$\frac{\dot{m}_0}{m_0} \neq \frac{\dot{m}_i}{m_i}, \quad \frac{\dot{m}_i}{m_i} \neq \frac{\dot{m}_j}{m_j} \quad (i, j = \overline{1, n}, \quad i \neq j). \tag{1}$$

In a relative coordinate system, the equations of motion of planets with isotropically varying masses may be written as [27–29]

$$\ddot{\boldsymbol{r}}_i = -f\frac{(m_0 + m_i)}{r_i^3}\boldsymbol{r}_i + f\sum_{j=1}^{n}{}'m_j\left(\frac{\boldsymbol{r}_j - \boldsymbol{r}_i}{r_{ij}^3} - \frac{\boldsymbol{r}_j}{r_j^3}\right) \quad (i, j = \overline{1, n}), \tag{2}$$

where r_{ij} are mutual distances between the centers of spherical bodies

$$r_{ij} = \sqrt{(x_j - x_i)^2 + (y_j - y_i)^2 + (z_j - z_i)^2} = r_{ji}, \tag{3}$$

f is the gravitational constant, $\boldsymbol{r}_i(x_i, y_i, z_i)$ is a radius-vector of the planet P_i, and the prime sign in summation means that $i \neq j$.

3   Equation of Motion in the Osculating Elements

3.1   Extraction of the Perturbing Function

Equations of motion (2) may be rewritten in the form

$$\ddot{\boldsymbol{r}}_i + f\frac{(m_0 + m_i)}{r_i^3}\boldsymbol{r}_i - \frac{\ddot{\gamma}_i}{\gamma_i}\boldsymbol{r}_i = \boldsymbol{F}_i, \quad \gamma_i = \frac{m_0(t_0) + m_i(t_0)}{m_0(t) + m_i(t)} = \gamma_i(t), \quad (4)$$

where t_0 is an initial instant of time, and

$$\boldsymbol{F}_i = grad_{r_i} W_i, \qquad W_i = W_{gi} + W_{ri}, \qquad (5)$$

$$W_{gi} = f\sum_{j=1}^{n}{}' m_j \left(\frac{1}{r_{ij}} - \frac{\boldsymbol{r}_i \cdot \boldsymbol{r}_j}{r_j^3} \right), \quad \boldsymbol{r}_{ij} = \boldsymbol{r}_j - \boldsymbol{r}_i, \quad W_{ri} = -\frac{\ddot{\gamma}_i}{2\gamma_i}r_i^2. \quad (6)$$

The equations of relative motion written in the form (4) are convenient for applying the perturbation theory developed for such non-stationary systems [6]. In the case under consideration the perturbing forces are given by the expressions (5), (6). Note that in the case of $\boldsymbol{F}_i = 0$ Eqs. (4) reduce to integrable differential equations describing unperturbed motion of the bodies along quasi-conic sections.

3.2   Differential Equations of Motion in Analogues of the Second System of Poincaré variables

For our purposes, analogues of the second system of Poincaré canonical elements given in the works [6, 28] are preferred

$$\Lambda_i, \quad \lambda_i, \quad \xi_i, \quad \eta_i, \quad p_i, \quad q_i, \qquad (7)$$

which are defined according to the formulas

$$\Lambda_i = \sqrt{\mu_{i0}}\sqrt{a_i},$$
$$\lambda_i = l_i + \pi_i, \qquad (8)$$

$$\xi_i = \sqrt{2\sqrt{\mu_{i0}}\sqrt{a_i}(1 - \sqrt{1 - e_i^2})}\cos\pi_i,$$
$$\eta_i = -\sqrt{2\sqrt{\mu_{i0}}\sqrt{a_i}(1 - \sqrt{1 - e_i^2})}\sin\pi_i, \qquad (9)$$

$$p_i = \sqrt{2\sqrt{\mu_{i0}}\sqrt{a_i}\sqrt{1 - e_i^2}(1 - \cos I_i)}\cos\Omega_i,$$
$$q_i = -\sqrt{2\sqrt{\mu_{i0}}\sqrt{a_i}\sqrt{1 - e_i^2}(1 - \cos I_i)}\sin\Omega_i, \qquad (10)$$

where

$$l_i = M_i = \tilde{n}_i[\phi_i(t) - \phi_i(\tau_i)], \qquad \pi_i = \Omega_i + \omega_i. \tag{11}$$

The differential equations of motion of n planets in the osculating analogues of the second system of Poincaré variables (8)–(11) have the canonical form

$$
\dot{\Lambda}_i = -\frac{\partial R_i^*}{\partial \lambda_i}, \quad \dot{\xi}_i = -\frac{\partial R_i^*}{\partial \eta_i}, \quad \dot{p}_i = -\frac{\partial R_i^*}{\partial q_i},
$$
$$
\dot{\lambda}_i = \frac{\partial R_i^*}{\partial \Lambda_i}, \quad \dot{\eta}_i = \frac{\partial R_i^*}{\partial \xi_i}, \quad \dot{q}_i = \frac{\partial R_i^*}{\partial p_i}, \tag{12}
$$

where the Hamiltonian functions are given by

$$R_i^* = -\frac{\mu_{i0}^2}{2\Lambda_i^2} \cdot \frac{1}{\gamma_i^2(t)} - W_i \left(t, \Lambda_i, \xi_i, p_i, \lambda_i, \eta_i, q_i\right). \tag{13}$$

The canonical equations of perturbed motion (12) are convenient for describing the dynamic evolution of planetary systems when the analogues of eccentricities e_i and analogues of the inclinations I_i of the orbital plane of the planets are sufficiently small

$$e_i << 1, \qquad I_i << 1 \quad (i = \overline{1, n}). \tag{14}$$

Let us rewrite the canonical equations of motion (12) as

$$
\dot{\lambda}_i = \frac{\partial R_i^*}{\partial \Lambda_i} = \frac{\mu_{i0}^2}{\gamma_i^2 \Lambda_i^3} - \frac{\partial W_i}{\partial \Lambda_i}, \qquad \dot{\Lambda}_i = \frac{\partial R_i^*}{\partial \lambda_i} = \frac{\partial W_i}{\partial \lambda_i},
$$

$$
\dot{\eta}_i = \frac{\partial R_i^*}{\partial \xi_i} = -\frac{\partial W_i}{\partial \xi_i}, \qquad \dot{\xi}_i = \frac{\partial R_i^*}{\partial \eta_i} = \frac{\partial W_i}{\partial \eta_i}, \tag{15}
$$

$$
\dot{q}_i = \frac{\partial R_i^*}{\partial p_i} = -\frac{\partial W_i}{\partial p_i}, \qquad \dot{p}_i = \frac{\partial R_i^*}{\partial q_i} = \frac{\partial W_i}{\partial q_i}.
$$

4   The Secular Part of the Main Part of the Perturbing Function

The secular part of the perturbing functions (13) has the form [29]

$$W_i^{(sec)} = W_{is}^{(sec)} + W_{ik}^{(sec)} + W_{ri}^{(sec)}. \tag{16}$$

Let us write the explicit form of the secular part of the perturbing function

$$
W_i^{(sec)} = f \sum_{s=1}^{i-1} m_s \Big(\frac{A_0^{is}}{2} + \Pi_{ii}^{is} \frac{\eta_i^2 + \xi_i^2}{2\Lambda_i} + \Pi_{is}^{is} \frac{\eta_i \eta_s + \xi_i \xi_s}{\sqrt{\Lambda_i \Lambda_s}} + \Pi_{ss}^{is} \frac{\eta_s^2 + \xi_s^2}{2\Lambda_s} -
$$

$$
-B_1^{is} \Big(\frac{p_i^2 + q_i^2}{8\Lambda_i} - \frac{p_i p_s + q_i q_s}{4\sqrt{\Lambda_i \Lambda_s}} + \frac{p_s^2 + q_s^2}{8\Lambda_s} \Big) \Big) + f \sum_{k=i+1}^{n} m_k \Big(\frac{A_0^{ik}}{2} + \Pi_{ii}^{ik} \frac{\eta_i^2 + \xi_i^2}{2\Lambda_i} +
$$

$$
+\Pi_{ik}^{ik} \frac{\eta_i \eta_k + \xi_i \xi_k}{\sqrt{\Lambda_i \Lambda_k}} + \Pi_{kk}^{ik} \frac{\eta_k^2 + \xi_k^2}{2\Lambda_k} - B_1^{ik} \Big(\frac{p_i^2 + q_i^2}{8\Lambda_i} - \frac{p_i p_k + q_i q_k}{4\sqrt{\Lambda_i \Lambda_k}} + \frac{p_k^2 + q_k^2}{8\Lambda_k} \Big) \Big) -
$$

$$
-\frac{\ddot{\gamma}_i \Lambda_i^4}{2\gamma_i \mu_{i0}^2} \Big(1 + \frac{3}{2\Lambda_i} (\xi_i^2 + \eta_i^2) \Big), \tag{17}
$$

where the following designations are accepted for the inner planets $(s < i)$

$$\Pi_{ii}^{is} = -\frac{3\alpha_{is}}{4}B_0^{is} - \frac{1}{2}B_1^{is} + \frac{15 + 6\alpha_{is}^2}{8}C_0^{is} - \frac{3\alpha_{is}}{2}C_1^{is} - \frac{9}{8}C_2^{is},$$

$$\alpha_{is} = \frac{\gamma_s a_s}{\gamma_i a_i} = \alpha_{is}(t) < 1, \tag{18}$$

$$\Pi_{is}^{is} = \frac{1}{8}\left(9B_0^{is} + B_2^{is}\right) - \frac{9\left(1 + \alpha_{is}^2\right)}{8\alpha_{is}}C_0^{is} + \frac{21}{16}C_1^{is} + \frac{3\left(1 + \alpha_{is}^2\right)}{8\alpha_{is}}C_2^{is} + \frac{3}{16}C_3^{is}, \tag{19}$$

$$\Pi_{ss}^{is} = -\frac{3}{4\alpha_{is}}B_0^{is} - \frac{1}{2}B_1^{is} + \frac{15\alpha_{is}^2 + 6}{8\alpha_{is}^2}C_0^{is} - \frac{3}{2\alpha_{is}}C_1^{is} - \frac{9}{8}C_2^{is}, \tag{20}$$

$(s < i).$

$$A_0^{is} = \frac{2}{\pi a_i \gamma_i} \int_0^{\pi} \frac{d\lambda}{\left(1 + \alpha_{is}^2 - 2\alpha_{is}\cos\lambda\right)^{1/2}}, \quad (s < i), \quad p = 0, 1, 2, 3, \tag{21}$$

$$B_p^{is} = \frac{2a_s\gamma_s}{\pi\left(a_i\gamma_i\right)^2} \int_0^{\pi} \frac{\cos(p\lambda)d\lambda}{\left(1 + \alpha_{is}^2 - 2\alpha_{is}\cos\lambda\right)^{3/2}}, \tag{22}$$

$$C_p^{is} = \frac{2\left(a_s\gamma_s\right)^2}{\pi\left(a_i\gamma_i\right)^3} \int_0^{\pi} \frac{\cos(p\lambda)d\lambda}{\left(1 + \alpha_{is}^2 - 2\alpha_{is}\cos\lambda\right)^{5/2}}.$$

For the outer planets $(i < k)$ the following notations are accepted

$$\Pi_{ii}^{ik} = -\frac{3\alpha_{ik}}{4}B_0^{ik} - \frac{1}{2}B_1^{ik} + \frac{15 + 6\alpha_{ik}^2}{8}C_0^{ik} - \frac{3\alpha_{ik}}{2}C_1^{ik} - \frac{9}{8}C_2^{ik}, \quad (i < k),$$

$$\alpha_{ik} = \frac{\gamma_i a_i}{\gamma_k a_k} = \alpha_{ik}(t) < 1, \tag{23}$$

$$\Pi_{ik}^{ik} = \frac{1}{8}\left(9B_0^{ik} + B_2^{ik}\right) - \frac{9\left(1 + \alpha_{ik}^2\right)}{8\alpha_{ik}}C_0^{ik} + \frac{21}{16}C_1^{ik} + \frac{3\left(1 + \alpha_{ik}^2\right)}{8\alpha_{ik}}C_2^{ik} + \frac{3}{16}C_3^{ik}, \tag{24}$$

$$\Pi_{kk}^{ik} = -\frac{3}{4\alpha_{ik}}B_0^{ik} - \frac{1}{2}B_1^{ik} + \frac{15\alpha_{ik}^2 + 6}{8\alpha_{ik}^2}C_0^{ik} - \frac{3}{2\alpha_{ik}}C_1^{ik} - \frac{9}{8}C_2^{ik}, \tag{25}$$

$$A_0^{ik} = \frac{2}{\pi a_k \gamma_k} \int_0^{\pi} \frac{d\lambda}{\left(1 + \alpha_{ik}^2 - 2\alpha_{ik}\cos\lambda\right)^{1/2}}, \quad (i < k), \quad p = 0, 1, 2, 3, \tag{26}$$

$$B_p^{ik} = \frac{2a_i\gamma_i}{\pi \left(a_k\gamma_k\right)^2} \int\limits_0^\pi \frac{\cos(p\lambda)d\lambda}{\left(1 + \alpha_{ik}^2 - 2\alpha_{ik}\cos\lambda\right)^{3/2}},$$

(27)

$$C_p^{ik} = \frac{2\left(a_i\gamma_i\right)^2}{\pi \left(a_k\gamma_k\right)^3} \int\limits_0^\pi \frac{\cos(p\lambda)d\lambda}{\left(1 + \alpha_{ik}^2 - 2\alpha_{ik}\cos\lambda\right)^{5/2}}.$$

Note that the Laplace coefficients A_0^{ij}, B_0^{ij}, B_1^{ij}, B_2^{ij}, C_0^{ij}, C_1^{ij}, C_2^{ij}, C_3^{ij} $(i \neq j)$ are interconnected by recursive relations.

5 Evolutionary Equations

5.1 Derivation of Evolution Equations

The evolutionary equations that determine the behavior of the orbital parameters over long time intervals are obtained from the equations of motion if instead of the perturbing functions W_i we substitute their secular part $W_i^{(sec)}$ according to (17).

The evolution equations have the form [29]

$$\dot{\xi}_i = f \sum_{s=1}^{i-1} m_s \left(\frac{\Pi_{ii}^{is}}{\Lambda_i}\eta_i + \frac{\Pi_{is}^{is}}{\sqrt{\Lambda_i\Lambda_s}}\eta_s \right) + f \sum_{k=i+1}^{n} m_k \left(\frac{\Pi_{kk}^{ik}}{\Lambda_i}\eta_i + \frac{\Pi_{ik}^{ik}}{\sqrt{\Lambda_i\Lambda_k}}\eta_k \right) - \frac{3\ddot{\gamma}_i\Lambda_i^3}{2\gamma_i\mu_{i0}^2}\eta_i,$$

(28)

$$\dot{\eta}_i = -f \sum_{s=1}^{i-1} m_s \left(\frac{\Pi_{ii}^{is}}{\Lambda_i}\xi_i + \frac{\Pi_{is}^{is}}{\sqrt{\Lambda_i\Lambda_s}}\xi_s \right) - f \sum_{k=i+1}^{n} m_k \left(\frac{\Pi_{kk}^{ik}}{\Lambda_i}\xi_i + \frac{\Pi_{ik}^{ik}}{\sqrt{\Lambda_i\Lambda_k}}\xi_k \right) + \frac{3\ddot{\gamma}_i\Lambda_i^3}{2\gamma_i\mu_{i0}^2}\xi_i,$$

(29)

$$\dot{p}_i = -f \sum_{s=1}^{i-1} m_s B_1^{is} \left(\frac{q_i}{4\Lambda_i} - \frac{q_s}{4\sqrt{\Lambda_i\Lambda_s}} \right) - f \sum_{k=i+1}^{n} m_k B_1^{ik} \left(\frac{q_i}{4\Lambda_i} - \frac{q_k}{4\sqrt{\Lambda_i\Lambda_k}} \right),$$ (30)

$$\dot{q}_i = f \sum_{s=1}^{i-1} m_s B_1^{is} \left(\frac{p_i}{4\Lambda_i} - \frac{p_s}{4\sqrt{\Lambda_i\Lambda_s}} \right) + f \sum_{k=i+1}^{n} m_k B_1^{ik} \left(\frac{p_i}{4\Lambda_i} - \frac{p_k}{4\sqrt{\Lambda_i\Lambda_k}} \right),$$ (31)

$$\dot{\lambda}_i = \frac{\mu_{i0}^2}{\gamma_i^2\Lambda_i^3} - \frac{\partial W_i^{(sec)}}{\partial \Lambda_i}, \quad \dot{\Lambda}_i = 0.$$ (32)

From the second equation of the system (32) we obtain

$$\Lambda_i = const$$ (33)

or

$$a_i = const.$$ (34)

Note that the first equation of system (32) is solved after integrating the Eqs. (28)–(31).

5.2   Transition to Dimensionless Variables

Let us rewrite the evolution equations for eccentric and oblique elements in the form

$$\dot{\xi_i} = f \sum_{s=1}^{i-1} m_s \left(\frac{\Pi_{ii}^{is}}{\Lambda_i} \eta_i + \frac{\Pi_{is}^{is}}{\sqrt{\Lambda_i \Lambda_s}} \eta_s \right) + f \sum_{k=i+1}^{n} m_k \left(\frac{\Pi_{kk}^{ik}}{\Lambda_i} \eta_i + \frac{\Pi_{ik}^{ik}}{\sqrt{\Lambda_i \Lambda_k}} \eta_k \right) - \frac{3\ddot{\gamma_i} \Lambda_i^3}{2\gamma_i \mu_{i0}^2} \eta_i,$$

(35)

$$\dot{\eta_i} = -f \sum_{s=1}^{i-1} m_s \left(\frac{\Pi_{ii}^{is}}{\Lambda_i} \xi_i + \frac{\Pi_{is}^{is}}{\sqrt{\Lambda_i \Lambda_s}} \xi_s \right) - f \sum_{k=i+1}^{n} m_k \left(\frac{\Pi_{kk}^{ik}}{\Lambda_i} \xi_i + \frac{\Pi_{ik}^{ik}}{\sqrt{\Lambda_i \Lambda_k}} \xi_k \right) + \frac{3\ddot{\gamma_i} \Lambda_i^3}{2\gamma_i \mu_{i0}^2} \xi_i,$$

(36)

$$\dot{p_i} = -f \sum_{s=1}^{i-1} m_s B_1^{is} \left(\frac{q_i}{4\Lambda_i} - \frac{q_s}{4\sqrt{\Lambda_i \Lambda_s}} \right) - f \sum_{k=i+1}^{n} m_k B_1^{ik} \left(\frac{q_i}{4\Lambda_i} - \frac{q_k}{4\sqrt{\Lambda_i \Lambda_k}} \right),$$

(37)

$$\dot{q_i} = f \sum_{s=1}^{i-1} m_s B_1^{is} \left(\frac{p_i}{4\Lambda_i} - \frac{p_s}{4\sqrt{\Lambda_i \Lambda_s}} \right) + f \sum_{k=i+1}^{n} m_k B_1^{ik} \left(\frac{p_i}{4\Lambda_i} - \frac{p_k}{4\sqrt{\Lambda_i \Lambda_k}} \right).$$

(38)

Physical units: t is measured in years, a_i are measured in astronomical units, m_i are measured in masses of the Sun. In the evolution Eqs. (35)–(38) we switch to dimensionless variables t^*, a_i^*, m_i^*,

$$t^* = \tau = \omega_1 t, \quad a_i^* = \frac{a_i}{a_1}, \quad m_i^* = \frac{m_i}{m_{00}},$$

(39)

$$\omega_1 = \frac{\sqrt{f m_{00}}}{a_1^{3/2}} = const, \quad T_1^t = \frac{1}{\omega_1} = \frac{1}{\sqrt{f m_{00}}} a_1^{3/2} = const,$$

(40)

$$m_{00} = m_0(t_0) = const, \quad a_1 = a_1(t_0) = const, \quad \frac{d}{d\tau} = ()',$$

(41)

$$a_i = a_1 a_i^*, \quad m_i = m_{00} m_i^*.$$

(42)

Then we obtain

$$\Lambda_i = \sqrt{f m_{00}} \sqrt{a_1} \Lambda_i^*, \quad \Lambda_i^* = \sqrt{\mu_{i0}^*} \sqrt{a_i^*}, \quad \mu_{i0}^* = 1 + \frac{m_{i0}}{m_{00}} = const,$$

(43)

$$\xi_i = \xi_i^* (f m_{00} a_1)^{1/4}, \quad \eta_i = \eta_i^* (f m_{00} a_1)^{1/4},$$

$$p_i = p_i^* (f m_{00} a_1)^{1/4}, \quad q_i = q_i^* (f m_{00} a_1)^{1/4}$$

(44)

$$\frac{3\ddot{\gamma_i} \Lambda_i^3}{2\gamma_i \mu_{i0}^2} = \omega_1 \frac{3\gamma_i'' \Lambda_i^{*3}}{2\gamma_i \mu_{i0}^{*2}}, \quad \frac{d^2}{d\tau^2} = ()''.$$

(45)

Thus, the dimensionless eccentric and oblique elements have the form

$$\xi_i^* = \sqrt{2\sqrt{\mu_{i0}^*}\sqrt{a_i^*}(1 - \sqrt{1 - e_i^2})} \cos \pi_i,$$
$$\eta_i^* = -\sqrt{2\sqrt{\mu_{i0}^*}\sqrt{a_i^*}(1 - \sqrt{1 - e_i^2})} \sin \pi_i, \tag{46}$$

$$p_i^* = \sqrt{2\sqrt{\mu_{i0}^*}\sqrt{a_i^*}\sqrt{1 - e_i^2}(1 - \cos I_i)} \cos \Omega_i,$$
$$q_i^* = -\sqrt{2\sqrt{\mu_{i0}^*}\sqrt{a_i^*}\sqrt{1 - e_i^2}(1 - \cos I_i)} \sin \Omega_i. \tag{47}$$

Using the introduced notations (39)–(42) and the relations (43)–(45), we can write down the evolution Eqs. (35)–(38) in dimensionless quantities

$$t^* = \tau, \ a_i^*, \ m_i^*, \tag{48}$$

$$\Lambda^* = const, \ \xi_i^*, \ \eta_i^*, \ p_i^*, \ q_i^*. \tag{49}$$

As a result, reducing the left and right sides of the Eqs. (35)–(38) by a common factor

$$\omega_1 (fm_{00}a_1)^{1/4} = const \tag{50}$$

we obtain the evolution Eqs. (35)–(38) in dimensionless quantities (48)–(49). For convenience of notation, we omit the symbol ($*$) and rewrite the Eqs. (35)–(38) in dimensionless variables (48)–(49). in the form

$$\xi_i' = \sum_{s=1}^{i-1} m_s \left(\frac{\Pi_{ii}^{is}}{\Lambda_i} \eta_i + \frac{\Pi_{is}^{is}}{\sqrt{\Lambda_i \Lambda_s}} \eta_s \right) + \sum_{k=i+1}^{n} m_k \left(\frac{\Pi_{kk}^{ik}}{\Lambda_i} \eta_i + \frac{\Pi_{ik}^{ik}}{\sqrt{\Lambda_i \Lambda_k}} \eta_k \right) - \frac{3\gamma_i''}{2\gamma_i} \frac{\Lambda_i^3}{\mu_{i0}^2} \eta_i, \tag{51}$$

$$\eta_i' = -\sum_{s=1}^{i-1} m_s \left(\frac{\Pi_{ii}^{is}}{\Lambda_i} \xi_i + \frac{\Pi_{is}^{is}}{\sqrt{\Lambda_i \Lambda_s}} \xi_s \right) - \sum_{k=i+1}^{n} m_k \left(\frac{\Pi_{kk}^{ik}}{\Lambda_i} \xi_i + \frac{\Pi_{ik}^{ik}}{\sqrt{\Lambda_i \Lambda_k}} \xi_k \right) + \frac{3\gamma_i''}{2\gamma_i} \frac{\Lambda_i^3}{\mu_{i0}^2} \xi_i, \tag{52}$$

$$p_i' = -\sum_{s=1}^{i-1} m_s B_1^{is} \left(\frac{q_i}{4\Lambda_i} - \frac{q_s}{4\sqrt{\Lambda_i \Lambda_s}} \right) - \sum_{k=i+1}^{n} m_k B_1^{ik} \left(\frac{q_i}{4\Lambda_i} - \frac{q_k}{4\sqrt{\Lambda_i \Lambda_k}} \right), \tag{53}$$

$$q_i' = \sum_{s=1}^{i-1} m_s B_1^{is} \left(\frac{p_i}{4\Lambda_i} - \frac{p_s}{4\sqrt{\Lambda_i \Lambda_s}} \right) + \sum_{k=i+1}^{n} m_k B_1^{ik} \left(\frac{p_i}{4\Lambda_i} - \frac{p_k}{4\sqrt{\Lambda_i \Lambda_k}} \right). \tag{54}$$

At the same time, the expressions

$$\Pi_{ii}^{is}, \ \Pi_{is}^{is}, \ \Pi_{kk}^{ik}, \ \Pi_{ik}^{ik} \tag{55}$$

in the Eqs. (51)–(54) and the Laplace coefficients keep their form, according to the formulas (18)–(22) and (23)–(27). They become dimensionless quantities.

6   The Algorithm of Calculations

In our model, it is more convenient to use analogues of the second system of Poincaré canonical elements [9] and to write the equations of motion in the form (15). The secular part of the perturbing functions W_i is defined in the form (16)–(20), (23)–(25) with the Laplace coefficients of the form (21), (22), (26), (27). The evolutionary equations are written in dimensionless variable in the form (51)–(54).

I. We define and study a system of 14 differential equations of the form (53)–(54). To do this, we perform the following steps:

a) Determine the type of change in the masses of the central star and planets (we consider dependencies accordingly Eddington-Jeans law)

$$m_0(t) = (\varepsilon_0(1 - n_0)(t - t_0) + m_{00}^{1-n_0})^{1/(1-n_0)} \quad (n_0 = 3),$$

$$m_i(t) = (\varepsilon_i(1 - n_i)(t - t_0) + m_{i0}^{1-n_i})^{1/(1-n_i)} \quad (n_i = 2, \quad i = \overline{1,7});$$

b) taking into account the type of functions α_i $(i = \overline{1,7})$ we compute the Laplace coefficients B_1^{ik} from (27) $(p = 1)$;

c) for each of the seven planets we build a system of two differential equations of the form (53)–(54) and add the initial conditions (47),

$$\sigma_1 = 1.374, \ \nu_1 = 10^{-5}, \ a_1 = 0.01154, \ e_1 = 0.00622, \\ \pi_1(t_0) = 21°, \ I_1(t_0) = 0.35°, \ \Omega_1(t_0) = 45° \tag{56}$$

for the first planet P_1. Similar initial conditions at the point $\tau = 0$ for the other six planets $P_2, ..., P_7$ can be taken from [1]. Adding these initial conditions to the system of differential equations, we obtain the required system (53)–(54);

d) using the numerical integration, we find the functions p_i, q_i, and then visualize the orbital elements (see Fig. 1)

$$\sin^2 I_j \approx \frac{p_j^2 + q_j^2}{\Lambda_j} \quad (j = \overline{1,7}). \tag{57}$$

II. We define and study a system of 14 differential equations of the form (51)–(52). To do this, we perform the following steps:

a) Using the results of step I a) we determine the Laplace coefficients $A_0^{ij}, B_0^{ij}, B_1^{ij}, B_2^{ij}, C_0^{ij}, C_1^{ij}, C_2^{ij}, C_3^{ij}$ from the system (21), (22), (26), (27);

b) on the next step we define functions Π_{ii}^{ij} from (18)–(20) and (23)–(25);

c) for each of the seven planets we build a system of two differential Eqs. (51)–(52), add the initial conditions (46) and other initial conditions from [1] at the point $\tau = 0$. Adding initial conditions to the system of differential equations, we obtain the required system (51)–(52);

d) using the numerical integration, we find the functions ξ_i, η_i, and then eccentric elements $e_i^2 \approx \frac{\xi_i^2 + \eta_i^2}{\Lambda_i}$ (Fig. 2), $\pi_i = -\arctan\frac{\eta_i}{\xi_i}$;

e) performing steps I d), II d) we can find and visualize the orbital elements $\omega_i = \pi_i - \Omega_i$.

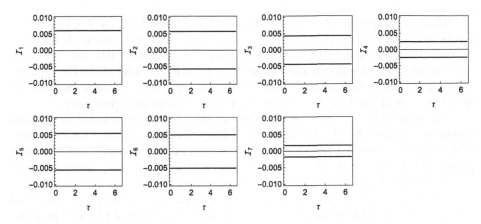

Fig. 1. Plots of the functions $\sin^2(I_i) = \frac{p_i^2 + q_i^2}{\Lambda_i}$ $(i = \overline{1,7})$ (see (57))

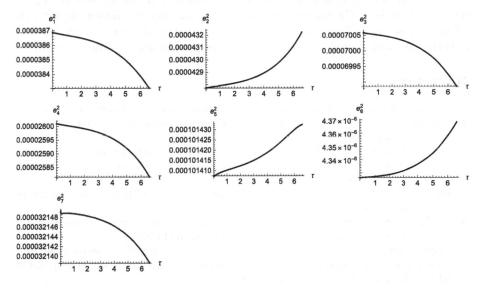

Fig. 2. Plots of the functions e_i^2 $(i = \overline{1,7})$

7   Conclusion

For the first time, a system of differential equations has been obtained that describes the motion of planets of variable masses around a central star in the TRAPPIST-1 system.

The evolution equations have been obtained in dimensionless variables, and their numerical solutions have been found in the case of the physical parameters corresponding to the TRAPPIST-1 system. The calculation time was chosen to be 2000 revolutions of the first planet.

Numerical experiment was carried out in three ways: first, the Laplace coefficients in elliptic functions were directly calculated; then (in the second and third cases) these coefficients were expanded into series up to the 4th and 2nd order of smallness. The results of the calculations practically coincided. The computation time has significantly decreased in the second case in comparison to the first one, and in the third case it has decreased even more significantly.

The graph (Fig. 1) shows the time dependence of the inclination of the planets orbits determined by the formulas (57). They change in a very narrow range which allows us to conclude that the system is moving in one plane (the Laplace plane). At the same time the changes of eccentricities of planets' orbits are more noticeable (see Fig. 2).

Note that choosing different laws of mass variations, one can find numerical solutions to the evolutionary equations for different values of physical parameters. Such simulation enables to investigate dynamical evolution of the exoplanetary system and to understand better an influence of the masses change on the system motion. The corresponding calculations may be carried out with the aid on any software but here we used the computer algebra system Wolfram Mathematica which enables to visualize easily the obtained results, as well.

References

1. NASA Exoplanet Exploration. http://exoplanets.nasa.gov/. Accessed 28 Apr 2022
2. Omarov, T.B. (ed.): Non-stationary Dynamical Problems in Astronomy. Nova Science Publ. Inc., New-York (2002)
3. Bekov, A.A., Omarov, T.B.: The theory of orbits in non-stationary stellar systems. Astron. Astrophys. Trans. **22**(2), 145–153 (2003). https://doi.org/10.1080/1055679031000084803
4. Eggleton, P.: Evolutionary Processes in Binary and Multiple Stars. Cambridge University Press, New York (2006)
5. Luk'yanov, L.G.: Dynamical evolution of stellar orbits in close binary systems with conservative mass transfer. Astron. Rep. **52**, 680–692 (2008)
6. Minglibayev, M.Z.: Dynamics of gravitating bodies with variable masses and sizes [Dinamika gravitiruyushchikh tel s peremennymi massami i razmerami]. LAMBERT Academic Publ, Saarbrucken (2012)
7. Berkovič, L.M.: Gylden-Meščerski problem. Celest. Mech. **24**, 407–429 (1981)
8. Omarov, T.B.: Two-body problem with corpuscular radiation. Sov. Astron. **7**, 707–714 (1963)
9. Hadjidemetriou, J.D.: Two-body problem with variable mass: a new approach. Icarus **2**, 440–451 (1963). https://doi.org/10.1016/0019-1035(63)90072-1
10. Veras, D., Hadjidemetriou, J.D., Tout, C.A.: An exoplanet's response to anisotropic stellar mass-loss during birth and death. Mon. Not. R. Astron. Soc. **435**(3), 2416–2430 (2013). https://doi.org/10.1093/mnras/stt1451
11. Minglibayev, M.Z., Mayemerova, G.M.: Evolution of the orbital-plane orientations in the two-protoplanet three-body problem with variable masses. Astron. Rep. **58**(9), 667–677 (2014). https://doi.org/10.1134/S1063772914090066
12. Prokopenya, A.N., Minglibayev, M.Z., Beketauov, B.A.: Secular perturbations of quasi-elliptic orbits in the restricted three-body problem with variable masses. Int. J. Non-Linear Mech. **73**, 58–63 (2015). https://doi.org/10.1016/j.ijnonlinmec.2014.11.007

13. Minglibayev, M.Z., Prokopenya, A.N., Mayemerova, G.M., Imanova, Z.U.: Three-body problem with variable masses that change anisotropically at different rates. Math. Comput. Sci. **11**, 383–391 (2017). https://doi.org/10.1007/s11786-017-0306-4

14. Prokopenya, A.N., Minglibayev, M.Z., Mayemerova, G.M., Imanova, Z.U.: Investigation of the restricted problem of three bodies of variable masses using computer algebra. Program. Comput. Softw. **43**(5), 289–293 (2017). https://doi.org/10.1134/S0361768817050061

15. Minglibayev, M., Prokopenya, A., Shomshekova, S.: Computing perturbations in the two-planetary three-body problem with masses varying non-isotropically at different rates. Math. Comput. Sci. **14**(2), 241–251 (2019). https://doi.org/10.1007/s11786-019-00437-0

16. Prokopenya, A.N., Minglibayev, M.Z., Mayemerova, G.M.: Symbolic calculations in studying the problem of three bodies with variable masses. Program. Comput. Softw. **40**(2), 79–85 (2014). https://doi.org/10.1134/S036176881402008X

17. Prokopenya, A.N., Minglibayev, M.Z., Shomshekova, S.A.: Applications of computer algebra in the study of the two-planet problem of three bodies with variable masses. Program. Comput. Softw. **45**(2), 73–80 (2019). https://doi.org/10.1134/S0361768819020087

18. Prokopenya, A., Minglibayev, M., Baisbayeva, O.: Analytical computations in studying translational-rotational motion of a non-stationary triaxial body in the central gravitational field. In: Boulier, F., England, M., Sadykov, T.M., Vorozhtsov, E.V. (eds.) CASC 2020. LNCS, vol. 12291, pp. 478–491. Springer, Cham (2020). https://doi.org/10.1007/978-3-030-60026-6_28

19. Murray, C.D., Dermott, S.F.: Solar System Dynamics. Cambridge University Press, New York (1999)

20. Lidov, M.L., Vashkov'yak, M.A.: On quasi-satellite orbits in a restricted elliptic three-body problem. Astron. Lett. **20**(5), 676–690 (1994)

21. Ford, E.B., Kozinsky, B., Rasio, F.A.: Secular evolution of hierarchical triple star systems. Astron. J. **535**, 385–401 (2000)

22. Perminov, A.S., Kuznetsov, E.D.: The implementation of Hori–Deprit method to the construction averaged planetary motion theory by means of computer algebra system Piranha. Math. Comput. Sci. **14**(2), 305–316 (2019). https://doi.org/10.1007/s11786-019-00441-4

23. Perminov, A., Kuznetsov, E.: The orbital evolution of the Sun–Jupiter–Saturn–Uranus–Neptune system on long time scales. Astrophys. Space Sci. **365**(8), 1–21 (2020). https://doi.org/10.1007/s10509-020-03855-w

24. Perminov, A.S., Kuznetsov, E.D.: Orbital evolution of the extrasolar planetary systems HD 39194, HD 141399, and HD 160691. Astron. Rep. **63**(10), 795–813 (2019). https://doi.org/10.1134/S1063772919090075

25. Gillon, M., et al.: Seven temperate terrestrial planets around the nearby ultracool dwarf star TRAPPIST-1. Nature **542**(7642), 456–460 (2017). https://doi.org/10.1038/nature21360

26. Shallue, C.J., Vanderburg, A.: Identifying exoplanets with deep learning: a five-planet resonant chain around Kepler-80 and an Eighth planet around Kepler-90. Astron. J. **155**(2), 94 (2018). https://doi.org/10.3847/1538-3881/aa9e09

27. Minglibayev, M. Zh., Kosherbayeva, A.B.: Differential equations of planetary systems. Rep. Nat. Acad. Sci. Repub. Kazakhstan **2**(330), 14–20 (2020). https://doi.org/10.32014/2020.2518-1483.26

28. Minglibayev, M.Z., Kosherbayeva, A.B.: Equations of planetary systems motion. News of the National Academy of Sciences of the Republic of Kazakhstan. Phys. Math. Ser. **6**(334), 53–60 (2020). https://doi.org/10.32014/2020.2518-1726.97
29. Prokopenya, A.N., Minglibayev, M.Z., Kosherbayeva, A.B.: Derivation of evolutionary equations in the many-body problem with isotropically varying masses using computer algebra. Program. Comput. Softw. **48**(2), 107–115 (2022). https://doi.org/10.1134/S0361768822020098
30. Wolfram, S.: An Elementary Introduction to the Wolfram Language, 2nd edn. Wolfram Media, New York (2016)

Transparent Checkpointing for Automatic Differentiation of Program Loops Through Expression Transformations

Michel Schanen[1,4], Sri Hari Krishna Narayanan[1,4(✉)],
Sarah Williamson[2,4], Valentin Churavy[3,4], William S. Moses[3,4],
and Ludger Paehler[3,4]

[1] Argonne National Laboratory, Lemont, IL 60439, USA
{mschanen,snarayan}@anl.gov
[2] Oden Institute for Computational Engineering and Sciences, University of Texas
at Austin, Austin, TX 78712, USA
swilliamson@utexas.edu
[3] MIT CSAIL, Cambridge, MA 02139, USA
{vchuravy,wmoses}@mit.edu
[4] Technical University of Munich, Munich 78712, Germany
ludger.paehler@tum.de

Abstract. Automatic differentiation (AutoDiff) in machine learning is
largely restricted to expressions used for neural networks (NN), with the
depth rarely exceeding a few tens of layers. Compared to NN, numerical simulations typically involve iterative algorithms like time steppers
that lead to millions of iterations. Even for modest-sized models, this may
yield infeasible memory requirements when applying the adjoint method,
also called backpropagation, to time-dependent problems. In this situation, checkpointing algorithms provide a trade-off between recomputation
and storage. This paper presents the package *Checkpointing.jl* that leverages expression transformations in the programming language Julia and
the package ChainRules.jl to automatically and transparently transform
loop iterations into differentiated loops. The user may choose between various checkpointing algorithm schemes and storage devices. We describe the
unique design of *Checkpointing.jl* and demonstrate its features on an automatically differentiated MPI implementation of Burgers' equation on the
Polaris cluster at the Argonne Leadership Computing Facility.

Keywords: Julia · Automatic differentiation · Checkpointing

1 Introduction

Automatic differentiation [8] (AutoDiff) is a technique for generating derivatives
of a given implemented function $y = f(x)$ with input $x \in \mathbb{R}^n$ and output $y \in \mathbb{R}^m$,
by differentiating the code at the statement level and applying the chain rule of
derivative calculus. The differentiated code is required in optimization, nonlinear
partial differential equations (PDE), sensitivity analysis, inverse problems, and

© The Author(s), under exclusive license to Springer Nature Switzerland AG 2023
J. Mikyška et al. (Eds.): ICCS 2023, LNCS 10475, pp. 483–497, 2023.
https://doi.org/10.1007/978-3-031-36024-4_37

machine learning. The associativity of the chain rule leads to two main modes of code differentiation: the forward mode and the reverse mode. The forward mode computes the Jacobian-vector product $\dot{y} = \nabla J(x) \cdot \dot{x}$ with ˙ denoting the tangents or directional derivatives. The reverse mode, also known as backpropagation in machine learning, computes the transposed Jacobian-vector product $\bar{x} = \bar{y} \cdot \nabla J(x)$, with ¯ denoting adjoints. Note that the adjoint of the input $\bar{x}$ is computed with respect to the adjoint of the output $\bar{y}$. This implies a data flow reversal throughout the entire program. During the forward run $y = f(x)$, all the *intermediate values* of x at each statement need to be stored for the reverse run $\bar{x} = \bar{y} \cdot \nabla J(x)$. This comes at a high cost of memory, increasing its complexity to at least the runtime complexity when assuming nonlinear functions f. The upside of the reverse mode is that the gradient of a scalar function f with $m = 1$ can be computed at $\mathcal{O}(1) \cdot cost(f)$ versus $\mathcal{O}(n) \cdot cost(f)$ for the forward mode. As a remedy, checkpointing in AutoDiff refers to a trade-off between recomputation and the memory requirement for storing the intermediate values.

In this paper, we will focus on the common pattern of time-stepping loops or iterative loops in general that appear in numerical simulations further explained in Sect. 1.1 and apply it to the Burgers' equation (see Fig. 1). For the first time, through expression transformations and code reflection in Julia, we make checkpointing for iterative loops in AutoDiff fully transparent to the user.

1.1 Adjoint Timestepping Checkpointing

Most numerical problems require the evaluation of nonlinear expressions, either due to direct nonlinear function expressions (e.g., polynomials, trigonometric functions, etc.) or due to the evaluation of conditional expressions (e.g., IF-ELSE). Furthermore, these expressions are found in iterative sequences, either as part of a time-stepping model or an iterative solver (or both). In reverse-mode AutoDiff, these variables are required in reverse order compared to the execution of the nonlinear primal model (see f and $\bar{f}$ in Fig. 2). Two extreme approaches exist to access these variables, either storing all (see Fig. 2) or recomputing all that are necessary. For complex models, neither of these approaches is practical. Checkpointing provides a computational solution that can help circumvent these issues by reducing the amount of storage at the expense of increased run time.

One well-known use is the computation of the so-called adjoint (gradient) of a model-data misfit (or cost) function, as is done in data assimilation based on gradient-based, PDE-constrained optimization. For example, the gradient of a cost function with respect to a very high-dimensional space of control variables via minimization of a Lagrangian,

$$\mathcal{L} = J - \sum_{t=1}^{t_f} \bar{\mathbf{x}}_t \left(\mathbf{x}_t - f(\mathbf{x}_{t-1}) \right), \tag{3}$$

where J is a previously defined cost function and, in general, requires knowledge of all forward steps. In this notation, $\mathbf{x}_t$ refers to the model state at time t, and f is a nonlinear model that steps the state from time $t - 1$ to time t. In this

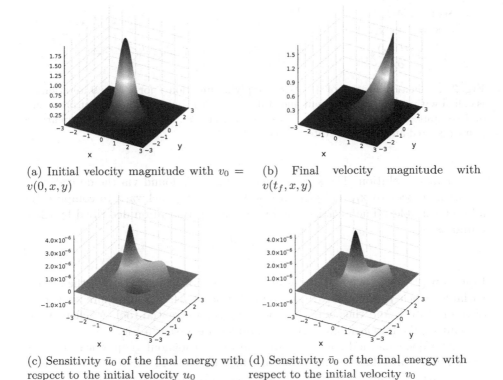

(a) Initial velocity magnitude with $v_0 = v(0, x, y)$

(b) Final velocity magnitude with $v(t_f, x, y)$

(c) Sensitivity $\bar{u}_0$ of the final energy with respect to the initial velocity u_0

(d) Sensitivity $\bar{v}_0$ of the final energy with respect to the initial velocity v_0

Fig. 1. Adjoint solution to Burgers' equation with $dx = 3e-2$, $dy = 3e-2$, $dt = 3e-3$, and $\nu = 1e-2$ on a grid $(Nx, Ny) = (1\,000, 1\,000)$ and 10 000 timesteps. This requires around $1000^2 \times 10\,000 \times 4$ fields $\times 8\text{B} = 320\text{GB}$ of memory to **store** all the intermediate timesteps for the adjoint computation. Our solution enables a user to transparently trade this high memory footprint for a runtime overhead of around $10--{-}12$ while reducing the footprint to 1.6GB

(and other examples), where the numerical state at each time step t may be of the order $10^5--{-}10^7$, and with iteration (i.e., time-stepping) loops of the order $10^4--{-}10^6$ keeping the required state in memory is not feasible. The solution is to use checkpointing. Instead of storing all system states during the forward pass, "checkpoints" at specified intervals are stored on disk, which can subsequently be restored to recompute future states.

Adjoint Method. The adjoint method aims to minimize the Lagrangian described in (3) to compute the adjoint variables, $\bar{\mathbf{x}}_t$. Say the cost function is given by $J(\mathbf{x}_{t_f})$, and we wish to know how J depends on the initial condition $\mathbf{x}_0$. This sensitivity is captured in $\bar{\mathbf{x}}_0$, the adjoint variable at the initial time. Taking a derivative of (3) with respect to $\bar{\mathbf{x}}_t$, one finds the first normal Eq. (1),

$$\mathbf{x}_{t+1} = f(\mathbf{x}_t) \tag{1}$$

$$\bar{\mathbf{x}}_t = \bar{f}(\mathbf{x}_t, \bar{\mathbf{x}}_{t+1})$$

$$= \frac{\partial f(\mathbf{x}_t)}{\partial \mathbf{x}_t} \bar{\mathbf{x}}_{t+1} \tag{2}$$

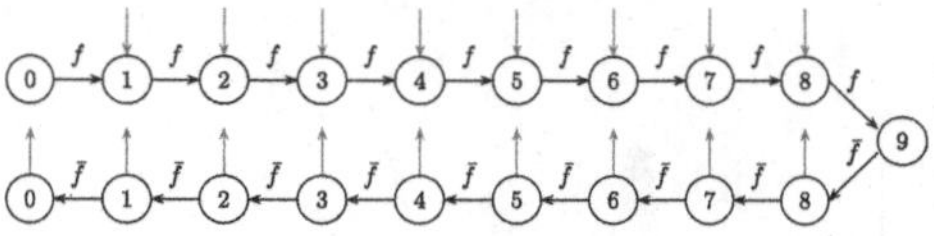

Fig. 2. Evaluation process of iteratively applying function f for $t = 1 : 9$ iterations, f is called with state x_t as the input and state x_{t+1} as the output. The adjoint function $\bar{f}$ of f computes state $\bar{x}_t$ with respect to state $\bar{x}_{t+1}$ and x_t. The red down and up arrows mark a stored and restored state, respectively.

the forward evolution. The second normal Eq. (2) is found via the derivative of (3) with respect to $\mathbf{x}_t$, and gives a rule for stepping backward to compute the adjoint variables. The initial value for the back-propagation described by (2) is found as

$$\bar{\mathbf{x}}_{t_f} = \frac{\partial J}{\partial \mathbf{x}_{t_f}}. \tag{4}$$

Equation (2) shows why all states are necessary for computation of the adjoint variables when f_t is nonlinear (i.e. computation of $f_t(x_t)$ will require knowledge of prior states), and thus why checkpointing is an essential tool. A schematic of computing the forward and backward problems is given in Fig. 2.

The adjoint method has many applications throughout geophysical sciences. Most notable are data assimilation, in which the cost function is a data misfit, and sensitivity analysis, where the cost function is a physical quantity of interest. In this paper, we employ the adjoint method for sensitivity analysis of solutions to the Burgers' equation.

1.2 Contribution

Checkpointing capability has been implemented in source transformation AutoDiff tools as well as popular differentiable programming frameworks for machine learning. In this work, we show how languages that support code reflection or metaprogramming can be leveraged to make checkpointing for AutoDiff of loops fully transparent to the user. While we use the programming language Julia, the various constraints and generalizations laid out in the design section Sect. 2 can be extrapolated to other programming languages. This significantly improves the user experience for inexperienced AutoDiff users who run into memory bottlenecks when differentiating their time-dependent numerical code.

We implemented our design in the software package *Checkpointing.jl*[1]. It currently supports

---

[1] https://github.com/Argonne-National-Laboratory/Checkpointing.jl.

- automated generation of the store and restore for the checkpointed object type,
- modular support of three checkpointing schemes: periodic, binomial, and online,
- and modular support of two storage devices Array and HDF5 files.

1.3   Use Case: Burgers' Equation

Checkpointing will be applied to the two-dimensional Burgers' equation[2]

$$\frac{\partial u}{\partial t} + u\frac{\partial u}{\partial x} + v\frac{\partial u}{\partial y} = \nu\nabla^2 u \tag{5}$$

$$\frac{\partial v}{\partial t} + u\frac{\partial v}{\partial x} + v\frac{\partial v}{\partial y} = \nu\nabla^2 v \tag{6}$$

where u and v represent the x and y velocities of a fluid and ν is the viscosity coefficient. The equation is solved on a square domain, $(x, y) \in [-L, L] \times [-L, L]$, with the initial velocities

$$u(0, x, y) = \exp\left(-x^2 - y^2\right), \quad v(0, x, y) = \exp\left(-x^2 - y^2\right),$$

and Dirichlet conditions on all four boundaries

$$u(t, x, -L) = u(t, x, L) = u(t, -L, y) = u(t, L, y) = 0.$$

An identical boundary condition is imposed on v.

To discretize the system, we use a centered finite difference scheme in space and an explicit forward Euler scheme in time.

Adjoint Example. Let N_x, N_y be the total number of grid points in x and y, respectively. Using the notation from Sect. 1.1 we define the cost function

$$J = \frac{1}{N_x \cdot N_y} \sum_{j=1}^{N_x} \sum_{k=1}^{N_y} \left(u(t_f, x_j, y_k)^2 + v(t_f, x_j, y_k)^2\right), \tag{7}$$

a measure of total kinetic energy in the system at the final time t_f. The interest then lies in computing

$$\bar{u}_0 = \frac{\partial J}{\partial u(0, x, y)}, \quad \bar{v}_0 = \frac{\partial J}{\partial v(0, x, y)}, \tag{8}$$

the sensitivity of the final energy with respect to the initial velocities. A schematic of computing the forward and backward problems is given in Fig. 2, and Eqs. 1, 2 demonstrate why all states are necessary for computation of the adjoint variables when f_t is nonlinear (i.e. computation of $f_t(x_t)$ will require knowledge of prior states), and thus why checkpointing is an essential tool.

---

[2] https://github.com/DJ4Earth/Burgers.jl.

2   Design of *Checkpointing.jl*

The goal of *Checkpointing.jl* is to implement a fully transparent and flexible solution for adjoint checkpointing in timestepping loops. This includes (1) the automated store and restore of the checkpointed variables, (2) support of multiple checkpointing schemes, and (3) support for various types of storage devices.

Model Object. To achieve this goal, we define a standard structure in Julia of such timestepped models. These codes use a context model object where the state of the current model is stored. This style of writing code is very common in Julia as it allows to dispatch methods on the object type using the language's multiple dispatch feature. In our case, the model type is `Burgers` as shown in Listing 1.1.

```
Burgers struct

1   mutable struct Burgers          8    µ::Float64
2     nextu::Matrix{Float64}        9    dx::Float64
3     nextv::Matrix{Float64}       10    dy::Float64
4     lastu::Matrix{Float64}       11    dt::Float64
5     lastv::Matrix{Float64}       12    tsteps::Int
6     Nx::Int                      13    ...
7     Ny::Int                      14  end
```

Listing 1.1. Model datatype that the timestepping loop will be dispatched on.

The model includes a field of type Matrix for u and v and for each a next and last storage place for the stencil where `next` is computed from `last`. In addition, it includes all the model parameters ν, dt, dx, dy, and the grid size N_x and N_y. Only the fields u and v need to be checkpointed. However, this requires the user to manually specify all the variables that are required in the adjoint computation. To alleviate this, we checkpoint the entire struct. This is an overestimation, but it enables us to automate the adjoint checkpointing, rendering it fully transparent. The assumption is that most of the memory required to store the struct is associated with variables required in the adjoint computation.

To store the checkpoints *Checkpointing.jl* currently implements two storage types. `ArrayStorage` $<:$ `AbstractStorage` is used to store the checkpoints in RAM whereas `HDF5Storage` $<:$ `AbstractStorage` is used to store them in an HDF5 file. For binary storage in a file we use Julia's built-in `Serialization` module to serialize the `Burgers` struct into disk-storable data. To extend *Checkpointing.jl* with additional storage devices, one can easily add another storage type derived from `AbstractStorage` and add an implementation of `getindex` and `setindex` method for the storage device, which allows the `[]` operator to be used for all stores and restores of a checkpoint with index i (see Listing 1.2).

```
getindex
1   # Array storage implementation of      8    # of right-hand read [] operator
2   # right-hand read [] operator          9    function Base.getindex(
3   Base.getindex(                         10       storage::HDF5Storage{MT}, i
4      storage::ArrayStorage{MT}, i        11    )::MT where {MT}
5   ) where {MT} = storage._storage[i]     12       blob = read(storage.fid["$i"])
6                                          13       return Serialize.deserialize(blob)
7   # HD5 storage storage implementation   14   end
```

Listing 1.2: Example of [] operator (`getindex`) for restoring `ArrayStorage` and `HD5Storage`

Loops. Relying on this abstraction, our timestepping loop is written as a `for` loop over the number of timesteps with an advance function and a halo exchange for the MPI implementation (see Listing 1.3). Note that the loop's body can be composed of any arbitrary code. In addition, *Checkpointing.jl* also supports `while` loops. It is important that the loop iterator bounds for the for loop and the variables in the evaluation of the while condition belong to the model object, here `burgers.tsteps`.

```
Final energy with final_energy
1   function final_energy(                 8        halo(burgers)
2      burgers::Burgers,                   9        copyto!(burgers.lastu, burgers.nextu)
3      scheme::Scheme,                    10        copyto!(burgers.lastv, burgers.nextv)
4   )                                     11      end
5   @checkpoint_struct scheme burgers     12      return energy(burgers)
6      for i in 1:burgers.tsteps          13   end
7         advance(burgers)
```

Listing 1.3. Timestepping loop implementation with a single time step (`advance`), halo exchange using MPI (`halo`), and field swaps with Julia's `copyto!` function.

Differentiation of Loops via Expression Transformations. In *Checkpointing.jl* we treat `for` and `while` loops as just another function that can be differentiated with the additional benefit of applying a checkpointing scheme that drastically reduces the memory footprint for storing the intermediate values. To achieve this we create a marco `@checkpoint_struct` that transforms *for* loops into function calls (see Listing 1.4). Using this macro as a decorator in Listing 1.3 allows the user to mark a loop to be differentiated using *Checkpointing.jl* by transforming it into a function call that is differentiated based on a rule defined in Sect. 2. In addition to this transformation, we make a copy of the original model object and create a *s*hadow copy that is used to store the adjoints of the adjoint evaluation.

```
┌─────────────────────────────────────────────────────────────────────────┐
│    checkpoint_struct macro                                                │
├─────────────────────────────────────────────────────────────────────────┤
│  1    macro checkpoint_struct(alg, model,   12          shadowmodel,      │
│  2      loop)                               13          range             │
│  3      if loop.head == :for               14        ) do $model          │
│  4        ex = quote                        15          $(loop.args[2])    │
│  5          shadowmodel = deepcopy($model)  16        end                 │
│  6          function range()                17      end                   │
│  7            $(loop.args[1])               18      elseif loop.head == :while │
│  8          end                             19        ...                 │
│  9          $model = checkpoint_struct_for( 20      end                   │
│ 10            $alg,                         21      esc(ex) # Return expression │
│ 11            $model,                       22    end                      │
└─────────────────────────────────────────────────────────────────────────┘
```

Listing 1.4: Loop transformation

Now, we must make the AutoDiff tool aware of how to differentiate the `checkpoint_struct_for` function call. Multiple efforts exist to standardize differentiation rules. Most AutoDiff tools differentiate the language's intrinsic operations like arithmetic operations (e.g., multiplication, addition) and certain special functions (e.g., cosine, sine). However, higher-level functions (e.g., linear solvers) or rarely used special functions like BesselK [1] are rarely supported out of the box and have to be defined as *external functions*. In Julia, the popular package ChainRules.jl [11] allows the specification of differentiation rules which AutoDiff tools may then rely on to apply the chain rule. That way, the differentiation rules do not have to be reimplemented for each AutoDiff tool. We refer the reader to the manual of ChainRules.jl for the details on defining such differentiation rules. In summary, it requires a user to define a rule for forward mode differentiation (`frule`) and a reverse mode differentiation rule (`rrule`). By defining those two rules, any combination of higher-order models using, for example, a forward over forward or forward over reverse model, may be generated by an AutoDiff tool. Our reverse rule is presented in Listing 1.5. ChainRules.jl implements joint reversal (Fig. 3) for external functions (see [8] for more details). The outer loop AutoDiff tool will execute the *augmented forward run* of the'Before" block (green) and store all intermediate values. When this tool hits the checkpointed loop it will apply our rule. The rule is composed of the forward run implementing the original function (orange). Then it defines a callback or pullback function that the outer AutoDiff tool will execute once it has executed the *reverse run* (blue) of the 'After' block. This pullback will set the adjoints of the time loop shadow model to zero and then copy the computed adjoints of the 'After' block into the starting adjoints or *seeds* of the time loop. Now, the augmented forward run (green) of the time loop will be executed in `checkpoint_struct_for`, followed by the reverse run (blue) based on the selected checkpointing scheme. After the adjoints are computed, they are again copied back into the respective seeds for the 'Before' block using `create_tangent`. Note that all other arguments of `checkpoint_struct_for` are passive and do not need to be differentiated. This is marked by `NoTangent()`.

```
      ChainRules.jl implementation

1     function ChainRulesCore.rrule(::typeof(Checkpointing.checkpoint_struct_for),
2         body::Function, alg::Scheme, model::MT, shadowmodel::MT,
3         range::Function) where {MT}
4       model_input = deepcopy(model)
5       for i in 1:alg.steps
6         body(model)
7       end
8       function checkpoint_struct_pullback(dmodel)
9         set_zero!(shadowmodel)
10        copyto!(shadowmodel, dmodel)
11        model = checkpoint_struct_for(body, alg, model_input, shadowmodel, range)
12        dshadowmodel = create_tangent(shadowmodel)
13        return NoTangent(), NoTangent(), NoTangent(), dshadowmodel, NoTangent(),
14            NoTangent()
15      end
16      return model, checkpoint_struct_pullback
17    end
```

Listing 1.5. Reverse rule for time loop

Such a differentiation rule may be defined for other differentiation rule systems that may be general or AutoDiff tool specific. ChainRules.jl covers the most general case, while other rule systems may include other attributes.

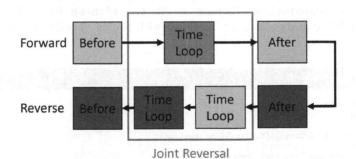

Fig. 3. Adjoining a time loop embedded into another code using ChainRules.jl. (green) denotes an *augmented forward* run where all the intermediate variables are stored. (blue) denotes a *reverse* run where the intermediate variables of the augmented forward run are used for the evaluation of the adjoints. (orange) is the original undifferentiated function evaluation.(Color figure online)

Modular Support of Schemes. The package currently provides three checkpointing schemes

- **Periodic Checkpointing**: For a computation consisting of l timesteps and c available checkpoints, the periodic checkpointing scheme that stores the input and the output of each $\lfloor \frac{l}{c} \rfloor$ iterations and restores them for computing the adjoint [3].
- **Binomial Checkpointing**: For a computation consisting of N time steps with the availability of c checkpoints, binomial checkpointing [2] gives a formulation for the minimal number of time steps $t(l,c)$, evaluated during the adjoint calculation with $t(l,c) = rl - \beta(c+1, r-1)$ where $\beta(c,r) = \binom{c+r}{c}$ and the repetition number r is the unique integer, such that $\beta(c, r-1) < l \leq \beta(c,r)$. We have ported the software `revolve` for providing an implementation of the binomial checkpointing algorithm.
- **Online Checkpointing**: In adaptive time-stepping procedures, the number of time steps, l, is not known a priori. Periodic checkpointing and binomial checkpointing are therefore not appropriate here. The online checkpointing scheme determines during the first forward integration where a checkpoint must be placed. Given the number of available checkpoints c, and the repetition number r, it is possible to determine the range of timesteps l for which the online checkpointing scheme generates an optimal schedule [10]. We have currently implemented the cases where $r = 1$ and $r = 2$.

Listing 1.6 gives an overview of the supported checkpointing schemes and storage devices. The created scheme `Scheme <: AbstractScheme` has to be passed to the macro `@checkpoint_struct` together with the checkpointed object.

```
    Example code
1   checkpoints = 50
2   tsteps = 10000
3   # Storage in RAM
4   storage = ArrayStorage{Burgers}(checkpoints)
5   # Storage to disk with HDF5
6   rank = MPI.Comm_rank(MPI.COMM_WORLD)
7   storage=HDF5Storage{Burgers}(snaps, filename="$rank.chkp")
8   # Storage to on-node SSD with HDF5
9   storage=HDF5Storage{Burgers}(snaps, filename="/local/scratch/$rank.chkp")
10  # Our three currently supported checkpointing schemes
11  scheme = Revolve{Burgers}(tsteps, snaps, verbose=1, storage=storage)
12  scheme = Periodic{Burgers}(tsteps, snaps, verbose=1, storage=storage)
13  # No tsteps needed for Online scheme!
14  scheme = Online_r2{Burgers}(snaps, verbose=1, storage=storage)
```

Listing 1.6. Example of checkpointing schemes and storage object instantiations based on the `Burgers` type

3   Implementation

Our implementation is available at [9]. It currently supports three checkpointing schemes (Periodic, Revolve, and Online_R2). It distinguishes between an outer AutoDiff tool for differentiating the code outside the loop and an inner AutoDiff tool that is used to differentiate the actual loop body. Both tools can be the same; however, the outer AutoDiff tool has to support differentiation rules through ChainRules.jl. Consider the Burgers' example, we apply the @checkpoint_struct to the timestepping loop and the code computes the energy with the energy function (Listing 1.3). The outside code uses the AutoDiff package Zygote.jl while the timestepping loop is differentiated with Enzyme.jl. The loop body consists of an advance function implementing one forward time step and halo implementing the halo exchange (see Sect. 3).

Enzyme [5–7] is an AutoDiff tool acting on the LLVM IR. It, therefore, supports C++ and Julia alike, with the Julia package Enzyme.jl providing Julia-specific support. The novel advantage of Enzyme is its optimization capabilities. AutoDiff tools are usually not integrated directly into a compiler, but use either language-inherent features like operator overloading or are implemented as a separate parsing and generation process before the code is passed to the compiled (source transformation). Enzyme, on the other hand, uses parts of the LLVM optimization pipeline, then differentiates the code, and finally, this IR is again optimized before the code is finally passed to the machine code generation. This three-stage process adds unique performance capabilities to Enzyme that other AutoDiff tools have trouble achieving.

In our example, we use Enzyme to differentiate the inner loop body. Any AutoDiff tool can be used here if it implements Jacobian-transposed vector products, which is the basic operation in the reverse mode of AutoDiff. Although Enzyme.jl does currently not support ChainRules.jl, it is not a requirement for the inner AutoDiff tool in *Checkpointing.jl*, and we can use it in our test case. A similar differentiation rule system is in development for Enzyme.

Zygote.jl [4] is an AutoDiff package originally designed for machine learning. As such, it lacks the support of mutation. This implies that in-place manipulation of array elements is impossible and requires a code to be written without any mutation, which our code outside the loop adheres to. Zygote.jl treats the underlying LLVM IR as static single assignment code and allows the compiler to highly optimize the generated differentiated code. However, due to its limitation to immutable code, it is not well suited for numerical simulations where in-place manipulation of values is common.

MPI. The halo function uses MPI send and receives to do the nearest neighbor halo exchange in all 4 directions of the 2D discretized Burgers' equation. The outside code uses MPI for the summation reduction of the local energy to the global energy of the velocity fields u and v. Enzyme.jl has intrinsic support of MPI, whereas Zygote is not aware of MPI. We added a ChainRules.jl rule for the MPI reduction that allows Zygote to differentiate through this method for the summation.

4 Results

Our experiments are conducted on an HPE Apollo 6500 Gen 10+ based system. Each node has a single 2.8 GHz AMD EPYC Milan 7543P 32-core CPU with 512 GB of DDR4 RAM and four Nvidia A100 GPUs connected via NVLink, a pair of local 1.6TB of SSDs in RAID0 as on-node scratch disks, and a pair of slingshot network adapters. For the timings, we used the Julia 1.8 built-in macro @time and the BenchmarkTools.jl provided @btime . To maximize the throughput of our code and avoid any overhead, we optimized it gradually based on PProf.jl results. An estimate of the total memory footprint was reported by the maximum resident memory through /usr/bin/time.

For the large-scale runs, we increase the grid size to (N_x, N_y) $= (10\,000, 10\,000)$. To achieve the same final state as in Fig. 1 the number of timesteps needs to be increased to 100 000. However, due to compute time limitations, we reduce this to 10 000. This has no effect on the overall claims of this paper. Other runtime parameters are $dx = 0.01$, $dy = 0.01$, $dt = 0.001$ with 100 ranks and a checkpoint size of 32MB.

The entire case of $10\,000 \times 10\,000$ grid points is partitioned among 100 MPI ranks. The goal of the increased resolution is to reduce the numerical error introduced by the sharp gradients at the shock boundary in Fig. 1. Each node has 32 cores, so we distribute the 100 MPI ranks over 4 nodes which in total have 2TB of RAM. Each rank gets a partition of the 10^6 points, which amounts to roughly 8MB. Each Burgers object includes 4 of these fields: nextu , nextv , lastu , and lastv . This gives us a total checkpoint size of 32MB, which agrees with our observed disk checkpoint file sizes. To store all 10 000 time steps, this would amount to 320 GB per process or 32 TB for all 100 processes. This implies that we cannot run our case without checkpointing at all because we have only 2 TB of RAM available. In Figure 4a and Fig. 4b, we compare the relative runtime and memory overhead of the adjoint computation compared to the primal evaluation of the final energy. In addition, we use checkpointing to RAM, to disk, and to a local on-node SSD drive.

The theoretical runtime is derived from the sum of additional forward steps that binomial checkpointing requires and the joint adjoint reversal that is implemented using ChainRules.jl (see Fig. 3). Joint reversal incurs a cost of at least a factor of 3 in integrating the loop function into ChainRules.jl. In addition, we add the forward steps necessary for binomial checkpointing. If the number of checkpoints equals the number of time steps *tsteps*, binomial checkpointing still executes *tsteps* forward steps. So in the most optimistic case, we end up with an overhead factor of 4. With fewer checkpoints than time steps, we can compute the required forward steps laid out in the binomial checkpointing analysis in [2]. The sum of all required steps gives us the theoretical overhead factor in Fig. 4a.

Each data point in the graph is computed from the average execution time of 3 separate runs. We did not obtain results for the 250 checkpoints data point of the "Node SSD" storage device due to instability with HDF5. In general, we are impacted by noise in our test runs. Due to compute time limitations, we are unable to provide a thorough statistical analysis of this noise.

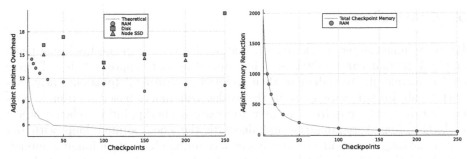

(a) The *adjoint runtime overhead* is the ratio between the total adjoint computation and the total forward computation of the original code $\frac{T_f}{T_r}$ for a given number of checkpoints. The theoretical overhead is based on an analytic formula in binomial checkpointing and assumes no memory latency. The random memory access of binomial checkpointing is expected to introduce an overhead using RAM, disk, and on-node SSD checkpointing.

(b) The memory reduction can be accurately predicted as the memory needed without checkpointing equal to the number of checkpoints times the object size divided by the number of checkpoints time memory times the object size $\approx \frac{tsteps}{snaps}$. Due to the object size being dominated by the TBR variables of u and v, the difference between theoretical and practical overhead is negligible.

Fig. 4. Results of adjoint runtime overhead (a) and memory consumption (b)

However, we can extract some patterns based on the results. First, we observe that RAM checkpointing yields the fastest results with an overhead of around 10–12 between 50 and 250 checkpoints. There is no substantial benefit to increasing the number of checkpoints beyond 50. Second, we have a general pattern from fast to slowest of RAM, on-node, and disk checkpointing, with disk checkpointing being the slowest and the one most impacted by noise since it is most affected by other jobs running on the system.

The memory reduction is the total number of time steps divided by the number of checkpoints. We measured 32 MB per checkpoint per process. So multiplying the number of checkpoints by 32 MB and by 100 processes gives the actual memory requirement ranging from 64 GB for 20 checkpoints up to 0.8 TB for 250 checkpoints. Due to MPI parallelism, this memory requirement is divided among 4 compute nodes. This is a dramatic reduction from the 32 TB required for storing all 10 000 time steps. Moreover, on-node SSD and disk checkpointing have the additional benefit of reducing the RAM overhead to zero, providing more RAM for the actual application. This allows for a decrease in the number of partitions and potentially reduces the required compute time spent on the run despite exhibiting a higher wall clock time. On-node SSD checkpointing provides an overhead of around 13 with more regular runtime results than disk checkpointing. Thus, it may provide the right compromise for this application.

5   Conclusion

We have implemented an extendable and flexible time-loop checkpointing package in Julia that can be integrated into any code that supports AutoDiff based on ChainRules.jl. According to their webpage, 6 AutoDiff tools currently support ChainRules.jl, with more in the works. Our macro-based solution is non-invasive and only requires the user to create a checkpointing scheme object with the desired parameters and decorate the checkpointed loop with our `@checkpoint_struct` macro. It relies on a common abstraction found in numerical simulations where models are encapsulated in a single context object based on a model type. Our results show the flexibility and performance of *Checkpointing.jl* illustrated by a canonical nonlinear PDE implementation of the Burgers' equation that runs on a state-of-the-art supercomputer with minimal development effort and without introducing any domain-specific language. It allows for fast testing of various checkpointing schemes and storage devices. The user may implement their own scheme or storage devices with a few lines of code without worrying about the technicalities of the underlying AutoDiff tool. In theory, such a package may be implemented in any programming language. However, the access to expression transformation in Julia reduces the complexity for both users and developers significantly, increases the modularity of the code, and avoids any restriction to a domain-specific language. All available storage options in *Checkpointing.jl* are currently implemented using synchronous reads and writes. Furthermore, although binomial checkpointing has a random access pattern, it does access the memory locations deterministically according to the binomial checkpointing algorithm. We will investigate the asynchronous prefetching of the next checkpoint concurrently with the adjoint computation relative to the last checkpoint.

Acknowledgements. We would like to thank Paul Hovland and Jan Hückelheim for their valuable suggestions and discussions. This work was funded and/or supported by NSF Cyberinfrastructure for Sustained Scientific Innovation (CSSI) award numbers: 2104068, 2103942, and 2103804, Argonne Leadership Computing Facility, which is a U.S. Department of Energy (DOE) Office of Science User Facility supported under Contract DE-AC02-06CH11357, DOE Computational Sciences Graduate Fellowship, NSF (grants OAC-1835443, AGS-1835860, and AGS-1835881), DARPA under agreement number HR0011-20-9-0016 (PaPPa), Schmidt Futures program, Paul G. Allen Family Foundation, Charles Trimble, Audi Environmental Foundation, DOE, National Nuclear Security Administration under Award Number DE-NA0003965, LANL grant 531711, and German Research Council (DFG) under grant agreement No. 326472365. Research was sponsored in part by the US Air Force Research Laboratory and the United States Air Force Artificial Intelligence Accelerator and was accomplished under Cooperative Agreement Number FA8750-19-2-1000. The views and conclusions contained in this document are those of the authors and should not be interpreted as representing the official policies, either expressed or implied, of the United States Air

Force or the U.S. Government. The U.S. Government is authorized to reproduce and distribute reprints for Government purposes notwithstanding any copyright notation herein. This material is based upon work supported by the DOE, Office of Science, Office of Advanced Scientific Computing Research.

References

1. Geoga, C.J., Marin, O., Schanen, M., Stein, M.L.: Fitting matérn smoothness parameters using automatic differentiation. Stat. Comput. **33**(2), 48 (2023). https://doi.org/10.1007/s11222-022-10127-w
2. Griewank, A., Walther, A.: Algorithm 799: revolve: an implementation of checkpointing for the reverse or adjoint mode of computational differentiation. ACM Trans. Math. Softw. **26**(1), 19–45 (2000). https://doi.org/10.1145/347837.347846
3. Griewank, A., Walther, A.: Evaluating Derivatives: Principles and Techniques of Algorithmic Differentiation. 2nd edn. No. 105 in Other Titles in Applied Mathematics. SIAM, Philadelphia (2008). http://bookstore.siam.org/ot105/
4. Innes, M.: Don't unroll adjoint: differentiating SSA-form programs (2018). https://doi.org/10.48550/ARXIV.1810.07951
5. Moses, W.S., et al.: Scalable automatic differentiation of multiple parallel paradigms through compiler augmentation. In: SC22: International Conference for High Performance Computing, Networking, Storage and Analysis, pp. 1–18. IEEE Computer Society, Los Alamitos (2022). https://doi.org/10.1109/SC41404.2022.00065
6. Moses, W., Churavy, V.: Instead of rewriting foreign code for machine learning, automatically synthesize fast gradients. In: Larochelle, H., Ranzato, M., Hadsell, R., Balcan, M.F., Lin, H. (eds.) Advances in Neural Information Processing Systems, vol. 33, pp. 12472–12485. Curran Associates, Inc. (2020). https://proceedings.neurips.cc/paper/2020/file/9332c513ef44b682e9347822c2e457ac-Paper.pdf
7. Moses, W.S., et al.: Reverse-mode automatic differentiation and optimization of GPU kernels via enzyme. In: Proceedings of the International Conference for High Performance Computing, Networking, Storage and Analysis. SC 2021, Association for Computing Machinery, New York (2021). https://doi.org/10.1145/3458817.3476165
8. Naumann, U.: The art of differentiating computer programs. Soc. Ind. Appl. Math. (2011). https://doi.org/10.1137/1.9781611972078
9. Schanen, M., Narayanan, S.H.K.: Argonne-National-Laboratory/Checkpointing.jl: v0.6.3 (2023). https://doi.org/10.5281/zenodo.7607916
10. Stumm, P., Walther, A.: New algorithms for optimal online checkpointing. SIAM J. Sci. Comput. **32**(2), 836–854 (2010). https://doi.org/10.1137/080742439
11. White, F.C., et al.: JuliaDiff/ChainRules.jl: v1.45.0 (2022). https://doi.org/10.5281/zenodo.7312560

Performance of Selected Nature-Inspired Metaheuristic Algorithms Used for Extreme Learning Machine

Karol Struniawski[1] , Ryszard Kozera[1,2] , and Aleksandra Konopka[1]([⊠])

[1] Institute of Information Technology, Warsaw University of Life Sciences - SGGW,
ul. Nowoursynowska 159, 02-776 Warsaw, Poland
{karol_struniawski,ryszard_kozera,aleksandra_konopka}@sggw.edu.pl
[2] School of Physics, Mathematics and Computing, The University of Western
Australia, 35 Stirling Highway, Perth, Crawley, WA 6009, Australia
ryszard.kozera@uwa.edu.au

Abstract. This work presents a research on Nature Inspired Metaheuristic Algorithms (MA) used as optimizers in training process of Machine Learning method called Extreme Learning Machine (ELM). We tested 19 MA optimizers measuring their performance directly on sample datasets. The impact of input parameters such as number of hidden layer units, optimization stopping conditions and population size on the accuracy results, training and prediction time is evaluated here. Significant differences in performance of applied methods and their parameters' values are detected. The most meaningful outcome of this paper shows that an increase of the number of MA iterations does not yield significant boost in accuracy with a huge increase in training time. Indeed a cap on number of MA iterations ranging from 1 to 5 is sufficient for analyzed machine learning tasks. In our research the best results are obtained for population size ranging between 50 and 100. Hybridized ELM outperforms classical implementation of ELM as higher accuracy is reached for the same number of neurons.

Keywords: Computational Optimization · Metaheuristic Algorithms · Bio-inspired computing · Extreme Learning Machine · Machine Learning

1 Introduction

Mathematical optimization algorithms play a vital role in many contemporary technology applications such as e.g. GPS or IT banking sector tools. In addition, the optimization driven behavior is also prevalent within all living organisms commonly relying on it while e.g. hunting or trying to move more efficiently. In fact, searching for the efficiency in nature can be a matter of life and death. Among all the latter is physically demonstrated by the reproduction capabilities. Thus, organisms that perform life activities more efficiently are better adopted

© The Author(s), under exclusive license to Springer Nature Switzerland AG 2023
J. Mikyška et al. (Eds.): ICCS 2023, LNCS 14075, pp. 498–512, 2023.
https://doi.org/10.1007/978-3-031-36024-4_38

to the environment and are more likely to pass these abilities to their offsprings by genes according to the Darwin's Theory [7].

Scientists attempt to describe "optimized" activities of organisms in terms of mathematical modeling [27] commonly called Metaheuristic Algorithms (MA). The combination of bio-inspired optimization algorithms with machine learning models may improve their performance [30]. Here one of such approaches is called Extreme Learning Machine (ELM) - the Machine Learning method with growing popularity since its formulation in 2004 [14].

Recently, a hybrid MA-ELM that combines ELM with Metaheuristic Algorithms (MA) is proposed and evaluated in practical applications. In doing so, Chia et al. [8] used particle swarm (PSO), moth-flame (MFO) and whale optimization algorithm (WOA). In other practical related context, Wu et al. [30] applied genetic algorithm (GA), ant colony optimization (ACO), cuckoo search algorithm (CSA) and flower pollination algorithm (FPA). The above research demonstrates superiority of hybridized ELM over the regular one. Nevertheless, most of the works in this topic deal with the practical applications of these methods. There is a shortage in literature on comprehensive comparison of metaheuristic algorithms used in ELMs. In this paper we evaluate hybrid ELM on MNIST handwritten and Wine Quality White datasets [9] for different MA. The comparison analysis for a separate set of parameters to investigate their impact on attained accuracy and registered computational time is also performed for each examined algorithm. The experiment is carried out on a single machine in MATLAB R2021b, Ryzen 9 3900X CPU, 64 GB RAM, GTX 1660TI GPU.

2   Extreme Learning Machine

Extreme Learning Machine (ELM) is a dense feed-forward neural network classifier and regressor introduced by Huang et al. in 2004 [14]. The network's topology consists of input layer, a single hidden-layer and an output layer of neurons. The numbers of selected neurons in input and output layer depends on the task characteristics. The number of hidden layer units requires an empirical determination as a consequence of the theoretical method scarcity permitting to determine upfront its optimal numbers controlling the topology of the ELM.

2.1   Classification

Input data regarding supervised classification task with N observations can be described as pairs of values $\{(x_i, t_i)\}_{i=1}^{N}$, where x_i is i-th vector of d features and t_i is i-th label of class to which selected x_i belongs. Here, $t_i = 0, \ldots, M - 1$, where M is the amount of distinctive classes in the classification task in question. Note here that for multiclass classification (when object belongs to more than one class) t_i is a vector. Based on the latter, matrix $X = (x_1, x_2, \ldots, x_N) \in \mathbb{M}_{d \times N}(\mathbb{R})$

is formed, where $x_i \in \mathbb{R}^d$ with vector $T = \{t_i\}_{i=1}^N$:

$$X = \begin{bmatrix} x_{11} & \cdots & x_{1N} \\ \vdots & \ddots & \vdots \\ x_{d1} & \cdots & x_{dN} \end{bmatrix} \quad \text{and} \quad T = \begin{bmatrix} t_1 \\ \vdots \\ t_N \end{bmatrix}.$$

The ELM input layer comprises of d neurons and its output layer consists of units number equal to M. As an output of the network, the corresponding N values $\{y_i\}_{i=1}^N$ are calculated forming the matrix $Y = (y_1, y_2, \ldots, y_N) \in \mathbb{M}_{N \times M}(\mathbb{R})$, where $y_i \in \mathbb{R}^M$. The recognition of a given input x_i is performed based on extracting the maximal value of y_i observed on the p-th index which assigns x_i to the p-th class. Thus, matrix Y is actually reformatted as N values $\{y_i\}_{i=1}^N$, where $y_i = [0, \ldots, \overset{p}{1}, \ldots, 0]$. Subsequently, one has to properly format T in order to facilitate comparison with Y. In the next step $1 - of - K$ scheme is applied to the vector T - see [6]. Such procedure is designed to reformat t_i as $\{t_{ij}\}_{j=0}^M$ that yields all values set to zero except one element at s-th index that in turn is set to one. Consequently, t_i can be written as $t_i = [0, \ldots, \overset{s}{1}, \ldots, 0]$, where s-th element indicates that i-th input vector of X affiliates to the s-th class. Correct classification is observed if and only if $s = p$ for a given input vector x_i.

Let L be a number of neurons in hidden layer that is chosen a priori. Weights between input and hidden layer determine the matrix $W \in \mathbb{M}_{d \times L}(\mathbb{R})$, where w_{ij} represent the weights associated with the connection of i-th input layer neuron with j-th in hidden layer (see left Eq. (1)). Bias connections are represented by a vector $b = \{b_i\}_{i=1}^N$. In learning process of ELM coefficients of W and b are computed using uniform distribution function $U(-1, 1)$. The outputs of hidden layer neurons are stored in matrix $H \in \mathbb{M}_{N \times L}(\mathbb{R})$ (see right Eq. (1)):

$$W = \begin{bmatrix} w_{11} & \cdots & w_{1L} \\ \vdots & \ddots & \vdots \\ w_{d1} & \cdots & w_{dL} \end{bmatrix}, H = \begin{bmatrix} f(\sum_{i=1}^d x_{i1} w_{i1} + b_1) & \cdots & f(\sum_{i=1}^d x_{i1} w_{iL} + b_1) \\ \vdots & \ddots & \vdots \\ f(\sum_{i=1}^d x_{iN} w_{i1} + b_N) & \cdots & f \sum_{i=1}^d x_{iN} w_{iL} + b_N) \end{bmatrix}.$$
(1)

The activation function $f : \mathbb{R} \to \mathbb{R}$ represents in our investigation a sigmoid function $f(x) = f_\alpha(x) = \frac{1}{1+e^{-\alpha x}}$, with $\alpha = 1$. The weights β between hidden and output layer can be computed upon solving the following equation $Y = H\beta$. The system cannot be directly solved since H with probability equal to 1 is irreversible and $\|H\beta - Y\| = 0$ (see Huang et al. [14]). We estimate β as a minimizer of mean residual square error:

$$\hat{\beta} = \underset{\beta}{argmin} \|H\beta - T\|^2 = H^\dagger T,$$
(2)

where $H^\dagger$ defines a Moore-Penrose generalized inverse of H [26]. The Pseudo-inverse of matrix $H^\dagger$ is uniquely determined and in the case of a non-singular matrix H it coincides with an ordinary inverse i.e. $H^\dagger = H^{-1}$. The matrix $H^\dagger$ gives solution $\hat{\beta}$ so that $H\hat{\beta}$ is close to Y in terms of mean square error (MSE).

Assigning random values to weights and bias between input and hidden ELM layer makes the network not susceptible to overtraining. Most importantly, the computed solution $\hat{\beta}$ is a global minimizer of (2). The latter contrasts with Multi-Layer Perceptron (MLP) supervised training procedure. Indeed Backpropagation Algorithm finds generically only a local minimizer of the given network's loss function that measures how well the neural network classifies the training data [12]. In addition, the optimal value of $\hat{\beta}$ is found here upon performing a non-iterative procedure in (2). The learning speed of ELM can be thousands times faster than other methods like MLP (see [15]).

3   Genetic Extreme Learning Machine

The original concept of ELM relies on selecting weights between input and hidden layer together with bias values as randomly generated. This principle has a remarkable advantage in terms of computational efficiency [15]. Still such randomness in weights generation in ELM can lead to the unstable performance [4]. The idea here is to somehow estimate weights and bias values in order to maximize the accuracy and stability of the model. A possible remedy to this problem is to combine the Genetic Algorithm (GA) (see [13]) with ELM to form the so-called hybridized Genetic Extreme Learning Machine (GELM). GAs are created as a computational representation of Darwinian evolution theories to search for the optimal solution of global non-linear optimization task by simulating the process of biological natural selection concept [16]. Our hope is that reflecting the natural processes of selection, crossover and mutation the fittest individuals are selected for reproduction that will provide better offspring in terms of improving an appropriate fitness evaluation function [5].

4   Nature-Inspired Metaheuristic Algorithms

In general, the constrained optimization problem can be formulated in terms of minimizing some objective function: $minimize f(x)$ with $x = (x_1, \ldots, x_n)$ admitted to fulfill either some equality(ies) and/or inequality(ies) [31]. All modern nature-inspired algorithms are called Metaheuristic Algorithms [17]. Up to now there is no commonly accepted definition of MA, but one can outline the following selected principles of MA adopted in the literature (see [24, 27, 31]): a strategy that the main aim is to guide the search process avoiding the disadvantages of iterative improvement allowing the local search to escape from local optima; starting to find solutions in more intelligent way than just providing random initial solutions; dealing with randomness in an biased form incorporating search experience (in a form of memory) to guide the search; in the simulation stage considered as a set of assumptions about the natural environment.

The search strategies of different MA are highly dependent on the philosophy of the metaheuristic itself. In this paper, as a comparison of the MA applied in ELM learning process, we use methods simulating behaviors of living organisms in terms of the following optimization processes: Artificial Ecosystem-based

Optimization (AEO) [35], Artificial Hummingbird Algorithm (AHA) [34], Artificial Rabbits Optimization (ARO) [29], African Vultures Optimization Algorithm (AVOA) [2], Coyote Optimization Algorithm (COA) [25], Dandelion Optimizer (DO) [32], Fast Cuckoo Search (FCS) [23], Gorilla Troops Optimizer (GTO) [3], Grey Wolf Optimizer (GWO) [20], Hybrid Grey Wolf and Cuckoo Search Optimization Algorithm (GWO-CS) [11], Improved Grey Wolf Optimizer (I-GWO) [21], Leader Harris Hawks Optimization (LHHO) [22], Mountain Gazelle Optimizer (MGO) [1], Manta Ray Foraging Optimization (MRFO) [36], Northern Goshawk Optimization (NGO) [10], Pelican Optimization Algorithm (POA) [28], Hybrid Particle Swarm Optimization and Gravitational Search Algorithm (PSOGSA) [19], Sea-horse Optimizer (SHO) [33] and lastly Salp Swarm Algorithm (SSA) [18].

5   Experiments and Results

The metaheuristic algorithms (briefly outlined in the previous section) used in this work have common prerequisites. In particular, from now on, the term MA directly refers to the algorithms exclusively used in this paper (see Sect. 4).

MA define a concept of population as a set S of S_n candidate solutions, where s_i, $i = 1, \ldots, S_n$ is a solution vector called also an individual and implement the concept of intelligent iterative ransacking search space taking as an input dimension of the vector $S_d = dim(s_i)$, number of population S_n and constraints applied to s_i. A termination condition for the algorithm and appropriate fitness function must be determined. As MA fall into iterative methods, in k-th iteration the set S^k called generation is produced with $s_i^k \in S^k$ representing generation's individual, where $k = 1, \ldots, k_n$. The output of MA $s^{min} = s_i^{k_n}$ yields a minimal value of a given fitness function in the last generation of the algorithm.

To integrate MA with ELM we first need to specify input parameters for MA. Analogously to GELM our aim is to evaluate optimal values of weights between input and hidden layer including bias. In fact, the output of the MA is a vector $s^{min} \in \mathbb{R}^{S_d}$, where $S_d = dN + N$. As a consequence we can reformat s^{min} properly constructing W and b:

$$W = \begin{bmatrix} s_{11}^{min} & \cdots & s_{1N}^{min} \\ \vdots & \ddots & \vdots \\ s_{d1}^{min} & \cdots & s_{dN}^{min} \end{bmatrix} \quad \text{and} \quad b = \begin{bmatrix} s_{dN+1}^{min} \\ \vdots \\ s_{dN+N}^{min} \end{bmatrix}.$$

Vector s^{min} forms the final, optimal value of $W^{s^{min}}$ and $b^{s^{min}}$. Fitness function g is prepared based on the response of the ELM network represented as Y_i^k for a given s_i^k (forming W_i^k and b_i^k) compared to the expected results T, $g(X, W_i^k, b_i^k, T) = \frac{1}{N} \sum_{j=1}^{N} (Y_{ij}^k - T_j)^2$, where $Y_i^k = H\beta$, $\beta = H^\dagger T$ and $H = f(X^T W_i^k + b_i^k)$ (see also (1)). The inequality constraints $-1 < s_{ij} < 1$ (with $j \in [1, \ldots S_d]$ for each $i \in [1, \ldots, S_n]$) enforce $s_i \in [-1; 1]^{S_d} \subseteq \mathbb{R}^{S_d}$. The impact of parameters S_n and the termination condition on the algorithm performance is investigated in this research. The admitted values for S_n are equal to 50, 100 or 200. It is noteworthy that lowering the vales of S_n led to unstable results, while higher S_n values resulted in impractically long evaluation times for our experiment. Two different approaches of selecting stopping conditions of MA are here considered. *First*, the stopping flag is activated once one of two conditions is fulfilled. More specifically, the upper limit on k iterations is a priori set (here $k_n = 10000$). In conjunction with the latter, the optimization procedure terminates once the following a posteriori condition is met $|g(X, W_i^k, b_i^k, T) - g(X, W_i^{k+1}, b_i^{k+1}, T)| < \varepsilon$ holding for longer than 200 iterations (here $\varepsilon = 0.0001$). In further presentation of calculation results the first variant of stopping condition is marked as *"Limit 0"*. *Second*, the impact of fixing ad hock an upper bound k on number of iterations is also analyzed here for $k_n = 1$, $k_n = 5$ and $k_n = 50$ that can be recognized in further considerations as *"Limit 1"*, *"Limit 5"* and *"Limit 50"*, respectively.

Another parameter taken also here into consideration is the number of neurons L in hidden layer of ELM. At this point one should mention a dilemma of evaluating results applying testing and training sets once MA is used for optimizing W and b. A core principle of the Machine Learning (ML) is to examine results returned by a selected method on data that cannot be used for training process. To enforce the latter the data is usually a priori divided into training and testing sets or alternatively one resorts to a cross-validation method [12]. Cross-validation is an iterative method that uses different portions of data to test and to train a model applying randomness. Thus, matrices W and b are optimized upon using MA on training data exclusively and cannot be specified as optimal on testing set. A similar approach should be adopted for β evaluation while computing weights between hidden and output layer of ELM. It is implicitly assumed here that dependencies for both training and testing sets are similar. Therefore the optimized W and b based on training set can equally successfully operate on testing set. The case of unbalanced number of observations obtained on training and testing sets deserves a short note. Indeed, should the latter occurs, the matrices W and b generated by MA on training set cannot be directly applied to estimate Y on testing set. In ML there exists an implicit assumption that the testing set should be essentially smaller than a training one. Typically, the proportion of observations abides from 9:1 to 7:3 ratio. Consequently, s^{min} is too large to be properly re-formatted to W and b which can still act on testing set. Assuming testing set contains N^{test} observations we solve this problem by taking $k = N^{test} \times d$ first elements of s^{min} transforming them into matrix $W_{N^{test} \times d}^{s^{min}}$ and last N^{test} elements of s^{min} creating vector $b^{s^{min}}$:

$$W^{s^{min}} = \begin{bmatrix} s_1^{min} & \cdots & s_N^{min} \\ \vdots & \ddots & \vdots \\ s_{k-N}^{min} & \cdots & s_k^{min} \end{bmatrix} \quad \text{and} \quad b^{s^{min}} = \begin{bmatrix} s_{S_d-N^{test}}^{min} \\ \vdots \\ s_{S_d}^{min} \end{bmatrix}.$$

The entire calculation process is presented in the flowchart (see Fig. 1).

First, we evaluate the model in question for a different number of neurons L in hidden layer, population size S_n and MA termination condition taken as *"Limit 0"*. Unfortunately, even for $S_n = 50$ and $L = 100$ computation time for

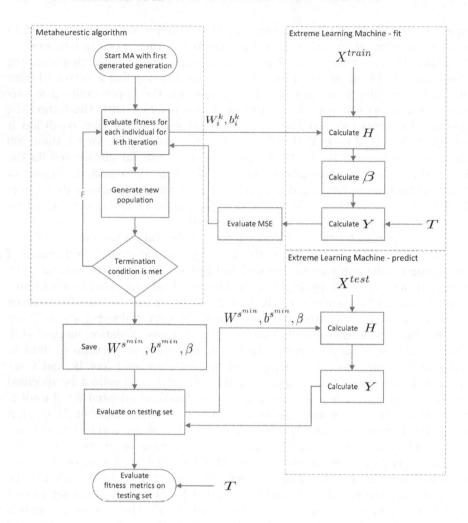

Fig. 1. Flowchart of MA-ELM training and testing process.

most of the methods exceeded a few hours. Then, a full comparison to the other Limits is impossible. Therefore we discarded *"Limit 0"* calculations and leave it for a future investigation. For our research two exemplary datasets are used. The first set is called MNIST handwritten digits. The dataset contains 60000 training and 10000 testing samples that are handwritten digits saved as greyscale images of size 28 × 28 pixels. Thus, input vectors' size is 784 after flatten operation is applied to the original image transforming image to the row by row vector. The dataset is a typical example of classification task where accuracy of the classifier is evaluated. The second set is called Wine Quality White which is composed of features describing chemical and physical parameters of white wine i.e. fixed acidity, volatile acidity, pH etc. Our task is to assign to a given wine sample the quality measure ranging from 0 to 10. Here classification performance is measured with MSE as we recognize differences between inappropriate class assignments i.e. attaching a wine which is truly categorized as 0 to class 9 is a graver mistake than assigning this wine to class 1. The dataset does not contain separate training and testing subsets and because of that to obtain the most meaningful results a 20% cross-validation method is applied that is repeated 50 times to properly estimate a statistically significant value of MSE and to discard randomness influence on the final results.

Table 1. Accuracy of ELM with selected MA applied for a given number of neurons in hidden layer, population size 50 and limit of iterations equal to 1 used directly as a classifier on MNIST handwritten digits dataset.

ACC [%]	100	200	300	400	500	600	700	800	900	1000
AEO	78.31	79.04	82.05	85.19	85.06	86.14	84.95	78.74	88.89	88.00
AHA	77.08	81.02	83.80	84.17	86.77	85.22	87.89	88.29	85.29	89.61
ARO	73.74	83.43	81.06	83.17	83.76	85.96	88.27	89.20	88.08	88.02
AVOA	68.50	82.11	84.06	81.66	84.46	85.78	86.55	88.85	89.84	89.60
COA	72.75	78.67	83.83	84.62	86.36	86.04	86.28	89.36	88.47	90.09
DO	73.48	81.41	83.61	80.31	87.55	81.95	89.22	86.20	87.54	85.98
FCS	72.89	80.43	80.94	86.15	85.77	87.17	86.76	89.25	88.02	88.80
GTO	78.54	81.73	78.35	84.67	75.38	88.85	84.99	80.26	83.8	84.41
GWO	69.42	74.79	81.01	83.28	83.95	88.52	87.58	85.05	88.65	90.08
GWO-CS	46.43	51.76	63.67	68.60	62.92	70.41	68.98	70.54	73.66	71.76
I-GWO	71.36	79.02	82.68	80.52	86.51	86.38	88.23	86.47	84.77	85.78
LHHO	72.48	81.02	80.57	87.39	84.42	86.93	85.53	87.43	87.88	87.84
MGO	71.59	78.65	78.06	80.13	86.27	86.93	86.93	88.36	88.42	88.92
MRFO	73.52	81.21	86.47	83.36	82.47	88.8	88.34	89.16	90.01	89.25
NGO	69.89	81.43	84.35	84.63	84.9	86.17	87.09	88.54	89.72	90.70
POA	69.81	81.06	83.16	84.73	85.89	87.31	87.83	88.09	87.36	89.60
PSOGSA	74.91	78.97	83.23	84.25	85.34	86.70	86.36	84.90	86.89	90.19
SHO	72.87	81.92	81.92	84.57	86.44	84.13	83.34	89.73	89.49	85.07
SSA	74.92	80.61	81.97	82.41	82.41	86.66	88.8	87.78	87.21	87.77

For MNIST similarly to the ELM (see Table 7) for MA-ELM we obtain better results upon increasing number of neurons (see Table 1). Then it was decided to fix L to the value of 1000 in further calculations of MA-ELM. In contrast, for Wine Quality White dataset the best results are observed for $L = 100$ (see Table 2), so this value will be fixed analyzing the dataset. Fitting process (see Table 3) for MNIST, $L = 1000$, $S_n = 50$ and selected limit of iterations combined with a method in question may take from a few seconds to some hours. The fastest methods turned out to be AVOA, DO, PSOGSA, SHO and SSA. In particular, the last one yields the most prominent results with 8s, 284s and 2784 s s fit time for limit k_n set to 1, 5 and 50, respectively. One can also notice that raising the number of iterations more or less linearly increases the resulting computation time. Differences in fitting time between various methods testify

Table 2. MSE of Extreme Learning Machine with selected Metaheuristic Algorithms applied for a given number of neurons in hidden layer, population size 50 and limit of iterations equal to 1 used directly as a classifier on Wine Quality White dataset.

MSE	100	200	300	400	500	600	700	800	900	1000
AEO	0.641	0.647	0.659	0.679	0.692	0.699	0.725	0.734	0.755	0.775
AHA	0.634	0.645	0.661	0.669	0.685	0.701	0.718	0.731	0.759	0.768
ARO	0.648	0.650	0.654	0.670	0.684	0.697	0.716	0.739	0.747	0.765
AVOA	0.645	0.647	0.663	0.675	0.688	0.706	0.722	0.739	0.753	0.776
COA	0.639	0.647	0.659	0.680	0.690	0.712	0.721	0.738	0.751	0.771
DO	0.653	0.652	0.661	0.681	0.692	0.708	0.725	0.743	0.754	0.768
FCS	0.647	0.650	0.659	0.677	0.694	0.707	0.727	0.738	0.762	0.761
GTO	0.650	0.650	0.659	0.661	0.674	0.719	0.712	0.723	0.740	0.756
GWO	0.648	0.647	0.661	0.678	0.690	0.709	0.719	0.741	0.757	0.774
GWO-CS	0.625	0.628	0.641	0.642	0.671	0.672	0.726	1.105	0.722	1.056
I-GWO	0.644	0.651	0.668	0.672	0.693	0.709	0.723	0.747	0.749	0.770
LHHO	0.630	0.638	0.660	0.676	0.695	0.681	0.722	0.727	0.758	0.794
MGO	0.635	0.637	0.656	0.668	0.695	0.694	0.721	0.726	0.726	0.752
MRFO	0.641	0.642	0.656	0.657	0.674	0.711	0.720	0.730	0.751	0.765
NGO	0.647	0.654	0.664	0.674	0.686	0.708	0.726	0.739	0.751	0.771
POA	0.656	0.651	0.664	0.676	0.691	0.704	0.724	0.745	0.756	0.773
PSOGSA	0.647	0.648	0.662	0.677	0.695	0.703	0.732	0.737	0.758	0.772
SHO	0.651	0.649	0.659	0.677	0.696	0.710	0.723	0.738	0.756	0.772
SSA	0.647	0.652	0.663	0.675	0.696	0.708	0.720	0.742	0.755	0.767

their practical applicability i.e. MGO needs fourfold more time for *"Limit 1"* to achieve comparable results with SSA. It should be emphasized here that there is no correlation between more computational time involved versus achieving better results. Indeed, a GWO method surpasses in terms of ACC the other methods that still need twice longer time to be executed. In previous section MA are defined as methods that tend to create the new generations with individuals characterized by lower fitness function value. Simultaneously, as we stated the low MSE value for a given individual being MA solution cannot be directly recognized as better in terms of accuracy upon applying on a testing set, because of the fact that in training and testing different subsets of dataset are used. Surprisingly, for many of the methods increasing number of iterations does not improve ACC (see Table 3). For most of them we observe a slight increase of ACC between *"Limit 1"* and *"Limit 5"*. The decrease of ACC between *"Limit 5"* and *"Limit 50"* was not expected. For some of the methods the lowest ACC is obtained for *"Limit 50"*. The classifier performance on Wine Quality White confirms results obtained on MNIST. For the majority of methods we do not observe significant changes of MSE. Setting k_n to higher value even increases MSE in the case of AHA, COA, GWO-CS, LHHO, MGO and PGOGSA. For the remaining methods change of k_n from 1 to 5 results in a slight decrease of

Table 3. Fit time and accuracy of Extreme Learning Machine with selected Metaheuristic Algorithms applied for $L = 1000$ neurons in hidden layer, population size $S_n = 50$ and $k_n = 1$, 5 or 50 used directly as a classifier on MNIST handwritten digits. For $L = 100$, $S_n = 50$, $k_n = 1$, 5 or 50 on Wine Quality White dataset.

Method / k_n	MNIST						Wine Quality White					
	Fit time [s]			ACC [%]			Fit time [s]			MSE		
	1	5	50	1	5	50	1	5	50	1	5	50
AEO	237	571	5697	88.00	86.79	84.95	0.6	2.4	21.6	0.641	0.635	0.641
AHA	155	312	3989	89.61	88.98	84.62	0.3	1.7	10.7	0.634	0.646	0.631
ARO	156	309	4028	88.02	88.15	81.87	0.3	1.4	11.0	0.648	0.637	0.636
AVOA	78	258	4050	89.60	84.94	77.36	0.1	1.0	10.9	0.645	0.629	0.634
COA	278	692	7274	90.09	82.7	88.83	0.6	3.2	26.6	0.639	0.643	0.624
DO	77	273	2666	85.98	88.46	86.9	0.1	1.3	11.5	0.653	0.649	0.650
FCS	233	564	5041	88.80	86.51	77.86	0.5	2.5	20.9	0.647	0.647	0.651
GTO	103	567	5062	84.41	89.99	89.89	0.5	2.6	20.5	0.650	0.637	0.633
GWO	250	277	2705	90.08	88.57	87.78	0.2	1.1	11.8	0.648	0.648	0.653
GWO-CS	270	311	3046	71.76	72.99	87.88	0.2	1.3	13.7	0.625	0.643	0.630
I-GWO	445	608	5466	85.78	88.99	86.67	0.6	3.2	24.7	0.644	0.643	0.635
LHHO	240	896	8879	87.84	89.87	76.34	0.6	3.6	36.7	0.630	0.635	0.623
MGO	236	1218	11486	88.92	90.55	89.89	1.0	5.6	52.6	0.635	0.633	0.640
MRFO	232	566	5069	89.25	88.14	83.28	0.5	2.9	21.3	0.641	0.642	0.643
NGO	141	566	5038	90.70	91.45	87.42	0.5	2.6	21.5	0.647	0.645	0.647
POA	212	556	4974	89.60	81.25	86.32	0.5	2.8	20.6	0.656	0.647	0.638
PSOGSA	82	507	4800	90.19	85.8	90.81	0.5	2.6	24.6	0.647	0.655	0.649
SHO	22	471	4152	85.07	87.37	72.69	0.5	2.4	17.0	0.651	0.644	0.634
SSA	8	284	2784	87.77	87.71	88.90	0.2	1.5	12.8	0.647	0.647	0.643

MSE. Methods AEO, AVOA, DO, FCS, GWO, MGO, MRFO and NGO can be characterized by decreasing MSE once k_n changes from 1 to 5. In terms of fitting time we observe more or less a linear growth of computational time when k_n is enlarged. In Table 4 we tested a different population size S_n. It shows that increasing S_n expands the fitting time, but to a lesser extent with exceptions like SSA method that needs almost $14\times$ computation time for $S_n = 100$ as compared to $S_n = 50$. Such tendency is similar for SHO method. Fitting time between $S_n = 100$ and $S_n = 200$ for most of the methods expand twice. In terms of accuracy, we do not observe a significant increase when population size is enlarged. The methods that are beneficial to this increase are DO, FCS, GTO, GWO-CS, I-GWO, LHHO, MGO, MRFO and SSO. Noticeably, for DO, FCS, I-GWO, MGO and MRFO the highest accuracy is registered for $S_n = 100$ and the lowest for $S_n = 200$. The observed dependencies on MNIST are even more visible for Wine Quality dataset. The lowest MSE for all methods is obtained for $S_n = 50$ and increases substantially when $S_n = 100$ or $S_n = 200$ is used.

For standard ELM classifier applied on MNIST handwritten dataset the highest accuracy 91.41% is generated for 4000 and 5000 neurons in hidden layer (see Table 7). In comparison, for MA-ELM the highest ACC of 91.45% is reached for NGO metaheuristic algorithm, 1000 neurons, limit of 5 iterations, population size

Table 4. Fit time and accuracy of Extreme Learning Machine with selected Meta-heuristic Algorithms applied for 1000 neurons in hidden layer, limit of iterations equal to 1 and a different population size $S_n = 50$, 100 or 200 used directly as a classifier on MNIST handwritten digits dataset and 100 neurons in hidden layer, limit of iterations equal to 1 and a different population size $S_n = 50$, 100 or 200 used directly as a classifier on Wine Quality White dataset.

| Method / S_n | MNIST | | | | | | Wine Quality White | | | | | |
| | Fit time [s] | | | ACC [%] | | | Fit time [s] | | | MSE | | |
	50	100	200	50	100	200	50	100	200	50	100	200
AEO	237	331	675	88.00	86.47	88.50	0.6	30	57	0.641	0.743	0.741
AHA	155	211	428	89.61	86.46	86.07	0.3	19	36	0.634	0.804	0.788
ARO	156	208	435	88.02	84.53	86.26	0.3	19	36	0.648	0.732	0.771
AVOA	78	104	215	89.60	88.13	88.87	0.1	9	18	0.645	0.745	0.712
COA	278	369	785	90.09	89.42	88.25	0.6	35	68	0.639	0.717	0.744
DO	77	102	209	85.98	89.93	87.88	0.1	9	17	0.653	0.704	0.734
FCS	233	313	654	88.80	90.06	88.67	0.5	29	57	0.647	0.737	0.711
GTO	103	321	649	84.41	80.56	89.26	0.5	28	54	0.650	0.804	0.716
GWO	250	115	231	90.08	88.99	88.85	0.2	11	21	0.648	0.750	0.731
GWO-CS	270	142	315	71.76	84.23	89.93	0.2	16	34	0.625	0.733	0.759
I-GWO	445	345	721	85.78	89.06	87.23	0.6	33	65	0.644	0.747	0.736
LHHO	240	422	845	87.84	88.39	89.68	0.6	39	71	0.630	0.715	0.749
MGO	236	657	1574	88.92	89.59	87.32	1.0	69	153	0.635	0.731	0.729
MRFO	232	318	673	89.25	91.43	82.92	0.5	30	60	0.641	0.774	0.788
NGO	141	312	651	90.70	86.34	88.46	0.5	29	56	0.647	0.737	0.749
POA	212	306	628	89.60	87.06	89.96	0.5	28	53	0.656	0.732	0.717
PSOGSA	82	296	996	90.19	87.65	88.91	0.5	42	140	0.647	0.664	0.747
SHO	22	286	593	85.07	88.07	89.39	0.5	28	56	0.651	0.717	0.743
SSA	8	110	229	87.77	90.69	88.18	0.2	10	20	0.647	0.687	0.746

Table 5. The 6 highest ACC for Extreme Learning Machine with selected Metaheuristic Algorithms used directly as a classifier on MNIST handwritten digits dataset.

Method	S_n	L	k_n	Fit MSE	Fit time [s]	Prediction time [s]	ACC [%]
NGO	50	1000	5	0.059	566	0.074	91.45
MRFO	100	1000	1	0.057	318	0.082	91.43
GTO	50	900	5	0.058	471	0.081	90.88
GTO	200	900	5	0.057	2928	0.069	90.85
SSA	50	900	5	0.062	236	0.065	90.81
PSOGSA	50	1000	50	0.060	4800	0.068	90.81

50 and 91.43% ACC for MRFO with 1000 neurons, limit of 1 iteration and population size 100 (see Table 5). Here we obtained comparable results of ELM and MA-ELM but for a different number of neurons. Training time, that is stated as a fit time for MA-ELM, is a lot longer than in case of typical ELM. Note here that in many practical application cases of ML a short prediction time is crucial. The time is extended when net is composed of more hidden layer units. Focusing on

Table 6. The 5 lowest MSE for Extreme Learning Machine with selected Metaheuristic Algorithms used directly as a classifier on Wine Quality White dataset.

Method	S_n	L	k_n	Fit MSE	Fit time [s]	Prediction time [s]	MSE
LHHO	50	100	50	0.417	36.706	0.008	0.623
COA	50	100	50	0.414	26.659	0.008	0.624
GWO-CS	50	100	1	0.425	0.285	0.005	0.625
GWO-CS	50	200	1	0.398	0.561	0.010	0.628
AVOA	50	100	5	0.421	1.073	0.008	0.629

Table 7. Extreme Learning Machine used directly as a classifier on MNIST handwritten digits dataset results.

Neurons	Fit Time [s]	Prediction Time [s]	ACC [%]
1000	3	0.069	88.69
2000	7	0.134	90.57
3000	14	0.256	91.26
4000	26	0.268	91.41
5000	45	0.412	91.41
6000	71	0.511	91.30
7000	115	0.527	90.58
8000	165	0.707	90.83

Table 8. Extreme Learning Machine used directly as a classifier on Wine Quality White results.

Neurons	Fit Time [s]	Prediction Time [s]	MSE
100	0.004	0.0003	0.646
200	0.008	0.0014	0.652
300	0.021	0.0032	0.660
400	0.028	0.0034	0.677
500	0.025	0.0025	0.691
600	0.040	0.0033	0.710
700	0.052	0.0037	0.721
800	0.053	0.0041	0.744
900	0.105	0.0061	0.757
1000	0.130	0.0084	0.770

prediction time we should compare nets with 1000 hidden layer neurons, then MA-ELM achieve ACC higher by 3pp (percentage points) over EML. Prediction time is highly dependent on L, then there is no difference of prediction time between MA-ELM and ELM. When we compare a similar ACC from the both methods (for MA-ELM $L = 1000$ and ELM $L = 4000$) prediction time for 1000 is 6 times shorter which can be very beneficial in models that require short classification time. Summing up the results obtained for the Wine Quality with ELM classifier applied leads to a rise of MSE when number of neurons in hidden layer increases (see Table 8). The lowest MSE=0.646 is generated for $L = 100$ with training time of the net equal to 0.004s and prediction coinciding with 0.0003s. According to

Table 6 which presents the top 5 lowest MSE across all exploited parameters' values of MA-ELM the best results are produced for $S_n = 50$, $L = 100$ and $k_n = 50$ for LHHO and COA methods. The lowest observed MSE=0.623 for MA-ELM is a better result than using core ELM method for which MSE=0.646 is detected. Both classifiers for this dataset have comparable prediction time as the best results are reached for the same value of L.

6   Conclusions

In this paper the concept of hybridized ELM with MA is introduced. Subsequently, the influence of the parameters' value selection on final results is examined. More precisely, the impact on the results of the number of neurons in hidden layer of ELM, the size of the population and the stopping conditions for MA are investigated. Based on this research we conclude that higher accuracy of the hybridized ELM can be detected even for lower number of neurons in hidden layer than in typical ELM. The latter leads to a significant fall in prediction time of the model. Surprisingly, the best results assessing MA termination condition of MA are registered as a hard limit of 5 iterations for MNIST handwritten and of 50 iterations for Wine Quality White dataset. Unfortunately, we were not able to examine termination condition of MA as a limit of 10000 iterations or changes of fitness less than $\varepsilon = 0.0001$ because of the computational complexity involved. This aspect should be further investigated. The population sizes examined in our study were set to 50, 100, and 200, as lower values led to unstable results, and higher values resulted in impractically long evaluation times for our experiment. In total 19 MA methods are tested and across all SSA and MRFO stand out for their high accuracy combined with shorter training times compared to other methods. Notwithstanding, one ought to emphasize that there is no method that yields excellent results on both datasets. It is worth noting that there is no direct correlation between increased computational time and improved results. The selection of the appropriate MA algorithm for a particular task should be based on comprehensive evaluation. However, in our work, we observed that certain algorithms exhibit a high computational complexity without a significant improvement in classification accuracy. Therefore, MGO, AEO, COA, and LHHO may not be suitable for hybridized ELM and can be discarded from further consideration.

References

1. Abdollahzadeh, B., Gharehchopogh, F.S., Khodadadi, N., Mirjalili, S.: Mountain gazelle optimizer: a new nature-inspired metaheuristic algorithm for global optimization problems. Adv. Eng. Softw. **174**, 103282 (2022). https://doi.org/10.1016/j.advengsoft.2022.103282
2. Abdollahzadeh, B., Gharehchopogh, F.S., Mirjalili, S.: African vultures optimization algorithm: a new nature-inspired metaheuristic algorithm for global optimization problems. Comput. Ind. Eng. **158**, 107408 (2021). https://doi.org/10.1016/j.cie.2021.107408

3. Abdollahzadeh, B., Soleimanian Gharehchopogh, F., Mirjalili, S.: Artificial gorilla troops optimizer: a new nature-inspired metaheuristic algorithm for global optimization problems. Int. J. Intell. Syst. **36**(10), 5887–5958 (2021). https://doi.org/10.1002/int.22535

4. Albadr, M.A.A., Tiun, S., AL-Dhief, F.T., Sammour, M.A.M.: Spoken language identification based on the enhanced self-adjusting extreme learning machine approach. PLOS One **13**(4), 1–27 (2018). https://doi.org/10.1371/journal.pone.0194770

5. Banzhaf, W., Francone, F.D., Keller, R.E., Nordin, P.: Genetic Programming: An Introduction: On the Automatic Evolution of Computer Programs and its Applications. Morgan Kaufmann Publishers Inc., Burlington (1998)

6. Bishop, C.M.: Pattern Recognition and Machine Learning. Springer, New York (2006)

7. Brooks, C.M.: The nature of life and the nature of death. J. UOEH **5**(2), 133–145 (1996)

8. Chia, M.Y., Huang, Y.F., Koo, C.H.: Swarm-based optimization as stochastic training strategy for estimation of reference evapotranspiration using extreme learning machine. Agric. Water Manag. **243**, 106447 (2021). https://doi.org/10.1016/j.agwat.2020.106447

9. Cortez, P., Cerdeira, A., Almeida, F., Matos, T., Reis, J.: Modeling wine preferences by data mining from physicochemical properties. Decis. Support Syst. **47**(4), 547–553 (2009). https://doi.org/10.1016/j.dss.2009.05.016

10. Dehghani, M., Hubalovsky, S., Trojovsky, P.: Northern goshawk optimization: a new swarm-based algorithm for solving optimization problems. IEEE Access **9**, 162059–162080 (2021). https://doi.org/10.1109/ACCESS.2021.3133286

11. Gupta, A.: Hybrid grey wolf and cuckoo search optimization algorithm (2023). https://www.mathworks.com/matlabcentral/fileexchange/69392-hybrid-grey-wolf-and-cuckoo-search-optimization-algorithm

12. Hassoun, M.H.: Fundamentals of Artificial Neural Networks, 1st edn. MIT Press, Cambridge (1995)

13. Holland, J.H.: Adaptation in Natural and Artificial Systems. University of Michigan Press, Ann Arbor (1975)

14. Huang, G.B., Zhu, Q.Y., Siew, C.: Extreme learning machine: a new learning scheme of feedforward neural networks. In: IEEE International Conference Neural Network, vol. 2, pp. 985–990 (2004). https://doi.org/10.1109/IJCNN.2004.1380068

15. Li, H.T., Chou, C.Y., Chen, Y.T., Wang, S.H., Wu, A.Y.: Robust and lightweight ensemble extreme learning machine engine based on eigenspace domain for compressed learning. IEEE TCAS-I **66**(12), 4699–4712 (2019). https://doi.org/10.1109/TCSI.2019.2940642

16. Liu, W.C., Chung, C.E.: Enhancing the predicting accuracy of the water stage using a physical-based model and an artificial neural network-genetic algorithm in a river system. Water **6**(6), 1642–1661 (2014). https://doi.org/10.3390/w6061642

17. Ireau, M., Potvin, J.Y.: Handbook of Metaheuristics, 2nd edn. Springer, New York (2010). https://doi.org/10.1007/978-1-4419-1665-5

18. Mirjalili, S., Gandomi, A.H., Mirjalili, S.Z., Saremi, S., Faris, H., Mirjalili, S.M.: Salp swarm algorithm: a bio-inspired optimizer for engineering design problems. Adv. Eng. Softw. **114**, 163–191 (2017). https://doi.org/10.1016/j.advengsoft.2017.07.002

19. Mirjalili, S., Hashim, S.Z.M.: A new hybrid PSOGSA algorithm for function optimization. In: ICCASM 2010, pp. 374–377 (2010). https://doi.org/10.1109/ICCIA.2010.6141614

20. Mirjalili, S., Mirjalili, S.M., Lewis, A.: Grey wolf optimizer. Adv. Eng. Softw. **69**, 46–61 (2014). https://doi.org/10.1016/j.advengsoft.2013.12.007

21. Nadimi-Shahraki, M.H., Taghian, S., Mirjalili, S.: An improved grey wolf optimizer for solving engineering problems. Expert Syst. Appl. **166**, 113917 (2021). https://doi.org/10.1016/j.eswa.2020.113917

22. Naik, M.K., Panda, R., Wunnava, A., Jena, B., Abraham, A.: A leader Harris Hawks optimization for 2-D Masi entropy-based multilevel image thresholding. Multimed. Tools. Appl. **80**(28), 35543–35583 (2021). https://doi.org/10.1007/s11042-020-10467-7

23. Naik, M.K., Swain, M., Panda, R., Abraham, A.: Novel square error minimization-based multilevel thresholding method for COVID-19 x-ray image analysis using fast cuckoo search. Int. J. Image Graph. (2022). https://doi.org/10.1142/s0219467824500049

24. Osman, I.H., Laporte, G.: Metaheuristics: a bibliography. Ann. Oper. Res. **63**, 511–623 (1996)

25. Pierezan, J., Dos Santos Coelho, L.: Coyote optimization algorithm: a new meta-heuristic for global optimization problems. In: IEEE-CEC 2018, pp. 1–8 (2018). https://doi.org/10.1109/CEC.2018.8477769

26. Rao, C.R., Mitra, S.K.: Generalized Inverse of Matrices and its Applications. Wiley, Hoboken (1971)

27. Stützle, T.: Local search algorithms for combinatorial problems - analysis, improvements, and new applications. In: DISKI (1999)

28. Trojovsky, P., Dehghani, M.: Pelican optimization algorithm: a novel nature-inspired algorithm for engineering applications. Sensors **22**(3), 855 (2022). https://doi.org/10.3390/s22030855

29. Wang, L., Cao, Q., Zhang, Z., Mirjalili, S., Zhao, W.: Artificial rabbits optimization: a new bio-inspired meta-heuristic algorithm for solving engineering optimization problems. Eng. Appl. Artif. Intell. **114**, 105082 (2022). https://doi.org/10.1016/j.engappai.2022.105082

30. Wu, L., Zhou, H., Ma, X., Fan, J., Zhang, F.: Daily reference evapotranspiration prediction based on hybridized extreme learning machine model with bio-inspired optimization algorithms: application in contrasting climates of China. J. Hydrol. **577**, 123960 (2019). https://doi.org/10.1016/j.jhydrol.2019.123960

31. Yang, X.S.: Nature-Inspired Metaheuristic Algorithms. Luniver Press, United Kingdom (2010)

32. Zhao, S., Zhang, T., Ma, S., Chen, M.: Dandelion optimizer: a nature-inspired metaheuristic algorithm for engineering applications. Eng. Appl. Artif. Intell. **114**, 105075 (2022). https://doi.org/10.1016/j.engappai.2022.105075

33. Zhao, S., Zhang, T., Ma, S., Wang, M.: Sea-horse optimizer: a novel nature-inspired meta-heuristic for global optimization problems. Appl. Intell. (2022). https://doi.org/10.1007/s10489-022-03994-3

34. Zhao, W., Wang, L., Mirjalili, S.: Artificial hummingbird algorithm: a new bio-inspired optimizer with its engineering applications. Comput. Meth. Appl. Mech. Eng. **388**(1), 114194 (2022). https://doi.org/10.1016/j.cma.2021.114194

35. Zhao, W., Wang, L., Zhang, Z.: Artificial ecosystem-based optimization: a novel nature-inspired meta-heuristic algorithm. Neural Comput. Appl. **32**(13), 9383–9425 (2019). https://doi.org/10.1007/s00521-019-04452-x

36. Zhao, W., Zhang, Z., Wang, L.: Manta ray foraging optimization: an effective bio-inspired optimizer for engineering applications. Eng. Appl. Artif. Intell. **87**, 103300 (2020). https://doi.org/10.1016/j.engappai.2019.103300

Simulation–Based Optimisation Model as an Element of a Digital Twin Concept for Supply Chain Inventory Control

Bożena Mielczarek(✉) ⓘ, Maja Gora, and Anna Dobrowolska ⓘ

Wrocław University of Science and Technology, 50-370 Wrocław, Poland
{bozena.mielczarek,anna.dobrowolska}@pwr.edu.pl

Abstract. Supply chain management is a critical success factor for many manufacturing companies. During the pandemic period, the problem of meeting delivery on time according to customer needs has intensified in many companies around the world. Companies would like to keep inventories at a level that ensures smooth order fulfilment while minimising their own costs. Determining the optimal parameters is, however, a major challenge for Supply Chain inventory policy (SC). Combining simulation methods with optimisation techniques offers a methodology for obtaining an acceptable solution and, at the same time, provides a high degree of flexibility in the formulation of assumptions and the possibility of improving the decision-making process with respect to risk management. In this paper we present a simulation-based optimisation model to improve the quality of inventory management decisions in SC design and planning. Finally, we refer to the benefits of implementing the model in the concept of digital twins.

Keywords: Discrete Simulation · Supply Chain · Digital Twin

1 Introduction

Supply chain (SC) management is the management of the flow of goods and services and includes all processes that transform raw materials into final products delivered to consumers. Supply chains can be external if they include processes for the flow of raw materials from suppliers through manufacturers to the final external consumer. Supply chains can also be internal, in terms of the processes of material flow from supplier to customer within an organisation. One of the key components of internal SC is an inventory management due to the high share of warehouse costs in the total cost of a product, considering the entire supply chain [1]. Companies therefore strive to have inventory levels set at optimal values to ensure smooth order fulfilment and keep costs in an acceptable range. Finding optimal parameters values is, however, severely hampered by the high level of supply and demand uncertainty, the large number of decision variables, various internal and external constraints, and, above all, the high variability observed dynamically in the company itself and its environment. In the search for the optimum values of the decision variables, optimisation methods are particularly

© The Author(s), under exclusive license to Springer Nature Switzerland AG 2023
J. Mikyška et al. (Eds.): ICCS 2023, LNCS 14075, pp. 513–527, 2023.
https://doi.org/10.1007/978-3-031-36024-4_39

helpful. However, its use is limited in cases where the degree of complexity of the problem and computational requirements make it impossible to find a solution using analytical methods. The combination of simulation methods with optimisation techniques makes it possible to obtain results that are impossible to achieve applying each of these approaches separately [2]. Simulation helps to dispense with the restrictive simplifying assumptions in the model and allows the inclusion of any number of parameters to achieve the required level of accuracy in the representation of the system under study. An important advantage of simulation is also its flexibility in mapping any number and various types of uncertainty and randomness.

The industry 4.0 philosophy is making more and more companies focus on increasing automation through the application of the Internet of Things, sensors, advanced communication systems, and the Digital Twin concept (DT) [3]. The use of simulation models in the form of DT provides technology that can lead to better decisions that result in more efficient systems, also with regard to external and internal supply chains.

The purpose of this paper is to contribute to existing work by presenting a simulation-based optimisation model to enhance the quality of inventory management decisions in SC design and planning, taking into account uncertainty and risk analysis. The paper will also propose the concept of using the simulation model as a DT to simulate the near future of SC to predict potential delays and calibrate the ordering and renewal procedures parameters based on risk analysis.

2 Decision Support in Supply Chain Management

In recent years, many manufacturing companies have begun to see supply chain management as a critical success factor. Organisations that are able to deliver products to the customer in the right quality, time, and cost win the competitive battle. The problem of timely delivery according to customer needs has become apparent in a pandemic period [4, 5]. Many of the risks associated with external supply chain conditions are beyond the control of companies, but they can mitigate risks in the operation of the internal supply chain, such as by properly managing the supply process based on the use of modern approaches to material inventory control.

The classic approach to inventory management uses a push approach, in which the flow of materials within an organisation is controlled by demand forecasts [6]. Inventory volumes are constant throughout the time covered by the forecast. The modern approach, the so-called pull approach, assumes that the flow of materials within the organisation is controlled by a signal received from the customer. The level of inventory in the company is held at a minimum and is replenished as demand arrives from the customer [6]. This approach is used in the concepts of Just-in-Time delivery and Lean Management, which have been increasingly used in organisations in recent years.

From the point of view of inventory management, the use of the pull approach compared to the classical approach brings many advantages, including [7] less storage, less use of warehouse space, reduction of material waste, reduction of costs. However, its use can also cause problems in the form of the risk of running out of stock, lack of control over the time frame, and more planning required to take into account risks from the environment. The pull approach requires the collection of information from customers and its processing in real time, and the adoption of a number of organisational solutions in the material flow process itself, as well as the use of a number of tools categorised as Industry 4.0 tools that increase the efficiency of information processing.

In terms of organisational solutions companies must:

- use a set of permanent suppliers based on selection criteria that primarily take into account the quality and reliability (timeliness) of deliveries,
- develop spatial arrangement of workstations and tools (equipment) that optimise flow efficiency,
- introduce automation of logistics processes,
- apply system/methods of process integration inside and outside the company,
- introduce a system for detecting and fixing problems inside the process,
- apply risk assessment in the process flow and from the environment.

The basic tools classified as Industry 4.0 tools that are applicable to inventory management are the following [8]:

- Internet of Things (IoT) [9, 10] that refers to the transmission and exchange between two or more objects different information through the Internet,
- strongly correlated to IoT, the OPC – UA (Open Platform Communications - Unified Architecture),
- Big Data, which helps to link (put in relation) large amounts of heterogeneous data in order to discover links between different phenomena and predict future ones,
- cloud computing identified as the provision of services from a supplier to the end customer through the use of an Internet network.

Simulation methods are widely used in supply chain inventory management to help optimise material flow processes or the spatial distribution of inventory. Discrete event simulation (DES), agent-based simulation, Monte Carlo simulation, and system dynamics (SD) are being applied [11], but of these, the most widely used approach is DES and SD [12]. When comparing the use of DES and SD models in logistics, the clear advantage of DES is seen in modelling supply chain processes. According to [12] DES is primarily used to address issues such as supply chain structure, supply chain integration, replenishment control policy, supply chain optimisation, cost reduction, system performance, inventory planning and management, demand planning and forecasting, production planning and scheduling.

3 Case Study

3.1 Problem Definition

This article deals with the delivery process of five products offered by the company. In addition to these five products, the company supplies customers with almost 200 other types of products, which the company manufactures on its own in Poland by importing materials from various countries. In the case of the five types of products under consideration, due to the high production costs in the country, the company decided to import them directly from the Chinese market. The process starts with the arrival of customer demand (Fig. 1). Once a demand request is received from a customer, the process of completing the accepted request begins. If there are enough products of a given type in the stock, the request is processed immediately. If there is an insufficient inventory level, the request waits for the stock level to be renewed. The inventory level is monitored on an ongoing basis. When the number of products in the stock falls to a certain minimum level (or below), an order is placed with a supplier in China. The production process at the Chinese manufacturer takes about 6 weeks (42 days) regardless of the quantity and type of products ordered. Finished products are imported by sea or air. The expected delivery time by air transport is 2 weeks, while the expected delivery time by sea is about 8 weeks. Due to the long distance and many stochastic factors that affect the timeliness of delivery, the risk of delay is quite significant. Once the order is received, there is a quality control that takes about a week. After this time, the product is ready for sale and is put into the stock. The general structure of the considered supply chain is given in Fig. 1.

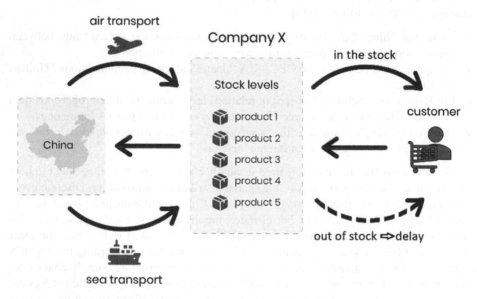

Fig. 1. General structure of the considered supply chain.

3.2 Methods

Discrete event simulation (DES) was used to simulate the supply chain from the moment of the arrival of the customer's request to the moment of the completion of customer service. Arena ® by Rockwell Automation software was used to build the simulation model. The simulation is carried out over a period of one year and starts with the set of initial inventory values. The goal of the simulation is to mimic the sales and ordering process of five types of products imported from China, taking into account the random values of parameters that affect the delivery and the dynamics of the process. One of the key decisions reproduced in the model is the choice of the mode of transportation:

by air or by sea. The decision made has a direct impact on the costs incurred by the company and on the average customer's waiting time. The verified simulation model was then used in a two-stage optimisation, the aim of which was to find such values of input parameters at which the company incurs low order processing costs and at the same time the customers waiting time is acceptable. In the first stage of optimisation, a series of simulation experiments was conducted to identify sets of values of boundary parameters in the vicinity of which a suboptimal solution might be found. The identified sets of input parameters were then entered into the OptQuest tool, which automatically searches for optimal solutions within the Arena simulation model, based on the set of baseline scenarios developed in the first phase of optimisation. The general scheme of simulation-based optimisation is shown in Fig. 2.

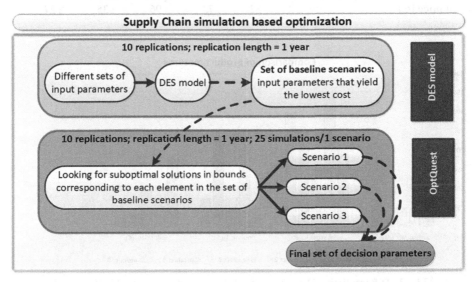

Fig. 2. Conceptual model of simulation-based optimisation.

3.3 Data and Input Parameters

The basic groups of input parameters, which were elaborated on the basis of real data, are shown in Table 1. Each of the five types of products has different characteristics. The most frequent demand requests are received for Product 4 and the fewest for Product 5. Within a single order, customers request the most units of Product 4 (from 1 to 235 with varying probability), while the lowest orders are for Product 1 (from 1 to 25 with varying probability). The price of individual products varies. Thus, the most expensive is Product 1 (€182/piece) and the cheapest is Product 5 (€27/piece). Transportation costs by air and sea are the highest for Product 3, while the lowest for Product 5. Differences between products are visualised in Fig. 3.

Table 1. Values of input parameters.

Parameter	Distribution	Prod.1	Prod.2	Prod.3	Prod.4	Prod.5
Arrival rate [days]	Poisson	67	48	52	22	72
Initial stock [units]		200	350	150	250	300
Demand volume [units]	Discrete	1–25	1–200	1–75	1–235	1–75
Cost of air transport [€/unit]		6	4.05	6.6	4.2	2.25
Cost of sea transport [€/unit]		2	1.35	2.2	1.4	0.75
Unit price [€]		182	78	96	35	27

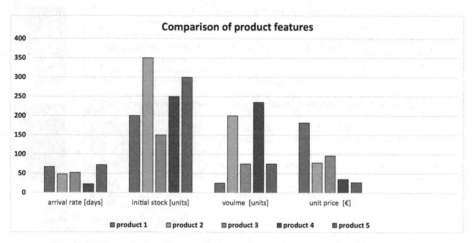

Fig. 3. Differentiating characteristics of the products considered in the model.

3.4 Assumptions

Several assumptions were formulated in the simulation model:

- The cost of transporting one unit of the product does not change and is independent of the number of ordered products of a given type.
- In the case of parallel execution of an order for several types of products, the transportation cost per one unit always remains the same.
- The production time of the products with the Chinese manufacturer does not depend on the number of ordered products.
- For the purpose of the simulation, it was assumed that the delivery time by sea transport is in the range of 8 – 12 weeks, while by air it takes 1.5 – 2.5 weeks.
- Inventory holding costs are not included in the simulation model.
- Lost sales are not included in the simulation model. All customer orders are processed from the time they are placed to delivery.

- The simulation ends after 365 days, and pending orders are not included in the final statistics.

Assumptions relating to the initial conditions of the simulation are also formulated. At the start of the simulation:

- the company is not fulfilling any previous demand requests from customers;
- no transportation is being carried out;
- there is a certain amount of each product in the warehouses.

3.5 Simulation Phase

Course of Simulation. In the simulation model, two types of entities are defined: demand generated by customers and orders placed by the company with a Chinese supplier.

The demand requests from customers arrive according to a Poisson distribution (Table 1) independently for five types of products. If products are in the stock when demand from customers arrives, they are released to customers without undue delay, and the stock is reduced accordingly. If the products are out of the stock (customer demand cannot be met in full), the customer waits for the stock to be renewed.

Stocks are continuously monitored. If the quantity of a product of a given type falls to a minimum level and, at the same time, an order has not been previously placed with the supplier for this product (no delivery is currently "on the way"), an order is placed with the supplier. The delivery process then begins. When the products are delivered to the company, the stock of the warehouse is increased, and demand requests from customers waiting in the queue are immediately fulfilled.

Mode of Transport. During the simulation, customer waiting times are continuously monitored. The decision to choose the mode of transportation (ship or plane) is made by comparing the current waiting time $QueueTime_i$ (i denotes customer, $i = 1, 2, ...$) of the customer who waits the longest in the queue with a critical value of the decision parameter $MaxWaitT$ (Eq. 1). The critical value reflects the company's preference for an acceptable maximum customer waiting time for a product.

$$QueueTime_i \geq MaxWaiT \tag{1}$$

When $QueueTime_i$ is greater than $MaxWaitT$, transport by air is selected, otherwise the sea route is chosen. Parameter $MaxWaitT$ is a decision variable that can be freely defined by the company and reflects customers' expectations regarding the quality of the service. Parameter $QueueTime_i$, on the other hand, is the current reading obtained during simulation. This parameter is continuously monitored and directly affects the choice of transport route.

Total Cost and Maximum Cost. During the simulation, the so-called cost of the frozen capital (in short, *TotalCost*) is calculated on the ongoing basis. This is the amount of money the company has to spend on fulfilling orders from the moment the order is placed with the supplier until the product is sold to the customer. The value of the *TotalCost* parameter changes over time. At the end of the simulation, the average value of this parameter is calculated for each replication (averaging over time). *TotalCost* consists of three components:

- *SCCost_{jk}* (Eq. 2): The total cost of the supply chain is calculated when a decision is made on the type of transport (air or sea) for the order *j* of the product *k*. The value of this element always adds to the current value of *TotalCost*.
- *PPCost_{jk}* (Eq. 3): The product purchase cost is calculated at the time the order *j* for the product *k* is placed with the supplier. It always increases the current value of *TotalCost*.
- *PSRev_{ik}* (Eq. 4): The product sale revenue relates to the sale of the product *k* to the customer *i*. It is calculated when the products are delivered to the customer. The value of this element always reduces the current value of *TotalCost*.

$$SCCost_{jk} = OrderVol_{jk} \cdot TransCost_k \tag{2}$$

$$PPCost_{jk} = OrderVol_{jk} \cdot UnitPrice_k \tag{3}$$

$$PSRev_{ik} = DemandVol_{jk} \cdot UnitPrice_k \tag{4}$$

where *i* denotes *Customer* ($i = 1, 2, ...$), *j* denotes *Order* ($j = 1,2,...$), and *k* denotes product type ($k = 1...5$).

OrderVol_{jk}: number of products of type *k* ordered from the supplier under the order *j*;

TransCost_k: the cost of transporting product *k* by sea or air, respectively (Table 1);

UnitPrice_k: the cost of purchasing one unit of product *k* from the manufacturer;

DemandVol_{ik}: the number of products of type *k* requested by the customer *i*;

In addition, at the end of the simulation the *MaxCost* parameter is also determined (Eq. 5). It informs what the highest amount of money the company had to spend at a certain point during the simulation to fulfil the order with the supplier.

$$MaxCost = max(TotalCost_t)_{t \in T} \tag{5}$$

where *t* denotes the simulation time advancing in steps at specific moments determined by events, from moment zero to the end *T* of replication.

Decision variables. The simulation model operates on 11 decision variables. These are:

- *ReorderPoint*: Minimum stock levels (5 variables) set separately for each product type. It defines the minimum level of the stock in the warehouse, below which the company places an order with the supplier to deliver the next batch of products.
- *OrderVolume*: The size of the order placed with the supplier (5 variables), set separately for each product type.
- *MaxWaitT*: The maximum acceptable waiting time for a customer (one variable).

Output variables. At the end of each simulation experiment, the following output variables are observed: the average cost of frozen capital (*TotalCost*), the maximum value of frozen capital (*MaxCost*), the average customer waiting time, the average number of customers waiting for products, and the number of times there were no products in stock when a new demand request from a customer arrived.

3.6 Verification and Validation

Each simulation experiment consists of 10 replications. The duration of one replication was 365 days. The simulation is run with the preset initial stock levels and with a one-month warm-up period. Key experts from the Company actively participated in the process of formulating assumptions for the model. The model was subjected to extensive verification and validation. As part of the verification, tests were conducted to verify internal consistency. These were visualisation tests, degeneration tests, extreme conditions tests, and many others. For validation purposes, two parameters were compared: the number of completed deliveries and the total volume of all deliveries. The results, shown in Table 2, confirm the credibility of the model.

Table 2. Validation of the simulation model (10 replications).

Parameter	Real system	Simulation	Half-width
Number of orders	19	20.1	6.9
Total volume ordered	5030	5106	1525.4
Product 1	270	206	
Product 2	1155	1250	
Product 3	675	850	
Product 4	2550	2500	
Product 5	380	300	

4   Results

4.1   Simulation-Based Optimisation Scenarios

Due to the large number of decision variables (11 variables; see Sect. 3.5), the optimisation was carried out in two phases. In the first phase, a series of simulation experiments were performed to determine potential sets of input parameters for further analysis using the OptQuest optimiser (Table 3). The objective function is to minimise the average value of the cost of the frozen capital (*TotalCost*). Three additional parameters are also taken into account: *MaxCost* (see Sect. 3.5), average customer waiting time *QueueTime* and average queue length of waiting customers *QueueLength*. In all experiments, the values of 10 decision variables (*ReorderPoint* and *OrderVolume*) are fixed, while the optimisation algorithm selects the value of the *MaxWaitT* variable moving with a step of 1, from the starting value of 1 to 70. Each experiment was performed with the same simulation parameters as in the base scenario. In Scenario 1, the baseline values of 10 decision variables are established. In Scenario 2, an equally low *ReorderPoint* for all products was introduced and, at the same time, *OrderVolume* values were raised slightly. In Scenario 3, the values of the 10 decision variables were set taking into account the characteristics of the products (see Table 1).

Table 3. Simulation–based optimisation scenarios.

Scenario		Decision variables		
		ReorderPoint [units]	*OrderVolume* [units]	*MaxWaitT* [days]
Sc1	Prod.1	15	100	1 – 70
	Prod.2	200	350	
	Prod.3	100	250	
	Prod.4	400	150	
	Prod.5	10	200	
Sc2	Prod.1	50	200	1 – 70
	Prod.2	50	400	
	Prod.3	50	150	
	Prod.4	50	400	
	Prod.5	50	300	
Sc3	Prod.1	100	200	1 – 70
	Prod.2	150	600	
	Prod.3	150	200	
	Prod.4	300	600	
	Prod.5	100	350	

4.2 Discussion

The optimisation results are shown in the Table 4.

The best results were obtained in Scenario 1. Although Scenario 2 provides a lower value of the objective function and a lower value of *MaxCost*, this is at the expense of high values of the parameters that describe the queue of waiting customers. This Scenario turns out to be immune to changes in the *MaxWaitT* parameter. The values of the variables relating to the cost, as well as the parameters describing the queues, change very little as the *MaxWaitT* parameter increases. Only setting *MaxWaitT* to 20 causes a noticeable change in values of the observed parameters. Scenario 3, on the other hand, turned out to be unfavourable in terms of all parameters. Figures 4 and 5 show the optimisation results for all *MaxWaitT* values obtained in the best Scenario 1.

Table 4. Scenario results: only the selection of results of the *MaxWaitT* parameter is displayed.

Scenario	*MaxWaitT*	Average *TotalCost*	*MaxCost*	Average Queue Waiting Time	Average Queue Length
Sc1	1	116.9	216.7	16.4	56.1
	30	112.5	215.3	19.4	76.9
	50	108.6	214.8	23.7	89.2
	70	104.5	215.4	32.7	116.4
Sc2	1	101.7	132.9	40.8	126.7
	10	101.7	132.9	40.9	126.8
	20	102.0	132.2	41.3	129.1
	30	101.9	131.5	41.3	129.1
Sc3	1	147.1	190.0	28.1	61.6
	30	143.8	185.2	37.2	80.7

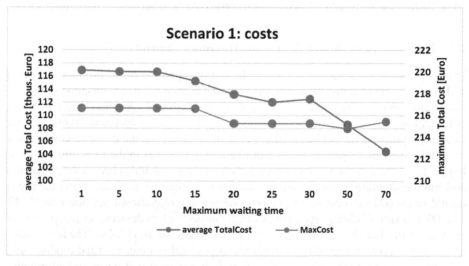

Fig. 4. Results of simulation-based optimisation according to Scenario 1: cost parameters.

The objective function reaches the lowest level for high values of *MaxWaitT* parameter. Analysis of the values of parameters that describe the queue allows us to select the *MaxWaitT* parameter = 50 days. At this value of the *MaxWaitT* variable, both cost parameters are low, while the parameters that describe the queue increase only slightly.

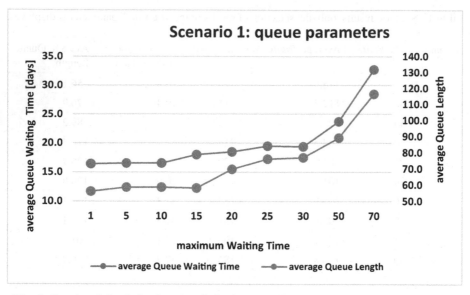

Fig. 5. Results of simulation-based optimisation according to Scenario 1: queue parameters.

4.3 The Concept of Digital Twin in SC Inventory Control

In the case study presented in the paper, the physical process of SC inventory planning and control is represented by a simulation model supported by optimisation tools. In a stable environment, the results of the applied approach that combines simulation with optimization, help to choose a solution that keeps costs at an acceptable level and at the same time provides the desired quality of services. However, the dynamic environment in which companies operate means that planned procedures in the near future may no longer be optimal. It is possible to use a cyclical (e.g. monthly) review of parameters and run the simulation again, but more and more companies are considering the use of digital twins (DT) to optimise warehouse design and operational performance [5, 13, 14]. DT is a virtual/ digital replica of physical entities such as devices, people, processes, or systems that help businesses make model-driven decisions, [13, 14]. The key concept of DT is the coexistence of three elements: a physical system, a virtual model, and a two-way connectivity that allows data to flow from real to virtual space, and information flow from virtual space to real system.

In the case study presented above, once DT is integrated with the physical system, real-system information on customer requests can be fed into the simulation model on an ongoing basis. The optimisation module finds the optimal solution with changed values of the input parameters, and new decisions and adjustments are transferred back to the physical system. As a result, new values of decision variables are determined, which are immediately implemented in SC inventory management procedures.

A simulation experiment was performed to illustrate the intercommunication between the real system and DT (see Figs. 6 and 7). After performing the optimisation described in Sect. 4.1, it was decided that the *MaxWaitT* decision variable should be 50 days. However, it was noticed that there was a slight increase in the arrival rate for Product 3, and the re-estimation of the inter-arrival rate for this product was performed. It turned out that after adjusting the value of the Arrival Rate parameter from 52 to 48 days, the value of the *MaxWaitT* decision variable should also change. Figures 6 and 7 compare the results of the two simulation-based optimisations. The values of the objective function in both optimisations change very similarly; that is, the lowest values of the *TotalCost* parameter are obtained at high values of *MaxWaitT*. However, the analysis of the parameters that describe the queue shows a deterioration in the customer waiting time and the length of the queue when the value of the parameter *MaxWaitT* is set to 50. A better choice would be to keep the *MaxWaitT* value less than 50 days.

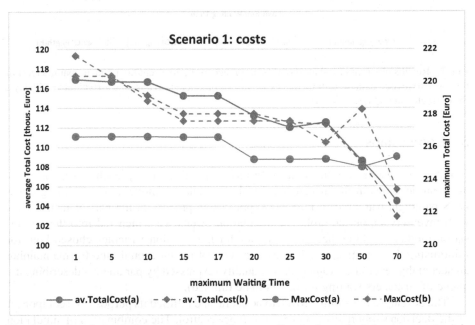

Fig. 6. Results of simulation-based optimisation according to Scenario 1 (cost parameters) in response to revised data describing the arrival rate of product 3; (a) baseline arrival rate = 52 days, (b) corrected arrival rate = 48 days.

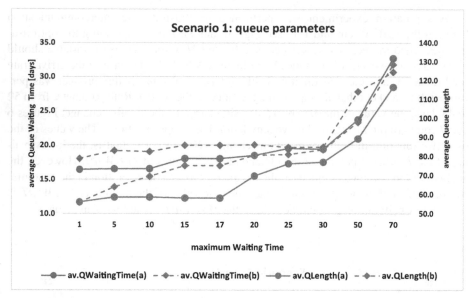

Fig. 7. Results of simulation-based optimisation according to Scenario 1 (queue parameters) in response to revised data describing the arrival rate of product 3; (a) baseline arrival rate = 52 days, (b) corrected arrival rate = 48 days.

5 Conclusions

In this paper, we discuss the simulation-based optimisation approach to support management decisions in SC inventory control. A simulation model built according to the DES paradigm was used to map the supply chain process in the part of it that deals with inventory supply control. The simulation results were then fed into the optimisation module to obtain acceptable values for the decision variables chosen both for minimising the objective function (average cost of frozen capital) and for maintaining an acceptable level of customer service quality (expressed by parameters describing the queue of customers waiting for orders to be fulfilled).

The developed algorithm proved to be useful and fulfilled its purpose as a component of the decision support system in SC inventory control. The combination of simulation and optimization can significantly improve the quality of SC managers' decisions, especially when the complexity of the problem, the need to consider many variables, and the accumulation of factors of a random nature make it impossible to obtain an acceptable solution by analytical methods. In addition, the use of a two-phase algorithm, discussed in the article in par. 4.1, reduces the time of performing calculations by preliminarily indicating the ranges for potential optimization. The duration of one full simulation run is less than a minute. Searching for the optimal solution within a predefined range therefore significantly reduces the duration of the entire study compared to full optimization process. From the company's point of view, any set of parameters that meets predefined quality requirements is satisfactory. Nevertheless, further research work will be aimed at an even more precise selection, through simulation experiments, of decision parameter ranges for further optimization.

The article also discusses the concept of using a simulation-based optimisation model as an element of DT and thus obtaining a significant impact on the efficiency of the supply chain.

References

1. Göçken, M., Boru, A., Dosdoğru, A.T., Geyik, F.: (R, s, S) inventory control policy and supplier selection in a two-echelon supply chain: An Optimization via Simulation approach. In: 2015 Winter Simulation Conference (WSC), pp. 2057–2067 (2015)
2. Ivanov, D., Dolgui, A., Sokolov, B., Ivanova, M.: Integrated simulation-optimization modeling framework of resilient design and planning of supply chain networks. IFAC-PapersOnLine. **55**, 2713–2718 (2022)
3. Kuhl, M.E., Bhisti, R., Bhattathiri, S.S., Li, M.P.: Warehouse Digital Twin: Simulation Modeling and Analysis Techniques. In: 2022 Winter Simulation Conference (WSC), pp. 2947–2956 (2022)
4. Swanson, D., Santamaria, L.: Pandemic supply chain research: a structured literature review and bibliometric network analysis. Logistics. **5**, 7 (2021)
5. Shen, Z., Sun, Y.: Strengthening supply chain resilience during COVID -19: A case study of JD .com. J. Oper. Manag. Oct18, 1–25 (2021)
6. Spearman, M., Zazanis, M.: Push and pull production systems: issues and comparisons. Oper. Res. **40**, 521–532 (1992)
7. Wee, H.-M., Peng, S.-Y., Yang, C.-C., Wee, P.K.: The Influence of Production Management Practices and Systems on Business Performance: From the Perspective of the Push-pull Production Systems. Oper. Supply Chain Manag. An Int. J. **2**(1) 11–23 (2019)
8. Gallo, T., Cagnetti, C., Silvestri, C., Ruggieri, A.: Industry 4.0 tools in lean production: A systematic literature review. Procedia Comput. Sci. **180**, 394–403 (2021)
9. Xu, F.: The Study of Just-in-time Inventory Management Based on the Perspective of the Internet of Things. In: The Annual International Conference on Social Science and Contemporary Humanity Development (SSCHD 2021). 238–243 (2021)
10. Xu, Y., Chen, M.: Improving just-in-time manufacturing operations by using internet of things based solutions. Procedia CIRP. **56**, 326–331 (2016)
11. Oliveira, J.B., Jin, M., Lima, R.S., Kobza, J.E., Montevechi, J.A.B.: The role of simulation and optimization methods in supply chain risk management: performance and review standpoints. Simul. Model. Pract. Theory. **92**, 17–44 (2019)
12. Tako, A.A., Robinson, S.: Comparing discrete-event simulation and system dynamics: users' perceptions. J. Oper. Res. Soc. **60**, 296–312 (2009)
13. Jiang, Y., Yin, S., Li, K., Luo, H., Kaynak, O.: Industrial applications of digital twins. Philos. Trans. R. Soc. A Math. Phys. Eng. Sci. **379**, 20200360 (2021)
14. Campos, J., López, J., Armesto, J., Seoane, A.: Automatic generation of digital twin industrial system from a high level specification. Proc. Manuf. **38**, 1095–1102 (2019)

Semi-supervised Learning Approach to Efficient Cut Selection in the Branch-and-Cut Framework

Jia He Sun[(✉)] and Salimur Choudhury

Queen's University, Kingston, ON, Canada
{20jhhs,s.choudhury}@queensu.ca

Abstract. Mixed integer programming (MIP) is an extremely versatile subclass of mathematical optimization problems. Applications of MIP are ubiquitous in our world today, ranging from scheduling to network design to production planning. The standard approach in state-of-the-art commercial solvers is called branch-and-cut. The selection of these cuts is an integral part of the branch-and-cut process as high-quality cuts can greatly increase solving efficiency. Currently, cut selection is decided by heuristics that both require expert knowledge and lack generalizability. In this paper, we propose an efficient and highly generalizable cut selection scheme based on semi-supervised learning. First, we design a cut evaluation metric that labels cuts based on whether they are efficient or not. Then, we train a deep learning classification model with unsupervised pre-training as a ranking function for cuts. In our evaluation, the proposed model outperforms standard heuristics and is comparable to existing machine learning approaches. Furthermore, the model is shown to be generalizable over both problem size and problem class.

Keywords: Machine learning · Semi-supervised learning · Mixed integer programming · Cutting planes

1 Introduction

MIP problems are linear programming (LP) problems with integrality constraints. This particular subclass of optimization problems can be applied to a plethora of industrial applications including but not limited to: scheduling [7], network design [8], and production planning [9]. Modern commercial MIP solvers take the branch-and-cut approach which is a combination of the branch-and-bound technique and the cutting planes technique [18]. The selection of solution variables for the branching and the selection of cuts are key decisions with a huge impact on the overall efficiency of the branch-and-cut algorithm [18]. Recently, there has been a surge in interest from the ML community in augmenting the branching process of the branch-and-cut framework [3–6,10].

Supported by NSERC.

© The Author(s), under exclusive license to Springer Nature Switzerland AG 2023
J. Mikyška et al. (Eds.): ICCS 2023, LNCS 14075, pp. 528–535, 2023.
https://doi.org/10.1007/978-3-031-36024-4_40

The cut selection process, the focus of this paper, has seen less focus from researchers aiming to integrate machine learning into the branch-and-cut framework. Tang et al. proposed a deep reinforcement learning (RL) formulation for intelligent adaptive cut selection for the cutting planes method [2]. Paulus et al. proposed an imitation-based learning model called "NeuralCut" based on a lookahead expert [13]. While these two works are in the same domain as our work, they are different in terms of target evaluation.

Huang et al. designed a multiple instance supervised machine learning model for cut selection in the branch-and-cut framework called "Cut Ranking" [1]. Huang et al.'s work is the most similar to the work proposed in this paper. However, not only do we propose a different labelling system for generating labelled cut data, but we also take a semi-supervised approach to the machine learning model as opposed to "Cut Ranking"'s completely supervised approach.

In this paper, we propose a generalizable and efficient cut selection scheme for the branch-and-cut framework. This selection scheme includes a cut classification system that differentiates efficient cuts from inefficient cuts and a semi-supervised machine learning model that learns to do the same.

2   Methods

For every MIP problem, and for each node of the branch-and-bound search tree, existing MIP solvers can generate a set of candidate cuts. The goal of our model is to select the most efficient cuts from this candidate set. In our work, the approach taken is multiple instance learning (MIL).

MIL is where the training data is generated based on bags of instances. This approach is chosen because individual cuts will have little measurable effect on the overall efficiency of the branch-and-cut framework, thus, cuts are grouped into bags and are evaluated at the bag level. Then, labels are assigned at the bag level. "Cut Ranking" takes the same approach for data generation [1].

To construct the bags of instances, consider a MIP problem P of the form:

$$max\{c^T x : Ax \leq b, x_j \in \mathbb{Z}, \forall j \in N_I\} \tag{1}$$

where $c, x \in \mathbb{R}^n, A \in \mathbb{R}^{m \times n}$, and $N_I \subseteq N = \{1, ..., n\}$. Let x_{LP} be an optimal solution to P's corresponding LP relaxation and let C be the candidate cut set generated by a solver. For each cut $c_i \in C$, it is of the form:

$$\alpha_i^T x \leq \beta_i \tag{2}$$

Let $f_{c_i} \in \mathbb{R}^l$ denote the feature vector of c_i. Let $B = \{B_1, ..., B_k\} \subseteq C$ be all bags of cuts sampled from C. Then, the feature vector of a bag B_u, denoted by f_{B_u}, is the average of the feature vectors f_{c_i} for all $c_i \in B_u$. That is, the feature vector of a bag of cuts is the average of the feature vectors of the cuts in the bag. Furthermore, $|B_u| \geq 0.1 \cdot |C|, \forall j \in \{1, ..., k\}$. In other words, the size of each sampled bag of cuts must be at least 10% of the size of the candidate cut set. This is to ensure that we do not have samples with not enough cuts to make a measurable difference in run time.

For each cut c_i, the features extracted are as follows:

1. cut coefficients features (4): maximum, minimum, mean, and standard deviation of cut coefficients α_i
2. objective function coefficients features (4): maximum, minimum, mean, and standard deviation of objective function coefficients that correspond to the non-zero cut coefficients
3. support
4. integral support
5. relative violation
6. distance
7. objective function parallelism
8. expected improvement

The first four features are basic structural data of the cut. The rest of the features are popular metrics for measuring the quality of cuts. The exact definitions of these features can be found in Wesselmann and Suhl's work [11].

2.1   Cut Evaluation and Data Labelling

For each MIP problem P, after we have sampled k bags from the generated candidate cut set C, every sampled bag B_u is evaluated by adding all cuts in B_u to P and running the solver. To evaluate the performance of each bag, the metric used in our scheme is normalized run time.

Let r_j be the run time of problem P with appended bag B_j for all $j \in \{1, ..., k\}$. Without loss of generality, assume B_v to be the bag with shortest run time and B_w be the bag with the longest run time. Then, the evaluation value assigned to each sampled bag B_u, normalized run time, is defined by:

$$r_j^* = 1 - \frac{r_j - r_m}{r_n - r_m} \tag{3}$$

In this format, for each MIP problem, the best performing bag will always be evaluated as 1 and the worst performing bag will always be evaluated as 0. The rest of the bags will have an evaluation of some value in $[0, 1]$.

After each bag of cuts has been evaluated, it will be given a discrete label. In our scheme, we will assign 1 to bags with normalized run time over λ_1 and assign 0 to bags with normalized run time under λ_2. λ_1 and λ_2 are both hyperparameters between 0 and 1 and $\lambda_1 > \lambda_2$. All other bags will not be labelled and consequently will not be used in the supervised training portion of the model.

Furthermore, notice that we choose to allow some data points to remain unlabelled. Intuitively, we are labelling the samples we know are good as 1 and the samples we know are bad as 0. The samples in the middle that could be good or could be bad are not labelled.

2.2 Unsupervised Pre-training

The proposed machine learning model is a semi-supervised deep learning model for tabular data which employs an unsupervised pre-training model. The reason that unsupervised pre-training is chosen is that we have an abundance of unlabelled data at our disposal. We implement the pre-training model used in TabNet, a deep tabular data learning model [17]. TabNet's pre-training model, similar to a denoising auto-encoder, is designed to predict missing feature values from corrupted feature input based on observed interdependencies.

2.3 Model Architecture

Table 1. Diagrams regarding the architecture and cut selection scheme

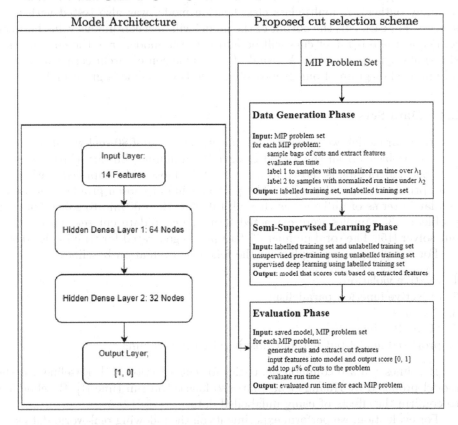

Following pre-training, the data will be pushed through a supervised classification model. The model consists of 4 fully connected layers with an input layer, an output layer, and 2 hidden dense layers of size 64 and 32 respectively. Since

the proposed model is a binary classification model we chose to use binary cross entropy as our loss function:

$$-\frac{1}{N}\sum_{i=1}^{N} y\log(p) + (1-y)\log(1-p) \qquad (4)$$

where N is output size, y is target value, and p is model output. For the same reasoning, we also choose to use sigmoid as our activation function:

$$\sigma(z) = \frac{1}{1+e^{-x}} \qquad (5)$$

A regression model was not chosen because, in our internal experiments, the binary classification model consistently outperformed it. We hypothesize that this is due to the relatively small amount of labelled data and the existence of many outliers. A multi-class classification model was also tested with little success. For each cut, the output of the model will be a continuous value between $[0, 1]$ and the top $\tau\%$ of cuts will be added to the model, τ is the cut selection threshold hyperparameter. A visual of our implemented architecture as well as a high-level diagram of our proposed cut selection scheme is given in Table 1.

2.4 Data Sets

To train our model, we procured a data set consisting of 80 real-world set partitioning problems from the Mixed Integer Programming Library (MIPLIB2017) [12]. Set partitioning is chosen as it is one of the most widely applied mathematical optimization problems [19]. The problems in our chosen problem set are all similar in terms of difficulty as they all take less than 30 min to solve. For each problem, 135 samples were extracted from the candidate cut set generated by our solver [15]. The total number of data points generated for training is 10,602.

For comparison, we implement the following evaluation baselines:

1. relative violation
2. objective function parallelism
3. distance
4. "Cut Ranking" [1]
5. proposed model but without unsupervised pre-training

Baselines 1–3 are common heuristics for cut selection [11]. Baseline 4 is the model proposed by Huang et al. that also focuses on run time [1]. Baseline 5 is to confirm the effects of using unlabelled data.

For evaluation, we perform experiments on the following real-world data sets:

1. 50 small set partitioning problems (different problems than the ones used in training) [12]
2. 50 large set partitioning problems [12]
3. 50 mixed integer knapsack problems [14]

4. 50 lot sizing problems [16]
5. 50 general MIP problems [12]

Data sets 1–2 are used to evaluate the performance of the model on similar problems that it was trained on as well as how well it generalizes in terms of problem size. Data sets 3–5 are used to evaluate the model's generalizability on different problem classes.

2.5 Hyperparameters

After tuning, the hyperparameters for data generation are $\lambda_1 = 0.7$ and $\lambda_2 = 0.45$. That is, cut samples with normalized run time over 0.7 are labelled 1, and cut samples with normalized run times under 0.45 are labelled 0. With these hyperparameters, the labelled samples total 5,020, and the unlabelled samples total 5,582. The cut selection threshold hyperparameter τ is set to be 0.7, that is, the top 30% of cuts are added to the model.

For model specific hyperparameters, dropout is set to be 0.01, the learning rate is set to be 0.0001, batch size is set to be 32, unsupervised pre-training is set for 512 epochs, and supervised training is also set for 512 epochs.

2.6 Implementation

The MIP solver used is the Coin-or Cut-and-Branch Solver [15]. The training data used in this study are openly available in Mendeley Data at DOI: 10.17632/thtz8h894m.1. The model implemented is based on Arık and Pfister's proposed model and is available here [17].

3 Results

In our experimentation, the cut selection scheme is implemented only at the root node of the search tree. In other words, cuts are only being added to the original MIP problem. Since each node of the branching search tree can be considered its own MIP problem, we believe that the results of our experimentation extend to all nodes of the branching search tree.

Table 2 displays the evaluation results of our proposed model compared against the selected baselines. First and foremost, the proposed model significantly outperforms all evaluated baselines on the data set it was trained on, set partitioning (small), achieving an average normalized run time of almost 50% better than the next highest baseline.

As for the other data sets, our proposed model is at worst comparable to "Cut Ranking" and the other heuristics. For the set partitioning (large) and the lot sizing data sets, our model outperforms all of the baselines by a comfortable margin. For mixed integer knapsack problems, the proposed model performs barely worse than "Cut Ranking". For the general MIP data set, our model is less efficient than "Cut Ranking" by a slight margin but outperforms the other baselines.

Table 2. Average normalized run time evaluation between proposed model, "Cut Ranking", and various heuristics (higher is better)

Problem Set	Proposed Model	Cut Ranking	Relative Violation	Distance	Parallelism	Without Pre-training
Set Partitioning (small)	**0.764**	0.515	0.476	0.372	0.351	0.676
Set Partitioning (large)	**0.749**	0.621	0.579	0.475	0.397	0.612
Mixed Integer Knapsack	0.991	**0.998**	0.974	0.762	0.764	0.975
Lot Sizing	**0.795**	0.698	0.667	0.197	0.740	0.596
General MIP	0.695	**0.729**	0.562	0.424	0.392	0.612

Lastly, we can see that our proposed model consistently outperformed its no pre-training counterpart. This confirms that the unsupervised pre-training portion of our model is indeed beneficial.

4   Conclusion

The backbone of modern state-of-the-art MIP solvers is the branch-and-cut framework. The selection of cutting planes to be implemented at each node of the branching search tree is an important task and, to tackle this, we proposed a semi-supervised deep learning based cut selection scheme. An unsupervised pre-training model is trained to reconstruct features based on inter-feature dependencies using unlabelled data. Then, the labelled data are trained using a standard binary classification approach. From our experiments on real-world MIP problem sets, we found that our model is not only comparable to current state-of-the-art approaches but also generalizable over both problem size and problem class.

Currently, both the branching process and the cutting process have received attention from the machine learning community. However, they are often considered completely separate. It can be interesting and fruitful to study the dependencies between variable/node selection in the branching process and cut selection in the cutting process.

References

1. Huang, Z., et al.: Learning to select cuts for efficient mixed-integer programming. Pattern Recogn. **123**, 108353 (2022)
2. Tang, Y., Agrawal, S., Faenza, Y.: Reinforcement learning for integer programming: Learning to cut. In: International Conference On Machine Learning, pp. 9367–9376. PMLR (2020)
3. Khalil, E.B., Dilkina, B., Nemhauser, G.L., Ahmed, S., Shao, Y.: Learning to Run Heuristics in Tree Search. In: Ijcai, pp. 659–666 (2017)

4. He, H., Daume III, H., Eisner, J.M.: Learning to search in branch and bound algorithms. In: Advances in Neural Information Processing Systems, vol. 27 (2014)
5. Balcan, M.F., Dick, T., Sandholm, T., Vitercik, E.: Learning to branch. In: International Conference On Machine Learning, pp. 344–353. PMLR (2018)
6. Khalil, E., Le Bodic, P., Song, L., Nemhauser, G., Dilkina, B.: Learning to branch in mixed integer programming. In: Proceedings of the AAAI Conference on Artificial Intelligence, Vol. 30, No. 1 (2016)
7. Pan, C.H.: A study of integer programming formulations for scheduling problems. Int. J. Syst. Sci. **28**(1), 33–41 (1997)
8. Guihaire, V., Hao, J.K.: Transit network design and scheduling: a global review. Transp. Res. Part A: Policy Pract. **42**(10), 1251–1273 (2008)
9. Díaz-Madroñero, M., Mula, J., Peidro, D.: A review of discrete-time optimization models for tactical production planning. Int. J. Prod. Res. **52**(17), 5171–5205 (2014)
10. Huang, L., et al.: Branch and bound in mixed integer linear programming problems: A survey of techniques and trends (2021). arXiv preprint arXiv:2111.06257
11. Wesselmann, F., Stuhl, U.: Implementing cutting plane management and selection techniques. In: Technical Report. University of Paderborn (2012)
12. Gleixner, A., et al.: MIPLIB 2017: data-driven compilation of the 6th mixed-integer programming library. Math. Program. Comput. **13**(3), 443–490 (2021)
13. Paulus, M.B., Zarpellon, G., Krause, A., Charlin, L., Maddison, C.: Learning to cut by looking ahead: Cutting plane selection via imitation learning. In: International Conference on Machine Learning, pp. 17584–17600 (2022). PMLR
14. Atamtürk, A.: On the facets of the mixed-integer knapsack polyhedron. Math. Program. **98**(1–3), 145–175 (2003)
15. Forrest, J. coin-or/Cbc: Release releases/2.10.8. Zenodo. https://doi.org/10.5281/zenodo.6522795 (2022)
16. Atamtürk, A., Munoz, J.C.: A study of the lot-sizing polytope. Math. Program. **99**(3), 443–465 (2004)
17. Arik, S.Ö., Pfister, T.: Tabnet: Attentive interpretable tabular learning. In: Proceedings of the AAAI Conference on Artificial Intelligence, Vol. 35, No. 8, pp. 6679–6687 (2021)
18. Mitchell, J.E.: Branch-and-cut algorithms for combinatorial optimization problems. Handbook Appl. Optimization **1**(1), 65–77 (2002)
19. Diaby, M.: Linear programming formulation of the set partitioning problem. Int. J. Oper. Res. **8**(4), 399–427 (2010)

Efficient Uncertainty Quantification Using Sequential Sampling-Based Neural Networks

Pavankumar Koratikere[1]([✉]), Leifur Leifsson[1] [iD], Slawomir Koziel[2,3] [iD], and Anna Pietrenko-Dabrowska[3] [iD]

[1] School of Aeronautics and Astronautics, Purdue University, West Lafayette, IN 47907, USA
{pkoratik,leifur}@purdue.edu
[2] Engineering Optimization and Modeling Center, Department of Engineering, Reykjavík University, Menntavegur 1, 102 Reykjavík, Iceland
koziel@ru.is
[3] Faculty of Electronics Telecommunications and Informatics, Gdansk University of Technology, Narutowicza 11/12, 80-233 Gdansk, Poland
anna.dabrowska@pg.edu.pl

Abstract. Uncertainty quantification (UQ) of an engineered system involves the identification of uncertainties, modeling of the uncertainties, and the forward propagation of the uncertainties through a system analysis model. In this work, a novel surrogate-based forward propagation algorithm for UQ is proposed. The proposed algorithm is a new and unique extension of the recent efficient global optimization using neural network (NN)-based prediction and uncertainty (EGONN) algorithm which was created for optimization. The proposed extended algorithm is specifically created for UQ and is called uqEGONN. The uqEGONN algorithm sequentially and simultaneously samples two NNs, one for the prediction of a nonlinear function and the other for the prediction uncertainty. The uqEGONN algorithm terminates based on the absolute relative changes in the summary statistics based on Monte Carlo simulations (MCS), or a given maximum number of sequential samples. The algorithm is demonstrated on the UQ of the Ishigami function. The results show that the proposed algorithm yields comparable results as MCS on the true function and those results are more accurate than the results obtained using space-filling Latin hypercube sampling to train the NNs.

Keywords: Uncertainty quantification · Monte Carlo simulation · efficient global optimization · neural networks · sequential sampling

1 Introduction

Uncertainty is ubiquitous in engineering design. For example, manufacturing processes create deviations from specifications, the system operating and loading conditions may vary, and some parameters are just inherently variable. Furthermore, the engineering models used in the design can be over simplified which

© The Author(s), under exclusive license to Springer Nature Switzerland AG 2023
J. Mikyška et al. (Eds.): ICCS 2023, LNCS 14075, pp. 536–547, 2023.
https://doi.org/10.1007/978-3-031-36024-4_41

introduces uncertainties. All of these can have a direct impact on the engineered system and its performance. Quantifying the impacts of uncertainties on the system is therefore an important part of the engineering design process.

Uncertainty quantification (UQ) of an engineering system can be divided into three major steps [25]. The first step involves identifying the types of uncertainties existing in the problem. For example, are there uncertainties in the system parameters, the design variables, or in the engineering model itself. The second step involves modeling the input parameter uncertainties. In particular, the uncertainties need to be defined in terms of the parameters and variables mathematically, for example using probability theory, and with respect to the quantities of interest (QoI), for example using the mean and standard deviation. The third step involves sampling the input uncertainties and propagating through the system model to yield the output probability distribution of the QoI and computing the associated statistics. This step often involves many evaluations of the QoI to obtain converged values of the statistics.

This work is focused on the third step in the UQ process described above, i.e., the forward propagation of the uncertainty through the system analysis, when dealing with computer simulations of the system. In particular, the goal of this work is to efficiently propagate uncertainties through simulations non-intrusively. In this work, it is assumed the system under consideration has a large number of uncertain parameters and each evaluation of system using the simulation requires the numerical solution of partial differential equations (PDEs). An example of such a system is the aerodynamic analysis of the flow past a civil transport aircraft at transonic speeds requiring a computational fluids dynamics (CFD) simulation which can take on the order of 24 h on a high-performance computing (HPC) system [10,17]. The key challenges of simulation-based UQ are (1) a large number of uncertain parameters, (2) time-consuming simulations, and (3) many model evaluations.

UQ is a large and active field of research. Methods for forward propagation of uncertainties are many. The four major classes of nonintrusive methods for forward propagation are perturbation methods [24], direct quadrature [2,18], polynomial chaos [12,19], and Monte Carlo simulation (MCS) [7,15]. Perturbation methods use a local Taylor series expansion of the functional output. These methods are limited to local modeling and need at least first-order derivative information. Direct quadrature uses numerical quadrature to evaluate the statistics. This method is limited to low-dimensional problems, although sparse grids enable partially alleviate this issue. Polynomial chaos represent uncertain parameters as a sum of orthogonal basis functions and can yield the statistics and the output distributions. This method is, however, limited to small number of dimensions. Monte Carlo methods approximate the statistics and output distributions using random sampling. These methods are easy to use and are independent of the problem dimension. However, a major weakness is their inefficiency, i.e., many samples are required to obtain converged values of the statistics.

In surrogate methods, the time-consuming simulations are replaced in the heavy computations of the QoIs with an approximation model, called a surrogate model or simply a surrogate, which is fast to evaluate [20]. This way the computational cost is shifted over to the creation of a surrogate model that can represent the true simulated response as a function of the parameters. In the context of simulation-based UQ, the surrogate needs to represent the uncertain output response of the simulation model in terms of the uncertain input parameter space. If that is possible, then the summary statistics and output distribution can be estimated using the aforementioned forward propagation methods.

Kriging is a widely used surrogate modeling method capable of approximating nonlinear responses [3]. An advantage of using kriging prediction is that it comes with its own prediction uncertainty. This enables the sequential (adaptive) sampling of the parameters space to enhance the kriging prediction surrogate. A widely used approach for sequential sampling is the efficient global optimization (EGO) algorithm [9]. A weakness of kriging is that it is limited to small data sets. Deep neural networks (DNNs) [6], on the other hand, scale more efficiently for large data sets [16,21] and while still being handle nonlinear responses. A major limitation, however, is that uncertainty estimates are not readily available for a single prediction [16], and it is necessary to make use of an ensemble of NNs with a range of predictions [5,14,23] or use dropout to represent model uncertainty [4] and these algorithms are computationally very intensive.

A recently created EGO algorithm with neural network (NN)-based prediction and uncertainty (called EGONN) partially alleviates some of these challenges [13]. The EGONN algorithm was created for unconstrained global optimization problems. In the algorithm, a NN model approximates a nonlinear high-dimensional objective function with initial samples and then proceeds to sequentially sample the design space and continuously update the NN-based prediction based on a predetermined computational budget or a termination condition is fulfilled. The update is based on an infill sample point determined by the prediction uncertainty of the NN model. In EGONN, a prediction uncertainty model is constructed using another NN that is trained on separate data set based on the current prediction NN model and sample from the true function. The prediction NN and the prediction uncertainty NN are used to maximize the expected improvement infill criterion to determine the next sample point.

In this paper, an extension to the EGONN algorithm to handle UQ problems is proposed. The goal is create an accurate global surrogate model of the true function that can be sampled quickly using MCS to compute the desired statistical information for the purpose of UQ. The proposed extension to EGONN is to model the spatial error of the prediction NN in order to construct the prediction uncertainty model. By maximizing the prediction error NN model, a new sample point is determined and appended to the current data set for training the prediction NN model. Since the prediction is being updated in each sequential sampling cycle it allows for termination of the UQ process based on convergence of the summary statistics. This is another new and unique feature of the proposed algorithm. This is achieved in the following way. In each iteration of the sequential sampling the current prediction NN is used to perform a MCS to yield the summary statistics and the change in the predicted mean and standard deviation with respect to the previous iteration is calculated. The algorithm terminates if the change in the statistics is below a pre-specified tolerance or a pre-specified maximum number of function evaluations has been reached. The proposed EGONN for UQ (uqEGONN) algorithm is demonstrated on two analytical test functions and compared against MCS of the true function.

The next section describes the proposed uqEGONN algorithm. The following section presents the numerical results of applying the proposed algorithm to an analytical test functions. Concluding remarks and possible next steps in this work are presented in the last section.

2 Methods

A surrogate-based forward propagation approach for simulation-based UQ is proposed. The proposed uqEGONN algorithm is given in Algorithm 1. Initially, two separate data sets, $(\mathbf{X}, \mathbf{Y})$ and $(\mathbf{X}, \mathbf{Y})_u$, are generated using design of experiments, such as space-filling Latin hypercube sampling (LHS). $(\mathbf{X}, \mathbf{Y})$ is for training a prediction model NN_y and $(\mathbf{X}, \mathbf{Y})_u$ is for training a prediction uncertainty model NN_u. The next steps in the algorithm comprise the sequential sampling. In the first step, the NN_y is fit to the current training data set $(\mathbf{X}, \mathbf{Y})$. In the second step, NN_y is evaluated at $\mathbf{X}$ and $\mathbf{X}_u$ to yield $\widehat{\mathbf{Y}}$ and $\widehat{\mathbf{Y}}_u$, respectively. This data is used to compute the spatial prediction errors $\mathbf{S} = \sqrt{(\mathbf{Y} - \widehat{\mathbf{Y}})^2}$ and $\mathbf{S}_u = \sqrt{(\mathbf{Y}_u - \widehat{\mathbf{Y}}_u)^2}$. The data is then appended as $\widetilde{\mathbf{X}} = \mathbf{X} \cup \mathbf{X}_u$, and $\widetilde{\mathbf{S}} = \mathbf{S} \cup \mathbf{S}_u$. The prediction uncertainty model NN_u is fit to the data set $(\widetilde{\mathbf{X}}, \widetilde{\mathbf{S}})$. A new sampling point $\mathbf{P}$ is now found by maximizing the prediction uncertainty model $\widehat{\mathbf{S}}(\mathbf{x}) = NN_u(\mathbf{x})$. The new sample point $\mathbf{P}$ is appended to $\mathbf{X}$ and the corresponding function value $y(\mathbf{P})$ is appended to $\mathbf{Y}$. The prediction model NN_y is used in a MCS to yield the summary statistics, the mean μ and standard deviation σ.

The algorithm terminates if the absolute relative change in the mean, calculated as $|\mu^{(i)} - \mu^{(i-1)}|/|\mu^{(0)}|$, is less than a predefined tolerance τ_μ, and the absolute relative change in the standard deviation, calculated as $|\sigma^{(i)} - \sigma^{(i-1)}|/|\sigma^{(0)}|$, is less than a predefined tolerance τ_σ, or if the number of sequential sample cycles exceeds N_{max}. In this work, the values of tolerance are set to $\tau_\mu = 0.005$ and $\tau_\sigma = 0.005$, and the maximum number of sequential samples is $N_{max} = 100$.

In this work, LHS is used to generate the initial sampling data set. The neural networks in uqEGONN are implemented within Tensorflow [1]. In this work, both NN_y and NN_u have two hidden layers and each with 8 neurons. The number of epochs is set to 10,000 and hyperbolic tan is used as activation function. All the NNs are trained using the Adam optimizer [11] with a learning rate of 0.001. For maximizing the infill criterion, differential evolution [22] is used with a population size of 210. The mutation and recombination is set to 0.8 and 0.9, respectively, with a maximum of 200 generations. MCS is performed using random sampling.

Algorithm 1. Uncertainty quantification with EGONN (uqEGONN)

Require: initial data sets $(\mathbf{X}, \mathbf{Y})$ and $(\mathbf{X}, \mathbf{Y})_u$
  repeat
    fit NN_y to data $(\mathbf{X}, \mathbf{Y})$
    use NN_y to get $\widehat{\mathbf{Y}}$ at $\mathbf{X}$ and $\widehat{\mathbf{Y}}_u$ at $\mathbf{X}_u$
    compute prediction errors: $\mathbf{S} \leftarrow \sqrt{(\mathbf{Y} - \widehat{\mathbf{Y}})^2}$ and $\mathbf{S}_u \leftarrow \sqrt{(\mathbf{Y}_u - \widehat{\mathbf{Y}}_u)^2}$
    combine data: $\widetilde{\mathbf{X}} \leftarrow \mathbf{X} \cup \mathbf{X}_u$, $\widetilde{\mathbf{S}} \leftarrow \mathbf{S} \cup \mathbf{S}_u$
    fit NN_u to data $(\widetilde{\mathbf{X}}, \widetilde{\mathbf{S}})$
    $\mathbf{P} \leftarrow \arg\max \widehat{\mathbf{S}}(\mathbf{x})$
    $\mathbf{X} \leftarrow \mathbf{X} \cup \mathbf{P}$
    $\mathbf{Y} \leftarrow \mathbf{Y} \cup y(\mathbf{P})$
    $\mu, \sigma \leftarrow$ Monte Carlo simulation using NN_y
  until convergence

---

3 Numerical Experiments

This section presents the numerical results of applying the proposed uqEGONN algorithm for the UQ of the Ishigami function [8], which is written as

$$Y = f(x_1, x_2, x_3) = \sin(x_1) + a\sin^2(x_2) + bx_3^4\sin(x_1), \tag{1}$$

where $a = 7$, $b = 0.1$, and $x_i \sim \mathcal{U}[-\pi, \pi] \; \forall \; i = 1, 2, 3$. The convergence of MCS on the true analytical function (1) is shown in Figs. 1(a) and (b). It can be seen that MCS needs around one million samples to reach a converged mean $\mu = 3.50$ and standard deviation $\sigma = 3.72$. Figure 1(b) shows the true output distribution.

    Figure 2 shows the convergence of the uqEGONN algorithm as a function of the sequential infill samples for the mean (Fig. 2(a)) and the standard deviation (Fig. 2(b)), as well as using NN modeling with LHS of the total number of samples, i.e., $n_s = n + n_i$, where n is the number of initial samples and n_i is the number of infill samples. It can be seen that the convergence history for both approaches exhibit oscillations which is due to the NN random fit. Also, the figures show that the sequential algorithm achieves more reduction in the both the mean and the standard deviation than the pure LHS approach.

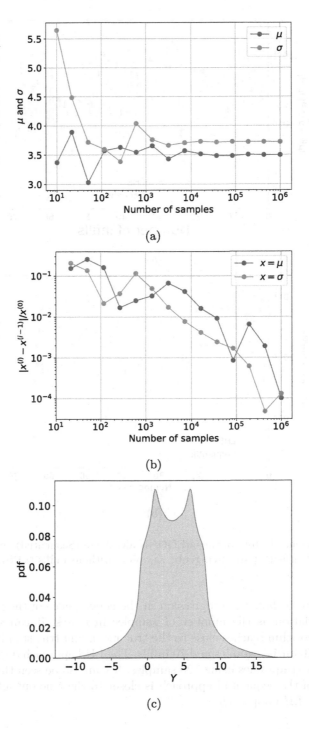

Fig. 1. Monte Carlo simulation convergence history of the true Ishigami function.

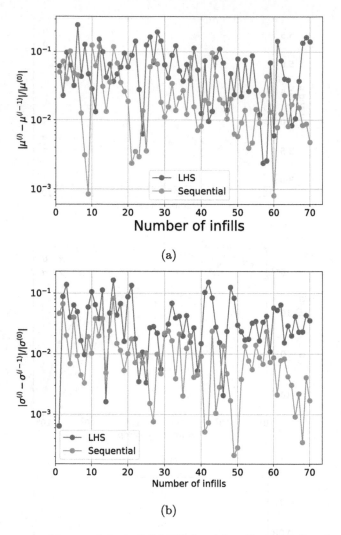

Fig. 2. Convergence history of the uqEGONN algorithm (Sequential) and using neural network modeling with Latin hypercube sampling without infills (LHS).

Figures 3 to 5 show the progression in the convergence of the mean and the standard deviation as the number of samples increases. It can seen that the sequential algorithm reaches close to the true mean and true standard deviation values using 50 initial samples and 70 infills. The LHS approach does not give get as close to the true values using 120 samples. It can also be seen that the output distribution of the sequential approach is closer to the true output distribution than the pure LHS approach.

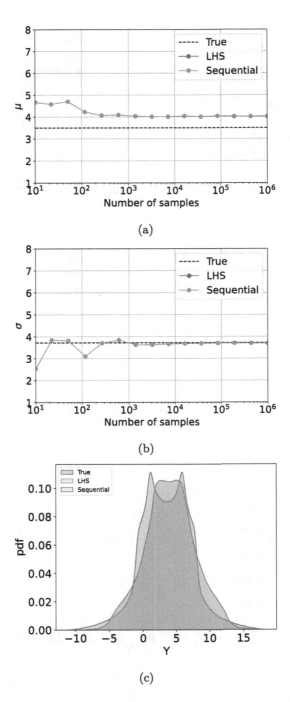

Fig. 3. Monte Carlo simulations on neural network models trained with the initial data sets ($n_s = 50$) using Latin hypercube sampling (LHS) and the sequential algorithm: (a) mean, (b) standard deviation, and (c) probability density functions of the output.

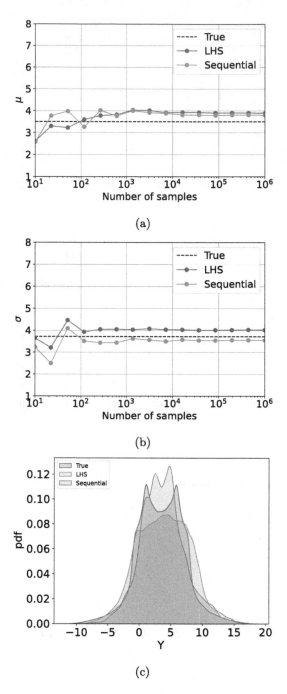

Fig. 4. Monte Carlo simulations on neural network models trained with the using Latin hypercube sampling (LHS) ($n_s = 85$) and the sequential algorithm ($n = 50$, $n_i = 35$): (a) mean, (b) standard deviation, and (c) probability density functions of the output.

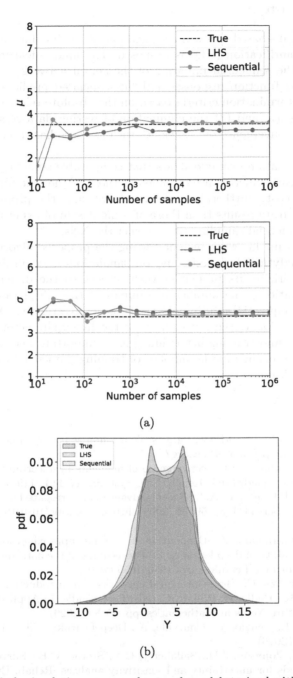

(a)

(b)

Fig. 5. Monte Carlo simulations on neural network models trained with the using Latin hypercube sampling (LHS) ($n_s = 120$) and the sequential algorithm ($n = 50$, $n_i = 70$): (a) mean, (b) standard deviation, and (c) probability density functions of the output.

4   Conclusion

A novel surrogate-based forward propagation algorithm for simulation-based uncertainty quantification (UQ) is proposed. The unique features of the algorithm are (1) the sequential updating of the neural network (NN) predictions of the nonlinear function responses and the associated prediction uncertainty, (2) automated termination criteria based on the absolute relative change in the summary statistics, and (3) the elimination of testing data sets and the arbitrary choice of convergence metric values such as those based on the root mean squared error.

The demonstration example shows that comparable summary statistics and probability density function are obtained as those of the true function at a low computational cost. Furthermore, it was shown that the proposed algorithm yields more accurate results than those of using design of experiments methods without the sequential sampling to construct the NNs.

Future steps in this work include testing the proposed algorithm on higher dimensional analytical functions, and on simulation-based problems with the goal of characterizing its properties with respect to the dimensionality. The effects of the ratio of the number of samples of the initial sampling plan to the number of sequential infill points needs to be investigated. Another important future direction is the automation of the hyperparameter tuning of the NN architecture and training algorithm. This is important because the optimal architecture may change as the number of training data points increase in the sequential sampling.

References

1. Abadi, M., et al.: TensorFlow: Large-scale machine learning on heterogeneous systems (2015). http://tensorflow.org/
2. Attar, P.J., Vedula, P.: On convergence of moments in uncertainty quantification based on direct quadrature. Reliab. Eng. Syst. Safety **111**, 119–125 (2013)
3. Forrester, A.I.J., Keane, A.J.: Recent advances in surrogate-based optimization. Prog. Aerosp. Sci. **41**(1–3), 50–79 (2009). https://doi.org/10.1016/j.paerosci.2008.11.001
4. Gal, Y., Ghahramani, Z.: Dropout as a bayesian approximation: Representing model uncertainty in deep learning. In: Proceedings of the 33rd International Conference on Machine Learning, pp. 1050–1059 (2016)
5. Goan, E., Fookes, C.: Bayesian neural networks: An introduction and survey. In: Mengersen, K., Pudlo, P., Robert, C. (eds) Case Studies in Applied Bayesian Data Science. Lecture Notes in Mathematics, pp. 45–87 (2020)
6. Goodfellow, I., Bengio, Y., Courville, A.: Deep Learning. The MIT Press, Cambridge, MA (2016)
7. Helton, J.C., Johnson, J.D., Sallaberry, C.J., Storlie, C.B.: Survey of sampling-based methods for uncertainty and sensitivity analysis. Reliab. Eng. Syst. Safety **91**(10–11), 1175–1209 (2006)
8. Ishigami, T., Homma, T.: An importance quantification technique in uncertainty analysis for computer models. In: [1990] Proceedings. In: First International Symposium on Uncertainty Modeling and Analysis, pp. 398–403. IEEE (1990)

9. Jones, D.R., Schonlau, M., Welch, W.J.: Efficient global optimization of expensive black-box functions. J. Global Optim. **13**(4), 455–492 (1998)
10. Kedward, L.J., Allen, C.B., Rendall, T.C.S.: Gradient-limiting shape control for efficient aerodynamic optimization. AIAA J. **58**(9), 3748–3764 (2020). https://doi.org/10.2514/1.J058977
11. Kingma, D.P., Ba, J.: Adam: A method for stochastic optimization. arXiv:1412.6980 (2014)
12. Knio, O.M., Le Maitre, O.: Uncertainty propagation in CFD using polynomial chaos decomposition. Fluid Dyn. Res. **38**(9), 616 (2006)
13. Koratikere, P., Leifsson, L.T., Barnet, L., Bryden, K.: Efficient global optimization algorithm using neural network-based prediction and uncertainty. In: AIAA SCITECH 2023 Forum, p. 2683 (2023)
14. Lampinen, J., Vehtari, A.: Bayesian approach for neural networks - review and case studies. Neural Netw. **14**(3), 257–274 (2001)
15. Landau, D., Binder, K.: A guide to Monte Carlo simulations in statistical physics. Cambridge University Press (2021)
16. Lim, Y.F., Ng, C.K., Vaitesswar, U.S., Hippalgaonkar, K.: Extrapolative Bayesian optimization with Gaussain process and neural network ensemble surrogate models. Adv. Intell. Syt. **3**, 2100101 (2021)
17. Lyu, Z., Kenway, G.K.W., Martins, J.R.R.A.: Aerodynamic shape optimization investigations of the common research model wing benchmark. AIAA J. **53**(4), 968–984 (2015). https://doi.org/10.2514/1.J053318
18. Monahan, J.F.: Numerical methods of statistics. Cambridge University Press (2011)
19. Najm, H.N.: Uncertainty propagation and polynomial chaos techniques in cfd. Annual Review of Fluid Mechanics (SAND2008-1167J) (2008)
20. Queipo, N.V., Haftka, R.T., Shyy, W., Goel, T., Vaidyanathan, R., Tucker, P.K.: Surrogate-based analysis and optimization. Prog. Aerosp. Sci. **21**(1), 1–28 (2005)
21. Snoek, J., et al.: Scalable bayesian optimization using deep neural networks. In: Proceedings of the 32nd International Conference on Machine Learning, pp. 2171–2180 (2015)
22. Storn, R., Price, K.: Differential evolution - a simple and efficient heuristic for global optimization over continuous spaces. J. Global Optim. **11**, 341–359 (1997)
23. Titterington, D.M.: Bayeisan methods for neural networks and related models. Stat. Sci. **19**(1), 128–139 (2004)
24. Wong, F.S.: First-order, second-moment methods. Comput. Struct. **20**(4), 779–791 (1985)
25. Yao, W., Chen, X., Luo, W., Van Tooren, M., Guo, J.: Review of uncertainty-based multidisciplinary design optimization methods for aerospace vehicles. Prog. Aerosp. Sci. **47**(6), 450–479 (2011)

Hierarchical Learning to Solve PDEs Using Physics-Informed Neural Networks

Jihun Han$^{(\boxtimes)}$ and Yoonsang Lee

Department of Mathematics, Dartmouth College, Hanover, NH 03755, USA
{jihun.han,yoonsang.lee}@dartmouth.edu

Abstract. The neural network-based approach to solving partial differential equations has attracted considerable attention. In training a neural network, the network learns global features corresponding to low-frequency components while high-frequency components are approximated at a much slower rate. For a class of equations in which the solution contains a wide range of scales, the network training process can suffer from slow convergence and low accuracy due to its inability to capture the high-frequency components. In this work, we propose a sequential training based on a hierarchy of networks to improve the convergence rate and accuracy of the neural network solution to partial differential equations. The proposed method comprises multi-training levels in which a newly introduced neural network is guided to learn the residual of the previous level approximation. We validate the efficiency and robustness of the proposed hierarchical approach through a suite of partial differential equations.

Keywords: hierarchical learning · scientific machine learning · physics-informed neural networks

1 Introduction

Many research efforts have focused on well-designed objective or loss functions to guide a neural network to approximate the solution of a PDE. An objective function measures how well a neural network satisfies the PDE, typically defined as the empirical mean of the residual by a neural network. Physics-informed neural networks (PINN) [10], and DGM [11] consider the direct PDE residual as the loss function so that the neural network satisfies the PDE in the domain.

In particular, PINN has flexibility in informing physical laws described in differential equations, and thus it has been employed in solving a wide range of PDEs. Despite its successful results in many applications, PINN suffers from a slow convergence rate and accuracy degradation for a certain class of PDEs. Such computational challenges are often inherent from the characteristics of the solution of a PDE, in particular when the solution involves a wide range of scales. The multiscale PDE problems arise in various scientific domains, such as fluid dynamics, quantum mechanics, or molecular dynamics. Standard methods, such as finite difference methods (FDM) or finite element methods (FEM), encounter

© The Author(s), under exclusive license to Springer Nature Switzerland AG 2023
J. Mikyška et al. (Eds.): ICCS 2023, LNCS 14075, pp. 548–562, 2023.
https://doi.org/10.1007/978-3-031-36024-4_42

an intractable computational complexity in resolving all relevant scales, numerical instabilities, or slow convergence in general. There have been significant efforts in developing efficient discretization methods for multiscale problems. As a representative example, the multigrid (MG) method [2] addresses the disparate convergence rates of different scale components through a hierarchical design of discretizations. The MG method captures the diverse target scale components of the solution from the collaboration of scale-corresponding grid approximations. The MG method achieves fast convergence as the method approximates all scale components corresponding to the grids. A hierarchical approach for multiscale problems has also been discussed in [6] for turbulent diffusion. Instead of using a fine resolution grid for whole domain at each level, the approach in [6] uses a local spatiotemporal domain. By designing a hierarchy that captures all possible scale ranges of the solution, the approach can capture the effective macroscopic behavior by a significant computational gain.

Neural network-based methods also face hurdles in approximating the multiscale solution of a PDE in that a neural network prefers low frequencies (F-Principle) [17]. Therefore, a standard network design can be ineffective in learning the high-frequency components. There are several recent research efforts to address the limitation in training high-frequencies by modifying the architecture or the ingredients of neural networks [3,12,15]. In particular, [15] proposed a neural network structure with Fourier feature embeddings to learn the multiscale solution efficiently. The embedding allows one to specify target characteristic frequencies of the neural network. The authors consider the multiple embeddings of inputs to simultaneously learn the diverse range of frequencies in the solution.

This work proposes hierarchical learning for solving PDEs to expedite the convergence through a sequential training of neural networks based on a hierarchy of networks. Using a set of networks of different target scales, where it is assumed that the sum of the networks can represent the solution of a PDE solution, we separate the training process so that each network can learn its corresponding scales without the training interruption from other networks. Once the training for a network finished, the training switches to the next network to correct the residual of the approximation up to the previous level. Among other methods to impose different target scales for networks, we test 1) standard MLPs of different complexity (number of layers and neurons) and 2) Fourier feature embedding. We emphasize that our proposed approach differs from other network design-based methods, such as [3,7,15], in that the focus of the hierarchical learning is the sequential training process. In our numerical experiments, it is shown that a sequential training process of networks with different target scales performs better than the training of the complex network that combines all networks of different levels at once. Thus, separating the training process for different target scales becomes a key ingredient of the proposed hierarchical learning. We believe the proposed hierarchical learning can apply to many neural network-based methods. In this study, we investigate the efficacy of the proposed hierarchical learning method in the framework of PINN along with other techniques to improve the convergence of PINN, such as the aforementioned adaptive weighting algorithms.

The rest of the paper is organized as follows. Section 2 reviews the PINN method and discusses the previous efforts to overcome the spectral barriers in training neural networks to solve PDEs. In Sect. 3, we propose the hierarchical learning methodology, while Sect. 4 provides numerical experiments validating the efficacy of the proposed method. We conclude this paper with discussions about the limitation and future directions of the current study in Sect. 5.

2 Physics-Informed Neural Networks

In this section, we summarize the standard Physics-informed Neural Networks (PINN) for a boundary value problem and discuss its variants to address the limitations of PINN. We consider the partial differential equation of unknown real-valued function u in a bounded domain $\Omega \subset \mathbb{R}^n$

$$\mathcal{N}[u](\boldsymbol{x}) = f(\boldsymbol{x}), \text{ and } \mathcal{B}[u](\boldsymbol{x}) = g(\boldsymbol{x}), \ \boldsymbol{x} \in \partial\Omega, \tag{1}$$

where $\mathcal{N}$ is a differential operator and $\mathcal{B}$ represents a boundary condition operator. General deep learning-based methods to solve Eq. (1) employ a neural network, $u(\boldsymbol{x}; \boldsymbol{\theta})$, to approximate the solution, and train the parameters $\boldsymbol{\theta}$ under the guidance of a loss function leading the neural network to satisfy Eq. (1). The PINN measures the direct PDE residuals in the loss function

$$\mathcal{L}(\boldsymbol{\theta}) = \lambda_\Omega \mathcal{L}_\Omega(\boldsymbol{\theta}) + \lambda_{\partial\Omega} \mathcal{L}_{\partial\Omega}(\boldsymbol{\theta}), \tag{2}$$

which consists of the interior and boundary loss terms

$$\mathcal{L}_\Omega(\boldsymbol{\theta}) = \sum_{i=1}^{N_r} \frac{|\mathcal{N}[u(\cdot; \boldsymbol{\theta})](\boldsymbol{x}_r^i) - f(\boldsymbol{x}_r^i)|^2}{N_r}, \quad \mathcal{L}_{\partial\Omega}(\boldsymbol{\theta}) = \sum_{i=1}^{N_b} \frac{|\mathcal{B}[u(\cdot; \boldsymbol{\theta})](\boldsymbol{x}_b^i) - g(\boldsymbol{x}_b^i)|^2}{N_b} \tag{3}$$

respectively. Here $\left\{\boldsymbol{x}_r^i\right\}_{i=1}^{N_r}$ and $\left\{\boldsymbol{x}_b^i\right\}_{i=1}^{N_b}$ are sampling points in the interior, and the boundary of the domain, respectively.

Despite the remarkable achievement in many applications, the PINN often struggles to learn the solutions of PDEs with either slow convergence or degraded accuracy. Recent works have endeavored to understand unfavorable training scenarios of neural networks and proposed alternative methodologies to overcome the limitations. One of the methods includes balancing different terms of the loss function in the context of multi-objective optimization discussed in Sect. 1 [8,14,16]. Another direction addresses the intrinsic behavior of training neural networks, which is specifically disadvantageous to learn functions involving diverse frequency spectrum [1,9,17,18].

The general learning process of neural networks has been studied from spectral analysis [1,9,17,18]. The F-principle [17] shows that the gradient-based training process has spectral bias as the neural networks tend to learn low frequencies while it requires a longer time to fit high frequencies. This phenomenon is a challenge in neural network-based methods to solve multiscale PDE problems that suffer from slow convergence or low accuracy. The networks miss the high-frequency components unless the training process is sufficiently long to learn high-frequency components. [3] proposed a neural network architecture with an input scaling treatment for converting the high-frequency components to low-frequency ones preferable to learning. The network is installed with a compact supported activation function and is effectively applied to multiscale applications [7,13].

Another work in [12] showed that a simple random Fourier feature embedding of inputs enables a standard MLP to learn high-frequency components more efficiently in applications of computer vision and graphics. Namely, the embedding corresponds to a map from the input $x \in \mathbb{R}^n$ to the $2m$-dimensional frequency domain as

$$x \in \mathbb{R}^n \quad \mapsto \quad \begin{bmatrix} a \odot \cos(B_\sigma x) \\ a \odot \sin(B_\sigma x) \end{bmatrix} \in \mathbb{R}^{2m}, \tag{4}$$

where $B_\sigma \in \mathbb{R}^{n \times m}$ is a random wave number matrix sampled from the Gaussian distribution $\mathcal{N}(0, \sigma^2)$ and $a \in \mathbb{R}^m$ is a scaling vector. The authors analyzed the effect of the embedding on the neural tangent kernel (NTK) of the standard MLP to attenuate the spectral bias using appropriate σ and a.

3   Hierarchical PINN

The methods discussed in the previous section focus on various strategies to improve the capability of a single neural network to learn a wide range of scales in the solution of a PDE. In the current study, we propose a hierarchical learning procedure of neural networks to represent the multiscale solution. The proposed method, which we call 'hierarchical Physics-informed neural network' (HiPINN), uses a sequence of neural networks to represent the multiscale solution of a PDE and trains them sequentially rather than simultaneously. Our hierarchical approach is motivated by the multigrid method that uses a hierarchy of different grid sizes to expedite the convergence of an iterative method to solve PDEs [2]. The rationale of the multigrid method is that a grid size has its characteristic scale with its corresponding convergence rate. The multigrid method achieves a fast convergence rate by capturing different scale components through variable grid sizes. The idea of the proposed hierarchical approach for PINN is to impose a hierarchy in training so that each network can capture its corresponding scales without the interruption in training other networks, which enables us to capture uniformly all possible ranges of scales.

3.1   Hierarchical Design of Networks

HiPINN employs a set of M neural networks $\{v_m(x; \theta_m)\}_{m=1}^M$ with a hierarchy to represent the PDE solution where M represents the number of levels for different characteristic scales. To mimic the hierarchy of the multigrid method, we consider two approaches in the current study. The first approach is the standard multiplayer perceptron (MLP) with various network sizes and complexity. We expect that a simple network will be enough to approximate for low variability components of the unknown solution, while high variability components require a more complicated network. Following this intuitive argument, we increase the complexity of networks by increasing the depth and width of each network. One issue of this approach is that it is unclear to cover a specific range of scales. Suppose two networks are significantly different in terms of complexity. In that case, we expect that the two networks will represent different scale components, but it is not clear whether there is a gap between them. In the study of the spectral bias of neural networks [9], it is shown that higher frequencies are significantly less robust than lower ones in the perturbation of the neural network parameters. This observation indicates that a limited volume in the parameter space is involved in expressing the high-frequency components. With the support of this observation, we consider the complexity of networks in composing the hierarchy. The schematic diagram of the hierarchical employment of the MLPs is displayed in Fig. 1.

Another approach to impose a hierarchy in the network is the Fourier feature embedding [12]. The structure of each network is identical through levels with the input embedding as Eq. (4). However, we vary them by increasing σ as the level (m) increases so that a high-level network represents high frequency or wavenumber behaviors compared to the ones captured by the low-level networks. As the network size does not change through the hierarchy, the Fourier embedding-based approach does not provide any computational efficiency in solving a low-level network compared to the hierarchy using the network complexity of MLP. However, the Fourier embedded hierarchy can specify the target characteristic scales through σ. In the multiscale approach using the Fourier embedding for PINN [15], a various range of σ values is incorporated to design a single network to target all possible ranges of scales in the solution. In terms of the network complexity, HiPINN does not necessarily use a network more complicated than the one used in [15]. The goal of HiPINN is to expedite the training process by dividing the training into specific scales instead of training all possible scales simultaneously. The schematic of a hierarchical neural network design using the Fourier feature embedding is shown in Fig. 2.

3.2 HiPINN Algorithm

The benefit of the proposed hierarchical learning method comes from the sequential training based on a hierarchy of networks. Using the neural networks with a hierarchy, the M-level HiPINN representation of the PDE solution is the sum of all neural networks, which is given as

$$u_M(\boldsymbol{x}) = \sum_{m=1}^{M} v_m(\boldsymbol{x}; \boldsymbol{\theta}_m). \tag{5}$$

Under this structure, the training of each level network is on the correction of the residual of the previous level solution representation. To add the $(M+1)$-th

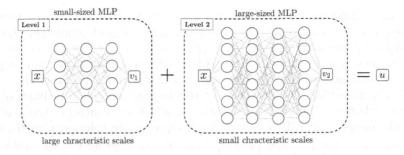

Fig. 1. Hierarchical composition of standard MLPs; small-sized MLP for capturing low variability components and large-sized MLP for high variability components.

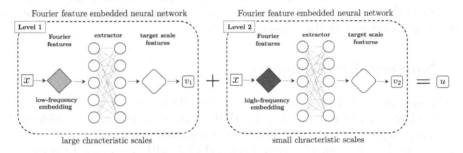

Fig. 2. Hierarchical composition of Fourier feature embedded neural networks; With the same network architecture, the target characteristic frequency is controlled by the Fourier feature embedding of inputs as Eq. (4).

level to u_M using v_{M+1}, the loss function $\mathcal{L}^{(M+1)}$ is

$$\mathcal{L}^{(M+1)}(\boldsymbol{\theta}_M) = \lambda_\Omega \mathcal{L}_\Omega^{(M+1)}(\boldsymbol{\theta}_{M+1}) + \lambda_{\partial\Omega} \mathcal{L}_{\partial\Omega}^{(M+1)}(\boldsymbol{\theta}_{M+1}), \qquad (6)$$

$$\mathcal{L}_\Omega^{(M+1)}(\boldsymbol{\theta}_{M+1}) = \frac{1}{N_r} \sum_{i=1}^{N_r} \left| \mathcal{N}[u_M + v_{M+1}(\cdot; \boldsymbol{\theta}_{M+1})](\boldsymbol{x}_r^i) - f(\boldsymbol{x}_r^i) \right|^2, \qquad (7)$$

$$\mathcal{L}_{\partial\Omega}^{(M+1)}(\boldsymbol{\theta}_{M+1}) = \frac{1}{N_b} \sum_{i=1}^{N_b} \left| \mathcal{B}[u_M + v_{M+1}(\cdot; \boldsymbol{\theta}_{M+1})](\boldsymbol{x}_b^i) - g(\boldsymbol{x}_b^i) \right|^2. \qquad (8)$$

We note that u_M is already approximated, and thus the training variable related to $\mathcal{L}^{(M+1)}$ is $\boldsymbol{\theta}_{M+1}$. If the differential operator and the boundary operator are linear, the $(M+1)$-th level training is equivalent to solving the original PDE operator using v_{M+1} for modified $f^{(M+1)}(\boldsymbol{x})$ and $g^{(M+1)}(\boldsymbol{x})$, which are given by

$$f^{(M+1)}(\boldsymbol{x}) = f(\boldsymbol{x}) - \mathcal{N}[u_M](\boldsymbol{x}) \text{ and } g^{(M+1)}(\boldsymbol{x}) = g(\boldsymbol{x}) - \mathcal{B}[u_M](\boldsymbol{x}), \qquad (9)$$

respectively. Therefore, the implementation for the linear case involves only marginal modification of the standard PINN method. When the differential operator $\mathcal{N}$ is nonlinear, the differential operator on v_{M+1} at the $(M+1)$-th level will be different from the original operator $\mathcal{N}$. However, the structure of the operator does not change over the level; it remains at minimizing the loss related to $\mathcal{N}[$'approximation up to the previous level' + 'current level network'] over the current level network. Thus HiPINN for a nonlinear problem requires only one implementation of a solver and uses it repeatedly for all levels.

We also note that HiPINN does not require any projection or interpolation operations between different level solutions, which are crucial in the algebraic multigrid method. In HiPINN, each level approximate solution uses the same sampling points $\left\{\boldsymbol{x}_r^i\right\}_{i=1}^{N_r}$ and $\left\{\boldsymbol{x}_b^i\right\}_{i=1}^{N_b}$. From the homogeneity of the problem to be solved at each level, it is straightforward to implement various types of cycles to iterate over different levels, such as V and W cycles [2]. The V cycle starts from a low resolution to a high resolution and iterates back to a low resolution. The W cycle repeats the V cycle to approximate scale components that are not sufficiently captured at the corresponding level.

4   Numerical Experiments

In this section, we validate the robustness and effectiveness of the proposed hierarchical learning methodology to solve PDEs through a suite of test problems. In all numerical experiments, we use the standard multilayer perceptrons (MLPs) and Fourier feature embedded neural networks [15] with the tanh activation function. In the Fourier feature embedding case, the architecture of each network is designed as follows in sequence; 1) multiple Fourier feature embeddings of input, each of embedding corresponding to the map in Eq. (4) with scaling vector $\boldsymbol{a} = \boldsymbol{1}$, 2) a multiscale feature extractor MLP common for each embedded feature, 3) a final linear layer passed by concatenated features extracted. In our numerical experiments, we consider the dimension of a Fourier feature embedding the same as that of the first hidden layer of the multiscale feature extractor. Moreover, we include a dense layer to pass the concatenated features, which performs better than direct linear mapping to the output in our experiments.

We train each neural network using the Adam optimizer [5] with $\beta_1 = 0.95$ and $\beta_2 = 0.95$, and all the trainable parameters are initialized from Glorot normal distribution [4]. Moreover, we employ the adaptive weights algorithm [16] in all experiments, updating the weights in every 100 gradient descent steps for computational efficiency. To validate the performance of the proposed method, we consider the standard training procedure using networks with and without a hierarchy. We note that the standard training using networks with a hierarchy uses the same overall network structure as in the proposed learning method. The difference is in the training process; the proposed method uses sequential training while the standard training uses simultaneous training (that is, the networks of different scales are trained at the same time). Except for the first test in which an exact solution is available, we obtain reference solutions using the FEM method with sufficiently large mesh sizes. We measure the accuracy of the network-based solutions $\tilde{u}$ using the relative $\mathcal{L}^2$-error, $\frac{\|\tilde{u}-u\|_{2,\Omega}}{\|u\|_{2,\Omega}}$. All benchmark losses are referred to the test losses computed on the grid points. We want to

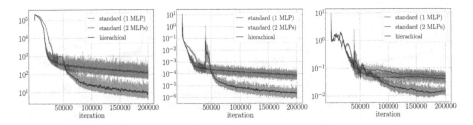

Fig. 3. Training procedures of MLPs in solving Eq. (10) by standard (single MLP: 3 hidden layers, 200 units, two MLPs: 2, 3 hidden layers, 200 units) and hierarchical (first level: 3 hidden layers, 200 units, second level: 5 hidden layers, 200 units) learning. (left) interior losses, (middle) boundary losses, (right) relative $\mathcal{L}^2$-errors. The hierarchical learning corresponds to second level initiation at 4×10^4 iterations.

note that the reference FEM simulation is much more efficient than the neural network-based methods in the tests we consider here, which are PDEs in the 2D space domain. The computational efficiency of the neural network-based methods comes in when the domain is in a high-dimensional space. Note that in a high-dimensional case, the mesh generation and solving its corresponding (nonlinear) system can be extremely slow compared to the Monte-Carlo-based training in the neural network approach [10].

4.1 Poisson Equation

As the first example, we consider the Poisson equation in the unit square $\Omega = [0,1]^2$ with a Dirichlet boundary condition,

$$\Delta u = f \ \text{ in } \ \Omega, \ \ u = g \ \text{ on } \ \partial\Omega. \tag{10}$$

Here, we choose the force term f and the boundary value g such that Eq. (10) has the exact solution $u(\boldsymbol{x}) = \sin(8\pi x_1^2 + 4\pi x_2)\sin(8\pi x_2^2 + 4\pi x_1)$. We consider two-level hierarchical learning with neural networks, $v(\boldsymbol{x}; \boldsymbol{\theta}_1)$ and $v(\boldsymbol{x}; \boldsymbol{\theta}_2)$, which are sequentially trained using the corresponding loss functions,

$$\mathcal{L}^{(1)}(\boldsymbol{\theta}_1) = \frac{\lambda_\Omega}{N_r} \sum_{i=1}^{N_r} \left| \Delta v(\boldsymbol{x}_r^i; \boldsymbol{\theta}_1) - f(\boldsymbol{x}_r^i) \right|^2 + \frac{\lambda_{\partial\Omega}}{N_b} \sum_{i=1}^{N_b} \left| v(\boldsymbol{x}_b^i; \boldsymbol{\theta}_1) - g(\boldsymbol{x}_b^i) \right|^2 \tag{11}$$

$$\mathcal{L}^{(2)}(\boldsymbol{\theta}_2) = \frac{\lambda_\Omega}{N_r} \sum_{i=1}^{N_r} \left| \left(\Delta v(\boldsymbol{x}_r^i; \boldsymbol{\theta}_1^*) + \Delta v(\boldsymbol{x}_r^i; \boldsymbol{\theta}_2) \right) - f(\boldsymbol{x}_r^i) \right|^2 \tag{12}$$

$$+ \frac{\lambda_{\partial\Omega}}{N_b} \sum_{i=1}^{N_b} \left| \left(v(\boldsymbol{x}_b^i; \boldsymbol{\theta}_1^*) + v(\boldsymbol{x}_b^i; \boldsymbol{\theta}_2) \right) - g(\boldsymbol{x}_b^i) \right|^2,$$

respectively. Here, $\boldsymbol{\theta}_1^*$ in Eq. (12) is the updated $\boldsymbol{\theta}_1$ at level 1 and is fixed at level 2. We note that as the differential (i.e., Laplacian) and boundary operators are linear, the second level PDE for the neural network $v(\boldsymbol{x}; \boldsymbol{\theta}_2)$ is also a Poisson

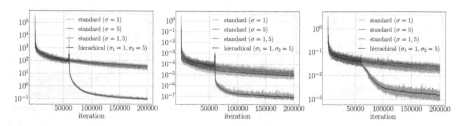

Fig. 4. Training procedures for Eq. (10) by standard and hierarchical learning. (left) interior losses, (middle) boundary losses, (right) relative $\mathcal{L}^2$-errors. The standard learning corresponds to single Fourier feature embedding with $\sigma = 1$, $\sigma = 5$, separately, and multiple embeddings $\sigma = 1, 5$. The hierarchical learning runs with single embedding $\sigma = 1$ at the first level, and $\sigma = 5$ at the second level, in sequence. The second level training is initiated at 6×10^4 iterations.

equation with a Dirichlet boundary condition with shifted force and boundary functions. We test our method using the standard MLPs, and Fourier feature embedded neural networks with training sample sizes $N_r = 400$ and $N_b = 400$.

First, we use the MLP with three hidden layers of dimension 200 at the first level and five hidden layers of dimension 200 at the second level. The hierarchical learning is compared with the standard learning (i.e., single-level hierarchy) with different sizes of MLPs, H numbers of hidden layers of dimension 200 for $H = 2, 3, \cdots, 8$, among which the MLP with $H = 3$ achieves the best performance in approximating the solution. We also test the standard learning with the sum of two MLPs with H_1 and H_2 numbers of hidden layers of dimension 200 for $H_1, H_2 = 2, 3, \cdots, 8$, $H_1 < H_2$, in which $\boldsymbol{\theta}_1$ and $\boldsymbol{\theta}_2$ are trained simultaneously. Among the combination of two MLPs, the networks with $H_1 = 2$ and $H_2 = 3$ perform the best approximation of the solution. Figure 3 shows the training procedures for 2×10^5 iterations. We observe that the correction of the hierarchical learning at the second level properly works to accelerate the convergence in two losses and achieve better approximation accuracy than standard learning with the single network or the sum of two networks. We want to emphasize that the importance of the sequential training process of the different level networks. In comparison with the proposed method and the two MLP case in which the total networks are the same, the hierarchical training (blue curve) shows better performance than the combined network case (green curve). This result shows that the training of each target scale network can be hindered by the training of other scale networks. For the case of the Fourier feature embedded neural networks, we use the same size neural networks at both levels, where each network has a different Fourier embedding. We use single Fourier feature embedding at each level with $\sigma = 1$ and $\sigma = 5$, respectively, in considering low target frequencies at the first level and relatively high frequencies at the next level. The rest of the network consists of the feature extractor with three hidden layers of dimension 200 followed by the last dense layer of dimension 200. To demonstrate the effectiveness of learning the diverse frequencies from low to

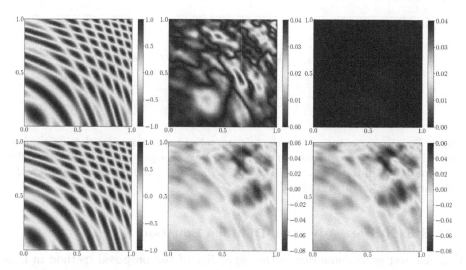

Fig. 5. first row: The numerical solutions of Eq. (10) by standard (multiple embeddings $\sigma = 1, 5$) and hierarchical (single embedding $\sigma_1 = 1$, $\sigma_2 = 5$ in sequence) learning. (left) the exact solution, (middle) the pointwise error corresponding to standard learning, (right) the pointwise error corresponding to proposed hierarchical learning. second row: the approximations at each level in the hierarchical learning. (left) the approximation $v(\cdot; \boldsymbol{\theta}_1^*)$ at the first level, (middle) the approximation $v(\cdot; \boldsymbol{\theta}_2^*)$ at the second level, (right) the target function for $v(\cdot; \boldsymbol{\theta}_2)$ at the second level, which is equal to $(u_{\text{exact}} - v(\cdot; \boldsymbol{\theta}_1^*))$.

high in sequence, we compare our method with the standard learning with the single embedding ($\sigma = 1$ and $\sigma = 5$) and multiple embeddings ($\sigma = 1, 5$) aiming to learn various frequencies simultaneously. As shown in Fig. 4, our method accelerates the convergence at the second level and achieves an accurate approximation (relative $\mathcal{L}^2$-error 1.33×10^{-3}) in comparison to the other experiments (best relative $\mathcal{L}^2$-error 1.65×10^{-2}). Particularly, in comparison with the multiple embedding (red curve) and the proposed hierarchical training (blue curve), where the overall network structure to represent the solution is comparable, the hierarchical training process shows a better performance than the simultaneous training process.

Figure 5 shows the point-wise errors of each level approximation in comparison with the standard learning method. Moreover, our method combined with Fourier feature embedding outperforms the performance of HiPINN using the standard MLP, as we can employ a neural network suitable for learning the target frequencies at each level. We address a question when it is appropriate to switch to the next level. Figure 6 presents the six training procedures of the Fourier feature embedded neural networks ($\sigma = 1$, $\sigma = 5$ in level sequence). The experiment shows that transition to the next level after 25000 iterations provide comparable overall accuracy using the two-level representation.

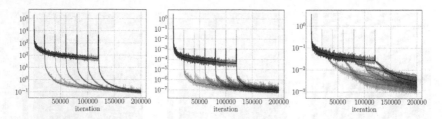

Fig. 6. Training procedures of Fourier feature embedded neural network (single embedding $\sigma_1 = 1$, $\sigma_2 = 5$ in sequence) in solving 2D Poisson equation, Eq. (10) with different second level initiations, $2n \times 10^4$, $n = 1, 2, 3, 4, 5, 6$. (left) interior losses, (middle) boundary losses, (right) relative $\mathcal{L}^2$-errors.

4.2   Steady-State Advection-Diffusion Equation

The last test is to demonstrate the capability of the proposed method in handling the multiscale behavior, which could arise from the intertwined consequences of both differential operator and force term. We consider the steady-state advection-diffusion equation with mixed Dirichlet and Neumann boundary conditions,

$$
\begin{aligned}
\boldsymbol{w} \cdot \nabla u - \nu \Delta u &= f \quad \text{in } \boldsymbol{x} \in [0,1]^2, \\
u(0, x_2) = g_1, u(1, x_2) &= g_2 \quad \text{for } x_2 \in [0,1], \\
\frac{\partial u}{\partial n} u(x_1, 0) = \frac{\partial u}{\partial n} u(x_1, 1) &= 0 \quad \text{for } x_1 \in [0,1].
\end{aligned}
\tag{13}
$$

We choose the diffusion coefficient $\nu = 0.01$, the force $f = \sin(4\pi x_2)$, Dirichlet boundary value, $g_1 = 0$ and $g_2 = 1$, and an incompressible velocity field $\boldsymbol{w}$,

$$
\boldsymbol{w}(\boldsymbol{x}) = (-5\sin(6\pi x_1)\cos(6\pi x_2), 5\cos(6\pi x_1)\sin(6\pi x_2))
\tag{14}
$$

We solve Eq. (13) using two levels, in which neural networks $v(\boldsymbol{x}; \boldsymbol{\theta}_1)$ and $v(\boldsymbol{x}; \boldsymbol{\theta}_2)$ are trained under the following loss functions

$$
\mathcal{L}^{(1)}(\boldsymbol{\theta}_1) = \frac{\lambda_\Omega}{N_r} \sum_{i=1}^{N_r} \left| \mathcal{R}(\boldsymbol{\theta}_1; \boldsymbol{x}_r^i) \right|^2 + \frac{\lambda_{\partial\Omega,1}}{N_{b,1}} \sum_{i=1}^{N_{b,1}} \left| \frac{\partial v}{\partial n}(\boldsymbol{x}_{b,1}^i; \boldsymbol{\theta}_1) \right|^2
\tag{15}
$$

$$
+ \frac{\lambda_{\partial\Omega,2}}{N_{b,2}} \sum_{i=1}^{N_{b,2}} \left| v(\boldsymbol{x}_{b,2}^i; \boldsymbol{\theta}_1) - g(\boldsymbol{x}_{b,2}^i) \right|^2,
$$

$$
\mathcal{L}^{(2)}(\boldsymbol{\theta}_2) = \frac{\lambda_\Omega}{N_r} \sum_{i=1}^{N_r} \left| \mathcal{R}(\boldsymbol{\theta}_2; \boldsymbol{x}_r^i) \right|^2 + \frac{\lambda_{\partial\Omega,1}}{N_{b,1}} \sum_{i=1}^{N_{b,1}} \left| \frac{\partial v}{\partial n}(\boldsymbol{x}_{b,1}^i; \boldsymbol{\theta}_1^*) + \frac{\partial v}{\partial n}(\boldsymbol{x}_{b,1}^i; \boldsymbol{\theta}_2) \right|^2
$$

$$
+ \frac{\lambda_{\partial\Omega,2}}{N_{b,2}} \sum_{i=1}^{N_{b,2}} \left| v(\boldsymbol{x}_{b,2}^i; \boldsymbol{\theta}_1^*) + v(\boldsymbol{x}_{b,2}^i; \boldsymbol{\theta}_2) - g(\boldsymbol{x}_{b,2}^i) \right|^2,
\tag{16}
$$

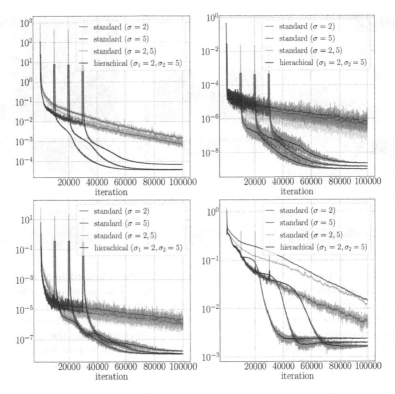

Fig. 7. Training procedures for steady-state advection-diffusion equation, Eq. (13), by standard learning and proposed hierarchical learning. (first row, left) interior losses, (first row, right) Dirichlet boundary losses, (second row, left) Neumann boundary losses, (second row, right) relative $\mathcal{L}^2$-errors. The standard learning corresponds to single Fourier feature embedding with $\sigma = 2$, $\sigma = 5$, separately, and multiple embeddings $\sigma = 2, 5$. The hierarchical learning runs with single embedding $\sigma = 2$ at the first level, and $\sigma = 5$ at the second level, in sequence. The second level training is initiated at $n \times 10^4$, $n = 1, 2, 3$, iterations. The detail approximations are also presented in Fig. 8.

where the residuals of PDE at each level are

$$\begin{aligned}
\mathcal{R}(\boldsymbol{\theta}_1; \boldsymbol{x}_r^i) &= \boldsymbol{w} \cdot \nabla v(\boldsymbol{x}_r^i; \boldsymbol{\theta}_1) - \nu \Delta v(\boldsymbol{x}_r^i; \boldsymbol{\theta}_1) - f(\boldsymbol{x}_r^i) \\
\mathcal{R}(\boldsymbol{\theta}_2; \boldsymbol{x}_r^i) &= \boldsymbol{w} \cdot \nabla v_c - \nu \Delta v_c - f(\boldsymbol{x}_r^i) \quad v_c = v(\boldsymbol{x}_r^i; \boldsymbol{\theta}_1^*) + v(\boldsymbol{x}_r^i; \boldsymbol{\theta}_2).
\end{aligned} \tag{17}$$

Here, $\boldsymbol{x}_{b,1}$ and $\boldsymbol{x}_{b,2}$ are sampling points for the Neumann and the Dirichlet boundary conditions, respectively, g is read as g_1 or g_2 depending on the location of $\boldsymbol{x}_{b,2}^i$, and $\boldsymbol{\theta}_1^*$ in Eq. (16) and Eq. (17) is the updated $\boldsymbol{\theta}_1$ at the first level and is fixed at the second level. Moreover, we also apply the adaptive weight algorithm [16] by treating the weight λ_Ω on boundary loss separately into two parts, $\lambda_{\partial\Omega,1}$ on the Neumann boundary loss and $\lambda_{\partial\Omega,2}$ on the Dirichlet boundary loss.

We apply the hierarchical learning method with the Fourier feature embedded neural networks. The exact size neural networks are considered at both

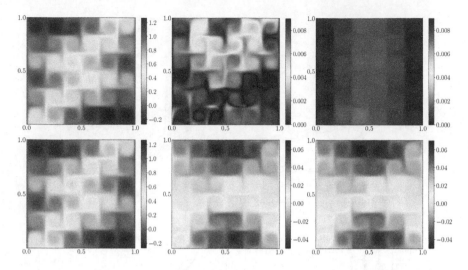

Fig. 8. first row: The numerical solutions of Eq. (13) by standard (single embedding $\sigma = 2$) and hierarchical (single embedding $\sigma_1 = 2$, $\sigma_2 = 5$ in sequence) learning. (left) the reference solution, (middle) the pointwise error corresponding to standard learning, (right) the pointwise error corresponding to proposed hierarchical learning. second row: the approximations at each level in the hierarchical learning. (left) the approximation $v(\cdot; \boldsymbol{\theta}_1^*)$ at first level, (middle) the approximation $v(\cdot; \boldsymbol{\theta}_2^*)$ at second level, (right) the target function for $v(\cdot; \boldsymbol{\theta}_2)$ at second level, which is equal to $(u_{\text{reference}} - v(\cdot; \boldsymbol{\theta}_1^*))$.

levels using different single embedding; $\sigma = 2$ at the first level and $\sigma = 5$ at the second level to learn low and high-frequency components. The rest of the network comprises the feature extractor with three hidden layers of dimension 200 followed by the last 200-dimensional dense layer. We train the neural networks over 1×10^5 iterations, where we switch to the second level at various instances (which are at $n \times 10^4$, $n = 1, 2, 3$, iterations).

We compare the hierarchical learning method with the standard learning approach using the same size neural network with different Fourier feature embeddings; single embedding using $\sigma = 2$ or $\sigma = 5$, and multiple embeddings using $\sigma = 2, 5$. Figure 7 shows the training procedures in terms of three losses and relative $\mathcal{L}^2$-errors. Among the standard learning experiments, $\sigma = 2$ embedding is suitable for this example as it has the most accurate approximation with a $\mathcal{L}^2$ error 4.76×10^{-3}. In comparison with the hierarchical learning method, hierarchical learning has the lowest error 2.41×10^{-3} when the second level training is triggered after 1×10^4 iterations of the first level. We also note that hierarchical learning converges after 4×10^4 iterations, which is 2.5 times faster than the other method. Figure 8 shows the numerical solutions from both standard learning and hierarchical learning for reference.

5   Discussions and Conclusions

Several research efforts have been focused on the design of a network to efficiently represent a PDE solution that contains a wide range of scales, including
hierarchical networks. However, the training process can suffer from slow convergence due to the interruptions of different scale components of the multiscale
network. Thus, the trainability of the network becomes an issue even though the
network can represent the multiscale solution if sufficiently trained. This study
proposed a hierarchical learning method to solve PDEs using neural networks,
which uses sequential training based on a hierarchy of networks. The rationale
of the sequential training is to focus on the target characteristic scales of each
network by training them separately rather than by training all networks simultaneously. Among several other methods to impose a hierarchy in the network,
we tested two methods; 1) multi-layer perceptrons (MLPs) with various network
complexities, and 2) Fourier feature embedded networks. The first approach has
a computational efficiency in solving a low complexity network while capturing
the low-frequency components of the solution. The second approach does not
provide any computational gain as each network at different levels has the same
complexity. Still, we can explicitly impose the range of scales of the solution
through the Fourier feature embedded layers. The proposed hierarchical learning method has been tested through a suite of numerical tests including the
advection-diffusion problem with a multiscale velocity field.

There are several issues to be addressed for the proposed hierarchical learning
method. It is unclear to see the connection between the network complexity
and its characteristic scales to represent a function. We have checked in our
numerical experiments that changing the complexity of a network will change
its corresponding scales. Still, we lack explicit and rigorous criteria to determine
the characteristic scales.

Also, we used the same network complexity for each level for the Fourier
feature embedding approach, assuming that the Fourier embedded layer will
determine its characteristic scales.

In applying the hierarchical learning to time dependent problems, there are
two approaches. One approach is to use a network to learn the spatiotemporal
scales at the same time. The other approach is to march the problem where
the spatial variations are learned through a network [10]. We are interested in
designing hierarchical networks to resolve multiscale behaviors in the temporal
domain, particularly to capture the long-time behavior of a dynamical system,
such as the climatology of geophysical fluid systems. Lastly, we have tested the
hierarchical learning method in the PINN framework in the current study. As
the overarching idea of the proposed method is in the efficient representation of
a multiscale function using hierarchical networks, we expect that the proposed
method can apply to other network-based methods for solving PDEs, which we
leave as future work.

Acknowledgements. YL is supported in part by NSF DMS-1912999 and ONR MURI
N00014-20-1-2595.

References

1. Arpit, D., et al.: A closer look at memorization in deep networks. In: International Conference on Machine Learning, pp. 233–242. PMLR (2017)
2. Briggs, W.L., Henson, V.E., McCormick, S.F.: A multigrid tutorial. SIAM (2000)
3. Cai, W., Xu, Z.Q.J.: Multi-scale deep neural networks for solving high dimensional pdes. arXiv preprint arXiv:1910.11710 (2019)
4. Glorot, X., Bengio, Y.: Understanding the difficulty of training deep feedforward neural networks. In: Proceedings of the Thirteenth International Conference on Artificial Intelligence and Statistics, pp. 249–256 (2010)
5. Kingma, D.P., Ba, J.: Adam: A method for stochastic optimization. arXiv preprint arXiv:1412.6980 (2014)
6. Lee, Y., Engquist, B.: Multiscale numerical methods for advection-diffusion in incompressible turbulent flow fields. J. Comput. Phys. **317**, 33–46 (2016)
7. Liu, Z., Cai, W., Xu, Z.Q.J.: Multi-scale deep neural network (mscalednn) for solving poisson-boltzmann equation in complex domains. arXiv preprint arXiv:2007.11207 (2020)
8. van der Meer, R., Oosterlee, C.W., Borovykh, A.: Optimally weighted loss functions for solving pdes with neural networks. J. Comput. Appl. Math. **405**, 113887 (2022)
9. Rahaman, N., et al.: On the spectral bias of neural networks. In: International Conference on Machine Learning, pp. 5301–5310. PMLR (2019)
10. Raissi, M., Perdikaris, P., Karniadakis, G.E.: Physics-informed neural networks: A deep learning framework for solving forward and inverse problems involving nonlinear partial differential equations. J. Comput. Phys. **378**, 686–707 (2019)
11. Sirignano, J., Spiliopoulos, K.: Dgm: A deep learning algorithm for solving partial differential equations. J. Comput. Phys. **375**, 1339–1364 (2018)
12. Tancik, M., et al.: Fourier features let networks learn high frequency functions in low dimensional domains. Adv. Neural. Inf. Process. Syst. **33**, 7537–7547 (2020)
13. Wang, B., Zhang, W., Cai, W.: Multi-scale deep neural network (mscalednn) methods for oscillatory stokes flows in complex domains. arXiv preprint arXiv:2009.12729 (2020)
14. Wang, S., Teng, Y., Perdikaris, P.: Understanding and mitigating gradient flow pathologies in physics-informed neural networks. SIAM J. Sci. Comput. **43**(5), A3055–A3081 (2021)
15. Wang, S., Wang, H., Perdikaris, P.: On the eigenvector bias of fourier feature networks: From regression to solving multi-scale pdes with physics-informed neural networks. Comput. Methods Appl. Mech. Engrg. **384**, 113938 (2021)
16. Wang, S., Yu, X., Perdikaris, P.: When and why pinns fail to train: A neural tangent kernel perspective. J. Comput. Phys. **449**, 110768 (2022)
17. Xu, Z.Q.J., Zhang, Y., Luo, T., Xiao, Y., Ma, Z.: Frequency principle: Fourier analysis sheds light on deep neural networks. Commun. Comput, Phys (2019)
18. Xu, Z.-Q.J., Zhang, Y., Xiao, Y.: Training behavior of deep neural network in frequency domain. In: Gedeon, T., Wong, K.W., Lee, M. (eds.) ICONIP 2019. LNCS, vol. 11953, pp. 264–274. Springer, Cham (2019). https://doi.org/10.1007/978-3-030-36708-4_22

SPMD-Based Neural Network Simulation with Golang

Daniela Kalwarowskyj(✉) and Erich Schikuta(ID)

Faculty of Computer Science, RG WST, University of Vienna,
Währingerstr. 29, Vienna 1090, Austria
dkalwarowskyj@yahoo.com, erich.schikuta@univie.ac.at

Abstract. This paper describes the design and implementation of parallel neural networks (PNNs) with the novel programming language Golang. We follow in our approach the classical Single-Program Multiple-Data (SPMD) model where a PNN is composed of several sequential neural networks, which are trained with a proportional share of the training dataset. We used for this purpose the MNIST dataset, which contains binary images of handwritten digits. Our analysis focusses on different activation functions and optimizations in the form of stochastic gradients and initialization of weights and biases. We conduct a thorough performance analysis, where network configurations and different performance factors are analyzed and interpreted. Golang and its inherent parallelization support proved very well for parallel neural network simulation by considerable decreased processing times compared to sequential variants.

Keywords: Backpropagation Neuronal Network Simulation · Parallel and Sequential Implementation · MNIST · Golang Programming Language

1 Introduction

When reading a letter our trained brain rarely has a problem to understand its meaning. Inspired by the way our nervous system perceives visual input, the idea emerged to write a mechanism that could "learn" and furthermore use this "knowledge" on unknown data. Learning is accomplished by repeating exercises and comparing results with given solutions. The neural network studied in this paper uses the MNIST dataset to train and test its capabilities. The actual learning is achieved by using backpropagation. In the course of our research, we concentrate on a single sequential feed forward neural network (SNN) and upgrade it into building multiple, parallel learning SNNs. Those parallel networks are then fused to one parallel neural network (PNN). These two types of networks are compared on their accuracy, confidence, computational performance and learning speed, which it takes those networks to learn the given task.

© The Author(s), under exclusive license to Springer Nature Switzerland AG 2023
J. Mikyška et al. (Eds.): ICCS 2023, LNCS 14075, pp. 563–570, 2023.
https://doi.org/10.1007/978-3-031-36024-4_43

The specific contribution of the paper is twofold: on the one hand, a thorough analysis of sequential and parallel implementations of feed forward neural network respective time, accuracy and confidence, and on the other hand, a feasibility study of Golang [8] and its tools for parallel simulation.

2   Related Work and Baseline Research

In the literature a huge number of papers on parallelizing neural networks can be found. An excellent source of references is the survey by Tal Ben-Nun and Torsten Hoefler [1]. However, only few research was done on using Golang in this endeavour.

In the course of our work on parallel and distributed systems [2,9,11] we developed several approaches for the parallelization of neural networks. In [6], two novel parallel training approaches were presented for face recognizing backpropagation neural networks. Further, we differentiate between topological data parallelism and structural data parallelism [10], where the latter is focus of the presented approach here.

3   Parallel Neuronal Networks

Go, often referred to as Golang [8], is a compiled, statically typed, open source programming language developed by a team at Google and released in November 2009. It is distributed under a BSD-style license, meaning that copying, modifying and redistributing is allowed under a few conditions.

Built-in support for concurrency is one of the most interesting aspects of Go, offering a great advantage over older languages like C++ or Java. One major component of Go's concurrency model are goroutines, which can be thought of as lightweight threads with a negligible overhead, as the cost of managing them is cheap compared to threads. If a goroutine blocks, the runtime automatically moves any blocking code away from being executed and executes some code that can run, leading to high-performance concurrency [8]. Communication between goroutines takes place over channels, which are derived from"Communicating Sequential Processes" found in [5]. A Channel can be used to send and receive messages from the type associated with it. Since receiving can only be done when something is being sent, channels can be used for synchronization, preventing race conditions by design.

Another difference to common object oriented programming languages can be found in Go's object oriented design. Its approach misses classes and type-based inheritance like subclassing, meaning that there is no type hierarchy. Instead, Go features polymorphism with interfaces and struct embedding and therefore encourages the composition over inheritance principle.

For the parallelization of neural network operations we apply the classical Single-Program Multiple-Data (SPMD) approach well known from high-performance computing [3]. It is a programming technique, where several tasks execute the same program but with different input data and the calculated output data is merged to a common result. Thus, based on the fundamentals of single feed forward neural network we generate multiple of these networks and set them up to work together in parallel manner.

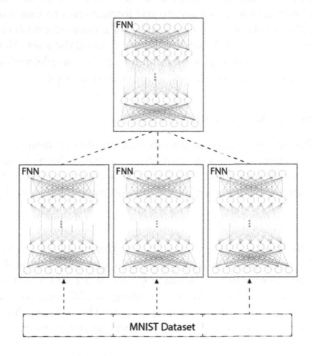

Fig. 1. Design of a Parallel Neural Network

The parallel-design is visualized in Fig. 1. On the bottom it shows the dataset which is divided into as many slices as there are networks, referred to as child-networks (CN). Each child-network learns only a slice of the dataset. Ultimately the results of all parallel child-networks are merged to one final parallel neural network (PNN). The combination of those CNs can be done in various ways. In the presented network the average of all weights, calculated by each parallel CN by a set number of epochs, is used for the PNNs weights. For the biases the same procedure is used, e.g. averaging all biases for the combined biases value.

4   Performance Evaluation

For our analysis, we use the MNIST dataset which holds handwritten numbers and allows supervised learning. Using this dataset the network learns to read handwritten digits. Since learning is achieved by repeating a task, the MNIST dataset has a "training-set of 60,000 examples, and a test-set of 10,000 examples" [7]. Each dataset is composed of an image-set and a label-set, which holds the information for the desired output and makes it possible to verify the networks output. All pictures are centered and uniform by 28×28 pixels. First, we start the training with the training-set. When the learning phase is over the network is supposed to be able to fulfill its task. To evaluate it's efficiency it is tested by running the neural network with the test-set since the samples of this set are still unknown. It is important to use foreign data to test a network since it is more qualified to show the generalization of a network and therefore its true efficiency. We are aware that MNIST is a rather small data set. However, it was chosen on purpose, because it is used in many similar parallelization approaches and allows therefore for relatively easy comparison of results.

4.1   Network Configurations

Number of Neurons. Choosing an efficient number of neurons is important, but it is hard to identify. There is no calculation which helps to define an effectively working number or range of neurons for a certain configuration of a neural network. Varying the number of neurons between 20 to 600 delivered great accuracy.

Number of Networks. To evaluate the performance of PNNs in terms of accuracy, PNNs with different amounts of CNs are composed and trained. The training runs over 20 epochs with a learning rate of 0.1 and a batchsize of 50. All CNs are built with one hidden layer consisting of 256 neurons. On the hidden layer the ReLU-function and on the output layer the Softmax-function is used. After every epoch, the networks are tested with the test-dataset. The results are visualized in Fig. 2.

Figure 2 illustrates a clear loss in accuracy of PNNs with a growing number of CNs. The 94.5% accuracy, for example, is reached by a PNN with 2 CNs after only one epoch, while a PNN with 30 CNs achieves that after 12 epochs.

Since the provided PNNs are built by using averaging of weights and biases it also seemed interesting to compare the average accuracy of the CNs with the resulting PNN, to grade the used combination function. The results are illustrated in Fig. 4.

It shows that the efficiency of an average function grows with the number of CNs. The first graph drawn with 2 CNs shows, that the resulting PNN is performing worse than the average of the CNs, it has been built from. By growing the number of CNs to 10, the average of CNs approximates towards the PNN. The last graph of this figure shows that a PNN composed of 20 CNs outperforms the average of its CNs after 200 epochs, and after 300 epochs levels with it. It has to be noted that the differences in accuracy are very small, as it is only a range of 0.1 to 0.2 percent. Overall it can be said that this combination function is working efficiently.

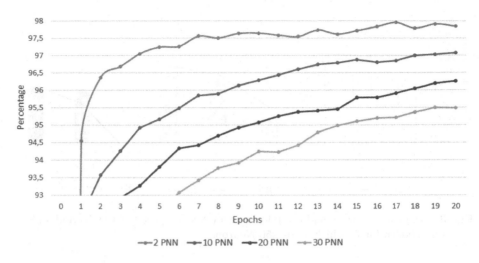

Fig. 2. Accuracy of PNNs, built with different amount of CNs, over 20 epochs

4.2 Comparing the Performances

Time. Time is the main reason to have a network working in parallel. To test the effect of parallelism on the time required to train a PNN, the provided neuronal network is tested on three systems. The first system is equipped with 4 physical and 4 logical cores, an Intel i7-3635QM processor working with a basic clock rate of 2.4 GHz, the second system holds 6 physical cores and 6 logical cores working with 2.9 GHz and an Intel i9-8950 HK processor and last the third system works with an AMD Ryzen Threadripper 1950X with 16 physical and 16 logical cores, which work with a clock rate of 3.4 GHz. The first, second and third systems are referred to as 4 core, 6 core and 16 core in the following.

In Fig. 5 the benefit in terms of time using parallelism is clearly visible. The results illustrated show the average time in seconds needed by each system for training a PNN consisting of one CN per goroutine.

The time in Fig. 5 starts on a high level and decreases with an increasing amount of goroutines for all three systems. Especially in the range of 1 to 4 goroutines, a formidable decrease in training time is visible and only starts to level out when reaching a systems physical core limitation. This means that the 4 core starts to level out after 4 goroutines, the 6 core after 6 goroutines and the 16 core after 16 goroutines, even though all systems support hyper threading. After reaching a systems core number the average time necessary for training a neural network decreases further with more goroutines. This should be due to the ability to work in parallel and in concurrency as one slot finishes and a waiting thread can start running immediately, without waiting for the rest of the running threads to be finished. All three systems show high time savings by parallelizing the neural networks. While time requirements decreased in every system, the actual time savings differ greatly as the 16 core system decreased 91% on average from 1 goroutine to 64 goroutines. In comparison, the 4 core

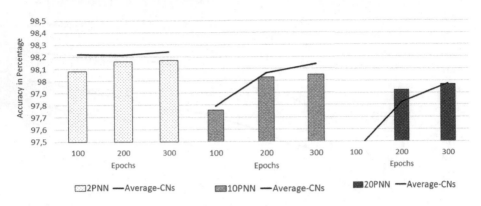

Fig. 3. Compare Accuracy and Confidence of a PNN composed of 10 CNs and an SNN with one Hidden Layer which holds 256 Neurons

system only took 65% less time. As the 16 core system is a lot more powerful than the 4 core system, it can perform an even greater parallel task and therefore displays a positive effect of parallelism upon time requirements.

Accuracy and Confidence of Networks. In this section the performance in terms of accuracy and confidence is compared between a PNNs and an SNN. For the test, illustrated by Fig. 3, both types of networks have been provided with the same random network to start their training. They have the exact same built, except that one is trained as SNN and the other is cloned 10 times to build a PNN with 10 CNs.

In Fig. 3 the SNN performs better than the PNN in both accuracy and confidence. While the SNNs accuracy and confidence overlap after 8 epochs, the PNN has a gap between both lines at all times. This concludes that the SNN is "sure" about its outputs, while the PNN is more volatile. The SNNs curve of confidence is a lot steeper than the PNNs and quickly approximates towards the curve of accuracy. Both curves of accuracy start off almost symmetric upwards the y-axis, but the PNN levels horizontally after about 90 percent while the SNN still rises until about 94 percent. After those points both accuracy curves run almost horizontally and in parallel towards the x-axis. The gap stays constantly until the end of the test. Even small changes within the range of 90 to 100 percent are to be interpreted as significant.

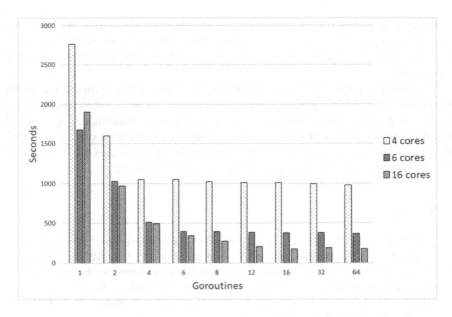

Fig. 4. Time in seconds, that was needed to train a PNN with a limited amount of one Goroutine per composed CN.

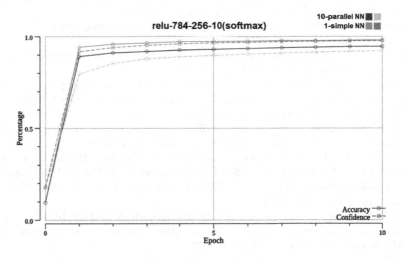

Fig. 5. Compare Accuracy and Confidence of a PNN composed of 10 CNs and an SNN with one Hidden Layer which holds 256 Neurons

5   Findings and Conclusion

This paper presents and analyses PNNs composed of several sequential neural networks. The PNNs are tested upon time and accuracy and compared to an SNN. Summing up, PNNs proved to be very time efficient but are still lacking in terms of accuracy. As there are plenty of other optimizations, e.g. adjusting learning rates [4], a PNN proved to be more time efficient than an SNN. However, until the issue of accuracy has been taken care of, the SNN surpasses the PNN in practice.

We close the paper with a final word on the feasibility of Golang for parallel neural network simulation: Data parallelism proved to be an efficient parallelization strategy. In combination with the programming language Go, a parallel neural network implementation is coded as fast as a sequential one, as no special efforts are necessary for concurrent programming thanks to Go's concurrency primitives, which offer a simple solution for multithreading.

References

1. Ben-Nun, T., Hoefler, T.: Demystifying parallel and distributed deep learning: An in-depth concurrency analysis. ACM Comput. Surv. (CSUR) **52**(4), 1–43 (2019)
2. Brezany, P., Mueck, T.A., Schikuta, E.: A software architecture for massively parallel input-output. In: Waśniewski, J., Dongarra, J., Madsen, K., Olesen, D. (eds.) PARA 1996. LNCS, vol. 1184, pp. 85–96. Springer, Heidelberg (1996). https://doi.org/10.1007/3-540-62095-8_10
3. Darema, F.: *The SPMD Model*: past, present and future. In: Cotronis, Y., Dongarra, J. (eds.) EuroPVM/MPI 2001. LNCS, vol. 2131, pp. 1–1. Springer, Heidelberg (2001). https://doi.org/10.1007/3-540-45417-9_1
4. Goyal, P., et al.: Accurate, large minibatch sgd: Training imagenet in 1 hour. arXiv preprint arXiv:1706.02677 (2017)
5. Hoare, C.A.R.: Communicating sequential processes. In: The origin of concurrent programming, pp. 413–443. Springer (1978)
6. Huqqani, A.A., Schikuta, E., Ye, S., Chen, P.: Multicore and gpu parallelization of neural networks for face recognition. Proc. Comput. Sci. **18**(Supplement C), 349–358 (2013): 2013 International Conference on Computational Science
7. LeCun, Y., Bottou, L., Bengio, Y., Haffner, P.: Gradient-based learning applied to document recognition. Proc. IEEE **86**(11), 2278–2324 (1998)
8. Meyerson, J.: The go programming language. IEEE Softw. **31**(5), 104–104 (2014)
9. Schikuta, E., Weishaupl, T.: N2grid: neural networks in the grid. In: 2004 IEEE International Joint Conference on Neural Networks (IEEE Cat. No.04CH37541), vol. 2, pp. 1409–1414 (2004)
10. Schikuta, E.: Structural data parallel neural network simulation. In: Proceedings of 11th Annual International Symposium on High Performance Computing Systems (HPCS 1997), Winnipeg, Canada (1997)
11. Schikuta, E., Fuerle, T., Wanek, H.: Vipios: The vienna parallel input/output system. In: Pritchard, D., Reeve, J. (eds.) Euro-Par 1998 Parallel Processing, pp. 953–958. Springer, Berlin (1998)

Low-Cost Behavioral Modeling of Antennas by Dimensionality Reduction and Domain Confinement

Slawomir Koziel[1,2]([✉]) [iD], Anna Pietrenko-Dabrowska[2] [iD], and Leifur Leifsson[3] [iD]

[1] Engineering Optimization and Modeling Center, Department of Engineering, Reykjavík University, Menntavegur 1, 102 Reykjavík, Iceland
koziel@ru.is
[2] Faculty of Electronics Telecommunications and Informatics, Gdansk University of Technology, Narutowicza 11/12, 80-233 Gdansk, Poland
anna.dabrowska@pg.edu.pl
[3] School of Aeronautics and Astronautics, Purdue University, West Lafayette, IN 47907, USA
leifur@purdue.edu

Abstract. Behavioral modeling has been rising in importance in modern antenna design. It is primarily employed to diminish the computational cost of procedures involving massive full-wave electromagnetic (EM) simulations. Cheaper alternative offer surrogate models, yet, setting up data-driven surrogates is impeded by, among others, the curse of dimensionality. This article introduces a novel approach to reduced-cost surrogate modeling of antenna structures, which focuses the modeling process on design space regions containing high-quality designs, identified by randomized pre-screening. A supplementary dimensionality reduction is applied via the spectral analysis of the random observable set. The reduction process identifies the most important directions from the standpoint of geometry parameter correlations, and spans the domain along a small subset thereof. As demonstrated, domain confinement as outlined above permits a dramatic improvement of surrogate accuracy in comparison to the state-of-the-art modeling approaches.

Keywords: Antenna design · surrogate modeling · behavioral modeling · dimensionality reduction · domain confinement · EM-driven design

1 Introduction

Contemporary antenna design largely relies on full-wave electromagnetic (EM) simulation tools, which are indispensable to account for mutual coupling effects [1]. Parameter tuning is increasingly often performed using rigorous numerical methods [2]. Yet, the major setback of EM-driven optimization is its high computational cost, often problematic for local tuning [3], and usually unmanageable for global search [4]. An extensive research has been devoted to expediting simulation-based design. Some of techniques attempt to lower the cost of direct EM-driven optimization (adjoint sensitivities [5],

© The Author(s), under exclusive license to Springer Nature Switzerland AG 2023
J. Mikyška et al. (Eds.): ICCS 2023, LNCS 14075, pp. 571–579, 2023.
https://doi.org/10.1007/978-3-031-36024-4_44

restricted Jacobian updates [6]). In surrogate modeling, expensive EM simulations are replaced by fast metamodels, which may be data-driven [7] or physics-based ones [8]. The former (kriging [9] or neural networks [10]) are significantly more popular due to their versatility and accessibility. The bottleneck of data-driven metamodels is the curse of dimensionality but also nonlinearity of high-frequency system responses. Physics-based modeling techniques (e.g., space mapping [11] are less prone to the mentioned difficulties. Other approaches capitalize on the distinctive shape of the system responses (e.g., feature-based technology [12]). Still, constructing accurate models is impeded by the problems related to dimensionality and extensive ranges of material and geometry parameters, the surrogate should be valid for to ensure its design utility.

In [13], a performance-driven methodology has been put forward, in which the modeling process is carried out in the section of the parameter space comprising designs of high quality w.r.t. the assumed figures of interest. Volume reduction permits radical improvement of the model predictive power [14]. From computational perspective, the limitation of this method is that the surrogate model domain is outlined using pre-optimized database designs, acquisition of which incurs considerable costs. In a recent alternative [15]; however, the domain is determined using a stochastic pre-selection procedure, thus no reference designs are employed.

This article proposes an advancement over the approach of [15], where the surrogate domain defined through pre-selection is further confined by means of the spectral analysis of the observable set. The final domain is spanned by a small number of eigenvectors of the covariance matrix of the observables. The outcome is a considerably improved accuracy and scalability of the model predictive power as a function of the training data set size. At the same time, design usability of the surrogate is retained.

2   Domain-Confined Modeling Using Pre-screening and Dimensionality Reduction

This section outlines the observable-based constrained modeling framework [15], being the core of our approach. Further, the inverse regression model employed for domain definition purposes is described, as well as dimensionality reduced domain.

2.1 Performance-Driven Modeling

Our approach capitalizes reference-design-free modeling technique [15], where the surrogate domain is constricted so that it encloses the parameter vectors of high quality as considered by the designer. Such a domain is considerably smaller volume-wise than the traditional domain. Thus, constructing the surrogate therein is significantly cheaper.

The notation utilized in the techniques [13] and [15] is summarized in Fig. 1. Observe that in any performance-driven technique, the modeling process is objective-oriented (i.e., focuses on the space of design objectives F), rather than the design space X.

Let us introduce the notion of design optimality [15] assessed with the use of the performance metric $U(\mathbf{x}, f)$, which $U(\mathbf{x}, f)$ appraises the quality of the vector x w.r.t. the vector of design objectives f. Consider an exemplary dual-band antenna and the design objectives defined as the operating frequencies $f_{0.1}$ and $f_{0.2}$. If the enhancement of antenna impedance matching over the fractional bandwidths B centered at both frequencies is of interest, then we may use $U(\mathbf{x}, \mathbf{f}) = U\left(\mathbf{x}, [f_{0.1} f_{0.2}]^T\right) = \max\{S_1(\mathbf{x}, \mathbf{f}), S_2(\mathbf{x}, \mathbf{f})\}$ (i.e., the merit function is defined as the maximum in-band reflection), with $S_k = \max\{f_{0.k}(1 - \beta/2) \leq f \leq f_{0.k}(1 + \beta/2) : |S_{11}(\mathbf{x}, f)|\}$, $k = 1, 2$, and f standing for the frequency.

Symbol	Description	Comments
$\mathbf{x} = [x_1 \ldots x_n]^T$	Vector of antenna parameters	Independent antenna dimensions to be tuned in the design process
$X = [\mathbf{l}\ \mathbf{u}]$	Conventional parameter space	$\mathbf{l} = [l_1 \ldots, l_n]^T$ and $\mathbf{u} = [u_1 \ldots, u_n]^T$ are lower and upper bounds on parameters, i.e., $l_k \leq x_k \leq u_k$ for $k = 1, \ldots, n$
$f_k, k = 1, \ldots, N$	Figures of interest	May include operating frequency (or frequencies), bandwidth, substrate permittivity, etc.
$\mathbf{f} = [f_1 \ldots f_N]^T$	Objective vector	Aggregates figures of interest pertinent to the antenna structure of interest
F	Objective space	Defined by ranges for figures of interest $f_{k.\min} \leq f_k^{(j)} \leq f_{k.\max}$, $k = 1, \ldots, N$, over which the surrogate is to be valid

Fig. 1. Notation of the nested kriging [13] and observable-based [15] modeling techniques.

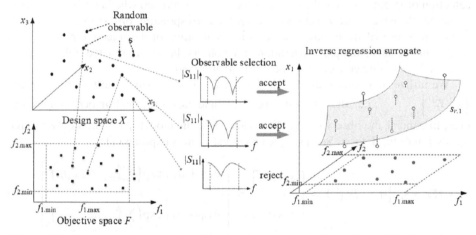

Fig. 2. (left) Observable generation for a dual-band antenna (X: 3-dimensional, F: 3-dimensional). (middle) Observable selection: samples of operating frequencies $\in F$ are kept, other are discarded. (right) Construction of an inverse model $s_r(\cdot)$ using the observable set $\{\mathbf{x}_r^{(j)}\}_{j = 1,\ldots,N_r}$, for a single component of $s_{r.j}$, Corresponding to parameter x_1 (shown as gray manifold).

Having defined $U(\mathbf{x}, f)$, the following minimization task is solved

$$\mathbf{x}^* = U_F(\mathbf{f}) = \arg \min_{\mathbf{x}} U(\mathbf{x}, \mathbf{f}) \tag{1}$$

where $\mathbf{x}^*$ refers to the optimal solution. We also define the N-dimensional manifold in the parameter space X comprising all the vectors rendered using (1) for all $\mathbf{f} \in F : U_F(F) =$

$\{U_F(\mathbf{f}) : \mathbf{f} \in F\}$. Setting up the surrogate model over the optimum design set suffices for ensuring the model design usability w.r.t. objective space F. Clearly, it is not possible to accurately identify $U_F(F)$ because this would require finding optimal solutions of (1) for all $f \in F$. Thus, its location within the design space X may be only roughly assessed. In the technique of [15], this is established by using random observables. As a consequence, the expenses associated with surrogate construction are low, whereas, the modeling accuracy is maintained at the same level as in the nested kriging [15]. Here, the aim is to improve the cost-efficiency even further by executing the modeling process in the domain of reduced dimensionality.

2.2 Domain Definition

In the reference-design-free constrained modeling [15], only a rough identification of $U_F(F)$ is possible, because the information pertaining to the location of the optimum design manifold encompassed in the observables is not precise. Thus, in [15], an auxiliary inverse regression surrogate is employed for the surrogate domain definition. This model is constructed as follows: over the design space X, the set of observables $\mathbf{x}_r^{(j)}, j = 1, 2, \dots,$ following a uniform probability distribution is gathered. Next, the respective responses are simulated, and the performance figure vectors $f_r^{(j)}$ are extracted therefrom. Next, a selection procedure is carried out: each vector $f_r^{(j)}$ is examined whether it belongs to the assumed objective space F (in such a case, the corresponding observable is accepted) or not (it is discarded). Figure 2 presents a visualization of the selection process for a dual-band antenna example. Generation of the observables is terminated if the assumed number N_r is collected (typically, between 50 and 100).

We utilize an auxiliary inverse model $s_r : F \rightarrow X$, which is built using the pairs: $\left\{ \mathbf{x}_r^{(j)} f_r^{(j)} \right\}_{j=1,\dots,N_r}$, i.e., the retained observables and the corresponding performance figures. In general, the observable quality is low, thus, the model s_r provides only a rough regression of the dataset. So, a simple analytical form is exploited [15]

$$s_r(\mathbf{f}) = \mathbf{s}_r\left(\left[f_1 \dots f_N\right]^T\right) = \begin{bmatrix} s_{r.1}(\mathbf{f}) \\ \dots \\ s_{r,n}(\mathbf{f}) \end{bmatrix} = \begin{bmatrix} a_{1.0} + a_{1.1}\exp\left(\sum_{k=1}^{N} a_{1.k+1}f_k\right) \\ \dots \\ a_{n.0} + a_{n.1}\exp\left(\sum_{k=1}^{N} a_{n.k+1}f_k\right) \end{bmatrix} \quad (2)$$

Despite featuring a small number of coefficients, the model (2) is sufficiently flexible. Identification of coefficients requires solving the weighted nonlinear regression task

$$\left[a_{j.0}\ a_{j.1} \dots a_{j.K+1} \right] = \arg\min_{[b_0 b_1 \dots b_{K+1}]} \sum_{k=1}^{N_r} w_k\left[s_{r.j}\left(f_r^{(k)}\right) - x_{r.j}^{(k)}\right]^2, \quad j = 1, \dots, n \quad (3)$$

where $\mathbf{x}_r^{(k)} = \left[x_{r:1}1^{(k)} \dots x_{r:n}^{(k)} \right]^T$, and the multipliers w_k serve to discriminate between the observables of different qualities. The latter requires defining auxiliary vectors $\mathbf{p}_r^{(j)} = \left[p_{r:1}^{(j)} \dots p_{r.N^{(j)}} \right]^T$ which are based on the evaluations of the design quality metric, and are derived from EM-evaluated antenna responses. The weights w_k are defined as $w_k = \left[\left(W - \max\{p_1(\mathbf{x}^{(i)}), \ldots, p_N(\mathbf{x}^{(j)})\}\right)/W\right]^2$, $k = 1, \ldots, N_r$., with $W = \max\{k = 1, \ldots, N_r, j = 1 \ldots, N : p_j^{(k)}\}$ being a normalization factor. The weighted regression is employed to ensure that better-quality vectors have a more profound effect on the regression surrogate, which is advantageous because they reside close to the optimum design set. The inverse surrogate for the exemplary dual-band antenna is shown graphically in Fig. 2. Surrogate $s_r(F)$ gives a rough appraisal of the manifold $U_F(F)$ of the optimal designs. The domain of the ultimate surrogate, rendering the antenna responses, is to encompass the entire $U_F(F)$, so we need to perform its orthogonal extension, as in [15].

First, a set of designs is allocated on $s_r(F)$ uniformly w.r.t. the objective space F on a rectangular grid $F_G \subset F$. The vector of objectives $\boldsymbol{f}_g = [f_{g.1} \ldots f_{g.n}]^T$ resides on the grid, if and only if $f_{g.k} = f_{k.\min} + (f_{k.\max} - f_{k.\min})m_k/(M-1)$, $k = 1, \ldots, N$, where $m_k \in \{0, 1, \ldots M - 1\}$, and M is grid density (its value is not critical, we set $M = 4$ or 5). The grid F_G encompasses M^N vectors in total. Thus, we have the points $\mathbf{x}_g^{(j)} = s_r\left(\mathbf{f}_g^{(j)}\right), j = 1, \ldots, M^N$, which render an approximate spatial allocation of the image $s_r(F)$ of the inverse model.

Symbol	Description	Default value and comments
N_r	Number of random observables	Typically set between 50 and 100. The value is not critical, but should be larger for higher-dimensional objective spaces
p	Surrogate domain dimensionality	Typically set to $p = 3$; the required value can be estimated by based on the analysis of the eigenvalues λ_k
N_B	Number of training data samples for model construction	Depends on required predictive power of the surrogate model. Typical values to ensure relative RMS error of a few percent are between 200 to 500 samples

Fig. 3. Control parameters of the introduced modeling procedure with reduced dimensionality.

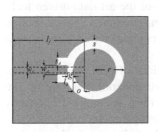

Substrate	ε_r – operating parameter, $h = 0.76$ mm
Design parameters[$]	$x = [l_f\, l_d\, w_d\, r\, s\, s_d\, o\, g]^T$
Other parameters[$]	w_f – adjusted for 50Ω line impedance
EM model	CST (~300,000 cells, sim. time 90 s)
Figures of interest	Center freq. f_0; substrate permittivity ε_r
Objective space	2.5 GHz $\leq f_0 \leq$ 6.5 GHz, 2.0 $\leq \varepsilon_r \leq$ 5.0
Design optimality	Minimum reflection at target operating frequency and substrate permittivity
Conventional parameter space X	$l = [22.0\ 3.5\ 0.3\ 6.5\ 3.0\ 0.5\ 3.5\ 0.2]^T$, $u = [27.0\ 8.0\ 2.3\ 16.0\ 7.0\ 5.5\ 6.0\ 2.3]^T$
What is modelled?	Reflection response S_{11}
	[$] Dimensions in mm.

Fig. 4. Ring-slot antenna [17], (b) details on antenna structure.

Next, the spectral analysis of the set $\left\{\mathbf{x}_g^{(j)}\right\}_{j=1,\ldots,M^N}$, is carried out so as to find the most important correlations between the coordinates of these designs. We define a covariance matrix $\mathbf{S}_g = \left[\sum_{j=1,\ldots,M^N}\left(\mathbf{x}_g^j - \mathbf{x}_m\right)\left(\mathbf{x}_g^{(j)} - \mathbf{x}_m\right)^T\right]/(M^N - 1)$, with, $\mathbf{x}_m =$

$[\sum_{j=1,...,M^N} \mathbf{x}_g^{(j)}]/M^N$ being the set's center. The principal components [16] of S_g are referred to as $\mathbf{v}_k$, $k = 1, ..., n$ (eigenvectors), and λ_k, $k = 1, ..., n$ (eigenvalues; listed in a descending sequence, i.e., $\lambda_1 \geq \lambda_2 \geq ... \geq \lambda_n \geq 0$). Typically, only the first few eigenvectors are meaningful. Thus, surrogate domain is delimited using them.

We use the following expansion $\mathbf{x}_g^{(j)} = \mathbf{x}_m + \sum_{k=1,...,n} b_{jk} \mathbf{v}_k$. We also have the center point: $\mathbf{x}_c = \mathbf{x}_m + [\mathbf{v}_1 \ \mathbf{v}_2 \ ... \ \mathbf{v}_n]\mathbf{b}_0$, where the entries of the vector $\mathbf{b}_0 = [b_{1.0} \ ... \ b_{n.0}]^T$ are: $b_{j.0} = (b_{j.\max} - b_{j.\min})/2$, $j = 1, ..., n$, and $b_{j\cdot\max} = \max\{k : b_{kj}\}$, $b_{j\cdot\min} = \min\{k : b_{kj}\}$, as well as the eigenvalue vector $\lambda_b = [\lambda_{b1} \ ... \ \lambda_{bn}]^T$, with $\lambda_{bj} = 0.5(b_{j\cdot\max} + b_{j,\min})$. The confined domain of reduced dimensionality, spanned by the vectors $\mathbf{v}_1$ through $\mathbf{v}_p$, is defined as

$$X_p = \left\{ \begin{array}{c} \mathbf{x} = \mathbf{x}_c + \sum_{k=1}^p (2\lambda_k - 1)\lambda_{b_k} \mathbf{v}_k \\ 0 \leq \lambda_k \leq 1, k = 1, ..., p \end{array} \right\} \tag{4}$$

In practice, the eigenvalues are quickly decreasing, so it suffices to exploit p directions, where p is much smaller than n (the design space dimensionality), which is beneficial from the point of view of the training data acquisition.

The control parameters of the developed modeling procedure along with the discussion of their recommended values are presented in Fig. 3. The procedure requires supplying the following input parameters: lower and upper bounds on parameters $\mathbf{l}$ and $\mathbf{u}$ delimiting the conventional parameter space X, as well as the bounds for operational conditions delimiting the space of design objectives F. Observe that in the presented approach, the final surrogate is built from both the training data set $\{\mathbf{x}_B^{(j)}\}$ and the observable data $\{\mathbf{x}_r^{(j)}\}$. This is to improve the overall modeling accuracy.

3 Results

This section demonstrates the performance of the introduced modeling methodology along with its design utility. Benchmarking against state-of-the-art data-driven techniques is provided as well. Figure 4 shows a ring-slot antenna employed as a verification case. Here, the aim is to build surrogate models representing complex reflection responses S_{11} versus frequency.

The surrogate models have been built using the technique of Sect. 2. The number of retained observables is set to $N_r = 50$, which required generating a total of 106 samples. The reduced dimensionality of the model domain was set to $p = 3$. To study the model scalability, the surrogates were rendered for training data sets of sizes $N_B = 50, 100, 200, 400$, and 800. The benchmark methods include: (i) kriging (in the standard space X); (ii) radial basis function (RBF) (within X); (iii) convolutional neural network (CNN) [18] (within X)(iv) Ensemble Learning [19] (in the confined domain X_S); (v) nested kriging [13] (within X_S); the cost of identifying the database designs is 864 EM simulations, (vi) reference-design-free constrained model [15], set up in domain X_S; the cost of generating the observables is added to the overall surrogate set up expenses, which is the same as for the method of Sect. 2. The predictive power of the surrogates is quantified using a relative RMS error, defined as $\|\mathbf{R}_s(\mathbf{x}) - \mathbf{R}_f(\mathbf{x})\| / \|\mathbf{R}_f(\mathbf{x})\|$, with $\mathbf{R}_s$

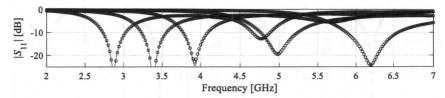

Fig. 5. Reflection responses at the selected test locations: EM model (—), and the introduced constrained surrogate with dimensionality reduction (o). The surrogate built with $N_B = 400$ data samples.

Table 1. Modeling results and benchmarking for ring-slot antenna

Modeling method		Number of training samples				
		50	100	200	400	800
Kriging	Modeling error	56.9%	50.8%	35.8%	31.5%	25.6%
	Modeling setup cost	50	100	200	400	800
RBF	Modeling error	61.0%	53.2%	37.9%	34.1%	27.2%
	Modeling setup cost	50	100	200	400	800
CNN	Modeling error	67.7%	58.8%	34.0%	22.3%	13.5%
	Modeling setup cost	50	100	200	400	800
Ensemble learning	Modeling error	73.8%	69.1%	63.9%	58.1%	55.8%
	Modeling setup cost	50	100	200	400	800
Nested kriging [13]	Modeling error	19.4%	12.9%	7.7%	5.1%	3.7%
	Modeling setup cost$	914	964	1,064	1,264	1,664
Constrained modeling [15]	Modeling error	13.4%	9.9%	6.9%	5.4%	4.4%
	Modeling setup cost#	156	206	306	506	906
This work	Modeling error	14.3%	10.1%	6.7%	3.5%	2.2%
	Modeling setup cost#	156	206	306	506	906

$ The cost includes acquisition of the reference designs.
# The cost includes generation of random observables, here, 106 simulations in total to yield N_r = 50 accepted samples

and R_f being the responses predicted by the surrogate and EM analysis. The error is computed using 100 randomly allocated test points.

Table 1 provides numerical results, and Fig. 5 shows the antenna responses for the selected test locations. The accuracy of our model is superior to all models built in the parameter space X, and also to nested kriging, which is mainly a result of dimensionality reduction and volume-wise confinement. Cost-efficiency of our method is better than that of nested kriging because no reference designs are required, thus the overall numbers of EM analyses is reduced by 80 percent for $N_B = 50$. For $N_B = 800$, the reduction is 45 percent. The efficiency of our method is identical to that of the procedure of [19].

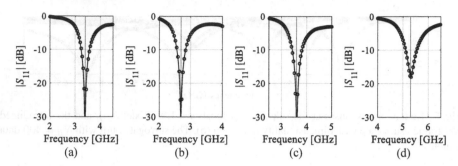

Fig. 6. Optimization using the proposed surrogate: surrogate prediction (o) and EM-evaluated characteristics (—) at the design yielded by optimizing the model. The responses evaluated for $F = [f_0 \ \varepsilon_r]$ (frequency in GHz): (a) $F = [3.4 \ 3.6]$, (b) $F = [2.7 \ 4.3]$, (c) $F = [3.6 \ 2.2]$, (d) $F = [5.3 \ 4.3]$. The intended operational frequency is shown with vertical line.

Figure 6 shows the results of applying the proposed surrogate (constructed using 800 training samples) to optimize a ring-slot antenna for a variety of operating frequencies and substrate permittivity ε_r. It can be observed that satisfactory designs are obtained for all assumed targets, which corroborates design utility of the models.

4 Conclusion

This article proposed a novel procedure for cost-efficient surrogate modeling of antenna structures. Our technique capitalizes on the performance-driven paradigm with the model domain established in the region containing high-quality designs. A supplementary dimensionality reduction is applied by spanning the domain along the most relevant directions that account for spatial orientation of the initially established region of interest. These mechanisms lead to a dramatic increase in the surrogate model predictive power without compromising its design usefulness. Both features have been conclusively corroborated through extensive benchmarking and application case studies.

Acknowledgement. The authors would like to thank Dassault Systemes, France, for making CST Microwave Studio available. This work is partially supported by the Icelandic Centre for Research (RANNIS) Grant 217771 and by National Science Centre of Poland Grant 2018/31/B/ST7/02369.

References

1. Wang, Y., Zhang, J., Peng, F., Wu, S.: A glasses frame antenna for the applications in internet of things. IEEE Internet of Things J. **6**(5), 8911–8918 (2019)
2. Tu, S., Cheng, Q.S., Zhang, Y., Bandler, J.W., Nikolova, N.K.: Space mapping optimization of handset antennas exploiting thin-wire models. IEEE Trans. Ant. Propag. **61**(7), 3797–3807 (2013)
3. Kolda, T.G., Lewis, R.M., Torczon, V.: Optimization by direct search: New perspectives on some classical and modern methods. SIAM Rev. **45**, 385–482 (2003)

4. Tang, M., Chen, X., Li, M., Ziolkowski, R.W.: Particle swarm optimized, 3-D-Printed, wideband, compact hemispherical antenna. IEEE Ant. Wireless Propag. Lett. **17**(11), 2031–2035 (2018)

5. Hassan, E., Noreland, D., Augustine, R., Wadbro, E., Berggren, M.: Topology optimization of planar antennas for wideband near-field coupling. IEEE Trans. Ant. Prop. **63**(9), 4208–4213 (2015)

6. Koziel, S., Pietrenko-Dabrowska, A.: Reduced-cost electromagnetic-driven optimization of antenna structures by means of trust-region gradient-search with sparse Jacobian updates. IET Microwaves Ant. Prop. **13**(10), 1646–1652 (2019)

7. Simpson, T.W., Pelplinski, J.D., Koch, P.N., Allen, J.K.: Metamodels for computer-based engineering design: Survey and recommendations. Eng. Comput. **17**, 129–150 (2001)

8. Cervantes-González, J.C., Rayas, J.E., López, C.A., Camacho-Pérez, J.R., Brito-Brito, Z., Chávez-Hurtado, J.L.: Space mapping optimization of handset antennas considering EM effects of mobile phone components and human body. Int. J. RF Microwave CAE **26**(2), 121–128 (2016)

9. Forrester, A.I.J., Keane, A.J.: Recent advances in surrogate-based optimization. Prog. Aerospace Sci. **45**, 50–79 (2009)

10. Kabir, H., Wang, Y., Yu, M., Zhang, Q.J.: Neural network inverse modeling and applications to microwave filter design. IEEE Trans. Microwave Theory Tech. **56**(4), 867–879 (2008)

11. Melgarejo, J.C., Ossorio, J., Cogollos, S., Guglielmi, M., Boria, V.E., Bandler, J.W.: On space mapping techniques for microwave filter tuning. IEEE Trans. Microw. Theory Tech. **67**(12), 4860–4870 (2019)

12. Koziel, S.: Fast simulation-driven antenna design using response-feature surrogates. Int. J. RF & Micr. CAE **25**(5), 394–402 (2015)

13. Koziel, S., Pietrenko-Dabrowska, A.: Performance-based nested surrogate modeling of antenna input characteristics. IEEE Trans. Ant. Prop. **67**(5), 2904–2912 (2019)

14. Koziel, S., Pietrenko-Dabrowska, A.: Performance-driven surrogate modeling of high-frequency structures. Springer, New York (2020)

15. Koziel, S., Pietrenko-Dabrowska, A.: Knowledge-based performance-driven modeling of antenna structures. Knowl, Based Syst. **237**, paper no. 107698 (2022)

16. Jolliffe, I.T.: Principal component analysis, 2nd edn. Springer, New York (2002)

17. Koziel, S., Pietrenko-Dabrowska, A.: Design-oriented modeling of antenna structures by means of two-level kriging with explicit dimensionality reduction. AEU Int. J. Electronics Comm. **127**, 1–12 (2020)

18. Mahouti, P.: Application of artificial intelligence algorithms on modeling of reflection phase characteristics of a nonuniform reflectarray element. Int. J. Numer. Model. **33** (2020)

19. Zhang, Y., Xu, X.: Solubility predictions through LSBoost for supercritical carbon dioxide in ionic liquids. New J. Chem. **44**, 20544–20567 (2020)

Real-Time Reconstruction of Complex Flow in Nanoporous Media: Linear vs Non-linear Decoding

Emmanuel Akeweje[1]([✉]) [iD], Andrey Olhin[2] [iD], Vsevolod Avilkin[2] [iD],
Aleksey Vishnyakov[3,4] [iD], and Maxim Panov[5]([✉]) [iD]

[1] Trinity College Dublin, Dublin, Ireland
eakeweje@gmail.com
[2] Skolkovo Institute of Science and Technology, Moscow, Russia
[3] Department of Physics, Moscow State University, Moscow, Russia
[4] Krestov Institute of Solutions Chemistry, Ivanovo, Russia
[5] Technology Innovation Institute, Abu Dhabi, UAE
maxim.panov@tii.ae

Abstract. Physical field reconstruction from limited real-time data is a topical inverse problem that attracts substantial research effort, and complex geometries present a formidable challenge. The paper describes a reconstruction of the velocity field of a steady fluid flow through a two-dimensional porous structure from the real-time gauge readings (that is, velocity values obtained at specific fixed locations). The dataset is composed of 300 Lattice-Boltzmann simulations of the flow with different boundary conditions. The number of the gauges and their locations are varied. Two reconstruction techniques are applied: neural network (NN) and linear least squares solver. The linear solver outperforms the NN in terms of both speed and precision. Sensor locations are optimized by Monte Carlo method. The porous structure is mapped onto a graph and the optimization is performed by Metropolis type node-to-node trial displacements of the gauges. With 100 gauges, the linear method enables reconstruction of the velocity field in a porous structure discretized on 256×256 2D grid with the normalized error of 0.57%.

Keywords: physical field reconstruction · fluid flow · porous materials · neural networks · MCMC

1 Introduction

In engineering computing, there are situations when the laws governing the behavior of a system are well known, and its behavior can be predicted with good accuracy, but the first-principle solution is complicated either by a lack of information on the initial/boundary conditions, or by computational expenses (for example, the response is needed within seconds, whereas calculations take

© The Author(s), under exclusive license to Springer Nature Switzerland AG 2023
J. Mikyška et al. (Eds.): ICCS 2023, LNCS 14075, pp. 580–594, 2023.
https://doi.org/10.1007/978-3-031-36024-4_45

hours). At the same time, experimental real-time monitoring using gauges is possible only in a limited number of locations and/or at select moments in time due to equipment-related or economic reasons. Real-time monitoring requires fast restoration of the values in the locations where direct observations are unavailable from the available readings. Modification of this problem is restoration of properties that are difficult to measure from the properties easy to monitor. An obvious extension is the optimization of the gauge locations.

Physical field restoration is a classical inverse problem: the direct problem can be solved using deterministic methods with a reasonable accuracy given the initial and/or boundary conditions, but condition the current gauge readings may correspond to are unknown. The problem may or may not be well-posed: that is, there is no guarantee that similar readings cannot be obtained with different set of conditions. The reconstruction problem is of a great practical importance: building monitoring systems, temperature tracking for processors and chips, water resource monitoring, and industrial processes control require reconstruction of velocity, pressure, temperature and magnetic fields [1–3]. A number of data-driven approaches to the physical field restoration problem have been developed over the last few years, most of them quite recently.

In many applications, field reconstruction tasks are approached using a low dimensional representation of the field [4–9]. This low-dimensional representation can be viewed as a tailored basis of the field such as proper orthogonal decomposition (POD) basis, Fourier basis and wavelet basis. For efficient reconstruction of thermal field, Li et al. [5] obtained low dimensional representations of the high dimensional temperature distribution state of the thermal maps via POD techniques and then implemented a greedy algorithm for optimal sensor placement. Willcox [8] extended the gappy POD method to handle unsteady flow reconstruction. Tan et al. [10] demonstrated the application of POD in an iterative procedure to reconstruct incomplete or inaccurate aerodynamic data.

The main challenge that POD methods face is the sensitivity of the reconstructions to the location of the sensors. Neural networks (NNs) promise more stable solutions in physical field reconstruction [11–14], although they are substantially more computationally expensive to train. Some of the NN-based schemes for related CFD problems directly take into account the physical laws that govern the system behaviour [15–17], the others do not [18–21]. Erichson et al. [11] proposed a simple shallow neural network based learning algorithm for reconstruction. The shallow neural network learned an end-to-end mapping between a set of sensor measurements and the high-dimensional fluid flow field from which the measurements were taken, without requiring special data preprocessing. More sophisticated NNs were employed to tackle the field reconstruction problem in [12]; the authors employed reshaping operations to allowed for the possibility of harnessing the performance of some state of the art convolutional image models as opposed to just the fully connected network architecture. Two of their models have architectures similar to that of U-Net; with one of them having Fourier Neural Operator layers. NN-based inversion methods are powerful for learning and are increasingly used field reconstruction [13, 22, 23].

Reconstruction of physical fields from limited sensor measurements from the system could be expressed as a map of any form, be it linear or non-linear. Clark et al. [24] approached the reconstruction problem with an algorithm based on linear maps and incorporated this algorithm in a greedy scheme for sensor placement. Thus, a field reconstruction technique strictly by linear map was introduced. This technique, even though yet to be explored in many applications, is advantageous for reconstruction problems in that it allows to simply determine the stability and optimality of a collection of (sensor) measurements, and also handy for building effective algorithms for optimal sensors placement. Despite the linearity constraint, this technique yields good reconstructions even with measurements from randomly placed sensors [24,25].

In this study, we consider data-driven techniques for reconstruction of steady flow in complex porous geometry from limited sensor readings and a dataset of velocity fields calculated for 300 different sets of boundary conditions. Most literature considered flow state reconstruction in simple geometries. Systems of complex morphology are more challenging to work with. The main **contributions** of this work are as follows.

- Using a least squares based linear solver and a neural network, we reconstruct the stationary state of the system from limited sensor measurements, see Sects. 3 and 4.
- We optimize the location of the sensors and determine the minimum number of sensors sufficient for the reconstruction with a given precision, see Sect. 5.

2   Fluid Flow in 2D Porous Medium

As a case study, we selected reconstruction of the velocity field of a 2D flow in a model porous media. The pore structure is generated using the Porespy package [26]. The algorithm generates random noise, then applies a gaussian blur to the noise. Parameter σ controlled by the desired level of sample blobiness as $\sigma = \bar{l}/(40 \cdot \text{blobiness})$, where $\bar{l}$ is mean value (in voxels) of the set width and height of the sample image. After that, we re-normalizing acquired data to uniform distribution. The resulting geometry is shown in Fig. 1. The binary structure is discretized on 256×256 voxels (each site is either filled or open), has a 0.72 porosity ratio and the blobiness of 1 in all direction.

On the north and south borders, periodic boundary conditions are applied. A constant pressure difference of 0.0005 (Lattice units) is maintained between the west and the east boundaries to establish a steady flow. The relaxation ratio is $\tau = 1.0$. The parameters are chosen so that the value of $u_x^{\max} < 0.2$ in the resulting velocity profile does not exceed the stability limits of the numerical simulation. In each simulated system, a constant velocity profile is maintained at the western boundary. The velocity profile is generated in a random fashion using harmonic function as $u_x(y) = \sum_i A_i \sin(k_i y + 2\pi\xi)$, where $A_i = a\left(1 + (k_i L_0)^2\right)^{-1/2}$ are the amplitudes of the modes, ξ is random number with uniform distribution $[0, 1]$, $k_i = \frac{1}{2}\frac{2\pi}{L}i$ is the wave number. The parameter L_0 is the characteristic length of the undulations. The sample velocity profiles are

Fig. 1. The pore structure used as a case study for this work.

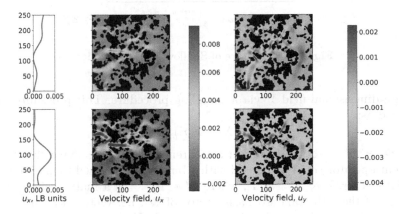

Fig. 2. The figure shows two examples of the velocity profile at the western boundary (left panels) and corresponding steady velocity fields in x (center panel) and y (right panel) dimensions

shown at Fig. 2. The velocity fields for each profile are generated by the LB simulations with BGK approximation of the Boltzmann equation for 2D, 9 discrete velocity model (D2Q9) [31]. Each simulation is carried out over 5000 iterations on average. The simulations are assumed to converge to the running average of the standard deviation of the energy over 500 iterative steps, decreased to $\epsilon = 10^{-4}$. In total, 300 random profiles are generated, and for each of them the steady state velocity fields are calculated. This dataset [32] is used for reconstruction of the velocity field from the gauge readings.

2.1   Mathematical Framework for the Velocity Field Reconstruction

The objective of velocity field reconstruction is to learn the relationship between gauge readings $X \in \mathbb{R}^{m \times 2}$ and velocity field data $Y \in \mathbb{R}^{n \times 2}$, with a constrain of limited gauge number $m \ll n$. For this study, each row of X and Y consist of velocity data in both x and y directions at a certain location (point) in the velocity field. The sensor measurements $\boldsymbol{x}_i \in X$, $1 \leq i \leq m$ are collected from

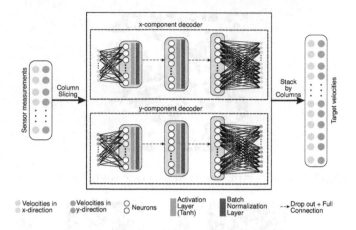

Fig. 3. Architecture of Shallow Neural Decoder.

the high-dimensional field Y via a sampling procedure. This procedure can be described as:

$$X = \Phi(Y),$$

where $\Phi: \mathbb{R}^{n \times 2} \to \mathbb{R}^{m \times 2}$ denotes a measurement operator. Typically, the measurement operator could be related with a binary matrix used to mask out some flow data. Therefore, $X = MY$, where $M \in \mathbb{R}^{m \times n}$ is a matrix whose m rows comprise of the j-th standard basis vector of $\mathbb{R}^n$, i.e. vectors e_j with j representing the location index of the measurement in Y.

The task at hand requires the construction of an inverse model, which generates the state Y from observations X, described as $Y = \Gamma(X)$, where $\Gamma: \mathbb{R}^{m \times 2} \to \mathbb{R}^{n \times 2}$ denotes the forward operator. This problem is frequently ill-posed, and we cannot invert the measuring operator Φ to get the forward operator Γ immediately. However, a function Ψ to approximate the forward operator Γ can be derived from a set of available sensor measurement and velocity field data, that is $\{X_i, Y_i\}_{i=1}^k$, where k is the number of flow snapshot available in the dataset. In particular, the objective is to learn a function $\Psi: X \to Y$ that maps a limited number of measurements to a predicted state Y,

$$\hat{Y} = \Psi(X), \tag{1}$$

so that the misfit is minimized, for instance in a Euclidean sense, over all sensor measurements $\|\Psi(X) - \Gamma(X)\|^2 < \epsilon$, where ϵ is a very small positive number. In this study, Ψ is taken to be a regression function: shallow neural network or linear solver.

3   Shallow Neural Decoder

We first consider estimating Ψ in Eq. (1) with neural networks that are generally labelled as universal estimators. Taking inspiration from the works of Erichson

Table 1. Hyperparameters for the optimal SNDs. There are no scheduler parameters for the SND for $m = 50$ sensors because the model was trained without scheduling the learning rate.

no. of sensors (m)	model hyperparameters			training hyperparameters				
	1st hidden layer size	2nd hidden layer size	output layer size	batch size	lr	betas	scheduler step	gamma
10	200	1000	10686	32	0.005	(0.9, 0.95)	50	0.5
50	200	1000	10646	32	0.001	(0.95, 0.95)	-	-
500	200	500	10196	16	0.05	(0.95, 0.99)	100	0.5
996	200	500	9700	32	0.01	(0.95, 0.95)	100	0.5

et al. [11], we developed a neural network architecture as in Fig. 3; a common architecture for decoders consists of layers of non-decreasing sizes, which increase the size of the representation from low-dimensional observations to the high-dimensional field in a continuous manner. Hence, it is named as Shallow Neural Decoders (SND). The SND consists of three hidden layers, activation layers to introduce non-linearity, batch normalization layers and drop out. For the activation function, we found Tanh to work better in our experiments, however other choices such as ReLU could produce more efficient result for other applications. The velocity of the flow in our system is two-dimensional, and for this reason we considered a separated-decoder architecture, such that the model learns the components of the target velocities independent of the other. Thus, the SND consists of x-component and y-component decoders as shown in Fig. 3.

Setting the number of sensors, m, to 10, 50, 500, and 996, we experimented with 3-layered SNDs. Adam optimization technique was used to obtain the model parameters which minimize the mean squared error loss function. Training the SNDs heavily depend on the hyperparameters, and to obtain the optimum SND, we repeatedly trained the SNDs with different combinations from a range (or set) of each of the hyperparameter: learning rate, betas, batch size, number of layers in hidden units, scheduler step, and scheduler gamma. The combinations of hyperparameters, which yield the optimum decoder for each number of sensor m is presented in Table 1.

Metrics. In this study, three metrics: the normalized error (NE), normalized fluctuation error (NFE) and coefficient of determination (also known as R^2), are employed to evaluate the performances models and algorithms. The metrics are defined as follow:

$$\text{NE}(Y, \hat{Y}) = \frac{\|Y - \hat{Y}\|_2}{\|Y\|_2}, \quad \text{NFE}(Y, \hat{Y}) = \frac{\|Y' - \hat{Y}'\|_2}{\|Y'\|_2} = \text{NE}(Y', \hat{Y}'),$$

$$R^2(Y, \hat{Y}) = 1 - \frac{\sum_{i=1}^{n}(Y_i - \hat{Y}_i)^2}{\sum_{i=1}^{n}(Y_i - \bar{Y})^2} = 1 - \frac{\|Y' - \hat{Y}'\|_2^2}{\|Y'\|_2^2} = 1 - \text{NFE}^2(Y, \hat{Y}),$$

where Y is the ground-truth velocity, $\hat{Y}$ is model prediction, $Y' = Y - \bar{Y}$, $\hat{Y}' = \hat{Y} - \bar{Y}$, and $\bar{Y}$ is the empirical mean. NE penalizes over-estimations and

Table 2. Performance of SNDs.

	Training set			Validation set			Testing set		
m	NE	NFE	R^2	NE	NFE	R^2	NE	NFE	R^2
10	0.326	0.546	0.702	0.303	0.490	0.760	0.274	0.489	0.761
50	0.146	0.243	0.941	0.148	0.240	0.942	0.148	0.264	0.930
500	0.033	0.055	0.997	0.039	0.064	0.996	0.032	0.058	0.997
996	0.051	0.085	0.993	0.054	0.087	0.992	0.057	0.102	0.990

under-estimations in model prediction. NFE eliminates the possible dominating empirical mean and focuses on the fluctuations in the field. Lastly, the coefficient of determination (R^2) checks how much of the variability in the dataset is explainable by a model. Using these metrics, the error in the flow state reconstruction can be quantified, and model performance is evaluate. We sought for models with the lower NEs, lower NFEs, and higher R^2 scores.

Results. The synthesized velocity dataset, as described in Sect. 2, was split into three sets such that 70% was used for training, 20% for validation and 10% for testing the models. As common with other neural networks, the validation set is to keep the model in check whilst training to prevent overfitting on the training set. We down-sampled (coarse-grained) each snapshots in the dataset to 128×128 for computational reasons, this resulted in having 10,696 velocity points in the "porous space". The performance of the SNDs given the number of sensors is presented in the Table 2. Although more sensor measurements generally produced better reconstruction, the results from the SND reconstructions reflects that reconstruction with 500 sensor measurement is better than that of 996 sensor measurements. This is an indication that for the system, there may be a sufficient number of sensor required for an efficient field reconstruction. With 500 sensor measurements, the SND produced reconstruction with R^2 scores of 0.997 on the testing set. Figure 4 is a sample of reconstruction produced by SND from the testing set.

4   Least Squares Linear Algorithm

A major concern with the neural decoders for flow state reconstruction tasks is the need to always retrain the decoders whenever the sensor positions are altered. Some optimization problems related to physical field reconstruction includes identification of the optimal positions to place sensors, and having knowledge of the number of sensors sufficient for such reconstructions. Retraining neural decoders is usually computational expensive. This concern motivated us to consider algorithms which requires less computation resources for modeling the relationship between sensor measurements and target velocities.

It is quite easy to spot the important role of least squares solutions in classical reconstruction methods such as POD. Without the least squares solutions, reconstructions with tailored basis will become intractable. However, by imposing a

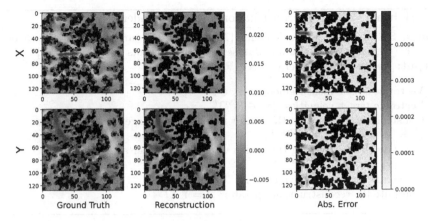

Fig. 4. SND reconstruction from 500 sensors. The green pixels represent the position of the sensors in the velocity field, the sensors were placed at randomly picked spots on walls of the porous medium for this experiment. (Color figure online)

linear relationship between sensor measurements and corresponding (missing) velocity information, the least square method alone can sufficiently reconstruct the state of the flow field. Hence, the name LS-decoder. The linear relationship in LS-decoder requires that the velocity data matrices, Y and X, be vectorized before imposing the linearity assumption. This algorithm is similar to the exact Dynamic Mode Decomposition where the Koopman operator could be approximated by least squares technique [27].

Let us consider a snapshot which consists of $m + n$ points such that there are m sensors which measures the velocity in both x- and y- directions, that is a total of $2m$ measurements would be recorded. To reconstruct the velocity field requires estimating missing $2n$ point velocities. Mathematically, these tall vectors are constructed from the velocity data matrices by vectorization:

$$\mathbf{x} = \text{vectorize}(X) \in \mathbb{R}^{2m}, \quad \mathbf{y} = \text{vectorize}(Y) \in \mathbb{R}^{2n}.$$

The linearity relationship then implies that $\mathbf{y} \approx \tilde{A}\mathbf{x}$ with $\tilde{A} \in \mathbb{R}^{2n \times 2m}$. Typically, $m \ll n$. Matrix $\tilde{A}$ is the (rectangular) matrix of importance coefficients a_{ij}. The importance coefficient, a_{ij}, is interpreted as the contribution of the sensor measurement indexed j in $\mathbf{x}$ to estimation of velocity at missing point indexed i in $\mathbf{y}$. Again, this relationship means that each velocity estimate is computed as a weighted sum of available sensor measurements. Typically, we expect that the proximity of the points at which velocity is been estimated to the gauges directly influence the importance of each gauge in the estimation. The coefficient matrix is computed strictly via data-driven approach. To obtain a linear operator which best fits the data, we adopt the method of snapshot method from DMD technique [27]. We structure the data in two snapshot matrices $\tilde{X}$ and $\tilde{Y}$.

Algorithm 1. Velocity reconstruction with LS-decoder

Inputs: Set of sensor measurements $\{X_i \in \mathbb{R}^{m \times 2}\}_{i=1}^{k}$ and set of target velocities $\{Y_i \in \mathbb{R}^{n \times 2}\}_{i=1}^{k}$ from the training dataset of flow snapshots $\{S_1, S_2, \cdots, S_k\}$
Output: Reconstruction operator $\tilde{A} \in \mathbb{R}^{2n \times 2m}$.

1. Vectorize sensor measurements: $\{\mathbf{x}_i \in \mathbb{R}^{2m}\}_{i=1}^{k}$.
2. Vectorize target velocities: $\{\mathbf{y}_i \in \mathbb{R}^{2n}\}_{i=1}^{k}$.
3. Construct $\tilde{X}$ by stacking the sensor measurements together as columns:
   $\tilde{X} = [\mathbf{x}_1 \ \mathbf{x}_2 \ \cdots \ \mathbf{x}_k] \in \mathbb{R}^{2m \times k}$.
4. Construct $\tilde{Y}$ by stacking the target velocities together as columns:
   $\tilde{Y} = [\mathbf{y}_1 \ \mathbf{y}_2 \ \cdots \ \mathbf{y}_k] \in \mathbb{R}^{2n \times k}$.
5. Construct the pseudo-inverse of $\tilde{X}$: $\tilde{X}^{\dagger}$.
6. Compute the reconstruction operator: $\tilde{A} = \tilde{Y}\tilde{X}^{\dagger}$.

That is, given that there are k training snapshots, then $\tilde{Y} \in \mathbb{R}^{2n \times k}, k \ll 2n$ has k columns, and each column has a length of $2n$ since $\mathbf{y} \in \mathbb{R}^n$.

Considering the training data and the linear relationship between sensor measurements and the (missing) velocity information, we can write a compact expression as below:

$$\begin{bmatrix} | & | & & | \\ \mathbf{y}_1 & \mathbf{y}_2 & \cdots & \mathbf{y}_k \\ | & | & & | \end{bmatrix} \approx \tilde{A} \begin{bmatrix} | & | & & | \\ \mathbf{x}_1 & \mathbf{x}_2 & \cdots & \mathbf{x}_k \\ | & | & & | \end{bmatrix}, \tag{2}$$

$$\tilde{Y} \approx \tilde{A}\tilde{X}. \tag{3}$$

This formulation makes it possible to compute the operator $\tilde{A}$ across the training set, and thus the operator can be used in reconstruction of the velocity field. Recall that the number of sensors available is very limited, thus the linear map is under-determined and has many solutions. However, given this new formulation in Eq. (3), we are interested in the "best fit" solution of

$$\min_{\tilde{A}} \|\tilde{Y} - \tilde{A}\tilde{X}\|_F,$$

in the Frobenius sense, which has the standard form $\tilde{A} = \tilde{Y}\tilde{X}^{\dagger}$ where $\tilde{X}^{\dagger}$ is the Moore-Penrose pseudo-inverse of $\tilde{X}$. The steps for implementing LS-decoder algorithm are highlighted in Algorithm 1.

Results. The dataset was separated into 2 sets: 70% for training and 30% for testing set. Validation set is not necessary for this algorithm since the model training is non-iterative. The performance of the reconstruction algorithm on these sets are recorded in Table 3. With 500 and 996 sensor, R^2 scores are almost perfect, and the errors are minimal. Figure 5 displays a sample reconstruction using the LS decoder with 500 sensor measurements. The absolute error in Fig. 5

Table 3. Performance of LS-decoder

	Training set			Testing set		
m	NE	NFE	R^2	NE	NFE	R^2
10	0.1124	0.1892	0.964	0.1264	0.2119	0.955
50	0.0371	0.0625	0.996	0.0412	0.0692	0.995
500	0.0017	0.028	1.000	0.0033	0.0055	1.000
996	0.0021	0.0035	1.000	0.0036	0.0060	1.000

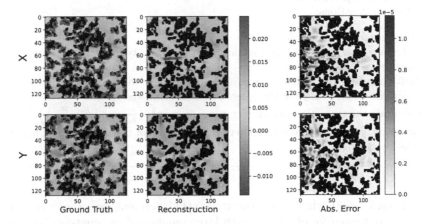

Fig. 5. LS-decoder reconstruction from 500 sensors. The green pixels represent the position of the sensors in the velocity field, the sensors were placed at randomly picked spots on walls of the porous medium for this experiment. (Color figure online)

reflects the precision of the reconstruction as the error are very low. LS-decoder also recorded impressive performances with just 10 and 50 sensors.

As with other deep learning models, back-propagation mechanism is employed to train the SNDs, that is to minimize the misfit between SNDs predictions and target velocities. Whereas for linear (LS decoder) algorithm, a standard minimum norm solution which is readily available is adopted. Obtaining the parameters of SNDs requires an iterative process of gradually approaching a potential solution which heavily depends on the choice of hyperparameters. The search for the optimal SNDs is computational expensive. The linear algorithm, on the other hand, does not involve such iterative process. The velocity field reconstruction with the linear algorithm requires lesser computation resources and time.

5   Monte Carlo Optimization of Gauge Placement

The ability to swiftly re-train the LS decoder and calculate the error allows for the optimization of sensor placement. The number of gauges in the system was

Algorithm 2. Metropolis Monte-Carlo algorithm

Inputs:

Initial gauge distribution.

Number of gauges $N_g \in [5, 10, 25, 50, 100, 200]$.

Number of iterations $N_{it} = 10000$.

Temperature parameter $T \in \{1e^{-5}, 5e^{-5}, 1e^{-4}, 5e^{-4}, 1e^{-3}, 5e^{-3}, 1e^{-2}, 5e^{-2}, 1e^{-1}\}$.

Output:

Optimal gauge positions.

Initialize the gauge distribution.

Cycle for N_{it} iterations:

1. Select a random gauge **A**.
2. Calculate an error of the approximation of the velocity field $\mathbf{E_A}$ for current gauge locations.
3. Shift the gauge to random unoccupied node **B** near the gauge **A**.
4. Calculate an error $\mathbf{E_B}$ for the gauge locations after the shift.
5. Calculate the transition probability $p_{A \to B} = \min\left(1, \frac{e_B}{e_A} \exp\left(\frac{E_B - E_A}{T}\right)\right)$.
6. Generate a uniform random number $u \in [0, 1]$. Accept the gauge shift if $u > p_{A \to B}$, reject otherwise.

varied from 5 to 500, which means up to 1000 independent gauge coordinates. MC methods are known for efficiency with large number of parameters. To optimize the gauge placement, the Canonical Metropolis MC method [28] is applied. The pore structure is presented as a graph network which was extracted via a marker-based segmentation algorithm known as the Sub-Network of an Over-segmented Watershed (SNOW) algorithm [29]. We implemented the SNOW algorithm made available in Porespy package [26] to extract the pore network of our 2D porous structure. In the initial configuration, the gauges are uniformly placed at the nodes, then the field is reconstructed and the normalized error (NE) calculated. The gauge locations are modified in a Markov stochastic process. At each step, we attempt a move of a randomly selected gauge located at node A to a neighboring node B along a randomly selected edge. The move is accepted with the probability of

$$p_{A \to B} = \min\left(1, \frac{e_B}{e_A} \exp\left(\frac{E_B - E_A}{T}\right)\right),$$

where E_A is the error of the approximation of the velocity field with the LS solver (of course, it is a function of coordinated of all gauges in the system), T is an optimization parameter (temperature), e_A and e_B are the number of edges that each node forms. If $T = 0$ only the moves leading to the improvement of approximation are accepted and thus the system never escapes any local minimum, $T \to \infty$ means any random displacement is accepted. The e_B/e_A factor compensates for the asymmetry of the transition matrix to make the

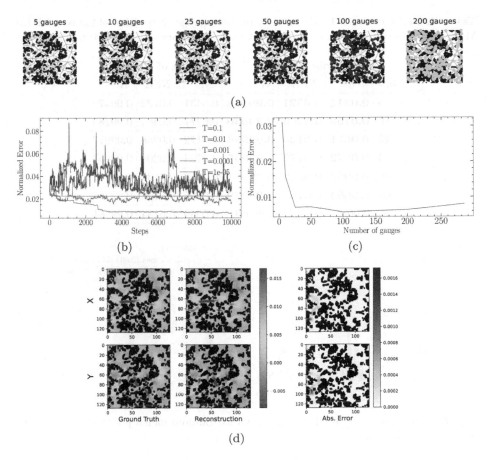

Fig. 6. (a) optimal locations of the gauges obtained in Monte Carlo optimization with $m = 5$ (red), $m = 10$(orange), $m = 25$ (yellow), $m = 50$ (green), $m = 100$ (blue) and $m = 200$ (purple) (b) dependence of the error on time for $m = 25$ and $T \in [1e^{-5}, 1e^{-4}, \ldots, 1e^{-1}]$ (c) dependence of the error on m for the optimal gauge placement (d) LS-decoder reconstruction from 25 sensors. The green pixels represent the position of the sensors in the velocity field, the sensors were placed optimally at the nodes of the graph network for this experiment. (Color figure online)

algorithm ergodic and obey the detailed balance [30]. The steps of the Metropolis algorithm are highlighted in Algorithm 2.

Figure 6(b) shows the optimization progress for Metropolis stochastic processes with $m = 25$ gauges and different values of T. Starting uniform gauge distribution is not an optimal one. For high T values (> 0.01) the acceptance rate is too high and the NE does not seem to decrease. If the T value is too low ($1e^{-5}$), acceptance rate is too low and the NE converges to a non-optimal local minimum. The optimal T for this algorithm and m value is close to $1e^{-4}$, which corresponds to acceptance probability equal to 56%. The fact that the system reaches the

Table 4. Performance of LS-decoder for optimal gauge disposition obtained with the Metropolis algorithm and picking the best gauge locations out of 10000 random ones

	Metropolis algorithm			Best out of 10000		
m	NE	NFE	R^2	NE	NFE	R^2
5	**0.0311**	**0.0521**	**0.99728**	0.0431	0.0722	0.99479
10	**0.0158**	**0.0265**	**0.99930**	0.0277	0.0464	0.99785
25	**0.0071**	**0.0120**	**0.99986**	0.0172	0.0288	0.99917
50	**0.0072**	**0.0122**	**0.99985**	0.0124	0.0208	0.99957
100	**0.0057**	**0.0096**	**0.99991**	0.0079	0.0133	0.99982
200	**0.0064**	**0.0107**	**0.99989**	0.0065	0.0109	0.99988
284	0.0080	0.0134	0.99981	0.0080	0.0134	0.99981

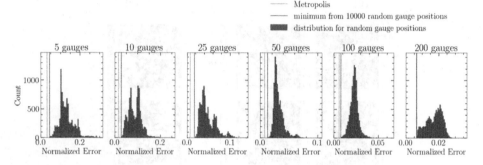

Fig. 7. Comparison of the NE distributions for random gauge position selection and result of MC simulations

"equilibrium" and high acceptance ratio for the attempted displacement moves means there many sets of gauge coordinates that allow approximately similar reconstruction quality. The coordinates corresponding to the lowest error are recorded as the optimal sensor locations, which are also shown in Fig. 6(a). The optimal locations are by no means counter-intuitive in this system: the gauges occupy the key "straights" and mostly located near the western border where the velocity profiles are set. As expected, the reconstruction quality improves with the number of gauges. Improvement is very fast when there are just a few gauges in the system, and slows down as the number of gauges reaches 25. As the number of gauges grows beyond the optimal number of 100, the NE rises slowly, which could be seen at Fig. 6(c) or in Table 4.

The Metropolis algorithm could be compared to the random selection of gauge positions. On the Fig. 7 it is shown that gauge positioning greatly affects the NE and the Metropolis algorithm gives better results than just selecting the best one from the random selection. It is especially reasonable for low number of gauges with a big variability in gauge positions.

6   Conclusion

In this work, we applied fast shallow decoders to reconstruction of a velocity field for complex steady flow through an irregular porous medium using a database of 300 simulated flows with different boundary conditions and gauge readings from a selected number of locations. Two decoders were applied: shallow nonlinear neural decoder and linear least square decoder. Despite a non-linearity of the Navier-Stokes equation that governs the flow, the linear least square solver outperformed the neural network in both precision and speed. The locations of the gauges are optimized with the Metropolis Monte-Carlo algorithm. It was found that 25 gauges are sufficient to reconstruct the velocity field in 256×256 2D grid with about 99% precision; taking precision with respect to NFE. The strategy developed here can be applied to the monitoring of water resources, pipe circuits at chemical plants or applied in anomaly detection.

Individual contributions: E.A.: NN and LS implementation, text; A.O.: LB simulations; V.A.: MC optimization; M.P.: conceptualization, ML methodology, text; A.V.: conceptualization, methodology, text. Authors declare no conflict of interests.

References

1. Zhou, H., et al.: An information-theoretic framework for optimal temperature sensor allocation and full-chip thermal monitoring. In: Proceedings of the 49th Annual Design Automation Conference 2012, pp. 642–647 (2012)
2. Reda, S., et al.: Improved thermal tracking for processors using hard and soft sensor allocation techniques. IEEE Trans. Comput. **60**(6), 841–851 (2011)
3. Ranieri, J., et al.: Near-optimal thermal monitoring framework for many-core systems-on-chip. IEEE Trans. Comput. **64**(11), 3197–3209 (2015)
4. Clenet, S., Henneron, T., Korecki, J.: Sensor placement for field reconstruction in rotating electrical machines. IEEE Trans. Magn. **57**(6), 1–4 (2021)
5. Li, B.J., Liu, H.R., Wang, R.Z.: Data-driven sensor placement for efficient thermal field reconstruction. Sci. China Technol. Sci. **64**(9), 1981–1994 (2021). https://doi.org/10.1007/s11431-020-1829-2
6. Chaturantabut, S., Sorensen, D.C.: Nonlinear model reduction via discrete empirical interpolation. SIAM J. Sci. Comput. **32**(5), 2737–2764 (2010)
7. Bui-Thanh, T., et al.: Aerodynamic data reconstruction and inverse design using proper orthogonal decomposition. AIAA J. **42**(8), 1505–1516 (2004)
8. Willcox, K.: Unsteady flow sensing and estimation via the Gappy proper orthogonal decomposition. Comput. Fluids **35**(2), 208–226 (2006)
9. Everson, R., Sirovich, L.: Karhunen-Loeve procedure for Gappy data. JOSA A **12**(8), 1657–1664 (1995)
10. Tan, B.T., Willcox, K.E., Damodaran, M.: Applications of proper orthogonal decomposition for inviscid transonic aerodynamics. AIAA J. 4213 (2003)
11. Erichson, N.B., Mathelin, L., Yao, Z., Brunton, S.L., Mahoney, M.W., Kutz, J.N.: Shallow neural networks for fluid flow reconstruction with limited sensors. Proc. Roy. Soc. A **476**(2238), 20200097 (2020)
12. Özbay, A.G., Laizet, S.: Deep learning fluid flow reconstruction around arbitrary two-dimensional objects from sparse sensors using conformal mappings. AIP Adv. **12**(4), 045126 (2022)

13. Yu, J., Hesthaven, J.S.: Flowfield reconstruction method using artificial neural network. AIAA J. **57**(2), 482–498 (2019)
14. Li, Y., Liu, Z.M., Wang, Y., Liu, Y., Xie Y.: Real-time physical field reconstruction for nanofluids convection using deep learning with auxiliary tasks. Numer. Heat Transfer Part A Appl. **83**(2), 213–236 (2023). https://doi.org/10.1080/10407782.2022.2091359
15. Raissi, M., Yazdani, A., Karniadakis, G.E.: Hidden fluid mechanics: learning velocity and pressure fields from flow visualizations. Science **367**, 1026–1030 (2020)
16. Raissi, M., Yazdani, A., Karniadakis, G.E.: Hidden fluid mechanics: a Navier-Stokes informed deep learning framework for assimilating flow visualization data (2018). arXiv preprint arXiv:1808.04327
17. Cai, S., Mao, Z., Wang, Z., Yin, M., Karniadakis, G.E.: Physics-informed neural networks (PINNs) for fluid mechanics: a review. Acta. Mech. Sin. **37**(12), 1727–1738 (2021)
18. Hennigh, O.: Lat-Net: compressing lattice Boltzmann flow simulations using deep neural networks (2017). arXiv preprint arXiv:1705.09036
19. Guo, X., Li, W., Iorio, F.: Convolutional neural networks for steady flow approximation. In: Proceedings of the 22nd ACM SIGKDD International Conference on Knowledge Discovery and Data Mining, pp. 481–490 (2016)
20. Sanchez-Gonzalez, A., Godwin, J., Pfaff, T., Ying, R., Leskovec, J., Battaglia, P.: Learning to simulate complex physics with graph networks. In: International Conference on Machine Learning, pp. 8459–8468. PMLR (2020)
21. Chen, J., Hachem, E., Viquerat, J.: Graph neural networks for laminar flow prediction around random two-dimensional shapes. Phys. Fluids **33**(12), 123607 (2021)
22. Fukami, K., Fukagata, K., Taira, K.: Super-resolution reconstruction of turbulent flows with machine learning. J. Fluid Mech. **870**, 106–120 (2019)
23. Carlberg, K.T., Jameson, A., Kochenderfer, M.J., Morton, J., Peng, L., Witherden, F.D.: Recovering missing CFD data for high-order discretizations using deep neural networks and dynamics learning. J. Comp. Phys. **395**, 105–124 (2019)
24. Clark, E., Askham, T., Brunton, S.L., Kutz, J.N.: Greedy sensor placement with cost constraints. IEEE Sens. J. **19**(7), 2642–2656 (2018)
25. Koo, B., Son, H., Kim, H., Jo, T., Yoon, J.Y.: Model-order reduction technique for temperature prediction and sensor placement in cylindrical steam reformer for HT-PEMFC. Appl. Therm. Eng. **173**, 115153 (2020)
26. Gostick, J.T., et al.: PoreSpy: a Python toolkit for quantitative analysis of porous media images. J. Open Source Softw. **4**(37), 1296 (2019). https://doi.org/10.21105/joss.01296
27. Proctor, J.L., Brunton, S.L., Kutz, J.N.: Dynamic mode decomposition with control. SIAM J. Appl. Dyn. Syst. **15**(1), 142–161 (2016)
28. Metropolis, N., Ulam, S.: The Monte Carlo method. J. Am. Stat. Assoc. **44**(247), 335–341 (1949)
29. Gostick, J.T.: Versatile and efficient pore network extraction method using marker-based watershed segmentation. Phys. Rev. E **96**(2), 023307 (2017)
30. Frenkel, D., Smit, B.: Understanding Molecular Simulation: From Algorithms to Applications, vol. 1. Elsevier (2001)
31. Krüger, T., Kusumaatmaja, H., Kuzmin, A., Shardt, O., Silva, G., Viggen, E.M.: The Lattice Boltzmann Method: Principles and Practice. Springer, Cham (2017). https://doi.org/10.1007/978-3-319-44649-3. ISBN 978-3-319-44649-3
32. Olhin, A.: Lattice Boltzmann velocity fields dataset. Mendeley Data **V2** (2023). https://doi.org/10.17632/kbrprbvtjw.2

Model of Perspective Distortions for a Vision Measuring System of Large-Diameter Bent Pipes

Krzysztof Borkowski[1]([⊠]) [iD], Dariusz Janecki[1] [iD], and Jarosław Zwierzchowski[2] [iD]

[1] Kielce University of Technology, al. Tysiaclecia P. P. 7, 25-314 Kielce, Poland
kborkowski@tu.kielce.pl
[2] Institute of Electronics, Lodz University of Technology, Al. Politechniki 10, 90-590 Lodz, Poland

Abstract. The measuring system described in the article was designed for measuring large and heavy bent pipes with diameters up to 1.2 m. Currently, measurements of large-diameter bent pipes are taken using either simple and inaccurate protractors or measuring systems requiring a 3D model of the pipe created in graphics software, an optical scanner and markers. This paper presents methods of modeling distortions for measuring system based on one camera that is easy to install in an industrial plant. Those models allow to perform measurement quickly at any position of the pipe on a large industrial measurement table. The paper describes the mathematical models of perspective projections of single- and double- bent pipes, as well as the method of bending angle determination and detection of straight sections of the bent pipe. As part of the research, measurement accuracy of the designed system model were described and confirmed.

Keywords: Modelling · Vision system · Pipes measurement · Perspective distortions

1 Introduction

Today's vision systems are widely used in manufacturing processes. Growing quality requirements entail the use of continuous control and correction of the production process parameters. Cameras are used as sensors in many applications [4,10]. With the cameras, it is possible to recognise, classify, separate objects and determine their direction and location in space [9]. The use of a vision system in production lines allows for automatic and continuous control of the dimensional and shape quality of the manufactured parts [1]. The measurement accuracy is obviously lower than the accuracy of measurements performed by metrological machines and devices [7]. However, features such as: ability to track moving objects, no need to change the orientation of objects in space, quick and non-contact measurement, high degree of automation, durability or failure-free operation are the advantages that distinguish the vision systems.

© The Author(s), under exclusive license to Springer Nature Switzerland AG 2023
J. Mikyška et al. (Eds.): ICCS 2023, LNCS 14475, pp. 595–609, 2023.
https://doi.org/10.1007/978-3-031-36024-4_46

In this paper, the models of perspective distortions are provided because those are important in measurement vision systems to obtain correct metric measurement values.

New, elements worth emphasizing in the paper are: (1) determination of a mathematical model for single- and double-bent pipes, (2) determination of the pipe bend radius and straight sections using mathematical methods.

In one of the first works [2] on the multi-camera measuring system of bent pipes, the accuracy of bending radius measurement was not lower than 0.5 mm. Moreover, several works related to optical control and measurement systems of bent pipes and cables have been published. Most of them determine the surfaces of the 3D model of pipes using passive multi-view systems. Therefore, the proposed system requires adding a pipe model created using graphic software to the measurement, and then, based on this model, measurement errors are estimated. The presented work does not provide for using a 3D model, and the mathematical model created is built on the basis of a perspective projection. Moreover, there is a tendency to increase the measurement accuracy by connecting an increasing number of cameras. Like in works [8,12], where a system consisting of 16 fixed cameras was used.

Work [11] presents a pipe measurement system using stereo-visions and a perspective model of a pipe to increase the accuracy of the spatial reconstruction of the bent pipe central axis. In work [6], an analysis of reconstruction errors for a 3D model of bent pipes is performed using a multi-camera system that is divided into pairs of stereo-vision cameras. Additionally, an illuminated work table was used in that case. Work [3] presents a system used for reconstructing the shape of a pipe with a constant diameter on the basis of photos taken with a three-view vision system. The work takes into account perspective distortions for each of the cameras, and the pipe's surface is described with the use of circles of the same diameter and centres on the centre line of the pipe. The mean square measurement error of the bending radius of 0.127 mm was obtained.

2   Description of the Proposed Measuring System

Due to the production costs, the pipe processing process requires shape and dimensional control in subsequent stages of production, especially after the following processes: bending, heat treatment, possible corrective works and in the preparation of the as-built report. Even small errors in the bending angle prevent correct installation of the pipe in the pipeline.

We assume that the measured object is a combination of cylindrical and toroidal surfaces with a constant cross-sectional radius. As a result of this assumption, it is possible to measure the parameters of an object based on the image obtained in the camera setting. It should be mentioned here that this is an approximation – in fact, the bend of the bent pipe consists of many smaller bends with different radii and, what is more, the pipe becomes oval (changing the shape from round to elliptical.

Parameters describing the double-bent pipe are given in Fig. 1. The developed system determines the following parameters for pipes bent in one plane:

The work also considers the measurement of pipes bent in planes inclined to each other at an angle ζ.

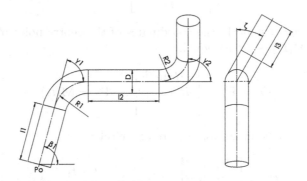

Fig. 1. Basic parameters of the bent pipe: bending angles γ_1, γ_2; pipe bending radii R_1, R_2; straight section lengths l_1, l_2, l_3; pipe diameter $D = 2r$; bending plane angles of deviation ζ

In the next section, we will show how to find those parameters for single and double bending.

3   Mathematical Models of Bent Pipes

3.1   Single-Bent Pipe Model

Let us begin with describing the parametric equations of the pipe surface centre line. We assume that the line consists of a straight line section $\mathbf{P_0P_1}$, an arc $\mathbf{P_1P_2}$ and another straight line section $\mathbf{P_2P_3}$. The position of the measured object is determined by:

- co-ordinates x, y, z end of the centreline $\mathbf{P_0} = [x_0 \ y_0 \ r]^T$,
- the slope of the centre line at point $\mathbf{P_0}$ to axis X: β_1.

One should note that the slope of the segment $\mathbf{P_2P_3}$ will be equal to:

$$\beta_2 = \beta_1 + \gamma, \ \gamma > 0. \tag{1}$$

Suppose the parameter t is the length of the centre line from the point $\mathbf{P_0}$ to the point selected on the line $\mathbf{P}$. We will denote:

$$t_1 = l_1, \quad t_2 = l_1 + R\gamma, \quad t_3 = t_2 + l_2. \tag{2}$$

So the parameters t_1, t_2, t_3 are the lengths of the line sections $\mathbf{P}_0\mathbf{P}_1$, $\mathbf{P}_1\mathbf{P}_2$, $\mathbf{P}_2\mathbf{P}_3$. The parametric equation of the section $\mathbf{P}_0\mathbf{P}_1$ can take the following form:

$$\mathbf{M}(t) = \begin{bmatrix} x_0 + t\cos\beta_1 \\ y_0 + t\sin\beta_1 \\ r \end{bmatrix}, \quad t \in [0, t_1]. \tag{3}$$

As the centre line is smooth, the coordinates of the centre point of the circle of which the arc $\mathbf{P}_1\mathbf{P}_2$ is a part equal:

$$\mathbf{O}_c = \mathbf{P}_1 + \begin{bmatrix} -R\sin\beta_1 \\ R\cos\beta_1 \\ r \end{bmatrix}, \quad \mathbf{P}_1 = \mathbf{M}(l_1) \tag{4}$$

and the parametric equation for the arc of a circle:

$$\mathbf{M}(t) = \mathbf{O}_c + \begin{bmatrix} R\sin\beta \\ -R\cos\beta \\ r \end{bmatrix}, \quad \beta = \beta_1 + \frac{t - t_1}{R}, \quad t \in [t_1, t_2]. \tag{5}$$

Similarly to (3), the parametric equation of the section $\mathbf{P}_2\mathbf{P}_3$ has the following form:

$$\mathbf{M}(t) = \mathbf{P}_2 + \begin{bmatrix} (t - t_2)\cos\beta_2 \\ (t - t_2)\sin\beta_2 \\ r \end{bmatrix}, \quad \mathbf{P}_2 = \mathbf{M}(t_2), \quad t \in [t_2, t_3]. \tag{6}$$

Figure 2b shows the projection of the centre line onto a plane for example parameter values: $x_0 = 100$, $y_0 = -30$, $\beta_1 = 150°$, $\gamma = 100°$, $R = 40$, $l_1 = 130$, $l_2 = 85$, $r = 15$.

Based on Eq. (3), (5) and (6) we can write a parametric equation for the entire surface. Suppose $\mathbf{M}(t)$ is the selected centre line point, and $\beta(t)$ is the angle of inclination of the tangent of this line to axis XY.

$$\beta(t) = \begin{cases} \beta_1 & \text{dla } t \in [0, t_1] \\ \beta_1 + \frac{t - t_1}{R_1} & \text{dla } t \in [t_1, t_2] \\ \beta_2 & \text{dla } t \in [t_2, t_3] \end{cases} \tag{7}$$

The cross-section of an object perpendicular to the centre line is described by the following equation:

$$\mathbf{T}(t) = \begin{bmatrix} \cos\beta(t) & -sin\beta(t) & 0 \\ \sin\beta(t) & cos\beta(t) & 0 \\ 0 & 0 & 1 \end{bmatrix} \begin{bmatrix} 0 \\ r\cos\alpha \\ r\sin\alpha \end{bmatrix} + \mathbf{M}(t), \tag{8}$$

where $\alpha \in [0, 2\pi]$, $\beta \in [0, t_3]$. The resulting equation is a parametric equation for the surface of a bent pipe. Figure 2a shows a pipe's surface generated using the previously selected example parameters. The measurement of the object parameters consists in analysing the edge of the object in the image captured by the camera. Based on the developed model, we will determine the parametric

(a)

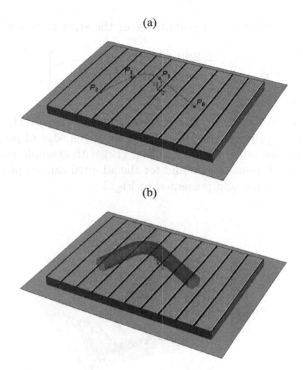

(b)

Fig. 2. (a) The centre line of a bent pipe (red) with points marking a straight line section $\mathbf{P}_0\mathbf{P}_1$, arc of a circle $\mathbf{P}_1\mathbf{P}_2$ and another straight line section $\mathbf{P}_2\mathbf{P}_3$ (b) Visualization of the pipe surface with its centreline (red) and the selected cross-section (blue). The reference coordinate system is also marked in the drawing (Color figure online)

equations of these edges. To begin with, we will designate points $\mathbf{T}$ on the object surface for which the straight line drawn through the camera point $\mathbf{C} = [0\ 0\ z_c]^T$ and point $\mathbf{T}$ will be tangent to the surface of the pipe. We will determine these points based on the condition:

$$\overrightarrow{\mathbf{CT}} \perp \overrightarrow{\mathbf{TM}}, \tag{9}$$

where $\mathbf{M}$ is the point of the centre line of the cross section passing through the point $\mathbf{T}$ from the condition:

$$(\mathbf{C} - \mathbf{T})^T(\mathbf{T} - \mathbf{M}) = 0 \tag{10}$$

we obtain (for better clarity parameter t was omitted):

$$(r - z_c)\sin\alpha + (m_y\cos\beta - m_x\sin\beta)\cos\alpha + r = 0, \tag{11}$$

where m_x, m_y are the corresponding elements of vector $\mathbf{M}$. The above equation has two solutions α_1, α_2 dependent on parameter t. It is understandable, since from the point of camera one can draw two tangents to the pipe surface in a fixed cross-section. Intersection points $\mathbf{CT}$ of the straight line with plane XY are equal

$$\mathbf{E}_{1,2} = \begin{bmatrix} \frac{rz_c \cos\alpha_{1,2}\sin\beta - m_x z_c}{r\sin\alpha_{1,2}} + r - z_c \\ \frac{rz_c \cos\alpha_{1,2}\cos\beta - m_y z_c}{r\sin\alpha_{1,2}} + r - z_c \\ 0 \end{bmatrix} . \tag{12}$$

It must be noted that in the above equations variables β, M_y, M_y are functions t, so $\mathbf{E}_1$ and $\mathbf{E}_2$ are parametric equations of the edge of projection of the pipe surface onto the table plane. Equation (12) with example pipe parameters adopted from Fig. 2b visualisation and for the adopted camera placement height $z_c = 130$ was crossed out and presented in Fig. 3.

Fig. 3. Visualization of the projection of the pipe surface onto the table plane

It should be emphasized that the edges of the cylindrical surface section projection form two parallel lines. It results from the fact that the set of straight tangents to the cylinder and passing through the selected point creates a plane. Whereas the edges of the torus fragment projection are not, as one might assume, sections of circles, but a curve described by quite complex equations.

3.2   Twice-Bent Pipe Model in One Plane

The surface of an ideal pipe was recorded using parametric equations describing spatial figures: a cylinder and a torus. First, we will consider the case where the centre line is in a plane parallel to the plane of the table. Let us begin by describing the parametric equations of the pipe surface centre line. We assume that the line consists of a straight line section $\mathbf{P}_0\mathbf{P}_1$, an arc $\mathbf{P}_1\mathbf{P}_2$, another straight section $\mathbf{P}_2\mathbf{P}_3$, another arc $\mathbf{P}_3\mathbf{P}_4$ and another straight section $\mathbf{P}_4\mathbf{P}_5$. The position of the measured object is determined by:

- coordinates of the end of the centre line $\mathbf{P}_0 = [x_0 \ y_0 \ r]^T$,
- the slope of the centre line at a point $\mathbf{P}_0$ to the axis $X : \beta_1$.

Let us presume the parameter t is the length of the centre line from the point $\mathbf{P}_0$ to the selected point on the centre line $\mathbf{M}$. We will denote:

$$
\begin{aligned}
t_1 &= l_1, \\
t_2 &= l_1 + R_1\gamma_1, \\
t_3 &= t_2 + l_2, \\
t_4 &= t_3 + R_2\gamma_2, \\
t_5 &= t_4 + l_3.
\end{aligned}
\tag{13}
$$

So the parameters t_1, t_2, t_3, t_4, t_5 are the lengths of the line segments $\mathbf{P}_0\mathbf{P}_1$, $\mathbf{P}_1\mathbf{P}_2$, $\mathbf{P}_2\mathbf{P}_3$, $\mathbf{P}_3\mathbf{P}_3$, $\mathbf{P}_4\mathbf{P}_5$. Parametric equation of the centreline for the first pipe section represented by segment $\mathbf{P}_0\mathbf{P}_1$ can be presented as follows:

$$
\mathbf{M}(t) = \mathbf{P}_0 + t \begin{bmatrix} \cos\beta_1 \\ \sin\beta_1 \\ 0 \end{bmatrix}, \ t \in [0, t_1].
\tag{14}
$$

As the centre line is smooth, the coordinates of the centre point of the circle of which the arc $\mathbf{P}_1\mathbf{P}_2$ equal:

$$
\mathbf{O}_1 = \mathbf{P}_1 + R_1 \begin{bmatrix} -\sin\beta_1 \\ \cos\beta_1 \\ 0 \end{bmatrix}, \ \mathbf{P}_1 = \mathbf{M}(t_1)
\tag{15}
$$

and the parametric equation of this arc of a circle:

$$
\mathbf{M}(t) = \mathbf{O}_1 + R_1 \begin{bmatrix} \sin\beta \\ -\cos\beta \\ 0 \end{bmatrix},
$$
$$
\beta = \beta_1 + (t - t_1)/R_1, \ t \in [t_1, t_2].
\tag{16}
$$

One should note that the slope of the segment $\mathbf{P}_2\mathbf{P}_3$ will be equal to:

$$
\beta_2 = \beta_1 + \gamma_1,
\tag{17}
$$

so similarly to (14), the parametric equation of the section $\mathbf{P}_2\mathbf{P}_3$ has the following form:

$$
\mathbf{M}(t) = \mathbf{P}_2 + (t - t_2) \begin{bmatrix} \cos\beta_2 \\ \sin\beta_2 \\ 0 \end{bmatrix},
$$
$$
\mathbf{P}_2 = \mathbf{M}(t_2), \ t \in [0, t_1].
\tag{18}
$$

Similarly to (15) and (16), the centre point and parametric equation of arc $\mathbf{P}_3\mathbf{P}_4$ take the following form:

$$
\mathbf{O}_2 = \mathbf{P}_3 + R_2 \begin{bmatrix} -\sin\beta_2 \\ \cos\beta_2 \\ 0 \end{bmatrix},
\tag{19}
$$
$$
\mathbf{P}_3 = \mathbf{M}(t_3),
$$

$$\mathbf{M}(t) = \mathbf{O}_2 + R_2 \begin{bmatrix} \sin\beta \\ -\cos\beta \\ 0 \end{bmatrix},$$ (20)

$$\beta = \beta_2 + (t - t_3)/R_2, \ t \in [t_3, t_4].$$

Segment slope $\mathbf{P}_4\mathbf{P}_5$ to the axis X is:

$$\beta_3 = \beta_2 + \gamma_2$$ (21)

and its parametric equation has the following form:

$$\mathbf{M}(t) = \mathbf{P}_4 + (t - t_4) \begin{bmatrix} \cos\beta_3 \\ \sin\beta_3 \\ 0 \end{bmatrix},$$ (22)

$$\mathbf{P}_4 = \mathbf{M}(t_4), \ t \in [t_4, t_5].$$

Figure 4 shows the visualisation of the centre line of the pipe placed on the measurement table for example parameter values: $x_0 = 120$, $y_0 = 10$, $\beta_1 = 150°$, $\gamma_1 = 100°$, $\gamma_2 = 75°$, $R_1 = 40$, $R_2 = 40$, $l_1 = 130$, $l_2 = 50$, $l_3 = 50$, $r = 15$.

The next section describes image processing algorithms, which are necessary for automatically detecting edges of the pipes on the measuring table.

4   Determining the Angle of Bending and Straight Sections of Pipes

There is assumed that the vision system correctly calculated the edge points of the measuring pipe. The obtained edge points of a selected pipe shown in Fig. 5 that are expressed in the measurement table coordinate system, will be used to determine the parameters of the bent pipe. This task was divided into three stages: determination of two pairs of parallel lines (projections of straight pipe sections), determination of curves being a projection of a torus section and matching pipe parameters using the presented analytical model.

An effective method to reduce errors is to match a straight line or curve with the obtained data. For this purpose, simple equations were determined using an algorithm based on the Hough Transform [5]. In the case of a straight line, we will define its position in polar coordinates using: the inclination angle θ and the distance from the origin of the coordinate system ρ. Through each given point on the plane (x, y) we can route infinite number of lines which parameters θ and ρ are related by the equation

$$x \cos\theta + y \sin\theta = \rho.$$ (23)

The graph of this equation for a single point is a sine curve. We assume angle $\theta \in [0, \pi]$ and distance $\rho \in [0, \rho_{max}]$, where value ρ_{max} is the distance from the origin of the coordinate system to the most distant point (for an image it is the diagonal length). Figure 6 shows the designated Hough plane $\Pi_h(\theta, \rho)$ with the

(a)

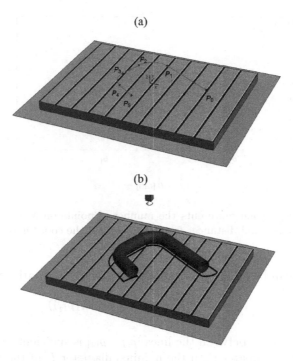

(b)

Fig. 4. (a) The centre line of a bent pipe (red) with points marking a straight line section $\mathbf{P}_0\mathbf{P}_1$, arc of a circle $\mathbf{P}_1\mathbf{P}_2$, straight line section $\mathbf{P}_2\mathbf{P}_3$, arc of a circle $\mathbf{P}_3\mathbf{P}_4$ and another straight line section $\mathbf{P}_4\mathbf{P}_5$, (b) Visualization of the projection of twice-bent pipe surface onto the table plane (Color figure online)

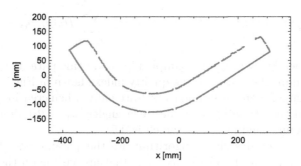

Fig. 5. The edge points of the recognized pipe converted to the measurement table coordinate system

step of discretisation $\Delta\theta = 0.1°$ and $\Delta\rho = 0.1$ mm. The parameters of the lines can be found by searching for local maxima, the value of which corresponds to the number of points lying on a given line. Note that the projection of the edge of a straight pipe section lying on the table is a pair of parallel lines, and therefore of the same inclination angle θ_i but different distances ρ_{i1} and ρ_{i2}. Such a pair

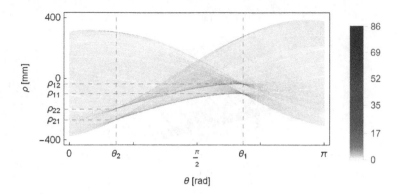

Fig. 6. The Hough plane represents the number of points on a straight line with an angle of inclination θ and distances from the origin of the coordinate system ρ

of simple lines can be found, for example, by solving the mathematical problem

$$\max_{\theta_i, \rho_{i_1}, \rho_{i_2}} \left(\Pi_h(\theta_i, \rho_{i_1}) + \Pi_h(\theta_i, \rho_{i_2}) \right) \tag{24}$$

assuming the distance between the lines $|\rho_{i1} - \rho_{i2}|$ is sufficient (e.g. $|\rho_{i1} - \rho_{i2}| > \delta\rho$, where $\delta\rho$ is not greater than the nominal diameter D of the tested pipes).

In this work, when searching for pairs of lines, points lying in the vicinity of the straight line in the distance η were additionally taken into account and match index was defined as the following function:

$$g(\theta_i) = \max_{\rho_{i_1}, \rho_{i_2}} \left(\sum_{j=-\eta}^{\eta} \Pi_h(\theta_i, \rho_{i_1} + j) + \sum_{j=-\eta}^{\eta} \Pi_h(\theta_i, \rho_{i_2} + j) \right) \tag{25}$$

Solving the dependency (25) for a sample pipe and using parameters $\eta = 0.1$ mm and $\delta\rho = 10.0$ mm, the course of the quality index shown in Fig. 7 was obtained. It has two maxima for angles θ_1 and θ_2. Additionally, when creating the algorithm it was assumed that the difference between angles $|\theta_1 - \theta_2|$ should be greater than the minimum bend angle $\delta\theta$, which is $0.5°$.

The result of the described algorithm are the parameters of two pairs of parallel lines $(\theta_1, \rho_{1_1}, \rho_{1_2})$ and $(\theta_2, \rho_{2_1}, \rho_{2_2})$ that are visible on the Hough plane and are marked in Fig. 6. The determined two pairs of parallel lines were marked on the edge points of the pipe in Fig. 8.

The next task is to determine the mean line for the surface of the straight sections of the pipe.

In order to determine the parameters of the pipe, an equation for middle lines of two cylindrical parts of the pipe. Let us consider a cylindrical section of a pipe with an inclination angle θ_1. We will three planes: the plane of the table π_1 and two planes tangent to the cylinder passing through the camera point π_2, π_3, and then a section perpendicular to the three described planes passing through the camera point $\mathbf{C}$ (Fig. 9).

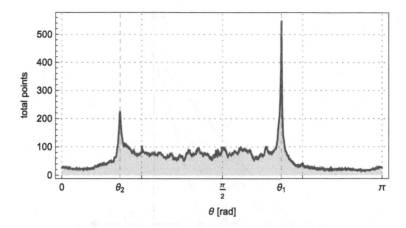

Fig. 7. Matching index for a pair of parallel lines inclined depending on angle θ

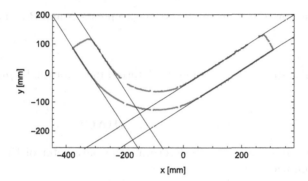

Fig. 8. Matching of the pairs of lines parallel to the edge points of the cylindrical sections of the bent pipe

Our goal is to determine the radius of the circle inscribed in the triangle ABC and the coordinate ρ_1 of the circle centre projection on the table plane, bearing in mind the coordinates of points $\mathbf{A}$ and $\mathbf{B}$ equal ρ_{11} and ρ_{12} respectively. We have:

$$r_1 = \frac{|\mathbf{AB}|z_c}{|\mathbf{AB}| + |\mathbf{BC}| + |\mathbf{CA}|}, \qquad (26)$$

where $|\mathbf{AB}| = |\rho_{i_1} - \rho_{i_2}|$, $|\mathbf{BC}| = \sqrt{z_c^2 + \rho_{1_2}^2}$, $|\mathbf{CA}| = \sqrt{z_c^2 + \rho_{1_1}^2}$. Similarly we will find the value of radius r_2. Of course, the radius of the pipe is actually the same at both ends. We can therefore assume:

$$r = \frac{r_1 + r_2}{2}. \qquad (27)$$

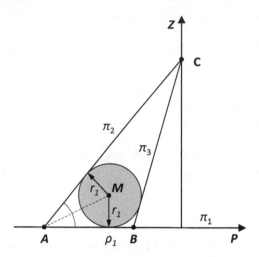

Fig. 9. Cross-section of the pipe with visible edge points **A** and **B** and a point on the centre line **M**

Figure 9 also shows the method of determining coordinate ρ_1. We have (I assume $r_1 = r$)

$$\rho_1 = \rho_{1_1} + r \cot(\frac{1}{2} \sphericalangle \mathbf{BAC}). \qquad (28)$$

And in the same way, we will determine the parameter of the mean line of the second cylindrical segment.

As a result of the calculations, we have obtained parameters of the straight middle parts of both cylindrical parts of pipes (θ_1, ρ_1) and (θ_2, ρ_1).

On the basis of the obtained values, we can determine angle γ:

$$\gamma = \begin{cases} |\beta_2 - \beta_1| & \text{for the bend turning to the left side} \\ \pi - |\beta_2 - \beta_1| & \text{for the bend turning to the right side} \end{cases} \qquad (29)$$

To determine the bend radius of pipe R, let us consider the projection of the edge of its bent fragment, which is described by the complex equations of analytical model (12). Note that the centre of torus fragment $\mathbf{O}_c$ is located on the bisector of an angle adjacent to γ between straight lines lying on the centre line of two cylindrical parts of the pipe and, at the same time, is offset from them to the distance of R. We will determine the point location $\mathbf{O}_c$ based on the real projection of the bent pipe edge.

Let us presume the distance between point $\mathbf{O}_c$ and the mean line, that is, value R falls within a certain range $[R_{min}, R_{max}]$. By making the interval discretisation by step δR we obtain values $\hat{R}_i$. Now, we define two accumulators A_{1i} and A_{2i} in the form of vectors with initial values equal to zero, indices of which are related to discrete values $\hat{R}_i$. For each edge point, we determine the distance from the two edge lines determined in the analytical model. If this distance is less than δP, we increase the value of accumulators A_{1i} or A_{2i} accordingly.

Finally, we designate index i and the corresponding value $R = \hat{R}_i$ for which sum:

$$\max_i = A_{1i} + A_{2i} \tag{30}$$

reaches its maximum.

The other parameters sought are the lengths of straight pipe sections l_1, l_2. By analysing the distance of the pipe edge points from the previously determined straight lines with parameters $(\theta_1, \rho_{11}, \rho_{12})$ and $(\theta_2, \rho_{21}, \rho_{22})$ we will define four points at both ends of the pipe profile. Based on these points, we will designate points on the centre line of the pipe $\mathbf{P}_0$, $\mathbf{P}_3$ which represent the position of the extreme cross sections of the pipe. Thus, the lengths of the straight pipe sections are, respectively:

$$l_1 = |\mathbf{P}_0\mathbf{P}_1|, \quad l_2 = |\mathbf{P}_2\mathbf{P}_3|. \tag{31}$$

The determined parameters are finally verified using the analytical model (12), by comparing the calculated outline of the pipe with the edges detected in the camera image (Fig. 10). A similar method is used to determine the bend angles for double-bent pipes.

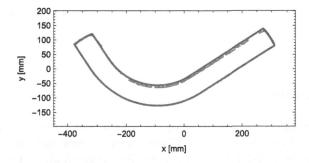

Fig. 10. Comparison of the pipe edge points (blue colour) to the perspective projection of the pipe calculated using the analytical model (red colour) (Color figure online)

5   Experimental Research

The measurement accuracy of the developed models was checked by comparing model results with results from a certified CMM Zeiss Prismo Navigator machine, designed for very accurate contact measurements. The measurement was taken along the outer and inner surface of the pipe, so that the outline of the pipe corresponded to its basic projection. The results of the accuracy are shown in Table 1. The CMM results are given in the first column and are considered a reference standard for the check. Then, a series of 14 measurements of the bent pipe were taken and the average value of these measurements is shown in the second column of Table 1. The last column presents the standard deviation of the measurements. From the industrial measurements point of view, the two most important parameters, i.e. the bend radius error and the bend angle error of the pipe, are respectively 0.85 mm, 0.34° and are similar to the results obtained in other scientific studies mentioned in the introduction of this paper. Moreover, they are accepted by the industry in pipelines components prefabrication.

Table 1. The results of the measuring system accuracy series of measurements of a reference pipe compared to the reference values measured by a Zeiss – Prismo Navigator CMM

	Reference measurement	Series of measurements	
		Average value	Standard deviation
Pipe diameter D [mm]	60.8	61.1	0.34
Bending angle [°]	89.9	89.8	0.34
Bending radius [mm]	238.6	238.4	0.85
Straight section length 1 [mm]	296.4	295.5	1.99
Straight section length 2 [mm]	100.9	101.7	1.58
Length of the centre line [mm]	771.5	770.8	1.68

6   Conclusions

The paper presents some aspects of optical measuring system intended for measuring large-diameter bent pipes. Measurement stages such as, mathematical models, the method of determining the bending angle of pipes and other parameters important for the pipeline industry are described. In particular, the focus was on mathematical models and analytical models of perspective projections for single- and double-bent pipes were developed. For pipes bent twice in one plane (torsional angle $\zeta = 0°$) taking into account the perspective deformations of the pipe contour in the developed algorithms positively influenced the increase in the accuracy and repeatability of the measurement results. Optical measurements with the use of a developed mathematical model for a single-bent pipe were compared with measurements results obtained on a certified CMM. The auxiliary measurements performed clearly confirm that the developed measurement system meets the assumptions as to the accuracy of the measurement and its repeatability. During the tests a measurement accuracy of 0.34° for the bending angle and 0.85 mm for the bend radius was achieved.

As part of further research, it is necessary to investigate the effect of the pipe flattening that occurs during the bending process on the final results.

The use of the developed vision measuring system can be used as a quality control tool for controlling products in subsequent stages of pipeline production, as well as a tool for preparation of as-built reports. The measurements data collected can also be used to correct the settings of process machines and contribute to the reduction of production costs.sts.

References

1. Borzykowski, J., Domańska, A.: Współczesna metrologia. WNT, Warszawa (2004)
2. Bösemann, W.: The optical tube measurement system olm photogrammetric methods used for industrial automation and process control. Int. Arch. Photogrammetry Remote Sens. **31**, 55–58 (1996)
3. Cheng, X., Sun, J., Zhou, F., Xie, Y.: Shape from apparent contours for bent pipes with constant diameter under perspective projection. Measurement **182**, 109787 (2021)
4. Giancola, S., Valenti, M., Sala, R.: A Survey on 3D Cameras: Metrological Comparison of Time-of-Flight, Structured-Light and Active Stereoscopy Technologies. Springer, Cham (2018). https://doi.org/10.1007/978-3-319-91761-0
5. Hart, P.E., Duda, R.: Use of the hough transformation to detect lines and curves in pictures. Commun. ACM **15**(1), 11–15 (1972)
6. Huang, H., Liu, J., Liu, S., Jin, P., Wu, T., Zhang, T.: Error analysis of a stereo-vision-based tube measurement system. Measurement **157**, 107659 (2020)
7. Isa, M.A., Sims-Waterhouse, D., Piano, S., Leach, R.: Volumetric error modelling of a stereo vision system for error correction in photogrammetric three-dimensional coordinate metrology. Precis. Eng. **64**, 188–199 (2020)
8. Jin, P., Liu, J., Liu, S., Wang, X.: A new multi-vision-based reconstruction algorithm for tube inspection. Int. J. Adv. Manuf. Technol. **93**(5), 2021–2035 (2017)
9. Ma, Y., Soatto, S., Košecká, J., Sastry, S.: An Invitation to 3-D Vision: From Images to Geometric Models, vol. 26. Springer, New York (2012). https://doi.org/10.1007/978-0-387-21779-6
10. Schwenke, H., Neuschaefer-Rube, U., Pfeifer, T., Kunzmann, H.: Optical methods for dimensional metrology in production engineering. CIrP Ann. **51**(2), 685–699 (2002). https://doi.org/10.1016/S0007-8506(07)61707-7. https://www.sciencedirect.com/science/article/pii/S0007850607617077
11. Sun, J., Zhang, Y., Cheng, X.: A high precision 3D reconstruction method for bend tube axis based on binocular stereo vision. Opt. Express **27**(3), 2292–2304 (2019)
12. Wang, X., Liu, J., Liu, S., Jin, P., Wu, T., Wang, Z.: Accurate radius measurement of multi-bend tubes based on stereo vision. Measurement **117**, 326–338 (2018)

Outlier Detection Under False Omission Rate Control

Adam Wawrzeńczyk[1] and Jan Mielniczuk[1,2]

[1] Institute of Computer Science, Polish Academy of Sciences, Jana Kazimierza 5,
Warsaw 01 -248, Poland
{adam.wawrzenczyk,jan.mielniczuk}@ipipan.waw.pl
[2] Faculty of Mathematics and Information Science, Warsaw University of
Technology, Koszykowa 75, Warsaw 00-662, Poland

Abstract. We argue that in many practical situations control of False
Omission Rate (FOR) or Bayesian False Omission Rate (BFOR) is of
primary importance. We develop and investigate such rule in the context
of outlier detection, and propose its empirical formulation for practical
use. We consider several score statistics used to detect outliers and study
how well the introduced method controls FOR in practice. It is shown by
analysis of several datasets that FOR control in contrast to FDR control
is inherently tied to performance of the score statistic employed on both
inlier and outlier data sets.

Keywords: False Omission Rate · Bayesian False Omission Rate ·
False Discovery Rate · outlier detection · fraud detection · one class
classification · truncation rule

1 Introduction

We consider the situation when a score statistic is learned on a random sample
of regular observations (inliers) and used to detect out of distribution observations (outliers) with an objective to control the percentage of undetected outliers
among the observations classified as inliers. Such a need arises in many practical
situations: imagine a scrutiny of possibly fraudulent transactions for which one
would like to detect all but a very small percent of frauds – such an approach
accounts for the fact that trying to detect all frauds will require very stringent
safety rules which would deter potential customers. Another example is development of a new test for a contagious disease (e.g. COVID-19), for which is vital to
ensure that randomly chosen person will not pass it *if infected* with large probability (see [14]). In such situations it is much more important to control False
Omission Rate (FOR, called also False Non-discovery Rate (FNR) [6]) than commonly used False Discovery Rate (FDR). FOR is defined as the expected value
of False Omission Proportion i.e. proportion of undetected outliers among observations classified as inliers, whereas FDR is the expected proportion of inliers
among observations deemed outliers. Obviously, FDR control in many situations

© The Author(s), under exclusive license to Springer Nature Switzerland AG 2023
J. Mikyška et al. (Eds.): ICCS 2023, LNCS 14075, pp. 610–625, 2023.
https://doi.org/10.1007/978-3-031-36024-4_47

has evident advantages but we argue that in numerous cases FOR control – or its Bayesian analogue defined below – is of main interest, and procedures which ensure it are worth studying.

Our main objective here is to develop a rule which approximately controls FOR and to investigate its properties both theoretically and by means of analysis of real data sets. The rule developed here is derived analogously to Benjamini-Hochberg rule [2] using Frequentist Bayes approach (see e.g. [6], Chap. 4). We also consider several methods scores for outlier detection and check how their choice influences control of FOR. Finally, we investigate ways of diminishing intrinsic variability of p-values due to the random split of the data set.

2   FOR Control Procedure

2.1   Preliminaries

Consider checking whether observations under study are outliers with the use of a specified score statistic $\hat{s}$ to test a null hypothesis $H_{0,i}$: i-th observation X_i is an inlier versus an alternative $H_{1,i}$: i-th observation is an outlier. We adopt throughout the convention that large values of $\hat{s}$ indicate outliers. It is known that when the cumulative distribution function (CDF) denoted by F of a test statistic is continuous, the distribution of the corresponding p-value equal to $1 - F(X_i)$, provided the null hypothesis is true, is uniform on $[0,1]$. This is the fundamental property used to bound Family Wise Error Rate (FWER) defined as probability of falsely rejecting at least one null, or False Discovery Rate (FDR) defined below, which is more easily controlled. For a discussion of numerous solutions to the problem from which Benjamini-Hochberg (BH) procedure is the most commonly used, see e.g. [5]). In [7] analysis of behavior of FOR for BH procedure is given. However, construction of rules controlling FOR remains, to the best of our knowledge, largely untreated. Imagine now that we have a sample of n observations generated by mixture of distribution of inliers (occurring with probability π) and outliers (occurring with probability $1 - \pi$), and denote by $p_1, \ldots, p_n$ corresponding p-values of a test under consideration. Then we can write

$$p_i \sim \pi U + (1 - \pi)F_1, \quad i = 1, \ldots, n, \tag{1}$$

where U stands for the distribution function of the uniform distribution $U[0,1]$: $U(t) = t$, F_1 is the cumulative distribution function of the p-values for outliers and „$\sim$" denotes „is distributed as". In the following we assume that mixing proportion π is known. This assumption is commonly met i.e. when prevalence of a certain disease can be precisely estimated based on independent data base.

We assume that n_0 observations are inliers (nulls) and $n_1 = n - n_0$ are outliers (non-nulls) and note that due to our mixture assumption (1) n_0 and n_1 are random and have Bernoulli distribution: $n_0 \sim Bin(n, \pi)$ and $n_1 \sim Bin(n, 1 - \pi)$. Consider a specific decision rule assigning each of n observations to inliers or outliers and denote by R the number of of rejected null hypotheses, by V the number of falsely rejected nulls and by Z the number of falsely not rejected alternatives. Note that $Z = n_1 - (R - V)$ and let NR be the number of not rejected items. We will consider threshold rules such that for any $p_i \leq u$ the corresponding null hypothesis $H_{0,i}$ is rejected i.e. i-th element is considered an outlier. Threshold u is assumed here to be a fixed, predetermined point. We will write $NR(u)$ for NR to underline the dependence on u. Let

$$\text{FOR} = \mathbb{E}\left(\frac{Z}{NR(u)}\ \mathbb{I}\{NR(u) > 0\}\right),$$

$$\overline{\text{FOR}} = \mathbb{E}\left(\frac{NR(u) - n\pi(1 - u)}{NR(u)}\ \mathbb{I}\{NR(u) > 0\}\right). \tag{2}$$

FOR stands for False Omission Rate and $\overline{\text{FOR}}$ is an estimable approximation of FOR obtained by replacing number of non-rejected nulls at the threshold u by its expected value $n\pi(1 - u)$. Our aim is to construct a decision rule which approximately controls FOR i.e. such that for any given $\alpha \in (0, 1)$ the inequality FOR $\leq \alpha$ is satisfied.

In the traditional setting one aims at controlling False Discovery Rate (FDR) at the level α, where FDR is defined as

$$\text{FDR} = \mathbb{E}\left(\frac{V}{R}\ \mathbb{I}\{R > 0\}\right). \tag{3}$$

We note that although it might appear at the first sight that controlling FOR defined in (2) is analogous to controlling FDR, this is not the case as the roles of inliers and outliers are not exchangeable. The difference is due to differences in distributions of p-values for false signals and false non-signals. Namely, we assume that the distribution of inliers'score $\hat{s}$ is known, and it follows that for a threshold u, the distribution of the p-value corresponding to false signal, i.e. inlier smaller than u is given by the uniform distribution on $[0, u]$, whereas for the false non-signal it pertains to unknown distribution F_1 and equals $F_1(s)/(1 - F_1(u))$ for $s \in [u, 1]$. As F_1 is unknown, in contrast to known (i.e. uniform) distribution of p-values for the inliers, the problem of control of FOR is considerably harder than the control of FDR. We would like the distribution of F_1 to be concentrated close to 0, but this may vary depending on the quality of the score function in general and its performance on the studied dataset in particular (see Fig. 5).

We also note that similarly to testing (where decrease of level of significance leads to smaller values of power), when FDR is controlled at the level α, FOR is uncontrolled and can attain any level less than proportion of outliers $1 - \pi$.

Example 1. Assume that distribution of the score statistic $\hat{s}$ for inliers is given by the standard normal distribution $N(0, 1)$ and outliers by $N(\theta, 1)$, where $\theta > 0$. We reject the null for large values of s. Then straightforward calculation show that the distribution function of p-value $(1 - \Phi)(s)$ for an outlier is given by $F_1(s) = 1 - \Phi\left(-\Phi^{-1}(s) - \theta\right) = \Phi\left(\Phi^{-1}(s) + \theta\right)$, where Φ is CDF of $N(0, 1)$. Using the formula (8) below, the values of FOR at threshold u_{FDR}^* corresponding to Benjamini-Hochberg procedure [2] or its modified version with Storey's correction [13] can be calculated, and are shown in Fig. 1. The figure shows that

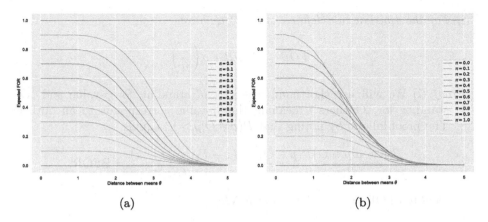

Fig. 1. Values of FOR when FDR is controlled against the mean distance θ between inliers and outliers (see text) (a) FDR $\leq \pi\alpha$ (Benjamini-Hochberg procedure) (b) FDR $\leq \alpha$ (Benjamini-Hochberg procedure with Storey's correction, [13]), $\alpha = 0.05$.

for small θ (when the inliers and outliers become less separated and threshold u_{FDR}^{*} becomes smaller), the value of FOR gets larger and approaches proportion $1 - \pi$ of inliers in the mixture.

We also introduce Bayesian False Omission Rate (BFOR)

$$\text{BFOR} = \frac{\mathbb{E}\,Z}{\mathbb{E}(NR(u))}, \tag{4}$$

following the analogous treatment of False Discovery Rates (see e.g. [6,7], Sect. 2.2). Efron [6] argues that from Bayesian view point, control of Bayesian False Discovery Rate – the quantity defined analogously to Eq. (4), but for false discoveries – is of main interest.

2.2 Control of FOR: Theoretical Results

We prove in Theorem 1 below that all introduced quantities are approximately equal for large sample sizes, namely

$$\text{FOR} \approx \text{BFOR}, \qquad \overline{\text{FOR}} \approx \text{BFOR}.$$

Let $G(t) = \pi U(t) + (1 - \pi)F_1(t)$ be a mixture distribution of p-values. We have

Theorem 1. *(i) Assume that considered decision rule rejects all null hypotheses with corresponding p-values smaller or equal u such that $0 < G(u) < 1$. Then*

$$FOR = BFOR \times (1 - (1 - G(u))^n) < BFOR.$$

A. Wawrzeńczyk and J. Mielniczuk

(ii) For the decision rule defined as in *(i)* we have

$$\overline{FOR} \leq BFOR + o\left(\frac{1}{n}\right).$$

Proof. (i) We will use shorthand $p \in O$ („p" standing for p-value and „O" standing for „Outliers") meaning that p-value corresponds to an outlier. The proof follows by noting that $P(p \in O|p > u)$ equals

$$\frac{P(p > u|p \in O)P(p \in O)}{P(p > u)} = \frac{(1 - F_1(u))(1 - \pi)}{1 - G(u)} = \text{BFOR}. \qquad (5)$$

Denote $\text{BFOR} = \gamma(u)$. Thus, given $NR(u)$,

$$Z|NR(u) \sim Bin(NR(u), \gamma(u)),$$

For $NR \neq 0$, using the formula for the expected value of the binomial, we have that

$$\mathbb{E}\left(\left.\frac{Z}{NR(u)}\right| NR(u)\right) = \frac{\mathbb{E}(Z|NR(u))}{NR(u)} = \frac{NR(u)\gamma(u)}{NR(u)} = \gamma(u) = \text{BFOR}$$

and thus

$$\begin{aligned} \text{FOR} &= \sum_{i>0} \mathbb{E}\left(\left.\frac{Z}{NR(u)}\right| NR(u) = i\right) P(NR(u) = i) \\ &= \sum_{i>0} \gamma(u)P(NR(u) = i) = \gamma(u)P(NR(u) \neq 0), \end{aligned}$$

which implies (i) as $NR(u) \sim Bin(n, 1 - G(u))$.
(ii) Observe that for $X \sim Bin(n, p)$ we have

$$\begin{aligned} \mathbb{E}\left(\frac{1}{X}\mathbb{I}\{X > 0\}\right) &\geq \mathbb{E}\left(\frac{1}{X+1}\right) - P(X = 0) \\ &= \frac{1}{p(n+1)}[1 - (1 - p)^{n+1}] - (1 - p)^n, \end{aligned} \qquad (6)$$

where the inequality follows from $\mathbb{E}\frac{1}{X+1}\mathbb{I}\{X = 0\} = P(X = 0)$ and the final equality above is proved in [4].

Thus for $X = NR(u) \sim Bin(n, 1 - G(u))$:

$$\mathbb{E}\left(\frac{NR(u) - n\pi(1 - u)}{NR(u)} \mathbb{I}\{NR(u) > 0\}\right)$$

$$= P(NR(u) > 0) - n\pi(1 - u) \times \mathbb{E}\left(\frac{1}{NR(u)} \mathbb{I}\{NR(u) > 0\}\right)$$

$$\leq 1 - G(u)^n - n\pi(1 - u) \times$$

$$\times \left[\frac{1}{(n+1)(1 - G(u))}(1 - G(u)^{n+1}) - G(u)^n\right] \tag{7}$$

$$= 1 - \frac{n\pi(1 - u)}{(n+1)(1 - G(u))} - G(u)^n(1 - n\pi(1 - u) -$$

$$- G(u)n\pi(1 - u))$$

$$= 1 - \frac{\pi(1 - u)}{1 - G(u)} + o\left(\frac{1}{n}\right) = \text{BFOR} + o\left(\frac{1}{n}\right),$$

where inequality follows from (6). The two first terms in penultimate expression above are equal to $\text{BFOR} + o(n^{-1})$ due to $n^{-1} - (n+1)^{-1} = (n(n+1))^{-1} = o(n^{-1})$ and all remaining terms are also $o(n^{-1})$.   ∎

We note that it follows from the proof that both $Z \sim Bin(n, (1 - \pi)(1 - F_1(u))$ and $NR(u) \sim Bin(n, 1 - G(u))$ are binomially distributed and thus we have

$$\frac{\mathbb{E}(Z)}{\mathbb{E}(NR(u))} = \frac{(1 - \pi)(1 - F_1(u))}{1 - G(u)} = \frac{1 - G(u) - \pi(1 - u)}{1 - G(u)} = P(p \in O | p > u). \tag{8}$$

Note that we assume in Theorem 1 that threshold u does not depend on data. We conjecture that in a general case, when threshold will be data-dependent, FOR, BFOR and $\overline{\text{FOR}}$ are also approximately equivalent.

Replacing FOR in the condition $\text{FOR} = \alpha$ by its approximation BFOR one obtains the following equality

$$\frac{1 - G(u) - \pi(1 - u)}{1 - G(u)} = \alpha, \tag{9}$$

or equivalently

$$1 - G(u) = \frac{\pi}{1 - \alpha} \times (1 - u). \tag{10}$$

Theorem 2. *Solution $u^* \in (0, 1)$ of (10) exists and is unique provided that (i) $G(\cdot)$ is strictly concave and (ii) $G'(1) \geq \pi/(1 - \alpha)$.*

Proof. Indeed, the condition (ii) is equivalent to the condition that the derivative of $1 - G(u)$ at 1 is not larger than the derivative of the line $(\pi/1 - \alpha) \times (1 - u)$ at 1. As $1 - G(u)$ is strictly convex it is enough to check that $1 - G(0) = 1 \geq \pi/(1 - \alpha)$. But this follows from (ii) since $1 > G'(1)$ as density $g(s) = G'(s)$ is strictly decreasing in view of strict concavity and $\int_0^1 g(s)\,ds = 1$. Uniqueness of the solution is due to the strict concavity of G.   ∎

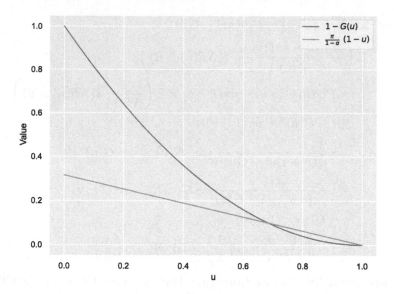

Fig. 2. Illustration of equation (10). Convex curve $1 - G(u)$ starts at 0 above the value of the line $(\pi/(1-\alpha)) \times (1-u)$ and intersects it at a point u^*.

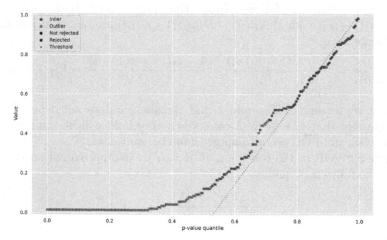

Fig. 3. Illustration of rule (11). The index of the smallest p-value which down-crosses the line $1 - (1 - \alpha)/\pi \times (1 - u)$ corresponds to the threshold in (11).

Theoretical solution is shown in Fig. 1. Thus we know that the truncation level u^* for such that $\text{BFOR} = \mathbb{E}(Z)/\mathbb{E}(NR(u^*)) = \alpha$ exists under above conditions. Note that the assumption that $G(\cdot)$ is strictly concave (or, equivalently, that $g(\cdot)$ is strictly decreasing) is natural in the considered context. Namely, it implies in the view of (1) that density f_1 of p-value distribution F_1 for outliers is strictly decreasing, and, consequently, it is more likely to obtain smaller p-values for outliers than larger ones.

2.3   FOR Control: Empirical Rule

Now we consider solution to the empirical counterpart of (10). Note that due to (8) BFOR is easily estimated, and we obtain the following rule: for a given $\alpha \in (0,1)$ find p-value $p_{(i^*)}$ such that

$$
p_{(i^*)} = \begin{cases} \min_{p_{(i)}} B & \text{if } B \neq \emptyset; \\ 1 & \text{otherwise,} \end{cases}
$$

$$
\text{where } B = \left\{ p_{(i)} : p_{(i)} \leq 1 - \left(1 - \frac{i}{n}\right) \frac{1-\alpha}{\pi} \right\}
$$

(11)

and „accept" (treat as inliers) all p-values strictly larger than $p_{(i^*)}$, where $p_{(1)} \leq p_{(2)} \leq \ldots \leq p_{(n)}$. This follows by plugging in empirical distribution $G_n(t) = \#\{i : p_i \leq t\}/n$ of all p-values for G in (10) and noting that $G_n\left(p_{(i)}\right) = \frac{i}{n}$. Note also that the threshold in (11) equals 1 for $i = n$ and is approximately equal to $1 - (1-\alpha)/\pi \leq 0$ (implied by condition (ii) of Theorem 1) for $i = 1$. Thus i^* is an index of the ordered p-value corresponding to the first moment when ordered p-values down-cross (cross from above to below) the line $1 - (1-u) \times (1-\alpha)/\pi$. This empirical rule is analogous (in a symmetric way) to Benjamini-Hochberg threshold construction for FDR control: starting from the largest p-values (as those are of interest when controlling not-rejected examples; this is an mirror image of Benjamini-Hochberg procedure starting from the smallest p-values) we look for the last (i.e. the smallest) index where FOR is still controlled, and use it as a threshold separating inliers from outliers.

2.4   Construction of p-values

We now discuss the framework in which p-values appearing in (11) are defined (*Multisplit* procedure). Note that as we do not know CDF of score statistic $\hat{s}$ for inliers, we can not compute p-value directly as $1 - F(X)$ and F needs to be estimated. We thus consider a sample $\mathcal{D} = \{X_1, \ldots, X_{2n}\}$ of size $2n$ consisting of inliers which will be split into training $\mathcal{D}^{train}$ and calibration $\mathcal{D}^{cal}$ samples consisting of n observation each. Moreover, let $\hat{s}$ will be a real-valued score statistic constructed to distinguish inliers from outliers. We adopt the convention that large values of $\hat{s}$ indicate a possible outlier. We consider the empirical distribution of $\hat{s}(X_i)$ for $X_i \in \mathcal{D}^{cal}$ as approximation of F and define p-value $\hat{p} = \hat{p}(X)$ of X as

$$
\hat{p} = \frac{\#\left\{X_i \in \mathcal{D}^{cal} : \hat{s}(X_i) \geq \hat{s}(X)\right\} + 1}{n + 1}.
$$

(12)

Consider the sample $S_1, \ldots, S_{n+1}$ consisting of observations $\hat{s}(X_i)$ for $X_i \in \mathcal{D}^{cal}$ augmented by $S = \hat{s}(X)$. When S corresponds to an inlier, observations $S_1, \ldots, S_{n+1}$ are equi-distributed and it follows that for continuous $\hat{s}(X)$, $\hat{p}(X)$ is uniformly distributed on $\{1/(n+1), \ldots, n/(n+1), 1\}$ given $\mathcal{D}^{cal}$, and thus $P(\hat{p}(X) \leq t|\mathcal{D}^{cal}) \leq t$ (see e.g. [1]) – this means that distribution of $p(X)$ is super-uniform.

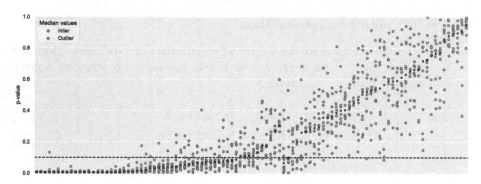

Fig. 4. p-value lottery for A^3 classifier on 100 first examples from *Tic-tac-toe* dataset; examples are sorted according to their class and median p-value. p-values for 10 random training-calibration splits are very unstable – median range width is 0.31, and maximal difference between minimal and maximal p-value for one of the examples exceeded 0.83. FOR control based a single split would not be reliable.

As the definition above depends on the training-calibration split we initially considered several versions of $\hat{p}$:

- p_{single}: one-split version defined above,
- p_{med}: median of $p_1, \ldots, p_k$ when $p_1, \ldots, p_k$ are p-values based on k random splits,
- $p_{2med} = 2 \times p_{med}$.

Definitions of p_{med} and p_{2med} are based on analogous proposals in variable selection and their purpose is to decrease variability incurred due to the random split; the phenomenon named p-value lottery (see [11]). Its occurrence is confirmed by Fig. 4 which shows substantial variability of p-values depending on a random split for A^3 classifier. The distribution of $p_{single}(X)$ is super-uniform and the the same is also true for p_{2med}, with the proof being analogous to that of Theorem 11.1 in [3]. p-value p_{med} is also considered as in practice p_{2med} is too conservative and thus inflates FDR in consequence. As our experiments confirm this, we focus on p_{med} only in the following.

3   Experimental Setting

We tested the proposed FOR control procedure as described in Sect. 2.3. We consider four different score functions to obtain the outlier scores:

- Isolation Forest [9] (abbreviated to *IForest*),
- Activation Anomaly Analysis A^3 [12] (neural network based model),
- Mahalanobis distance [10] based score (abbreviated as *Mahalanobis*),
- Empirical Cumulative distribution based Outlier Detection *ECOD* [8], as well as its variant applying ECOD to PCA-transformed data (abbreviated as *ECOD+PCA*).

Table 1. Dataset summary

Dataset	Samples	Features	Inlier rate π	Dataset	Samples	Features	Inlier rate π
Abalone	4177	8	0.34	Madelon	2600	500	0.50
Arrhythmia	452	279	0.54	Musk	6598	166	0.85
Banknote-auth	1372	4	0.56	Optdigits	5620	64	0.20
Breast-w	699	9	0.66	Pendigits	10992	16	0.20
Dermatology	366	34	0.31	Satimage	6430	36	0.24
Diabetes	768	8	0.65	Segment	2310	19	0.29
Fertility	100	9	0.88	Seismic-bumps	210	7	0.33
Gas-drift	13910	128	0.51	Semeion	1593	256	0.20
Glass	214	9	0.68	Sonar	208	60	0.47
Haberman	306	3	0.74	Spambase	4601	57	0.61
Heart-statlog	270	13	0.56	Tic-tac-toe	958	9	0.65
Ionosphere	351	34	0.64	Vehicle	846	18	0.26
Isolet	7797	617	0.27	Waveform-5000	5000	40	0.34
Jm1	10885	21	0.81	Wdbc	569	30	0.63
Kc1	2109	21	0.85	Yeast	1484	8	0.16

For each score function, the p-values are obtained from scores using *Multisplit* procedure (Sect. 2.4), and control procedures (e.g. FOR control procedure) were applied; number of random splits in Multisplit procedure was set as $k = 10$. Each experiment used 60% of the inliers for training+calibration, and the remaining inliers and all of the outliers as the test set. For each test case, we repeated the entire process (starting from the training+calibration / test split) 20 times. For the control level we used $\alpha = 0.1$, which is a common value considered in literature. Code implementing the FOR control procedure, all tested methods and experiments is available publicly on GitHub[1].

Tests were conducted on 30 datasets constructed from real-world classification data. One of the classes (or several relatively similar ones) was selected as the inlier class, while other classes were considered as outliers. Basic summary of the datasets is presented in Table 1. For details on dataset construction, as well as their visualizations, we refer to the GitHub repository[2].

4   Results

Table 2 aggregates FOR mean values (and their standard errors) for the proposed FOR control procedure. FOR is controlled by at least one method on 20 datasets, but there are only 3 datasets where the same holds true for all methods at once. Mean FOR value was below 2α for at least one classifier in 29 out of 30 cases (except *Yeast* dataset). Even though A^3 controlled FOR on the largest number of datasets, we will concentrate on *IForest* due to the higher consistency of its

---

[1] https://github.com/wawrzenczyka/FOR-CTL.
[2] https://github.com/wawrzenczyka/FOR-CTL-datasets.

Table 2. FOR values and standard errors under FOR control on tested datasets, for level $\alpha = 0.1$. Magenta „✓" denotes FOR $\leq \alpha$; black „✓" denote weaker FOR $\leq 2\alpha$.

Dataset	IForest	A^3	Mahalanobis	ECOD	ECOD + PCA
Musk	0.000 ± 0.000 ✓✓	0.137 ± 0.005 ✓	0.415 ± 0.040	0.000 ± 0.000 ✓✓	0.135 ± 0.011 ✓
Seismic-bumps	0.061 ± 0.018 ✓✓	0.093 ± 0.028 ✓✓	0.056 ± 0.018 ✓✓	0.048 ± 0.016 ✓✓	0.099 ± 0.025 ✓✓
Ionosphere	0.067 ± 0.014 ✓✓	0.107 ± 0.016 ✓	0.082 ± 0.011 ✓✓	0.081 ± 0.016 ✓✓	0.140 ± 0.009 ✓
Tic-tac-toe	0.075 ± 0.008 ✓✓	0.061 ± 0.004 ✓✓	0.140 ± 0.023 ✓	0.117 ± 0.012 ✓	0.272 ± 0.030
Breast-w	0.076 ± 0.005 ✓✓	0.083 ± 0.003 ✓✓	0.086 ± 0.004 ✓✓	0.057 ± 0.005 ✓✓	0.095 ± 0.004 ✓✓
Isolet	0.078 ± 0.006 ✓✓	0.047 ± 0.014 ✓✓	0.282 ± 0.006	0.090 ± 0.009 ✓✓	0.363 ± 0.004
Dermatology	0.078 ± 0.013 ✓✓	0.037 ± 0.006 ✓✓	0.049 ± 0.014 ✓✓	0.052 ± 0.008 ✓✓	0.206 ± 0.022
Semeion	0.078 ± 0.014 ✓✓	0.074 ± 0.012 ✓✓	0.049 ± 0.012 ✓✓	0.118 ± 0.016 ✓	0.000 ± 0.000 ✓✓
Banknote-auth	0.079 ± 0.009 ✓✓	0.086 ± 0.029 ✓✓	0.078 ± 0.003 ✓✓	0.069 ± 0.012 ✓✓	0.085 ± 0.006 ✓✓
Pendigits	0.091 ± 0.005 ✓✓	0.091 ± 0.004 ✓✓	0.108 ± 0.006 ✓	0.100 ± 0.012 ✓✓	0.099 ± 0.010 ✓✓
Satimage	0.094 ± 0.004 ✓✓	0.000 ± 0.000 ✓✓	0.084 ± 0.005 ✓✓	0.129 ± 0.010 ✓	0.105 ± 0.007 ✓
Segment	0.094 ± 0.007 ✓✓	0.317 ± 0.082	0.109 ± 0.009 ✓	0.085 ± 0.005 ✓✓	0.151 ± 0.017 ✓
Kc1	0.094 ± 0.008 ✓✓	0.091 ± 0.008 ✓✓	0.089 ± 0.009 ✓✓	0.170 ± 0.022 ✓	0.129 ± 0.007 ✓
Wdbc	0.097 ± 0.009 ✓✓	0.088 ± 0.010 ✓✓	0.100 ± 0.006 ✓✓	0.097 ± 0.010 ✓✓	0.137 ± 0.007 ✓
Optdigits	0.101 ± 0.010 ✓	0.074 ± 0.011 ✓✓	0.083 ± 0.008 ✓✓	0.160 ± 0.036 ✓	0.188 ± 0.012 ✓
Gas-drift	0.114 ± 0.010 ✓	0.000 ± 0.000 ✓✓	0.146 ± 0.011 ✓	0.088 ± 0.014 ✓✓	0.143 ± 0.013 ✓
Spambase	0.122 ± 0.021 ✓	0.139 ± 0.008 ✓	0.140 ± 0.017 ✓	0.217 ± 0.056	0.173 ± 0.025 ✓
Vehicle	0.127 ± 0.024 ✓	0.000 ± 0.000 ✓✓	0.073 ± 0.009 ✓✓	0.160 ± 0.045 ✓	0.162 ± 0.014 ✓
Glass	0.147 ± 0.030 ✓	0.163 ± 0.025 ✓	0.121 ± 0.019 ✓	0.186 ± 0.039 ✓	0.156 ± 0.018 ✓
Heart-statlog	0.153 ± 0.020 ✓	0.277 ± 0.035	0.170 ± 0.021 ✓	0.140 ± 0.026 ✓	0.246 ± 0.050
Diabetes	0.179 ± 0.017 ✓	0.151 ± 0.021 ✓	0.240 ± 0.030	0.271 ± 0.079	0.133 ± 0.023 ✓
Waveform-5000	0.204 ± 0.020	0.264 ± 0.076	0.161 ± 0.041 ✓	0.142 ± 0.024 ✓	0.355 ± 0.046
Abalone	0.206 ± 0.008	0.093 ± 0.014 ✓✓	0.170 ± 0.015 ✓	0.302 ± 0.030	0.167 ± 0.017 ✓
Arrhythmia	0.214 ± 0.037	0.187 ± 0.076 ✓	0.197 ± 0.030 ✓	0.300 ± 0.058	0.292 ± 0.017
Fertility	0.219 ± 0.024	0.121 ± 0.016 ✓	0.218 ± 0.024	0.229 ± 0.036	0.223 ± 0.022
Yeast	0.228 ± 0.069	0.299 ± 0.088	0.226 ± 0.078	0.301 ± 0.084	0.251 ± 0.092
Jm1	0.237 ± 0.008	0.000 ± 0.000 ✓✓	0.189 ± 0.022 ✓	0.312 ± 0.027	0.242 ± 0.020
Haberman	0.293 ± 0.035	0.472 ± 0.057	0.366 ± 0.058	0.153 ± 0.033 ✓	0.259 ± 0.039
Sonar	0.431 ± 0.078	0.125 ± 0.047 ✓	0.252 ± 0.052	0.175 ± 0.083 ✓	0.506 ± 0.030
Madelon	0.748 ± 0.013	0.495 ± 0.010	0.722 ± 0.008	0.100 ± 0.069 ✓✓	0.150 ± 0.082 ✓

results. *IForest* managed to control FOR $\leq \alpha$ in 14 cases, $FOR \leq 2\alpha$ in 7 additional ones, and failed to keep FOR below 2α on the remaining 9 datasets.

Figure 5 illustrates 2-dimensional t-SNE representation of the data (1st column), distributions of the obtained p-values (2nd column) and FOR control procedure visualization (3rd column) for three of the datasets. *Tic-tac-toe* is an example of an easy dataset: we can see that the data forms distinct, separate clusters (Fig. 5a) and therefore there is a clear difference in inlier and outlier p-value distributions (Fig. 5b), as well as nearly perfectly uniform inlier distribution – *IForest* captured the inlier distribution really well. In that case, FOR control procedure has no issues capturing the clean portion of inliers (Fig. 5c) with the occasional outlier samples allowed by the α parameter.

Vehicle dataset in the second row is of medium difficulty: inlier and outlier samples (Fig. 5d) are a lot more difficult to separate. Note that the outlier p-value distribution (Fig. 5e) is shifted right, towards higher values, and inlier distribution is not as regular as in previous case. Though this example is significantly harder, we can see that in this particular case FOR control (Fig. 5f) divided the examples perfectly – though multiple outlier samples are extremely close to the threshold and might be incorrectly undetected with a small variations in their p-values. That leads to FOR for this dataset in Table 2 being slightly higher than desired.

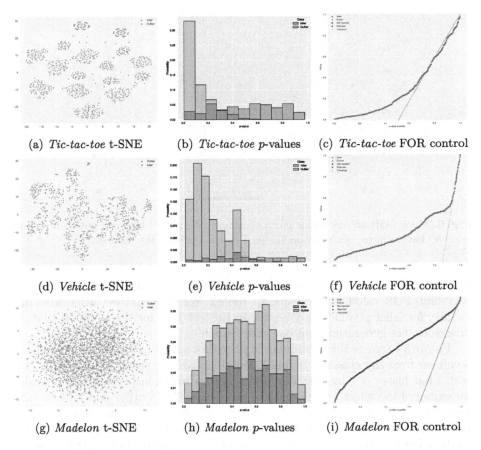

(a) *Tic-tac-toe* t-SNE   (b) *Tic-tac-toe* p-values   (c) *Tic-tac-toe* FOR control

(d) *Vehicle* t-SNE   (e) *Vehicle* p-values   (f) *Vehicle* FOR control

(g) *Madelon* t-SNE   (h) *Madelon* p-values   (i) *Madelon* FOR control

Fig. 5. FOR control for datasets with varying difficulty, based on Isolation Forest scores. Red (green) dots correspond to inliers (outliers). (Color figure online)

Madelon dataset, on the other hand, was selected as an hard problem example. T-SNE visualization (Fig. 5g) does not capture any visible outlier characteristics, which suggests that the relationships in the data are complex. As a consequence, p-value distributions obtained from *IForest* scores (Fig. 5h) are extremely similar between outliers and inliers; note that the inlier p-value density is slightly bell shaped and thus deviates from the uniform, moreover the outliers p-value density does not decrease. As the proposed procedure assumes those properties, their lack has a profound impact on the FOR control (Fig. 5i). Lower than expected (when uniformity holds) number of inlier examples with high p-values causes outliers with high p-values to take their place; this results

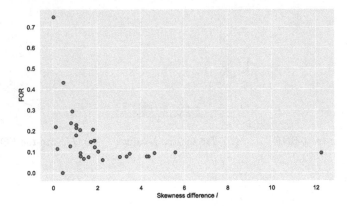

Fig. 6. Mean FOR values versus mean skewness difference I between outlier and inlier p-value distributions; each dot on the plot corresponds to one dataset.

in the dramatic omission of dominant part of outlier samples, which results in a very high FOR value, as presented in Table 2. We note that deviation from uniformity for inlier p-values may be due to the fact that for this synthetic dataset inliers are not generated from one distribution.

Figure 5 suggests that obtaining high quality scores (and, as a result, reliable p-values) from the classifier is fundamental in order to ensure a good FOR control. That makes p-value distribution properties worth inspecting. In particular, we explored the effect of difference in skewnesses I of outlier and inlier p-value distribution, given by formula $I = Skew_{OUT} - Skew_{IN}$, on the empirical FOR value. We expect inlier distribution to be uniform (so $Skew_{IN}$ should be close to 0), whereas probability mass for the outlier distribution should be concentrated on small p-values (resulting in large positive values of $Skew_{OUT}$), which should mean that for a p-value distributions satisfying the imposed assumptions, I should be both positive and relatively large. Indeed, as illustrated in the Fig. 6, datasets where I is large are also the ones where the proposed procedure works really well; on the other hand, when I falls below 2, FOR control becomes unreliable. This emphasizes the dependency of FOR control on outlier score quality – we can control FOR only if outlier scores make that possible.

Figure 7 visualizes relationship between FOR and FDR when FOR is controlled on all sets. Observe that good control of FOR doesn't imply low FDR value (and vice versa, see Fig. 1). Moreover, this holds irrespectively of the scoring method – even when a given method controls FOR at a given level, its FDR might remain high.

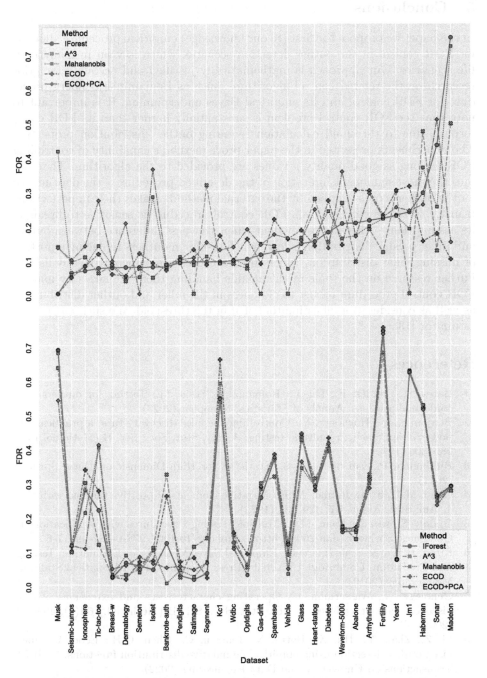

Fig. 7. FOR values for all methods and data sets (upper panel) with the corresponding values of FDR (lower panel).

5   Conclusions

In the paper we propose the first (to our knowledge) empirical procedure allowing for FOR control in the outlier detection scenario, which is vital in many real-life scenarios. Our approach is mathematically justified and accounts for prior research on the related control algorithms, such as Benjamini-Hochberg procedure for FDR control and its empirical Bayes underpinning. It is important to note that the FOR control problem is substantially harder than its FDR counterpart, due to threshold calculation requiring outlier distribution properties. The experiments presented in the paper prove method's capability of controlling FOR as long as good quality p-values are provided to the algorithm. That ties into the most significant limitation of the described procedure – the dependence on the outlier scores supplied by the external methods makes their imperfections transfer to the researched task. FOR control procedure is sensitive to breaking its assumptions – this is most evident when skewness difference between outlier and inlier p-value distribution is low, which results in outliers replacing a portion of missing inlier distribution, which in turn causes their uncontrolled omission. Futher research on the topic might include handling those scenarios by making FOR control procedure robust to closeness in the inlier and outlier distribution, as well as consideration of estimators of π in the threshold rule such as Storey's estimator [13].

References

1. Bates, S., Candès, E., Lei, L., Romano, Y., Sesia, M.: Testing for outliers with conformal p-values. Annals of Statistics, to appear (2023)
2. Benjamini, Y., Hochberg, Y.: Controlling the false discovery rate: a practical and powerful approach to multiple testing. J. Roy. Stat. Soc.: Ser. B (Methodol.) **57**, 289–300 (1995)
3. Bühlmann, P., van de Geer, S.: Statistics for High-Dimensional Data. Springer (2011)
4. Chao, M.T., Strawderman, W.E.: Negative moments of positive random variables. J. Am. Stat. Assoc. **67**, 429–431 (1972)
5. Dudoit, S., van der Laan, M.J.: Multiple Testing Procedures with Applications to Genomics. Springer (Jan 2008). https://doi.org/10.1007/978-0-387-49317-6
6. Efron, B.: Large-Scale Inference: Empirical Bayes Methods for Estimation, Testing, and Prediction. Cambridge University Press, Institute of Mathematical Statistics Monographs (2010)
7. Genovese, C., Wasserman, L.: Operating characteristics and extensions of false discovery rate procedures. J. Roy. Stat. Soc.: Ser. B (Methodol.) **64**(3), 499–517 (2002)
8. Li, Z., Zhao, Y., Hu, X., Botta, N., Ionescu, C., Chen, G.H.: Ecod: Unsupervised outlier detection using empirical cumulative distribution functions. In: IEEE Transactions on Knowledge and Data Engineering (2022)
9. Liu, F.T., Ting, K.M., Zhou, Z.H.: Isolation forest. In: 2008 Eighth IEEE International Conference on Data Mining, pp. 413–422 (2008)
10. Liu, R.Y., Parelius, J.M., Singh, K.: Multivariate analysis by data depth: descriptive statistics, graphics and inference. Ann. Stat. **27**(3), 783–858 (1999)

11. Meinshausen, N., Meier, L., Bühlmann, P.: P-values for high-dimensional regression. J. Am. Stat. Assoc. **104**, 1671–1681 (2008)
12. Sperl, P., Schulze, J.-P., Böttinger, K.: Activation anomaly analysis. In: Hutter, F., Kersting, K., Lijffijt, J., Valera, I. (eds.) ECML PKDD 2020. LNCS (LNAI), vol. 12458, pp. 69–84. Springer, Cham (2021). https://doi.org/10.1007/978-3-030-67661-2_5
13. Storey, J.: Direct approach to false discovery rates. J. Royal Stat. Society. Series B (Methodological) **64**, 479–498 (2002)
14. Takahashi, H., Ichinose, N., Yasusei, O.: False-negative rate of sars-cov-2 rt-pcr tests and its relationship to test timing and illness severity. IdCases 28 (2022)

Minimal Path Delay Leading Zero Counters on Xilinx FPGAs

Gregory Morse[1]([✉])[iD], Tamás Kozsik[1][iD], and Péter Rakyta[2][iD]

[1] Department of Programming Languages and Compilers, Eötvös Loránd University, Budapest, Hungary
morse@inf.elte.hu
[2] Department of Physics of Complex Systems, Eötvös Loránd University, Budapest, Hungary

Abstract. We present an improved efficiency Leading Zero Counter for Xilinx FPGAs which improves the path delay while maintaining the resource usage, along with generalizing the scheme to variants whose inputs are of any size. We also show how the Ultrascale architecture also allows for better Intellectual Property solutions of certain forms of this circuit with its newly introduced logic elements. We also present a detailed framework that could be the basis for a methodology to measure results of small-scale circuit designs synthesized via high-level synthesis tools. Our result shows that very high frequencies are achievable with our design, especially at sizes where common applications like floating point addition would require them. For 16, 32 and 64-bit, our real-world build results show a 6%, 14% and 19% path delay improvement respectively, enough of an improvement for large scale designs to have the possibility to operate close to the maximum FPGA supported frequency.

Keywords: FPGA · MaxCompiler · Leading Zero Counters · High-Level Synthesis · Vivado

1 Introduction

Leading Zero Counters (LZC [4]) are of importance for various bit-level tasks, most notably floating point addition and subtraction [10,12,18] such as in the IEEE-754 standard. This is due to an effective subtraction when the signs are of opposite polarity with addition, or identical polarity with subtraction. Rather than the leading bit being one more, one less or identical to the original mantissa size (adjusted for its guard and round bits), which are also the three cases where rounding will occur, there also are all the cases where the result has some number of less digits up to the mantissa size plus the guard and round bit or be an all zero result. Since the exponent will need to be adjusted as well as the zero case handled, all of this information is essential. Detecting the all zero case can be merged into the logic of the leading zero detect as its intermediate information must be propagated anyway.

In fact, a traditional clever use of floating point units (FPUs) addition/subtraction unit has been using the normalization process post-subtraction with

© The Author(s), under exclusive license to Springer Nature Switzerland AG 2023
J. Mikyška et al. (Eds.): ICCS 2023, LNCS 14075, pp. 626–640, 2023.
https://doi.org/10.1007/978-3-031-36024-4_48

custom byte-packing [2], by inserting an integer in the mantissa m and setting the exponent to $e = b - 1$ where b is the mantissa size of the data type (e.g. $b = 24$ for float32, $b = 53$ for float64). The floating point value is $m * 2^e$. There is always an implied 2^e added to the stored mantissa, in the IEEE standard variants. Then by subtracting the exponent part 2^{b-1}, the exponent becomes the leading zero count. For more details, see the well known C implementations of *Find the integer log base 2 of an integer with an 64-bit IEEE float* which is a prime use case [2]. As an example, a double-precision number would set the 12-bits representing the exponent to $1023 + 53 - 1 = 0x433$ where 1023 is the exponent bias and the 52-bits of mantissa to the value to determine the LZC. By subtracting 2^{53-1}, it effectively subtracts out the implied one at the 53rd position, and requires re-normalizing, which places the leading zeros as the exponent. Merely subtracting the bias or 1023 from the 12-bits of the exponent, yields the LZC.

More formally, we define the LZC-n for bit-vector $X_{n..1}$ as an ordered pair (V, C) where

$$V = \bigwedge_{K=n}^{1} \overline{X_k} = \overline{X_n} \wedge \overline{X_{n-1}} \wedge \cdots \wedge \overline{X_1} \qquad (1)$$

is the all-zero signal and

$$z(i,j) = \left(\bigvee_{k=n-2^{i+j}}^{n-2^{i+j+1}} X_k \right) \vee \left(\bigwedge_{k=n-2^{i+j+1}}^{n-2^{i+j+2}} \overline{X_k} \wedge z(i, j+2) \right) \qquad (2)$$

$$C = \prod_{i=0}^{\lceil \log_2 n \rceil - 1} \left(V \vee \left(\bigwedge_{k=n}^{n-2^i} \overline{X_k} \wedge z(i, 0) \right) \right) \qquad (3)$$

represents the leading zero count as a bit-string (which is built via the concatenation operator $\|$) in Boolean algebra as an infinite recurring relationship (where $\vee$ and $\wedge$ are logical OR and logical AND respectively). In our notation, a bar above represents a logical negation and $\lceil x \rceil$ is the ceiling operation of rounding x up to the nearest integer. In the special case that X contains all zeros, then V and all bits of C are set to 1. In some contexts, such as one-hot decoding, these artifacts may be unnecessary. However as will be seen, their intermediate computation is convenient regardless.

Although traditionally a focus on power is prevalent, we have chosen to focus on performance, then area and only minimize power if it does not effect performance or area. As higher area allows more concurrency and thus more performance, our justification for high-performance computing (HPC) is due to work in the area of Quantum simulation. But an investigation into the latest offerings for HPC in Ultrascale and Vivado is thus forthcoming.

Our contribution is thus a more general framework which uses careful and precise integration of a more modular framework, which yields a better result. Expert re-synthesis of integrated units of a modular design, can unsurprisingly yield a better state-of-the-art result. The exact ideas and optimizations used are important in a broader range of circuits.

1.1 Efficient Logic Block Usage on Xilinx FPGAs

This discussion is in part intended to draw attention to the complexity and multi-layered and even ambiguous approach the tools take to configuring the more advanced FPGA features. Various settings being utilized at various stages inclusively or exclusively with in some cases profound effects on the output make Vivado an expert tool on a per platform or even per module basis. It intends to draw attention to the reader all attributes and settings which are relevant or worth of consideration in this specific context.

Xilinx FPGAs have long utilized Look-Up Table (LUT) 6's with 6 input signals and one output signal [1] providing a realization of a generic 6 input binary function $f(x_0, x_1, x_2, x_3, x_4, x_5)$ where $x_k \in \{0,1\} \forall k \in \{0, \ldots, 5\}$, thus appropriate to realize high performance signal processing implementations [8]. 4 LUT-6 appear in a single slice in the architecture. A LUT-6 can also be thought of as a 4-to-1 multiplexer where 4 inputs are selected from based on the other 2 inputs as

$$f(x_0, x_1, x_2, x_3, x_4, x_5) = \begin{cases} x_0 & \text{if } \overline{x_4} \wedge \overline{x_5} \\ x_1 & \text{if } \overline{x_4} \wedge x_5 \\ x_2 & \text{if } x_4 \wedge \overline{x_5} \\ x_3 & \text{otherwise} \end{cases} \tag{4}$$

where the number encoded by the 2 binary inputs can be used to govern one of the remaining 4 inputs onto the output. However modern Xilinx 7-series and Ultrascale FPGAs also offer additional features. The *LUT N:M* (LUTNM) VHDL feature allows combining logical LUTs into a single physical LUT with multiple outputs (explicitly specified by VHDL property *LUTNM* and its hierarchical counterpart *HLUTNM*). This is ubiquitous as its presence is in both the logic slices (*SLICEL*) and the memory slices (*SLICEM*) the latter of which can be repurposed as Shift-Register LUTs (SRLs). It will occur by inference during synthesis if the no LUT combining option (*-no_lc*) is absent as well as during placement when no physical synthesis in placer (*-no_psip*) is absent [15]. Since the LUT-6 are actually implemented as an element called a *LUT6-2* as seen in Fig. 1, the flexibility of $N \leq 5$ common inputs mapped to $M = 2$ outputs has allowed increased efficiency involved with common design patterns. In practice this means a realization of two 5-input LUTs with common inputs and having 2 separate outputs [13], as the 6-th input is used to multiplex between the 2 outputs. Therefore, any LUT-2 and LUT-3 without common inputs can be easily combined into a single instance of LUT6-2. Likewise two LUT-3 which share a single common input would also be combine-able. In 1, x_5 is the 6th input used for multiplexing, while $x_0, \ldots, x_4$ are the 5-common inputs to the two LUT-5s. As can two LUT-5 with 5 common inputs, etc. Formally we have the two LUT-5s with outputs $f_2(x_0, \ldots, x_4), f_3(x_0, \ldots, x_4)$ (see Fig. 1) but the latter output is replaced with

$$f_1(x_0, \ldots, x_5) = \begin{cases} f_3(x_0, \ldots, x_4) & \text{if } x_5 \\ f_2(x_0, \ldots, x_4) & \text{otherwise} \end{cases} . \tag{5}$$

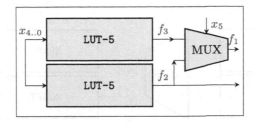

Fig. 1. LUT6-2 in Xilinx Ultrascale Configurable Logic Blocks (CLBs).

This physical realization turns out to be very convenient for logic circuits like LZCs. For clarity, notation $LUT-k$ where $2 \leq k \leq 6$ represents a logical LUT, while $LUT6 - 2$ represents the physical element. The second feature of modern Xilinx FPGAs is the existence of two extra multiplexing units in all of the slices called a $MUXF7$ and $MUXF8$ as shown in Fig. 2. The figure shows its name usage context but these elements can also function as a generic multiplexer of two LUTs or two MUXF7s). On the left, is the optimal CLB arrangement of a 7-bit multiplexer with the MUXF7 preceding the output, while on the right is the optimal CLB arrangement of an 8-bit multiplexer, from which these units receive their name. Inputs and outputs are indicated by arrows. These elements allow further multiplexing of the outputs of the LUT6-2s. Conceptually, $MUXF7$ and $MUXF8$ allow 2 LUT-6s to be turned into an 8 to 1 multiplexer or 16 to 1 multiplexer, respectively [3]. However, they need not be strictly used for this specific purpose. At this point we notice the presence of a $MUXF9$ unit in the design, however, it is not inferred during the synthesis process and the direct utilization of this element might cause undesirable path delays (unless under specific circumstances), so we do not consider this element in this work. $MUXF7$ and $MUXF8$ are explicitly specified via the $MUXF\_MAPPING$ VHDL property, and can be converted to LUT-3s via the -$muxf\_remap$ option in the design optimization phase. (The 3-input $MUXF7$ can be also considered as a 3-input LUT having a configuration space of size $2^3 = 8$. [15]) A limitation of the outlined design is that the LUT6-2 pairs in a slice connect to one of the two multiplexed inputs of these multiplexer units, but not to the selector signal (see x_6/x_7 on the left/right side of Fig. 2.). This limitation makes the $MUXF7$ and $MUXF8$ units furthermore only optimal in specific contexts, depending upon the routing intricacies. Xilinx provides the Vivado Design Suite [16] to perform synthesis and implementation for its various FPGA models. The synthesis translates the VHDL into an internal model format of LUT, MUXF7, MUXF8 and optionally LUTNM units. Post-synthesis, the optional *optimize design* stage is typically invoked to start the implementation process. It can for example, reduce logic which is unnecessarily cascaded. The -*remap* option (and its related -*aggressive_remap* and -*resynth_remap*) could have a significant effect on the final circuit, often adding additional LUTs which would reduce path delay. Nevertheless, this option for simple measurements can be disabled. The physical optimizations, power optimizations and routing remain to achieve an implementation whose bitstream can be utilized.

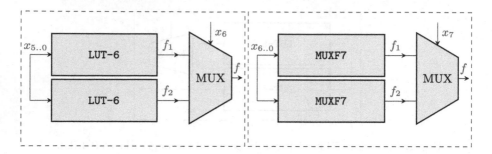

Fig. 2. MUXF7 and MUXF8 in Xilinx CLBs when used as a 7 and 8 bit multiplexer.

2   High-Level-Synthesis (HLS) Tools Background

For this project, rather than writing VHSIC Hardware Description Language (VHDL, where VHSIC is Very High-Speed Integrated Circuit), we used the MaxCompiler Data-Flow Engine (DFE) framework to provide a high-level Java description describing the calculations in a high-level extended variant of the Java programming language.

MaxCompiler uses a kernel design methodology, where each kernel is buffered by pulling input from a data stream and pushing output onto a data stream. It provides IP to integrate the PCIExpress subsystem for input and output seamlessly as data streams which can feed into the various kernels. MaxCompiler ultimately invokes Vivado to synthesize and implement the design. MaxCompiler has been used for large-scale research involving trigger algorithms [11], Complex FIR Filters [14] and even stencil computing [5] and most recently for quantum computer simulation [9]. But the idea of creating state-of-the-art circuits at a small scale has not been investigated. Although MaxCompiler supports at a kernel-level IP integrated solutions, it does not offer such functionality at a circuit level within a kernel.

The main feature needed to implementation efficient designs is part of the aptly named optimization library in MaxCompiler. The pipelining register disable and pipelining register enable features which effect a stack of states describing how operations pipeline which are constructed during any Java code in the current scope. Manually pipelining a signal s is easily done via explicit pipelining. By default, high level tools often pipeline every logic at the operator level which is not efficient for constructing efficient LUT-mapped circuitry. For example consider a bitwise AND: A & B will be stored in a register by default. This is equivalent to the behavior in no pipelining mode of pipeine(A & B).

There is furthermore far less research into state-of-the-art circuits which are constructed via high-level synthesis tools as opposed to those based upon hand-crafted VHDL.

3   Description of the LZC Design and Implementation

There are two typical design patterns when synthesizing logic circuits, one is the cascade, sequential arrangement, and the other is a tree, or parallel arrangement. The timing properties, and amount of resources are very much dependent upon this choice, and there is usually a point at which performance versus resources becomes a critical deciding factor. The interesting case, is when one aspect can be improved without worsening the other.

Several recent LZC designs have been published. The first introduced a leading-zero anticipation framework to handle carry-lookahead operations [4]. This was improved upon to use complex gates to improve the speed [6]. There was also a design which focused on modularity rather than efficiency [7]. Vivado High-Level Synthesis itself also provides a default implementation based on the C instrinic _builtin_clz which are generic but not optimal, and further leave the all-zero case as undefined [16].

We offer an improvement over the method which Zahir, et al. introduced [17]. Their method used an LZC-8 consisting of 3 LUTs cascading into a LUT6-2 as a primitive for larger LZC units. By removing the cascaded LUT structure of their 8-bit LZC primitive which although necessary for a 7 or 8 bit LZC, turns out to be logic expandable and reducible into the 16-bit layer, the path delay can be significantly improved. This in turn allows for meeting timing at higher frequencies. Furthermore, by careful use of LUT combining, rather than a 3-level cascade with 10 LUTs, we achieve 6 LUTs cascading into 4 LUTs, preserving the LUT usage.

Furthermore we generalize the solution to any bit size, and not just even powers of two. Floating point implementations for various mantissa sizes can be further optimized to its precise LZC bit-optimized variant. Even at common sizes of IEEE-754 single and double precision simple implementations require LZC of 26 and 55 while an extended precision x86 long double requires 68, depending on implementation details regarding rounding strategies. In an FPGA, there is no native or built-in FPU, so this functionality cannot be replicated via the method mentioned in the introduction.

Now we note several special base cases of LZCs: (i) LZC for 1-bit is trivial and requires no LUTs just the constants, the original signal or its inversion (which count as 1 LUT only if a final signal). (ii) The 2-bit variant is similarly trivial requiring a 2-LUT but these signals can be merely propagated to the next level if part of a bigger LZC. (iii) For 3 or 4 bits, it will require 2 LUTs. (iv) For 5 or 6 bits, 3 LUTs. (v) And finally for 7 or 8 bits, 4 LUTs like in the prior method. (vi) Various occurrences where the formulae simplify down to a single signal, or even two or in some cases 3 signals may allow merging of signals into later layers, but such fine-tuned optimization should be reserved for Intellectual Property (IP)-level solutions, and we do not concern ourselves with this.

For convenience, we label a k-bit LZC unit ($k >= 2$) with LZC$-k$. The primary design issue in creating larger LZC units is to solve the odd/even parity computation of the result in a LZC-15/16 unit (i.e LZC with 15 and 16 bits) whose inner intermediate values we refer to as leading-parity-one/two/three

(*LP1, LP2, LP3*) along with one to compute the all-zero indicator V (labeled by *LP4*) and whose truth table is defined in Table 1. X6-X1 represent the 6 most significant bits in little-Endian order. Note that *LP3* is not present for LZC-4 or less while LP2 is not present for LZC-2 or less. The input value of X represents "don't care" or any value. These values are cascaded into a second stage of LUTs provided in later equations. We also provide the formulae which are straight-forward derivations from Eqs. (1) and (3), and we have left them in their form which uses the minimal possible Boolean operations:

$$LP1 = \overline{X_6} \wedge \left(X_5 \vee \left(\overline{X_4} \wedge (X_3 \vee \overline{X_2})\right)\right) \tag{6}$$

$$LP2 = \overline{X_6} \wedge \overline{X_5} \wedge \left(X_4 \vee X_3 \vee \left(\overline{X_2} \wedge \overline{X_1}\right)\right) \tag{7}$$

$$LP3 = \overline{X_6} \wedge \overline{X_5} \wedge \overline{X_4} \wedge \overline{X_3} \tag{8}$$

$$LP4 = \overline{X_6} \wedge \overline{X_5} \wedge \overline{X_4} \wedge \overline{X_3} \wedge \overline{X_2} \wedge \overline{X_1} \tag{9}$$

A general 16-bit function takes at least a two-stage *LUT* cascade on a LUT-6-based architecture, which turns to be both sufficient and efficient in the case of LZC-15/16, due to the regularity and evenness of the calculation. A possible realization of a LZC-15/16 is to cascade two LZC-8 sub-units of Ref. [17]. Determining the parity with these LZC-8 sub-units would require 4 intermediate parity signals, and 3 all-zero signals. Since this approach would not properly fit the Ultrascale CLB architecture, the solution is to introduce a specialized strategy by utilizing a special signal splitting along groups of 6, 6 and 5 bit-slices of the input X for *LP1* and the all-zero checks (see Fig. 3 for details). In 3, combined LUT6s are colored dark. The high (H) 6 bits of the input X are elaborated by the upper half of the design, while the LUTs in the bottom row operates on 5-bit slices of the input X denoted by low (L) and lower low (LL) labels. The design can be adopted for LZC-9 up to LZC-14 implementations with signal reductions as described in the text. Note that $LP1_{LL}$ is computed by the LP1 truth Table 1 by setting $X_6 = 0$. The *LP1* computation need not consider the least significant bits (X_{11}, X_5, X_1) in the presence of the all-zero signal (reducing the bit-slices effectively to 5, 5 and 4 bits). This observation reduces the final computation of *LP1* to the combinations of 5 signals compared to 7 signals if maintaining the approach of Ref. [17]. To minimize the path delay for LZC-15/16, we use an LZC-8-Intermediate, i.e. a non-cascaded design of 2 LZC-8 units. The concept of the improved circuit is outlined in Fig. 3. In contrast to the LZC-8-Intermediate used by 15 and 16-bit inputs, the 9 to 14-bit input X is partitioned into two 8-bit segments (high and low) and transferred into LZC-8-High and LZC-8-Low components. LZC-8-High refer to the top half of Fig. 3 while LZC-8-Low indicate the high unit duplicated with $X_{8..1}$ substituted for $X_{16..9}$, having a symmetrical structure in this case. In order to fit the Xilinx CLB architecture, the LZC-8-High and LZC-8-Low units are utilized in the following way: 6 signals are received from LZC-8-High ($LP1_H$, $LP2_H$, $LP3_H$, $LP4_H$, X_9, X_{10} in Fig. 3) and from Low units ($LP1_H$, $LP2_H$, $LP3_H$, $LP4_H$, X_9, X_{10} in Fig. 3). In case the final LZC unit is less than 2, 4 or 6 bit-wide (corresponding to LZC-9 up to LZC-14), the lower part of the design returns 1,

Table 1. Boolean Logic Mappings used by LZC-8-Intermediate results.

X6	X5	X4	X3	X2	X1	LP3	LP2	LP1	LP4
1	X	X	X	X	X	0	0	0	0
0	1	X	X	X	X	0	0	1	0
0	0	1	X	X	X	0	1	0	0
0	0	0	1	X	X	0	1	1	0
0	0	0	0	1	X	1	0	0	0
0	0	0	0	0	1	1	0	1	0
0	0	0	0	0	0	1	1	1	1

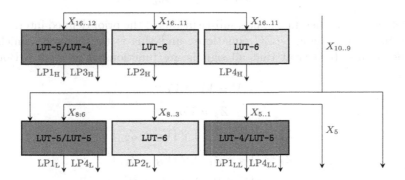

Fig. 3. Fully Parallel LZC-8-Intermediate circuit for LZC-15/16.

2 or 3 signals, respectively. Now we switch to a more generic notation, as after the base case, the units can be combined recursively to form larger units. In general, for LZC-k where $k = 8q$ or $k = 8q - 1$ ($q \geq 2$, $q \in \mathcal{Z}$), we have the following equations which define temporary logic values (signals) V_H, Z_{0_H}, Z_{1_H}, Z_{2_H}, Z_{1_L}, Z_{2_L} to be used after (where the equations for $V_L, Z_{2_L}, Z_{1_L}, Z_{0_L}$ are equivalent to the high equations for LZC-9 up to LZC-14):

$$
\begin{aligned}
V_H &= \text{LP4}_H \wedge \overline{X_{10}} \wedge \overline{X_9}, \\
Z_{2_H} &= \text{LP3}_H, \\
Z_{1_H} &= \text{LP2}_H, \\
Z_{0_H} &= \left(\text{LP1}_H \wedge \overline{\text{LP4}_H}\right) \vee \left(\text{LP4}_H \wedge \overline{X_{10}}\right), \\
Z_{2_L} &= \text{LP4}_L \wedge \overline{X_5}. \\
Z_{1_L} &= \text{LP2}_L.
\end{aligned}
\tag{10}
$$

Observe that intermediary V_H specifically has 3 signals involved in its computation, which will reduce possibility for LUT combining on the second stage, but we have compensated for this with an efficient first stage. It will be combined with at most 2 or 3 further signals to compute the LZC-16 unit signals. This combined usage can be represented by on-chip multiplexers. V_L and Z_{0_H}, Z_{0_L}

are not needed in further processing as will be described shortly. However, Z_{0_H} is used in each LZC ranging from 9-bits up to 14-bits but not necessary in the LZC-15/16.

Generalizing this scheme to arbitrary LZC sizes where $8 \leq (k \mod 16) \leq 14$, one requires a simple fallback strategy where if $LP3_L$ is not present, $LP2_L$ is used. If $LP2_L$ is not present then $LP1_L$ is used while both $LP1_L$ and $LP4_L$ are always present. When $1 \leq (k \mod 16) \leq 7$ the same fallback strategy is used for $LP3_H$ and $LP2_H$, and X_{10} and X_9 can be removed or set to zero if they are not present. Furthermore, in this uneven case when $k \mod 16 < 8$, we can assume when no low pair is present that:

$$V_L = Z_{2_L} = Z_{1_L} = Z_{0_L} = 1 . \tag{11}$$

The correct result then just makes substitutions of the prior defined intermediate values into the standard LZC equations, and this is the point at which the cascade occurs. It is clear that all signals are functions of at most 6 Boolean signals:

$$V = V_H \wedge V_L \tag{12}$$

$$Z_3 = V_H \tag{13}$$

$$Z_2 = (\overline{V_H} \wedge Z_{2_H}) \vee (V_H \wedge Z_{2_L}) \tag{14}$$

$$Z_1 = (\overline{V_H} \wedge Z_{1_H}) \vee (V_H \wedge Z_{1_L}) \tag{15}$$

$$Z_0 = (\overline{V_H} \wedge Z_{0_H}) \vee (V_H \wedge Z_{0_L}) \tag{16}$$

For LZC-15 and LZC-16, the LUT reduction modifications require the following substitutions defining signals V and Z_0:

$$V = LP4_H \wedge LP4_L \wedge LP4_{LL} \tag{17}$$

$$Z_0 = LP4_H \wedge ((LP4_L \wedge LP1_{LL}) \vee (\overline{LP4_L} \wedge LP1_L)) \vee (\overline{LP4_H} \wedge LP1_H) \tag{18}$$

The multiplexer usage possibility is based on the simple relationship e.g.:

$$(\overline{V_H} \wedge Z_{0_H}) \vee (V_H \wedge Z_{0_L}) == \begin{cases} Z_{0_L} & \text{if } V_H \\ Z_{0_H} & \text{otherwise} \end{cases} \tag{19}$$

(while programmers might be more familiar with $V_H ? Z_{0_L} : Z_{0_H}$).

The same technique as the first set of equations is then generalized and repeated for the larger building blocks from LZC $-k$ with $k > 16$ and beyond, including inferring the low-signals as all true when there is an odd number of signal groups leaving one without a pairing. Since signals created by Z_1 and Z_2 as well as Z_3 and V are candidates for LUT combining into LUT6-2 units or alternatively into MUXF7 and MUXF8 units, the latter stages also yield an area minimized solution. The final result can be returned as

$$(V, C) = \left(V, \overset{0}{\underset{k=\log_2 n-1}{||}} Z_k \right) . \tag{20}$$

Direct propagation of 3 signals from the LZC-8 stage to the LZC-16 stage creates an additional constraint for the routing part of the implementation. To address this, we can also allow Vivado synthesis flexibility in optimizing the logic as it sees fit. Since synthesis strategies for performance, area and power exist separately, this is preferred approach. Otherwise we can use VHDL *KEEP* directives on LUT output signals to prevent any sort of combining at synthesis time. This is as opposed to the VHDL *DONT_TOUCH* attribute which is similar but also applied during implementation having the unfortunate side-effect of preventing LUTNM combining during placement. However, the optimize design step may no longer preserve the original plan without this attribute.

So the 16-bit LZC achieves the smaller path delay without any additional LUT cost. At most $\lceil \frac{k}{16} \rceil$ additional LUTs are used while the theoretical path delay is reduced by $\lceil \log_2 \frac{k}{8} \rceil / (\lceil \log_2 \frac{k}{8} \rceil + 1)$ percent. We also indicate that this design is suitable for older architectures like Virtex 7 or any other which support the LUT6-2 units.

Finally, we also present a MUXF7/MUXF8 specific IP-level solution for an LZC-8 which does not cascade LUTs. Although the circuit would not infer from high-level tools, its existence is noteworthy. Namely *LP2*, *LP3* are computed by Eqs. (7) and (8). On the other hand, *LP1*, *LP4* and *V* are computed by:

$$LP4 = \begin{cases} 0 & \text{if } X_7 \\ \overline{X_6} \wedge \overline{X_5} \wedge \overline{X_4} \wedge \overline{X_3} \wedge \overline{X_2} \wedge \overline{X_1} & \text{otherwise} \end{cases},$$

$$V = \begin{cases} 0 & \text{if } X_8 \\ LP4 & \text{otherwise} \end{cases}, \tag{21}$$

$$LP1 = \begin{cases} 0 & \text{if } X_8 \\ X_7 \vee LP1(X_{6..1}) & \text{otherwise} \end{cases}$$

This straight-forward method uses 4 LUTs, and for *LP1* an additional single MUXF7, while for *V*, both a MUXF7 and a MUXF8. All the signals enter and leave the slice only one time, providing minimal routing delay, and only the delay of the MUXF7 and MUXF8 units themselves.

3.1 Demonstration of the Design on a Small-Sized Example

As an example to show the main computational stages in the outlined procedure, consider the LZC-16 of the number 2 which is "00000000 . 00000010b". It is clear that $(V, C) = (0, 14)$. Computing the LZC-8-Intermediate values shows that $LP1_H, LP2_H, LP3_H, LP4_H, LP1_L, LP2_L, LP3_L, LP4_L$ will be 1 while $LP1_{LL}, LP4_{LL}$ are both 0. This implies that V, Z_1, Z_2, Z_3 are 1 while Z_0 is 0. We can calculate C from concatenated binaries Z_i as: $C = 2^3 Z_3 + 2^2 Z_2 + 2^1 Z_1 + 2^0 Z_0 = 14$, as expected.

If we used an LZC-8-High and LZC-8-Low in this example, instead of LZC-8-Intermediate, then the values turn out to be the same, except X_1, X_2, X_8, X_9 along with $LP1_H, LP4_H, LP1_L$ are needed to compute Z_0 since $LP1_{LL}, LP4_{LL}$ are not present.

4   Results and Discussion

For our experiments and design we target the Ultrascale+ architecture and specifically Alveo U250 FPGAs. We targeted a 650MHz clock frequency, the highest that MaxCompiler can synthesize with the Mixed-Mode Clock Manager (MMCM) oscillators/frequency synthesizers on the FPGA, from the "freerun" clock which enters the chip through an IO port at 300MHz. We only use the Super Logic Region (SLR) of SLR1 as the PCIExpress ports are bound there by MaxCompiler (though the Xilinx floorplan does indicate SLR0 as the true point of entry). Our MaxCompiler version is 2021.1 which works alongside Vivado 2020.1. The Vivado implementation was based upon the versatile "Performance_ExplorePostRoutePhysOpt" strategy. We utilized a global clock buffer for the clock enable (CE) signal of the kernels which uses a BUFGCE unit on the Ultrascale device.

Comparing to high-level MaxCompiler provided machinery was determined to be inappropriate as it would require a combination of a leading one detector (via the simple two's complement property leading1detect(x)=-x&x where here a bitwise AND is used) and a one-hot decoder which generates an $O(n^2)$ VHDL algorithm, giving high area and power, and degraded performance due to presence of addition (as -x=~ x+1), fanout and congestion. However, the one-hot decoder did not allow disabling pipeline register insertion, so we were unable to make a meaningful comparison without modifying the VHDL and we thereby opted not to compare it. However the combination of a leading-one-detector and a one-hot decoder as an alternative means of computing LZC, is convenient and perfectly fine in some scenarios.

We chose an synthesis strategy optimized for performance based on Vivado's "Flow_PerfOptimized_high", and hence will have a resulting non-optimal LUT count, but optimal path delay and ability to synthesize at higher frequencies. For purpose of comparison, we used sizes the prior algorithm supported rather than the minor differences of the optimal generalized size variants.

Our results will appear different from those of the referenced paper because they targeted Virtex 7 with a 28nm process. While here we target Ultrascale+ with a 16nm process. There is a common practical rule that assumes roughly 3 levels of cascaded LUTs can meet timing regardless of the frequency in most situations. This has been applied for barrel shifters and other various circuits that require cascading for efficiency. At 16nm, several more layers of cascading seem to be possible, but it will start to place difficult timing constraints on the signals involved and influence their allowed spatial arrangements. So for an actual implementation, for very large LZC implementations (such as LZC-68) some adjustment of lower frequency or inserting pipeline registers into the algorithm after some number of levels becomes strictly necessary.

We collected the results using a hand designed Tool Command Language (TCL) script which integrates with Vivado. The input stream registers are easily found via regular expressions and by traversing output pins, nets and cells, the circuit can be recursively discovered in a depth-first search (DFS) manner. The specific slices and Basic ELements (BELs) are tracked during this traversal. Then the "get_timing_paths" command measures the worst timing path amongst the starting and terminating elements of the realized circuit (from register output pin to register input pin). Power was measured hierarchically via "report_power". When a build succeeds, this script is automatically run based on the saved checkpoint preserved by setting MaxCompiler not to clean the build folder ("clean_build_directory" set to "false").

The explanation for the LUT counts being high requires a deeper understanding of Vivado synthesis. Although it can merge or split apart Boolean functions, when it sees a function of more than 6 arguments, how it choose to cascade or plan them is according to its own algorithms. At the lowest-level, VHDL can describe individual LUTs, while at the highest level it describes mere logic which infers some sort of LUT structure. The Vivado design methodology makes it reasonable to trust their automatic synthesis choices as long as the design is constrained in appropriate areas. Our area of interest is primarily performance, though optimization for power is also possible. We do not expect too much beyond some reasonable heuristics as the circuit satisfiability problem (circuit-SAT) is NP-complete, only here we have 6-input gates rather than binary gates. Tools are beginning to incorporate machine learning (ML) for improved heuristic inference.

Table 2. Performance Results for various LZC sizes.

LZC bitwidth	LUTs(LUTNMs /MUXF7/MUXF8)	Slices	Power (mW)	Delay (ns)	Freq. (MHz)
8 new/old [17]	4 (1)	2	10	0.808	600
16 old [17]	12 (1)	3	13	1.016	650
32 old [17]	29 (1)	7	11	1.226	650
64 old [17]	58 (2)	14	11	1.69	470
16 new	11 (3)	3	10	0.952	500
32 new	27 (5)	10	13	1.142	600
64 new	67 (1)	16	15	1.429	650
8	5 (1)	2	13	0.772	610
16	10 (4)	5	10	0.988	510
32	27 (0)	7	12	1.052	650
64	56 (0/8/0)	15	20	1.363	650

We note several things about the results in Table 2. The number of Logical LUTs introduced is LUTs plus LUTNMs. The data was gathered by compiling 2 independent but identical circuits so LUTs and slices have been ceiling divided by 2. The Zahir, et al. [17] scheme is labeled "old", the proposed scheme synthesized with *KEEP* to preserve LUT output signals is labeled "new" while the unlabeled does not use this attribute. We refer interested readers to [17] for comparison to the other algorithms from [4, 6, 7, 16]. First, the number of slices for two identical circuits can have overlap, but slices are not considered a useful modern metric in large scale designs as the tools are not directly optimizing for minimal slices. The power measurement in Vivado is likely inaccurate and not very detailed for specific circuits. We therefore measured the power to the whole MaxCompiler kernel core in milli-Watts, the finest granularity and resolution offered by the tool. The kernel has some registers, counters and control mechanisms which make this an impure result, however it is a fixed cost across all kernels of identical input/output sizes.

The results are all for builds at 650 MHz which although not the $f_{\max}$ of 725 MHz., is the highest frequency at which MaxCompiler can configure an MMCM unit. Such excellent performance shows that the LZC is unlikely to be a limiting factor in the designs which utilize it at least through to 64 bits. We don't consider overall efficiency metrics such as the power-delay-area product (PDAP), as our focus was on highest build speed, and path delay while other metrics requires uniform build speeds or consideration of frequency weighting. The frequency and power have a simple linear relationship.

Furthermore, at very high frequencies, deeper circuits can in some cases perform better as the path delay effects the setup (which balances the clock skew against the path delay, clock uncertainty and setup time) as well as the hold (balancing path delay against clock skew, uncertainty and hold time) slacks for the registers which ultimately latch the result signals. For example at a clock speed of 650MHz, the clock period is $\frac{10^3}{650\mathrm{MHz}} = 1.538$ nanoseconds (while Vivado timing scores are reported in picoseconds), which although an upper bound on path delay to achieve a build at this frequency, needs to account for the setup and hold slack in full.

5   Conclusion

The shortcomings of the proposed technique is less modularity, more complicated logic, and that certain non-power of two LZC sizes may have further unique optimizations which require complicated generation algorithms or on a case-by-case basis analysis to determine or achieve. The ideas presented are enough to find such simplifications. However, if willing to implement a more complicated and general framework, our design allows higher frequency builds, and savings in LUT and routing resources and more optimal IP. Future works can use the ideas presented to improve circuits beyond the trivially equivalent leading/trailing-one-counter, but counting bits set (sometimes called *popcount*), checking for powers of two, or rounding up to the nearest power of two, etc.

We have given a survey of the synthesis features for a specific modern tool and architectures where it comes to optimizing a small but important circuit as it relates to HPC. It shows that a generic design with a reasonable signal layout will synthesize by sophisticated tools with success in likely any optimization scenario when the logic is minimally constrained. However, due to the various complications of the physicality and details of setup and hold, trying multiple options may be necessary as the minimal path delay is not always the build that ultimately succeeds.

We furthermore carefully provided details towards a research methodology for designing small-scale circuits with HLS tools, understanding the ways of constraining the underlying build tool, as well as measuring and collecting data points. This methodology is geared specifically toward achieving high frequency and minimal path delay builds.

Although the synthesis and implementation process is complicated and depends on a broad set of constraints and target goals from performance to power to resource usage and chip area, the design here should from all perspectives have better versatility and be applicable for maximizing performance and/or minimizing area and power. The fact that LZC-16 and higher circuits builds at the highest synthesize-able frequency of 650MHz from high-level language tools, is indicative of both the flexibility of modern high-level synthesis tools like MaxCompiler, and the diverse features of modern FPGA architectures like Ultrascale.

Acknowledgements. This research was supported by the Ministry of Culture and Innovation and the National Research, Development and Innovation Office within the Quantum Information National Laboratory of Hungary (Grant No. 2022-2.1.1-NL-2022-00004), by the ÚNKP-22-5 New National Excellence Program of the Ministry for Culture and Innovation from the source of the National Research, Development and Innovation Fund. RP. acknowledge support from the Hungarian Academy of Sciences through the Bolyai János Stipendium (BO/00571/22/11) as well.

References

1. Anderson, J.H., Wang, Q.: Area-efficient FPGA logic elements: architecture and synthesis. In: 16th Asia and South Pacific Design Automation Conference (ASP-DAC 2011), pp. 369–375 (2011). https://doi.org/10.1109/ASPDAC.2011.5722215
2. Anderson, S.E.: Bit twiddling hacks (2005). http://graphics.stanford.edu/seander/bithacks.html
3. Chapman, K.: Multiplexer design techniques for Datapath performance with minimized routing resources. Xilinx All Programmable **1**, 1–32 (2014)
4. Dimitrakopoulos, G., Galanopoulos, K., Mavrokefalidis, C., Nikolos, D.: Low-power leading-zero counting and anticipation logic for high-speed floating point units. IEEE Trans. Very Large Scale Integr. Syst. **16**(7), 837–850 (2008). https://doi.org/10.1109/TVLSI.2008.2000458
5. Dohi, K., Okina, K., Soejima, R., Shibata, Y., Oguri, K.: Performance modeling of stencil computing on a stream-based FPGA accelerator for efficient design space exploration. IEICE Trans. Inf. Syst. **98**(2), 298–308 (2015)
6. Miao, J., Li, S.: A design for high speed leading-zero counter. In: 2017 IEEE International Symposium on Consumer Electronics (ISCE), pp. 22–23 (2017). https://doi.org/10.1109/ISCE.2017.8355536
7. Milenković, N.Z., Stanković, V.V., Milić, M.L.: Modular design of fast leading zeros counting circuit. J. Electr. Eng. **66**(6), 329–333 (2015). https://doi.org/10.2478/jee-2015-0054

8. Nadjia, A., Mohamed, A.: Efficient implementation of AES S-box in LUT-6 FPGAs. In: 2015 4th International Conference on Electrical Engineering (ICEE), pp. 1–4 (2015). https://doi.org/10.1109/INTEE.2015.7416679
9. Rakyta, P., Morse, G., Nádori, J., Majnay-Takács, Z., Mencer, O., Zimborás, Z.: Highly optimized quantum circuits synthesized via data-flow engines (2022). arXiv preprint arXiv:2211.07685
10. Srivastava, P., Chung, E., Ozana, S.: Asynchronous floating-point adders and communication protocols: a survey. Electronics 9(10), 1687 (2020). https://doi.org/10.3390/electronics9101687, https://www.mdpi.com/2079-9292/9/10/1687
11. Summers, S., Rose, A., Sanders, P.: Using MaxCompiler for the high level synthesis of trigger algorithms. J. Instrumentation 12(02), C02015 (2017). https://doi.org/10.1088/1748-0221/12/02/C02015
12. Suzuki, H., Morinaka, H., Makino, H., Nakase, Y., Mashiko, K., Sumi, T.: Leading-zero anticipatory logic for high-speed floating point addition. IEEE J. Solid-State Circ. 31(8), 1157–1164 (1996). https://doi.org/10.1109/4.508263
13. Walters, E.G.: Array multipliers for high throughput in xilinx FPGAs with 6-input LUTs. Computers 5(4), 20 (2016). https://doi.org/10.3390/computers5040020, https://www.mdpi.com/2073-431X/5/4/20
14. Wang, H., Gante, J., Zhang, M., Falcão, G., Sousa, L., Sinnen, O.: High-level designs of complex FIR filters on FPGAs for the SKA. In: 2016 IEEE 18th International Conference on High Performance Computing and Communications; IEEE 14th International Conference on Smart City; IEEE 2nd International Conference on Data Science and Systems (HPCC/SmartCity/DSS), pp. 797–804 (2016). https://doi.org/10.1109/HPCC-SmartCity-DSS.2016.0115
15. Xilinx, I.: Ultrascale architecture configurable logic block user guide (2023). https://docs.xilinx.com/v/u/en-US/ug574-ultrascale-clb
16. Xilinx, I.: Vivado design suite user guide: High level synthesis UG902 (v2020.1) (2023). https://docs.xilinx.com/v/u/en-US/ug902-vivado-high-level-synthesis
17. Zahir, A., Ullah, A., Reviriego, P., Hassnain, S.R.U.: Efficient leading zero count (LZC) implementations for xilinx FPGAs. IEEE Embed. Syst. Lett. 14(1), 35–38 (2022). https://doi.org/10.1109/LES.2021.3101688
18. Zhang, H., Chen, D., Ko, S.B.: High performance and energy efficient single-precision and double-precision merged floating-point adder on FPGA. IET Comput. Digit. Tech. 12(1), 20–29 (2018). https://doi.org/10.1049/iet-cdt.2016.0200, https://ietresearch.onlinelibrary.wiley.com/doi/abs/10.1049/iet-cdt.2016.0200

Reduction of the Computational Cost of Tuning Methodology of a Simulator of a Physical System

Mariano Trigila[1]([⊠]) [iD], Adriana Gaudiani[2] [iD], Alvaro Wong[3] [iD], Dolores Rexachs[3] [iD], and Emilio Luque[3] [iD]

[1] Facultad de Ingeniería y Ciencias Agrarias, Pontificia Universidad Católica Argentina, Ciudad Autónoma de Buenos Aires, Argentina
`mariano_trigila@uca.edu.ar`
[2] Instituto de Ciencias, Universidad Nacional de General Sarmiento, Buenos Aires, Argentina
`agaudiani@campus.ungs.edu.ar`
[3] Departamento de Arquitectura de Computadores y Sistemas Operativos, Universidad Autònoma de Barcelona, 08193 Bellaterra (Barcelona), Spain
`alvaro.wong@uab.cat, {dolores.rexachs,emilio.Luque}@uab.es`

Abstract. We propose a methodology for calibrating a physical system simulator and whose computational model represents its events in time series. The methodology reduces the search space of the fit parameters by exploring a database that contains stored historical events and their corresponding simulator fit parameters. We carry out the symbolic representation of the time series using ordinal patterns to classify the series, which allows us to search and compare by similarity on the stored data of the series represented. This classification strategy allows us to speed up the parameter search process, reduce the computational cost of the adjustment process and consequently improve energy cost savings. The experiences showed a reduction in the computational cost of 29% compared with our tuning methodology proposed in previous research.

Keywords: Parametric simulation · Tuning methodology · Ordinal pattern · Time series classification · Data driven

1 Introduction

Computer simulators are software components developed from the implementation of a model that represents a real system whose behavior is interesting to study or predict future events. As the simulator evolves over time, it tends to lose its calibration, since the parameters that define the system depend on physical magnitudes that define the real system and these change over time. A simulator is out of calibration when it produces output data that differs from the observed data, which is the data that is| measured in the physical system, and the difference between both data exceeds a predetermined limit of error [10–12].

The calibration of a simulator is the process by which the search for the set of parameters close to the optimum is carried out through parametric simulations. The

© The Author(s), under exclusive license to Springer Nature Switzerland AG 2023
J. Mikyška et al. (Eds.): ICCS 2023, LNCS 14075, pp. 641–651, 2023.
https://doi.org/10.1007/978-3-031-36024-4_49

search for the values of the adjustment parameter is carried out over a very large search space given the large number of different values and the number of parameters that can be associated with the model that represents the physical system. Consequently, the simulator will be executed with each set of parameters (simulation scenario), generating the simulated data set for each of the scenarios [11, 12]. As can be seen, the calibration process takes a large amount of time and requires a high cost of computing resources, which consequently results in a process with a high energy cost. In our research, we propose a novel methodology to reduce the computational cost of the adjustment and calibration process of a physical system simulator.

In this article we propose a methodology that offers us a more efficient way to find the parameters' values that best fit the observed data. We improve the efficiency of the search algorithms by reducing the search space of the adjusted set of parameters. Consequently, we can achieve considerable savings in computing resources. We use the strategy of storing the data defining a particular event in each moment in the past, when we had detected a lack of precision in the simulation. We also store the set of adjusted parameters found for tuning the simulation at that moment. In this way, we can use the stored events in the future when a disruption event occurs, and we can find those similar events that can potentially tune the simulator to the current event using a low computational cost method.

Our methodology fits into the data assimilation paradigm, since it combines observations of reality data and predictions of the states of the model parameters [8]. It also fits into the data driven paradigm in that measurement data is used to improve model accuracy and runtime, while the computational output produced by the model helps drive the measurement process itself [9]. We implement an ordinal pattern [1–4] approach to efficiently find similar events stored in the past events database. In this work, we use a riverbed computational model that represents its events of interest through time series.

In Fig. 1 you can see the entities that interact with the "Fit Model" component: "Real System" represents the real physical system and it is this which provides the observed data. "Model Simulator" represents the simulator, which is fed with the simulation scenarios and produces the output data, and "Past Events Store" represents the data store where the events that have occurred are recorded.

Carrying out the experiences and using our proposed methodology, we achieved savings in the use of computational resources of 29% compared to the successive steps methodology SSM [12]. Compared to the number of simulation runs we needed with the SSM methodology, we saved on average 1044 simulator runs, which leads directly to energy cost savings. It was also possible to speed up the search process for the adjustment parameters.

This article is organized as follows: In Sect. 2 we describe the simulation model and the simulation domain characterization, and in Sect. 3 we describe the proposed methodology, which includes, in subsections, the explanation of symbolic representation and the search temporal subsequences by similarity. In Sect. 4 we explain the experimental environment and the results of this experimentation. The conclusions are presented in Sect. 5.

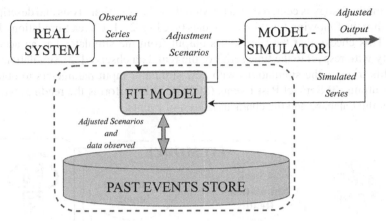

Fig. 1. Fit Model Contextual Diagram.

2 The Simulation Model and the Simulation Domain Characterization

We use a simulation model developed by the National Water Institute of Argentina (INA), which calculates the translation of waves in rivers and canals [11]. The simulator output data to be considered in this work are the heights of the river at 15 measurement stations, that is to say, physical points of the riverbed where the height of the river is measured. The simulator fit parameters that have the greatest impact on the output data are the roughness coefficients, the "Manning coefficient" and the "Manning edge" [10]. The simulation model discretizes the riverbed in 76 sections, and for each section, supplying the Manning coefficient values will be required. For a complete run of the simulator, a set of adjustment parameters must be provided, which is made up of all the Manning values for both edge and plain for each of the 76 sections. For each section, riverbed and plain subsections are considered at both ends of the section, and it is necessary to have parameter values in each of these subsections.

The simulator provides the simulated time series, which are the outputs it produces in each complete run, for each station, for the configured scenario. This series contains the daily height of the river for the simulated period.

There is also a time series set of observed data, which are the real records taken daily of the height of the river in 15 monitoring stations for a period of 11 years. With this data set, the output of the simulator will be validated at each station.

3 Proposed Methodology

We propose a calibration methodology that reduces the search space of the fit parameters by using the exploration of a database that contains stored historical events of the real system, the historical events of the simulated system and the history of the best fit parameters of the simulator.

644   M. Trigila et al.

The methodology is centered on a Fit Model, which for its analysis and description is divided into the following functional modules (see Fig. 2): 1) Check Simulation - Reality (CSR): This checks that the time series output from the simulator has an acceptable similarity with respect to the time series of the real data observed. 2) Simulations Shoot (SS): This triggers the simulations with new simulator input parameters to obtain the best-fit outputs. 3) Set/Get Past Events (SGPE): This performs the reading, writing of events in the database and the characterization of events.

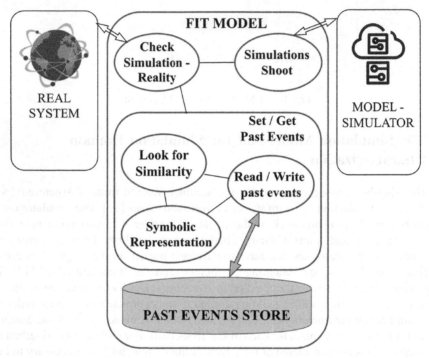

Fig. 2. Fit Model Detail - Functional Modules.

In the following subsections, we carry out a detailed explanation of each of the functional modules including the innovative concepts that are used in the methodology.

3.1   Check Simulation - Reality

The CSR functional module has the purpose of measuring and controlling the difference between the output values produced by the simulator for a simulation scenario and the actual measured values observed at station k.

We define S as the set of the complete scenarios of the simulator, S_k as the set of scenarios for station k and $\widehat{s_K} \subseteq S_k \subseteq S$ as the best fit scenario for measuring station k.

Using a divergence index implemented with the mean square error estimator (RMSE), we determine the best scenario $\widehat{s_K}$ by comparing the simulated data set SD with the observed data set OD.

$$DI_{k,d}^t = RSME_{k,d}^t = \sqrt[2]{\frac{\sum_{i=0}^{i=d-1}\left(H_k^{OD,t} - H_k^{SD,t}\right)_i^2}{d}} \tag{1}$$

The $DI_{k,d}^t$ index is the RMSE error of the series of simulated river heights $H_k^{SD,t}$ in relation to the observed series of river heights $H_k^{OD,t}$, at a measuring station k, for a value in time t, for a subsequence of the series of size d.

We define acceptable difference AD as a reference value, provided by the engineers who modeled the river and use the simulator selected for this work, and its decimal value can be between $0 \leq AD \leq 1$. AD as will be the reference value to determine if $DI_{k,d}^t$ is acceptable or not. When AD is close to 0, it will indicate that the scenario is a good fit, otherwise it will indicate that it is a bad scenario.

If the $DI_{k,d}^t$ index is greater than the reference value AD, then a difference has been found between the observed and simulated series. This indicates that an adjustment of the simulator must be carried out by requesting Simulations Shoot to execute the simulator for a new set of parameters.

If the $DI_{k,d}^t$ index is less than or equal to the reference value AD, then the difference between the observed and simulated series is acceptable, therefore a scenario $\widehat{s_K}$ has been found. In this case, Set/Get Past Events is requested to store $\widehat{s_K}$, the observed height subsequence $H_{k,d}^{OD,t}$ of size d, the simulated subsequence $H_{k,d}^{SD,t}$ of size d and the time t.

3.2 Simulations Shoot

The Simulations Shoot functional module has the purpose of requesting its execution to the simulator by supplying the values of the scenario set S that is desired to be tested in the simulation process. Scenario set S is composed of the adjustment parameters $Sc_m \subseteq S$, one for each section. m is the section number and its value is between $1 \leq m \leq 76$. The scenarios Sc_m that are supplied to the simulator contain the mp_m that is Manning value of plain and mc_m that represents the Manning value of the channel. We define a scenario Sc_m for a section m, subdivided into three subsections, like a 3-tupla (2).

$$Sc_m = (mp_m, mc_m, mp_m) \tag{2}$$

Simulator execution is triggered with the set of adjustment parameters S provided by the CSR module after requesting its execution.

The simulation response, which is the simulated data series, is sent as a response to the CSR module for evaluation.

3.3 Set/Get Past Events

Functional module Set/Get Past Events performs the management of writing, reading the database and it handles characterization and symbolic representation of the subsequences of time series that are stored. Receives from RSC requests to retrieve past events from the database based on a current event or to update the database with new events that are not yet registered.

For better analysis and description, we divide it into the following functional sub-modules: Symbolic representation (SR), Look for similarity (LFS) y Read/Write past events (RWPE). In the following subsections we explain this in detail.

3.3.1 Symbolic Representation

The SR module has the purpose of mapping the subsequences of time series of the observed heights $H_{k,d}^{OD,t}$ or simulated heights $H_{k,d}^{OS,t}$ to symbolic representations. It also performs the unmapping of symbolic representations to subsequences of time series Fig. 3.

The methods of symbolic representations of time series discretize the series and transform it into a sequence of symbols making, in our case, the search for series of similar events from the past, stored in the database, more efficient, thus accelerating the process of search by comparison. We will use the symbolic representation methodology of Bandt and Pompe (2002) [4], hereinafter BP, to represent a subsequence of values of a time series into an array of values belonging to an alphabet Θ which will represent the characterization of the subsequence, where φ is a symbol, $\varphi \in \Theta$, and $0 \leq \varphi \leq d, d (d \in \mathbb{N})$, d is the number of values in the subsequence is the number of values that make up each subsequence and this is called the embedding dimension, which is usually set between $3 \leq d \leq 7$.

In the BP approach, the concept of ordinal pattern (OP) (also called permutation pattern) is defined [1–4]. Given a time series X_t, It is divided into subsequences of consecutive values $(x_t, x_{t+\tau}, ..., x_{t+\tau d})$; $d(d \in \mathbb{N})$; y $\tau(\tau \in \mathbb{N})$ the time between consecutive points and this is called the embedding delay, usually set to 1 [1]. An ordinal pattern π_t is a sequence of symbols φ of dimension d which is associated with each of the time series subsequences. The OP is a sequence of symbols $\varphi \in \mathbb{N}$, where each symbol inside the pattern indicates the permutations that must be applied to the element of the temporal subsequence in order obtain a set of d dimension, with the increasing order of the values of the temporal subsequence. To broaden any knowledge, we recommend the bibliography where the subject is explained in detail [1, 2].

We then use, in our methodology, the BP approach through a mapping function M, which transforms a time series subsequence $sX_{t\alpha}$ into an ordinal pattern $\pi_{t\alpha}$, for time series X_t with $(sX_{t\alpha} \subseteq X_t)$, for a moment in time $t\alpha$, for an interval of time between consecutive points $\tau(\tau \in \mathbb{N})$, for a number of values $d(d \in \mathbb{N})$ of values that make up the subsequence [5, 6]. All the above can be summarized in (3).

$$M(sX_{t\alpha}, t\alpha, \tau, d) \mapsto \pi_{t\alpha} \tag{3}$$

To solve the mapping M, we developed a software program that performs the transformation using "ordpy" [3], which is a software package developed in python that implements permutation entropy and several of the main methods related to the Bandt and Pompe framework.

To get a subsequence or a list of subsequences that have a certain ordinal pattern $\pi_{t\alpha}$, we use the unmapping function U. To solve the function unmapping U, we develop a

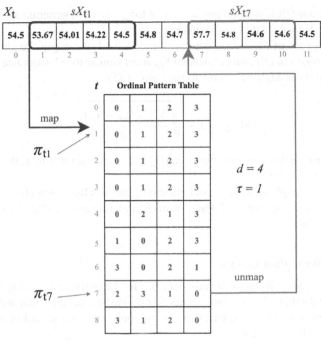

Fig. 3. Map/Unmap Ordinal Pattern.

software program that searches the database of stored past events and returns a list L of the subsequences which have the same ordinal pattern $\pi_{t\alpha}$. That is, given a $\pi_{t\alpha}$ we get a list L where $sX_{t\alpha} \subseteq L$ (4).

$$U(\pi_{t\alpha}) \mapsto L \tag{4}$$

3.3.2 Look for Similarity

The LFS functional module has the purpose of searching for subsequences of similar time series.

From the mapping of each subsequence, we can identify its associated ordinal pattern and classify the temporal sequences [6, 7]. Therefore, we can relate a set of temporary subsequences to an OP that will be the class to which the subsequence belongs. Thus, we can indirectly compare two time subsequences by comparing their OP. Two temporal subsequences with the same OP belong to the same class, that is, it indicates a certain similarity between the subsequences, but it does not necessarily indicate that they are the same. With a similarity index implemented with the root mean square error estimator (RMSE), the similarity between two temporal subsequences that have the same OP is quantified.

Let us see how a similarity search develops:

Given the subsequence $sX_{t\alpha}$, from Eq. (3), we obtain the ordinal pattern OP $\pi_{t\alpha}$. Then we access the database of past events and look for the temporary subsequences that have the same OP. If we find any $\pi_{t\beta}$ equal to $\pi_{t\alpha}$, then we generate a list of temporal subsequences that are related to the pattern $\pi_{t\beta}$.

Then, with each of the subsequences found in list L we apply the similarity index $SI_{\alpha,\beta}$ to determine what the subsequence $sX_{t\beta}$ most similar to subsequence $sX_{t\alpha}$ is. That is to say, if the index $SI_{\alpha,\beta}$ is closest to the value 1 (5).

$$SI_{\alpha,\beta} = 1 - \sqrt[2]{\frac{\sum_{i=0}^{i=d}(x_\alpha - x_\beta)_i^2}{d}} \qquad (5)$$

The similarity index can vary between $0 \leq\leq 1$, the closer it is to 1, the more similar the subsequences will be.

As previously explained, the search method generates list L, which is obtained from the data previously stored in PES (Past Events Store), so that $L \subseteq PES$. The PES module is described below.

3.3.3 Read/Write Past Events

The functional module Read/Write past events is for the purpose of storing and obtaining the data in the database PE S (Past Events Store). Past events are recorded in the PE S, which can then be used in similar current situations. They are stored in a table whose record has the following structure:

t: The time the event occurs.

sX_t : The subsequence of the time series X_t.

π_t: The ordinal pattern related to the subsequence sX_t.

Sc: The full set of simulator best fit parameters associated with the moment t.

4   Experience Environment and Experimental Results

The experiments that we present in this article were carried out focusing on a single measurement station. The station on which the experiments were carried out has the real name of "Itabaite" and its mnemonic in the simulator is "ITAE" and represents a city located in the domain of the river OD_K. So, for the station $k = "ITAE"$ we have a series of observed data SD_K with a quantity of 5533 records of daily measurements of water height H_k^{OD} between the dates $09/04/2010 \leq t \leq 09/04/2010$. The OD_K series has a maximum height of 58.3, a minimum height of 52.3 and a standard deviation of 0.93.

To carry out the experiments, the observed data was partitioned OD_K at 70%–30% (see Fig. 4). 70% corresponds to the reference series, which are the first records of the observed data series that goes from $t_1 = 01/08/1994$ to $t_{70} = 31/12/2005$.

These observed data $OD_k^{t_1,t_{70}}$ are stored in the PES database, together with the corresponding simulated data $SD_k^{t_1,t_{70}}$, with their associated ordinal patterns $\pi_k^{t_1,t_{70}}$ (see Fig. 5, the histogram of π) and their corresponding best-fit parameters $Sc_m^{t_1,t_{70}}$. Therefore, with these data we built the database, PES, in a startup/initial state, to search for similarities.

The process of carrying out the experiment consisted of adjusting the simulator every 5 days, the number of days is 1500 for the period of t_{71} to t_{100}, therefore 300 adjustments should be made. If we consider that the SSM methodology has an average execution

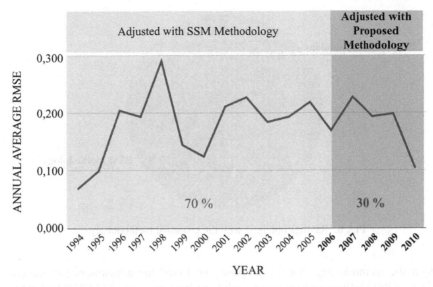

Fig. 4. Annualized mean square error between the simulated series with adjusted parameters and the observed series.

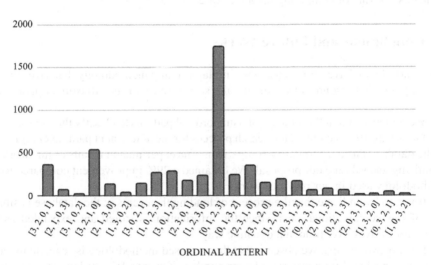

Fig. 5. Histogram of ordinal patterns.

of 12 calls to the simulator for each adjustment, then SSM on average would call the simulator 3600 times to perform the 1500-day adjustment.

To determine the efficiency of our methodology, we measured the number of times the correct fit parameters were obtained for a current event by accessing the PES database and carrying out a similarity search. If the adjustment parameters were not found in PES, then we proceeded to obtain it using the SSM methodology, making parametric simulations. In Fig. 6 the goodness of the proposed methodology can be observed.

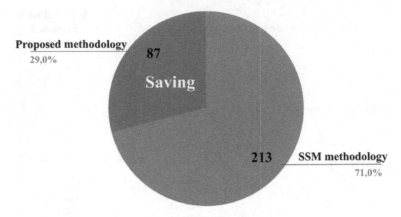

Fig. 6. Savings obtained with the proposed methodology.

With the methodology that we propose, we found the adjustment parameters in PES in 87 (29%) adjustment processes, while in the remaining 213 (71%) adjustment processes, we obtained the adjustment parameters with the SSM methodology. This indicates a saving of computing resources of 29%.

5   Conclusions and Future Works

In this article, we have proposed a new simulator tuning methodology that saves the use of computational resources by reducing the search space of the adjustment parameters set.

We understand that the strategy of using ordinal patterns to classify the time series in a database greatly accelerated the search process for the adjustment parameter registered in the database through similarity comparison. The experimental results of this work are promising and validate our proposal. We obtained a 29% improvement compared to the methodology used in our previous work.

It is good to indicate that the process of loading the events in the database is simple and of low computational cost, since the information is stored and expanded as the simulator evolves over time in a dynamic way.

It is noteworthy that we observe that the proposed methodology is scalable in terms of the potential ability to process a larger number of events. We attribute this quality to the classification approach of the time sub-series using the symbolic representation.

There are several directions for future research, some of which could be: experimenting with much more stored event information, investigating the relationships between ordinal patterns and the probability of occurrence of sequence patterns of ordinal patterns over time, and we are working to tune more domain points in the same system calibration process. Also, a future direction in which we are focused is to extrapolate our methodology to other simulation models of physical systems that need to find behavior patterns and that also need to tune their simulator automatically. The level of abstraction achieved with our methodology allows us to infer that extrapolations to other physical simulators are highly feasible, thus expanding its field of application. The motivation is broad and calls us to continue expanding our knowledge about this methodology.

Acknowledgments. This research has been supported by the Agencia Estatal de Investigacion (AEI), Spain and the Fondo Europeo de Desarrollo Regional (FEDER) UE, under contract PID2020-112496GB-I00 and partially funded by the Fundacion Escuelas Universitarias Gimbernat (EUG).

References

1. Zanin, M., Olivares, F.: Ordinal patterns-based methodologies for distinguishing chaos from noise in discrete time series. Commun. Phys. **4**(1), 190 (2021)
2. Bariviera, A.F., Guercio, M.B., Martinez, L.B., Rosso, O.A.: Libor at crossroads: stochastic switching detection using information theory quantifiers. Chaos, Solitons Fractals **88**, 172–182 (2016)
3. Pessa, A.A., Ribeiro, H.V.: ordpy: a python package for data analysis with permutation entropy and ordinal network methods. Chaos Interdisc. J. Nonlinear Sci. **31**(6), 063110 (2021)
4. Bandt, C., Pompe, B.: Permutation entropy: a natural complexity measure for time series. Phys. Rev. Lett. **88**(17), 174102 (2002)
5. Mohr, M., Wilhelm, F., Hartwig, M., Möller, R., Keller, K.: New approaches in ordinal pattern representations for multivariate time series. In: The Thirty-Third International Flairs Conference, pp. 124–129 (2020)
6. Sinn, M., Keller, K., Chen, B.: Segmentation and classification of time series using ordinal pattern distributions. Eur. Phys. J. Spec. Top. **222**, 587–598 (2013)
7. Cuesta-Frau, D.: Using the information provided by forbidden ordinal patterns in permutation entropy to reinforce time series discrimination capabilities. Entropy **22**(5), 494 (2020)
8. Reich, S.: Data assimilation: the Schrödinger perspective. Acta Numer. **28**, 635–711 (2019)
9. Fujimoto, R., et al.: Dynamic data driven application systems: research challenges and opportunities. In: Winter Simulation Conference (WSC), pp. 664–678. IEEE (2018)
10. Berends, K.D., Warmink, J.J., Hulscher, S.J.M.H.: Efficient uncertainty quantification for impact analysis of human interventions in rivers. Environ. Model. Softw. **107**, 50–58 (2018)
11. Gaudiani, A., Wong, A., Luque, E., Rexachs, D.: A computational methodology applied to optimize the performance of a river model under uncertainty conditions. J Supercomputing **79**, 4737–4759 (2023). https://doi.org/10.1007/s11227-022-04816-6
12. Trigila, M., Gaudiani, A., Luque, E.: Agile tuning method in successive steps for a river flow simulator. In: Shi, Y., et al. (eds.) ICCS 2018. LNCS, vol. 10862, pp. 639–646. Springer, Cham (2018). https://doi.org/10.1007/978-3-319-93713-7_60

A Hypergraph Model and Associated Optimization Strategies for Path Length-Driven Netlist Partitioning

Julien Rodriguez[1,2]($\boxtimes$), François Galea[1], François Pellegrini[2], and Lilia Zaourar[1]

[1] Université Paris-Saclay, CEA List, 91120 Palaiseau, France
{julien.rodriguez,francois.galea,lilia.zaourar}@cea.fr
[2] Université de Bordeaux & INRIA, 33405 Talence, France
francois.pellegrini@u-bordeaux.fr

Abstract. Prototyping large circuits on multi-FPGA platforms requires to partition the circuits into sub-circuits, each to be mapped in a given single FPGA. While most existing partitioning algorithms focus on minimizing cut size, the main issue is not to map long paths across multiple FPGAs, as it may cause an increase in critical path length. To address this problem, we propose a new hypergraph model, for which we design algorithms for initial partitioning and partition refinement. We integrate these algorithm in a multilevel framework, combined with existing min-cut solvers, to tackle simultaneously both path length and cut size objectives. We observe a significant reduction in critical path degradation, by 12%–40%, at the cost of a moderate increase in cut size, compared to path-agnostic min-cut methods.

Keywords: hypergraph partitioning · critical path · VLSI · FPGA · fast prototyping

1 Introduction

The typical hardware design flow comprises several steps, such as prototyping, verification, floor planning, placement and routing, possibly on very large logic circuits. The methods which perform these steps often take advantage of *netlist partitioning*, which enables divide-and-conquer approaches to separate circuits into parts of smaller size that are easier to handle. Thus, reduce as much as possible the work on the global circuit. This study focuses on netlist partitioning for prototyping on multi-FPGA platforms, for circuits that do not fit into a single FPGA. In this case, the circuit is divided into several parts, one for each FPGA. Valid partitions must respect capacity constraints on each FPGA and their interconnections. In addition, it should mitigate a possible increase in the signal propagation delay of the longest combinatorial path, known as the *critical path*. Indeed, in synchronous circuits, critical path length determines the maximum frequency at which the circuit may operate. Mapping long paths across

© The Author(s), under exclusive license to Springer Nature Switzerland AG 2023
J. Mikyška et al. (Eds.): ICCS 2023, LNCS 14075, pp. 652–660, 2023.
https://doi.org/10.1007/978-3-031-36024-4_50

several FPGAs is likely to degrade the critical path. During the last 30 years, several hypergraph partitioning tools have been developed [4,9,14]. These tools use a multi-level framework, that consists of three phases: *coarsening*, *initial partitioning*, and *refinement*. The coarsening phase recursively uses a clustering method to transform the hypergraph into smaller ones, that possess the same topological properties. Then, an initial partitioning is computed on the smallest coarsened hypergraph. Finally, for each coarsening level, the solution for the coarser level is prolonged to the finer level, and then refined using a local refinement algorithm. A survey on hypergraph partitioning has been recently issued [5]. Common metrics to measure the quality of hypergraph partitions are f_c, which sums the number of cut hyperedges, and f_λ, which sums the number of connected parts, minus one, for each hyperedge, and which represents the amount of information that needs to be transferred across parts. However, all of these tools, while commonly used in production chains, have not been designed to minimize the critical path length, as shown in [2]. This is why several authors proposed pre- and/or post-processing steps to reduce the degradation of cut paths [2,12]. However, the main objective addressed by these works is to reduce the cut along critical paths as much as possible, not considering the critical path's degradation resulting from subsequent mapping onto a non-fully connected target topology. This paper proposes a cost function to minimize path cost degradation during partitioning, based on a modeling of circuits as *red-black hypergraphs*. We also propose a weighting scheme to drive *min-cut* tools as well as a partitioning scheme comprising several algorithms. It includes an adaptation of a refinement algorithm that aims at preventing cuts across the longest paths while still trying to minimize cut size. The remainder of the paper is organized as follows. Section 2 presents the notations, definitions, and previous work on time-driven partitioning. Section 3 presents our approach, relying on a new initial partitioning algorithm and an extension of the well-known Fiduccia-Mattheyses (FM) algorithm [8] to minimize path cost. Simulation and results are presented in Sect. 4. We conclude in Sect. 5.

2 Preliminaries

2.1 Notations and Definitions

Oriented hypergraphs are a generalization of oriented graphs in which the notion of arc is extended to that of hyperarc, which can connect one or more source vertices to one or more sink vertices. In our context, we consider only hyperarcs that comprise a single source vertex. Let $\mathcal{H} \overset{\text{def}}{=} (\mathcal{V}, \mathcal{A}, \mathcal{W}_v, \mathcal{W}_a)$ be a directed hypergraph, defined by a set of vertices $\mathcal{V}$ and a set of hyperarcs $\mathcal{A}$, with a vertex weight function $\mathcal{W}_v : \mathcal{V} \to \mathbb{R}^+$ and a hyperarc weight function $\mathcal{W}_a : \mathcal{A} \to \mathbb{R}^+$. Every hyperarc $a \in \mathcal{A}$ is a subset of vertex set $\mathcal{V}$: $a \subseteq \mathcal{V}$. Let $s^+(a)$ be the source vertex set of hyperarc a, and $s^-(a)$ its sink (destination) vertex set. We consider here that each hyperarc has a single source, so $\forall a, |s^+(a)| = 1$. As hyperarcs connect vertices, let $\Gamma(v)$ be the set of neighbor vertices of vertex v, and $\Gamma^-(v) \subseteq \Gamma(v)$

and $\Gamma^+(v) \subseteq \Gamma(v)$ the sets of its inbound and outbound neighbors, respectively. In the model we propose, hypergraphs that model circuits are represented as sets of interconnected Directed Acyclic Hypergraphs (DAHs), associated with a red-black vertex coloring scheme. Red vertices correspond to I/O (Input/Output) ports and registers, and black vertices to combinatorial components. Let $\mathcal{V}^R \subset \mathcal{V}$ and $\mathcal{V}^B \subset \mathcal{V}$ be the red and black vertex subsets of $\mathcal{V}$, such that $\mathcal{V}^R \cap \mathcal{V}^B = \emptyset$ and $\mathcal{V}^R \cup \mathcal{V}^B = \mathcal{V}$. A hypergraph or sub-hypergraph $\mathcal{H}$ is a DAH iff its red vertices $v_R \in \mathcal{V}^R$ are either only sources or sinks (*i.e.*, $\Gamma^-(v_R) = \emptyset$ or $\Gamma^+(v_R) = \emptyset$), and no cycle path connects a vertex to itself. Using this definition, we can represent circuit hypergraphs as *red-black hypergraphs*, which are sets of DAHs that share their red vertices. Let $\mathbf{H}(\mathbf{V}, \mathbf{A}) \overset{\text{def}}{=} \{\mathcal{H}_i, i \in \{1 \dots n\}\}$ be a *red-black hypergraph*, such that every $\mathcal{H}_i$ is a DAH and an edge-induced sub-hypergraph of $\mathbf{H}$. In our model, the paths in $\mathbf{H}$ to consider when addressing the objective of minimizing *path-cost* degradation during partitioning are only the paths interconnecting red vertices, as these red-red paths represent register-to-register paths. Since only red vertices are shared between DAHs in $\mathbf{H}$, any red-red path only exists within a single DAH and can never span across several DAHs.

Let us define $\mathbf{P}$ as the set of red-red paths in $\mathbf{H}$. From these paths and a function $d_{max}(u, v)$ which computes the maximum distance between vertices u and v of some DAH $\mathcal{H}$, we can now define the longest path distance for $\mathcal{H}$ as: $d_{max}(\mathcal{H}) \overset{\text{def}}{=} \max(d_{max}(u,v) | u, v \in \mathcal{H})$ and, by extension, for $\mathbf{H}$, as: $d_{max}(\mathbf{H}) \overset{\text{def}}{=} \max(d_{max}(\mathcal{H}) | \mathcal{H} \in \mathbf{H})$.

A k-partition Π of $\mathbf{H}$ is a splitting of $\mathbf{V}$ into k vertex subsets $\pi_1 .. \pi_k$, called parts, such that: (i) all parts π_i, given a capacity bound C_i, respect the capacity constraint: $\sum_{v \in \pi_i} \mathcal{W}_v(v) \leq C_i$; (ii) all parts are pairwise disjoint: $\forall i \neq j, \pi_i \cap \pi_j = \emptyset$; (iii) the union of all parts is equal to $\mathbf{V}$: $\bigcup_i \pi_i = \mathbf{V}$. For a given partition Π of $\mathbf{H}$, the connectivity $\lambda_\Pi(a)$ of some hyperarc $a \in \mathcal{A}$ is the number of parts connected by a. If $\lambda_\Pi(a) > 1$, then a is said to be cut; otherwise it is entirely contained within a single part and is not cut. The cut of partition Π is the set $\omega(\Pi)$ of cut hyperarcs, such as $\omega(\Pi) \overset{\text{def}}{=} \{a \in \mathcal{A}, \lambda_\Pi(a) > 1\}$. The cut size is defined as $f_c \overset{\text{def}}{=} \sum_{a \in \omega(\Pi)} \mathcal{W}_a(a)$. The *connectivity-minus-one* cost function f_λ of some partitioned hypergraph $\mathbf{H}^\Pi$ can then be defined as: $f_\lambda = \sum_{a \in \mathbf{A}} (\lambda_\Pi(a) - 1)$. Consequently, in our model, the distance between two vertices u and v may increase during partitioning due to the additional cost of routing paths between two (or more) parts. Let D_{ij} be the penalty associated with parts i and j such that if u is in part i and v is in part j, then: $d_{max}^\Pi(u, v) \geq d_{max}(u, v) + D_{ij}$. The objective function f_p can therefore be defined as the minimization of the longest path of $\mathbf{H}$ subject to partition Π: $f_p = \min \ d_{max}(\mathbf{H}^\Pi)$.

2.2   Related Work

Many previous works exploit existing *min-cut* partitioning tools, used as black boxes, and try to account for additional constraints. For instance, [2] presents a multi-objective approach based on HMETIS. In [1] is proposed a pre-processing coupled with a recursive bi-partitioning algorithm using HMETIS. They use path length values a hyperarc weights, to channel the partitioner towards minimizing path cost. Since path length values can change as the result of a bi-partitioning step, hyperarc weights are reevaluated after each of them to identify critical hyperarcs. Reference [6] compares a classic method using HMETIS for partitioning, followed by a placement algorithm, with a derived approach consisting in placing and routing during the partitioning step. More recently, [10] performs some pre- and post-processing on the hypergraph to capture the critical path minimization objective within the cut-size metric, using HMETIS as a partitioning tool. However, minimizing cut size is often not the most relevant objective, and biasing *min-cut* cost functions to take *path-cost* minimization into account does not allow to handle properly platform topologies.

3 Contributions

Our first contribution consists of two initial partitioning algorithms, called DBFS and DDFS. They are based on breadth-first-search and depth-first-search methods, respectively. Both are driven by *vertex criticality*, *i.e.*, the length of the longest path traversing a vertex. These algorithms are combined with some cut minimization tools to achieve both objectives of preserving critical paths and minimizing the cut. Our second contribution consists of an extension of the FM heuristic [8], called DKFM. It aims to perform moves that minimize the cost function f_p, taking into account the target topology. The objective is to use DKFM as a post-processing cut minimization tool, for instance as a refinement method within a multilevel framework.

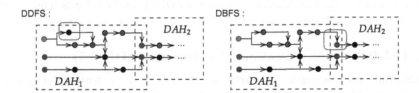

Fig. 1. A hint on DDFS vs. DBFS. All hatched nodes will be placed in the same part. DBFS avoids multiple cuts along paths within a DAH, while DDFS does not (see the red circle in the DDFS example). However, if a critical path starts in DAH_2 from a sink of DAH_1, DBFS can produce a cut along this critical path, while DDFS cannot (see the red circle in the DBFS example). (Color figure online)

3.1 Initial Partitioning Based on Breadth-First Search Driven by Vertex Criticality (DBFS)

This first initial partitioning algorithm is based on an extended search along the vertices of **H**, driven by criticality. It groups neighboring vertices of each vertex v according to their criticality, to avoid multiple cuts along a long path. In our context, the multiple-cut constraint is also important because we need to avoid cutting the same path multiple times. Indeed, the cutting cost D_{ij} is often larger than the path length. However, many paths can be shorter than the longest path before partitioning. Therefore, our algorithm considers the criticality of each vertex v in addition to the criticality of its neighbors. To select vertices to consider, we use their in-degree value, decremented for all outgoing neighbors of v when visiting v. This gives us a topology-driven hypergraph traversal, considering vertex criticality. This algorithm allows us to perform a walk along the topological order in the current DAH, by selecting, at each step, a neighbor of maximum criticality. This choice allows us to favor the grouping, within the same part, of neighboring vertices with high criticality. As long as the size constraint is respected, each selected vertex will be placed in the same part. An example can be found in Fig. 1.

3.2   Initial Partitioning Based on Depth-First Search and Vertex Criticality (DDFS)

This second initial partitioning method performs a depth-first traversal driven by the criticality of the vertices. The main difference with the previous method is that all vertices will be inserted in the same priority queue, regardless of whether they are red or black. It enables a more compact visit order concerning the criticality value, but does not consider the topological structure of DAHs in **H**. The main idea behind this method is to be able to pack interconnected critical paths of several DAHs in the traversal order. If the topology allows for it, the most critical neighboring paths will be placed in the same part, as long as the size constraint is respected. An example can be found in Fig. 1. However, DDFS may induce cuts within the DAHs and a possible path cost degradation for paths of smaller criticality. The relative efficiency of DBFS and DDFS is likely to vary, depending on circuit topologies and the distributions of path lengths.

3.3   FM-Based Local Optimization Heuristic (DKFM)

A commonly used heuristic in partitioning tools is the algorithm proposed by Fiduccia and Mattheyses (FM) [8]. It is based on *local search* to move vertices across parts so as to reduce the cut of balanced hypergraph bipartitions. FM computes a move gain for each vertex and performs a single move at each step.

   We propose an extension of this algorithm to minimize critical path degradation during partitioning. Let Π be a partition of **H**; the gain function for some vertex v and some candidate partition Π' in which v may be moved to part π_k is defined as: $gain(\Pi, \Pi', \mathbf{H}) \stackrel{\text{def}}{=} d_{max}^{\Pi}(\mathbf{H}) - d_{max}^{\Pi'}(\mathbf{H})$.

4   Experimental Results

4.1   Methdology

The hypergraphs of our benchmark are taken from the Titan23 [11] and ITC99 [7] benchmarks, and *Chipyard* [3]. We also defined two target topologies composed of four elements. Test Architecture 1 connects its four FPGAs as a cycle

$\pi_0, \pi_1, \pi_2, \pi_3$, while Architecture 2 is a chain $\pi_0, \pi_1, \pi_2, \pi_3$. In order to evaluate the DBFS and DDFS algorithms, we first generated an initial partition and then ran KKAHYPAR to refine it, using the objective function f_λ. To test the DKFM refinement algorithm, we first ran PAToH-S, KHMETIS, and KKAHYPAR on the weighted hypergraph instances xxx_w, with our weighting scheme driven by vertex-hyperarc criticality. Then, we applied our DKFM algorithm as post-processing to these outputs, to study the *path-cost* improvement and possible degradation of the *connectivity-minus-one* metric induced by our refinement pass. We chose KHMETIS over HMETIS, used in previous works, because we get better results in the majority of cases for both target topologies. We select PAToH in its "speedy" (S) version to see whether it could produce acceptable solutions in combination with DKFM, because computation time is a critical aspect of industrial-size circuit prototyping.

4.2 Initial Partitioning Algorithms

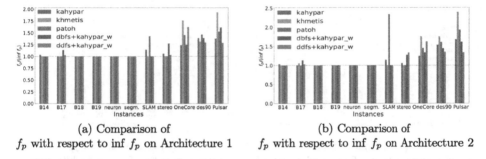

(a) Comparison of
f_p with respect to inf f_p on Architecture 1

(b) Comparison of
f_p with respect to inf f_p on Architecture 2

Fig. 2. Comparison of f_p on Architecture 1 (a) and Architecture 2 (b).

Figures 2a and 2b evidence that in many cases, initial partitioning algorithms DBFS and DDFS, combined with our weighting scheme, can influence a cut minimization partitioning tool to improve path cost. However, for some instances (*b17*, *OneCore*), the algorithms perturb the cut minimization tools and produce worse results. Combined with a cut minimization tool, these two algorithms yield an average improvement in 18% (DBFS) and 45% (DDFS) of the instances when mapping onto Architecture 1, and an average improvement in 36% of the instances on Architecture 2, for both algorithms. Compared to the cut minimization approach, DBFS improves the path cost metric by 8% and 7% on the two architectures, respectively, while DBFS improves path cost by 7% and 17%, respectively. Moreover, in many instances, the produced path cost is equal to the lowest possible cost, and it is impossible to improve it further. It shows that both algorithms indeed prevented critical paths from being cut. We already see a gain at this scale, which is encouraging for testing these methods on more significant cases. The quality of the partitions produced by the two algorithms varies, depending on the instances. We suppose this somewhat reflects the topological properties of the instances (e.g., whether the black vertices are heavily interconnected at each level). Further analysis are required to determine relevant criteria. Our two algorithms differ in the way vertices are visited: DDFS (depth-first) is more likely to cluster long paths in the same part, while DBFS (breadth-first) is more likely to cluster together vertices on the same level with respect to source red vertices. We assume that these algorithms succeed in preventing the same path from being cut multiple times, by packing together the neighborhood of black vertices within DAHs. It allows them to produce results that are topologically compatible with the different hardware topologies. Notably, DBFS is adapted to linear (streamlined) hardware topologies, where vertices belonging to the same levels of computation must be packed together.

4.3 Refinement Algorithm

Figures 3a and 3b display path cost values for the smallest possible value of f_p (that is, the one obtained when no edge is cut).

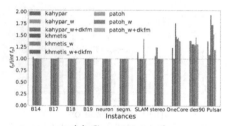 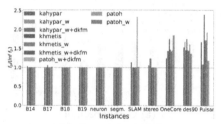

(a) Comparison of
f_p with respect to inf f_p on Architecture 1, with 20% DKFM moves and $\rho=0$

(b) Comparison of f_p
with respect to inf f_p on Architecture 2, with 20% DKFM moves and $\rho=0$

Fig. 3. Comparison of f_p on Architecture 1 (a), and Architecture 2 (b).

Figure 3a and 3b evidence an improvement in maximum path cost on all instances, ranging between 27% and 54%, depending on the architectures. Interestingly, despite using the fast mode, PATOH-S +DKFM provides similar results. In some cases, DKFM does not improve the solutions produced by the cut minimization tools or can even produce worse solutions (e.g., for *stereo_vision*). We attribute this behavior to difficulties experienced by the DKFM algorithm to get out of the local minima computed by the min-cut algorithms, all the more in the case of weighted cut minimization. This, already partially addresses the path cost minimization objective. On the other hand, DKFM can improve a weighted cut minimization solution (e.g., for *des90*) on both architectures. Finally, in instances not already of minimal path cost, DKFM improves the path cost by 12% to 17% for Architecture 1, and by 12% to 40% for Architecture 2. At this scale, we already see a gain from a solution close to a local minimum, which is encouraging to test DKFM on larger target topologies and with a dedicated cut minimization algorithm that allows more movement. Indeed, Architecture 2 evidences the drawbacks of plain partitioners when communication costs vary significantly between different parts. DKFM is more likely to benefit to partitions in which vertices have been mapped onto remote parts instead of closer ones. In this case, moving a few vertices to the right parts can make a huge difference. It is especially true for case *SLAM_spheric*, for which the improvement is of more than 50% (highly influencing the computation of the average gain over all instances for Architecture 2).

5   Conclusion and Future Work

This paper presents the red-black oriented hypergraph model and its associated red-red path cost metric. Our results show that the DBFS and DDFS algorithms are relevant and complementary initial partitioning methods, depending on circuit instances and their underlying topologies. DKFM results show an average improvement in critical path length ranging between 12% and 40%, compared to the initial min-cut solutions. These algorithms seem to be a good approach to improve performance of multi-FPGA prototyping. However, these methods can degrade cut size by a factor from 1.2 up to 5.0. At this stage, only the DKFM refinement algorithm takes into account the topology of the target hardware and correctly manages outliers. Hence, it may not be easy to improve incrementally and locally on results provided by the architecture-unaware algorithms DBFS and DDFS. It would be interesting to work on the placement of vertices according to the target topology from the initial partitioning phase, as SCOTCH [13] does for unoriented graphs.

References

1. Ababei, C., Bazargan, K.: Timing minimization by statistical timing hMetis-based partitioning. In: 16th International Conference on VLSI Design (2003). Proceedings
2. Ababei, C., Navaratnasothie, S., Bazargan, K., Karypis, G.: Multi-objective circuit partitioning for cutsize and path-based delay minimization. In: International Conference on Computer Aided Design (ICCAD) (2002)
3. Amid, A., et al.: Chipyard: integrated design, simulation, and implementation framework for custom SoCs. IEEE Micro **40**(4), 10–21 (2020)
4. Çatalyürek, Ü., Aykanat, C.: PaToH. Springer, US, Boston, MA (2011)
5. Çatalyürek, U., et al.: More recent advances in (hyper) graph partitioning. ACM Comput. Surv. **55**(12), 1–38 (2023)
6. Chen, M.H., Chang, Y.W., Wang, J.J.: Performance-driven simultaneous partitioning and routing for multi-FPGA systems. In: 2021 58th ACM/IEEE Design Automation Conference (DAC) (2021)
7. Corno, F., Reorda, M., Squillero, G.: RT-level ITC'99 benchmarks and first ATPG results. IEEE Des. Test Comput. **17**(3), 44–53 (2000)
8. Fiduccia, C., Mattheyses, R.: A Linear-Time Heuristic for Improving Network Partitions. In: 19th Design Automation Conference (DAC) (1982). https://doi.org/10.1109/DAC.1982.1585498

9. Karypis, G., Kumar, V.: Hmetis: a hypergraph partitioning package. ACM Trans. Archit. Code Optim. (1998)
10. Liou, S.H., Liu, S., Sun, R., Chen, H.M.: Timing Driven Partition for Multi-FPGA Systems with TDM Awareness. Ass. Comp. Mach, New York, USA (2020)
11. Murray, K.E., Whitty, S., Liu, S., Luu, J., Betz, V.: Timing-driven titan: enabling large benchmarks and exploring the gap between academic and commercial CAD. ACM Trans. Reconf. Technol. Syst. 8(2), 1–8 (2015)
12. Ou, S.L., Pedram, M.: Timing-driven partitioning using iterative quadratic programming
13. Pellegrini, F., Roman, J.: Scotch: a software package for static mapping by dual recursive bipartitioning of process and architecture graphs. In: Liddell, H., Colbrook, A., Hertzberger, B., Sloot, P. (eds.) HPCN-Europe 1996. LNCS, vol. 1067, pp. 493–498. Springer, Heidelberg (1996). https://doi.org/10.1007/3-540-61142-8_588
14. Schlag, S.: High-Quality Hypergraph Partitioning. Ph.D. thesis, Karlsruhe Institute of Technology, Germany (2020)

Graph TopoFilter: A Method for Noisy Labels Detection for Graph-Structured Classes

Artur Budniak[✉][ID] and Tomasz Kajdanowicz[ID]

Department of Computational Intelligence, Wrocław University of Science and Technology, Wrocław, Poland
{artur.budniak,tomasz.kajdanowicz}@pwr.edu.pl

Abstract. Detection of incorrectly conducted failure repairs is not a trivial task for companies manufacturing big volumes of goods. Extensive data sets of service calls are periodically updated and subject matters experts would not be efficient in manual annotating of the data. Symptoms described in free text form might be caused by different components - not necessarily by the most obvious. Classes are imbalanced due to different time to failure of particular components and thus actions taken for some rare failures might be noted as incorrect ones. The presented problem is similar to the problem of learning in a presence of noisy labels, which are caused by human errors, variation of annotator to annotator perception, faults made by annotating algorithms or by other reasons. There are multiple techniques to prevent neural networks from overfitting to the noisy data, but to our best knowledge none of them considers relationships between classes, which is crucial in engineering systems built from multiple components connected in a specific way. A novel approach of selecting clean data samples in an unsupervised manner is presented in this paper. It is based on a topological approach exploring the deep representation of the features in the hidden space, enriched with knowledge graphs reflecting the structure of the classes. We present the case study of the algorithm utilized for service calls data set for home appliances.

Keywords: noisy labels · knowledge graph · natural language processing

1 Introduction

Deep neural networks are capable of fitting to the training set when it is tagged with true or entirely reshuffled labels [13], which is an impressive property, but also a drawback, when real data sets are considered. A case of a not correct label for given features is noted as a noisy label. Four sources of noisy labels have been indicated by Frenay et al. [4]: poor quality of input data, human mistakes, annotator to annotator variation of perception and data or communication problems. Lower accuracy of a model and longer training time may be

Supported by Whirlpool Company Polska Sp. z o. o.

© The Author(s), under exclusive license to Springer Nature Switzerland AG 2023
J. Mikyška et al. (Eds.): ICCS 2023, LNCS 14075, pp. 661–667, 2023.
https://doi.org/10.1007/978-3-031-36024-4_51

caused by the assumption that data set is tagged perfectly and when no strategy for the noise is planned [13]. As a consequence of above, many techniques have been developed to improve generalization of classifiers trained on the real data. However, relationships between classes have not been evaluated in the literature. To benefit from knowing the class structure, we modified the TopoFilter algorithm [10], which is based on the topology of deep representation of features. The algorithm works as follows. First, k-Nearest Neighbors (kNN) graph G of data points from all classes is created. Then nodes of a given class are selected by breaking edges to other class nodes. As a result a sub-graph G_i is set. Samples are considered clean, if they belong to the largest connected component Q_i spread within G_i. For the hidden representation points lying close to the other class the TopoFilter might be too rigorous. Data points of that class i may be not included in Q_i because of points from class j in their neighborhood that break edges. It is correct for independent classes, but for related ones (components in engineering systems) it results in removing too many points. A knowledge graph KG representing the structure of classes prevents our algorithm from deleting the points and thus helps to keep more information from the data set. The graph denotes arbitrarily understood similarity between classes or known by subject matter experts connections (e.g. distance between weather stations [7], current flow between points in power grids [3] or photovoltaic cells placement on farms [2]).

2   Related Work

There are a few techniques to handle the presence of noisy labels and one of them is the sample selection. It may be performed by one neural network (MentorNet) to train the other network (StudenNet) by preparing curriculum of more and more complex data points [6]. Instead of selecting for curriculum samples with small loss and adding samples with gradually larger loss during the training phase at constant rate, MentorNet adjusts that process according to StudentNet feedback. Two networks can also work together in an architecture of co-teaching [5], which limits the flow of the error caused by noisy labels due to the different learning rates of the two networks. For every mini-batch the networks select small-loss instances for each other. Later on the method was improved [12] by passing between the two networks only those samples that the networks disagree about in predictions. In contrast to this approach, the joint agreements between two models are used for training in Joint Agreement Method [9]. It is less probable for the two models to fit to the noisy labels at the same time having different learning rates. A deep representation of data points derived from network weights may help to find noisy labels, assuming that samples from the same class are similar [8], which is proposed to be verified by a new version of Local Outlier Factor algorithm (pcLOF). Another approaches use kNN graph construction over points in hidden space. NGC (Noisy Graph Cleaning) propagates their labels [11], while TopoFilter [10], in order to find clean samples, builds the largest connected component within given class. In addition, some points are removed, if in their neighborhood there are too many neighbors from other classes.

3 Problem Definition

Notation follows the survey [1]. A data set with noisy labels is denoted as

$$D = \{x_i, y_i\}_{i=1}^{|D|} \tag{1}$$

where x_i is a feature vector, $y_i \in Y$ is an observed label and $|D|$ is the number of entities in the data set D. A true label $\widehat{y}_i \in Y$ is not known. A noise transition matrix presents the probability for each true label $\widehat{y}_i$ to be marked as a noisy, observed label y_i:

$$p(y = j|x_i, \widehat{y}_i = c) = \eta_{jc}(x_i) \tag{2}$$

$$\sum_{j \in Y} \eta_{jc}(x_i) = 1 \tag{3}$$

Hidden representation of the feature vector x_i, denoted as $x_i{}^h$, is learned during the training. In the initial phase of the training it is learned on all samples, including the noisy ones. After sample selection, it is adjusted on samples considered clean. The structure of the classes is represented by a knowledge graph KG. Class labels are nodes and relationships between them are edges of that graph. Cycles are allowed.

The aim of the task is to split the data set D into two subsets containing true ($\widehat{y}_i \in Y$) and false ($Y \setminus \widehat{y}_i$) labels in unsupervised manner. This is a binary classification of samples within each class.

4 The Algorithm

As noted before, the algorithm is an extension of TopoFilter [10]. For first m epochs a model is trained on a whole training set. After that the clean samples are selected before the training in each epoch and weights are updated on clean data set up to N epochs. In contrast to approach mentioned above, hidden representations $x_i{}^h$ of clean samples in class i are not only nodes from the largest connected component (lcc) of kNN sub-graph G_i, which is made by removing edges connected to nodes from other classes in G, but the lcc is constructed on nodes from class i and k_{hops} neighbor classes in accordance with the knowledge graph KG. Then neighbor nodes are removed and remaining ones are considered clean. That allows points laying among those from related classes to be kept. The algorithm is presented in pseudo-code below and depicts the differences to the original TopoFilter:

1. Noisy training data S, milestone m, training epochs N, number of classes Γ, number of neighbors k_{NN}, ~~filtering parameter ζ~~, [ADDED: k_{hops}]

2. Output: Collected clean data C
3. Initialize $C \leftarrow \emptyset, \widehat{S} \leftarrow \emptyset$
4. for t=1, $\cdots$, N do
5.   Train network on $\widehat{S}$
6.   if $t \geq m$ then
7.   Extract feature vectors x^h from training data S
8.   Compute k-NN graph G over x^h
9.   for i=1, $\cdots$, Γ do
10.   Construct subgraph Gi by selecting feature vectors x_i^h from i-th class and removing all edges associated with x_j^h ~~for $j \neq i$~~
                [ADDED: if $d(i,j)_{KG} > k_{hops}$]
11.   Compute the largest connected component Qi over Gi.
                [ADDED: remove x_j^h for $j \neq i$]
12. $C \leftarrow C \cup Qi$
13.   end for
14.   ~~Find outliers O within C based on ζ -- filtering; update $C \leftarrow C\backslash O$~~
15. $\widehat{S} \leftarrow C$
16.   end if
17. end for

The implementation is available at https://github.com/ArturBudniak/Graph_TopoFilter.

5   Case Study

5.1   Data Set

The data set is the Service Calls Register of Wall Oven products in Whirlpool Company. For each symptom description x_i claimed by a customer there is a label y_i denoting the component replaced by a serviceman. After removing rows with missing values it contains 79469 rows with 95 unique labels. Classes having less than 100 instances or those not possible to be represented as nodes (e.g. "customer refused repair", "general safety question") are also removed. As a result 46 classes and 62312 rows are selected. For the algorithm validation the following components have been selected for manual annotation by subject matter experts: hinges, lamp, door lock, cooling fan, product fuse and thermostat. There are in total 7207 clean samples. We assume one failure at the time. To keep the company's internal data unrevealed the data sets are not published.

5.2   Knowledge Graph

All engineering systems, including home appliances, may be represented by graphs, where nodes indicate components and edges the relationship between them. Edge direction is not utilized, although it might be helpful for engineers. For the same purpose edge types may also be included to distinguish between mechanical, electrical, functional or any other sort of interface. Type of edge is not used by our algorithm.

5.3   Experiments

The experiments compare two algorithms for unsupervised learning - TopoFilter [10] (baseline) and our modified version - Graph TopoFilter. The task is to classify service calls as clean or noisy samples. The factors in the experiments are k_{hops} and size of the data set $|D|$. Three runs are performed for each setting. The other parameters are found experimentally and kept constant. The neural network has input layer with https://tfhub.dev/google/nnlm-en-dim50/ 2 embedding, two hidden layers with ReLU activation function, each of 200 neurons and a softmax layer. Optimizer is set to adam, loss function - CategoricalCrossentropy. Batch size is set to 512, number of epochs N is 20, milestone m is 15, number of neighbors k_{NN} for kNN graph is 13 and ζ is 0 ($\zeta > 0$ removes most external points in lcc). Parameter k_{hops} is set to 1 and 2 (for 0 the new algorithm is the same as TopoFilter). From the whole data set 20%, 40%, 60%, 80% and 100% of rows are taken randomly for training. Due to the data set imbalance the weighted F1-score metric was chosen to rate the algorithm performance. The values and the standard deviation over three trials are presented in Table 1. Precision and recall are shown in Fig. 1 and Fig. 2. Higher weighted F1-score is reported for the Graph TopoFilter compared to the original TopoFilter [10]. The smaller the size of the data set is, the bigger the difference is noted. Precision drops for the new algorithm. However, recall obtains much

Table 1. Weighted F1-score for clean labels classification.

Algorithm	Data set size				
	20%	40%	60%	80%	100%
TopoFilter $k_{hops} = 0$	0.70 ± 0.03	0.77 ± 0.01	0.80 ± 0.01	0.80 ± 0.01	0.82 ± 0.01
Graph TopoFilter $k_{hops} = 1$	0.87 ± 0.01	0.88 ± 0.01	0.89 ± 0.01	0.88 ± 0.01	0.89 ± 0.01
Graph TopoFilter $k_{hops} = 2$	0.89 ± 0.01	0.89 ± 0.01	0.89 ± 0.01	0.89 ± 0.01	0.89 ± 0.01

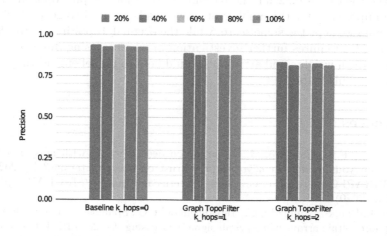

Fig. 1. Precision per algorithm and data set size.

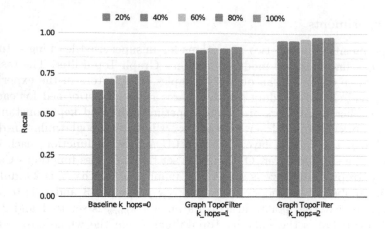

Fig. 2. Recall per algorithm and data set size.

higher values. It is worth noting that the impact of the data set size is less visible for Graph TopoFilter. The new algorithm needs only 20% of the data to achieve better results than the original one for the whole data set.

6   Summary

The algorithm presented in this paper selects clean samples from a text data set containing service calls in unsupervised learning. It is based on TopoFilter [10] that annotates too many samples as noisy because of the nature of failures and their multiple possible root causes. Our approach takes additional information about the structure of the classes and improves noisy labels detection. Case study experiments conducted on the whole data set showed that weighted F1-score increased from 0.82 for the baseline to 0.89 for Graph TopoFilter. What is more impressive, for the number of service calls reduced to 20% (to simulate rare claims) we kept the same score, while the original TopoFilter obtained only 0.70. The idea requires further investigation of whether using directed edges or adding to them weights and attributes would help to model engineering systems behavior in fault states.

References

1. Cordeiro, F.R., Carneiro, G.: A survey on deep learning with noisy labels: how to train your model when you cannot trust on the annotations? In: 2020 33rd SIBGRAPI Conference on Graphics, Patterns and Images (SIBGRAPI), pp. 9–16. IEEE (2020)
2. Fan, J., Rao, S., Muniraju, G., Tepedelenlioglu, C., Spanias, A.: Fault classification in photovoltaic arrays using graph signal processing. In: 2020 IEEE Conference on Industrial Cyberphysical Systems (ICPS), vol. 1, pp. 315–319. IEEE (2020)

3. de Freitas, J.T., Coelho, F.G.F.: Fault localization method for power distribution systems based on gated graph neural networks. Electr. Eng., 1–8 (2021)
4. Frénay, B., Verleysen, M.: Classification in the presence of label noise: a survey. IEEE Trans. Neural Netw. Learn. Syst. **25**(5), 845–869 (2013)
5. Han, B., et al.: Co-teaching: robust training of deep neural networks with extremely noisy labels. arXiv preprint arXiv:1804.06872 (2018)
6. Jiang, L., Zhou, Z., Leung, T., Li, L.J., Fei-Fei, L.: MentorNet: learning data-driven curriculum for very deep neural networks on corrupted labels. In: International Conference on Machine Learning, pp. 2304–2313. PMLR (2018)
7. Owerko, D., Gama, F., Ribeiro, A.: Predicting power outages using graph neural networks. In: 2018 IEEE Global Conference on Signal and Information Processing (GlobalSIP), pp. 743–747. IEEE (2018)
8. Wang, Y., et al.: Iterative learning with open-set noisy labels. In: Proceedings of the IEEE Conference on Computer Vision and Pattern Recognition, pp. 8688–8696 (2018)
9. Wei, H., Feng, L., Chen, X., An, B.: Combating noisy labels by agreement: a joint training method with co-regularization. In: Proceedings of the IEEE/CVF Conference on Computer Vision and Pattern Recognition, pp. 13726–13735 (2020)
10. Wu, P., Zheng, S., Goswami, M., Metaxas, D., Chen, C.: A topological filter for learning with label noise. Adv. Neural. Inf. Process. Syst. **33**, 21382–21393 (2020)
11. Wu, Z.F., Wei, T., Jiang, J., Mao, C., Tang, M., Li, Y.F.: NGC: a unified framework for learning with open-world noisy data. In: Proceedings of the IEEE/CVF International Conference on Computer Vision, pp. 62–71 (2021)
12. Yu, X., Han, B., Yao, J., Niu, G., Tsang, I., Sugiyama, M.: How does disagreement help generalization against label corruption? In: International Conference on Machine Learning, pp. 7164–7173. PMLR (2019)
13. Zhang, C., Bengio, S., Hardt, M., Recht, B., Vinyals, O.: Understanding deep learning (still) requires rethinking generalization. Commun. ACM **64**(3), 107–115 (2021)

Dynamic Data Replication for Short Time-to-Completion in a Data Grid

Ralf Vamosi[ID] and Erich Schikuta[✉][ID]

Faculty of Computer Science, University of Vienna, Vienna, Austria
{ralf.vamosi,erich.schikuta}@univie.ac.at

Abstract. Science collaborations use computer grids to run expensive computational tasks on large data sets. Tasks as jobs across the network demand data and thereby workload management and data allocation to maintain the computational workflow. Data allocation includes data placement with different replication factors (multiplicity) of data.

The proposed data replication & allocation model can place multitudes of subsets of a data population in a distributed system, such as a computer cluster or computer grid. A stochastic simulation with a data and computing example from the ATLAS Physics Collaboration shows its potential usability in one of the largest Computing Grids. This paper showcases data allocation with different replica factors and various numbers of subsets to improve the overall situation in a computer network.

Keywords: data replication · data placement · data partitioning · data-intensive computing

1 Introduction

Data replication is essential for quicker response or time-to-completion in computer clusters and data grids. Replication involves generating copies of data on different nodes or systems in the network. Data parallelization reduces the time required to read and complete computations, balances the processing load, and improves system resilience. The base case has a replication factor of 1, meaning that only one copy of the data is stored in the system. This is often referred to as a single-replica or non-replicated setup. Other cases are replication factors of 2 or more. A fraction of an integer means that a data subset is replicated.

The motivating use case for the presented method is the ATLAS physics collaboration. The scientific collaboration stores data as files across a worldwide computing grid with 160 geographically distant sites. The data is organized into labeled data sets or file collections that can change over time, and the files are the primary units for computational jobs.

Delay or latency in computer networks, especially across wide area networks (WAN), can cause delays in workflow, and how data replication can improve the situation by promoting the local processing of data sets. However, jobs must be allocated to appropriate sites with the necessary resources to support the

© The Author(s), under exclusive license to Springer Nature Switzerland AG 2023
J. Mikyška et al. (Eds.): ICCS 2023, LNCS 14075, pp. 668–675, 2023.
https://doi.org/10.1007/978-3-031-36024-4_52

required computation. Load balancing ensures that jobs are placed at sites that can support the necessary computations. The site with input data is preferred as the target to fully utilize processing resources and reduce delays.

In the same way, data sets cannot be randomly placed on sites or nodes due to storage limitations leading to two major points: Every time a job needs to run, it may transfer files as its input and thereby use a system bottleneck, the interlinking network (WAN) between sites. The network consists of non-uniform sites providing resources with considerable differences. Furthermore, network connections vary from site to site due to cost and provision limitations.

2   State of the Art

Data replication has a long history. The technique is referred to with multiple terms and seen from different points of view. The main reasons for replication are data backup and data parallelization. It was investigated when distributed databases were used, so parallelization was to be utilized. Data replication is NP-complete due to the combinatorial explosion in the selection process. Research on data replication presents various models and algorithms for file placement and replication in distributed systems. The models focus on factors such as storage capacity, network bandwidth constraints, load balance, reliability, and cost trade-offs. They also use different approaches, including deterministic and stochastic, and hybrid models that combine both. Many software architectures have been proposed for parallel systems [2]. Mathematical models like game theoretic and Markov decision process models also make file placement decisions.

The paper of [3] investigates the mathematical modeling of a network of nodes with replicated data files, focusing on transactions involving multiple files. The authors consider two types of transactions, namely, query and update traffic, and use a linear cost model to illustrate optimal file assignment to nodes. They present bounds on the number of file copies required in the network based on the query, update traffic, and derive a test to determine the optimal configuration.

An algorithm for dynamic replication of a data object in distributed database systems that adapts to changes in the read-write pattern, continuously moving the replication policy toward an optimal one, is presented in [7]. The algorithm can be integrated with concurrency control and recovery mechanisms and provide a lower bound on the performance of any dynamic replication algorithm.

The paper of [4] proposes replication management services and protocols for efficient and fast access to large and widely distributed data in Data Grids. The cost model for replication decisions consider factors such as run-time read/write statistics, response time, bandwidth, and replica size. The system evaluates the costs and performance gains of creating each replica and organizes them in hierarchical and flat topologies to minimize inter-replica communication costs.

The paper from [5] summarizes research on genetic algorithms (GA) to optimize data replication in large distributed systems and reduce network traffic delays. The authors compare their GA approach to a greedy method in a static

scenario with constant read/write demands and also propose a hybrid GA approach to handle changing patterns. The experiments are conducted with up to
25 sites and 90 data objects.

The paper of [1] explores the energy efficiency and bandwidth consumption of
data replication in cloud computing data centers, considering the QoS improvement achieved by reducing communication delays. A mathematical model and
extensive simulations are used to evaluate the performance and the trade-offs.

The paper from [6] proposes an approach based on evolutionary clustering to
improve data allocation in distributed systems. Popularity information allocates
data items to storage nodes, including a stochastic search with a fast clustering
method. The method improves data allocation without changing the number
of replicas in the global file system while considering storage constraints and
improving data access performance.

3   Optimization Method

The proposed heuristic optimization solves data replication and partitioning by
inferring from a large data population into smaller subsets for replication and
placement. Major factors in parallel computing are data available on multiple
nodes, the different types of data, and the read frequencies. Read patterns introduce some degree of unpredictability in the job execution process among sites.
The uncertain nature is modeled as probability mass functions (PMF).

Our model shows how patterns in data are used to fulfill replication for
a distributed workflow. For an overall improvement, average values are taken
from network connections. Dynamic replication is supported because runs can
be repeated as often as necessary for any data share and replication factor. The
method returns a tuple of subsets replication runs. The optimization process
aims to minimize the target cost/loss by conditionally finding subsets in the
data population:

$$\underset{A}{argmin}\ cost_{network}(A)$$

$$s.t. \tag{1}$$

$$A^T \times w_{point} \leq w_{site}$$

where the parameter A defines the objects or points per subset $A_1, ..., A_N$. A
is a slim matrix with one row per data point and one column per subset. A
'1' indicates one replica (row) in a subset (column). Multiples are possible for
higher replication factors. The network cost $cost_{network}(A)$ will be introduced
next. w_{point} is a column vector representing the sizes for data points, and w_{site}
is a column vector holding the maximum sizes for the final N subsets.

During search steps, the method keeps the parameters if it results in improvement and repeats until no further satisfaction can be achieved. The effort to
improve continuously increases as better data sampling becomes harder on a

higher baseline. **Algorithm 1** outlines this process as pseudo-code: The function setSystemParameters() sets the values for system parameters and variables before each run, including network variables, site variables, datasets, and job sets. *setupClustering()* setups and selects the clustering method. Due to space limitations, we are unable to provide any discussion. Next-neighbor is the base method. *setSeeds()* memorizes the seed positions and partially alternates them. The cost function *calculateCost()* depends on its parameters, system variables, and meta parameters. Its parameters are the data set X and a tuple of data placement at each site or node denoted as $A = (A_1, ..., A_N)$. Some A_i may be empty. The method adds subsets $A' = A'_1, ..., A'_N$ for each replication to the allocated data $A = A_1, ..., A_N$. The final data allocation always depends on the replication case. An optimization yields the initial data placement A merged with A', where the tuples are joined element-wise (denoted by operator $+$), incorporating constraints for capacities. The free space is $w' \leftarrow w_i - |A_i|$ for the subset with index i depending on the maximum size w_i and the data A_i already placed at the i'th site or node. Depending on the cost, the update process continues and returns the dominating solution stored as A''.

Algorithm 1. Data sampling algorithm with clustering for data replication. Input parameters are data set X, subsets of data placement $A = A_1$ to A_N, maximum subset sizes $w = w_1$ to w_N, and the number of clusters k.

```
function GETSUBSETS(X, A, w, k)
    for i in 1 to N do
        w' ← w_i − |A_i|
        k' ← GetNumberOfSeeds(w', i, k)
        setSeeds(X, k')
        A'_i ← getSubsetComplementary(X, A_i, w')
    end for
    return A' = A'_1, ..., A'_N
end function
setSystemParameters()
cost ← calculateCost(X, A)
counter ← 0
while termination condition not met do
    setupClustering(counter)
    A' ← getSubsets(X, A, w, k)
    cost' ← calculateCost(X, A + A')
    if cost' < cost then
        A'' ← A'
        cost ← cost'
    else
        counter ← counter + 1
    end if
end while
return A'', cost
```

Within processes performing the algorithm, the stochastic search of subsets results better over time, and the best solution converges to the optimum. However, it will generally not reach the global optimum since of the large non-convex search space. A mechanism intended to prevent a process from getting stuck is to monitor and count the failures to improve the solution. If stuck in a local minimum, it cannot proceed, and the parameters are then randomized again to start over. The data sampling is based on clustering, whose parameters are type, weights, and number of clusters. Type is, e.g., soft or hard clustering. The parameters alternate over time. The cluster seeds may sometimes spread if the solution becomes better this way. An opposite case would be, for example, one cluster for a subset that stays as one cluster in the final solution.

In the context of data replication and data placement, the primary factor is the data transfer time, which will be selected as the target cost/loss. The $cost_{network}(A)$ penalizes external data transfers from site to site.

$$cost_{network}(A) = \sum_{j \in \{jobs\}} cost_{network}(A, j) \tag{2}$$

where A is the allocation parameter as before. j is a job from the job batch (validation). $cost_{network}(A, j)$ denotes an affine approximation dependent on the non-local files for job j given A. The value per file depends on the file size and the bandwidth between the source and the target.

4   Justification and Evaluation

Resources, such as computing resources, storage, and network bandwidth, are randomly chosen for each simulation run. Random selection means for storage, for instance, that each node gets a random value within a range to store files. At the end of value sampling for parameters, the situation must be such that the sites or targets (subsets from the data perspective) provide room for all replicas. Additionally, there are several job types and several data types to represent the real-world scenario. This shall represent the essential types or classes if there are more than that. The simulation runs multiple times to average out uncertainties on parameters and variables. The results ensure a fair evaluation without picking random or biased values in the first place. The **model** of the parallel system consists of a network with sites, jobs as basic processing units, and data points representing datasets or collections of files:

Network connections of sites are each sampled before a simulation run relative to the maximal connection bandwidth:
$bandwidth_i \in [0.1, ..., 1]$ for all sites $1, ..., N$

Sites or nodes hold data and process jobs on the data. Each site has several different resource types, such as high-mem (high memory) or GPU. A site with small storage must rely more on others to provide files to be processed. The model mimics that each node handles different job types differently quickly. The number of jobs is expressed as a value in the probability mass functions depending on type and size, such as in Fig. 2 (Fig. 1).

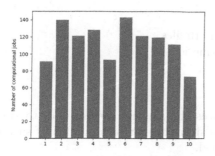 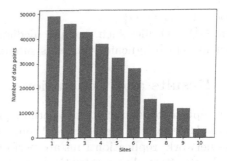

(a) Summed up running jobs per site. (b) Stored data points per site.

Fig. 1. Random sample for 10 sites (ordered by storage capacities) with jobs (left) and data stored (right).

$resourcetypes = \{A, B, C, D\}$: $\forall t \in resourcetypes : Pr(t) \in [0, 1]$, normalize such that $\sum_{i \in resourcetypes} Pr(i) = 1$ (PMF)

Jobs reads a data point of the same type:

$type \in \{A, B, C, D\}$ The location of a running job depends on its type and the provided capacities of the same type on sites based on probability mass functions (PMF) looking like the one in Fig. 2 (left). A job prefers local data since the workload balancer would rather choose a site with input data. This rate is built into the model with a 50 percent suppression of the branching factor of N (N sites). In other words, jobs aim at internal data with double chances.

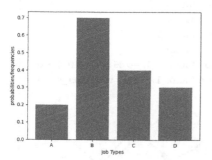 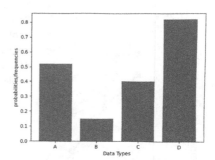

(a) Job probabilities of one site. (b) Probabilities of global data types.

Fig. 2. A random sample of independent probabilities of local job types (left) versus global data types (right).

Data points are collections of files called datasets in the ATLAS use case. Such a data set comprises 1 to 100 files in the model to reflect dynamic dataset sizes. Some popularity is assigned to datasets since there are differences in how likely a data set will be read. Data sets are from 100 users. Those users are in classes of job types and of activeness.

$type \in \{A, B, C, D\}$
$size \in [1, 100]$ files such that the median is 10 files.
user_id, which indicates the users submitting jobs.

5   Results and Conclusions

For replication factor 1, the method places the single copies of data to sites. Each data point is placed once. From this situation, for each number of sites in the network, replication runs are performed.

In the **first demonstration**, replicated data is added on all sites. Figure 3 shows replication of different shares of the global data and for a different number of sites or nodes in the network. A replication factor of, e.g., 1.4 means replication of 2 on a 40 percent data share. The **second demonstration** includes the most extensive five sites providing free space for replicating data. The **third demonstration** covers feeding only the variable of *user_id* as data into the replication method. As in the Sect. 4 on data described, there are 100 users, of which about ten are very active. These are like power users submitting lots of jobs. Users further submit jobs onto one or two data types. The results of the second and third cases are shown in Fig. 3.

Each run in the evaluation is repeated several times. System parameters for each run differ significantly, so the fluctuations from run to run are large. Many more simulation runs (per replication, per number of sites) would be necessary to obtain smooth curves.

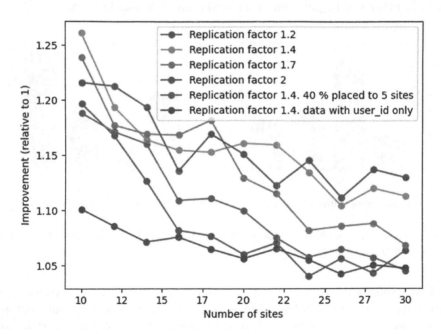

Fig. 3. Relative improvements on network metric depending on replication factors and the number of sites in the network.

6 Conclusion and Future Work

The method looks for the best cluster parameters on a 10 % sub batch and then uses the near-optimal solution on the full data. We observe only several percentage points decrease from a 100k batch to the entire data population of 1M. On a high-performance computer, applying the optimization would be feasible on a data batch of 1 million data points from 10 million data points in total. Complexity also depends on the dimensionality of data and the number of clustering types besides general model parameters and population size.

It is further shown that even on a weak predictor such as users, the method finds proper subsets to get closer to a near-optimum data replication and placement.

References

1. Boru, D., Kliazovich, D., Granelli, F., Bouvry, P., Zomaya, A.Y.: Energy-efficient data replication in cloud computing datacenters. Cluster Comput. **18**(1), 385–402 (2015). https://doi.org/10.1007/s10586-014-0404-x
2. Brezany, P., Mueck, T.A., Schikuta, E.: A software architecture for massively parallel input-output. In: Waśniewski, J., Dongarra, J., Madsen, K., Olesen, D. (eds.) PARA 1996. LNCS, vol. 1184, pp. 85–96. Springer, Heidelberg (1996). https://doi.org/10.1007/3-540-62095-8_10
3. Casey, R.G.: Allocation of copies of a file in an information network. In: Proceedings of the May 16–18, 1972, Spring Joint Computer Conference, pp. 617–625 (1971)
4. Lamehamedi, H., Szymanski, B., Shentu, Z., Deelman, E.: Data replication strategies in grid environments. In: Fifth International Conference on Algorithms and Architectures for Parallel Processing, 2002, Proceedings, pp. 378–383. IEEE (2002)
5. Loukopoulos, T., Ahmad, I.: Static and adaptive distributed data replication using genetic algorithms. J. Parallel Distrib. Comput. **64**(11), 1270–1285 (2004)
6. Vamosi, R., Lassnig, M., Schikuta, E.: Data allocation based on evolutionary data popularity clustering. In: Shi, Y., et al. (eds.) ICCS 2018. LNCS, vol. 10860, pp. 153–166. Springer, Cham (2018). https://doi.org/10.1007/978-3-319-93698-7_12
7. Wolfson, O., Jajodia, S., Huang, Y.: An adaptive data replication algorithm. ACM Trans. Database Syst. (TODS) **22**(2), 255–314 (1997)

Detection of Anomalous Days in Energy Demand Using Leading Point Multi-regression Model

Krzysztof Karpio(✉) ⓘ and Piotr Łukasiewicz ⓘ

Institute of Information Technology, Warsaw University of Life-SGGW, Nowoursynowska 159, 02-787 Warsaw, Poland
{krzysztof_karpio,piotr_lukasiewicz}@sggw.edu.pl

Abstract. Leading Point Multi-Regression Model was utilized to detect days with abnormal energy consumption profiles. They were identified based on the statistical analysis of relative errors of the model. Two ranges of error values were identified: above 4.98% and 3.88%–4.98%. All days with anomalous energy consumption profiles were identified as major religious holidays in Poland: Easter, All Saints, and Christmas Eve, as well as days related to a celebration of the new year: New Year's Eve and New Year.

Keywords: Leading Point Multi-Regression Model · energy consumption · untypical daily profiles

1 Introduction

In recent years, new methods of electricity consumption analysis have been proposed. They go beyond standard methods such as Holt-Winters or ARIMA [1, 2]. More advanced methods, such as nonlinear machine learning (ML), have been utilized in order to forecast power consumption. Among those methods are KNN (K-nearest neighbors) [3], SVM (support vector machine) [4], GBM (gradient boosting machine) [5], RF (random forest) [6], and ANN (artificial neural networks) [7]. An analysis of energy consumption in order to identify sources of energy demand and other factors influencing the consumption, including weather, season, and economics gained importance [8]. One of the most important aspects of the analysis of energy consumption is peak identification. Artificial neural networks were used in [7], while extended CART trees and the K-Nearest Neighbors classifiers were utilized in [3]. In turn, generalized combined additive models and deep ANN were adopted in [9]. Another important issue in a modeling of the power demand is an identification of outliers [10]. In [11] the hybrid model combining Long Short Term Memory (LSTM) and the K-means algorithm was used, while in [12], the authors used a combination of the deep learning model Transformer and a clustering approach based on K-means. Other advanced methods were described in [13, 14].

In this paper, we deal with a detection of outliers in terms of daily energy consumption profiles. In contrast to the majority of studies [14, 15], we do not start with the possible

© The Author(s), under exclusive license to Springer Nature Switzerland AG 2023
J. Mikyška et al. (Eds.): ICCS 2023, LNCS 14075, pp. 676–684, 2023.
https://doi.org/10.1007/978-3-031-36024-4_53

reasons for an untypical profile and do not verify them. We use the Leading Points Multi-Regression model (LPMR) [16], which is solely based on the energy consumption during a few chosen hours. It does not use any other variables, and it's precision is very high, regardless of other factors, like weather conditions, season. In order to detect outliers, we used the error measure of the model and limit values of errors that were defined precisely.

2　Leading Points Multi-regression Model

The purpose of LPMR is to model hourly energy demands for the whole day using only a few variables, such as energy consumption at chosen hours. However, in this work, we use the model for other purposes: to detect untypical days from the energy consumption point of view. The model's details are presented and discussed in [16].

2.1　Data

These studies were carried out based on the data regarding total electricity consumption in the Polish power system [17]. The consumption was denoted in MWh on an hourly basis from 1 Jan 2008 to 31 Dec 2020. The data being analyzed contained 4,749 days, which corresponded to 113,976 h. While our model was solely based on the energy consumption, additional factors were used in the discussion of the results to distinguish separate days and hours, such as a day of the week, specific dates, working hours, etc. The data set was divided into two subsets: the training set and the testing set, consisting of 1583 and 3166 days, respectively.

2.2　Model, Errors and Variable Selection

Data is analyzed on the daily basis: 24 time series of hourly electricity consumptions. We use 24 variables: $E(h_m) = (E_1(h_m), \ldots, E_i(h_m), \ldots, E_N(h_m))$, where $h_m \in \{h_1, h_2, \ldots, h_{24}\}$, $E_i(h_m)$ is an electricity consumption at hour h_m in the i-th day, and N is the total number of analyzed days. The energy consumption is described by the multiple equation linear regression model of the form:

$$E(h_p) = a_{0p} + a_{1p}E(h_1) + a_{2p}E(h_2) + \cdots + a_{kp}E(h_k) + \xi_p \tag{1}$$

where $p \in \{1, \ldots, 24\} \setminus \{1, \ldots, k\}$ and $a_{0p}, a_{1p}, \ldots, a_{kp}$ are model parameters. The number of equations is related to the number of describing variables. In the case of k variables, the model consists of $24 - k$ equations.

　　Model (1) takes electricity consumptions at hours $h_1, h_2, \ldots, h_k$ and use them to describe electricity consumptions at the remained hours. The selection of variables is based on the analysis of the random components ξ_p. For each model equation, one calculates the standard deviation of the residuals:

$$\sigma(h_p) = \sqrt{\frac{1}{N-k-1}\sum_{i=1}^{N}\left(E_i(h_p) - \hat{E}_i(h_p)\right)^2} \tag{2}$$

where, $\widehat{E}_i(h_p)$ denotes theoretical value and $E_i(h_p) - \widehat{E}_i(h_p) = \xi_p$. The quality of the model regressions is measured by means of the relative standard deviation:

$$v(h_p) = \sqrt{\frac{1}{N-k-1} \sum_{i=1}^{N} \left(E_i(h_p) - \widehat{E}_i(h_p)\right)^2} / \overline{E(h_p)} \tag{3}$$

where, standard deviation of residuals is divided by the mean electricity consumption. The quality of the whole model (1) is measured based on the mean values of the measure (4) calculated for all $24 - k$ equations in the model:

$$MRSD = \frac{1}{24-k} \sum_{p=1}^{24-k} v(h_p) \tag{4}$$

The independent variables are selected by the algorithm in successive steps. In each step, one describing variable $E(h_i)$ is chosen, the new model is built and its precision is evaluated by the $\sigma(h_p)$ and $MRSD$ measures. The procedure can be stopped after reaching a desired precision. An algorithm of variable selection is described precisely in [16]. During the model's construction, we observed a strong decrease of errors in steps 1–4. Already in step two, the error decreased below 3%, and in step four, it was below 2%. Subsequent declines are not so significant.

2.3 Application of the Model

We concluded that four variables are sufficient to describe a data with reasonable quality, $MRSD = 1.74\%$. They corresponded to energy consumptions at hours: 14, 20, 2, 18. The model quality was evaluated for each data point (hour) and for every day in the testing data set, using the relative measure:

$$RSD(i) = \sqrt{\frac{1}{20} \sum_{p=5}^{24} \left(E_i(h_p) - \widehat{E}_i(h_p)\right)^2} / \frac{1}{20} \sum_{p=5}^{24} E_i(h_p) \tag{5}$$

where in the sum, the following hours are omitted: from h_1 to h_4, $i = 1, 2, ..., N$, and N denotes the number of analyzed days. For illustration of the model precision one shows empirical and theoretical daily time series for selected six days in Fig. 1.

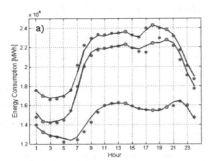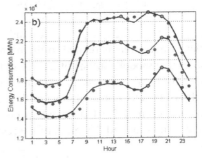

Fig. 1. Empirical (red dots) and theoretical (blue lines) time series for six selected days. From the top: a) 2016-01-14, 2016-02-22, 2016-07-17; b) 2017-12-14, 2017-09-12, 2017-10-08. Green dots denote independent variables of the model.

For the days presented, the error was between 0.0118 and 0.0178. The model exhibits very good agreement with the data; the mean *RSD* is 0.0175 and for about 90% of days it does not exceed 0.0251.

3   Analysis of Anomalyous Profiles

3.1   Daily Errors of the Model

In Fig. 2, logarithms of the daily *RSD* errors are shown for the whole testing data set. They do not exhibit any trend and are symmetric around the mean value.

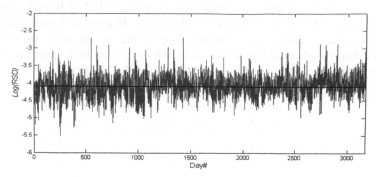

Fig. 2. Logarithms of the daily errors of the model for testing the data set. The mean value of the errors is indicated by the horizontal line at -4.102.

A comparison of the distribution of log-*RSD* with a normal distribution is shown in Fig. 3a) in vertical log scale. The Q-Q plot in Fig. 3b) demonstrates a good agreement between empirical and Gaussian distributions.

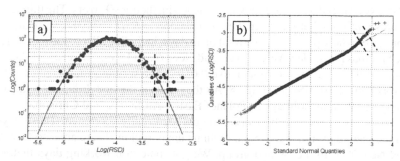

Fig. 3. The compatibility of error and normal distributions: a) distributions of log-errors with normal curve on a logarithmic vertical scale; b) comparison of empirical and theoretical quantiles. The black dashed lines indicate threshold values of errors.

However, there is a visible deviation at the right tail, the number of counts is higher than for the theoretical distribution. Those days are anomalous, characterized by their untypical daily profiles of energy consumption. Moreover, the distortions from normal distribution may indicate the existence of additional factors influencing data apart from statistical fluctuations.

3.2 Identification of Unusual Daily Energy Consumption Profiles

We identified two ranges of big values of relative errors, based on the analysis of their distribution.

(1) $0.0498 < RSD \ (-3 < \log(RSD))$ and
(2) $0.0388 < RSD \leq 0.0498 \ (-3.25 < \log(RSD) \leq -3)$.

Boundaries between ranges are indicated by vertical dashed lines in Fig. 4. We observe deviations from the normal distributions in both ranges. However, in the first range, the normal distribution is negligible, while in the second range, we may expect some days distributed according to Gaussian. All days from both ranges are listed in Table 1. Weekdays are numbered from 1 (Monday) to 7 (Sunday).

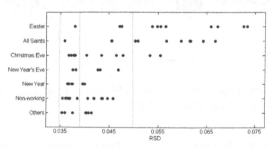

Fig. 4. Identified days corresponding to the biggest errors. Three ranges of error values are indicated by dashed lines.

All days discussed below are shown in Fig. 4 with their *RSD* errors. The third group of days with slightly smaller errors, in the range $0.0353 < RSD \leq 0.0388$, were also added to the plot. The whole plot contains a total of 45 days, including all the: Easters, All Saints Days, Christmas Eves, New Year's Eves, and New Year's Days. There are 11 days in the first range with the greatest errors, above 0.0498. The second group contains 19 days with errors between 0.0388 and 0.0498. All the points in the first group are related to the three biggest religious holidays in Poland: Easter, All Saints, and Christmas Eve. The last one is working day, but the working hours are usually shortened. Those days also exist in the second group, predominantly in its upper region. There are also New Year's Eve and New Year in the second group. We got some non-working days that are not holidays but are not random, e.g., a day before Christmas Eve, which was a Sunday; a second day of Christmas, which is a paid holiday in Poland; or Easter Monday. Figures 5 and 6 show daily profiles for days with the greatest values of errors, which are All Saints, Easter, New Year's, and Christmas Eve. Untypical profiles are accompanied by profiles for adjacent or corresponding days. Solid lines denote theoretical values, and green dots correspond to the four independent variables.

Table 1. Days from both ranges of the big measure values (see text).

No	Date (Weekday)	RSD	Description	No	Date (Weekday)	RSD	Description
1	2016-03-27 (7)	0.0672	Easter	16	2013-12-25 (3)	0.0458	Non-working
2	2013-11-01 (5)	0.0667	All Saints	17	2018-11-01 (4)	0.0456	All Saints
3	2019-04-21 (7)	0.0658	Easter	18	2012-11-01 (4)	0.0455	All Saints
4	2019-11-01 (5)	0.0567	All Saints	19	2018-12-23 (7)	0.0437	Non-working
5	2020-04-12 (7)	0.0566	Easter	20	2014-12-24 (3)	0.0435	Christmas Eve
6	2012-04-08 (7)	0.0555	Easter	21	2018-12-31 (1)	0.0432	New Year's Eve
7	2019-12-24 (2)	0.0555	Christmas Eve	22	2013-12-31 (2)	0.0427	New Year's Eve
8	2014-04-20 (7)	0.0548	Easter	23	2012-12-23 (7)	0.0420	Non-working
9	2013-12-24 (2)	0.0533	Christmas Eve	24	2012-04-09 (1)	0.0419	Non-working
10	2015-11-01 (7)	0.0509	All Saints	25	2020-04-06 (1)	0.0414	Other
11	2014-11-01 (6)	0.0504	All Saints	26	2017-10-05 (4)	0.0408	Other
12	2015-12-24 (4)	0.0479	Christmas Eve	27	2016-12-24 (6)	0.0404	Christmas Eve
13	2015-04-05 (7)	0.0472	Easter	28	2020-04-07 (2)	0.0404	Other
14	2019-12-31 (2)	0.0469	New Year's Eve	29	2019-12-20 (5)	0.0402	Other
15	2020-12-24 (4)	0.0465	Christmas Eve	30	2019-01-01 (2)	0.0400	New Year

We limit a discussion to only the days mentioned above. The daily profiles for each type of day are very similar to one another, so the presented profiles in Figs. 5, 6 can be treated as representative. (1) Easter: when compared to other Sundays, a profile is more flattened. There is a clear maximum between 9:00 and 11:00, followed by a long slow decrease. (2) All Saints: is compared to adjacent days. The first maximum is moved to the left; we also observe a significantly more flattened profile before 17.

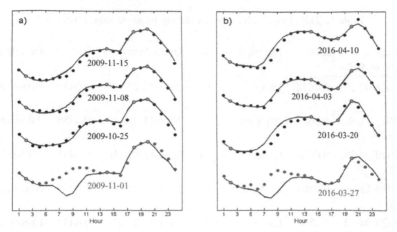

Fig. 5. Daily energy consumption profiles for a) All Saints (red dots $RSD = 0.0642$) and adjacent days, b) Easters (red dots $RSD = 0.0672$) and other Sundays.

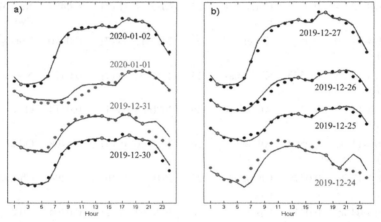

Fig. 6. Daily energy consumption profiles: a) New Year's Eve (red dots, $RSD = 0.0469$) and New Year (red dots, $RSD = 0,0375$) and adjacent days; b) Christmas Eve (red dots, $RSD = 0.0555$) and the following days.

(3) Christmas Eve: compared to the next three days. The analyzed profiles exhibit clear peaks around 10–12 and 17, followed by an anomalous drop of convex shape. (4) New Year's Eve: compared to the profiles for Dec 30th and Jan 2nd. Profile is in general similar to them. However, we observe a different drop in consumption after around 18–19 in both cases. (5) New Year: compared to the profiles for Dec 30th and Jan 2nd. There is a semi-flat shape with no maximum before 17 o'clock, completely different than for other days. For all the days in the upper range, the social factors related to the Easter, All Saints, and Christmas Eve holidays exist and significantly influence energy consumption profiles. Due to the common celebration of those holidays in Poland, one can assume that these factors are related to the short-term, intensive migration of people.

4 Summary

The LPMR-based method of identification of days with untypical daily profiles of energy consumption was presented herein. The following conclusions have been drawn out: (1) Analyzed data could be described with high precision using four independent variables; (2) the distribution of model's errors follow a Gaussian distribution with a high accuracy; (3) days with untypical energy consumption profiles were precisely defined, as days with errors deviating from Gaussian distribution; (4) untypical days were identified as the major religious holidays in Poland: Easter, All Saints, and Christmas Eve; (5) there were also New Year's Eve and New Year identified in the range of smaller errors; and (6) the main factor causing anomalies in the daily profiles was related to the short-term migration of people. The future research will focus on the quantitative description of anomalies in the daily energy consumption profiles as well as the investigation of their reasons. The studies presented herein can be easily extended to other countries and regions. This subject of studies is of great interest because faults in the energy consumption predictions cause non optimal energy production.

References

1. Chicco, G., Mazza, A.: Load profiling revisited: prosumer profiling for local energy markets. In: Pinto, T., Vale, Z., Windergrean, S. (eds.) Local Electricity Markets, Cambridge, MA, USA, pp. 215–242 (2021)
2. Karpio, K., Łukasiewicz, P., Nafkha, R.: Regression technique for electricity load modeling and outlined data points explanation. In: Pejaś, J., El Fray, I., Hyla, T., Kacprzyk, J. (eds.) ACS 2018. AISC, vol. 889, pp. 56–67. Springer, Cham (2019). https://doi.org/10.1007/978-3-030-03314-9_5
3. Gajowniczek, K., Nafkha, R., Ząbkowski, T.: Seasonal peak demand classification with machine learning techniques. In: International Conference on Applied Mathematics & Computer Science (ICAMCS), Paris, France, pp. 101–1014 (2018)
4. Niu, D., Wang, Y., Wu, D.D.: Power load forecasting using support vector machine and ant colony optimization. Expert Syst. Appl. **37**, 2531–2539 (2010)
5. Massaoudi, M., Refaat, S.S., Chihi, I., Trabelsi, M., Oueslati, F.S., Abu-Rub, H.: A novel stacked generalization ensemble-based hybrid lgbm-xgb-mlp model for short-term load forecasting. Energy **214**, 118874 (2021)
6. Huang, N., Lu, G., Xu, D.: A permutation importance-based feature selection method for short-term electricity load forecasting using random forest. Energies **9**, 767 (2016)
7. Gajowniczek, K., Nafkha, R., Ząbkowski, T.: Electricity peak demand classification with artificial neural networks. In: Federated Conference on Computer Science and Information Systems (FedCSIS), Prague, Czech Republic, pp. 307–315 (2017)
8. Hong, T., Fan, S.: Probabilistic electric load forecasting: a tutorial review. Int. J. Forecast. **32**, 914–938 (2016)
9. Berrisch, J., Narajewski, M., Ziel, F.: High-resolution peak demand estimation using generalized additive models and deep neural networks. Energy AI **13**, 100236 (2023)
10. Berthold, M.R., Borgelt, C., Höppner, F., Klawonn, F., Silipo, R.: Guide to Intelligent Data Science: How to Intelligently Make Use of Real Data. Springer, Cham (2020). https://doi.org/10.1007/978-1-84882-260-3

11. Chahla, C., Snoussi, H., Merghem, L., Esseghir, M.: A novel approach for anomaly detection in power consumption data. In: De Marsico, M., Sanniti di Baja, G., Fred, A. (eds.) International Conference on Pattern Recognition Applications and Methods - ICPRAM, 2019, Prague, Czech Republic, 19–21 February 2019, vol. 1, pp. 483–490 (2019)
12. Zhang, J., Zhang, H., Ding, S., Zhang, X.: Power consumption predicting and anomaly detection based on transformer and k-means. Front. Energy Res. **9**, 779587 (2021)
13. Fu, T., Zhou, H., Ma, X., Hou, Z.J., Wu, D.: Predicting peak day and peak hour of electricity demand with ensemble machine learning. Front. Energy Res. **10**, 944804 (2022)
14. Zhang, W., Dong, X., Li, H., et al.: Unsupervised detection of abnormal electricity consumption behavior based on feature engineering. IEEE Access **8**, 55483–55500 (2020)
15. Hu, M., Ji, Z., Yan, K., et al.: Detecting anomalies in time series data via a meta-feature based approach. IEEE Access **6**, 27760–27776 (2018)
16. Karpio, K., Łukasiewicz, P., Nafkha, R.: New method of modeling daily energy consumption. Energies **16**(5), 2095 (2023)
17. Polskie Sieci Elektroenergetyczne. http://www.pse.pl. Accessed 15 Jan 2023

Correction to: Coupling Between a Finite Element Model of Coronary Artery Mechanics and a Microscale Agent-Based Model of Smooth Muscle Cells Through Trilinear Interpolation

Aleksei Fotin⬚ and Pavel Zun⬚

Correction to:
Chapter 20 in: J. Mikyška et al. (Eds.):
Computational Science – ICCS 2023, LNCS 14075,
https://doi.org/10.1007/978-3-031-36024-4_20

The original version of the book was inadvertently published with incorrect author names. The given names and the last names of both authors were mixed-up. Instead of "Fotin Aleksei and Zun Pavel" it should read "Aleksei Fotin and Pavel Zun". This has been corrected.

The updated version of this chapter can be found at
https://doi.org/10.1007/978-3-031-36024-4_20

© The Author(s), under exclusive license to Springer Nature Switzerland AG 2023
J. Mikyška et al. (Eds.): ICCS 2023, LNCS 14075, p. C1, 2023.
https://doi.org/10.1007/978-3-031-36024-4_54

Author Index

A

Akeweje, Emmanuel 580
Alba, Emilio 228
Alves, Domingos 170, 300
Ameijeiras-Rodriguez, Carolina 313
Andrade Martins, Pedro Emilio 170
Assis, Laura S. 90, 105
Asteriou, Konstantinos 323
Avilkin, Vsevolod 580

B

Baazaoui, Hajer 185
Bączkiewicz, Aleksandra 48
Baglamis, Selami 345
Bänziger, Rolf B. 243
Basukoski, Artie 243
Belhadj, Kamel 123
Bernardi, Filipe Andrade 170, 300
Borkowski, Krzysztof 595
Boterman, Willem 74
Bouhamoum, Redouane 185
Bouraoui, Chokri 123
Bródka, Piotr 33
Budniak, Artur 661

C

Cardoso, Douglas O. 90, 105
Cassão, Victor 300
Chang, Qiong 454
Chaussalet, Thierry 243
Chichurin, Alexander 469
Choudhury, Salimur 528
Churavy, Valentin 483
Colombo Filho, Márcio Eloi 170
Czarnowski, Ireneusz 17

D

de Andrade Mioto, Ana Clara 170, 300
Dębski, Roman 395
der Heijden, Maartje van 345
Dignum, Eric 74

D (cont.)

Diniz, Jose Miguel 313
Dobrowolska, Anna 513
Dreżewski, Rafał 395
Dul, Magdalena 138

F

Félix, Têmis Maria 170
Filippi-Mazzola, Edoardo 337
Flache, Andreas 74
Fotin, Aleksei 258
Freitas, Alberto 313
Funkner, Anastasia 213

G

Galea, François 652
Gaudiani, Adriana 641
Gora, Maja 513
Grzeszczyk, Michal K. 138
Guedria, Najeh Ben 123
Gupta, Richa 185

H

Han, Jihun 548
Hanawa, Toshihiro 378
Harzallah, Omar Anis 123
Hashimoto, Atsushi 410
Helali, Ali 123
Huaman, Israel 270, 286

I

Ida, Akihiro 378
Idoumghar, Lhassane 123
Imura, Naoto 454

J

Janecki, Dariusz 595
Jankowski, Jarosław 33
Jedrzejowicz, Joanna 3
Jedrzejowicz, Piotr 3
Jerez, José M. 228

© The Editor(s) (if applicable) and The Author(s), under exclusive license
to Springer Nature Switzerland AG 2023
J. Mikyška et al. (Eds.): ICCS 2023, LNCS 14075, pp. 685–687, 2023.
https://doi.org/10.1007/978-3-031-36024-4

K

Kajdanowicz, Tomasz 661
Kalwarowskyj, Daniela 563
Karpio, Krzysztof 676
Kawai, Masatoshi 378
Kharlunin, Aleksandr 286
Konopka, Aleksandra 498
Koratikere, Pavankumar 425, 536
Kosherbayeva, Aiken 469
Kovalchuk, Sergey 213
Kozera, Ryszard 439, 498
Koziel, Slawomir 363, 425, 536, 571
Kozsik, Tamás 626
Krawczyk, Przemek M. 345
Kritski, Afrânio 300
Król, Robert 48

L

Lamine, Sana Ben Abdallah Ben 250
Lee, Yoonsang 548
Lees, Mike 74
Leifsson, Leifur 363, 425, 536, 571
Leonenko, Vasiliy 270, 286
Lesiński, Wojciech 162
Lima, Vinícius Costa 170, 300
Lima, Willian P. C. 90, 105
López-García, Guillermo 228
Łukasiewicz, Piotr 676
Luque, Emilio 641

M

Marques, Lucas C. 105
Martin, Kyle 153
Masmoudi, Maroua 185
Massie, Stewart 153
Mehrotra, Deepti 185
Mesa, Héctor 228
Michalski, Radosław 33
Miedema, Daniël M. 345
Mielczarek, Bożena 513
Mielniczuk, Jan 610
Minglibayev, Mukhtar 469
Miotti, Pietro 337
Miyoshi, Newton Shydeo Brandão 300
Moreno-Barea, Francisco J. 228
Morse, Gregory 626
Moses, William S. 483

N

Nakajima, Kengo 378
Narayanan, Sri Hari Krishna 483
Nishinari, Katsuhiro 454
Noakes, Lyle 439
Nojszewska, Ewelina 138
Nowacki, Jerzy P. 200

O

Olhin, Andrey 580

P

Paehler, Ludger 483
Panov, Maxim 580
Pellegrini, François 652
Pieliński, Bartosz 59
Pietrenko-Dabrowska, Anna 363, 425, 536, 571
Pivkin, Igor V. 337
Pogrebnoi, Dmitrii 213
Prokopenya, Alexander 469
Przybyszewski, Andrzej W. 200, 278

R

Radaoui, Marouane 250
Rakyta, Péter 626
Rb-Silva, Rita 313
Rexachs, Dolores 641
Ribelles, Nuria 228
Rodriguez, Julien 652
Rudnicki, Witold R. 162

S

Saha, Joyaditya 345
Sahatova, Kseniya 286
Saputa, Karol 59
Schanen, Michel 483
Schikuta, Erich 563, 668
Schreiber, Hanna 59
Seweryn, Karolina 59
Sheraton, Vivek M 345
Silva, Guilherme G. V. L. 90
Sitarz, Konrad 200
Sitek, Arkadiusz 138
Śledzianowski, Albert 200
Soares, Giovane Thomazini 300
Souza, Julio 313
Spieker, Christian J. 323
Stępień, Stanisław 33

Struniawski, Karol   498
Sun, Jia He   528
Sysko-Romańczuk, Sylwia   59
Szarmach, Marta   17

T
Trigila, Mariano   641

U
Upadhyay, Ashish   153

V
Vamosi, Ralf   668
van Gent, Démi   345
Veredas, Francisco J.   228
Vermeulen, Louis   345
Vishnyakov, Aleksey   580

W
Wątróbski, Jarosław   48
Wawrzeńczyk, Adam   610
Wichrowska, Aleksandra   59
Wijekoon, Anjana   153
Williamson, Sarah   483
Wiratunga, Nirmalie   153
Wit, Ernst C.   337
Wong, Alvaro   641
Wróblewska, Anna   59

Z
Zaourar, Lilia   652
Zav́odszky, Gaбor   323
Zehner, Paul   410
Zghal, Hajer Baazaoui   250
Zha, Aolong   454
Zun, Pavel   258
Zwierzchowski, Jarosław   595

Printed in the United States
by Baker & Taylor Publisher Services

Printed in the United States
by Baker & Taylor Publisher Services